The Encyclopedia of
WORLD AIR POWER

The Encyclopedia of
WORLD AIR POWER

Consultant Editor Bill Gunston

TEMPLE PRESS
AEROSPACE

Published by Temple Press
an imprint of Newnes Books
Bridge House, 69 London Road, Twickenham,
Middlesex, TW1 3SB
and distributed for them by
Hamlyn Distribution Services
Rushden, Northants, England

Produced by Stan Morse
Aerospace Publishing Limited
179 Dalling Road
London W6 0ES

First published 1980
Second edition 1986

ISBN 0 600 35165 3

Printed in Italy

Picture acknowledgements

The publishers would like to thank the following individuals and organisations for their help
in supplying photographs for this book.

Page 8: Austin J. Brown/Peter R. Foster. **9:** Saab/Klaus Niska. **10:** Dassault-Breguet. **11:** British Aerospace/Dassault-Breguet. **12:** Dassault-Breguet/Aérospatiale. **13:** Vought Corporation/Aeritalia. **15:** Lockheed. **16:** Peter R. Foster/Philip Chinnery. **17:** McDonnell Douglas/McDonnell Douglas. **18:** Saab. **19:** Peter R. Foster. **20:** McDonnell Douglas/Malcolm English. **21:** Robbie Shaw. **22:** Westland Helicopters/Bob Munro. **23:** Peter R. Foster/Peter R. Foster. **27:** US Navy. **28:** General Dynamics. **29:** Lockheed. **30:** Israel Aircraft Industries. **31:** British Aerospace/US Navy. **32:** Lockheed/British Aerospace. **33:** Westland Helicopters/British Aerospace. **34:** Israeli Defence Force/Aermacchi. **35:** Paul A. Jackson. **36:** Austin J. Brown. **38:** Dassault-Breguet/Aermacchi. **39:** Dornier. **40:** Dornier/Austin J. Brown/Fokker. **42:** Herman Potgieter/Herman Potgieter/Fokker. **44:** British Aerospace/Gamma. **45:** Lockheed. **48:** Boeing/Lockheed. **49:** McDonnell Douglas/US Air Force. **50:** British Aerospace. **51:** Lindsay Peacock. **52:** US Air Froce/Northrop. **53:** Vietnamese Embassy/Peter R. Foster. **54:** Peter R. Foster/Chris Pocock. **55:** US Air Force/US Air Force. **56:** US Navy/Robert L. Lawson. **57:** McDonnell Douglas. **58:** Sikorsky. **59:** FMA. **60:** Lockheed/Aermacchi. **61:** EMBRAER. **63:** Cessna. **65:** British Aerospace. **66:** Dassault-Breguet. **68:** Philip Chinnery. **70:** Aermacchi/Aermacchi. **73:** Aérospatiale. **74:** Peter Steinemann. **75:** Aérospatiale. **78:** Aérospatiale/Aérospatiale. **79:** Agusta. **81:** Agusta/Agusta. **87:** Herman Potgieter/Herman Potgieter. **91:** Peter R. Foster. **92:** Bell Helicopter Textron /Bell Helicopter Textron. **93:** Bell Helicopter Textron. **94:** Peter R. Foster. **96:** US Air Force. **97:** David Oliver. **98:** Boeing. **99:** Lindsay Peacock. **100:** Boeing. **102:** US Navy/Boeing Vertol. **103:** Robbie Shaw/Malcolm English. **104:** Peter R. Foster. **105:** Bill Gunston/British Aerospace. **106:** British Aerospace. **107:** British Aerospace. **108:** British Aerospace. **110:** Royal Navy/MoD. **112:** Jon Lake. **114:** RAF Germany. **115:** British Aerospace. **116:** MoD. **117:** British Aerospace/MoD. **118:** Britten-Norman. **119:** Peter R. Foster/CASA. **121:** Cessna. **122:** US Air Force/US Air Force. **123:** Cessna. **125:** Dassault-Breguet. **126:** Stan Morse. **127:** Dassault-Breguet. **128:** Herman Potgieter. **130:** Peter R. Foster/Peter R. Foster. **131:** Dassault-Breguet/Dassault-Breguet. **134:** Dassault-Breguet. **135:** Dassault-Breguet. **136:** MoD. **138:** Peter R. Foster. **139:** US Air Force/Peter R. Foster. **140:** Dornier. **141:** US Navy.

142: EMBRAER. **143:** EMBRAER/EMBRAER. **144:** ENAER. **145:** US Air Force. **148:** Fokker. **149:** Government Aircraft Factories. **150:** Gates/Peter R. Foster. **151:** US Air Force. **154:** US Air Force. **155:** US Air Force/Grumman. **156:** Grumman. **157:** Grumman/Grumman. **158:** Robert L. Lawson/US Navy. **159:** Robert L. Lawson/Grumman. **160:** US Navy. **161:** US Navy. **164:** Grumman. **165:** US Air Force. **168:** US Air Force. **169:** Hughes. **170:** Hughes/Hughes. **171:** Hughes. **176:** US Navy/Paul A. Jackson. **177:** Israel Aircraft Industries. **178:** Israeli Defence Force. **179:** Israel Aircraft Industries. **183:** Lockheed/US Air Force. **186:** MoD. **187:** US Air Force. **188:** US Air Force. **189:** Peter R. Foster. **190:** Peter R. Foster. **191:** US Navy. **192:** Lockheed/Lockheed/Peter R. Foster. **193:** Lockheed/Lockheed/Lockheed. **194:** Lockheed/Lockheed. **195:** Bob Munro. **196:** Peter R. Foster/Bob Munro. **197:** US Air Force. **198:** US Navy/McDonnell Douglas. **199:** US Air Force. **200:** McDonnell Douglas/MoD. **201:** MoD. **202:** Bob Munro. **203:** Peter R. Foster. **204:** US Air Force. **205:** US Air Force. **208:** McDonnell Douglas/McDonnell Douglas. **209:** McDonnell Douglas/McDonnell Douglas. **210:** Messerschmitt-Bölkow-Blohm. **219:** US Navy. **222:** Lindsay Peacock. **225:** Mitsubishi/Mitsubishi. **228:** Austin J. Brown. **230:** Rockwell. **231:** Peter R. Foster/Paul A. Jackson. **232:** Northrop/Northrop. **233:** US Air Force/Panavia. **234:** RAF News. **238:** Bob Munro. **239:** Pilatus. **240:** Denis Hughes. **245:** Robin/US Air Force. **246:** US Air Force. **248:** Bob Munro. **249:** Saab. **250:** Saab/Saab. **251:** Saab/Saab. **254:** British Aerospace/Jon Lake. **256:** British Aerospace. **257:** Lindsay Peacock. **258:** Shin Meiwa/Shorts/Shorts. **259:** SIAI. **260:** SIAI. **261:** Philip Chinnery/US Air Force/Peter R. Foster. **262:** US Navy. **263:** US Army/US Navy/US Navy. **264:** US Air Force/US Air Force. **269:** Austin J. Brown. **276:** Transall. **277:** Swedish Air Force/Swedish Air Force. **280:** US Air Force. **281:** MoD. **285:** US Navy. **286:** Valmet. **287:** US Navy/Vought Corporation. **288:** US Air Force/Vought Corporation. **289:** Vought Corporation. **290:** Westland Helicopters. **291:** Westland Helicopters. **294:** COI/Westland Helicopters. **295:** Westland Helicopters (three). **296:** Westland Helicopters/Westland Helicopters. **297:** No. 28 Sqn RAF/Lieutenant K. P. White. **298:** Westland Helicopters. **300:** MoD. **302:** McDonnell Douglas. **311:** Royal Navy. **315:** US Navy.

The World's
Air Forces

Western Europe

Within NATO modernization is well under way, with Belgium and Denmark receiving General Dynamics F-16 Fighting Falcons from the SABCA production line and the Netherlands and Norway F-16s from the Fokker line. Germany, Italy and the United Kingdom are now receiving the Panavia Tornado. Turkey has ordered the F-16, as has Greece who will also receive the Dassault Mirage 2000. Spain has selected the Northrop F-18A for her re-equipment programme. Only Portugal is not buying new aircraft, but continues to re-equip the Força Aérea Portuguesa with reconditioned LTV A-7P Corsairs.

Outside NATO France is receiving the Mirage 2000 in large numbers and Sweden is developing the Saab Gripen. Austria has still to decide on a replacement for her ageing Saab 105s and Finland is receiving more ex-Swedish air force Drakens. Switzerland has finally retired her Venoms and is procuring more Northrop F-5E/Fs, but has yet to decide on a Hawk/Alpha Jet type to replace the Vampire.

Belgium, Norway and the Netherlands had all withdrawn their Starfighters by the end of 1984; West Germany and Denmark will have followed by 1986. The day of the Starfighter is nearly over, with the F-16 now appearing all over Europe. With regard to rotary-wing aircraft, France, West Germany, Great Britain, Italy and Spain all have, or are developing, TOW armed anti-tank helicopters to counter the threat from the massive Warsaw Pact armoured forces.

Austria Österreichische Luftstreitkrafte

The OLk is administered by the Austrian army and has three main units: Fliegerregiment I at Tulln-Langenlebarn has Hubschraubergeschwader I with 1 Staffel operating the Agusta-Bell AB 206A in training and SAR roles, 2 Staffel operating the Bell OH-58B for observation and reconnaissance and 3 Staffel with 12 Agusta-Bell AB 212s for utility and transport duties. Also under FlgRgt-I control is the Flachenfliegerstaffel with Pilatus PC-6B Turbo Porters and Shorts Skyvans in the transport role. Fliegerregiment II at Graz-Thalerhof has the Uberwachungsgeschwader with 2 Staffel operating the Saab 105 in the air defence role and Hubschraubergeschwader II with 1 Staffel operating the Aérospatiale SA.319B Alouette III at Aigen Im Ennstal. Fliegerregiment III at Linz-Horsching has the Jagdbombergeschwader with 1 and 3 Staffeln providing ground attack support for the army with their Saab 105s. Also with FlgRgt III is Hubschraubergeschwader III with 1 Staffel flying the Agusta-Bell AB 212 and 2 Staffel operating the Alouette II.

Training is carried out by the Pilotenschule des Bundesheeres at Zeltweg with primary training on the Pilatus PC-7 Turbo Trainers of the Ubungsstaffel and basic training following on the Saab 105s of the Dusenflugstaffel. Equipment of the OLk includes the few survivors of 24 Cessna O-1A/E Bird Dogs which are used in a liaison role, and only a few Saab 91 Safirs remain in use at Zeltweg following receipt of the first six of 10 PC-7s in early 1984. Other fixed-wing assets include 11 Pilatus PC-6B Turbo Porters, two Shorts Skyvans and about 30 Saab 1050Es. This last type is in need of replacement but no decision on a successor

has yet been finalized.

Helicopters include 12 AB 206As, 23 AB 212s, 12 SA.313 Alouette IIs, 12 SA.319Bs and 12 OH-58Bs. The 19 surviving Agusta-Bell AB 204B were withdrawn from use by the end of 1981; these are being refurbished and at least five are back in service.

Austria has great need of helicopters for use in the mountains which cover most of the country. Bell JetRangers form a large part of the fleet, including both Agusta-Bell AB 206As and Bell OH-58B Kiowas.

Belgium Force Aérienne Belge/Belgische Luchtmacht

The FAB/BLu is currently upgrading its aircraft equipment as part of its commitment to NATO and the combined Anglo-German-US-Dutch-Belgian 2 ATAF to which the FAB/BLu Commandement de la Force Aérienne Tactique is allocated.

Commandement de la Force Aérienne Tactique has four wings. 1 Wing de Chasse Tous-Temps at Beauvechain consists of 349 Squadron and 350 Squadron, and 10 Wing de Chasseurs Bombardiers at Kleine Brogel comprises 23 Squadron and 31 Squadron. In early 1985 they received the last of 94 General Dynamics F-16A and 20 F-16B Fighting Falcons between them. Of the total received by the end of 1984, nine F-16As and one F-16B had been written off in accidents, but 44 more F-16As were ordered in 1983 for delivery from 1987 onwards. The only other jet equipment in front line service is the Dassault Mirage 5. 2 Wing Tactique at Florennes has 2 Squadron flying the Mirage 5BA strike aircraft and 42 Squadron with the Mirage 5BR reconnaissance version as its two components. 3 Wing Tactique at Bierset has 1 Squadron with the Mirage 5BA and 8 Squadron with both the Mirage 5BA and the Mirage 5BD trainer version. Mirage procurement totalled 63 5BAs, 16 5BDs and 27 5BRs, which by late 1984 had been reduced by attrition to 42, 13 and 20 respectively. 1 Squadron and 2 Squadron are both thought to operate two Mirage 5BDs in addition to their Mirage 5BAs.

The Groupement de Transport consists of 15 Wing at Brussels/Melsbroek with two component squadrons. These are 20 Squadron operating 12 C-130H Hercules, and 21 Squadron with two ex-Sabena Boeing 727-29Cs, two Dassault Falcon 20s, three Hawker Siddeley HS.748s and five Swearingen Merlin IIIAs. Training is carried out by the Groupement d'Instruction et Entrainement and has subordinate units at two bases. The Ecole de Pilotage Elementaire at Gossoncourt, also known as 5 Squadron, fulfils the primary flying training role with 29 SIAI-Marchetti SF.260MBs. The second unit is the Opleiding en Training Wing at Brustem which has three squadrons, Nos 7 and 11 with 32 Dassault-Breguet/Dornier Alpha Jets, and 9 Squadron operating the 22 survivors of 50 Aérospatiale CM-170 Magisters. 9 Squadron also loans Magisters out to act as currency trainers for aircrew who are on a ground tour and most FAB/BLu bases have a station flight of two Magisters. The Belgian Air Cadets have many gliders and a few Piper Super Cubs in use as tugs.

Force Navale Belge/Belgische Zeemacht

The joint air force/navy helicopters unit is 40 Squadron based at Coxyde. In use for SAR are four Westland Sea King Mk.48s, and there are three SA.316B Alouette IIIs which also deploy aboard the support vessels Zinnia and Godetia for liaison duties. Two old Sikorsky S-58s are still in use, although they must be nearly due for withdrawal.

Aviation Légère de la Force Terrestre/Licht Vliegwezen van het Landmacht

The army currently has four squadrons. The main base is Brasschaat where 15 Squadron operates six Britten-Norman BN-2A Islanders and about 18 Alouette IIs. Similarly equipped is 16 Squadron at Butzweilerhof. 17 Squadron based at Werl has 18 Alouette IIs and 18 Squadron also has 18 Alouette IIs but is based at Merzbruck.

Belgium relies heavily on the General Dynamics F-16 Fighting Falcon for its tactical air assets. 114 have been delivered, with another 44 on order. Several have been lost.

Denmark Flyvevabnet

Denmark was one of the founder members of NATO in 1949 and has the primary role of supporting Allied Air Forces Baltic approaches, with a secondary internal role of fishery protection for and communications with and within Greenland.

Flyvertaktisk Kommando has six Eskadrilles under its control. Esk-723 at Aalborg, and Esk-727 and Esk-730 at Skrydstrup operate the General Dynamics F-16 Fighting Falcon, with a total of 44 F-16As and 10 F-16Bs delivered from the SABCA production line. Esk-727 and Esk-730 had previously operated the North American F-100 Super Sabre but all surviving aircraft were withdrawn in 1982-83. Esk-726 at Aalborg is now the only Lockheed/Canadair Starfighter operator following the conversion of Esk-723 to the F-16. The RDAF acquired a mixed bag of 51 Starfighters, but following the withdrawal of the Canadair-built aircraft in early 1984 Esk-726 has in service 18 F-104Gs and three TF-104Gs. These are scheduled to be withdrawn in 1986 when 12 F-16C/D Fighting Falcons will start to

replace them. The final two combat units are Esk-725 and Esk-729, both based at Karup. Esk-725 operates the Saab A35XD Draken in the interceptor/strike role and Esk-729 the S35XD Draken in the tactical reconnaissance role. Both units have a few SK35XD Draken two-seat trainers. A total of 51 Drakens were delivered during 1968-76, and of these 16 A35XDs, 18 S35Xs and 9 SX35XDs remain in service.

Flyvematerial Kommando is responsible for the transport and helicopter squadrons. Esk-721 and Esk-722 are based at Vaerlose and the former operates three Lockheed C-130H Hercules, three Grumman Gulfstream IIIs and three Saab MFI-17 Supporters (RDAF designation T-17). The RDAF received 23 T-17s and, in addition to the three with Esk-722, the Base Flights at Aalborg, Skrydstrup and Karup have seven between them. The remaining 13 are with the Flyveskolen at Avno. Esk-722 also operates eight Sikorsky S-61As which are used for SAR duties and detachments are operated at Aalborg, Skrydstrup and in Greenland. A Gulfstream III is also permanently detached to Narsarssuaq in Greenland. The Flyveskolen carries out initial pilot selection and training on the T-17 Supporter. Successful pilots then go to Vance AFB in the United States for basic training on the Cessna A-37B, followed by advanced training at Williams AFB on the Northrop T-38A Talon. Weapons training is also carried out at Williams AFB.

Haerens Flyvetjaeneste

The HF's aircraft are based at Vandel, but would deploy around the country in support of army units in times of conflict. The main type in the inventory is the Hughes H.500M, of which 14 are in use. The helicopters carry TOW anti-tank missiles, but are also deployed in observation roles. Following initial training in T-17s, at Avno, continuation training in the same type follows at Vandel. Basic helicopter training is carried out at Fort Rucker in the United States, with advanced training following on the Hughes H.500M at Vandel.

The Saab Draken soldiers on with the Royal Danish Air Force in both the interceptor (illustrated) and reconnaissance roles.

Sovaernets Flyvetjaeneste

Eight Westland Lynx HAS.80s delivered in 1980-81 have replaced the Aérospatiale Alouette IIIs which had all been withdrawn by 1982. The Lynx are deployed aboard five 'Hvidbjornen' class fishery protection frigates. Initial naval pilot training is undertaken at Avno with the Flyveskolen, and continued on the Hughes TH-55 and Bell UH-1 at Fort Rucker in the United States.

Eire Irish Air Corps

The Headquarters of the IAC, HQ Air Corps Group, is co-located with HQ Irish Army at Parkgate. There are five direct reporting wings within the IAC, of which two are administrative and engineering wings and have no active aircraft.

No. 1 Support Wing at Baldonnel/Casement is the main aircraft operating unit with four squadrons. They comprise the Helicopter Squadron (consisting of a SAR flight and a support flight) which operates seven Aérospatiale SA.316B Alouette IIIs; the Light Strike Squadron with six Aérospatiale CM-170-2

Super Magisters which can carry machine-guns, rocket pods and other underwing stores; the Transport and Training Squadron operating a BAe 125-700B and a Beech Super King Air 200; and the Maritime Squadron with two Beech Super King Air 200s. No. 2 Support Wing at Gormanston operates the Reims-Cessna FR.172; eight FR.172Hs were delivered in 1972 and one FR.172P in 1982. They all have underwing hardpoints capable of taking rocket pods for ground attack roles, in addition to their more usual border surveillance and liaison tasks.

The Training Wing at Baldonnel/Casement has four component units, two of which, the General Training Depot and the Apprentice School, have no active aircraft; the latter has a collection of withdrawn IAC airframes plus an ex-AdlA Magister and an ex-Zaïre AF SF.260MC for training. The basic flying training

school operates eight SIAI SF.260-WE trainers which can carry machine-gun pods for weapons training and light strike. The advanced flying school borrows SF.260WEs and Magisters from other units when required, but also has two Aérospatiale SA.342L

Eire's only jet combat aircraft are six Aérospatiale Magisters, based at Baldonnel for light strike duties.

Gazelles. In prospect for the IAC are five Aérospatiale SA.365F Dauphin IIs, ordered in 1983 for delivery in 1985.

Finland Suomen Ilmavoimat

Finland has ties with both East and West, and consequently has equipment supplied from both sides of the Iron Curtain. Mil helicopters have figured in the Finnish inventory for some years, including the Mi-4 'Hound' shown here. These are thought to be no longer in service.

Under the 1947 Neutrality Treaty with the UK and the USSR, Finland is limited to sixty combat aircraft and no bombers.

Havittajalentolaivue 11 (HavLLv 11 – 11th fighter squadron) at Rovaniemi inside the Arctic Circle is part

of the Lapin Lennosto (Lapland Wing) and operates the Saab Draken and a flight of BAe Hawk T.51s. The Finnish Draken order consisted of six refurbished J35BS, 12 new J35XSs assembled in Finland, six ex-Swedish J35Fs and three SK35C trainers, all delivered

between 1972 and 1976. These were followed in 1984 by two more SK35C trainers and 18 J35F interceptors in 1985. 50 BAe Hawk T.51s have been delivered during 1980-84, all except the first four being assembled locally by Valmet. HavLLv 21 is part of the Satakunnan Lennosto (Satakunta Wing), based at Tampere/Pirkkala, and fulfils the advanced and weapons training role with the Hawk T.51 and Fouga Magister. The unit received 80 Magisters, 18 built in France and 62 licence-built by Valmet, but only a few now remain in service; all should have been withdrawn by the end of 1985. Some of the Saab J35F Drakens due for delivery in 1985 are going to be attached to HavLLv 21. HavLLv 31

at Kuopio Rissala is part of the Karjalan Lennosto (Karelian Wing) and operates about 15 Mikoyan MiG-21Fs and a flight of Hawk T-51s. Finland requested equipment with MiG-23s but was only allowed MiG-21bis all-weather interceptors, of which 30 were delivered during 1978-80; for operational conversion two MiG-21UTIs, two MiG-21USs and two MiG-21UMs are in use.

Research and development is carried out by the Tiedustelulento-läivue (TiedLLv) at Jyväskylä/Tikkakoski. The TiedLLv operate four MiG-21Fs in the tactical reconnaissance role and also a Piper PA.31 Navajo, a PA.28R Arrow, a few CM.170 Magisters and Hawk T.51s. The transport squadron is the

Kuljetuslentoläivue (KuljLLv) based at Utti with a detachment at Malmi and the associated helicopter flight (Helikopterilentue) also at Utti. With the KuljLLv are six PA.31 Navajo Chieftains, six PA.28R Arrows (though most of these are loaned to other units for liaison use), four PA.28-201RT Arrows, one Fokker Friendship F.27-400 and two F.27-100s, and three Gates Learjet 35As. Helicopters in service include about 11 Mil Mi-8s (though some five of these are operated by the Border Police) and three Hughes 500Cs. Arrival of the Learjet 35As for use in the target towing, survey and limited ECM role should have resulted in withdrawal of the two ageing Ilyushin

Il-28s. Training is carried out by the Ilmasotakoulu (IlmaSK – air academy) at Kauhava. Primary training is completed on the Valmet L-70 Vinka, of which 30 have been delivered; this has resulted in the surviving Saab 91 Safirs being withdrawn and sold on the civil market. The IlmaSK is also receiving Hawk T.51s to replace its CM.170 Magisters. The final unit that operates aircraft is the Koelentue (flight test unit) at Kuorevesi.

In prospect for the Ilmavoimat are more MiG-21bis interceptors, which were reportedly ordered in early 1985; their receipt should result in withdrawal of the MiG-21Fs and early Drakens.

France L'Armée de l'Air

The AdlA is divided into seven major commands which are as follows:

Commandement des Forces Aériennes Stratégiques (CoFAS) consists of four squadrons, each with four Dassault Mirage IVAs allocated to the 91 and 94 Escadres de Bombardement in the nuclear strike role, and 11 Boeing C-135F tankers in three squadrons. The Mirage IVAs are deployed as follows: EB 1/91 'Gascogne' at Mont-de-Marsan, EB 2/91 'Bretagne' at Cazaux, EB 1/94 'Guyenne' at Avord and EB 2/94 'Marne' at St-Dizier. In addition to the Mirage IVAs above, four are with the Centre d'Instruction de FAS 328 (CIFAS 328 'Acquitaine') at Bordeaux. Another 18 Mirage IVAs are in the process of being converted into Mirage IVPs with improved avionics and the ability to carry the new Aérospatiale ASMP tactical nuclear weapon. The C-135F tankers are deployed as follows: ERV 1/93 'Aunis' at Istres, ERV 2/93 'Sologne' at Avord and ERV 3/93 'Landes' at Mont-de-Marsan. Finally, as mentioned above, CIFAS 328 carries out the strategic training role with four Mirage IVAs, about 12 Mirage IIIBs, about 10 Nord N.2501 Noratlas and a few Dassault-Breguet/Dornier Alpha Jets.

Force Aérienne Tactique (FATAC) is responsible for the tactical nuclear, reconnaissance and ground attack roles, and consists of six wings. Escadre de Chasse has as its component squadrons EC 1/3 'Navarre' and EC 2/3 'Champagne' with the Mirage IIIE and EC 3/3 'Ardennes' with the SEPECAT Jaguar A/E. This wing, based at Nancy, is assigned to the defence suppression role with Martel and other weapons being used. EC 4 at Luxeuil has two component squadrons, EC 1/4 'Dauphine' and EC 2/4 'La Fayette', both flying the Mirage IIIE. EC 7 has three squadrons based at Saint Dizier, these being EC 1/7 'Provence', EC 2/7 'Argonne' (the Jaguar operational conversion unit) and EC 3/7 'Languedoc', plus a fourth squadron, EC 4/7 'Limousin', at Istres; all operate the Jaguar A/E. EC 11 has three squadrons based at Toul: EC 1/11 'Roussillon', EC 2/11

'Vosges' and EC 3/11 'Corse', plus a fourth squadron, EC 4/11 'Jura' at Bordeaux; again, all operate the Jaguar. EC 13 at Colmar has three component squadrons, EC 1/13 'Artois' with Mirage IIIEs, along with EC 2/13 'Alpes' and EC 3/13 'Auvergne' both with Mirage 5Fs. Finally, ER 33 at Strasbourg is the tactical reconnaissance wing with three squadrons. ER 1/33 'Belfort' still operates the Mirage IIIR but is shortly to receive the Mirage F.1CR, ER 2/33 'Savoie' now fully equipped with the Mirage F.1CR and ER 3/33 'Moselle' with the Mirage IIIRD. Each wing also has an SLVSV (Section de Liaison et de Vol Sans Visibilité) operating a few CM-170 Magisters and MH.1521 Broussards. The AdlA received 160 Jaguar As and 40 Jaguar Es, but at least 20 and possibly more have been lost in accidents. About 180 Mirage IIIEs were delivered but only about half that number remain in use. The Mirage IIIRs and Mirage IIIRDs, of which respectively 50 and 20 were delivered, were scheduled to be replaced by Dassault Mirage F.1CRs by the end of 1986. 43 Mirage F.1CRs are on order with well over half of this total already delivered, although re-equipment of ER 3/33 is now in some doubt following reduction of the Mirage F.1CR order from 62 to 43. Of the 50 Mirage 5Fs acquired, eight

were sold to Chile in 1983. The latter are being replaced by eight new aircraft, four of which had been noted by the end of 1984. Long term re-equipment plans included the delivery of 85 Mirage 2000Ns (nuclear attack version), of which two prototypes were flying in 1984, and transfer of ex-CAFDA Mirage F.1Cs following that command's re-equipment with the Mirage 2000C.

Commandement Air des Forces de la Défense Aérienne (CAFDA) has five component wings. First is EC 2 with three squadrons based at Dijon. EC 1/2 'Cigognes' converted to the Mirage 2000C during 1984 and will be followed by the Mirage IIIE equipped EC 3/2 'Alsace' during 1985-86. ECT 2/2 'Cote d'Or', an Escadre de Chasse de Transformation, has a mixture of Mirage IIIB/BE/Rs but will eventually re-equip with Mirage 2000B trainers. EC 5 at Orange has three component squadrons, EC 1/5 'Vendée' and EC 2/5 'Ile de France' with the Mirage F.1C interceptor and ECT 3/5 'Comtat Venaissin' with the Mirage F.1B trainer. EC 10 at Creil has three component squadrons but the base is due to go into a reserve status in 1985. EC 1/10 'Valois' with Mirage F.1Cs is due to transfer to Reims and become ECTT 1/30 'Loire'. EC 2/10 'Seine' with the Mirage IIIC will disband, but EC 3/10 'Vexin', also with

The Dassault-Breguet Mirage F.1 forms the bulk of France's air defence, but will be replaced by the Mirage 2000. These carry the Matra Magic missile.

the Mirage IIIC but permanently detached to the former French colony of Djibouti, may retain her aircraft for some time to come. EC 12 at Cambrai is also equipped with Mirage F.1Cs, and EC 1/12 'Cambresis', EC 2/12 'Picardie' and EC 3/12 'Cornouaille' all operate these aircraft. At Reims, ECTT 30 (the TT stands for *Tous Temps*/all-weather, which is a historic title as the whole of CAFDA now has all-weather capability) currently has two component Mirage F.1C units, ECTT 2/30 'Normandie-Niemen' and ECTT 3/30 'Lorraine'. Of the 20 Mirage F.1B trainers, 87 Mirage F.1Cs (less six converted to F.1C-200s) and 83 Mirage F.1C-200 interceptors delivered to CAFDA, at least 15 have been lost in accidents. It is planned that all its Mirage F.1s will be replaced by the Mirage 2000B/C. A total procurement of 158 is envisaged and deliveries to EC 1/2 began in 1984.

Commandement du Transport Aérien Militaire (CoTAM) provides transport support. Its VIP and long-range transport duties are the responsibility of Escadre de Transport 60 at Villacoublay. ET 1/60 operates six Dassault Falcon 20s

SEPECAT Jaguars carry the brunt of tactical strike work. Lacking the LRMTS of their UK cousins, the French Jaguars are nevertheless highly capable.

and one Falcon 50; two Aérospatiale SA.330 Pumas, one SA.365 Dauphin II and an SE.210 Caravelle. ET 1/60 is also known as GLAM (Groupe de Liaisons Aériennes Ministerielles). ET 3/60 'Esterel' operates two Douglas DC-8-55Fs and three DC-8-72CFs. ET 61 at Orleans has three component squadrons, ET 1/61 'Touraine', ET 2/61 'Franche-Comte' and ET 3/61 'Poitou' all operating the Transall C-160F. Of the 49 on inventory, two are used by the French postal service. ET 2/63 at Toulouse is the last operator of the N.2501 Noratlas in any quantity. Also at Toulouse is EI 1/63 (Escadron d'Instruction 1/63) with two de Havilland Canada DHC-6 Twin Otters and loaned C-160F Transalls. EI 1/63 is also known as CIET – Centre d'Instruction des Equipages de Transport. ET 64 at Evreux is in the process of retiring the N.2501 Noratlas and ET 1/64 'Béarn' and ET 2/64 'Anjou' have received nearly all of the 25 Transall NGs on order. ET 65 (also known as GAEL – Groupe Aérien d'Entrainement de Liaison) carries out the light transport OCU duties as well as providing aircraft for liaison. ET 1/65 'Vendome' operates about twelve Nord 262D Frégates along with a Falcon 20. ET 2/65 'Rambouillet' has about 20 MS.760 Paris and a handful of MH.1521 Broussards.

EH 67 (Escadre de Hélicoptères) contains five helicopter squadrons which are deployed at various bases. EH 1/67 'Pyrénées' is at Cazaux with Pumas, Alouette IIs and Alouette IIIs. EH 2/67 'Valmy' is at Metz with Alouette IIs and Alouette IIIs. EH 3/67 'Paris' at Villacoublay has Alouette IIs and Alouette IIIs. EH 4/67 'Durance' is at Apt with Pumas and Alouette IIs, and EH 5/67 'Alpilles' at Istres has Pumas, Alouette IIs and Alouette IIIs. CIEH 341 (Centre d'Instruction des Equipages d'Hélicoptères) carries out helicopter training on the Puma, Alouette II and Alouette III at Chambéry. The helicopter force comprises about 60 Alouette IIIs, 30 Alouette IIs and 36 SA.330 Pumas. Other communication and liaison squadrons are Escadrons de Transport et d'Entrainment 41, 43 and 44. ETE 41 'Verdun' at Metz operates the

MS.760 Paris, MH.1521 Broussard and 262D Frégate. ETE 43 'Medoc' at Bordeaux operates the Paris, Broussard and two EMBRAER EMB.121 Xingu. ETE 44 'Mistral' at Aix operates the Paris, Broussard and four EMB.121 Xingu. The final liaison unit is Escadron de Transport et de Sauvetage 1/44 (ETS 1/44) 'Solenzara' with two Pumas and two Alouette IIIs for SAR duties at Solenzara, Corsica.

The following overseas based squadrons (ETOM – Escadron de Transport Outre-Mer) are also attached to CoTAM: ETOM-50 at St. Denis on Réunion Island with two Alouette IIs and three C-160Fs. ETOM-52 at Noumea-Tontouta with two DHC-6 Twin Otters and two Pumas. ETOM-55 at Dakar, Senegal with a C-160F and two Alouette IIs. ETOM-58 at Point-à-Pitre with two C-160Fs, three Alouette IIs and four Pumas. ETOM-82 at Papeete-Faaa, Tahiti, with three Caravelles, one Twin Otter and three newly-delivered Aérospatiale AS.332B Super Pumas, and ETOM-88 in Djibouti with two Broussards, four Alouette IIs and a couple of N.2501 Noratlas. The final unit within CoTAM is GAM 56 (Groupe Aérien Mixte) at Evreux with two Twin Otters, two C-160Fs and two Pumas, which are all operated on behalf of the French Secret Service.

Commandement des Ecoles de l'Armée de l'Air (CEAA) is responsible for all AdlA pilot and navigator training. Initial training is carried out at Clermont-Ferrand/Aulnat with Groupement Ecole 313. EFIPN 2/313 (École de Formation Initiale du Personnel Navigant) operates about 20 Mudry CAP 10 and CAP 10Bs for basic grading. Also at Aulnat is EFM 1/313 (Escadron de Formation des Moniteurs) which trains flying instructors on Magisters. From Aulnat prospective officer pilots move to Groupement d'Instruction 312 at Salon de Provence, where more Mudry CAP 10Bs are based along with CM.170 Magisters for initial jet training. In a parallel scheme, NCO pilots receive training at Cognac with GE 315, again with a mixture of Mudry CAP 10, CAP 10B and CM.170 Magisters. From GI 312 and GE 315 future fast jet pilots are transferred to GE 314 at Tours which currently has 65 Alpha Jets, and successful pilots leave here to begin weapons train-

ing with EC 8, which has two component units, EC 1/8 'Saintonge' and EC 2/8 'Nice', both equipped with the Alpha Jet. From EC 8 pilots pass to active combat units for operational conversion. Within GI 312 is the AdlA aerobatic team, the 'Patrouille de France' with Alpha Jets. Future transport pilots go to GE 319 at Avord, where the majority of 25 EMB.121 Xingus are now in use. Deliveries to AdlA of a total 168 Alpha Jets is almost complete. About 300 CM.170 Magisters are in use and will remain active until the 1990s at least. Just entering service with GE 315 at Cognac is the Aérospatiale Epsilon primary/basic trainer, of which planned procurement totals 152, and up to 26 Aérospatiale AS.355N Ecureuils are also on order.

The final command is Commandement des Transmissions de l'Armée de l'Air (CTAA), which has Escadron Électronique 51 'Aubrac' at Evreux with a DC-8-55. EE 54 'Dunkerque' at Metz has a handful of ECM-equipped N.2501 Noratlas. Escadrille de Calibration 57 'Commercy', which carries out navaid checking, is based at Villacoublay with a single Falcon 20. Escadron de Convoyage 70 at Châteaudun ferries crews and equipment with Nord 262D Frégates, MH.1521 Broussards and the MS.760 Paris. CEAM – Centre d'Expériences Aériennes Militaires at Mont-de-Marsan, with detachments at Bretigny and Istres, carries out trials of new aircraft. CEV – Centre d'Essais en Vol at Bretigny is the AdlA research organization and operates a wide variety of research, test and communications aircraft. Detachments are at Istres and Cazaux, the test pilots' school being at the former. Several base gliding squadrons also exist with Jodel DR.140 tugs.

Aéronautique Navale

Aéronautique Navale, or Aéronavale as it is usually abbreviated, is an integral part of the Marine Nationale and responsible for anti-submarine, anti-shipping and air defence protection of French coastal shipping lanes and ships at sea. The Marine Nationale has two aircraft carriers capable of taking strike aircraft, the *Clemenceau* and *Foch*. In addition to the helicopter carrier *Jeanne d'Arc*, a further 20 destroyers are able to carry helicopters.

The first-line squadrons are all called Flotilles and carry an 'F' suffix to their squadron number. 4F at Lann-Bihoué and 6F at Nimes-Garons both operate the Breguet 1050 Alizé in the ASW role. 4F and 6F both have a complement of eight aircraft and are normally carrier-based. The total inventory of Alizés is now about 37 aircraft. 11F and 14F at Landivisiau and 17F at Hyères are the first-line Dassault Super Etendard units with 66 aircraft which are normally carrier-based. 12F at Landivisiau is the last LTV F-8E (FN) Crusader operator with some 20 aircraft. 16F, also at Landivisiau, is equipped with 10 Etendard IVPs and four Etendard IVMPs for tactical reconnaissance. 21F and 22F at Nimes-Garons and 23F and 24F at Lann-Bihoué operate the Breguet 1150 Atlantic for maritime reconnaissance, ASW and anti-shipping duties in the Mediterranean, and have about 30 aircraft in use. Two were converted into the Atlantic Nouvelle Génération prototypes and an eventual procure-

Now beginning to re-equip many units of the Armée de l'Air, the Mirage 2000 is available in both single- and two-seat versions.

ment of 42 of this type is envisaged. 31F at St-Mandrier, along with 34F and 35F at Lanvéoc-Poulmic, are the first-line Westland Lynx operators, with 24 Lynx HAS.2(FN) and 14 Lynx HAS.4(FN) in service. The type is frequently deployed aboard destroyers and the *Jeanne d'Arc*. The final two flotilles are 32F at Lanvéoc-Poulmic and 33F at St-Mandrier operating the SA.321 Super Frelon. Approximately 32 SA.321LG (ASW) and SA.321G (cargo) aircraft are in use.

The support and training units, or Escadrilles de Servitude as they are known, all carry the letter 'S' suffix to their squadron number. 2S at Lann-Bihoué and 3S at Hyères both operate four Nord 262A Frégates and four Piper PA.31 Navajos. In addition, 3S has three Dassault Falcon 10MERs for instrument and radar training. 9S at Noumea-Tontouta, in New Caledonia, and 12S at Papeete-Faaa,Tahiti, have five Falcon 20H Gardians for communications and SAR duties. 9S has two and 12S the other three Gardiens. 20S at Saint Raphaël is the helicopter test squadron and operates examples of the Super Frelon, Alouette II, Alouette III and Lynx. 22S at Lanvéoc-Poulmic and 23S at St-Mandrier operate the Alouette II and Alouette III in the training, liaison, SAR and carrier plane-guard roles. A few Alouette IIs and about 15 Alouette IIIs remain in service. 50S (known formerly as the Section d'Avions Légers), operates 10 examples of the Socata Rallye 100S for initial pilot selection training. 52S at Lann-Bihoué uses the EMB.121 Xingu in the multi-engine training role; of the 16 Xingu delivered nearly all are with 52S, which also has two Piper PA.31 Navajos. 55S at Aspretto, Corsica, carries out co-pilot training for the Breguet Atlantic using the Nord 262 Frégate, with about four aircraft in use. 56S at Nimes-Garons operates 12 Frégates for flight engineer and navigator training, and 57S at Landivisiau has three Falcon 10MERs and about 10 MS.760 Paris for continuation training. 59S carries out deck landing training at Hyères and has some 15 Fouga CM.175 Zéphyrs with arrester hooks, 12 Etendard IVMs and eight Alizés. Before joining 59S, pupils have been trained by the Armée de l'Air.

Other units include 11S at Le Bourget, which carries out communications duties with a Douglas

DC-6, an EMB.121 Xingu, a Nord 262 and a Navajo; the SIV (Section d'Instruction au Vol) at Saint Raphaël with at least six CAP 10Bs; the ERC (Escadrille de Réception et de Convoyage) at Cuers with a single Xingu; the SES (Section Expérimentation et Soutien) at Saint Raphaël with an N.2504 Noratlas, a Xingu, two Alizés loaned from 59S and six Rallye 100STs operated on behalf of the local naval college for air experience flights. Finally, DCAN (Direction des Constructions et Armes Navales) at Cuers operates two Robin HR.100/250s, a Broussard and two Agusta-Bell AB. 47G-2 helicopters. The Aéronavale has a requirement for 22 AS.350B Ecureuils to replace its Alouette IIs and IIIs.

Aviation Légère de l'Armée de Terre

The ALAT comes under the administrative control of Commandement ALAT (COMALAT) with operational units being assigned to three Commandements d'Aviation Légère de Corps d'Armée (COMALCA) which consist of COMALCA 1 at Metz controlling 1 RHC, 3 RHC and 11 GHL; COMALCA 2 at Baden-Oos controlling 2 RHC and 12 GHL; and COMALCA 3 at Camp des Loges, Germain-en-Laye, controlling 6 RHC, 13 GHL, 2 GHL and 3 GHL.

The major helicopter operating units are the RHC (Régiment d'Hélicoptères de Combat) which should consist of 1 and 2 EHL (Escadrille d'Hélicoptères Légers) each with 10 SA.341F Gazelles for liaison and forward observation; 3, 4 and 5 EHA (Escadrille d'Hélicoptères Antichar) each with 10 SA.341M/ SA.342M Gazelles with HOT for anti-tank duties; and 6 and 7 EHM

(Escadrille d'Hélicoptères de Manoeuvre) each with 10 SA.330 Pumas for assault and transport duties. 1 RHC at Camp La Horie, Phalsbourg and 2 RHC at Trier are both fully equipped as above, but 3 RHC at Etain-Rouves only has 1 EHL and 3 and 4 EHA are still equipped with Alouette IIIs. 4 RHC is the 'shadow' designation of the EAALAT (École d'Application d'ALAT) at Luc/Le Cannet which carries out instrument training and tactical flying with 12 Alouette IIs, 15 SA.341F Gazelles, a few SA.341M Gazelles and 24 Pumas. Eight of the Pumas are detached with the GMEA (Groupe de Manoeuvre d'Ecole d'Application) at Aix/Les Milles. 5 RHC is based at Pau in the south of France and has 1 and 2 EHL with SA.341F Gazelles; 3 EHA with Alouette IIIs and 4 and 5 EHM with SA.330 Pumas. Finally, 6 RHC at Compiègne only has 1 EHL with SA.341F Gazelles, 2 EHA with SA.342M Gazelles and 3 EHM with SA.330 Pumas.

The other major units are the GHLs (Groupes d'Hélicoptères Légers) which carry out liaison duties with between 10 and 30 helicopters on strength. 11 GHL at Nancy/Essey and 12 GHL at Trier both have a complement of 30 aircraft. Each unit has three EHLs, one equipped with 10 SA.341F Gazelles, one with 10 Alouette IIs and one with 10 Alouette IIIs. 13 GHL at Les Mureaux has 1 EHL with 10 Alouette IIs and 2 EHL with 10 SA.341F Gazelles. The final four GHLs are 2 GHL at Lille/Lesquin, 3 GHL at Rennes/St-Valery, 4 GHL at Bordeaux/St-Medard and 5 GHL at Lyons/Corbas, each with 1 EH (Escadrille d'Hélicoptères) with 10 Alouette IIs. In addition, 5 GHL has detachments at Gap Tallard with Alouette IIs and at Grenoble with Alouette IIs and Alouette IIIs. 11 to 13 GHLs are GHL de Corps d'Armée and 2 to 5 GHLs are GHL de Zone de Défense and are eventually planned to each receive 10 SA.341F Gazelles in addition to the Alouette IIs.

The training system starts at the ESALAT (Ecole de Spécialisation de l'ALAT) at Dax where at least 60 Alouette IIs and 12 SA.341F Gazelles are used for training. At the GALSTA (Groupement ALAT de la

Section Technique de l'Armée de Terre) at Valence, experienced co-pilots may qualify as captains on the SA.330 Puma. In use here are Alouette II, Alouette III, SA.341F Gazelles and SA.330 Pumas, but only a few of each type.

Smaller units include EALAT 1A (Escadrille ALAT de l'Armée) at Baden-Oos with SA.341F Gazelles; DALAT (Detachment ALAT Djibouti) with Alouette IIIs and SA.330 Pumas, five of each type; EEABC (Escadrille de l'Ecole de l'Armée Blindée et de la Cavalerie) at Samur with Alouette IIs; EEAI (Escadrille de l'Ecole d'Application de l'Infanterie) at Montpellier with Alouette IIs; EDCM (Escadrille de la Direction Centrale du Matériel) at Bourges with Gazelles, Pumas, Alouette IIs and Broussards; and associated EERGM (Escadrille de l'ERGM ALAT) at Montaubin with a few Gazelles and Broussards. Finally, each of the operational RHCs has a liaison flight. 1, 2, 3, 5 and 6 GSALAT (Groupement de Soutien de l'ALAT) are each equipped with a Broussard at least.

The ALAT received 166 SA.341F Gazelles of which 110 are in the process of being modified to SA.341M standard to carry the Euromissile HOT ATM. In addition, 158 SA.342M Gazelles are required but so far only about half that number are in service. A total of 130 SA.330 Pumas are in service, with final procurement currently set at 138. There are 70 SA.316B Alouette IIIs still in service, these being armed with the SS.11 anti-tank missile. The final helicopter in ALAT service is the Alouette II, of which 140 SE.3130 Alouette IIs and 50 SA.318C Alouette Astazous were delivered; of these some are now being replaced by SA.341F Gazelles. A few MH.1521M Broussards remain in use for communications duties, and these are the only fixed-wing aircraft operated by the ALAT since withdrawal of the Cessna 0-1 Bird Dogs.

Gendarmerie Nationale

The Gendarmerie has an air element which has close links with the ALAT, to the extent that were general mobilisation to occur the Gendarmerie Nationale would take-over the military police role. In use are six Cessna U-206s, 27 Alouette IIs, 10 Alouette IIIs, and seven AS.350B Ecureuils against orders for 12 and there is a planned requirement for 30 to replace the Alouette IIs.

The home-grown Dassault Breguet Super Etendard is the major fixed-wing aircraft of the Aéronavale, although the Breguet Alizé and Vought Crusader still serve in small numbers. The prime task is maritime strike, armed with Exocets.

The army has a sizeable force of Aérospatiale-built helicopters. The Gazelle is the main anti-armour helicopter, armed with HOT missiles. These will serve until the new Eurocopter HAC is available.

Greece Polemiki Aeroporia

Although part of NATO, the continued poor relations of Greece with Turkey and the United States, and the often stated view that Turkey not the Warsaw Pact is the main threat to security, means that her future membership of NATO is in some doubt.

The PA consists of three commands. Firstly, the Taktiki Aeroporiki Dynamis (tactical air force) with headquarters at Larissa has control of seven combat wings or *pterix*. 110 Pterix Mahis has three component *moira* (squadrons): 337 Moira Anagaitiseos (interceptor) operating the McDonnell Douglas F-4E Phantom, 347 Moira Dioseos with LTV TA/A-7H Corsairs, and 348 Moira Taktikis Anagnoriseos (tactical reconnaissance) with RF-4E Phantoms and Republic RF-84F Thunderflashes. 111 Pterix Mahis at Nea Ankhialos has 341 Moira 'Arrow' for day fighter duties with the Northrop F-5A/B and 349 Moira Dioseos Bombardismoy 'Phoenix' also with F-5A/Bs for ground attack. 113 Pterix Mahis at Thessaloniki-Mikra has 343 Moira Dioseos Bombardismoy for ground attack. 114 Pterix Mahis at Tanagra operates the Dassault Mirage F.1CG in the air superiority role. Both her component squadrons, 334 Moira Pandos Kairou (all weather) and 342 Moira fly the Mirage F.1CG. 115 Pterix Mahis at Souda has 340 Moira DB and 345 Moira DB operating the LTV TA/A-7H Corsair. 116 Pterix

Mahis is the Starfighter wing and both 335 Moira 'Tiger' and 336 Moira 'Olympus' fly TF/F-104G versions of the aircraft. Final combat wing is 117 Pterix Mahis at Andravida with 338 and 339 Moira. Aircraft available to the tactical air force are believed to total about 63 LTV A-7H/TA-7H Corsairs, 61 F-4E/RF-4E Phantom IIs, 50+ F-5A/B/RF-5As, a few RF-84F Thunderflashes, about 70 F-104G/TF-104G Starfighters and 34 Mirage F.1CGs.

The second command is Diokisi Aeroporikoy Ythikoy (air material command) which controls the transport and ASW squadrons by 112 Pterix Mahis at Elefsis. 353 Moira Nautikis Synergasias fulfils the ASW role with 12 Grumman HU-16B Albatrosses. 354 Moira Taktikon Metaforon carries out the medium-range transport work with about 20 Nord N.2501 Noratlas. Also with 354 Moira are about 12 C-47 Dakotas. 355 Moira operates six NAMC YS-11 transports and 11 Canadair CL-215 water-bomber amphibians. 356 Moira TM operates twelve Lockheed C-130H Hercules. Also in use with the DAY are at least nine and probably more Dornier Do 28D-2 Skyservants which have recently been sold to Greece from West German air force stocks and are operated by an unknown unit. Another unknown unit operates five Meridionali-Vertol CH-47C Chinooks.

Diokisi Aeroporikoy Ekpitheyseos (air training command) has its HQ at Dekelia (Tatoi). It includes 359 Moira Elikopteron, operating at least 16 Agusta-Bell AB.205As, three AB.206As, three AB.212s and a Grumman Gulfstream I. Pilot instruction begins at the Ethniki Aeroporia Acadymia (national air academy) at Dekilia with 360 Moira operating 21 Cessna T-41Ds. From there, successful students go to 120

Pterix at Kalamata and initially 361 Moira Ekpitheyseos with the Cessna T-37C, followed by 362 Moira Ekpitheyseos with Rockwell T-2E Buckeyes for advanced training, then 363 Moira Ekpitheyseos with T-2Es for weapons training. DAE has about 25 T-37Cs and some 35 T-2Es. Also with DAE is the Moira Aeroporkis Exipiretisis Dimosion Ypiresion, which undertakes crop spraying on behalf of the Ministry of Agriculture with a few Bell 47Gs, Grumman G-164 AgCats and PZL M-18 Dromaders.

Greece still uses the Lockheed T-33A in fairly large numbers to equip communications and/or target towing flights. Procurement plans cover replacement of the Northrop F-5 and Lockheed Starfighter by 40 General Dynamics F-16 Fighting Falcons and a similar number of Dassault Mirage 2000s.

Elliniki Pterix Naftica

The naval air wing was formed in 1976 with the receipt of four Aérospatiale SA.319B Alouette IIIs for shipboard ASW and SAR opera-

Over 60 Vought Corsair IIs are used for tactical strike duties, including several two-seat TA-7H aircraft. These two-seaters retain full combat capability, the only penalty being extra weight.

tions, and from 1979-84 11 AB.212 ASWs were acquired. One destroyer can operate the Alouette III and two frigates can each operate two AB.212s.

Elliniki Aeroporia Stratou

The Greek army air wing, which has its main base at Megara, has expanded dramatically since 1975 when it was equipped initially with 15 AB.205As. Subsequent helicopter procurements have included 35 AB.205As and 49 Bell UH-1Hs, most of which remain in service. Transport capability is provided by five Meridionali-Vertol CH-47C Chinooks. For VIP use a Beech C-12A, Bell 212 and a Bell 206 Jet-Ranger are in service. Five Bell 47Gs are thought to remain in use for training. Some 25 Cessna U-17As and three Rockwell Commanders carry out communications and liaison duties.

Italy Aeronautica Militare Italiana

Italy is part of NATO and commits her combat aircraft to 5 ATAF in support of NATO's southern flank. The AMI has one main command, Comando Nationale della Difesa Aerea, which also incorporates the old transport, training and rescue commands. The only other command is Comando Logistico, responsible for maintenance and logistics. AMI control is exercised through two regional operation centres, 1 ROC at Montevanda, near Milan, and 3 ROC at Martina Franca, near Taranto. All flying units are attached to one of the two regional operation centres. The basic AMI unit is the gruppo (squadron) which usually has a parent stormo (wing).

The air defence and strike role is fulfilled by the Aeritalia/Lockheed F-104S Starfighter which equips the following units: 4 Stormo Amendo d'Aosta/9 Gruppo Caccia Intercettori at Grossetto under 1 ROC. 5 Stormo Giuseppe Cenni/23 Gruppo Caccia Intercettori Ognitempo and 102 Gruppo Caccia Bombardieri at Rimini Miramare under 1 ROC. 9

Stormo Francesco Baracca/10 Gruppo CI at Capua Grazzanise under 3 ROC. 36 Stormo Helmut Seidl/12 Gruppo at Gioia del Colle under 3 ROC. 51 Stormo Ferruccio Serafini with 22 Gruppo CIO and 155 Gruppo CB at Treviso Istrana under 1 ROC. 53 Stormo Gugliemo Chiarini with 21 Gruppo CI at Novara Cameri under 1 ROC. A newly formed unit is 37 Stormo/18 Gruppo CIO at Trapani Birgi which became operational late in 1984. 4, 9 and 53 Stormo have a sole interception role hence the gruppo designation Caccia Intercettori (fighter interceptor), whereas 5, 36 and 51 Stormo each have two gruppi, one for interception and the other a Caccia Bombardieri (fighter bomber) for the strike role. The AMI has a force of about 165 F-104S Starfighters for operational use. Each of the above wings has a liaison unit, namely 604, 605, 609, 636 and 653 Squadriglie Collegamenti which operate a few Aermacchi MB.326 and SIAI-Marchetti S.208Ms. 653 Squadriglia Collegamenti is also responsible for helicopter base rescue at Cameri, equipped with a few Agusta-Bell AB.204Bs.

The AMI is also an operator of the Fiat-assembled F-104G and TF-104G Starfighter, with a combined total of about 70 in service. Remaining TF/F-104G Starfighter units are 3 Stormo Carlo Emanuele Buscaglia with 28 and 132 Gruppo Caccia Ricognizione at Verona Villafranca

Italy is one of the last major users of the Starfighter, most of which are the locally-built Aeritalia F-104S. This aircraft is a two-seat TF-104G which serves with 20° Gruppo, the OCU.

under 1 ROC and 20 Gruppo Addestramento Operatino Autonomo. 20 Gruppo has no parent stormo and operates as the operational conversion unit with all the surviving TF-104G Starfighters and a few single-seat F-104Gs. The F-104G and partially the F-104S are being replaced in the attack role by 100 Panavia Tornado strike aircraft and 12 dual control trainers. Over half have now been delivered and equip 6 Stormo Alfredo Fusca/154 Gruppo CB at Brescia Ghedi under 1 ROC and 36 Stormo Helmut Seidl/156 Gruppo CB at Gioia del Colle under 3 ROC. 51 Stormo/155 Gruppo CB will also convert from the Aeritalia F-104S to the Tornado within the next two years. 3 Stormo and 6 Stormo both

have their respective Squadriglie Collegamenti 603 SC and 606 SC with a mix of MB.326s and S.208Ms.

For light strike and tactical reconnaissance duties, the Fiat G.91 still serves in considerable numbers and equips the following units: 2 Stormo Mario d'Agostini with 14 Gruppo Caccia Bombardieri/Ricognitori and 103 Gruppo CBR at Treviso San Angelo under 1 ROC control. 8 Stormo Gino Priolo with 101 Gruppo CBR at Cervia San Giorgio under 1 ROC. 32 Stormo Armando Boetto with 13 Gruppo CB at Brindisi under 3 ROC. About half of the 125 production G.91/A/R/PAN survive and are with 2 Stormo. The G.91PAN were aerobatic aircraft operated by the *Frece Tricolori*, until replaced by the Aermacchi MB.339, and then passed on to 2 Stormo. Most of 67 Aeritalia (Fiat) G.91Y twin-jet aircraft are still in service with 8 and 32 Stormo. Additionally, 2, 8 and 32 Stormo have their respective Squadriglie Collegamenti 602, 608 and 632 SC which operate MB.326s, S.208Ms and a few loaned Fiat G.91Ts. On order as a replacement for the Fiat G.91 and some of the strike Starfighters is the Aeritalia/EMBRAER AMX, of which 187 are required. The final air defence unit is 313 Gruppo Autonomo Addestramento Acrobatico or, as it is more familiarly known, the *Frecce Tricolori*. The national aerobatic team with its MB.339s also has an operational ground attack and anti-helicopter role in times of crisis.

Part of 1 ROC is 14 Stormo Radiomisure based at Pratica di Mare with two component gruppi. 71 Gruppo Guerra Elettronica is used to give ECM training and operates six Piaggio PD.808GE executive jets and one Aeritalia G.222VS (Versione Speciale). The second unit is 8 Gruppo Sorveglianza Elettronica with four PD.808GMs, three MB.339AS, a few EC-47 Dakotas and four G.222RMs for navaid calibration and monitoring of the transmissions made by 71 Gruppo GE aircraft. 14 Stormo is to expand when the remaining ten transport examples of the PD.808 are converted to PD.808GEs and PD.808RMs, five of each.

The main tactical transport unit is 46 Aerobrigata Trasporti Medi based at Pisa San Giusto under 1 ROC. The unit has three component gruppi: 2 Gruppo TM and 98 Gruppo TM flying the G.222 and 50 Gruppo TM operating 12 Lockheed C-130H Hercules. A total of 44 G.222s were ordered, of which six are G.222VS and four G.222RMS; deliveries should be completed in 1985. 646 Squadriglia Collegamenti flies loaned Piaggio P.166Ms for continuation training. The VIP transport unit is 31 Stormo Carmelo Raiti based at Roma Ciampino. It has two component gruppi, 93 Gruppo Elicotteri which is the helicopter unit operating two Agusta-Sikorsky AS-61A4s, an Agusta A.109A and a few AB.47Gs and AB.204Bs. 306 Gruppo Trasporti Speciali has two Douglas DC-9-32s and 10 PD.808s for VIP transport duties. Two Dassault Falcon 50s and a Grumman Gulfstream III are on order to release the PD.808s to 14 Stormo RM and the Douglas DC-9-32s to general transport duties.

Search and rescue duties are carried out by 15 Stormo Stefano Cagna which is headquartered at Roma Ciampino but maintains the following detachments: 82 Centro SAR at Trapani Birgi, 83 Centro SAR and 84 Centro SAR at Brindisi, and 85 Gruppo SAR at Ciampino, all operating the Agusta-Sikorsky HH-3F Pelican of which 20 were delivered in 1977-80. Shorter range SAR duties are carried out by AB.204Bs at Amendola, Cameri, Decimomannu, Grosseto, Istrana and Villafranca. The AB.204Bs are being replaced currently by 21 AB.212s. The AB.212 is also in use at the Reparto Sperimentale e di Standardizzazione del Tiro Aereo (air weapons training installation) at Decimo-

mannu where three more are used for SAR duties by 670 Squadriglia Soccorso e Collegamenti along with four MB.326s. Two final autonomous units are 303 Gruppo Volo Autonomo at Guidonia and 512 Squadriglia Collegamenti at Bari Palese operating P.166Ms, S.208Ms, MB.326s and loaned G.92s.

Training command (Comando Generale delle Scuole) has several flying training schools under its command. Initially students go to the SVBAE (Scuolo Volo Basico e Avancato ad Elica) at Latina which has 33 SIAI-Marchetti SF.260AMs. Also in use for twin-engine conversion of naval pilots are 12 P.166Ms. The SF.260AMs are operated by 207 Gruppo/Scuola di Volo di Primo Periodo and the P.166Ms by Scuola Plurimotori. From the SVBAE successful students progress to the SVBIA (Scuolo Volo Basico Iniziale Aviogetti) at Lecce-Galatina with 212, 213 and 214 Gruppi operating the MB.339A. Delivery of 81 MB.339As is now nearly complete. Of the total 15 are with the *Frecce Tricolori* and three with 14 Stormo/8 Gruppo SE. The SBVIA is thought to still operate a few MB.326s, the remainder having been passed to the various squadriglie collegamenti. Final stage of training is the SVBAA (Scuola Volo Basico Avanzato Aviogetti) at Foggia Amendola with 201 and 204 Gruppi operating the G.91T, and about 80 remain in service. The successor to the G.91T has yet to be decided, though a two-seat Aeritalia/EMBRAER AMX is a possible contender. Helicopter pilots are trained for all three services at the SVE (Scuola Volo Elicotteri) at Frosinone using AB.47Gs and AB.204Bs. The Starfighter OCU is 20 Gruppo, but Tornado pilots go to the Tri-national Tornado Training Establishment at Cottesmore, UK, where seven AMI Tornados are currently based. The gliding school for military personnel is the CVV (Centro Volo a Vela) at Guidonia with a detachment at Riete, operating Calif A.21, Twin Astir, Blanik and Canguro gliders with about 10 S.208M tugs. Final unit of the AMI is the RSV (Reparto Sperimentale Volo) at Practica di Mare, the AMI test and development unit which operates a variety of in service types.

Another paramilitary organizations is the Guardia di Finanza which use a variety of helicopters

with military serials for frontier patrol, anti-smuggling operations, fishery protection, customs surveillance and protection of historical sites. In use are about 50 AB.47G/Js and 70 BredaNardi (Hughes) NH-500, operating from about 15 bases throughout Italy. Secondly, the Corpo delle Guardie Pubblica Sicurezza, which was demilitarized in 1981 and became the Polizia di Stato, operates about 13 AB.206A-1s, three Partenavia P.64B-200 Oscars, nine A.109A Hirundos, two Bell 212s and four AB.212s.

Aviazione per la Marina Militare Italiana

The Marinavia is a helicopter equipped force of five ASW squadrons supported by AMI operated fixed-wing aircraft. The helicopters deploy aboard the helicopter carrier *Vittorio Veneto* and two cruisers, four destroyers and nine frigates. In addition, another helicopter carrier, the *Garibaldi*, is nearing completion. This vessel also incorporates a ski-jump for STOVL operations, despite the fact that the AMI or Marinavia have no aircraft in service or as yet on order that can make use of the ramp. Two more destroyers and seven frigates are also under construction. Helicopter units are under the control of 6 Reparto Elicotteri MM with its HQ in Rome. The three major helicopter bases at Luni-La Spezia, Taranto-Grottaglie and Catania-Fontanarossa are known as MARISTAELI (Stazione Elicotteri della Marina Militare). At MARISTAELI Luni are based 1 Gruppo Elicotteri (1 GRUPELICOT) operating the Agusta-Sikorsky SH-3D and 5 GRUPELICOT with the AB.212ASW. Based at MARISTAELI Catania is 2 GRUPELICOT with AB.204AS and 3 GRUPELICOT with SH-3Ds, and at MARISTAELI Grottaglie is 4 GRUPELICOT with AB.212ASW. Marinavia's helicopter force comprises about 25 AB.204As used in training and utility roles, 53 AB.212ASWs and 33 SH-3D Sea Kings. A handful of AB.47Js may still remain in use for communications and liaison.

The fixed-wing force of Breguet 1150 Atlantics is under the control of the AMIs Inspettorato per la Marina Militare and comprises 18 aircraft operated by 30 Stormo Valerio Scarabellotto/86 Gruppo

Italy uses the home-produced Agusta-Sikorsky S.61 for search and rescue duties. These are based at Rome but operate detachments around the coast.

Antisomergibili at Gagliari Elmas and 41 Stormo Athos Ammannato/88 Gruppo AS at Catania Sigonella.

Aviazione Leggera del l'Esercito Italiano

The ALE is a predominantly helicopter equipped force acting in support of the Italian Army. Major units in the ALE are the four wings or Raggrupamente ALE (RALE). Each have component squadrons known as Gruppi Squadroni (GS) which are also designated by role, as GSALE indicating GS Aerei Leggeri ed Elicotteri (light aeroplane and helicopter); GSETM, GS Elicotteri Trasporto Medio (medium transport helicopter); GSEM, GS Elicotteri Multiruolo (multi-role helicopter); and GSERI, GS Elicotteri da Ricognizione (reconnaissance helicopter).

The first wing, 1 RALE 'Antares', is under Stato Maggiore Esercito (Army high command) direct control and has all its units based at Viterbo. The component units are 11 GSETM with 111 and 112 Squadroni operating the Meridionali-Vertol CH-47C Chinook, 12 GSETM with 121 and 122 Squadroni operating the CH-47C Chinook, 51 GSEM with 511 and 513 Squadroni operating the AB.204B, 512 Squadrone with the AB.205A and 514 Squadrone the AB.212. 3 RALE 'Aldebaran' is attached to III Corpo d'Armata and has as subordinate units 23 GSALE with 231 Squadrone operating the SIAI-Marchetti SM.1019 and 423 Squadrone operating the AB.206 at Milano-Bresso, and 53 GSEM at Padova with 521 and 522 Squadroni operating the AB.205A. 4 RALE 'Altar', which is attached to IV Corpo d'Armata, has the 24 GSALE with 241 Squadrone operating SM.1019ES and 440 Squadrone with the AB.206 at Bolzano. 44 GSERI at Venaria with 441 and 442 Squadroni also has the AB.206. 54 GSEM with 541, 542, 543 and 544 Squadroni at Bolzano and 545 Squadrone at Aosta all operate the AB.205A. 5 RALE 'Rigel' is attached to V Corps d'Armata and has 25 GSALE with 251 Squadrone operating the SM.1019E and 425 Squadrone operating the AB.206A at Vittorio Veneto and 55 GSEM with 551, 552, 553 and 554 Squadroni at Casarsa della Delizia under its control. The four independent gruppi squadroni attached to divisioni (46, 47, 48 and 49 GSERI) will be disbanded during 1985 and their AB.206As distributed amongst 1, 3, 4 and 5 RALE.

Other independent gruppi squadroni are as follows: 20 GSALE consisting of 201 Squadrone with SM.1019Es and 42 Squadrone with AB.206A at Salerno-Pontecagnano attached to 10 Commiliter. 21 GSALE at Cagliari with 211 Squadrone operating SM.1019Es, 421 Squadrone operating AB.206As and 521 Squadrone operating AB.205As attached to Commiliter Sardegna.

27 GSALE at Firenze attached to 7 Commiliter with 271 Squadrone operating SM.1019Es and 427 Squadrone operating AB.206As. 28 GSALE at Rome-Urbe attached to 8 Commiliter with 281 Squadrone operating SM.1019Es and 428 Squadrone operating AB.206As. 30 GSALE at Catania attached to 11 Commiliter with 301 Squadrone operating the SM.1019Es and 398 Squadrone operating the AB.206A. 26 GSALE at Pisa attached to Brigata Paracadutisti 'Folgore' with 426 Squadrone operating the AB.206A and 526 Squadrone operating the AB.205A. Finally, 13 GRACO (13 Gruppo Acquisizione Obbiettivi – range squadron) at Verona-Boscomantico attached to 3za Briga-

ta Missili 'Aquileia' with 398 Squadrone operating the SM.1019E and 598 Squadrone operating the AB.206A.

There are four ALE maintenance units known as RRALE – Reparto Riparazioni ALE. 1 RRALE is at Bracciano, 2 RRALE is at Bologna, 3 RRALE is at Orio al Serio and 4 RRALE is at Viterbo. The final ALE unit is the CALE (Centro ALE) at Viterbo under direct high command control with various types on charge for use by its four component units, namely the Sezione Addestramento Volo Aerei (fixed wing training unit), SAV Elicotteri (helicopter training unit), Reparto Mezzi Aerei (tactical training unit) and SA Impiego Operatino (operational trials

unit).

Aircraft operated by the ALE's units include five A.109As, a few AB.47Gs and AB.204s, 90 AB.205s, 150 AB.206s, 18 AB.212s, a few Bell OH-13Hs, about 30 Cessna 0-1E Bird Dogs, 27 CH-47C Chinooks and some 80 SM.1019Es. On order for the ALE are 71 Agusta A.129 Mangusta anti-tank helicopters.

A para-military force with close Army connections is the Carabinieri, which carries out anti-terrorist operations in the north, anti-bandit operations on Sardinia and anti-Mafia operations in the south and on Sicily. Nine AB.205As, 43 AB.206A/Bs, 12 A.109As and a few AB.47Js are in use.

Malta Armed Forces of Malta, Helicopter Flight

The helicopter flight received three Agusta-Bell 47G-2s and a Bell 47G-2 as a gift from West Germany in 1971. A gift from Libya in 1973 of an Agusta-Bell 206B JetRanger resulted in a strengthening of ties be-

tween the two countries and Malta later received from Libya three Aérospatiale SA.316B Alouette IIIs, and two Libyan Arab Air Force Aérospatiale SA.321M Super Frelons were loaned to the Helicopter

Flight. Following a disagreement between the two countries in 1980 the Super Frelons were returned to Libya, but the Alouette IIIs have remained in storage at Luqa since then. In 1983 the Italians loaned the

Helicopter Flight two Agusta-Bell 204Bs for SAR duties and this arrangement continues to date; this may change, however, with recent renewed friendship between Malta and Libya.

NATO/Luxembourg NATO Airborne Early Warning Force/NAEWF

The NAEWF was formed in 1982 to operate an eventual total of 18 Boeing E-3A Sentry AWACS aircraft. The first NATO E-3A was delivered to its base at Geilenkirchen

in West Germany in early 1982 and the final aircraft will arrive in 1985. Three squadrons, each with six aircraft, will be formed and up to six aircraft will be assigned to forward

operating locations/bases at Orland (Norway), Konya (Turkey), Previza (Greece) and Trapani (Italy). All NATO member countries contribute to the programme, but as NATO

does not have national status the aircraft carry Luxembourg's lion insignia and registration prefix (LX-), followed by US registration prefix (N), then their USAF serial number.

Netherlands Koninklijke Luchtmacht

Like Belgium, the Royal Netherlands Air Force forms part of NATO's 2nd ATAF and is also re-equipping with the General Dynamics F-16 Fighting Falcon. The Dutch aircraft are being licence-built by Fokker and will eventually replace the KLu's Lockheed TF/F-104G Starfighters and Northrop NF-5A/Bs. A total of 201 F-16A/B Fighting Falcons have been ordered, of which about 112 had been delivered by the end of 1984. All are scheduled for delivery by 1992; of those which entered service, about 10 have been lost in accidents. The first F-16s began to equip the TCA (Tactical Conversion Unit) at Leeuwarden in 1979, with 322 Squadron and 323 Squadron, both also at Leeuwarden, becoming operational in 1981 and 1982 respectively. Conversion of the Volkel-based units is now nearly complete with 311 Squadron replacing its Star-fighters with Fighting Falcons in 1982 and becoming operational in 1983. 306 Squadron, also at Volkel, had received all its RF-16A reconnaissance aircraft by the summer of 1984 and was working towards operational status. The final Volkel based unit, 312 Squadron, relinquished its Starfighters in the middle of 1984 and has now started receiving Fighting Falcons. Some of the KLu Starfighters were sold to Greece (10) and Turkey (53) and the surviving aircraft are currently at

Ypenburg in storage awaiting disposal. The last two operators of the Starfighter were the CAV (Conversie Afdeling) and the UFO (Uit-faserings-Onderdeek) both at Volkel, the last KLu Starfighter flight taking place on 26 November 1984. Also under Dutch control is the 32nd TFS of the USAFE at Soesterberg, with a squadron of McDonnell Douglas F-15C/D Eagles for use in the air defence role.

In use for ground attack is the Canadair-Northrop NF-5A and NF-5B. 313 and 315 Squadrons at Twenthe, along with 314 Squadron at Eindhoven and 315 at Gilze Rijen, operate about 90 of these aircraft. All are due to be replaced by the General Dynamics F-16 Fighting Falcon, with 315 Squadron scheduled to begin conversion in late 1985. The KLu only has one transport unit, 334 Squadron based at Soesterberg with nine Fokker F.27-300M Troopships and three Fokker F.27-100 Friendships. Two other F.27s are in use, these being F.27-200MPA maritime surveillance aircraft operated by 336 Squadron based at Hato in the Netherlands Antilles. The Testgroep KLu at Twenthe carries out all the KLu research and development work and relinquished its single NF-5A for an F-16A in 1984. Finally, the SAR flight at Leeuwarden operates four Aérospatiale Alouette IIIs. Pilot training for the KLu was carried out in Canada for many years but the training programme has now been transferred to the USA.

Group Lichte Vliegtuigen/GpL V

Theoretically, the GpLV is part of the KLu but is under army control and flies exclusively for the army.
298 Squadron at Soesterberg and 300 Squadron at Deelen operate

some 63 SE.3160 Alouette IIIs. 300 Squadron also acts as an operational training unit for both GpLV and navy pilots. Finally, 299 Squadron, also at Deelen, operates 30 MBB BO 105Cs. Initial training for GpLV and navy aircrew is carried out at Fort Tucker in the USA.

Marine Luchtvaartdienst/MLD

As the air arm of the Dutch navy, the MLD provides maritime patrol aircraft and ship based helicopters primarily for ASW duties. The helicopters are also tasked with providing transport for Dutch marine commandos.

The MLD now has only one fixed-wing type in its inventory, the Lockheed P-3C Orions of 320 Squadron based at Valkenburg. All 13 Orions are in service, delivered between July 1982 and September 1984 to replace Lockheed SP-2H Neptunes. 321 Squadron, also at Valkenburg, operated the Breguet SP-13A Atlantic until the end of 1984. Its six aircraft are all in store at Valken-

burg awaiting a possible sale, but could re-enter MLD service if needed. The MLD has now standardized its helicopter fleet on the Westland Lynx. 7 Squadron at de Kooij has five UH-14A Lynx for SAR, utility and commando transport. 860 Squadron, also at de Kooij, has 17 SH-14B/C Lynx in service for use on 'van Speijk' and 'Kortenaer' class frigates for ASW duties. The Lynx replaced the ageing Westland Wasps which were sold to Indonesia. 2 Squadron at Valkenburg has no helicopters of its own, but borrows from other units to carry out its role as the Operational Conversion Unit. A long-term requirement for the MLD is a heavy ASW helicopter, of which eight are required from the late 1980s.

The Netherlands maintains a heavy maritime commitment, using 13 Lockheed P-3 Orions for ASW duties in the North Sea and Atlantic Ocean. One is permanently based on Iceland for Atlantic patrol.

Norway Kongelige Norske Luftforsvaret/KNL

Norway abandoned its traditional policy of neutrality after World War II and is now a member of NATO. Her air force is committed to NATO's Allied Forces Northern Europe (AF-North) in the case of a major conflict. The KNL is divided into two air commands – Luftkommando Nord-Norge and Luftkommando Sor-Norge.

Luftkommando Nord-Norge consists of Skvadron 330 (Skv 330) with 'A' Flight at Bodo and 'B' Flight at Banak both operating the Westland Sea King Mk.43, Skv 331 at Bodo with the General Dynamics F-16A/B Fighting Falcon, Skv 333 at Andova operating the Lockheed P-3B Orion and Skv 719, also at Bodo, has Bell UH-1Bs and the de Havilland Canada DHC-6 Twin Otter. Luftkommando Sor-Norge consists of Skv 330 'C' Flight (at Orland) and 'D' Flight (Stavanger-Sola) with the Sea King Mk.43, Skv 332 (Rygge) F-16A/B Fighting Falcon, Skv 334

(Bodo) F-16A/B Fighting Falcon, Skv 335 (Oslo-Gardermoan) Lockheed C-130H Hercules and Dassault Mystère Falcon 20, Skv 336 (Tygge) Northrop F-5A/B, Skv 337 (Bardufoss) Lynx Mk.86, Skv 338 (Orland) Northrop F-5A/B, Skv 339 (Bardufoss) and Skv 720 (Rygge) both have Bell UH-1Bs. Skv 336 has converted to the F-16A/B Fighting Falcon in late 1984.

Apart from initial pilot selection and grading by the primary flying school on the Saab-MFI 15 Safari at Trondheim-Vaernes, all pilot training is undertaken in the United States. The KNL provide the Luftvernartilleriet (field artillery observer service) on behalf of the army and operates Cessna O-1A Bird Dogs from Trondheim-Vaernes. The Skv 337 Lynx Mk.86s at Bardufoss are operated on behalf of the Kystvakt (coastguard) and carry their markings.

Along with Belgium, Denmark and the Netherlands, Norway is an operator of the F-16A/B Fighting Falcon and had 69 in service at the end of 1984. In addition, 24 F-16C/D Fighting Falcons are on order for delivery in 1990-91. The KNL still operates the Northrop F-5A/B/RF-5A and had 88 on inventory prior to the transfer of 11 F-5As to the Turkish air force in 1983. Skv 338 is the sole remaining F-5 operator, but Skv

718 may re-form in 1985 and use F-5 stocks to equip as an operational conversion unit. Other KNL equipment includes seven P-3B Orions for ASW and maritime patrol; six C-130H Hercules for heavy transport, with lighter duties carried out by three DHC-6 Twin Otters. For the VIP transport and ECM role there are three Mystère Falcon 20s. Helicopters in service consist of 10 Sea King Mk.43s for SAR duties, plus six Lynx Mk.86s delivered in 1981 for SAR, fishery protection and oil rig security duties. Final helicopter in use is the utility UH-1B of which 25 remain. There are about 23 Cessna O-1A Bird Dogs which the KNL

The F-16A/B carries the Norwegian air defence with the aid of a squadron of Northrop F-5s. F-16C/Ds are on order, but these will not be delivered until around 1990. The F-5s will soldier on for some years yet.

operate for the army, and some 16 SAAB-MFI 15-200A Safaris which replaced the Saab 91 Safir in the training role. The KNL also has in storage about 20 Canadair CF-104D/G/RF-104G Starfighters, but their fate is so far undecided.

Portugal Fôrça Aérea Portuguesa

Portugal is a member of NATO but current budgetary restrictions have resulted in few new arms purchases for the FAP.

The main strike element of the FAP is Esquadra de Ataque 302 at BA-5 (Base Aérea-5 or Air Base 5), Monte Real. 20 ex-United States Navy LTV A-7A Corsairs were refurbished as A-7Ps and delivered to the FAP from 1981 onwards, and one USN TA-7C trainer is also on loan to the FAP for conversion training. A second batch of 24 A-7Ps and six TA-7P trainers were ordered in 1983, and the first three were delivered in October 1984. The other attack capabilities of the FAP are provided by Esquadra de Ataque 301 at BA-6 Montijo and Esquadra de Ataque 303 at BA-4 Lajes. Both units operate ex-Luftwaffe Fiat G.91s. At least 73 G.91R-3/-4s have been received, plus some 35 further examples for spares recovery. For conversion to the type six Fiat G.91T-1s were followed by at least five more operational examples and 10 for spares recovery. It is believed that 40 G.91R-3/R-4s and eight G.91T-1s are in service. The FAP has had no real air defence capability since Esquadra de Caca 201 retired its North American F-86F Sabres in 1980. It has since received 12 Northrop T-38A Talon supersonic trainers, all of which are in service and based at BA-5, Monte Real. Limited air defence capability is provided by arming the A-7s and G-91Rs with AIM-9 Sidewinders.

Main transport capability is provided by Esquadra de Transport 501 at Montijo with five Lockheed C-

130H Hercules. Lighter transport duties are carried out with 23 CASA C-212s. Of these eight C-212As equip Esquadra de Transport 502 at BA-3, Tancos; six C-212As are with Esquadra de Transport 503 at BA-4, Lajes; five C-212As are with the operational conversion unit, Esquadra de Instruçao 111 at BA-3, Tancos; and the remaining four aircraft are C-212B survey aircraft which are with Esquadra de Reconnecimento at BA-1, Sintra. At least two of the C-212Bs are fitted with magnetic anomaly detectors, and two of the Esq.502 aircraft have been recently converted into ECM trainers. Esq.111 also operates about six Aérospatiale Alouette IIIs for training; Esquadra de Transport 552, also at BA-3, Tancos, has around 15 Alouette IIIs and the final Alouette III operator is Esquadra de Transport 551 at BA-6, Montijo, also with about 15. At least two Alouette II are operated by the Guardia Nacional (national guard), which

may receive up to six more. The other helicopter in FAP service is the Aérospatiale SA.330C Puma; about 12 are in use with Esquadra de Busca e Salvamento 751 at BA-6, Montijo, and Esquadra de Busca e Salvamento 752 at BA-4, Lajes, for search and rescue duties. Four of the Pumas are fitted with maritime surveillance radar and others are used in exercises with Portuguese army units.

Communications and liaison duties are carried out by Esquadra de Ligacao 701 at BA-2, Ota, and Esquadra de Ligacao 702 at AM-2 (Aerodromo de Manobra-2, or Transit Airfield 2), Aveiro, operating the Reims-Cessna FTB-337G; 16 have armament capability, eight are photo-reconnaissance versions and eight trainers, and the majority are thought to remain in service. Training duties are carried out by the Escola de Instruçao Elementar de Pilotagem at BA-2, Ota. Primary training is carried out at Ota by Es-

Ex-US Navy Vought A-7s have been supplied to Portugal to form the country's main strike element. These can be equipped with Sidewinders to give a limited air defence capability. Other strike assets are the ex-Luftwaffe Aeritalia G91s.

quadra de Instruçao 1010 on the DHC-1 Chipmunk, with about 40 remaining of 10 built by de Havilland Canada and 66 by OGMA. Basic training is carried out by Esquadra de Pilotagem 102 at BA-1, Sintra, which has about 24 Cessna T-37Cs; Esq.102 also provides pilots and aircraft for the national display team, the *Asas de Portugal*. Conversion from the T-37 to 'fast jets' is undertaken by Esquadra de Pilotagem 103 at BA-5, Monte Real, which operates about 18 Lockheed T-33As. As noted earlier, transport and helicopter pilots go to Esq.111 and final jet training is carried out by Esq.201 with the T-38A.

Spain Ejercito del Aire

Spain has recently joined NATO and now has one of the larger European air arms. The EdA is split into three groups, which are as follows:

The Fuerza Aérea consisting of Mando Aéreo de Combate (MACOM) which has three component wings: Ala de Caza 11 at Manises AFB with 111 and 112 Escuadrones flying the Dassault Mirage, and operating about 18 Mirage IIIEE interceptors and six Mirage IIIED two-seat trainers. Ala 11 will be the first unit to receive the McDonnell-Douglas F-18 Hornet, of which 60 single-seat F-18As and 12 two-seat F-18Bs are on order, and will move to Zaragoza to convert to the F-18. Ala de Caza 12 at Torrejon AFB has two component units, 121 and 122 Escuadrones, both flying the McDonnell Douglas F-4C Phantom II. About 34 F-4Cs are in use, and Ala 12 also has four RF-4C Phantoms which are operated at wing level and are not attached to either squadron. Ala de Caza 14 at Los Llanos AFB has two component units, 141 and 142 Escuadrones flying the Dassault Mirage F.1CE/F.1BE, and is currently equipped with about 42 F.1CE interceptors and five F.1BE trainers.

Mando Aéreo Tactico (MATAC) has two component wings: Ala de Ataque 21 at Moron AFB with 211 and 212 Escuadrones operating some 31 Northrop SF-5A/SRF-5As; in addition each squadron also has two SF-5Bs. Ala de Patrulla 22 at Jerez AFB has all six EdA Lockheed P-3A Orions in service with 221 Escuadron. It has been reported that some may be exchanged for aircraft with updated avionics and ASW equipment. Also with MATAC is 407 Escuadrilla operating a small number of Dornier Do 27s and the CASA licence-built equivalent, the C-127, in communications and liaison roles.

Mando Aéreo de Transporte (MATRA) has three component wings. Ala de Transporte 31 at Zaragoza AFB has Escuadrones 311 and 312 operating the Lockheed KC/C-130H Hercules, 311 Esc. having five C-130Hs and 312 Esc. one C-130H and five KC-130H tankers. Also attached to Ala 31 at wing level are a handful of CASA C-127s and Dornier Do 27s for liaison duties. Ala de Transporte 35 is at Getafe AFB with component units 351 and 352 Escuadrones operating the CASA C-212 Aviocar. The EdA has received a total of 82 C-212 Aviocars of several variants. From this number 351 and 352 Escuadrones each have 12 on strength. Also with Ala 35 is 353 Escuadron operating nine ex-USAF C-7B Caribou. Ala de Transporte 37 at Villanubla AFB has subordinate units 371 and 372 Escuadrones flying the de Havilland Canada DHC-4/C-7 Caribou. 371 Esc. has two DHC-4s and eight ex-USAF C-7Bs and 372 Esc. has ten DHC-4s on strength.

Mando Aéreo de Canarias (MACAN) has one component wing, Ala 46 with two subordinate escuadrones: 461 Escuadron which operates a few Aviocars and 462 Escuadron which uses 22 Dassault Mirage

F.1EEs in the interception role. Also at Gando with MACAN is 802 Escuadron, a SAR unit operating three Fokker F.27-200MPAs and two or three Aérospatiale AS.332B Super Pumas.

The Logistica Aérea consists of Mando de Personal (MAPER) and has several component units. Escuela Militar de Paracaidismo 'Mendez Parada' at Alcantrilla airfield consists of 721 Escuadron flying about eight Aviocars for parachute training. Escuela de Reactores at Talavera la Real AFB has 731 and 732 Escuadrones flying 22 SF-5Bs for operational jet conversion use. Escuela Militar de Transporte y Transito Aéreos at Matacan AFB has two training units, 744 and 745 Escuadrones, each with about six C-212 Aviocars. 744 Escuadron also has two Piper PA.31 Navajos on strength for liaison duties. Helicopter training for all three services is carried out by the Ala de Ensenanza 78 at Granada AFB with two component escuadrones. 782 Escuadron operates 17 Hughes H.269A-1s and possibly a few surviving Bell OH-13H/AB-47Gs. 783 Escuadron operates the six remaining EdA Agusta-Bell AB.205As and seven Bell UH-1Hs. The Academia del Aire at San Javier AFB has three component escuadrones. 791 with 25 Beech T-34A Mentors and about 12 Beech Bonanza F.33Cs, 792 Escuadron which has about six C-212 Aviocars, and 793 Escuadron operating 40 CASA C.101EB Aviojets. 41 Grupo at Zaragoza AFB also has two component escuadrones, 411 and 412, operating the C.101EB Aviojet. A total of 88 production Aviojets were received, from which one has been lost. Discounting the 40 with 793 Esc., the remainder are split between 411 and 412 Escuadrones. 42 Grupo at Getafe AFB has two component escuadrones: 421 Escuadron operating 14 Bonanza F.33Cs and 423 Escuadron with six Piper PA.23-25 Aztecs and six Beech 95 Baron B.55s. Also part of Logistica Aérea are the Mando de Material (MAMAT) and the Direccion de Infraestructura Aérea (DINFA), neither of which have flying units, but the MAMAT controls the following aircraft maintenance units. Ala de Logistica 51 at Cuatro Vientos, AdL 52 at Tablada and AdL 53 at Albacete.

The third group is the Cuartel

General del Ejercito del Aire (CGEA) which has its flying operations controlled by Agrupacion de Cuartel General del Ejercito del Aire and consists of the following units. 401 Escuadron at Madrid's International Airport, Barajas, has four Dassault Falcon 20s for navaid checking and VIP duties, two McDonnell Douglas DC-8-52s and a Falcon 50 for long-range VIP duties. 402 Escuadron at Cuatro Vientos military airfield has one SA.330C, two SA.330H and three SA.330J Pumas, and two AS.332B Super Pumas. 403 Escuadron, also at Cuatro Vientos, operates a few C-127s, Dornier Do 27s and about six C-212B Aviocars. 43 Grupo at Torrejon AFB has two component units, 431 and 432 Escuadrones both flying the Canadair CL-215 water bomber with a total of 14 in service. Both escuadrones also have a few C-127s and Dornier Do 27s. 406 Escuadron at Torrejon carries out test flying for the Instituto Nacional de Tecnica Aeronautica (INTA) with two prototype Aviocars and four prototype Aviojets. As well as 802 Esc. at Gando, SAR is provided by 801 Esc. at Son San Juan AFB, Palma, on the island of Majorca and 803 Esc. at Cuatro Vientos military airfield. Nine C-212S1 Aviocars have been delivered for this purpose along with two converted C-212AAs. The Aviocars are with 801 and 803 Escuadrones. 803 Esc. also operates six SA.319B Alouette IIIs. All three SAR squadrons operate the AS.332B Super Puma, of which 12 have been delivered, including two

121 and 122 Escuadrones at Torrejon fly elderly McDonnell F-4C Phantoms. These are mainly used for strike duties but could be called upon for air defence if needed.

for VIP use with 402 Escuadron.

The ENA or Escuela Nacional de Aeronautica is a State-run civil training centre for commercial pilots which has 12 Beech 95 Baron B.55s and 10 Beech B.65 King Air C.90s, both ex-EdA, plus two King Air A.100s, and 24 Beech Bonanza F.33As which still retain their EdA serials.

Guardia Civil

The para-military civil guard operates two types of helicopter, the MBB/CASA BO 105 and the MBB-Kawasaki BK 117. Of 20 BO 105s on order for the Guardia Civil at least 13 have been delivered; half of the total will be CASA-assembled. Four BK 117s are on order and deliveries began in 1984.

Arma Aérea de la Armada

The naval air wing provides fixed-wing aircraft and helicopters for the navy. The flagship is the escort carrier *Dedalo* which, scheduled for retirement in 1983-84, was due to be replaced by the 15,000-ton carrier *Principe de Asturias*, which has a 12° ramp for STOVL operations and can

Escuadrilla 008 at Rota operates the AV-8S Matador from the carrier *Principe de Asturias*.

carry eight Harriers and 14 ASW helicopters. Six destroyers, three LSTs and an attack transport also have the ability to carry helicopters.

The Arma Aérea de la Armada currently has six active units as follows: Escuadrilla 001 at Rota carrying out training and liaison duties operates a mix of Bell 47G, 47G-5 and HTL-4s, and Agusta-Bell AB.47G-2, AB.47G-2A and AB.47G-2A-1 helicopters to a total of 10. Escuadrilla 003 is based at Rota and operates the Agusta-Bell AB.212AS. The first four were received in 1974 and the unit now uses about 13 in the ASW, ECM and anti-shipping roles. One of the helicopters is equipped in a dedicated ECM configuration. They are deployed aboard the *Dédalo*, as well as the *Galicia* attack transport. Escuadrilla 004 is based at Rota and operates two Piper PA-24-260 Comanches, two PA-30-160 Twin Comanches and two Cessna 550 Citation IIs in the instrument training and search/observation roles. Escuadrilla 005 is based at Rota but maintains a detachment at Cartagena and also deploys aboard the *Dedalo*. It operates

14 Sikorsky SH-3 Sea Kings delivered in SH-3D, SH-3G and SH-3H variants, but all of these aircraft have since been modified to SH-3H standard. Escuadrilla 006 is based at El Ferrol and operates 11 Hughes H.369M (ASW)s which deploy aboard the helicopter platforms of the navy's destroyers. Escuadrilla 008 based at Rota operates the nine AV-8A and two TAV-8A Matadors. Escuadrilla 002 and 007 are at present disbanded. On order are 12 McDonnell Douglas AV-8B Harrier IIs and six Sikorsky SH-60B Seahawks.

Fuerzas Aeromoviles del Ejército de Tierra (FAMET)

Formed in 1965, the FAMET has recently been undergoing a dramatic expansion; its current order of battle follows:

BHELA I (Batallon de Helicopteros de Ataque I) based at Almagro Air Base with 28 CASA/MBB BO 105ATHs, 12 BO 105GSHs and two BO 105LOHs deployed in two Compania de Helicopteros. UHELMA II (Unidad de Helicopteros de Maniob-

ra II) based at Betera Air Base with two component units. Seccion de Reconocimiento with four BO 105 LOHs and two BO 105GSHs, and Compania de Transporta Medio with 12 Bell UH-1Hs. UHELMA III based at Agoncillo Air Base with the same component units and strengths as UHELMA II. UHELMA IV based at El Copero Air Base with the same component units and strengths as UHELMA II, with the addition of four Agusta-Bell AB.412s to the Compania de Transporte Medio. BHELTRA V (Batallon de Helicopteros de Transporte V) based at Los Remedios Air Base, Colmenar, with two component units. Compania de Transporte Pesado has 12 Boeing CH-47 Chinooks and Compania de Enlace y Transporte Medio six Bell OH-58Bs and six UH-1Hs. CEFAMET (Centro de Ensenanza de las FAMET – Training Centre) based at Los Remedios Air Base, Colmenar with 10 BO 105Cs, six OH-58Bs, one Agusta-Bell AB.206, three Bell UH-1Cs and several UH-1Hs. Plana Mayor (FAMAT Headquarters liaison flight), also at Colmenar with two

Agusta-Bell AB.212s (VIP), four AB.206s and some UH-1Hs. The aircraft maintenance unit is JEMAN (Jefatura de Mantenimiento) at Colmenar and third line servicing is also carried out at Colmenar by the SEHEL (Servicio de Helicopteros).

Aircraft in service consist of the following: three Bell UH-1C Iroquois, used mainly in training duties, 57 UH-1Hs, 12 OH-58B Kiowas, five Agusta-Bell AB.206As, and six AB.212s that include two in VIP configuration. Heavy transport duties are carried out by 12 Boeing CH-47C Chinooks. The most common aircraft in FAMET service is the MBB BO 105, with 11 ex-West German army test and evaluation MBB BO 105Cs in use for training, and CASA has assembled 19 BO 105GSHs (armed gunship version), 29 BO 105ATHs (armed anti-tank helicopters) and 14 BO 105LOHs (light observation helicopters). On order for the FAMET are 12 Boeing CH-47D Chinooks and 28 Agusta-Bell AB.412s (four in VIP configuration), with delivery of both types scheduled to begin in 1985.

Sweden Flygvapen

The Swedes have followed a policy of neutrality since 1915 but have always maintained well equipped armed forces to defend her neutrality.

The Flygvapen is divided into seven regional military commands (Milos-Militaromraden), four of them air defence wings which are as follows: Milo Sodra (southern command) has its HQ at Kristianstad and controls two air defence wings, F10 (also known as Skanska Flygflottiljen F10 and also designated F10/Sektor Sodra) at Angelholm with four component squadrons. 1, 2 and 3 Jaktflygdivisionen have Saab J35F Drakens in the air defence role and the Helikoptergrupp with Agusta-Bell AB.204Bs (HKP3B) for SAR and liaison duties. The second wing is F17 (Blekinge Flygflottilj F17) at Ronneby and comprising 1 Jaktflygdivisionen with Saab JA37 Viggens for air defence and 2 Spaningsflygdivisionen with SF/SH37 Viggens for reconnaissance duties. Also with F17 is a Helikoptergrupp with AB.204Bs and Boeing Vertol Model 107-II-4s (HKP4A). Milo Mellertsa covers the Baltic east coast from Stockholm and controls two air defence wings. F13 (Bravalla Flygflottilj F13) at Norrkoping-Bravalla has 1 Spaningsflygdivisionen with SF/SH-37 Viggens and 2 Jaktflygdivisionen with JA37 Viggens. Also part of F13 is F13M at Malmslatt consisting of the Malflygdivisionen operating Saab J32 Lansens for target towing and the Transportflygdivisionen with Aérospatiale SE.210 Caravelles (TP79). The second wing is F16 (Upplands Flygflottilj F16) which is also designated F16/Sektor Mellertsa as the sector station. F16 has 1, 2 and 3 Jaktflygdivisionen with the J35F Draken for air defence and 2 Divisionen also has some

SK35C Draken trainers. There is also a Helikoptergrupp with a few AB.204Bs and 5 (GTU/LA) Divisionen operating Saab 105s (SK60) for pilot and navigator instruction. The GTU carries out weapons training. Milo Nedre Norrlands extends up to the Norwegian border and controls one air defence wing F4 (Jamtlands Flygflottilj, also designated F4/Sektor Nedre Norrlands) as the sector station and is based at Froson with 1 Jaktflygdivisionen operating J35D Drakens and 2 Jaktflygdivisionen with JA37 Viggens. Also here is a Helikoptergrupp with AB.204Bs. Milo Ovre Norrlands controls Wing F21 (Norrbottens Flygflottilj F21) at Kallax which is also designated F21/Sektor Ovre Norrlands as the sector station. 1 Spaningsflygdivisionen operates SF/SH37 Viggens, 2 Jaktflygdivisionen J35D Drakens, 3 Jaktflygdivisionen JA37 Viggens and the Helikoptergrupp AB.204Bs and the Model 107-II-4.

The other Flygvapen combat aircraft are formed into the Forster Flygeskader as a tactical attack force. Wings which form part of this unit are F6 (Vastgota Flygflottilj F6) at Karlsborg with 1 and 2 Attackflygdivisionen operating AJ37 Viggens and a Helikoptergrupp with Aérospatiale SE.3130 Alouette IIs (HKP2); F7 (Skaraborgs Flygflottilj F7) at Satenas with 1 and 2 Attackflygdivisionen operating AJ37 Viggens, the Transportflygdivisionen with Lockheed C-130E/H Hercules and a Helikoptergrupp with SE.3130 Alouette IIs; F15 (Halsinge Flygflottilj F15) at Soderhamn with 1 and 2 Attackflygdivisionen operating AJ37 Viggen. 2 Divisionen also has all the Flygvapen SK37 Viggen trainers and there is also a Helikoptergrupp with Model 107-II-4s.

Flying training is carried out by the F5 wing (Krigsflygskolan – central flying school) based at Ljungbyhed. 1 Divisionen operates the Scottish Aviation Bulldog (SK61) in the primary training role and 2 and 3 Divisionen use the Saab 105 (SK60) for basic and advanced training. Weapons training is carried out by 5 Divisionen of F16 and oper-

ational conversion is carried out by 2 Divisionen of F16 for Draken pilots and 2 Divisionen of F15 for Viggen pilots. F14 at Halmstad is a non-flying training wing which consists of the air force cadet college (FOHS), the ground defence training school (BBS), the communications school (FSS) and the air force technical training school. F18 at Tullinge is another non-flying training wing and consists of the communications technical training school (FTTS) and the air force air defence control and reporting school (STRILS). Also here is a communications flight with SK50s, SK60s and a TP88. Also detached here is 1 Divisionen of F16 with J35F Drakens, but this will disband in 1985 and be followed by F18 in 1986.

Aircraft in service with the Flygvapen include a few J32D Lansens for target towing and some J32E Lansens for ECM duties. The Flygvapen retains about 50 J35D and 150 J35F Drakens in the interceptor role and about 25 SK35C Draken two-seat trainers, but these last are scheduled to be withdrawn in 1985. Viggen totals, of which about 20 have been lost in accidents, comprise seven prototypes and the following production aircraft: 110 AJ37s for the strike role, 149 JA37 interceptors (deliveries not yet complete), 18 SK37 Viggen trainers, plus 26 SH37 and 27 SF37 reconnaissance aircraft. Most of the Saab SK50 Safirs have been withdrawn from use but a

Sporting a white air defence scheme, this JA37 Viggen carries AIM-9P Sidewinder and BAe Sky Flash missiles. The Viggen in its various forms equips most of the Flygvapen's tactical units.

few of the 90 received remain in service for communications and liaison duties. Most of the 152 Saab 105 trainers that were produced, and which have the Flygvapen designations SK60A/B and C, continue in use. Of the 58 Scottish Aviation Bulldogs received (designated SK61 in Flygvapnet service), a few have been written off and the rest have been modified to SK61A/B/D or E standard. Two Cessna 404 Titans are in service for communications and VIP transport duties with F17 and the other with F21, and a Swearingen Metro III was leased in late 1984 and is operated by F18 at Tullinge.

Helicopters received by the Flygvapen include 12 SE.3130 Alouette IIs designated HKP2, seven AB.204Bs and 10 Model 107-II-4s. Most of these helicopters are still in service but four Model 107-II-4s will be transferred to the navy and converted for ASW duties.

In prospect for the Flygvapen are about six MBB BO 105Cs for SAR duties as replacements for the HKP2s and HKP3Bs. More long term is the order for 115 Saab JAS39 Gripens for interceptor/attack and reconnaissance duties and 25 SK39

Grippen two-seat trainers. The first prototype should fly in 1987 and they will initially replace J35F Drakens and early production Viggens, but an eventual total procurement of 350 is envisaged to completely replace the Viggen.

The Forsvarets Materielverk or FMV is the Government body responsible for all testing, supply, care and administration of all defence equipment and is, in fact, the owner of all the aircraft and helicopters. The test centre at Prov Malmen as well as operating Flygvapen aircraft for test and research studies also has two Rockwell Sabre 65s (TP86) in service.

Armeflygkar

The helicopters and fixed-wing aircraft of the army at present have no anti-tank capability and are mainly used for observation and liaison. The main unit is the Norrbottens Armeflygbattaljon AF1 based at Borden, with a subordinate unit Flygenheten which was formed on 1 July 1984 and consists of 1 Transporthelikopterkompani (1 TPHKPKOMP) with AB.206As (HKP6C) and the Skolavdelningen with Hughes 300Cs (HKP5B). Forming in 1985 is 3 Pansarvarnshkpkomp (3 PVHKPKOMP) with 20 BO 105CBs for anti-tank duties.

The other unit is the Armeflygskolan (Armeflyg-2) at Skavsta which has moved to Malmen during 1985. The subordinate unit is the Flygenheten which consists of the following training units: Flygskola 1 with Hughes 300Cs, Flygskola 2 with the Scottish Aviation Bulldog series 101 (FPL61C) and the Flygkompani with AB.204s, AB.206s, Dornier Do 27s (FPL53), and Scottish Aviation Bulldogs. Armeflygkar aircraft totals include 14 AB.204s, 21 AB.206s, 10 Hughes 300Cs, three Do 27s and 17 Bulldogs.

Marinen

The first naval squadron was formed in 1957 and today three units exist: 1 Helikopterdivisionen at Berga operating Boeing Vertol and Kawasaki Vertol 107-IIs, AB.206As and a Cessna 404 Titan. 2 Helikopterdivisionen at Save has SE.3130 Alouette IIs, Kawasaki Vertol 107-IIs and AB.206As. 3 Helikopterdivisionen was formed at Kallinge in late 1984/ early 1985 with Kawasaki Vertol 107-IIs. Marinen aircraft totals include nine SE.3130 Alouette IIs, seven AB.206As, one Cessna 404 Titan, four Boeing Vertol 107-II-5s, and eight Kawasaki Vertol 107-IIs.

Switzerland Kommand der Flieger und Fliegerabwehrtruppen

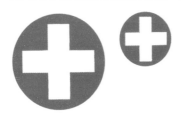

Switzerland is a neutral country with a very small peacetime army and air force but with a highly trained militia which would expand the armed forces rapidly to meet any threat to the nation's policy of neutrality.

The air force is organised in three commands, Flugwaffenbrigade 31 controls the flying units; Flugplatzbrigade 32 provides military logistic support and Flugabwehrbrigade 33 controls the anti-aircraft units with SAMs and AA guns. Additionally, Flugabwehr-Park 35 carries out airframe and avionics maintenance. The regular combat component of Flugwaffenbrigade 31 is called the Uberwachungageschwader (surveillance wing) and consists of Flugplatzregiment 1 with a detachment of Flieger Staffel 10 (FlSt 10) Dassault Mirage IIIRS at Sion. Flugplatzregiment 2 has its headquarters at Interlaken controlling FlSt 1 with Northrop F-5E/Fs at Payerne; FlSt 11 with F-5E/Fs at Meiringen; FlSt 17 at Payerne with the Mirage IIIS and another detachment of FlSt 10 with the Mirage IIIS at Payerne. Flugplatzregiment 3 with FlSt 16 at Buochs has the Mirage IIIS and another detachment of FlSt 10 Mirage IIIRS at Buochs. FlSt 18 at Alpnach has F-5E/Fs and there is a helicopter SAR service at Alpnach with Aérospatiale Alouette IIIs. It is thought that FlSt 18 is normally

based at Dubendorf but would deploy to Alpnach in times of crisis.

The reserve elements of the air force comprise:
FlSt 2 Hunter F.58s, probably based at Interlaken
FlSt 3 Hunter F.58s, based at Ambri
FlSt 4 Hunter F.58s, probably based at Meiringen
FlSt 5 Hunter F.58s, probably based at Meiringen
FlSt 7 Hunter F.58s, possibly based at Sion
FlSt 8 Northrop F-5E/Fs at an unknown base
FlSt 9 has disbanded following withdrawal of the Venom
FlSt 12 has disbanded following withdrawal of the Venom
FlSt 13 Northrop F-5E/Fs based at Emmen
FlSt 14 has disbanded following withdrawal of the Venom
FlSt 15 Hunter F.58s based at Interlaken
FlSt 19 Northrop F-5Es based at Alpnach or possibly Sion
FlSt 20 Hunter F.58s based at Mollis
FlSt 21 Hunter F.58s based at Turtmann
ECM FlSt 24 Hunter T.68s at an as yet unknown location

Other known bases are Airolo, Belp, Chur, Grenchen, Locarno, Turtig and Turtmann, but those at which units are located is a source of endless confusion; it is thought that larger bases are often used by reserve squadrons for maintenance and training, but each unit would deploy to one of the many auxiliary bases within Switzerland.

Transport duties are carried out by Leichte Fliegerstaffel 1 to 7 using the Alouette I, Alouette III, Dornier Do 27 and Pilatus PC-6B Turbo-Porter. The other transport unit is the Transportfliegerkorps at Duben-

dorf with three Beech Twin Bonanzas. Initial training is carried out by Piloten Rekrutenschule 42/242 at Magadino with the Pilatus PC-7 Turbo-Trainer, advanced instruction following on the PC-7 and de Havilland Vampire T.55. The Zeilfliegerkorps provides aircraft for target towing duties, these comprising 23 EKWC C.3605s and a few Vampire FB.6s.

Aircraft in service include about 15 SE.3130 Alouette IIs, 78 SE.3160 Alouette IIIs, 23 examples of the ancient EKW C.3605, one Dornier Do 27H-2, six Pilatus PC-6B Turbo-Porters, 12 converted PC-6A Porters, three Beech Twin Bonanza E.50s, and 62 Pilatus P-3s, though many of these may now be retired as all 40 Pilatus PC-7 Turbo-Trainers had been delivered by the end of 1983. Approximately 60 Vampire FB.6s had uprated avionics fitted for use in the training role, and most of these plus 37 Vampire T.55s are still in service. The air force has four examples of the Dassault Mirage IIIBS, the last two able to carry armament, 18 Mirage IIIRS for reconnaissance and 30 Mirage IIIS in-

Northrop F-5Es stand alongside Dassault Breguet Mirage IIIs in the defence of Switzerland. Many aircraft are hangared within mountains, using runways on valley floors. A new fighter is required soon to replace the ageing Mirages.

terceptors. Of the 95 F-5Es and 12 F-5F trainers in service, 84 of the F-5Es were assembled by FFA at Emmen. The final aircraft remaining in Swiss air force service is the Hawker Hunter, there being a total of about 130 Hunter F.58/F.58As, and seven T.68s. These latter aircraft have recently been fitted with radar warning receivers, chaff and flare dispensers and other ECM equipment. All the Hunters are now concentrated in ECM Fliegerstaffel 24. On order for the SAFAAC are three Sikorsky S-70A-4 Blackhawks for delivery in 1986 against a requirement for 40 assault helicopters. Also in prospect is a replacement for the Mirage III, Hunter and Vampire, which will probably be fulfilled by an advanced combat aircraft such as the F-16/Mirage 2000/ JAS 39 for interception and Hawk/ Alpha Jet for ground attack.

Turkey Turk Hava Kuvvetleri/THK

Turkey is part of NATO and allocates her combat forces to NATO's 6 ATAF. The THK combat element is divided into two tactical air forces which consist of the following units:

1st Tactical Air Force with HQ at 1st Air Base, Eskisehir, consists of the 1st TAF training and com-

munications flight with Douglas C-47s, Bell UH-1Hs and Lockheed T-33As. Also at Eskisehir are 111 and 112 Squadrons operating McDonnell Douglas F-4E Phantom IIs, 113 Squadron with the RF-4E Phantom and Simsek Flight as an operational training unit with F-4E Phantoms. At 4 AB, Murted, 141 and 142 Squadrons fly the Lockheed TF/F-104G Starfighter and Sahin Flight acts as the OTU; there is also a base flight with T-33As. At 6 AB, Bandirma, 161 Squadron operates the TF/F-104G and 162 Squadron uses the Northrop F/RF-5A/B, as does the Safak Flight OTU. The base flight has T-33As, UH-1Hs and Sikorsky UH-19Bs. 9 AB, Balikesir, is the

final part of the 1st TAF and has 191 and 192 Squadrons operating the F-104S/TF-104G Starfighter, 193 Squadron using the F-104G and a base flight with T-33As and UH-1Hs. 2nd Tactical Air Force with HQ at 8th Air Base, Diyarbakir, consists of 2nd TAF training and communications flights with C-47s, T-33As and UH-1Hs. Also at Diyarbakir are 181 and 182 Squadrons operating the North American F-100D/F Super Sabre and 184 Squadron with the RF-5A/F-5B. At 5 AB, Merzifon, 151 and 152 Squadrons operate the F-5A/B and the base flight uses T-33As and UH-1Hs. Finally, 7 AB, Erhac, has 171, 172 and 173 Squadrons operating the F-4E Phantom II

and a base flight with T-33As and UH-1Hs.

Air Transport Force has two main bases, 12 AB, Erkilet, where there is 221 Squadron with 20 ex-Luftwaffe Transall C-160Ds and 222 Squadron with seven Lockheed C-130H Hercules and several C-47 Dakotas. The second base is at Etimesgut, where the Etimesgut Transport Group has 223 Squadron operating the C-47 Dakota in the ECM/SAR/Elint/ calibration roles and 224 Squadron with C-47s, Vickers Viscounts and UH-1Hs. Training Command has its headquarters at Gaziemir, where the training and communications flight is based with C-47s and UH-1Hs. It has training units deployed at four other airfields: at 2 AB, Cigli, 123 Squadron with Beech T-34As and Cessna T-41Ds is based, but the aircraft are usually detached to Cumaovassi where initial training can be carried out away from the busy circuit at Cigli. 121 Squadron with Northrop T-38A Talons and T-33A/ANs and 122 Squadron with Cessna T-37Bs continue the training process at Cigli, and 124 Squadron trains instructors with aircraft borrowed from 121 and 123 Squadrons. Weapons training is carried out at 3 AB, Konya by 131 and 132 Squadrons with the North American F-100D/F Super Sabres, and there is also a base flight with T-33As and UH-1Hs. The Air Force Academy in Istanbul has a training communications and target towing flight based at Yesilkoy with C-47s and UH-19Bs.

Aircraft in THK service include approximately 12 Beech T-34 Mentors, 24 Bell UH-1H, the survivors of 54 Canadair T-33ANs, 30 Cessna T-37Cs and 20 T-41Ds, 50 Douglas C-47s, about 200 Lockheed F-104G/TF-104G and F-104S Starfighters from many sources, 80 Lockheed T-33As, some 95 McDonnell Douglas F-4E/RF-4E Phantom IIs, with 15 more F-4E due for delivery, about 95 North American F-5A/-5B/RF-5As, 100 North American F-100C/D/F Super Sabres, 30 Northrop T-38A Talons, five Sikorsky UH-19Bs and

three Vickers Viscounts.

Turk Cumhuriyet Bahrya

Although the THK is responsible for the ASW aircraft based at Topel at the edge of the Black Sea, they operate with joint navy-THK crews. Eight Grumman S-2As, two TS-2As and 12 S-2E Trackers are in service; helicopters consist of three Agusta-Bell AB.204AS, and six AB.212ASWs which deploy aboard

two frigates.

Turk Kara Kuvvetleri

Army aviation is a large element with about 350 aircraft. The main army base is the Central Army Aviation Establishment at Guverncinlik/Ankara. In service are about 20 AB.204Bs, 100 AB.205As, 20 UH-1Bs, over 50 Bell OH-13H/TH-13Ts, 50 Cessna O-1E Bird Dogs, 20 Cessna U-17As, 20 Dornier Do 28D

Five squadrons operate the McDonnell Douglas F-4E Phantom for attack and fighter duties. A further unit has RF-4E reconnaissance aircraft.

Skyservants, four Cessna 421 Golden Eagles, 10 Dornier Do 27s, two DHC U-6A Beavers and 30 Hughes H.300Cs. In addition to these 40 UH-1Hs and 40 AB.205As are in the process of delivery, and six Bell AH-1Ss and three AB.212s are on order.

United Kingdom Royal Air Force

Coming under the control of the Ministry of Defence (Air), the RAF is divided into three commands: Strike Command, Support Command and RAF Germany.

Strike Command is divided into 1 Group, 11 Group, 18 Group, RAF Cyprus, RAF Norway and RAF Hong Kong. 1 Group, HQ Upavon, Wiltshire, is assigned to SACEUR (NATO Supreme Allied Command Europe) and ACE/MF (Allied Command Europe/Mobile Force). It consists of the following units: 1 Sqn and 233 Operational Conversion Unit at Wittering, 1417 Flight in Belize and 1453 Flight at Port Stanley operating the BAe Harrier GR.3 strike aircraft and the Harrier T.4 trainer. About 87 Harrier GR.3/T.4s are in service and four Harrier GR.3s are on order as replacements for losses in the Falklands conflict. Also on order are 60 McDonnell Douglas/BAe AV-8B(K) Harrier GR.5s. The other two Harrier units are 3 and 4 Sqns at Gütersloh with RAF Germany. 6, 41 and 54 Sqns at Coltishall, along with 226 OCU at Lossiemouth, operate the SEPECAT Jaguar GR.1/T.2 strike aircraft. The RAF loaned 18 Jaguars to the Indian Air Force, but only a few were returned to service after completion of the loan. With the re-equipping of RAF Germany (RAFG) Jaguar units with the Tornado, surplus Jaguars have been returned to the UK for storage. About 120 Jaguar GR.1/T.2s remain in use. Jaguar units are 2 Sqn at Laarbruch, and 14 and 17 Sqns at Brüggen, all with RAFG.

7 Sqn and 240 OCU at Odiham, along with 1310 Flight at Port Stanley, operate the Boeing Chinook HC.1. Of the 33 Chinooks received three were lost in the Falklands conflict, and an additional order for eight was placed which is currently being fulfilled. 18 Sqn with RAFG at Gütersloh is the only other operator of this medium-lift transport helicopter. The Panavia Tornado is entering service in large numbers and 1 Group has five Tornado GR.1 units under its control. They are the Tri-national Tornado Training Establishment (TTTE) at Cottesmore for pilot training, the Tactical Weapons Conversion Unit (TWCU with 'shadow' designation 45 Sqn) at Honington for weapons training, 9 Sqn also at Honington, and 27 and 617 Sqns at Marham. There are 220 Tornado GR.1s on order, including about 40 dual-control Tornado GR.1Ts. About 170 Tornados are currently in service with the above units and also 15, 16, 20 and 31 Sqns with RAFG. 17 Sqn with RAFG also began conversion to the Tornado at the end of 1984. 10 Sqn at Brize Norton operates 13 BAC VC10s in the strategic transport role. The Lockheed Hercules is in service with the Lyneham Tactical Wing which consists of 24, 30, 47 and 70 Sqns plus 242 OCU. There are 62 Hercules in service, one of them having been converted as a Hercules W.2 for weather research. Of the remaining 61, 29 are being converted to Hercules C.3 standard with an extended fuselage, 16 others are Hercules C.1Ps with the ability to receive fuel by inflight refuelling and six aircraft are Hercules C.1(K)s which can provide inflight refuelling via a hosedrum unit.

32 Sqn at Northolt is the VIP and communications unit with 12 BAe 125 CC.1/CC.2/CC.3s, four Gazelle HT.3/HCC.4s and four Hawker Siddeley Andover C.1/CC.2s. 33 Sqn and 240 OCU at Odiham, 1563 Flight in Belize and 230 Sqn with

RAFG at Gütersloh operate the Westland Puma HC.1 of which 44 are in service. For inflight refuelling 55 and 57 Sqns plus 232 OCU at Marham operate the Handley Page Victor K.2 with 22 aircraft converted from the strategic nuclear role in the 1970s. The Victors are now nearing the end of their useful life and 101 Sqn at Brize Norton is equipping with five VC10 K.2s and four Super VC10 K.3s that have been converted to tanker aircraft at the BAe plant at Filton. Additional inflight refuelling capacity is to be provided by six ex-British Airways and four ex-Pan American World Airways Lockheed L-1011 TriStar 500s. All have been delivered to the RAF for operation by 216 Sqn but some are being used as pure transport aircraft prior to receiving their inflight refuelling equipment. 115 Sqn at Benson operates eight Andover C.1/E.3/E.3As in the navaid checking and aeromedical evacuation roles.

72 Sqn at Belfast-Aldergrove operates about 20 Wessex HC.2s in the tactical transport role. 241 OCU at Brize Norton is the strategic transport conversion unit and bor-

The Panavia Tornado is now rapidly re-equipping the RAF's strike and bomber squadrons, both at home and in Germany. Tornado F.Mk 2s will shortly begin re-equipping air defence squadrons.

rows VC10s from 10 Sqn and 101 Sqn and TriStars from 216 Sqn on an 'as needed' basis. The unit has been evaluating two BAe 146 C.1s for possible use with The Queen's Flight. One has been returned to the manufacturer and the other is still in service. Tactical weapons training is also one of 1 Group's responsibilities and 1 Tactical Weapons Unit (1 TWU) at Brawdy with 'shadow' designations 79 and 234 Sqns and 2 TWU at Chivenor with 'shadow' designations 63 and 151 Sqns operate the BAe Hawk T.1, except for 79 Sqn which still has the Hunter F.6A/FGA.9/T.7. The RAF received 175 Hawk T.1s; 34 are with 1 TWU, 51 with 2 TWU, and the others are with Support Command with 4 FTS and the RAF Display Team, the 'Red Arrows'. The Andover Training Flight at Benson borrows aircraft from 115 Sqn as required to produce crews for 115 Sqn, 32 Sqn and The Queen's Flight; the last is also based at Benson and has two Wessex HCC.4s and three Andover CC.2s. Following successful evaluation of the BAe 146 C.1 by 241 OCU an order has been placed for two of these aircraft for Queen's Flight use.

Second group within Strike Command is 11 Group with its HQ at Bentley Priory, Middlesex. 11 Group is responsible for air defence of the UK and comprises the following units: 5 and 11 Sqns together with the Lightning Training Flight and the Lightning Augmentation Flight at Binbrook operate all the remaining BAC Lightnings. Just under 70 Lightning F.3/F.6/T.5s remain at Binbrook, but at any one time a third are in storage as aircraft are rotated to keep their hours even. The Lightnings are due for withdrawal by the end of 1986. Replacing the Lightning, and to some extent the Phantom, will be the Panavia Tornado F.2. A total of 165 of the interceptor version are on order and deliveries to the reformed 229 OCU at Coningsby began in late 1984. The Tornado F.2s will be based at Leeming, Coningsby, Lossiemouth and Wattisham.

In addition to the Tornado F.2s, all of the 1 and 2 TWU Hawk T.1s are being modified to Hawk T.1A standard, giving the ability to carry the AIM-9 Sidewinder for limited use in the air defence role. Still in the air defence use is the Phantom. The Royal Navy received 48 McDonnell Douglas Phantom FG.1s for aircraft carrier operations and about 35 of these remain in RAF service with 43 and 111 Sqns at Leuchars. The RAF's 91 Phantom FGR.2s equip 29 Sqn and 228 OCU at Coningsby, 56 Sqn at Wattisham and 23 Sqn at Port Stanley in 11 Group and, additionally, 19 and 92 Sqns at Wildenrath with RAFG. Following the Falklands conflict and subsequent transfer of 23 Sqn from Wattisham to Port Stanley, 15 ex-US Navy F-4J Phantoms were bought and refurbished and all entered service as F-4J(UK)s with 74 Sqn at Wattisham. For airborne early warning, 8 Sqn at Lossiemouth still operates the Shackleton AEW.2. Six are still in service and development problems with the BAe Nimrod AEW.3 mean they will remain in service during 1985. Eleven Nimrod AEW.3s are on order and will eventually be based at Waddington. Also with 11 Group are 25 and 85 Sqns operating the Bloodhound SAM and all RAF Regiment Rapier SAM units.

18 Group with headquarters at Northwood, Middlesex is assigned to SACLANT (Supreme Allied Command Atlantic) with the following units: 42, 120, 201 and 206 Sqns with 236 OCU operating the Nimrod MR.2 for maritime surveillance, anti-submarine and anti-shipping operations. Except for 42 Sqn and the OCU at St Mawgan, all the other units are based at Kinloss. Of the 46 Nimrod MR.1s acquired by the RAF, 34 have been converted to Nimrod MR.2s, one was lost in an accident and the other eleven are being converted to Nimrod AEW.3s. For Elint 51 Sqn at Wyton operates three Nimrod R.1s. The BAC Canberra is still in service and 100 Sqn provides target facilities and calibration aircraft, operating about 22 BAC Canberra B.2/E.15/T.4/TT.18s. 360 Sqn uses 12 Canberra T.17s in the ECM role, 1 PRU carries out survey duties with five Canberra PR.9s, and 231 OCU operates 10 Canberra B.2/T.4 in the operational conversion role.

In addition to the Wessex HC.2s that are with 72 Sqn at Belfast-

Aldergrove, Wessex HC.2 units under direct Strike Command control are 28 Sqn, the sole unit of RAF Hong Kong at Sek Kong and 84 Sqn, the sole unit of RAF Cyprus at Akrotiri. 84 Sqn consists of 'A' Flight which carries out SAR duties and 'B' Flight which supports the United Nations peace-keeping forces. The remainder of the Wessex HC.2s are operated by 22 Sqn and the SARTF under 18 Group control providing SAR cover. 22 Sqn has its headquarters at Finningley with SAR detachments at Chivenor, Leuchars, Valley, Leconfield and Manston; the Search and Rescue Training Flight (SARTF) is at Valley. The RAF has a total of about 60 Wessex HC.2s, of which about 20 are with 22 Sqn and the SARTF. Additional SAR aircraft are provided by 202 Sqn, again with headquarters at Finningley, operating the Sea King HAR.3 of which 16 were delivered during 1977-80. Three additional aircraft are on order. 202 Sqn has detachments at Boulmer, Brawdy, Lossiemouth and Coltishall. Additionally, the Sea King Training Flight is at Culdrose and 1564 Flight at Port Stanley. Final unit with 18 Group is the Electronic Warfare Avionics Unit at Wyton, which carries out evaluation of ECM equipment and borrows aircraft from 360 and 51 Sqns when required. The final unit under direct Strike Command control is RAF Norway with the C-in-C AFNE Flight at Fornebou operating one Andover C.1 for the Commander in Chief Air Forces Northern Europe.

RAF Germany is effectively another command under direct Ministry of Defence (Air) control. In times of conflict it would be assigned to NATO's 2 ATAF. RAFG consists of 3 and 4 Sqns at Gütersloh operating the Harrier GR.3/T.4; 2 Sqn at Laarbruch with Jaguar GR.1/T.2s in the tactical reconnaissance role; and 14 and 17 Sqn at Brüggen operating the Jaguar GR.1T.2 in the tactical strike role, but 17 Sqn began conversion to the Tornado GR.1 in late 1984. 15 and 16 Sqns at Laarbruch together with 20 and 31 Sqns at Brüggen operate the Panavia Tornado GR.1. 19 and 92 Sqns at Wildenrath operate the McDonnell Douglas Phantom FGR.2 in the air defence role. 18 Sqn and 230 Sqn both at Gütersloh provide tactical helicopter support with Chinook HC.1s and Puma HC.1s respectively. 60 Sqn at Wildenrath carries out communications duties with seven Hunting Pembroke C.1s. Final unit with RAF Germany is the Berlin Station Flight with two DHC-1 Chipmunk T.10s at Gatow for communications duties.

Third and final command is Support Command, which controls a wide variety of training schools and maintenance units. Initial pilot selection is carried out at Swinderby by the Flying Selection Sqn on the Chipmunk T.10 for non-graduate students; about 10 Chipmunk T.10s are in service. University graduates go to the RAF College, Cranwell, and are instructed on 43 BAC Jet Provost T.5As. From Swinderby non-graduate pilots go either to 1 Flying Training School (1 FTS) at Linton-on-Ouse which has 33 Jet Provost T.3As and 17 Jet Provost T.5As, or 7 FTS at Church Fenton with 28 Jet Provost T.3As and 10 Jet Provost T.5As. After graduation, pilots are graded into three groups. The fast jet pilots are sent to 4 FTS at Valley where they are instructed on 65 BAe Hawk T.1s before going on to 1 or 2 TWU for weapons training and

then to an OCU for final operational conversion. Pilots for multi-engine aircraft go to 6 FTS at Finningley where they receive their first multi-engine instruction on the Jetstream T.1. Helicopter pilots go to 2 FTS at Shawbury for instruction on Gazelle HT.3s and Wessex HC.2s, and where 22 Gazelle HT.3s and four Gazelle HT.2s are in use. Instructors are trained by the Central Flying School which has Bulldog T.1s and Jet Provost T.5As at Leeming with 3 FTS, Hawk T.1s at Valley with 4 FTS, and Gazelle HT.3s at Shawbury with 2 FTS. Finally, the 'Red Arrows' Display Team at Scampton with 10 Hawk T.1s is also part of the CFS. Also at Shawbury is the Central Air Traffic Control School which has 12 Jet Provost T.4s for air traffic controller training. All other aircrew training is carried out by 6 FTS at Finningley which provides instruction for navigators, air electronics officers and air engineers on 13 Jet Provost T.5s and 20 Dominie T.1s. About 200 Jet Provost T.3/T.4/T.5s remain in service, but in March 1985 the EMBRAER/Shorts Tucano was announced as winner of the competition for a replacement. The initial contract will be for 130 aircraft for delivery during 1986-91.

Maintenance units within Support Command comprise RAF Engineering Wing at St Athan, Jaguar MU at Abingdon (which now also overhauls Hawks and Hunters) and 431 MU at Bruggen. At Shawbury there is an aircraft storage unit. Aircraft technicians are trained either at 1 School of Technical Training at Halton or 2 SOTT at Cosford. A new technical training unit is being formed at Scampton and the RAF Technical College is based at Henlow.

Also within Support Command are the University Air Squadrons (UAS) and Air Experience Flights (AEF). The UAS all operate the Bulldog T.1 and are as follows: Aberdeen and St. Andrews UAS at Leuchars, Birmingham UAS at Cosford, Bristol UAS at Filton, Cambridge UAS at Teversham, East Lowlands UAS at Turnhouse, East Midlands UAS at Newton, Strathclyde UAS at Glasgow, Liverpool UAS at Woodvale with Manchester UAS, London UAS at Abingdon with Oxford UAS, Northumbrian UAS at Leeming, Queens UAS at Sydenham, Southampton UAS at Hurn, Wales UAS at St. Athan and Yorkshire UAS at Finningley. There are 126 Bulldogs in service, including 10 CFS aircraft. The UAS are used to give university students basic flying training. The AEFs are used to give ATC and CCF cadets air experience using the DHC-1 Chipmunk T.10. 1 AEF is at Manston, 2 AEF at Hurn, 3 AEF at Filton, 4

AEF at Exeter, 5 AEF at Teversham, 6 AEF at Abingdon, 7 AEF at Newton, 8 AEF at Shawbury, 9 AEF at Finningley, 10 AEF at Woodvale, 11 AEF at Leeming and 12 AEF at Turnhouse. Over 50 DHC-1 Chipmunk T.10s are still in service. Also within Support Command are the Central Gliding School at Syerston and 28 volunteer gliding schools around the country. They operate 50 Slingsby T.61A Venture T.2s, two Schempp-Hirth Janus Cs, five Sleicher ASK-21 Vanguards and 10 Sleicher ASW-19 Valiants. Just being delivered are 100 Grob G-103 Viking T.1s.

The Ministry of Defence also controls several research and test organisations such as the Aeroplane & Armament Experimental Establishment (A&AEE) at Boscombe Down with a wide range of service, special and manufacturers aircraft, Royal Aircraft Establishment (RAE) with bases at Thurleigh (signals and radar research), Farnborough (radar and signals), West Freugh (weapons testing), Llanbedr and Aberporth (missile testing), all of which have a variety of aircraft, and finally the Empire Test Pilots School (ETPS) at Boscombe Down with various aircraft.

Fleet Air Arm/Royal Navy

The Fleet Air Arm is primarily a helicopter operator with Sea Harriers being the only fixed-wing combat aircraft. Following an impressive performance in the Falklands conflict, combat losses are not just being replaced but additional aircraft are being ordered.

The three main aircraft carriers in the Royal Navy are HMS Invincible, Illustrious and Ark Royal. These each carry a squadron of Sea Harriers, a squadron of Sea Kings and two or three AEW Sea Kings. 800 and 801 Sqns are attached to Illustrious and Invincible respectively and 802 Sqn is due to form shortly following the commissioning of Ark Royal. All operate Sea Harriers and are based at Yeovilton when not deployed on board ship. Also at Yeovilton is 899 Sqn, the Sea Harrier training unit. A total of 34 BAe Sea Harrier FRS.1s were delivered initially followed by three Harrier T.4N two-seat trainers. A further 14 Sea Harrier FRS.1s are on order; 10 Sea Harrier FRS.1s and a Harrier T.4N have been lost in accidents or combat to date. The Westland Sea King is also in service with the Fleet Air Arm in large numbers and is

used mainly in the anti-submarine role. The survivors of early Sea King HAS.1s (delivered from 1969) and Sea King HAS.2s (1976-79) are now being converted to Sea King HAS.5 standard. Together with 17 new built HAS.5s (delivered 1980-82) and 13 HAS.5s now on order, they will provide the Fleet Air Arm with a total of about 85 Sea Kings. 814 Sqn is allocated to *Illustrious* and 820 Sqn to *Invincible*; both are based at Culdrose. Also at Culdrose are 706, 810, 824, 826 and 849 Sqns which all operate the Sea King. 706 Sqn is the Sea King training unit; 810 Sqn acts as the operational conversion unit and may be allocated to *Ark Royal* when she is commissioned. 824 Sqn provides ASW cover for Royal Fleet Auxilliary (RFA) vessels, but currently has only two aircraft on strength as 826 Sqn has gained several of 824's aircraft to give her a complement of 14, five of which are permanently detached to give ASW cover around the Falkland Islands. Final and latest Culdrose-based Sea King unit is 849 Sqn, which will eventually receive eight AEW.2s for use in the airborne early warning role. Final Sea King unit is 819 Sqn based at Prestwick, still operating HAS.2As in the SAR role with six aircraft usually on strength.

For destroyer- and frigate-based anti-submarine operations, the Westland Lynx is gradually replacing the Westland Wasp HAS.1. The Lynx HAS.2 is currently deployed aboard two 'County' class and nine Type 42 destroyers, five Type 21, four Type 22, all Exocet-equipped 'Leander' and half the 'Broad Beamed Leander' class frigates. 815 Squadron at Portland is responsible for all the Lynx ship's flights and, in addition, has a Headquarters Flight with six Lynx HAS.2/3s. 702 Sqn is the Lynx training unit, also based at Portland, which operates nine Lynx HAS.2s and three HAS.3s. A total of 60 Lynx HAS.2s and 20 HAS.3s with uprated engines have been delivered. The Westland Wasp HAS.1 continues in service with 829 Sqn at Portland and, in addition to the Headquarters Flight, still provides ASW cover to one Type 21, half the 'Broad Beamed Leander' class, all the active Ikara-equipped 'Leander' class and the 'Rothesay' class frigates, survey vessels and, of course, HMS *Endurance*. Of the 98 Wasps received only about 42 are still active, but six others are held in store and could be returned to service if required.

In addition to ASW and Fleet defence, the other main Fleet Air Arm role is tactical assault. For this the

FAA provides transport helicopters for use by Royal Marine Commando units. 845, 846 and 707 Sqn, all based at Yeovilton, currently fulfil this role. 846 Sqn was the first unit to receive the Sea King HC.4. A total of 15 HC.4s were ordered, but three were lost during the Falklands conflict. A further eight aircraft are being delivered with eight more on order. 707 Sqn is the operational training unit with a mix of Westland Wessex HU.5s and Sea King HC.4s. Wessex HU.5s remain in use with 707 Sqn (six), 771 Sqn (six for SAR), 772 Sqn (seven for SAR) and with 845 Squadron (20).

Training for FAA pilots begins with RAF-operated but Fleet Air Arm-owned Scottish Aviation Bulldog T.1s at the Royal Navy Elementary Flying Training School, Leeming, though the aircraft are often detached to Topcliffe. Successful students then progress to either 705 Sqn at Culdrose which has 20 Westland Gazelle HT.2s for initial helicopter pilot training, or go through the RAF jet training course and onto Sea Harriers. Operational conversion is carried out by 810 Sqn for the Sea King, 815 Sqn for the Lynx, 899 Sqn for the Sea Harrier and 829 Sqn for the Wasp. 702 Sqn and 706 Sqn provide systems training for Lynx and Sea King crews respectively and, finally, 750 Sqn at Culdrose operates 16 Scottish Aviation Jetstream T.2s in the observer training role; four Jetstream T.3s are on order for 750 Sqn. Two Station Flights are currently active, one at Culdrose with four de Havilland Sea Devon C.20s and two de Havilland Canada DHC-1 Chipmunk T.10s. The other is at Yeovilton with four Sea Heron C.1s, one Heron C.4 and two DHC-1 Chipmunk T.10s. Another support flying unit is the Britannia Royal Naval College's air experience flight, with 11 DHC-1 Chipmunk T.10s based at Roborough, Plymouth.

The Air Engineering School is at Lee-on-Solent and has numerous instructional airframes. The main maintenance facilities are at Fleetlands and Wroughton, but second-line work is undertaken by the Naval Aircraft Support Units at Culdrose and Yeovilton. Aircraft handling and towing is taught by the School of Aircraft Handling at Culdrose with several instructional airframes. Target facilities are provided by two units. Firstly, the Fleet Target Group at Portland operating Northrop MQM-36A and MQM-74A Drones, and secondly by the civilian-operated Fleet Requirements and Aircraft Director Unit (FRADU) at Yeovilton. The FRADU currently

operates 17 BAC Canberra TT.18/T.22s and 28 Hawker Siddeley Hunter GA.11/T.7/T.8s for target towing, radar interception duties and target facilities. Up to eight Dassault Falcon 20s are being modified by Flight Refuelling Ltd., and the first had entered service early in 1985 as a long term Hunter and Canberra replacement.

Army Air Corps

The AAC, which has its HQ at Middle Wallop, is currently re-equipping and upgrading its aircraft. The TOW-armed Westland Lynx AH.1 has totally replaced the Westland Scout AH.1 within the BAOR (British Army of the Rhine). The Westland Gazelle is also in service in large numbers for reconnaissance and liaison duties.

The BAOR has three Armoured Divisions on its strength and each division has its own Army Air Corps Regiment. They are as follows: 1 Regiment AAC based at Hildesheim with 651 and 652 Sqns each operating eight Lynx AH.1s and four Gazelle AH.1s and 661 Sqn with 12 Gazelle AH.1s. 651 and 652 Sqns operate in the attack/anti-tank roles and 661 Sqn is a reconnaissance unit. Similarly, 3 Regiment AAC at Soest has 653 and 662 Sqns operating a mix of Lynx AH.1s and Gazelle AH.1s and 663 Sqn operating solely Gazelle AH.1s. Finally, 4 Regiment AAC at Detmold has 654 and 659 Sqns with a mix of Lynx AH.1s and Gazelle AH.1s and 669 Sqn solely with Gazelles. Additional units within BAOR are 664 Sqn based at Minden with a few Gazelle AH.1s as the BAOR Headquarters Flight, 7 Flight at Berlin-Gatow with three Gazelle AH.1s for communications duties, and 12 Flight at Wildenrath with the same role and complement.

Further anti-tank units are based within the UK and consist of 655 Sqn at Belfast-Aldergrove with a detachment at Ballykelly operating Gazelle AH.1s and eight Lynx AH.1s, 656 Sqn at Netheravon operating about four Gazelle AH.1s and eight Lynx AH.1s, 657 Sqn at Oakington with the standard complement of eight Lynx AH.1s and four Gazelle AH.1s, and finally 658 Sqn at Netheravon which operates about eight Scout AH.1s and four Gazelle AH.1s. Part of 658 Sqn is 6 Flight which operates the Gazelles and 8 Flight which operates some of the Scout AH.1s Communications

units within the UK include 2 Flight at Netheravon and 3 Flight at Topcliffe, each with six Gazelle AH.1s, Beaver Flight at Aldergrove with five de Havilland Canada DHC.2 Beaver AL.1s and Beaver Flight at Middle Wallop with two Beaver AL.1s.

The AAC still has a few units spread around the world: 660 Sqn operates about 12 Scout AH.1s from its main base at Sek Kong, Hong Kong and a detached flight in Brunei. 16 Flight is based at Dhekelia, Cyprus with four Aérospatiale Alouette AH.2s for communication duties. 25 Flight is at Belize International Airport with four Gazelle AH.1s for support and communication duties. 29 Flight at Suffield, Canada, has five Gazelle AH.1s and one DHC.2 Beaver AL.1 operating in the cold weather training role. United Nations Flight Cyprus (UNFICYP), with four Alouette AH.2s at Nicosia, support the UN peace-keeping forces on the island. Finally, the Garrison Air Sqn at Port Stanley in the Falkland Islands has four Gazelle AH.1s and eight Scout AH.2s to support Army units on the Falklands. Two other units warrant a mention, namely 3 Commando Brigade Air Sqn (3 CBAS) at Yeovilton with 12 Gazelle AH.1s and four Lynx AH.1s which operates in support of the Royal Marine Commando units. Finally, 7 Regiment AAC is the parent unit for 656, 657 and 658 Sqns and late in 1984 formed a Headquarters Flight with four Agusta A.109As. Two are new aircraft but the other two are ex-Argentine Army aircraft captured during the Falklands conflict. These three squadrons are believed to be available for Special Forces use within the United Kingdom and 658 Sqn also supports the newly formed 5th Airborne Brigade.

Pilot training commences at Middle Wallop on the DHC-1 Chipmunk T.10 of the Basic Fixed Wing Flight. Pilots then proceed to the Advanced Rotary Wing Sqn (ARWS) which has 26 Gazelle AH.1s for basic and advanced helicopter training. From there, successful pilots go to either the Lynx Conversion Flight with 10 Lynx AH.1s or the Scout Conversion Flight with five Scout AH.1s. Gazelle pilots go straight to their active units following an extended course with the ARWS. The Air Engineering Training Wing at Middle Wallop uses a number of redundant airframes for ground instruction in en-

gineering. A Forward Air Control Flight uses a Chipmunk T.10 to train forward air controllers. Also at Middle Wallop is the Development and Trials Sqn which operates single examples of the Scout AH.1, Lynx AH.1 and Gazelle AH.1 for evaluation and development. In the future the Territorial Army is to receive

helicopters, with 666 Sqn being due to form shortly at Netheravon with Westland Scout AH.1s.

Aircraft equipping the AAC include nine Alouette AH.2s, and only about eight DHC-2 Beavers remain in service, though others are stored at RAF Shawbury or are in use as instructional airframes. About 21

DHC.1 Chipmunk T.10s serve as primary trainers. The AAC has 189 Gazelle AH.1s, and deliveries of the Lynx AH.1 now total 113 from orders for 114. The final outstanding aircraft is being converted to interim Lynx AH.5 standard with uprated systems and avionics. A further interim Lynx AH.5 and nine

production versions are currently on order. The Scout AH.1 is being withdrawn from service and only about 40 are in use, although others are held in storage and could be returned to service if required.

West Germany Luftwaffe der Deutschen Bundesrepublik

The Luftwaffe is fully committed to NATO's 2 ATAF and 4 ATAF Air Forces and is currently replacing the Lockheed F-104G Starfighter with the Panavia Tornado.

All combat units are assigned to the Taktische Luftflotten Kommando and divided administratively into four components: 1 Luftwaffe Division at Messstetten, 2 Lw Div at Birkenfeld, Heinrich-Hertz-Kaserne, 3 Lw Div at Kalkar and 4 Lw Div at Aurich, Bluecher-Kaserne. The Tornado is now entering Luftwaffe service in large numbers with a total of about 110 delivered against the 172 strike and 54 trainer versions on order. Initial conversion to the Tornado is carried out at the Tri-national Tornado Training Establishment (TTTE) at RAF Cottesmore, UK, where 26 Luftwaffe Tornados are currently based, along with RAF and AMI examples. From the TTTE, pilots go to Jagdbombergeschwader 38 (Jbg-38) at Jever where 20 Tornados are used for weapons training. The first operational wing, Jbg-31 at Norvenich, has received nearly 40 aircraft and Jbg-32 at Lechfeld has begun to equip with the type. A few other Tornados are used by Erprobungstelle 61 (Est-61) for flight test work. Jbg-32 also has seven HFB.320 Hansa Jets for ECM duties. Eventually Jbg-33 at Buchel and Jbg-34 at Memmingen will also receive the Tornado, but currently they are still operating the Lockheed TF/F-104G Starfighter, and each should still have about 60 on strength. The combined air force and navy procurement of the Starfighter F-104F/F-104G/TF-104G totalled 916 of which nearly 300 were lost in accidents. A total of 66 were transferred to Taiwan from the Starfighter School at Luke AFB in Arizona (2 Ausbildungsgruppe) when it disbanded in 1983, and many have also been transferred to Greece and Turkey.

The tactical reconnaissance role is fulfilled by Aufklärungsgeschwader 51 (Akg-51) at Bremgarten and Akg-52 at Leck, operating the McDonnell Douglas RF-4E Phantom II. Current allocations are two with Est-61 and one maintenance airframe, with 79 contributed between Akg-51 and Akg-52. The second Phantom purchase covered 175 F-4F Phantom IIs for fighter-bomber and air defence use. The two fighter-bomber units are Jbg-35 at Pferdsfeld and Jbg-36 at Rheine-Hopsten. The two air defence units are Jagdgeschwader 71 (Jg-71) at Wittmundhaven and Jg-74 at Neuberg. Of the above total 13 F-4Fs have been lost in accidents. Ten F-

4E Phantom IIs are based with 3 Ausbgp at George AFB in the USA for training, and although finished in full USAF paint schemes they are West German owned.

Deliveries of the Dornier Alpha Jet are now complete with 173 in service and equipping Jbg-41 at Husum, Jbg-43 at Oldenburg and Jbg-49 at Fürstenfeldbruck, each wing having 51 aircraft on strength. Jbg-41 and Jbg-43 are dedicated to the ground attack role with Jbg-49 being unusual in having three component squadrons, unlike Jbg-41 and Jbg-43 which only have two. One of Jbg-49's squadrons is dedicated to ground attack, the second is the Alpha Jet conversion unit and the third operates about 40 surviving Piaggio P.149Ds and a few Dornier Do 28D-2s for pilot screening and initial navigator training respectively. Finally, 18 Alpha Jets are based with the Luftwaffenkommando Beja in Portugal for weapons training. In a crisis this unit would become Jbg-44 and deploy to Leipheim.

The Lufttransportkommando is responsible for controlling the air transport support and has four wings and a VIP transport squadron under its control. Lufttransportgeschwader 61 (Ltg-61) at Landsberg has a squadron of Transall C-160Ds and two squadrons of Bell UH-1Ds, plus a few Dornier Do 28D2 Skyservants for liaison duties. Ltg-62 at Wunstorf has a squadron of C-160Ds but is also the transport operational conversion unit and has an additional squadron of Skyservants. Ltg-63 at Hohn has two squadrons of C-160Ds and a few Skyservants. Hubschraubertransportgeschwader 64 (Htg-64) at Ahlhorn is a helicopter wing with three squadrons of UH-1D Iroquois. In addition, Ltg-61 provides rescue cover at Bremgarten, Goose Bay (Canada), Manching, Neuberg and Pferdsfeld, each with detachments of two UH-1Ds. Similarly, Htg-64 provides cover for Beja, Hopsten, Jever and Norvenich. Additionally, Htg-64 has 12 UH-1Ds at Fassberg with the Hubschrauberführerschule for pilot training. Of the 115 C-160Ds received 20 were sold to Turkey, five were withdrawn from use and one crashed. The Luftwaffe received 138 UH-1D Iroquois and most are still in service. A total of 125 Dornier Do 28D-2 Skyservants were received, including 20 operated by the navy, but budget restrictions have resulted in 43 being withdrawn from use and at least half of these have been sold to Greece and Turkey. In addition to the transport wings, all the Luftwaffe combat wings have verbindungsschwarm (liaison flights) with a few Skyservants. The final unit within the transport command is the VIP transport squadron, the Flugbereitschaftstaffel (FBS). Operating

Luftwaffe Tornados are now an important part of NATO strategy. They carry different weapon loads to RAF aircraft, including the lethal MW-1.

from Cologne/Bonn, it carries out its role with four Boeing 707-307Cs, three Lockheed JetStars, six HFB-320 Hansa Jets, three VFW-614s, six Do 28D-2 Skyservants and four UH-1D Iroquois. The HFB-320s and JetStars are to be replaced by seven Canadair CL-601 Challengers ordered in 1984.

As mentioned earlier, pilot selection is carried out by Jbg-49 and successful candidates then pass to 1 Ausbildungsgruppe (1 Ausbgp) at Sheppard AFB, Texas, where basic training is carried out on the Cessna T-37B and advanced training on the Northrop T-38A Talon. The Luftwaffe initially purchased 47 T-37Bs and 46 T-38As and most are still in service; all have standard USAF training aircraft colour schemes. As noted before, Phantom pilots then go to 3 Ausbgp and to the 3rd squadron of Jbg-36 for acclimatisation to European flying conditions.

Other miscellaneous units are the MilGeoAmt which carries out survey duties from Jever with two BAC Canberras; and the Schiessplatz-

175 F-4F Phantoms were purchased, roughly split between air defence and strike duties. A further 79 RF-4Es are used for reconnaissance. This F-4F is seen prior to refuelling from a USAF Boeing KC-135.

staffeln which provides target towing facilities at Mönchengladbach with four civil registered IAI Westwinds maintained by RFB, at Husum with 24 Fiat G.91s maintained by Condor Flugdienst, and at Lübeck with 16 Rockwell OV-10 Broncos again maintained by Condor Flugdienst. The major ground training technical establishments which use aircraft for instruction are Tslw-1 at Kaufbeuren and Tslw-3 at Fassberg.

Marineflieger

The Marineflieger is tasked with land and ship-based ASW, plus shore-based ASW, attack and defence of the Baltic Sea area.

Four naval air wings are currently active. Marinefliegergeschwader

1 (Mfg-1) at Schleswig/Jagel is the first of two wings that will receive the Panavia Tornado. Mfg-1 now has her full complement of 48 Tornadoes from a total of 96 on order for the navy and comprising 86 strike and 10 dual-control aircraft. Mfg-2 at Eggebek still operates the Lockheed TF/F-104G Starfighter but will eventually receive the other 48 Tornadoes on order for the navy. Mfg-2 has two squadrons of Starfighters, one of which is a reconnaissance unit with RF-104Gs, and these have a total of about 60 aircraft. Mfg-3 at Nordholz has three component squadrons, two of them flying the Breguet 1150 Atlantic. Of 19 Atlantics on inventory, four have been converted for Elint use, leaving 15 for ASW duties. The third squadron of Mfg-3 operates 12 Westland Lynx Mk. 88s for shipboard ASW duties on 'Bremen' class frigates. Two more Lynx were ordered in 1984 for delivery in 1986. Mfg-5 at Kiel/Holtenau comprises one squadron of 20 Dornier Do 28D-2 Skyservants and a second unit with 22 Westland Sea King Mk. 41s. Both types are used for SAR and the Skyservants also have a utility function. Technical training is carried out by the Marineflieger Lehrgruppe at Westerland/Sylt with a few redundant airframes.

Heeresfliegertruppen

The Heeresfliegertruppen is basically divided into three Heeresfliegerkommandos with a few independent units. Its order of battle is as follows:
HeeresFliegerKommando 1 (HFK-1) at Münster/Handorf controls the following units SSHFK-1 (StabStaffelHFK-1) with three MBB BO 105VBHs for HQ staff transport and HFVS-101 (HFlVerbindungsStaffel-101) with 12 BO 105VBHs for HQ communications, both based at Rheine-Bentlage. LHFTR-10 (LiechtesHFlTransportRegiment-10) with 48 UH-1D Iroquois in two squadrons plus five BO 105VBHs for light transport duties based at Fassberg. MHFTR-15 (MittlersHFlTrsRgt-15) with 32 VFW-Sikorsky CH-53Gs and five BO 105VBH in two squadrons for medium airlift capability, based at Rheine-Bentlage. PAR-16 (PanzerAbwehrRegiment-16) with 56 BO 105PAH-1s and five BO 105VBHs in two squadrons for anti-tank duties based at Celle. For divisional use, four HeeresFliegerStaffel (HFS) each with 10 Alouette IIs are in use for liaison, anti-tank duties, transport and airlift of infantry sections with anti-tank weapons; with HFK-1 they are HFS-1 at Hildesheim, HFS-3 and HFS-11 at Rotenburg and HFS-7 at Celle. Similarly, HFK-2 at Laupheim has SSHFK-2 and HFVS-201 at Laupheim with BO 105VBHs, LHFTR-20 at Neuhausen with UH-1Ds and BO 105VBHs, MHFTR-25 at Laupheim with CH-53Gs and BO 105VBHs, PAR-26 at Roth with BO 105PAH-1s and BO 105VBHs, and finally HFS-4 at Straubing, HFS-8 at Penzing and HFS-10 at Neuhausen with BO 105VBHs instead of the Alouette IIs. Finally, HFK-3 at Mendig has SSHFK-3 and HFVS-301 based there with BO 105VBHs. LHFTR-30 at Niederstetten with UH-1Ds and BO 105VBHs and, additionally, LHFTR-30 loans UH-1Ds to 1/FliegerAbteing 301 (1/FA-301) which is a rapid reaction force and would form part of NATO's ACE mobile force. MHFTR-35 at Mendig with CH-53Gs and Alouette IIs in place of 105VBHs. PAR-36 at Fritzlar has BO 105PAH-1s and BO 105VBHs. Finally, HFS-2 at Fritzlar, HFS-5 at Mendig and HFS-12 at Niederstetten Alouette IIs.
Outside the corps structure is HeeresFliegerRegiment-6 (HFR-6) at Itzehoe with three squadrons, one with BO 105VBHs, one with UH-1Ds and one with BO 105PAH-1s, assigned for rapid deployment to support Norway. Army pilots are trained initially in the United States but carry out weapons training with the HerresfliegerWaffenschule (HFWS) at Buckeburg on CH-53Gs, UH-1Ds, Alouette IIs and BO 105PAH-1s. Instructor training is also carried out here. Heeresfliegertruppen equipment includes about 190 of the 204 UH-1Ds that were built under licence by Dornier, and there are about 150 of the 280 Alouette IIs that were received. Most of the balance has been transferred to the Bundesgrenschutz, as well as sold to Turkey, Portugal and for civil use. A total of 100 BO 105VBH unarmed utility helicopters and 212 BO 105 PAH-1 anti-tank versions have been delivered and there are about 107 VFW-Sikorsky CH-53Gs in service. A replacement for the BO 105PAH-1 is in prospect in the shape of the Eurocopter (MBB/Aérospatiale) PAH-2, of which first deliveries are scheduled for 1993. This would allow some BO 105PAH-1s to be converted to BO 105VBHs to complete replacement of the Alouette II.

Eastern Europe

The countries behind the Iron Curtain are, of course, dominated by the Soviet Union, which supplies virtually all their combat equipment. The MiG-21 is still the major fighter of these countries, although token MiG-23 units have been equipped to form a sharp edge to the air defence forces. The Su-25 'Frogfoot' is now being supplied for close support duties, with Czechoslovakia and Hungary being the first confirmed recipients. These are replacing the earlier Su-7 'Fitter'. A notable exception to the Soviet-supplied rule is the Orao, designed and constructed jointly between Soko in Jugoslavia and CNIAR in Romania. In the Soviet Union itself, the air forces are under major re-equipment, with the new fighter trio (MiG-29 'Fulcrum', MiG-31 'Foxhound' and Su-27 'Flanker') now beginning to enter service. The bomber force will be greatly enhanced by the massive supersonic 'Blackjack'.

Albania   Albanian People's Army Air Force

The most introspective European nation, Communist Albania severed diplomatic relations with the USSR in 1961 (and was then expelled from the Warsaw Pact), suffering in 1978 an end to the liaison with China which had brought Shenyang-built combat aircraft. The front-line force, whose serviceability is open to question comprises 20 Chinese-produced MiG-15s, 35 Shenyang F-4s (MiG-17s), 30 F-6s (MiG-19s) and 20 F-7s (MiG-21s). In the interception role these are armed with Chinese SB06 versions of the AA-2 'Atoll' AAM. Transports comprise 10 Pinkiang C-5s (Antonov An-2s), three Ilyushin Il-14s and possibly still a few Lisunov Li-2s (Douglas DC-3s). The sole helicopter type appears to be the Pinkiang H-4 (Mil Mi-4), of which about 20 were received. In the training role are the Yakovlev Yak-11, Nanchang BT-5 (Yak-18) and MiG-15UTI. The army-operated SAM force has CSA-1 Chinese versions of the SA-2 'Guideline'. Much, if not all, equipment is in need of replacement, but Albania has no obvious sources of supply or finance.

Bulgaria   Balgarski Vizdusny Vojski

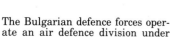

The Bulgarian defence forces operate an air defence division under direct command from Moscow, plus a tactical air force which is part of the army. Stationed at bases around Sofia, Plovdiv and Yambol, the former has its HQ at Sliven and a force of some 280 SA-2 'Guideline', SA-3 'Goa' and SA-4 'Ganef' SAMs. The tactical air force (HQ: Plovdiv) has a single division of three regiments, each with some four squadrons. Air superiority over the battlefield is entrusted to 80 Mikoyan-Gurevich MiG-21 'Fishbed-Cs and -Fs' in six squadrons, plus one token unit with 12 MiG-23MF 'Flogger-Bs'. The fighter-bomber force is of five MiG-17 squadrons, supported by two more with the same aircraft for reconnaissance, or some 100 MiG-17s in all. A helicopter regiment with 10 Mil Mi-2s, 40 Mi-4s and ten Mi-8s was expanded early in the 1980s with the arrival of 12 Mi-24 'Hind' gunships. Transports are based at Sofia/Vrajdebna and comprise 10 An-24RVs, a dwindling number of Ilyushin Il-14s and two VIP aircraft: a Tupolev Tu-134A and a Yakovlev Yak-40. At Bozhurishte a flying school operates the usual Warsaw Pact fleet of Yak-11/18s, Aero L-29s and MiG-15UTIs, students then converting to operational types via the MiG-21U and MiG-23U. There is also a naval air arm responsible for patrols of the Black Sea coast with three Mil Mi-14s, assisted by six Mi-2s and a similar number of Mi-4s.

Czechoslovakia   Ceskoslovenske Letectvo

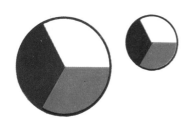

Two air armies comprise the CL, of which the 7th is assigned to air defence and the 10th to tactical support of ground forces. All units are,

Seventy Sukhoi Su-7 'Fitters' form the bulk of the Czech ground attack force, assisted by MiG-21 'Fishbeds' and MiG-23 'Floggers'. The 'Fitters' are now obsolete and are likely to be replaced in the near future.

of course, dedicated to the Warsaw Pact. The 7th air army has two regional divisions, each with three fighter regiments (comprizing three squadrons of 12 aircraft each) or a total establishment of 216. These are mostly all-weather Mikoyan-Gurevich MiG-21 'Fishbed-Ds, -Fs and -Js', but a squadron or two of MiG-23MF 'Flogger-Bs' is also in

service and a few locally-built S.107s (MiG-21F day fighters) may still remain. The 10th air army has four fighter-bomber regiments and three recce squadrons, the latter with 20 MiG-21RFs and 30 camera-equipped Aero L-39 Albatros. Attack aircraft are 40 MiG-17 'Frescos', 40 MiG-21 'Fishbed-Fs and -Js', 70 Sukhoi Su-7 'Fitter-As' and a

squadron of MiG-23BN 'Flogger-Fs', although unconfirmed reports claim a token number of Su-20 'Fitters' also to be in operation. A combat helicopter force includes 12 Mil Mi-24 'Hind' gunships and some of the 70 Mil Mi-4s and 20 Mi-8s in service. Czech-built Avia 14Ts and Let L.410 Turbolets provide transport support, assisted by eight Antonov An-24s,

five An-26s and the ubiquitous An-2, while VIPs fly in three Tupolev Tu-154Bs, four Tu-134As, eight Yakovlev Yak-40s and an Ilyushin Il-62. Main trainers are naturally local Aero L-29s and L-39s, aided by the MiG-15bis, MiG-15UTI, Yak-11/C.11, Zlin 43 and 526, and MiG-21U, MiG-23U and Su-7U conversion models.

East Germany   Luftstreitkräfte und Luftverteidigung (LSK/LV)

Although the Soviet Union attaches great importance to its Central Front with NATO and bases some 1,200 modern combat aircraft in East Germany, the local Air Attack Force and Air Defence Command operates only about 350 advanced front-line aircraft. From a joint HQ at Eggersdorf, LSK/LV controls six interceptor, seven SAM and two radar regiments organized in two geographical commands: 1st Air Defence Division at Cottbus, in the South, and the northerly 3rd ADD at Neubrandenburg. All-weather models of the Mikoyan-Gurevich MiG-21, numbering 250 'Fishbed-D, -F and -Js', form the backbone of the manned interceptor force, supported by some 30 sites for SA-2 'Guideline' and SA-3 'Goa' SAMs. MiG regiments are based at Cottbus, Drewitz and Marxwalde in the 1 ADD area; and Neubrandenburg and Peenemünde in 3 ADD. Soviet MiG-23 'Flogger-B' interceptors arrived in East Germany during 1973 and some have reportedly transferred to

the local air force. Attack-model MiG-23BN 'Flogger-Fs' equip a squadron or two of six assigned to tactical operations in support of the army, the remainder having 35-40 MiG-17 'Frescos' and a dozen MiG-21RF 'Fishbed-H' recce aircraft. The transport wing at Dresden flies Antonov An-2s and An-26s on general duties, plus a VIP force of five Tupolev Tu-134As and seven Let L.410 Turbolets. A transport

helicopter wing at Brandenburg-Briest is equipped with Mil Mi-8s and a combat heli wing at Basepohl flies Mi-8 'Hip-Fs' and about 30 Mi-24 'Hind' gunships. There are 80 Mi-8s in total, supported by a few older Mi-2s and Mi-4s, but the naval helicopter wing at Parow augmented its 'Hip-Cs' from 1982 onwards by Mi-17 'Haze' specialized ASW models. Pilot training is undertaken at Bautzen on the Aero L-29 Delfin and

In common with most countries under Soviet influence, the principal transport aircraft is the Antonov An-26 'Curl'. These are employed alongside the ubiquitous An-2 'Colt'.

L-39 Albatros, supported by Yakovlev Yak-11/18s, Zlin 226s, MiG-15UTIs, MiG-21Us and MiG-23Us. Miscellaneous types include three An-12 Pchelkas and five ECM-configured VEB/Ilyushin Il-14s.

Jugoslavia

A European Communist nation outside the Warsaw Pact, Jugoslavia maintains relations with the Eastern and Western Blocs, although at present most of its 300 combat aircraft are of Soviet or local origin. The JRV received Republic F-84s and North American F-86s from the US during the 1950s and has recently shown interest in the Northrop F-20 Tigershark. Two air corps with HQs at Zagreb and Zemun each have interceptor and ground attack divisions, the former operating 100 MiG-21 'Fishbed-Cs, -Ds and -Js', supported by eight battalions of SA-2 and six of SA-3 SAMs. Attack units are equipped with 100 locally-produced SOKO J-1 Jastrebs and about 20 SOKO P-2 Kraguj COIN lightplanes, the tactical reconnaissance role being fulfilled by RJ-1 versions of the Jastreb and possibly a few remaining Lockheed RT-33As. Modernization of this tactical force is now beginning with the SOKO

Orao (alias IAR-93) developed in conjunction with Romania. About 100 are understood to be required by the JRV. A cosmopolitan transport force flies 12 Antonov An-12s, 10 An-26s, two Douglas DC-6Bs, six Yakovlev Yak-40s, a Dassault Falcon 50, two LearJet 25Bs, a Dornier Do 128D-2, ten Ilyushin Il-14s and nine Pilatus Turbo-Porters. Some indigenous UTVA-66s are used for liaison and three of four Canadair CL-215 'water-bombers' remain.

Leading the helicopter force are most of the 153 SA 341H Gazelles and 100 SA 342Ls acquired from Aérospatiale and mostly built in Jugoslavia; 15 Alouette IIIs; five Agusta-Bell 205As; 30 Mil Mi-8s; and a few Kamov Ka-25s and Mi-4s. Gazelles armed with AT-3 ATMs and SA-7s used as AAMs are known as the 'Partizan.' Local flavour in the training role is provided by at least 30 basic UTVA-75s, 60 advanced SOKO G-2A Galebs and 20 similar

Jugoslavia is another nation which has links with both East and West, while also maintaining an indigenous aviation industry. This Yakovlev Yak-40 is used for general and VIP transport.

JT-1 Jastrebs, while a few T-33s, MiG-15UTIs and MiG-21Us also serve. The swept-wing SOKO G-4 Super Galeb has recently entered service as a basic/advanced jet trainer with light attack capability.

Hungary Magyar Légierö

Financial constraints result in Hun-

gary having the smallest Warsaw Pact air arm, despite which the mistrustful USSR would not, in any event, permit it to return to the size attained before the 1956 uprising. Soviet forces stationed in the country (with HQ at Tokol) have 200 combat aircraft, but the ML appears to have recently lost its fighter-bomber element of 50 Mikoyan-Gurevich MiG-17 'Frescos' and Sukhoi Su-7 'Fitter-As', although some MiG-21F 'Fishbed-C' day-

fighters could be used in their stead. No more modern replacements having been received, the ML is essentially an air defence force. The remaining air division includes two interceptor regiments equipped with up to 120 MiG-21 'Fishbed-Ds and -Fs' and a squadron or two of more capable MiG-23MF 'Flogger-Bs'. Newest combat equipment is the Mil Mi-24 gunship, 20 of which augment 15 Mi-4s and 30 Mi-8s are assigned to army and general support work. Up

to 25 Kamov Ka-26s are in service, mainly with the Danube River Guard. As normal within the Warsaw Pact, there is little use for large transport capacity, so one regiment has a squadron of 12 Antonov An-26s and a handful of An-2s, An-24s and Ilyushin Il-14s. The training organization flies the usual mixture of Aero L-29s, Yak-11/18s, MiG-15UTIs and MiG-21Us, while a few Ilyushin Il-28 Beagles are still used for target-towing, ECM and survey.

Poland Polskie Wojska Lotnicze

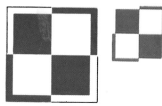

Second in size only to the Soviet air forces within the Warsaw Pact, the PWL operates almost 700 combat aircraft, primarily in the air defence role, alongside a further 350 of the Soviet forces based in Poland. There are 17 PWL fixed-wing air regiments assigned to four divisions, supported by nine regiments with 400 SA-2 'Guidelines' and SA-3 'Goas' at 50 sites. Eleven of these manned interceptor regiments (totalling 33 squadrons) are equipped with 350 Mikoyan-Gurevich MiG-21 'Fishbeds' and 50 MiG-23MF 'Flogger-Bs', forming Obrona Przeciwlot-

nocza, the air defence organization. The remaining six flying regiments are assigned to tactical air support of the army, mainly with 40 Sukhoi Su-7 'Fitter-As' and 80 locally-built LIM-6 versions of the MiG-17. However, one regiment has 40 variable-geometry Sukhoi Su-20 'Fitter-Cs' and another is dedicated to interdiction roles with a similar number of MiG-23BN 'Flogger-Hs'. Tactical reconnaissance is assigned to 35 MiG-21RF 'Fishbed-Hs' and 15 LIM-6s; while a few converted Ilyushin Il-14 transports and Il-28 light bombers are used for ECM and Elint. A small fixed-wing transport force flies a dozen Antonov An-26s and locally-built An-2s, plus two VIP Tupolev Tu-134As and an Il-18. Poland also produces Mil Mi-2 helicopters, 150 of which are in service, including some armed with AT-2 and AT-3s for anti-tank operations in conjunction with 30 Mi-8s and a dozen Mi-24 gunships, plus eight

utility Kamov Ka-26s. The indigenous TS-11 Iskra is used by the flying school at Deblin, aided by the TS-8 Bies, LIM-1 (MiG-15UTI), MiG-21U and Su-7U, plus a few PZL-104 Wilgas. A naval air arm (Morskie Lotnictwo Wojskowe) has

Poland's sizeable air arm uses around 40 of the Sukhoi Su-7 for ground attack, alongside the swing-wing Su-20.

an attack regiment of 40 LIM-5/6s, plus five recce Il-28s and a force of 30 or so Mi-2, Mi-4 and Mi-8 helicopters.

Romania Fortele Aeriene ale Republicii Socialiste Romania

Situated on the Warsaw Pact's southern flank, Romania is allowed remarkable latitude by the USSR – even to the extent of failing to com-

ply with arms expenditure targets. There are 300 combat aircraft in the FARSR's two tactical air divisions assigned to the Bucharest and Cluj military districts. (Iasi military district is supported by Soviet fighters on mobilization.) A separate air defence division is equipped with 108 SA-2 'Guideline' SAMs at 20 sites, while 150 Mikoyan-Gurevich MiG-21 'Fishbed-Cs, -Ds and -Js' and about 50 recently-delivered MiG-23 'Floggers' provide the manned interceptor component. Modernization of the close-support force with its MiG-

17s and Sukhoi Su-7s is being achieved by continuing deliveries of the IAR.93. After protracted development in association with SOKO of Jugoslavia (by which it is known as the Orao), the IAR.93A entered service in the early 1980s, a total of 20 now being followed by 165 IAR.93Bs with afterburning turbojets. Two transport regiments have 10 Antonov An-24s, five An-26s and a few Ilyushin Il-14s and An-2s, while three An-30s have a survey role and ten Ilyushin Il-28s are used for recce and ECM. Helicop-

ters include 45 Alouette IIIs and 15 Pumas built under licence from Aérospatiale, plus 10 Mil Mi-4s, 25 Mi-8s and some Mi-2s. For pilot training, the present course employing 40 IAR-823s, 50 Aero L-29s and 30 MiG-15UTIs will be replaced by a new sequence using the local IAR-825TP Triumpf turboprop (15 delivered for evaluation; 85 on option) and the prospective IAR-99 Soim light jet. Type conversion follows on MiG-21Us, while miscellaneous types used are the IAR-28MA, Yak-18 and Morava L-200.

Soviet Union Soviet Air Armies

What is known for convenience in the West as the 'Soviet Air Force' comprises three separate armed services (in addition to the army and navy). These are the Strategic Rocket Forces (*Raketnyye Voyska Strategicheskovo Naznachiya*) of long-range missiles; Troops of Air Defence (*Voyska Protivovozdushnoy Oborony – Voyska PVO*) with SAMs and manned interceptors; and a more traditional air force (*Voyenno-Vozdushnyye Sily*). The latter has three components: Long-Range Aviation (*Dal'nyaya Aviatsiya*), which is the strategic bomber force;

Frontal Aviation (*Frontovaya Aviatsiya*), the tactical force; and Transport Aviation (*Voyenno-Transportnaya Aviatsiya*).

In association with a re-grouping of forces in 1981, there has been some merging of the above elements so that the peacetime air structure more closely resembles the mobilization configuration. A new command, Headquarters of the Air Armies of the Soviet Union, now co-ordinates strategic and tactical components through five subordinate Theatres of Military Operations. Each TMO has an Air Army (the basic level of command in wartime), comprising strategic bomber, air defence and transport components, plus between three and five Air Forces of the Military District – formerly known as Tactical Air Armies of Frontal Aviation.

Air Defence Forces, with some 500,000 personnel, added the army SAM units to its own sizeable missile element in 1980, although these

are separate from the aircraft essential component, *Aviatsiya PVO*. A new capability is being gained by APVO's 1,250 aircraft with the continuing retirement of some 500 Sukhoi Su-15 'Flagons' and Yakovlev Yak-28 'Firebars' and their replacement by

Now constituting the majority of the Soviet fighter forces, the MiG-23 'Flogger' is replacing the MiG-21 in front-line units. These aircraft are MiG-23MF 'Flogger-Gs', optimized for air-to-air fighting and distinguished by having a shorter dorsal fin.

aircraft armed with look-down/shoot-down radar and AAMs. At present forming the backbone of the force, and with limited look-down ability are 450 Mikoyan-Gurevich MiG-23 'Floggers' and 250 MiG-25 'Foxbats', these now being augmented by the first of 600 more modern aircraft of three types. The MiG-31 'Foxhound' with its pulse-doppler radar entered service in 1983 as a development of the 'Foxbat', while the first MiG-29 'Fulcrum' regiment formed in 1984, the type being a close parallel to the McDonnell Douglas F-18 Hornet. Expected soon, the Sukhoi Su-27 'Flanker' is seen as the equivalent of the McDonnell Douglas F-15 Eagle. Working in conjunction with these potent new interceptors will be the Ilyushin Il-76 'Mainstay' airborne early warning aircraft which has just entered service as a replacement for the Tupolev Tu-126 'Moss'. By the end of the 1980s, 30 Il-76s are expected to have been built.

The assets of Frontal Aviation are now assigned to Air Forces of Military Districts, between three and five of which would come under a theatre commander in wartime. This tactical arm has added a considerable medium-range offensive element in recent years and now comprises 2,850 aircraft (90 per cent of them modern types) with the Western Theatre facing NATO's central Front; 1,250 in the South-Western Theatre; 850 in the Southern; 1,800 in the Far East; and a small number in the North-Western. Heading the interdictor force are over 400 Sukhoi Su-24 'Fencers' with accurate all-weather delivery capability and a range sufficient to reach East Anglia from their home bases. In addition to those *Aviatsiya PVO* interceptors, the tactical units have their own air superiority aircraft for gaining ascendancy over the battlefield. This is primarily the task of 900 MiG-23 'Flogger-Bs and -Gs', although a dwindling number of MiG-21s may be assigned when not engaged in air support. Main air support equipment, however, comprises 800 Sukhoi Su-17 'Fitters' and 650 MiG-27 versions of the 'Flogger' which may be used in close proximity to the army's front line. MiG-21 'Fishbed-Hs' and MiG-25 'Foxbat-Bs and -Ds' meet the reconnaissance need, while aircraft as diverse as the Antonov An-12 'Cub-C' and Yak-28 'Brewer-E' have been converted for ECM and Elint roles. Since 1982 the war in Afghanistan has provided opportunity to evaluate the new Su-25 'Frogfoot' aircraft which is dedicated to close support as a parallel to the USAF's Fairchild A-10 Thunderbolt II. Su-25s are entering service in ever-increasing numbers, although little has been heard recently of a battlefield surveillance aircraft

code-named 'Ram-M' which was under development in the late-1970s to undertake duties similar to the Lockheed TR-1.

Since 1981 units of Long-Range Aviation have been subordinated to theatre commanders and thus available for intervention at a tactical level, if required. Currently there are 45 Myasischev M-4 'Bisons', 115 Tu-95 'Bears', 315 Tu-16 'Badgers', 140 Tu-22 'Blinders' and 140 Tu-22M 'Backfires' in service, of which the last-mentioned presents the greatest threat. Its armament includes the AS-4 'Kitchen' nuclear cruise missile, soon to be replaced by the AS-X-15. The Tu-95 in its 'Bear-H' form has recently been put back into production to carry the same weapon, and earlier models are being converted to 'Bear-G' standard for the AS-4. Almost all the Tu-22 'Blinders' in use for medium-range work also carry AS-4s. Theatre nuclear attack still provides a useful role for the 'Badger', which is armed with AS-6 'Kingfish' or AS-5 'Kelts', although some perform a reconnaissance task. In fact, Long-Range Aviation has about 500 recce, electronic warfare and tanker aircraft for support. Around 1987 its assets will be greatly augmented with first deliveries of the Tupolev 'Blackjack' intercontinental bomber, capable of reaching any target in the USA without inflight refuelling. 'Blackjack' is larger than the Rockwell B-1 and will be able to carry the AS-X-15 and later types of cruise missile. Providing backing for manned bombers are the Strategic Rocket Forces with silos holding 520 SS-11s, 60 SS-13s, 150 SS-17s, 308 SS-18s and 360 SS-19s, the older weapons shortly to be replaced by SS-X-24 and SS-X-25 missiles.

The strategically important Transport Aviation has 600 four-engined aircraft and great quantities of smaller machines, yet even so can call in an emergency on the 1,400 medium and heavy types held by Aeroflot. Modernization of the fleet is under way, with some 300 Il-76 'Candids' delivered so far as replacements for the An-12 'Cub', of which there are about 350 remaining. The four-jet 'Candid' will carry twice its predecessor's load over five times the distance, but the heavyweight of Transport Aviation is the An-22 'Cock'. Fifty of this type are shared with Aeroflot and are the only Soviet aircraft able to move a T62 tank. Within the next two or three years even they will be eclipsed by the An-124, the proto-

type of which is currently the largest aircraft flying in the world.

Aviatsiya Voyenno-Morskoy Flot

Soviet Naval Aviation is a sizeable force of about 1,500 aircraft assigned to the Baltic (275 aircraft), Black Sea (405), Northern (425) and Pacific (440) Fleets. Its most profound change in recent years has involved the commissioning between 1976 and 1984 of four Kiev-class ASW carriers, each carrying a dozen Yak-38 'Forgers'. These rudimentary shipboard VTOL attack aircraft are technologically far in arrears of the BAe Sea Harrier, but may gain operational experience to assist design of more advanced aircraft. In prospect is a full-sized aircraft carrier with a complement of new catapult-launched fighters. ASW duties from ship and shore are undertaken by 180 Kamov Ka-25 'Hormones', which have recently begun to give way to Ka-27 'Helixes'. AVMF is the only navy to possess heavy bombers, over 300 of which fly from coastal bases for maritime strike. There are 110 Tu-22M 'Backfires' armed with AS-4 cruise missiles; and 250 Tu-16 'Badgers' carrying AS-2 'Kippers' and AS-5 'Kelts' for the tactical role (occasionally replaced by two AS-6 'Kingfish'.) In addition, 35 smaller Sukhoi Su-17 'Fitter-Cs' operate in

The Tu-22M/26 'Backfire-B' forms the backbone of the Soviet strategic and maritime attack force, although this is to be supplemented by the 'Blackjack'.

the Baltic with AS-7 'Kerrys'. Fixed-wing ASW and recce aircraft comprise 60 Il-38 'Mays', 80 Beriev Be-12 'Mail' amphibians, 95 Tu-142 'Bears' and 40 Tu-22 'Blinders', assisted by shore-based Mil Mi-14 'Haze' helicopters.

Sukhopputnyye Voyska

The Soviet Army recently formed an aviation branch from the rotary-wing component of Frontal Aviation, although this has merely served to formalize an existing situation. Helicopters are assigned at various levels of the army command for gunship, anti-tank, assault, re-supply, heavy-lift and observation duties. There are 1,600 Mi-8 'Hips' and a growing number of improved Mi-17 'Hip-Hs', over 1,000 Mi-24 'Hinds', many of them 'Hind-D' and subsequent strongly-armed types, and 400 heavy-lift Mi-6 'Hooks'. The even larger Mi-26 'Halo' is appearing in military insignia, while at the lighter end of the spectrum the next few years will see service-entry of the Mi-28 'Havoc' (a parallel to the Hughes AH-64 Apache attack helicopter) and a new air-to-air combat machine code-named 'Hokum'.

The AV-MF maintains many Tu-16 'Badgers' for maritime strike and for various reconnaissance and ECM duties. This is a 'Badger-F', identified by two Elint pods carried under the wings, with other equipment in the fuselage.

North Africa and the Middle East

Egypt has now fully joined the Western sphere of influence with the arrival of her General Dynamics F-16A/Bs and Lockheed C-130H Hercules, along with continuing supplies of French Dassault Mirage 5SDE, Gazelle helicopters, licence-built production of Dassault-Breguet Alpha Jets and supplies from Canada of the DHC-5D Buffalo. In the Western Sahara portion of Morocco the Polisario guerrillas are still active with Algerian and Libyan assistance but no major clashes have occurred recently. Libya is still rapidly expanding her armed forces with Soviet equipment and technicians, and though her borders with Egypt and the Sudan are now relatively quiet she invaded Chad in 1983. Libya only recently

agreed to withdraw following lengthy negotiations with France, who deployed forces into Chad to counter the Libyan advance. Libya continues to claim territorial water rights about the Gulf of Sirte but two Sukhoi Su-22s that tried to enforce this claim against two US Navy carrier-borne Grumman F-14A Tomcats were quickly shot down.

In the Middle East, Israel is involved with producing the Lavi multi-role fighter to maintain its air defences, while Saudi Arabia has ordered a large batch of Tornados. Iraq and Syria both receive help from Moscow, while Iran is trying to keep its American supplied aircraft in the air to continue the war with Iraq.

Algeria Al Quwwat al Jawwiya al Jaza'Eriya

After receiving independence from France in 1962 Algeria received only two Beech C-45s as the ex-Colonial package, but began to receive Soviet aid later that year. By late 1965 20 MiG-15s, 40 MiG-17s, 24 Ilyushin Il-28s, 12 C-11s (Czech-built Yak-11s), 12 Gomhuriah trainers, eight Ilyushin Il-14s, eight Antonov An-12s, 18 Mil Mi-4s and six MiG-21s had been received. The surviving MiG-15bis remain in use as armed trainers. Around 60 MiG-17s were eventually delivered and about a third of them remain in the fighter-bomber role, being replaced gradually by MiG-23BMs of which 40 have been received. As many as 95 MiG-21F/MFs were supplied to equip three squadrons, most of which are still in use in the air defence role today, but have been supplemented by a squadron of 18 MiG-25s. Four MiG-25R reconnaissance aircraft are also in use, but doubts exist as to whether the MiG-25s are crewed by Algerians or Russian advisers. In the strike role are about 20 Sukhoi Su-7BMs and Su-20s, the surviving Ilyushin Il-28s having been relegated to a secondary role. Reports of MiG-19s in service are unconfirmed. In the light strike role are two squadrons operating the survivors of 28 ex-Luftwaffe Fouga CM.170 Magisters refurbished by France and delivered from 1970 onwards. An order has been placed recently for 12 BAe Hawks for delivery in 1985.

The transport fleet has expanded greatly over recent years with the arrival of 20 Lockheed C-130H Hercules, the latest of which were delivered as recently as October 1984. Some Hercules are in a grey and white civil colour scheme, but all are believed to be operated by the air force. The Hercules joined eight Antonov An-12s and three Fokker F.27-400M Friendships (the other eight having been transferred to Air Algérie). The Il-14s were withdrawn some years ago. For training, six Beech T-34C-1s, four Beech B24R Sierras, several MiG-21Us, two MiG-23Us, four MiG-25Us, a few C-11s, Gomhuriahs and some Su-7Vs are in use. Operating in the liaison and VIP field are a Beech Queen Air 65, three Beech King Air 90s and three Beech Super King Air 200s. In government use are three Grumman Gulfstream IIIs, a Grumman Gulfstream II, an Ilyushin Il-18, two Dassault Falcon 20s and three Beech Super King Air 200s.

Helicopters in use include up to 40 Mil Mi-24s, five Aérospatiale SA.330 Pumas, seven Hughes 269As, 28 Mil Mi-4s, at least four Mil Mi-6s, 12 Mil Mi-8s and possibly two Aérospatiale SE.3130 Alouette IIs.

Egypt Al Quwwat al Jawwiya Ilmisriya

Following the severing of ties with the Soviet Union in the mid-seventies and Camp David Peace Agreement with Israel in 1979, Egypt has become the recipient of mainly Western equipment. However, an agreement with the People's Republic of China in 1976 has resulted in Egypt receiving quantities of Shenyang F-6s (Chinese built MiG-19) in return for supplying the PRC with MiG-23s for possible copy and production.

The air defence role is fulfilled by several types, including the Dassault Mirage 5SDE/SSE/SDE2 of which 32/22/16 respectively were acquired between 1975 and 1983. The two Mirage wings are based at Tanta and Genaclis, the former also being home to the operational conversion unit equipped with six Mirage 5DDs, and at the latter is based a reconnaissance unit with six Mirage 5DRs. The General Dynamics F-16 is also now in use, with 34 F-16As and six F-16Bs delivered by the end of 1983 to Nos 72 and 74 Sqns of the 232nd Tactical Fighter Brigade at Inchas, replacing MiG-21s. The MiG-21 still serves in the air defence role, about 100 being operational from over 400 MiG-21F, MiG-21PFS, MiG-21PFM, MiG-21FL, MiG-21MF aircraft supplied, along with about 207 MiG-21RF reconnaissance versions and a few MiG-21U two-seat trainers. Following the agreement with China, approximately 30 Xian F-7s (Chinese built MiG-21) were delivered in 1980 and a further 80 were reportedly being assembled at Genaclis in 1983 with Chinese assistance. Known MiG-21/F-7 bases are Wadi Natrun North, Fayid (F-7), Gebel El Basur, Al Mansurah and Mersa Matruh. The MiG-21RFs operate from Genaclis. All remaining MiG-23s are handed to the US and China for evaluation.

Several types of fighter-bomber are operated, including 35 McDonnell Douglas F-4E Phantom IIs delivered from 1979 onward to 222 Tactical Fighter Brigade at Cairo West, but at least two have been lost in accidents and overall serviceability was believed to be very low. An attempt to sell them to Turkey in 1983 fell through, but continued US maintenance assistance has reportedly improved airworthiness. The Shenyang F-6 (Chinese built MiG-19) is also used. In 1979 40 were delivered, followed by more aircraft for Egyptian assembly in 1983, and the total in service could exceed 80. At least one wing of these is based at Beni Suef and another unit may be at Bilbeis with the Air Academy. Approximately 120 MiG-17Fs were received from the Soviet Union and about half of them remain in use at Almaza, Draw and the Academy at Bilbeis. Other Soviet aid included quantities of Sukhoi Su-7s and Su-20s; these were to have undergone a series of refurbishment schemes that were subsequently cancelled, so doubts exist as to whether many of these remain in use. A more recent ground attack aircraft now entering service is the Dassault-Breguet/Dornier-Helwan Alpha Jet MS2, of which 15 are in the process of being delivered. The first four are French built, the following 11 being assembled at Helwan. A few of approximately 24 Tupolev Tu-22 'Badgers' are still in service in the long-range strike role.

The transport fleet now relies heavily on Western types with 23 Lockheed C-130H Hercules delivered between 1976 and 1982. At least one C-130H has been lost and two are reported to be EC-130H ECM/airborne command post versions. Ten DHC-5D Buffaloes were delivered in 1982, though one was lost during 1984. These have replaced the Soviet supplied Ilyushin Il-14Ms, Antonov An-12Bs and An-24s although a few An-12Bs may remain operational. A handful of Antonov An-2s and one Fairchild C-123B Provider are also in service. VIP duties are performed by a government unit, based at Cairo West along with all the other transport aircraft, operating a Boeing 707-366C, a 737-266 and two Dassault Falcon 20s.

In the training role, as well as the Mirage 5SDD, MiG-21U and F-16B mentioned earlier, are the survivors from 200 Helwan Gomhuriahs based at Bilbeis with the Air Academy for basic training. Dassault-Breguet/Dornier-Helwan Alpha Jet MS1s, of which 30 are on order, are being delivered to Bilbeis and will replace the Aero L-29 Delfin at El Minya for jet conversion; four are being French-built and the remainder assembled at Helwan. A few MiG-15UTIs remain in use and tactical training is carried out on the MiG-17F at Draw, though Bilbeis also has a few F-6, MiG-17 and Su-7s for some initial conversion duties. A few PZL-104 Wilgas are in service for liaison duties, but all the Fournier RF-5Bs have been withdrawn. Navigators are trained at Bilbeis on a few surviving Ilyushin Il-14s. A small number of Helwan HA.200 Al Kahiras are in use for training, along with a few Zlin Z.526s and Ilyushin Il-28s.

At least two types of Soviet helicopters remain in service, including an estimated 50 Mil Mi-8s from an original complement of about 80 and six Mil Mi-6s for heavy lift duties. Some Mil Mi-4s may also still be in use. Egypt has been a large purchaser of the Westland Commando since 1974, with five Mk.1s, 17 Mk.2s, two Mk.2Bs (VIP) and four Mk.2Es. The latest were delivered in 1980 and other recent helicopter deliveries include two Agusta-Sikorsky S-61s (1983), 20 Hiller UH-12Es (1982), and 15 Meridionali-Vertol CH-47Cs delivered to Kom Awshim from 1981 onwards. The most numerous helicopter in EAF service is the Aérospatiale SA.342 Gazelle of which four SA.342Ks and at least 60 SA.342Ls were delivered. The latest order, placed in 1981, covers 36 SA.342Ls of which 30 will be assembled by Helwan.

On order for the EAF are 16 Dassault Mirage 2000EM (reportedly increased to 40) and four Mirage 2000BM trainers, with deliveries expected to begin in 1985. A further 40 General Dynamics F-16s are scheduled for delivery from 1986 onwards; they are reported to comprise 34 F-16C and six F-16D, but one F-16B was noted under construction in late 1984. For training use 30 EMBRAER EMB-312 Tucanos are being procured as a bridge between the Gomhuriah and the Alpha Jet MS1. The four Grumman E2C Hawkeyes scheduled for late 1985 delivery will give Egypt her first airborne early warning capability.

Naval aviation consists of only one ASW unit, based at Alexandria, with six Westland Sea King Mk.47s.

Following Egypt's split from the Soviet camp, it has received 40 F-16s for air defence duties. These serve alongside MiG-21s and Mirage 5s.

A long war of attrition, begun in September 1980 with an assault by Iraq, has further weakened an air arm already undermined by the revolution of February 1979. International embargoes – not least of which is that imposed by the US – have made the acquisition of spares difficult, so that many aircraft not lost in battle remain unserviceable. An air arm built-up to major proportions by the former Shah, largely with US assistance and equipment, is now unable to guarantee the integrity of home airspace or launch a decisive air offensive against Iraq, its lack of equipment being compounded by faulty tactics and inadequately-trained personnel.

Principal interceptor of the IRIAF is the Grumman F-14A Tomcat and its associated Hughes AIM-54A Phoenix AAM – the importance to US policy of the Shah being indicated by the fact that Iran is the only export customer for this sophisticated team. However, no Phoenix had been fired by late 1985, and a handful of operable Tomcats from 80 delivered appear to have been used more for airborne early warning than interception (attempting to replace the seven Boeing E-3A Sentry AWACS cancelled after the revolution). The aircraft operate with one squadron which has detachments at Mehrabad, Bushehr and Shiraz.

Having received 32 McDonnell Douglas F-4D, 177 F-4E and 16 RF-4E Phantoms for interception, attack and reconnaissance, the IRIAF has only some 45 serviceable, although this figure will decline with the ending in 1984 of clandestine Israeli spares supplies. Few, if any, units have been disbanded so that 13 under-strength squadrons fly the surviving F-4s from Mehrabad (two squadrons), Tabriz (two),

Hamadan (two), Dezful, Bushehr (two), Shiraz (two), Bandar Abbas and Chabhar. In future, some may be operated from the new air base being built at Aghajari/Umidiyeh (north of the major oil terminal at Kharg Island).

Only 55 Northrop F-5E Tiger IIs were available in mid-1984 from 140 delivered together with 28 F-5F trainers. These are stationed at Mehrabad (two squadrons), Tabriz (two), Hamadan, Dezful (two) and Isfahan. The IRIAF has exhausted supplies of precision attack weapons such as Hughes AGM-54A Maverick ASMs and laser-guided bombs supplied with Phantoms and Tigers. Two of the original six Lockheed P-3F Orions remain in service for maritime patrol operations from Bandar Abbas.

The reduction of combat strength to 20 per cent of its pre-revolutionary level has been matched by a fall in the number of transports to 25. At the main base of Mehrabad four Boeing 747s remain from 16, and sorties are undertaken by some of the 14 Boeing 707s, including four with inflight refuelling tanker capability and a VIP aircraft. The Lockheed Hercules force, once numbering 21 C-130Es and 43 C-130Hs, has been cut to a dozen or so with units at Mehrabad (two squadrons), Shiraz (two) and Chabahar, plus a detachment at Tabriz. A few Fokker F.27 Friendships fly from Shiraz, but the communications fleet of two Lockheed JetStars, three Rockwell Commander 681Bs and three Dassault

Falcon 20Fs may be unserviceable.

A modest IRIAF helicopter fleet of two Meridionali-Vertol CH-47C Chinooks, 30 Bell 214As, two Agusta-Sikorsky S-61A-4s and two Agusta-Bell JetRangers has likewise been decimated. As its main primary trainer, the air force operates from Mehrabad the survivors of 49 Beech Bonanzas. Basic instruction is now undertaken on 36 Pilatus PC-7 Turbo trainers supplied in 1983-84 as a replacement for facilities once provided at USAF schools, and even a few Lockheed T-33As are to be seen at Mehrabad providing advanced flying tuition.

Islamic Republic of Iran Navy

According to a former C-in-C, who defected to the West in 1984, the naval air arm has lost 80 per cent of the 50 helicopters with which it began the Gulf War. The main anti-submarine type is the Agusta-Sikorsky SH-3D Sea King, of which 15 were in service, while in the mine-sweeping role two Sikorsky RH-53Ds (from six) operate from the sole naval helicopter base at Bandar Abbas. There may also be a few Agusta-Bell AB.205As, AB.206A JetRangers and AB.212s left.

The fixed-wing communications fleet is based at Mehrabad and includes some of the original four F.27 Friendships. At one time there were also four Falcon 20Es, three Rockwell Commander 500Ss and seven Shrike Commanders.

Operating from Bandar Abbas, Lockheed P-3 Orions are still active on maritime patrol duties. The Orion force has suffered the same fate as other US-supplied types. Six were originally supplied.

Islamic Republic of Iran Army

The immense helicopter force which would have been of value to Iran in its conflict with neighbouring Iraq has been mostly lost in combat, rendered unserviceable or cancelled before delivery. Expansion of army aviation began in the mid-1970s, augmenting the original 25 Bell 205As and 23 Bell 206s with 20 AB.205s and 91 AB.206s for light transport and liaison respectively. The transport fleet was further boosted by 287 Bell 214s, although co-production in Iran of a further 400 was a victim of the revolution.

In furtherance of its aim of acquiring a 'sky cavalry' force, Iran bought 202 Bell AH-1J SeaCobra attack helicopters armed with Hughes TOW anti-tank missiles. For medium lift support, orders were placed for 90 Boeing-Vertol CH-47C Chinooks, of which only 65 have been received, including 45 built by Meridionali. Of the total force, some 150 helicopters remain, including only a few AH-1Js. Fixed-wing aircraft comprise two Falcon 20Es, two Friendships (one with target-towing equipment), five Shrike Commanders and 45 Cessnas, few of which are operable.

Iraq Al Quwwat al Jawwiya al Iraqiya

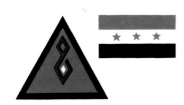

Locked in combat with Iran since its invasion of September 1980, Iraq has been obtaining arms from various sources to sustain the war effort and dilute the 95 per cent Soviet Bloc content of a decade ago. Diplomatic links were restored with the US in 1984, and there are suggestions of further large contracts for Western aerospace firms when the war is over, although France has gained the status of a major supplier during hostilities. The USSR replaced losses on a one-for-one basis, later pledging additional equipment. In the interceptor role are about ten Mikoyan-Gurevich MiG-25 'Foxbat-As' and 20 MiG-23 'Flog-

ger-Es', backed by over 150 MiG-21 'Fishbeds'. Some 50 are 'Fishbed-Fs' and 40 are 'Fishbed-Js', but the 80 'Fishbed-Cs' in the inventory are Chinese-built Shenyang F-7s delivered from 1982. 'Fishbeds' are equipped with AA-2 'Atoll' and MATRA R.550 Magic AAMs. Like the USSR, Iraq has a separate air defence organisation which includes army-operated SA-2, SA-3 and SA-6 SAMs.

Limited action over Iran has been seen by about nine Tupolev Tu-22 'Blinder' supersonic bombers which spearhead the attack force, these dropping conventional bombs despite the AS-4 'Kitchen' long-range ASMs with which they were delivered in 1974. Eight older Tupolev Tu-16 'Badger' bombers may still operate. There is a sizeable attack force, the oldest members of which are 80 Sukhoi Su-7 'Fitter-As', 30 MiG-17 'Frescos', 20 Hawker Siddeley Hunters and 40 or so Shenyang F-6 'Farmer-C' variants of the MiG-19 – possibly all relegated to training. More modern equipment

comprises 40 Su-17/20 'Fitter C/Js', 70 MiG-23 'Flogger-Fs' and 89 multirole Dassault Mirage F1s. The first 36 Mirages (30 F1EQs and six trainer F1BQs) were delivered from January 1981 onwards, and followed in 1983 by 23 F1EQ-200s and six F1BQ-200s, both with inflight refuelling probes. In October 1984 delivery began of 24 F1EQ5-200s and compatible trainers fitted with

Iraq relies heavily on Soviet equipment, although some French aircraft are also used. The transport fleet uses Antonov An-12s (illustrated) and Ilyushin Il-76s.

Agave radar (in place of Cyrano) and Aérospatiale AM-39 Exocet anti-ship missiles. These aircraft will be capable of continuing the naval missile attacks begun by Iraq's 11 Aérospatiale Super Frelons and intensified after five Dassault Super Etendards were diverted from the French navy in 1983. Eight MiG-25R 'Foxbat-Bs' are used for reconnaissance.

In support roles are transport aircraft and trainers, the former comprising six Antonov An-12 'Cubs', 24 An-24 'Cokes', a couple of An-26 'Curls', a few ageing Ilyushin Il-14 'Crates' and some 10 An-2 'Colt' biplanes. Twenty CASA C-212

Aviocars are reported to be on order and various civil aircraft of Iraqi Airways (Super Hercules, Il-76s, Tu-124s and executive jets) assist the air force. Six Gates Learjet 35As are assigned to survey, target-towing and VIP duties. The large training fleet has 40 Swiss FFA AS.202/18S Bravos and 52 Pilatus PC-7 Turbo Trainers, plus 48 Czech Aero L-39 Albatros for pilot instruction, and is adding 40 (with 40 more on option) Brazilian EMBRAER EMB-312 Tucanos from Egyptian licence

manufacture. Other training is undertaken on a few Aero L-29 Delfins, MiG-17s and MiG-15UTI 'Midgets'.

An important role is played in the war by helicopters, the Aérospatiale stable being represented by 30 Alouette IIIs (AS-11 or AS-12 ASMs), 50 or so SA.342L Gazelles (Euromissile HOT ATMs) and 20 SA.330 Pumas. Over 50 Mil Mi-8 'Hips' and a dozen Mi-24 'Hind' gunships from the USSR are supported by 20 Mi-4 'Hounds' and 15

heavy-lift Mi-6 'Hooks', while 40 MBB BO 105s (with 32 more reportedly on order) are also armed with HOTs. Late in 1983, Italy received naval helicopter orders believed to be for five Agusta-Bell 212s and three Agusta A.109s configured for anti-shipping attack, as a follow-on to deliveries of six Agusta-Sikorsky AS-61TS variants. Helicopter training is provided by 30 Hughes 300Cs delivered in 1983, accompanied by a similar number of 500Ds.

Israel La Tsvah Hagana Le Israel/Heyl Ha'Avir

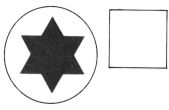

Despite incurring the wrath of its American backer on several occasions, and being subject to more than one remarkably brief arms embargo, Israel continues to receive massive US aid, amounting to one quarter of that supplied by Washington to the entire globe. This includes, for the 1985 fiscal year, a free military equipment grant of $1,400 million, while total yearly defence expenditure by Israel is some $4,000 million. (The exact sum is difficult to compute as, because of rampant inflation, the defence budget is set for three-month periods only).

War and tension between Israel and its Arab neighbours has produced in the IDF/AF one of the world's most efficient air arms. Spearheading the all-weather interceptor force from December 1976 onwards is the McDonnell Douglas F-15 Eagle, supplied in batches of 25, 15 and (ordered in 1982) 11. Eagle weaponry includes the AIM-9L Sidewinder and AIM-7M Sparrow short- and medium-range AAMs, in addition to earlier marks of these missiles from the US. They have been responsible for Eagles destroying over 50 opposing aircraft for no loss to themselves. Indirectly sharing in these victories are the four Grumman E-2C Hawkeye airborne surveillance and control systems delivered from December 1977. Hawkeyes have proved vital in coordinating defensive and offensive operations and are supported in a most competent intelligence-gathering and ECM force by two Grumman EV-1E Mohawks, four Beech RC-12Ds and a few Beech RU-21s, amongst others.

In the multirole category are the General Dynamics F-16 Fighting Falcon and locally-built IAI Kfir. An initial 67 F-16As and eight F-16B trainers was supplied from January 1980 for three squadrons, but when a further 75 F-16C/Ds were ordered, to be supplied from 1987, the alternate General Electric F110 turbofan engine was specified in place of the P & W F100. About 200 Kfirs are in service as interceptors and fighter-bombers, following the sale to Argentina of the surviving Dassault-built Mirage IIICJs and unlicensed IAI copies known as Neshers. The Kfir has a US J79 engine in place of the French Atar and provision for optional canards in its Kfir C2 form. An uprated engine and improved avionics are to be found in the current Kfir C7 variant, while there are also TC2 and TC7 trainer equivalents.

Some 120 McDonnell Douglas F-4E Phantoms and a dozen reconnaissance RF-4Es remain from 204/18 deliveries in 1969-78, the former now largely relegated to attack with precision weapons such as the Shrike, Maverick and (soon) local Gabriel 3A/S missiles, plus laser-guided bombs. Close support is still provided by just over 100 McDonnell Douglas A-4 Skyhawks, aided by 20 TA-4 trainers, the original 268 or so having been reduced by sales and storage, as well as battle losses. Replacement of Kfir and Skyhawk will be undertaken by 300 IAI Lavi attack aircraft due to be delivered from 1989 onwards after prototype first flight in February 1986. The Lavi programme involves a crippling $1,700 development cost, while the individual aircraft price has at least doubled from the planned $10m.

The IDF/AF transport fleet includes a dozen or so Boeing 707s which have both civil and military identities, including five equipped for inflight refuelling and a similar number assigned to gather electronic intelligence in peacetime and provide stand-off jamming during a

conflict. More regular tactical transport facilities are provided by 12 Lockheed C-130E and eight C-130H Hercules, a further two being KC-130H tankers. A squadron of venerable Douglas C-47 Dakotas continues to serve, as do 14 Dornier Do 28Bs and three IAI Westwind jets. Three more Westwinds, fitted with appropriate radar, are assigned to naval surveillance.

Helicopter support for the army is available in a variety of forms, and in the attack or gunship role are around 35 Bell AH-1G/Q/S Cobras and 32 (plus six more planned) Hughes 500MDs armed with the TOW anti-tank missile. For rapid deployment of troops and equipment there are 12 Aérospatiale Super Frelons (converted to General Electric T58 turboshaft power) and 30 Sikorsky S-65s (from 33 CH-53Ds and two ex-Austrian S-65Os re-

Symbolic of Israel's determination, this IAI Kfir carries the Rafael Python 3 missile. Following numerous arms embargoes, Israel is turning to its own industry to produce its weapons.

ceived). Light transports are 25 Bell 212s, while over 30 Bell OH-58A Kiowas are used for liaison and observation.

The training organisation includes a few Piper Super Cubs for grading and some 70 Fouga Magisters (many licence-produced) which are being converted to IAI Tzoukit standard, involving a complete rebuild with extensive modifications. Twenty Beech Queen Airs are used for twin-engine training. Liaison aircraft include 25 Cessna U206s and a few Model 172s, 180s and 185s. The IDF/AF makes considerable use of RPVs, including the IAI Scout and Tadiran Mastiff for reconnaissance.

Jordan Al Quwwat al Jawwiya Almalakia al Urduniya

Jordan is in the latter stages of an improvement programme for its air force, involving the replacement of Lockheed Starfighters, Northrop F-5As and Hawker Hunters by Das-

sault Mirage F1s and Northrop F-5Es. An Arab country with Western leanings, Jordan relies on Europe and the US for many of its aircraft and missiles, although the Jewish lobby in Washington ensures that it does not receive unlimited military support. Having failed to obtain General Dynamics F-16s, the RJAF turned to France for a Starfighter replacement and, with Saudi help, bought 17 Mirage F1CJ interceptors, 17 multirole F1EJs and two F1BJ trainers in 1979, then added a further ten aircraft in 1983. No. 25 Sqn Azraq received its F1CJs in

1981, and No. 1 took delivery of F1EJs two years later.

No. 2 Sqn flies the few remaining F-5As from Mafraq in the advanced training role, these surviving from 30 (and four F-5Bs) passed on by Iran. Beginning in 1975, 55 improved F-5E Tiger IIs and eight F-5F trainers were supplied to four squadrons, including Nos 9 and 17 at Prince Hassan AB, and No. 11. The F-5 is used for interception and ground attack, employing AIM-9J and AIM-9P Sidewinder AAMs in the former role. An order has been placed for Hughes Maverick ASMs,

allegedly the AGM-65D imaging infra-red model, and deliveries are taking place in 1985 of 24 Bell AH-1S Cobra attack helicopters armed with Hughes BGM-71 TOWs for anti-tank operations.

Static defence of airfields and radar sites is provided by 14 batteries totalling 532 Raytheon MIM-23B Improved Hawk SAMs delivered in 1977-78 as part of a $540 million package including 100 GE Vulcan AA guns. When, in 1981, the US proved reluctant to deliver more Hawks in mobile form, Iraq underwrote a $200m order for 20 batteries of Soviet SA-8 'Gecko' SAMs, ZSU-23-4 cannon and SA-7 'Grail' infantry SAMs.

Most transports are based at Amman/Marka, where No. 3 Sqn has two Lockheed C-130Bs and three C-130H Hercules, plus the surviving two of four CASA C-212 Aviocars. Three Antonov An-12 'Cubs' are also flown, these apparently donated by Iraq and used for collecting Iraqi Shenyang F-6 fighters from China. Helicopter units are No. 5 Sqn in the training role at Mafraq with eight Hughes 500Ds, and No. 7 at Amman with 18 Sikorsky S-76s (plus a few

remaining Aérospatiale Alouette IIIs).

Pilot training begins on the survivors of 22 BAe Bulldogs at the RJAF Academy, Mafraq, and continues at the same base with No. 6 Sqn's 13 Cessna T-37B/Cs. The advanced stage is conducted on No. 2 Sqn F-5A/Bs. Several civil registered jets are used on official business, including King Hussein's Boeing 727.

The most potent aircraft in the Jordanian air force is the Dassault-Breguet Mirage F1. Thirty-six were supplied, of which 17 are dual-role F.1EJ fighter-bombers.

Kuwait   Kuwait Air Force

Non-aligned apart from its support for Iraq in the Gulf War, Kuwait buys arms from both West and East, the latter – in the form of the USSR – having delivered or promised SA-6, SA-7 and SA-8 SAMs, as well as FROG 7 unguided SSMs. Ironically, 1973 border clashes with Iraq prompted large-scale orders for the equipment now in the KAF's front line, led by 16 Dassault Mirage F1CK interceptors and a single F1BK trainer from 18/2 bought. These were delivered in 1976-77 together with MATRA R.530 and R.550 Magic AAMs, while a dozen more Mirage F1s ordered in 1983 are

expected to have the new MATRA ARMAT anti-radiation ASM included in their weapon options. The Mirages replace little-used BAe Lightnings which have spent the time since 1977 in storage.

Close support is provided by some of the 30 McDonnell Douglas A-4KU and six trainer TA-4KU Skyhawks supplied in 1977-78 and based on the A-4M and TA-4J models respectively. Mirages and Skyhawks have provision for inflight refuelling, although the KAF has no known tanker aircraft. In training roles, but with light attack potential, are the remaining nine of 12 BAe Strikemaster Mk 83s and a dozen BAe Hawk Mk 64s bought in October 1983.

An armed role is also foreseen for some of the French helicopters supplied from 1975 onwards. These include over 30 Aérospatiale SA.342K Gazelles, half of them carrying Euromissile HOT ATMs, with the rest assigned to observation; nine SA.330H Pumas for army support;

and six AS.332F Super Pumas with Aérospatiale AM-39 Exocets for anti-shipping attack. France has also (in 1983) received an order to improve the air defence reporting system, which includes links with Raytheon MIM-23B Improved HAWK SAMs supplied by the US. A data transfer system will also be established with Boeing E-3A Sentry AWACS aircraft operating in Saudi Arabia, and may lead to wider

Kuwait's nine Strikemaster Mk 83s are mainly used for training, but their light attack capability can be called upon.

defence co-operation between the more moderate Arab nations.

Transport requirements are modest, being met by two McDonnell Douglas C-9Ks (DC-9 Srs 32s) and four Lockheed L-100-30 Super Hercules.

Lebanon   Al Quwwat al Jawwiya al Lubnamia

Having survived civil war, Syrian

occupation and Israeli invasion, shattered Lebanon is attempting to return to normality with armed forces strong enough to maintain the rule of law. The LAF has taken little part in the recent fighting, except for a brief appearance in September 1983 to support the army against attacks by militia forces. This resulted in a HS Hunter and a BAe Bulldog being shot down.

In theory, the LAF's principal interceptor is the Dassault Mirage III,

but nine IIIELs and a single IIIBL trainer have been in storage for some time. Ground attack duties are assigned to the surviving 10 Hunters (including two T.66 trainers), while five Fouga Magisters may also be armed for similar operations if required. Five of the six Bulldog Srs 126 primary trainers remain and appear to have an operational role of forward air control and reconnaissance.

Army support has been improved

with delivery of additional helicopters since 1980, augmenting two Aérospatiale Alouette IIs, a dozen Alouette IIIs and six Agusta-Bell 212s. The new equipment comprises six more AB.212s (two of which have been lost), ten Aérospatiale SA.342L Gazelles (armed with SS.11 or SS.12 ASMs, plus a 20-mm/0.79-in cannon) and six SA.330L Pumas. Light transport is provided by a DH Dove and a Rockwell Turbo Commander 690A.

Libya   Al Quwwat al Jawwiya al Libiyya

Massive Soviet aid and imported equipment from France, along with Soviet, Syrian, North Korean and Pakistani personnel to operate the more sophisticated items, has resulted in Libya having an air force far larger than necessary for her defensive needs. In recent years Libyan forces have been involved in border actions against Egypt and the Sudan. They also invaded Northern Chad in 1983 and have only recently agreed to withdraw, following

pressure and an armed French presence in Chad.

Air defence is carried out by 16 Dassault Mirage F.1EDs, delivered in 1977 to Okba Ben Nafi with six Mirage F.1BD two-seat trainers. In 1979 15 more Mirage F.1EDs were noted at Bordeaux, but delivery of these and a further 26 ordered in 1981 is still unconfirmed. Joining the Mirage F.1EDs are about 60 MiG-25 'Foxbats' operated by three squadrons, and at least 160 MiG-23 'Floggers' based at Okba Ben Nafi and Gemal Abdul Nasser. Strategic reconnaissance is carried out by about five MiG-25R 'Foxbats' delivered

Libya operates a force of over 60 MiG-25 'Foxbats'. Most are 'Foxbat-A' interceptors, such as this example photographed from a US Navy Tomcat. Others are 'Foxbat-C' trainers and 'Foxbat-B/D' reconnaissance platforms.

Lebanon (continued)

in 1978 (probably Soviet piloted) and the survivors from ten Mirage 5DRs. The fighter-bomber role is fulfilled by several types, including 100 Sukhoi Su-20/22s, 20 MiG-23BNs and the survivors from 32 Dassault Mirage 5DEs and 53 Dassault Mirage 5Ds both ordered in 1970 and delivered in 1971-74, and 16 Dassault Mirage F.1ADs ordered in 1974 with delivery starting in 1977. Also in use in this role is the MiG-21, of which 92 were received. About 12 Tupolev Tu-22 'Blinders' for use in bomber and reconnaissance roles were delivered in the mid-1970s and most are thought to remain in service.

The operational conversion units are equipped with 15 Dassault Mirage 5DDs, six Mirage F.1BDs, 20 MiG-23U 'Floggers', five MiG-25U 'Foxbats', several MiG-21Us and a few Tupolev Tu-22U 'Blinders'. Ad-

vanced training is carried out on the Aero L-39Z and the Soko G-2A Galeb, many of which have been delivered. The Magisters are believed to have been withdrawn from use. The primary training role is fulfilled mainly by the SIAI-Marchetti SF.260WL Warrior, of which some 140 were delivered between 1977-79 to Kufra and Sebha; many of these were assembled in Libya. Also in use are four Socata Rallye 180E and four Rallye 235GT.

In use for transport duties are 20 Aeritalia G.222Ls delivered from 1981 onwards. Of eight Lockheed C-130H Hercules (delivered 1970-71), one was lost supporting pro-Amin forces at Entebbe, Uganda in 1979; a further eight ordered in 1973 were embargoed by the US and are still in store with Lockheed at Marietta, Georgia, despite full payment by Libya. At least two Antonov An-26s

are in service and six LET L-410UVP Turbolets were delivered in 1984; a total of 10 of this last type are believed to be in use. A few remaining Douglas C-47 Dakotas may also remain active. Libyan Arab Airlines and Jamahiriya Air Cargo are believed to carry out military transport and VIP duties using eight Boeing 707s, nine Ilyushin Il-76s, one Canadair CL-44, three Grumman Gulfstream IIs and two Lockheed JetStars.

Few helicopters are in service when set against the large numbers of combat aircraft received, and they comprise 20 Meridionali-Vertol CH-47Cs, of which 14 are in army aviation use, four Aérospatiale SA.316B Alouette IIIs, 12 Mil Mi-8s, two Agusta A.109 Hirundos, around 25 Mil Mi-24s and one Agusta-Sikorsky S-61D for VIP duties. One Aérospatiale SA.330 Puma has also

been noted, but there is no evidence of the arrival of 40 Aérospatiale SA.342 Gazelles ordered in 1980. The navy operates a squadron of Mil Mi-14 'Haze'.

The army aviation unit was initially equipped with more than 10 ex-Italian Cessna O-1 Bird Dogs and at least 10 Agusta-Bell 47G-4As. These were followed by 14 SA.316B Alouette IIIs of which three were loaned to the Malta Defence Force and not returned. As noted earlier, fourteen CH-47C were delivered but one was lost in 1981. Seven Aérospatiale SA.321G Super Frelons were delivered in the early 1970s and six SA.321GM Super Frelons in 1980-81; most should still be in service. Five Agusta-Bell 206A JetRangers were delivered but one was presented to the Malta Defence Force in 1976. Two Agusta-Bell 212s are used as VIP transports.

Morocco Al Quwwat al Jawwiya al Malakiya Marakishiya

The air force was founded in 1956 following independence from France and initial equipment included six Morane-Saulnier MS.733 Alcyon, four ex-Iraqi Hawker Sea Furies, one de Havilland Heron, one Bell 47G, three MH.1521M Broussards and two Beech Twin Bonanzas. The Soviet Union began to supply aid in 1961; 12 MiG-17Fs and two MiG-15UTIs were received, but had been withdrawn from use by 1966. The French continued to supply aircraft during the early 1960s, with eight more MH.1521 Broussards, 12 North American T-6G Texans (with more in 1968) and 25 T-28 Fennecs. In 1962 eight Fouga CM.170R Magisters were delivered, followed by 24 ex-Luftwaffe which were refurbished before delivery. Expansion really started in 1975 with an order for 25 Dassault Mirage F-1CHs and 25 F.1EHs in 1977. All are in service except for about six which have been lost. The US supplied 18 Northrop F-5As, four F-5Bs and two RF-5As in 1966 to equip two squad-

rons at Kenitra. At least two more ex-USAF F-5As were delivered along with around six ex-Iranian F-5As. Very few of these F-5s are believed to be still in service. More Northrop F-5s were delivered from 1981 onwards against an offer of 20 F-5Es and four F-5Fs; confirmation has still to be received of delivery of four of the F-5Es. With withdrawal of the T-6Gs and Fennecs from the light attack role, the Fouga CM.170 Magisters continue to soldier on and were joined by 24 Dassault-Breguet/Dornier Alpha Jets in 1979; confirmation that all 24 were delivered is still awaited. These were followed by six refurbished ex-US Marine Corps Rockwell OV-10As in 1981.

The helicopter fleet also expanded to meet the guerrilla threat with the arrival of 34 Aérospatiale SA.330 Pumas between 1974-77, 48 Agusta-Bell 205As, five Agusta-Bell AB.206Bs and five Agusta-Bell 212s. Six Meridionali-Vertol CH-47Cs were delivered in 1979 followed by six more in 1982; a further six are on order. Aérospatiale SA.342L Gazelles were delivered during 1982 and reports suggest that five more AB.212s and AB.206s are on order.

The transport fleet is now heavily dependent on the Lockheed C-130H Hercules of which 17 have been delivered and at least two lost. Inflight refuelling capability is provided by two KC-130H Hercules and two

The ubiquitous Lockheed C-130 Hercules forms the backbone of the Moroccan transport fleet. Fifteen aircraft are on strength with two more equipped for inflight-refuelling.

more are on order. In the liaison role eight Socata MS.885 Rallye and two MS.893E Commodores are in use. For communications and VIP duties six Beech King Air A100, two Beech Super King Air 200s, one Grumman Gulfstream II and two Boeing 707s are in use, one of the 707s being equipped for inflight refuelling.

For training 10 FFA AS.202/18 Bravos, 12 Beech T-34C-1 Mentors and two Mudry CAP.10Bs delivered in 1983 are in use. An indigenous design, the AMIN Gepal V is undergoing tests with a view to 20 being acquired. Two Dassault Falcon 20s are operated in the ECM role and

two Dornier Do 28D-2s are used for maritime surveillance; unconfirmed reports suggest that these are to be supplemented by six Do 128s.

The Gendarmerie is a paramilitary force and operates six SA.330 Pumas, six SA.342K Gazelles, two SA.315B Lamas, two SA.316B Alouette IIIs, and two Aérospatiale SA.365N Dauphins delivered in 1983.

Oman Al Quwwat al Jawwiya al Sultanat Oman

Oman retains close ties with the UK to the extent that two-thirds of its air force officers are British and the Commander is seconded from the RAF. Leading the combat force are 19 single-seat SEPECAT Jaguar S attack aircraft and four Jaguar B trainers, half of which operate with No. 8 Sqn at Thumrait in the air defence role with AIM-9P Sidewinder AAMs. Those more recently delivered to No. 20 Sqn are reported to

have anti-shipping tasks with Aérospatiale AM-39 Exocet missiles in the strategically-placed Straits of Hormuz. An initial 12 Jaguars were delivered in 1977-78, followed by a similar number in 1983, plus a single ex-RAF T.2 which had previously been loaned to India.

For ground attack No. 6 Sqn at Thumrait has 17 HS Hunters of various types (four of them trainers) surviving from 31 F.6s, FR.10s,

SEPECAT Jaguars are the sharp end of the Omani air force, although Panavia Tornado ADV interceptors will shortly take over the air defence role. The Jaguars have a major anti-shipping role, carrying Exocet missiles.

F.73/73As and T.66s donated by Jordan and two T.67s from Kuwait. Two SAM squadrons (Nos 10 and 12) have BAe Rapiers with all-weather capability provided by Blindfire radars. These SAMs were delivered from December 1976 as part of a defence package that included two Marconi S600 surveillance radars and communications equipment. Once used for light attack missions against rebels, until the end of the Dhofar emergency in 1976, 11 survivors of 24 BAC Strikemaster Mk 82/82As are now assigned to basic training with No. 1 Sqn at Masirah. Armament may also be added to the seven remaining BN-2A Defenders of No. 5 Sqn.

Defenders normally operate from Salalah and/or Seeb with the rest of the SOAF transport force which comprises 15 Short Skyvan 3Ms with No. 2 Sqn (also assigned to SAR and to be fitted with Racal ASR 360 radars under a 1984 contract), No. 4 Sqn with three BAe One Elevens and three Lockheed C-130H Hercules, and a solitary DHC-5D Buffalo. Royal Flight aircraft are civil-registered and comprise a VC10, Gulfstream II, Falcon 20E, two AS 202 Bravos, three Bell 212s and a Bell 205A-1.

The helicopter fleet lost some of its number to ground fire during the Dhofar operations, leaving 13 of the original 31 Agusta-Bell 205A-1s with No. 3 Sqn at Salalah, or No. 14 at Seeb for SAR together with six newly-arrived radar equipped Bell 214STs and four older AB 206 Jet-Rangers; No. 3 Sqn also has five Bell 214Bs. Two Aérospatiale AS.332C Super Pumas are used as VIP transports. The para-military Oman Police Air Wing uses two Learjets for communications, backed by a utility fleet of three Buffaloes, two Dornier Do 228s, four Bell 205As, two Bell 206Bs, a 206L LongRanger, one Bell 222 and a Hughes 369D.

Qatar Qatar Emiri Air Force

A substantial expansion programme was begun by the QEAF in 1980, replacing its three remaining HS Hunters with new and more advanced aircraft. The greatest step forward came in 1984 when deliveries began – after training in France – of 12 Dassault Mirage F1EDAs and two F1DDA trainers from an order placed four years earlier. SAM defences comprise a unit of BAe Rapiers, bought to augment Short Tigercats. France has also been responsible for providing advanced trainers in the form of six Dassault-Breguet Alpha Jets, supplied from October 1980. The Hunters, which may remain for instructional duties, comprise two FGA.78s and a single T.79 two-seat trainer.

The first aircraft delivered to the QEAF, in 1968, were two Westland Whirlwinds, and Qatar has retained its close links with this UK company with orders for three Lynx HC.28s and 12 Commando versions of the Sea King. The first batch of four, delivered in 1975-76, contained three utility Mk 2As and a VIP Mk 2C, while in 1982-84, eight armed Mk 3s were supplied. France's contribution to the helicopter fleet was made in 1983 in the form of a small number of navalised Aérospatiale AS.332F Super Pumas and SA.342L Gazelles. In the para-military role are two Westland-built SA.341 Gazelles flown by the Qatar Police. Official communications flights are made in a

civil-registered Boeing 727 and a 707 (the latter for the use of Shaikh Khalifa bin Hamad al-Thani) but there are no military fixed-wing transports.

Qatar operates 12 Westland Commandos, showing the close links this country has with the United Kingdom. One of the Commandos is equipped for VIP transport.

Saudi Arabia Al Quwwat al Jawwiya Assa'udiya

Home of Islam and leader of the Western-looking Arab states, Saudi Arabia continues to expand its armoury for the protection of oil reserves, and assists neighbours in the procurement of some military equipment for their own use. Britain is well represented through its operation of the training programme and support of around 25 BAC Lightnings remaining from five F.52s, two T.54s, 34 F.53s and six T.55s, mostly delivered in 1968-69 and now operated from Tabuk by No. 2 Sqn. Configured for ground attack as well as interception with Firestreak and Red Top AAMs, the Lightnings were largely replaced in 1982-83 by 47 McDonnell Douglas F-15C Eagles and 15 F-15D trainers flown by No. 5 Sqn at Taif, No. 13 at Dhahran and one other at Khamis Mushayt. Armed with AIM-7F Sparrow and AIM-9L Sidewinder AAMs for interception, the Eagles are protected against pre-emptive attack by hardened shelters.

Saudi Arabia announced in 1985 that it would buy 48 Tornado IDS and 24 Tornado ADV, along with 30 BAe Hawks, in the face of stiff competition from Dassault-Breguet's Mirage 2000/Alpha Jet package.

Ground-attack in support of the army is provided by 74 Northrop F-5E Tiger IIs (plus 20 F-5B and 25 F-5F trainers) operated by Nos 3, 7, 10 and 15 Sqns at Taif, Dhahran, Taif and Khamis Mushayt respectively, while 10 RF-5E Tigereyes look after tactical reconnaissance. No. 15 is the type OCU and No. 7 is an armament-practice unit, although all aircraft can carry AIM-9P Sidewinder or MATRA R.550 Magic AAMs as an alternative to their attack equipment, which includes Hughes AGM-65 Maverick and Texas Instruments AGM-45 Shrike ASMs. The survivors of 47 BAC Strikemaster Mk 80/80As have light attack roles in addition to their prime task of basic instruction (No. 9 Sqn) and weapons training (No. 11 Sqn) within the King Faisal Air Academy at Riyadh. The 200-hour Strikemaster course is prefixed by 25 hours of grading on a dozen Reims-Cessna 172s of No. 8 Sqn at the same base.

Air defences include army-operated Raytheon MIM-23B Improved Hawk SAMs and Thomson-CSF/MATRA Shahine SAMs (development of the latter having been paid-for by Saudi Arabia), but the greatest advance in this area is now at hand with the imminent delivery of five Boeing E-3A Sentries and 10 KE-3 tankers, all powered by CFM 56 turbofans. Giving the Saudis one of the world's most sophisticated surveillance systems, they will eventually be based at Al Kharj, where a major control centre for air defence is being built, linked to the air force HQs of neighbouring countries.

Transport support is mostly entrusted to the Lockheed Hercules, of which 48 have been received for Nos 4 and 16 Sqns at Jeddah and the Royal Flight (No. 1 Sqn) at Riyadh – the last-mentioned with two VIP VC-130Hs and two 'flying hospital' variants. No. 1 also operates two Lockheed Jet Stars, two Gates Lear Jets, a Gulfstream III, three Agusta-Sikorsky AS-61A-4s and a Bell 212, as well as some civil airliners. Eight KC-130H tankers were in the Hercules deliveries. Other helicopters include around 15 Agusta-Bell 205As, 25 AB.206s and 29 AB.212s shared by Nos 12 and 14 Sqns at Taif for light transport, liaison and SAR. No. 4 Sqn has 18 larger Kawasaki-Vertol KV-107-IIs in both fire-fighting and SAR configurations.

Saudi Arabia has been a good market for British weapons. These Strikemasters were supplied for basic and weapons training.

As main supplier to the Saudi navy, France has received contracts for warships and Naval Crotale ship-to-air missiles. In 1984, production began of 24 Aérospatiale SA.365F Dauphin helicopters, of which 20 are equipped with deck-landing equipment and Aérospatiale AS 15TT anti-ship missiles.

Syria Al Quwwat al Jawwiya al Arabia as-Suriya

Although the material damage has been made good, the SAF is still suffering the effects of having lost 83 fighters (mostly MiG-21 'Fishbeds' and MiG-23 'Floggers') to Israel during the Lebanon conflict of 1982, and being unable to substantiate a single air-to-air claim in return. Finance for the Syrian stance as Israel's principal opponent is provided by several Arab nations which contribute $1,300 million per year towards the $2,500 million military budget. As the main interceptor in a separate Air Defence Command, the MiG-21 is represented by three generations of this ubiquitous fighter in 'Fish-

bed-D, -J and -L' forms, supported by 80 MiG-23 'Flogger-Es' and 30 MiG-25 'Foxbat-As'. In a move to avert a further catastrophic defeat for Syria, the USSR has delivered AA-6 and AA-7 AAMs and is implied by US intelligence reports to have supplied MiG-31 'Foxhounds' with AA-9 missiles; however, these may be operated by Soviet personnel, like the 48 SA-5 long-range SAMs also added to the defence network in 1983. Syria has also received SA-11 SAMs to add to its 100 missile sites with SA-2, SA-3 and SA-6 weapons.

A sizeable attack force flies 60 MiG-23 'Flogger-Fs' and 48 Sukhoi Su-22 'Fitter-Js', augmented by 80 older MiG-17F 'Frescos' and a few Su-7 'Fitters'. Six Antonov An-12 'Cubs', four Ilyushin Il-18 'Coots' and six Il-14 'Crates' form the transport force, but are assisted by civil-registered equipment comprising two An-24 'Cokes', two An-26 'Curls', seven Yak-40 'Codlings' and two Dassault Falcon 20Fs. For training, there are the Yakovlev Yak-11

'Moose', Yak-18 'Max' and SIAT Flamingo at the primary stage, plus 60 Aero L-29 Delfins and 40 L-39 Albatros for basic training. Jet type-conversion is via the MiG-15UTI 'Midget' to two-seat models of operational types.

Syria's main assault and utility helicopter is the Mil Mi-8 'Hip', assisted by Mi-24 'Hinds' supplied from 1980 onwards. There are about 100 Mi-8s and 35 Mi-24s, together

Aérospatiale Gazelles are operated in the anti-armour role. This example was captured by the Israelis in the Bekaa.

with 20 Mi-2 'Hoplites', 10 Mi-4 'Hounds' and 10 Mi-6 'Hooks'. Up to 50 Aérospatiale SA.342L Gazelles have been delivered for observation and anti-tank roles (with Euromissile HOT ATMs), and five Kamov Ka-25 'Hormones' fly coastal patrol sorties.

Tunisia Al Quwwat al Jawwiya al Djoumhouria at-Tunisia

Tunisia achieved independence in 1956 but the air force was not formed until 1960 and an order was placed for 15 Saab 91D Safirs; delivered in

1960-61, they have since been withdrawn from use. France supplied two Aérospatiale SE.3130 Alouette IIs in 1963 and later deliveries brought the number in service to eight, plus eight Aérospatiale SE.3160/SA.316B Alouette IIIs. Further helicopter deliveries have included 18 Agusta-Bell 205s, (delivered 1980-81) and six Bell 205A-1s (1980). An Aérospatiale AS.350 Ecureuil is on order.

Fixed-wing aircraft strength expanded in 1963 with the arrival of

three Dassault Flamant transports and 12 North American T-6G Texans, since withdrawn. Eight Aermacchi MB.326B were acquired from Italy in 1965 for training and 1969 saw the formation of the first fighter squadron with the arrival of 12 North American F-86F Sabres; formerly in Japanese service, these had been returned to the US and were then passed on to Tunisia. The F-86Fs have since been relegated to the training role. Further military contracts with Italy provided 12

SIAI-Marchetti SF.260WTs in 1974 to replace the Safirs, followed by nine SIAI-Marchetti SF.260CTs in 1978 and two SIAI-Marchetti SF.208As in 1979. An order for four two-seat Aermacchi MB.326LTs and seven single-seat MB.326KTs was fulfilled in 1977. On order are six Northrop F-5E and six F-5F Tiger IIs, along with a Lockheed C-130 Hercules. Two F-5Fs and a single F-5E were delivered via Prestwick and Alconbury during November 1984.

United Arab Emirates United Arab Emirates Air Force

Of the seven Gulf states which comprise the UAE, Abu Dhabi is by far the largest, and accordingly contributes some 80 per cent of the combined military budget. Dubai operates a small air arm, while the minor members of the group are Sharjah, Ajman, Fujairah, Ras al Khaimah and Umm al Qaiwain. The UAE is sub-divided into three defence districts: Western Command in Abu Dhabi; Central Command in Dubai; and Northern Command, which contains the remaining area. Despite the notion of unified defence, states retain some autonomy in procurement, and aircraft may be seen wearing Abu Dhabi or Dubai insignia as an alternative to that of the Union. The ADAF component of the UAEAF has a well-equipped combat arm comprising 14 Dassault Mirage IIIEAD interceptors and three Mirage 5DAD trainers of I Shaheen Sqn at Al Dhafra/Maqatra; and 12 ground attack Mirage 5ADs and three tactical reconnaissance 5RADS with II Shaheen Sqn at the same base. These will be replaced from 1986 onwards with the first of 12 Mirage 2000AD fighters, three reconnaissance 2000RADs and three 2000DAD trainers, while a

further 18 Mirage 2000s will be added later. Assisting with ground attack, if required, are 16 BAe Hawk T.63s recently supplied to III Shaheen Sqn at Sharjah to replace eight HS Hunters given to Somalia in 1983 (with four surplus BN Islanders).

Transport units operate from Bateem with four Lockheed C-130H Hercules, four DHC-5D Buffaloes, four CASA C-212 Aviocars, 12 Aérospatiale SA.330 Pumas, two VIP SA.332L Super Pumas, and seven Alouette IIIs under the name of 'C-130 Squadron', 'Buffalo Squadron', etc. From 1982, the Puma Squadron has operated eight navalised AS.332F Super Pumas which are believed to have an armament of AM 39 Exocet and AS 15TT anti-ship missiles. Ten more helicopters

of the Gazelle Squadron (Aérospatiale SA.342J) fly from al Dhafra alongside the 14 Pilatus PC-7 Turbo Trainers of the Training Squadron.

The Dubai Air Wing contribution is based at the local airport and al Dhafra, its most potent elements being four Aermacchi MB.326K COIN light jets, backed by an armed trainer fleet with two MB.326LDs, eight BAe Hawk T.61s, a single SIAI-Marchetti SF-260W Warrior and two (with two more on option) Aermacchi MB.339s. It will be noted that there is little commonality with the ADAF apart from the joint order for Hawks as the UAEAF's standard jet trainer – although purchase of similar MB.339s appears to introduce unnecessary servicing complexity. Basic training takes place on four turboprop-powered SF-

So far two Aermacchi M.B.339s have been supplied to the Dubai air force. The same country also operates the very similar (in role) BAe Hawk. Most aircraft have a secondary strike capability.

260TPs.

Two Lockheed Super Hercules and an Aeritalia G.222 constitute the Dubai medium transport force, assisted by lighter equipment in the form of a BN-2T Turbo Islander. Some 23 helicopters are in government service, about half assigned to the UAE or local Police. There are four Bell 205A-1 Iroquois, three 206A JetRangers, a 206L LongRanger, four police 212s, four 214Bs, and seven MBB BO 105s (four of them police-operated). The only fixed-wing aircraft in police use is a Cessna 182N Skylane.

Yemen Yemen Arab Republic Air Force

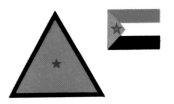

North Yemen well illustrates the changing nature of Middle East

alliances with its swing out of the Soviet orbit in the late-1970s and rapid return to links with Moscow. A move of Soviet aid to South Yemen (with which there have been brief conflicts, despite long-term plans for unification) resulted in Saudi Arabia giving four Northrop F-5B trainers and funding 10 F-5E Tiger II fighter-bombers and two Lockheed C-130H Hercules transports. However, a 1979 rapprochement

with the USSR resulted in early delivery of 30 MiG-21 'Fishbeds' to Hodeida, 20 Cuban-flown Su-22 'Fitter-Js' to Sana'a and the return to serviceability of 20 MiG-17F 'Frescos' at Taiz. F-5Es and Su-22s have air defence roles with AIM-9 Sidewinder and AA-2 'Atoll' AAMs respectively.

In addition to the pair of Hercules, the transport force has two An-26 'Curls', four Il-14 'Crates' and two

Short Skyvans plus, in civil markings, three An-24 'Cokes', an Il-18 'Coot' and four Douglas Dakotas. Three second-hand Fokker F.27 Friendships are said to have been acquired in 1984. Major helicopter is the Mi-8 'Hip' with 20 in use, and there are also two AB.204 Iroquois, six AB.212s, six AB,206 JetRangers and two Mi-4 'Hounds'. Training is undertaken on the Yak-11 'Moose' and MiG-15UTI 'Midget'.

South Yemen Air Force of the South Yemen People's Republic

South Yemen has relied in recent years totally on the Soviet Union for its military aircraft, and this is reflected in the inventory. Bombers are the elderly Ilyushin Il-28 'Beagle', of which a dozen are thought to be still in service. Attack is handled by a mixed force of Sukhoi Su-22 'Fitters' and MiG-17s, ably

assisted by over a dozen Mil Mi-24 'Hind' helicopters. MiG-21 'Fishbeds' are the most common combat type, with three air defence squadrons and one attack squadron being equipped with the type. Several of these aircraft are reputed to be flown by Soviet, East German or Cuban crew.

The recent civil war in the country between the Marxist government and even more extreme rebels has left the state of the armed forces as something of a mystery.

Africa

Only South Africa can claim to have an effective independent air force in the whole of sub-Saharan Africa. The other nations all have less than 50 combat aircraft each, with the exception of Angola and Ethiopia which rely heavily on Soviet advisers, along with Cubans and East Germans, to keep their aircraft airworthy.

South Africa continues to receive arms from France and Israel, the United Nations embargo having little effect other than creating a problem in finding a Shackleton replacement. The recent ceasefire agreements with Mozambique and Angola will reduce SAAF running costs.

Zimbabwe has ceased to field an effective air force since inde-

pendence and appears to be unable to rebuild it with the mass exodus of white aircrew and engineers from the service.

Internal disputes still trouble many of the countries, with guerrillas being active in Mozambique, Zimbabwe, Mauretania, Ethiopia and Angola; coups or attempted coups have recently taken place in Kenya, Nigeria and Ghana. The Libyan invasion and subsequent withdrawal from Chad also succeeded in heightening tensions in this area. The droughts and famine across the countries closest to the Sahara have also reduced the amount of money available for defence spending, as nations struggle to feed their ever-growing populations.

Angola Fôrça Aérea Populaire de Angola/Difesa Anti-Aviões

The FAPA/DAA was formed early in 1976 by the Marxist MPLA govern-

ment during the civil war which broke out following the granting of independence from Portugal in 1975. Initial equipment was eight MiG-21MFs which were flown from Brazzaville in the Congo by Cuban-trained Portuguese and Cuban mercenaries. Further Soviet equipment was received and is now believed to consist of a single squadron of MiG-17Fs with about 25 aircraft and a single squadron of MiG-21MF/bis with around 30 aircraft. Both units are based near Lubango and are be-

lieved to be flown by Cuban and East German pilots in missions against Unita guerrillas. Unita has made numerous claims of having shot down several MiG-17/21s and two MiG-21s are known to have been shot down by the South African air force.

Maritime surveillance is carried out by a Fokker F.27-200MPA delivered in 1980 and based at Luanda. Transport capability is provided by four ex-Air Algérie Nord 262As, one Fairchild FH-227B, one Tupolev Tu-

134A, 10 Antonov An-2s and at least two Antonov An-26s. Pilot training is carried out on 12 Pilatus PC-7 Turbo Trainers, four PC-6B1-H2 Turbo Porters and approximately three MiG-15UTIs. Helicopters in use comprise several Mil Mi-8s and between 20 and 30 Aérospatiale/ICA Brasov Alouette IIIs. The government operates 18 Britten-Norman BN-2A-21 Islanders, one Dassault Falcon 20E, one Rockwell Commander 690A, one Yak-40FG, four Antonov An-26s and several Cessna 172s.

Benin Forces Armées Populaires du Benin

Benin, or Dahomey as it was former-

ly called, achieved independence in 1960 and received the usual aid from France in the shape of one Agusta-Bell 47G, one C-47 Dakota and three Max Holste MH.1521 Broussards. Although delivery of the Broussards was reported as 1964, it is believed they were in use from 1961. An Aero Commander 500B was also delivered in 1961 and a Cessna 337D in 1970. Further C-47 Dakotas were acquired, one in 1965, two in 1971, two in 1972 and finally three former

Aéronavale examples in 1982. It is doubtful if more than a handful of the Dakotas remain in use and one of the aircraft delivered in 1982 has gone to Air Benin. Two Fokker F.27-600s were delivered in 1978 but transferred to Air Benin at the end of that year upon the arrival of two Antonov An-26s. Two helicopters reported but not confirmed as being in use are a Kamov Ka-26 and an Alouette II. Known bases are Abourney, Cotonou, Natitingou and Save.

Benin still operates a handful of the venerable Douglas C-47. These are used for general transport duties.

Bophuthatswana

Bophuthatswana Defence Force

Following independence from South Africa in 1977, a small defence force was formed and had acquired two Helio H295 Couriers, two Partenavia P.68B/Cs, two Aérospatiale SA.316B Alouette IIIs and an AS.355B Ecureuil 2 by 1984.

Botswana

Botswana Defence Force

A former British protectorate, granted independence in 1976, Botswana formed an aviation unit in 1977 in an attempt to prevent the then increasing number of cross-border raids by Rhodesian troops on anti-guerrilla missions. Initial equipment was a Britten-Norman BN-2A Defender delivered in 1977;

one more followed in 1977, but was impounded in Nigeria and subsequently destroyed in 1978. Two more Defenders arrived in 1978 and the final two in 1979, resulting in a total strength of five aircraft. Two Short Skyvan 3Ms were delivered in 1979 along with two Cessna A.152s; one of the latter was sold in 1980.

Botswana (continued)

Basic training is carried out on five surviving BAe Bulldog Series 120-1210s ordered and delivered in 1980. There is some doubt as to whether a DHC-6-300M was delivered in 1982 or merely on a demonstration tour. Two Pilatus Britten-Norman BN-3 Trislanders were delivered in 1984. Squadron Z1 operates the Skyvans, Bulldogs and two Defenders; Squadron Z2 operates the Cessna A152 and one Defender; and Squadron Z3 operates two further Defenders.

Burkina-Faso Force Aérienne de Burkina-Faso

This country, known previously as Haute-Volta and for a short period as Bourkina-Fasso, is one of the world's poorest nations. Upper Volta was granted independence from France in 1960 but only began formation of an air force in 1964. Initial equipment was an ex-Armée de l'Air C-47 and two MH.1521 Broussards, a further Broussard arriving in 1967 and a second C-47 in 1971; all were based at Ouagadougou. A Cessna 337D was delivered prior to 1971 and a Reims-Cessna F 337E in 1971. Further communications and transport types were an Aero Commander 500B and two Nord 262Cs delivered in 1974, one HS.748 in 1977 and a second in 1981 and a Reims-Cessna F 172N delivered in 1977. Helicopters were first provided in the shape of two Alouette IIIs but have since been supplemented by two SA.365N Dauphins. An unconfirmed report suggests that MiG-21s may have been received during late 1984. The two main bases are at Bobo-Dioulasso and Ougadougou.

Burundi Force Armée du Burundi

Though granted independence by Belgium in 1962, it was 1966 before an army air wing was established with the arrival of a Dornier Do 27Q-4 and two second-hand SA.318B Alouette IIIs; these were followed by two new SA.316Bs in 1967. France supplied three ex-Armée de l'Air C-47 Dakotas in 1969 but these were subsequently transferred to Air Burundi. Later equipment included three Reims-Cessna FRA 150Ls, delivered during the 1970s, and three SIAI-Marchetti SF.260Ws and three SF.260Cs in 1981. An order for four SIAI-Marchetti SF.260TPs was placed in 1982 followed by an order for two Aérospatiale SA.342L Gazelles in 1983. Confirmation of delivery of these orders is still required.

Cameroun L'Armée de l'Air du Cameroun

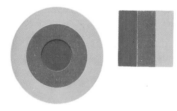

Cameroun gained her independence from France in 1960 and received the standard French ex-colonial package of a Douglas C-47 Dakota, three MH.1521M Broussards and an Agusta-Bell 47G in 1961; these were followed by five Dassault M.D.315 Flamants between 1962 and 1967, but all have now been withdrawn or returned to the Armée de l'Air. More C-47s comprised two in 1965 and two in 1966, but at least two have been withdrawn and one transferred to the government flight. A Dornier Do 28A-1 and Do 28B-1 were delivered in 1964, the former later transferred to government use. An Aérospatiale SA.318E Alouette II was delivered in 1968, later followed by four others, but at least one has been lost. A Mil Mi-4 delivered in the early seventies was withdrawn during 1975.

Transport strength was increased with the receipt of two Lockheed C-130H Hercules in 1977, a C-130H-30 in 1982, two DHC-4A Caribou in 1971 plus two DHC-5D Buffaloes in 1982 and two more in 1984. Two Hawker Siddeley HS.748s delivered in 1977 were transferred to government use in 1979. Liaison aircraft include two Piper PA-23 Aztecs acquired in 1975 and three Dornier Do 128s in 1982. More recent helicopter deliveries have included two Aérospatiale SA.319B Alouette IIIs in 1978 and four SA.342L Gazelles in 1981-82.

The initial jet equipment was four second-hand Fouga CM.170R Magisters refurbished by Aérospatiale and delivered in 1973, followed by two more in 1978-79; four were returned to Aérospatiale in exchange for four newly refurbished aircraft in 1982-83. Six Dassault-Breguet/Dornier Alpha Jet MS2s ordered in 1981 were delivered in 1984.

Cape Verde

Forza Aérea Caboverdaine

This small country was granted independence from Portugal in 1975 and the only air force aircraft noted so far have been two Antonov An-26 transports reported in 1982. The Cape Verde Islands plan eventually to unite with Guinea-Bissau.

Central African Republic Force Aérienne Centrafricaine

The FAC was formed during 1961 with the usual French gift of an Agusta-Bell 47G, a C-47 Dakota and two MH.1521 Broussards. By the end of 1966 four Dakotas had been delivered; one was written off in 1966 and one withdrawn from use in 1975. Five more Broussards were received by 1965 and, with the exception of one lost in 1963, all should still be in service; a further Broussard delivered in 1972 was written off in 1974. A C-54B is also in use.

For VIP use, a Douglas C-54 Skymaster was delivered in 1968 (sold in 1979), a Falcon 20 in 1969 (sold in 1976) and a Caravelle III in 1981. Other VIP-use aircraft include the civil registered SN.601 Corvette, Beech Baron 58, and two Socata Rallye 235Es.

In 1970 10 Aermacchi AL.60C/Fs were delivered. An order was placed in 1983 for 12 FMA IA.58 Pucarás and one Aérospatiale AS.350 but delivery of these has still to be confirmed. The main base is at Bangui-M'Poko.

Two Douglas C-47s are thought to still be in service with the Central African Republic.

Chad Escadrille Nationale Tchadienne

Chad was another French colony to

receive the standard package of three MH. 1521 Broussard, one C-47 Dakota and one Agusta-Bell 47G in 1961 following independence in 1960. Between 1961 and 1972 a further 10 C-47 Dakotas were received, though at least two were written off. Two more Broussards were in use by early 1965, but two have been lost and all may now be withdrawn from use. To expand the transport fleet, six Douglas C-54/DC-4s were delivered during 1973-

78 and at least three and possibly four are still in use. Also received were five Reims-Cessna FTB 337s in 1971 and two Pilatus PC-6B Turbo Porters in 1976; more recent expansion, in 1983, saw the arrival of two C-130As (one ex-Royal Australian Air Force and the other from the US Air National Guard) and one CASA C-212.

The only attack aircraft were six Douglas AD-4 Skyraiders, four delivered in 1976 and two in 1977; they

were used operationally against rebel forces in the north of the country but are believed to have now been withdrawn from use. Other equipment may include two SA.341 Gazelles, one received in 1974 and one in 1975; four SA.330 Pumas were also received from the French army in 1980 and some Alouette II/IIIs may have been handed over at the same time. The government uses a civil-registered Caravelle 6R for VIP transport.

Ciskei Ciskei Defence Force

Following the grant of independence by South Africa, the tribal homeland of Ciskei has formed a small defence

force. With the exception of a Jet Commander (delivered in 1983 but sold in 1984), all carry South African

civil registrations and consist of two Short Skyvans, two Pilatus Britten-Norman BN-2A-21 Islanders, one

MBB BO 105 and one MBB/Kawasaki BK117. Also reportedly in use for pilot training are six Mooney 201s.

Comores Aviation Militaire Comorienne

This small group of islands off the Madagascar coast declared independence in 1975 and has since formed a token air force. A Cessna 402B was acquired in 1976, followed by a Douglas C-47 Dakota which has since been withdrawn from use. Some confusion exists over a quantity of SIAI-Marchetti SF.260W and SF.260C allegedly ordered by Com-

ores, but at least 31 actually turned up in Rhodesia. It is thought that three SF.260Ws were the total quantity of this type acquired by Comores but two of these were sold in 1983.

Congo Republic Force Aérienne Congolaise

A former French colony, granted independence in 1960, the Congo Republic established an air force in 1961 with the arrival of one Douglas C-47 Dakota, two MH.1521M Broussards and an Agusta-Bell 47G. Two more Broussards were delivered during 1963-65, but two were later written off and the other two transferred to the Ivory Coast air force. Two more C-47 Dakotas were delivered during 1969-70 but one was subsequently written off. The C-47

Dakotas were supplemented by a Nord 262C and a Nord N.2501 Noratlas between 1975 and 1981.

Three Aérospatiale SA.318C Alouette IIs and an SE.3160 Alouette III were acquired in 1971 to replace the Agusta-Bell 47G, but one Alouette II and the Alouette III have since been lost in accidents. An Aérospatiale SA.330 Puma was delivered in 1974. As well as French aid, the Soviet Union has supplied equipment in the shape of six

Ilyushin Il-14s (one of which has been written off), three Antonov An-24Vs and two Antonov An-24RVs. Cuban assistance came in the form of eight MiG-17s and a MiG-15UTI. Unconfirmed reports suggest that a further 22 MiG-17s and four Aero L-39 Albatros trainers have also been received. The air force headquarters is at Brazzaville. The government operates a Fokker F.28-1000 Fellowship and an Aérospatiale SN.601 Corvette.

Djibouti Force Aérienne du Djibouti

Granted independence from France in 1977, Djibouti began formation of her air arm in the same year with the arrival of a single Nord N.2501 Noratlas on loan from the Armée de l'Air; it was replaced by two similar aircraft in 1978 (again on loan), both returned in 1979. Djibouti received an N.2501 as a gift in 1978; since twice replaced, it is still in use today and is the force's only transport air-

craft. An ex-Armée de l'Air Aérospatiale SE.3130 Alouette II was received in 1979 and remains in use, now supplemented by two Aérospatiale AS.355F Ecureuil 2s delivered in 1983. A Socata Rallye 235GT and a Cessna U206G Stationair were received in 1980 for communications duties, and the government operates a Dassault Falcon 20 which was presented in 1980 as a gift from Iraq.

Ethiopia Ye Ityopya Ayer Hayl

Following the arrival of Soviet equipment in 1977, which came about because the Soviet Union switched support from Somalia to Ethiopia during the Ogaden Desert dispute, this air arm is now primarily Soviet equipped. Fighter-bomber strength is believed to consist of 20 MiG-17Fs, 60 MiG-21MFs and 24 MiG-23s. Eritrean guerrillas re-

gularly claim to shoot down EAF aircraft and presumably the USSR provides attrition replacements. Aircrew training is believed to take place in the Soviet Union, with only operational conversion on MiG-21Us and MiG-23Us being carried out within Ethiopia. Some of the MiGs operate from Dire Dawa. Unconfirmed reports suggest a quantity of Sukhoi Su-20s may be in use, probably with non-Ethiopian crews.

The survivors of earlier Western equipment, comprising four BAe Canberra B.52s (delivered in 1969), 36 North American F-86Fs (1960), 10 Northrop F-5As and two F-5Bs (1966), plus three more F-5As (1972) and three ex-Iranian F-5As (1975), are all believed to have been withdrawn. Some doubt now exists as to

whether eight of 14 F-5Es on order were delivered. Other unserviceable Western equipment should include the survivors from 10 C-119G/Ks, 13 C-47 Dakotas, two C-54s, three de Havilland Doves, six Agusta-Bell AB.204Bs, five Alouette IIIs, 16 T-

Since political re-alignment with Moscow in 1977, Ethiopia has received a steady supply of Soviet materiel, including MiG-23s. Western-supplied aircraft, such as this Northrop F-5A, are thought to be no longer in service.

33As, 47 Saab Safirs and 30 T-28A/Ds.

Helicopters in use consist of 10 Aérospatiale/ICA-Brasov IAR-316 Alouette IIIs, one ICA-Brasov IAR-330 Puma, a quantity of Mil Mi-6 and Mil Mi-8s. Deliveries began in 1980 of at least 20 Mil Mi-24 gunships for anti-guerrilla operations. Transport capability is provided mainly by the Antonov An-12, of which six were delivered (but at least one has been lost), two Antonov An-26s, a Yak-40, three DHC-6 Twin Otters (delivered to the army in 1976) and an Ilyushin Il-14.

Gabon Forces Aériennes Gabonaises

Formerly part of French Equatorial Africa, Gabon achieved independence in 1960 and received from the Armée de l'Air during 1961 one Douglas C-47 Dakota, three Holste MH.1521M Broussards and a Bell 47G; three more C-47s followed during 1965-69. One C-47 was sold subsequently and all may now be withdrawn from use. A further Broussard was delivered (1964) but one was sold (1973) and the arrival of two Cessna 337s (1969) and a Reims-Cessna F 152 (1976) may have resulted in the Broussards being withdrawn. The Bell 47G was sold in 1970 following arrival of the first of five Aérospatiale Alouette IIIs. Four of these are still in use

along with the survivors from seven Aérospatiale SA.330 Pumas delivered between 1971-80. A more recent helicopter delivery is an Aérospatiale AS.350B Ecureuil which was noted during 1984.

The FAG ordered its first combat equipment during 1975 in the shape of three Dassault Mirage 5G, two Mirage 5RG (reconnaissance) and two Mirage 5DG (trainers). The two Mirage 5RG were completed but not delivered and the others were delivered to M'Vengue in 1978. A Mirage 5G and 5DG were lost in 1981 and an order for six more was placed in 1983. Four Mirage 5G-IIs and two Mirage 5DGs were delivered in 1984 to this order and were reportedly followed by two more Mirage 5G-IIs and a Mirage 5DG.

Pilot training is carried out on four Beech T-34C-1s delivered in 1982. Transport requirements are met by three Nord 262Cs, four Lockheed Hercules of various marks, two EMBRAER EMB-110P1s and one EMB-111 maritime patrol aircraft. VIP duties are carried out by a Grumman Gulfstream III, an earlier

Gulfstream II having been lost in 1980. The government use a civil-registered Dassault Falcon 20, a Piper PA-31 Navajo Chieftain, a Fokker F.28-1000 Fellowship and a Douglas DC-8-73CF.

The second arm of the FAG is La Guarde Presidentielle, which received from the Armée de l'Air four Douglas AD-4 Skyraiders (1976) and three (1977); at least two of these have been sold recently which

Gabon operates four Mirage 5DG trainers alongside the Mirage 5G and 5RG fleet. These represent the combat jet fleet.

may indicate that all have been withdrawn from use. Four T-6 Texans and six ex-Austrian air force Fouga Magisters and an EMBRAER EMB.110P Bandeirante are also in use. La Guarde Presidentielle is based at Libreville/Leon M'Ba.

Ghana Ghana Air Force

Formed in 1959, the Ghana AF was used mainly for transport, survey and policing duties until the light strike role was added in 1978 with receipt of the Aermacchi MB.326KB. Modernisation of the transport fleet began in 1973 with eight BN-2A-9 Islanders for 3 Sqn at Takoradi, followed by six Short Skyvan 3Ms (1974) for 1 Sqn, also at

Takoradi. Three Fokker F.27-400Ms and two Fokker F.27-600s were delivered during 1975 and are based at Accra. The arrival of these aircraft allowed Herons, DHC-2 Beavers, DHC-3 Otters and DHC-4 Caribou to be withdrawn from use. In addition to the transport role, the Islanders can carry out aerial survey, air ambulance, coastal patrol and paratroop training. The Skyvans also undertake SAR, coastal patrol and casevac, while the F.27s also have a survey and SAR role.

Four Alouette IIIs and two Bell 212s were delivered during 1972 and these helicopters operate out of Accra on policing, communications and VIP duties. Basic training is carried out on the survivors of 13 BAe Bulldogs delivered during

1973-76, replacing Chipmunks at Takoradi. An order for eight SIAI SF.260TPs should now have been fulfilled to provide an interim training stage between the Bulldogs and the Aermacchi MB.326F. Seven MB.326Fs were delivered to Tamale during 1965, followed by two attrition replacements in 1972. The combat squadron at Tamale shares the MB.326Fs with the training unit, but also have six Aermacchi MB.326KBs which were delivered in 1978.

Ghana's combat force consists entirely of the Aermacchi M.B.326. Both single seat and two-seat versions are used, with some of the former also operating in the training role.

Guinea-Bissau Air Force of Guinea-Bissau

This small Marxist state gained independence from Portugal in September 1974. The initial equipment of her air arm was a few T-6 Texans, C-47 Dakotas, Dornier Do 27s and Alouette III helicopters. A small army air wing was formed using two Alouette IIIs and two Dornier Do 27s, all of which are believed to be in use. Soviet assistance re-

sulted in the formation of a fighter squadron at Bissalanca with eight MiG-17s and two MiG-15UTIs, and a single Mil Mi-8 helicopter was also supplied. In 1978 France provided a Reims-Cessna FTB 337 for coastal patrol and more recently an Alouette II. A Yak-40 and Dassault Falcon 20 are used for VIP purposes and unconfirmed reports suggest a

quantity of Aero L-39s and Antonov An-26s are in use. The main base is at Bissau-Bissalanca.

Guinea Republic Force Aérienne de Guinée

This small West African republic has for many years been under the influence of the Soviet Union and, therefore, little is known of her air force since the breakaway from French influence in 1958. The Soviet Union is believed to have supplied 10 MiG-17s and two MiG-15UTIs as main jet equipment, along with three Aero L-29 Delfins for jet training. A report of three MiG-21PFMs

being in service has yet to be confirmed. Transport capability is in the form of two Ilyushin Il-18s, four Il-14s, four Antonov An-14s, two An-12s and possibly four ex-Air Guinea An-24Vs. A government operated Yak-40 is also available. Basic training is carried out using Yak-18s, seven of which were acquired. Known helicopters in use are one Alouette III, two Romanian assem-

bled SA.330 Pumas, one SA.342M Gazelle and possibly a few Mil Mi-4s. Known bases are Boke-Baralande, Conakry-G'Bessia, Kankan-Diankana, Kissidougou, Labe-Tata, N'Zerekore and Siguiri.

Ivory Coast Force Aérienne de Côte d'Ivoire

Independence was gained from France in 1960, but a current bilateral defence agreement results in her air force still having a distinctly French look about it. Initial equipment delivered from France comprised three C-47 Dakotas received in 1961, none of which are now in service, seven MH.1521 Broussards (between 1962-65), and an Agusta-Bell 47G (1962) which was damaged beyond repair at the end of 1963. The helicopter element is provided by the two Alouette IIs received in 1963, followed by four Alouette IIIs, two of which have been lost, four SA.330 Pumas and, finally, four SA.365C Dauphins delivered in 1983-84. For liaison the force received three Reims-Cessna F 337Es

and three F 150Hs to replace the Broussards, though only two of the F 150Hs now remain in use.

The three C-47s of the Groupe Aérienne de Transport et de Liaison were supplemented by two Fokker F.27 Friendships in 1971, a further example in 1979 and five Fokker F.28 Fellowships in 1978-79. Four of the F.28s and all three F.27s were sold to Air Ivoire during 1979 leaving one F.28 in Presidential use. Light transport/VIP duties were carried out initially by a Beech UC-45 and an Aero Commander 500B; they

have been replaced by a Cessna 421 and a Beech Super King Air 200. The Escadrille Presidentielle operates an F.28, a Gulfstream II and a Gulfstream III. One of the Dauphin 2s and the Cessna 421 and King Air 200 are at times loaned for Presidential duties. Pilot training is carried out on six Beech F33C Bonanzas by the Escadrille d'Ecolage.

The first combat aircraft were six Alpha Jets received in 1980. These are operated by the Escadrille de Chasse and an attrition replacement was received during 1984.

Kenya Kenya '82 Air Force

The effects of the unsuccessful military coup by Kenyan Air Force non-commissioned officers and subsequent army takeover of the Air Force appear to be minimal, apart from the changed title of Kenya '82 Air Force. Primary fighter equipment comprises nine Northrop F-5Es and two F-5Fs; 10 F-5Es and two F-5Fs were delivered in late 1978,

but one F-5E and both F-5Fs were written off and two F-5Fs were delivered as attrition replacements in 1982. All the F-5s are based at Nanyuki. For training and light attack duties 12 BAe Hawk T.52s received in 1980-81 are in use along with five survivors of six BAe Strikemaster Mk.87s delivered in 1981. Basic training is provided by the 12 survivors of 14 BAe Bulldogs delivered in 1972; all are based at Eastleigh with the Strikemasters.

Transport and liaison is the task of six DHC-5D Buffalo, eight Dornier Do 28D-2 Skyservant, four DHC-4 Caribou and two Piper PA.31 Navajo aircraft. The helicopter squadron operates one Aérospatiale SA.342K Gazelle, 13 Aérospatiale/IAR SA.330 Pumas, two

Hughes 500D Scouts, 15 500M Scouts and 15 500MD Defenders. The Police Air Wing operates six Cessna 310s, three Cessna 402s, one Cessna 404, one Cessna 185D, three Cessna U-206Fs and a Bell 206L.

Kenya has a mixed transport fleet centred around six DHC-5s and four DHC-4s. The Dornier Do-28D Skyservant (illustrated) provides light transport, its STOL performance and toughness proving extremely useful in the bush.

Lesotho Police Mobile Unit Air Wing

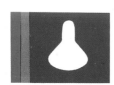

Lesotho, or Basutoland as it was known formerly, gained independence in 1966 but the Police Mobile Unit Air Wing was not formed until

1978. Initial equipment was two Short Skyvans, delivered in 1979, which are still in use. The first helicopters comprised two MBB BO 150s, received in 1978-80 (one sold 1984), followed by a Soloy-Bell 47G and a Westland-Bell 47G-3B-1 (1981), Mil Mi-2 (1983) and two Agusta-Bell 412s (1984). The last two aircraft carry 'LPF' prefixed to their serials, suggesting use by a different unit. Further liaison types acquired were a Dornier Do 27A-4, Do 28A-1 and Beech 58P (1980) and a Cessna 182Q (1981).

Liberia Liberian Army Air Reconnaissance Unit

The army formed an Air Wing in 1971 with a Cessna 180E, a 150K and a 337F. These were all civilianised during 1974 and transferred to the Justice Air Wing in 1977, where they remain today. The Air Reconnaissance Unit was formed in 1976 using a Cessna U207, (written off 1980) and two Cessna 172s which are believed to remain in use. An Executive Air Wing operates a Cessna 402B in civil markings. An IAI Arava was delivered to the ARU during 1984 and three more are reportedly on order.

Madagascar Armée de l'Air du Madagascar

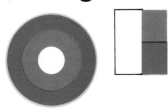

Another former French colony, Madagascar was granted independence in 1958 and known for a period as Malagasy Republic before readopting the name Madagascar. It received the French ex-colonial package of one Douglas C-47 Dakota, three MH.1521M Broussards and an Agusta-Bell 47G in 1961. A Dassault Flamant had been received in 1958. Subsequent deliveries from France resulted in a total of 10 Douglas C-47 Dakotas, 11 Broussards, one Agusta-Bell 47G, one SE.3130 Alouette II, two SE.3160 Alouette III and six Flamants being

received by 1974. Except for about six C-47s, all of these aircraft have been returned to France, withdrawn from service or written off.

Communications and liaison aircraft delivered more recently are three Reims-Cessna F 337E/F Super Skymasters (1971), a Piper PA-27 Aztec (1974), a Britten-Norman BN-2A-21 Defender (1974), plus a Cessna 310R and four 172M Skyhawks for training (1975). Transport aircraft include a BAe HS 748-360, two

Yak-40s, and at least one Antonov An-26 delivered in 1980. Two Mil Mi-8 helicopters form the total rotary-wing complement of the air force. A new base at Tamatave received eight MiG-21FLs via North Korea from 1978 onwards and may also have four MiG-17Fs for conversion training.

The national airline, Air Madagascar, operates six DHC-6 Twin Otters which are loaned to the air force as necessary.

Mali Republic Force Aérienne de la République du Mali

Independence was granted from France during 1960. Unusually, only one Broussard was handed over from the Armée de l'Air during 1961, although two C-47 Dakotas followed in 1969. The main source of aid was the Soviet Union who provided five MiG-17s and a single MiG-15UTI in 1967. Unconfirmed reports suggest they may have been replaced by eight MiG-19s and 14 MiG-21s. In addition to the French

Dakotas, transport aircraft were acquired in the shape of Antonov An-2 and An-24, Ilyushin Il-14 and Yakovlev Yak-12. The Dakotas were replaced by at least two Antonov An-26s in 1983 and the Il-14 and Yak-12 are believed to have been retired. An Aérospatiale SN.601 Corvette is in use for Government VIP flights. A few Mil Mi-4s and a single Mil Mi-8 are the only helicopters used. Pilot training is carried

out on six Yak-11/18s, but another unconfirmed report suggests that six L-29 Delfins are in use.

Malawi Malawi Army Air Wing

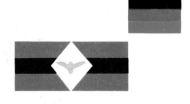

This former British protectorate became independent in 1964 and the

MAAW received its first aircraft in 1976; its main base is at Blantyre. Initial equipment consisted of two Dornier Do 28D-2s based at Zomba, complemented by an Aérospatiale SA.330 Puma and SA.316B Alouette III in 1977.

These were followed by an order for six refurbished ex-Belgian air force Dornier Do 27As, but only three have been confirmed as received (1979). Six more Dornier Do 28Ds were ordered but only four of these have been noted in service. Of

two SA.330 Pumas received (1980), one was lost in an accident. A BAe HS.125-700B is in service for VIP use. The Malawi Youth Pioneers provide para-military flying training with a Cessna A150, three 172Ms, one 182P and a 402B.

The Police Air Wing operates a Cessna 185A and four Short Skyvan 3s. At least one Pilatus Britten-Norman BN-2T Turbo-Islander was delivered during 1984 as a replacement for two BN-2A Defenders lost in accidents.

Malawi has no combat aircraft as such, concerning itself only with general duties. Six Dornier Do-28Ds are used.

Mauritania Force Aérienne Islamique de Mauritanie

This former French protectorate and colony achieved independence in 1960 and over the next few years received four MH.1521 Broussards and at least five C-47s. All but one of the Broussards and all the C-47s had been lost or withdrawn from use by the end of 1983. Modest expansion

began in the early 1970s with procurement of one Aermacchi AL-60 and two Reims-Cessna F 337Gs in 1973. Two C-54s delivered in 1974 have since been withdrawn from use. The first attack and counter-insurgency element was formed in late 1975 to counter a growing threat from Algerian-backed Polisario guerrillas in the former Spanish Sahara. Two Short Skyvan 3Ms were obtained (late 1975) plus two FTB-337s and six Britten-Norman BN-2A Defenders. To supplement the Skyvans in the transport role two DHC-5D Buffaloes were acquired (1977), one lost subsequently in an accident. Three Defenders delivered in 1978 gave a total comple-

ment of nine, but at least three of these have been lost to date. The latest arrivals were two Piper PA.31T Cheyenne IIs (1981), for use on anti-smuggling/coastal patrol duties. Known bases are Bir

Six Britten-Norman Defenders were supplied for transport and light counter-insurgency work.

Moghreim, F'Derik, Nouadhibou and Nouakchott.

Mozambique Fôrça Populare Aérea de Libertaçao de Moçambique

A former Portuguese territory,

Mozambique gained independence in 1975 when power was handed over to the Marxist Frelimo organisation and, apart from ex-Portuguese aircraft, the FPALM has been equipped solely from the Soviet Union. The FAP left behind three Auster D5/160s, six Douglas C-47 Dakotas, seven Nord N.2501 Noratlas, 10 North American T-6G Texans, four Piper Cherokee Six 300s and at least eight SA.316B Alouette

IIIs. All remain in use except for the Austers, and at least four of the Alouette IIIs which were 'liberated' by Rhodesian troops during anti-guerrilla operations into Mozambique.

Initial Soviet aid came in the shape of seven Zlin 326 basic trainers and eight MiG-21s delivered to Nacala in 1977 and supplemented by 35 Cuban-crewed MiG-21s at Nacala, Tete and Sofala in 1978. It is

doubtful if national crews fly any of the MiG-21s. Subsequently 23 MiG-17Fs were delivered to Maputo along with two MiG-15UTI for conversion training. One MiG-17 pilot defected with his aircraft to South Africa in 1981 and at least one other has crashed. Transport aircraft include five Antonov An-26s, a civil registered Tupolev Tu-134A and helicopter support is provided by at least eight Mil Mi-8s.

Niger Force Aérienne du Niger

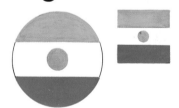

Another former French colony, Niger gained independence in 1960 and received an ex-Armée de l'Air C-47 Dakota in 1961. Further aircraft from the same source included two MH.1521 Broussards (1964), plus two more Broussards and a C-47 (1965). The C-47 complement was brought up to four aircraft in 1969 but by 1980 all had been disposed of

or withdrawn. Further communication types added to the inventory included an Aero Commander 500B in 1964 (sold 1983), a Beech Model 18 in 1965 (sold 1975) and three Cessna 337s during 1969-71, one since sold. The C-47s in the transport fleet were supplemented by four ex-Luftwaffe N.2501 Noratlas in 1969 plus one from the Armée de l'Air in 1977. All

but one are still believed to be active. A Douglas DC-4 was acquired from Aéropostale and it may still be in use. The West German government donated three Dornier Do 28Ds in 1979 and the latest acquisition for this air arm was two Lockheed C-130H Hercules in 1980. The Niger government operates a Boeing 737-2N9C.

Nigeria Federal Nigerian Air Force

Despite the military coup in late 1983 and continuing financial problems, the NAF continues to expand. Combat aircraft include the survivors of 26 MiG-21MFs based at

The Nigerian Fokker F.27 fleet has been severely depleted by attrition and sales. One operates in the transport role, while two others patrol the coast.

Kano, and all Il-28s and MiG-17Fs have been withdrawn from use. Operational conversion takes place at unit level, initially with MiG-15UTIs and then MiG-21Us. Two of each type were delivered. Production of BAe Jaguars is now well in hand against a Nigerian order for 13 single-seat and five two-seat aircraft; deliveries were scheduled for 1985, following completion of crew training. Light attack capability is provided by 12 Aermacchi MB.339As (delivered in 1984) and 12 Dornier-built Alpha Jets (1981-82). In peace time the Alpha Jets have an advanced training role, re-placing Aero L-29s. A second batch of 12 Alpha Jets was ordered in 1983 to equip an attack squadron but deliveries have yet to start.

The base at Murtala Muhammed Airport, Lagos, is the centre of the NAF transport operations. Based here are the six Lockheed C-130Hs delivered 1975-76, but three new C-130Hs were at Marietta in October 1984 awaiting delivery. Also due for delivery are five Aeritalia G.222s which were nearing completion in late 1984. Six Fokker F.27 Friendships were delivered to the NAF in 1972, but four have been sold to Nigeria Airways and a fifth written off. Two F.27-200MPAs are on order for maritime surveillance and were due for delivery by the end of 1984. The government uses civil-registered transports for VIP use, these including a Boeing 707-3F9C and a 727-2N6, Fokker F.28-1000 Fellowship, G.159 Gulfstream II and a BAe HS.125-600B.

Communications and liaison roles are fulfilled by 20 Dornier Do 28D Skyservants, five Do 128s and eight turboprop Do 128-6s which are currently being delivered and will be based at Kaduna. Earlier Dornier Do 27A-4s and Dornier Do 28Bs have all been withdrawn. Helicopter equipment comprises four MBB BO105CBs, 20 BO105Ds and 15 SA.330 Pumas. Five Boeing CH-47C Chinooks were ordered during 1983 and should now be in service, as should 14 Hughes 300Cs ordered in 1982, but neither have yet been confirmed. Basic training is carried out on the survivors of 37 BAe Bulldogs delivered during 1974-82 to replace Piaggio P.149Ds.

The Nigerian Navy ordered three Westland Lynx Mk.89s in late 1982 and following the completion of training, being carried out in the UK during 1984, all three were scheduled for delivery.

Rwanda Force Aérienne Rwandaise

This former Belgian colony was granted independence in 1962. Six Aermacchi MB.326s and three Aermacchi AM-3Cs were ordered in 1972 but only the three latter aircraft were delivered and these were returned to the manufacturer after a short period of use. Three refurbished Fouga CM.170 Magisters were ordered in 1975 but were never delivered.

With the failure to acquire fixed wing aircraft five Aérospatiale SA.316B Alouette III helicopters acquired during the 1970s and six Aérospatiale SA.342M Gazelles delivered in 1983 are in use. Two Socata R235 Guerriers are on order. An ex-Armée de l'Air Nord N.2501 Noratlas is in service, and the government operates an Aérospatiale Caravelle III. In addition the army, known as the Garde Nationale, operates two Douglas C-47 Dakotas, one Dornier Do 27, two Britten-Norman BN-2A Islanders and two Aérospatiale SE.3160 Alouette IIIs. The main base is Kagali International Airport.

Senegambia Armée de l'Air du Senegambia

This former French colony integrated her armed forces with those of the Gambia under Senegalese control during 1981. Like all former French colonies, initial equipment came from France in the form of seven ex-Armée de l'Air C-47s, five MH.1521 Broussards and two Agusta-Bell 47G helicopters delivered between 1961 and 1972. Both the AB-47Gs had been written off by the end of 1964 and were replaced by five Aérospatiale Alouette IIs delivered during 1965-72. One of these was damaged during 1972 and sold. A Reims-Cessna F 337 was delivered during 1971.

First jet equipment arrived during the mid-seventies in the shape of five ex-Brazilian Air Force Fouga CM.170 Magisters delivered after refurbishment by Aérospatiale; one was lost during 1976. An Aérospatiale SA.341 Gazelle was delivered during 1973, the helicopter strength being upgraded in 1976 by four Aérospatiale SA.330 Pumas, but two of these have since been lost. The air transport capability was increased from late 1977 by receipt of the first of six Fokker F.27-400Ms, the last of them delivered in 1979. These appear to have supplemented the airlift capability rather than replaced the Dakotas. A DHC-6-300M with COIN capability was received during 1983, and delivered during 1982 were two Rallye 110STs and two Rallye 235As. On order and due for delivery in 1984 were two Guerrier 235s. Known bases are Bel-Air, Dakar-Yoff and Tambacounda Ziguinchor.

Sierra Leone

Sierra Leone Defence Force

Although Sierra Leone was granted independence in 1961, the formation of an air arm was not announced until 1973 when two Saab MFI-15s were delivered. Three Hughes 369s were also received but all the surviving aircraft (one MFI-15 and two Hughes 369s) were sold in 1978. The sole aircraft on the inventory of this air arm is an MBB BO 105C which is used as a Presidential transport.

Somalia Dayuuradaha Xoogga Dalka Somaliyeed

Following the 1977 rift with the Soviet Union, which gave its support to Ethiopia when Somalia invaded the Ogaden region of Ethiopia in 1977, Somalia has recently become dependent on the West for military aid.

Initial Russian military aid provided 54 MiG-15UTIs and MiG-17Fs, but only two squadrons each flying 10 MiG-17Fs and a few MiG-15UTI for operational conversion now remain. These were later supplemented by at least 10 MiG-21Fs and a few MiG-27Us in 1974. The surviving MiGs are believed to be based at Hargeisa though unserviceability and attrition losses probably mean very few are airworthy. More recent fighter-bomber equipment has reportedly included 30 Shenyang F-6s (Chinese-built MiG-19), though confirmation of these is still awaited, and at least eight ex-Abu Dhabi Hawker Siddeley Hunters delivered during 1983. The Hunters are based at Mogadishu. Light bomber capability is provided by the survivors of about 10 Ilyushin Il-28s received in 1974-75, though some doubt exists if any are airworthy.

Soviet helicopters included Mil Mi-4s and Mi-8s. The Mi-4s were noted withdrawn from use at Moga-dishu but at least two Mil Mi-8s are still in service. More recent equipment includes at least one Italian army surplus Agusta-Bell 204 and four new AB.212s delivered in 1981. Transport equipment, based at Mogadishu, consists of two Aeritalia G222s delivered in 1980 and four ex-Abu Dhabi Britten-Norman BN-2A Islanders delivered during 1983. An order for six CASA C-212 Aviocars was placed during 1984.

Trainers total six SIAI SF.260WS aircraft and an order for six SIAI S.211s was placed during 1984. Two Reims-Cessna FRA 150Ls delivered in 1973 are still used for liaison. Aircraft operated for government use consist of a single Piaggio P.166DL-3 used by the ministry of air defence, three P.166DL-3s used by the ministry of transport and two Dornier Do 28D-1s used by the police air wing.

South Africa Suid-Afrikaanse Lugmag

Despite the United Nations-backed embargo, South Africa still receives some support from friendly nations and has developed its own aviation and armament industries to a point of near self-sufficiency. The South African air force is one of three services which form the South African defence force and is divided into six commands comprising an Air Logistics Command and five operational commands as follows:

Strike Command, with its headquarters at Waterkloof, controls five fighter, bomber and reconnaissance squadrons along with seven Active Citizen Force reserve units. These comprise No. 1 Sqn at Hoedspruit with Dassault Mirage F.1AZs, of which 32 were received from 1976 onwards for use in the ground attack role. All are with No. 1 Sqn except at least one lost in an accident. No. 2 Sqn, also at Hoedspruit, serves in the fighter-bomber and reconnaissance roles using Dassault Mirage IIIs, comprising the IIICZ (15), IIIRZ (4), IIIR2Z (4) and IIIBZ (3). No. 3 Sqn at Waterkloof operates the Dassault Mirage F.1CZ, of which 15 were obtained from France, deliveries commencing in 1974. Two lost in a collision during 1979 were replaced clandestinely by France with two identically serialled aircraft. No. 12 Sqn at Waterkloof operates the survivors from six BAC Canberra B(I).12s and three Canberra T.4s ordered in 1962. No. 24 Sqn, also at Waterkloof, operates the Hawker Siddeley Buccaneer S.50 of which 16 were delivered in 1965-66. An option for a further 14 was blocked by Britain and accidents have reduced the number of operational aircraft to six. The Active Citizen Force squadrons assigned to strike command are equipped mainly with the Atlas Impala. Procurement covered 151 Impala Is (based on the Aermacchi MB.326M), all except the first sixteen being built or assembled in South Africa. These were followed by 100 Impala IIs based on the MB.326K single-seat light attack aircraft. ACF squadrons comprise No. 4 at Lanseria, No. 5 at Durban and No. 8 at Bloemfontein with Impala IIs and a few Impala Is for instructional use. No. 6 Sqn at Port Elizabeth, No. 7 at Ysterplaat and No. 40 at Waterkloof have Impala Is and, finally, No. 41 Sqn at Lauseria operates the Atlas-Macchi C4M Kudu (of which 40 were acquired) in a training role.

Maritime Command with headquarters at Cape Town is responsible for ASW, maritime surveillance, patrol and naval co-operation tasks and has four component squadrons. No. 22 Sqn based at Ysterplaat operates the 10 survivors from 16 Westland Wasp HAS.1s delivered; two more were embargoed by the UK before delivery. The Wasps deploy to South African navy frigates and destroyers and often wear navy/Vloot markings but are technically under air force control. A few Alouette III are also used. No. 25 Sqn, also based at Ysterplaat, operates six Douglas C-47 Dakotas for transport and target towing. No. 27 Sqn based at D.F. Malan Airport has 19 of the 20 Piaggio P.166S delivered from Italy during 1973-74. They are used for inshore SAR, patrol and general transport duties. No. 35 Sqn, also at D.F. Malan Airport, has operated the Avro Shackleton MR.3 since 1957; eight were acquired but one was written off in 1963 and the re-

Six Buccaneers are in use with No. 24 Sqn for strike duties. Sixteen were originally supplied, but these have suffered from lack of spares and accidents. They have seen action in Angola.

maining aircraft were due for withdrawal in 1984.

Air Transport Command has its headquarters at Waterkloof and has three component squadrons. No. 21 Sqn based at Waterkloof operates only two BAe HS.125-600Bs, as earlier aircraft have been written off or sold. The sole Vickers Viscount 781 D was transferred to No. 44 Squadron during late 1983. No. 28 Sqn, also at Waterkloof, operates nine Transall C.160Z transports and seven Lockheed C-130B Hercules for long range operations. No. 44 Sqn at Zwartkop operates the solitary Viscount, five Douglas DC-4/C-54 Skymasters and about 10 Douglas C-47 Dakotas. A total of 89 C-47s have seen service with the SAAF but only 25 are believed to be left in use with No. 44 Sqn, No. 86 Advanced Flying School and the Air Navigation School.

Light Aircraft Command with its headquarters at Zwartkop now has seven component squadrons as No. 11 Sqn at Potchefstroom with Cessna 185s was disbanded in 1983-84. The remaining seven are No. 15 Sqn at Durban with a mix of Aérospatiale Alouette III, Super Frelon and Puma helicopters. No. 16 Sqn based at Port Elizabeth operates the Alouette III, as does No. 17 Sqn at Zwartkop. No. 19 Sqn at Zwartkop operates the Puma, No. 30 Sqn at Ysterplaat the Puma and Super Frelon and No. 31 Sqn at Hoedspruit the Alouette III and Puma. No. 42 Sqn based at Potchefstroom operates the Bosbok. At least 112 SE.3160/SA.316B Alouette IIIs have been received, though less than that number are now in service; more than thirty were loaned to Rhodesia and the survivors were returned prior to independence. At least 68 SA.330F/J/L Pumas have been delivered, most remaining in service, and 16 SA.321H/L Super Frelons were delivered from 1967 onwards but at least one has been written off. Either

39 or 40 Atlas-Macchi AM.3C Bosboks were acquired, all except the first eight built in South Africa.

Training Command has its headquarters at Dunnottar, its component units including No. 84 Advanced Flying School based at Potchefstroom with Cessna 185s, Bosboks and Kudus. No. 85 AFS at Pietersburg has the Mirage IIIDZ (3), Mirage IIID2Z (11), Mirage IIIEZ (16) and a few Impala Is and IIs. No. 86 AFS at Bloemfontein uses Douglas C-47 Dakotas and also acts as the multi-engine conversion unit. No. 87 AFS is based at Ysterplaat and borrows P.166S and Shackleton MR.3 as required for training. The Central Flying School based at Dunnottar has about 50 North American T-6 Harvards and is home also to the Impala I-equipped 'Silver Falcons'

Principal strike aircraft in the Angolan bush war is the Atlas Impala Mk II, a licence-built version of the Aermacchi M.B.326K.

display team. The Flying Training School is based at Laangebaanweg with Impala Is, as is the Air Navigation School equipped with C-47 Dakotas.

The Test Flight and Development Centre is based at Waterkloof and operates a few Vampire T.55s in the test role. In addition, the 13 Air Commando squadrons (Nos 101 to 112 and 114) are available for liaison and AOP duties in an emergency using civil registered light aircraft. A further Citizen Force Light Aircraft Squadron is with the South-West Africa Territory Force in Namibia.

Sudan   Silakh al Jawwiya as Sudaniya

The Sudanese air force was formed with British assistance and primarily British equipment in 1955, but a subsequent left-wing coup in 1969

resulted in an influx of Soviet and Chinese equipment. The situation reversed again in the late 1970s with the expulsion of Soviet advisers and Sudan returned to the Western sphere of influence. A few of the 12 BAC Jet Provost T.52s are believed to be still held in long term store, but the three surviving BAC Jet Provost T.55s from a batch of five ordered in 1967 are used for close air support. These were joined by three BAe Strikemaster Mk.90s in late 1983 and a further three are on order.

Soviet equipment still in use includes the survivors from 16 MiG-21PFMs delivered in the early 1970s and approximately 30 MiG-17PFs (actually Chinese-built Shenyang F.5s) delivered in 1970. The first sign of Western fighter equipment was the delivery of two Northrop F-5Fs in 1982, followed by two Northrop F-5Es (in 1984) of 10 which are on order. All Soviet supplied Mil Mi-8 helicopters have now been withdrawn following the arrival of 12 Agusta-Bell 212s from 1982 on-

wards, one of the latter written off in 1983. Reports of 20 MBB BO 105s being in service are unconfirmed.

The Khartoum-based transport unit uses six Lockheed C-130H Hercules (delivered 1978) and four DHC-5D Buffaloes (1978-79). Soviet supplied Antonov An-24s are all now in storage at Khartoum. Six CASA C-212s were ordered in 1984. The Police air wing operate two Beech King Air 90s and a Socata Rallye 235E, and the government uses a Dassault Falcon 20F.

Tanzania   Jeshi la Wanaanchi la Tanzania

The Tanzanian Defence Force was established after Pemba, Zanzibar and Tanganyika had united as the state of Tanzania in 1964. Initial equipment, from West Germany, consisted of eight Piaggio P.149Ds, and orders were placed for eight Dornier Do 28As and six N.2501 Noratlas. West Germany withdrew her training mission in early 1965 following Tanzania's diplomatic recognition of East Germany after only

the P.149Ds had been delivered. Canada then assumed the role left vacant by West Germany which resulted in the delivery of 12 DHC-4 Caribous, six DHC-2 Beavers, eight DHC-3 Otters and six DHC-5D Buffaloes. The DHC-5Ds were delivered

This Fokker F.28 Fellowship is used by the Tanzanian government for VIP duties alongside a BAe HS 748.

from 1979 onwards and the other Canadian aircraft have all been withdrawn from use.

Tanzania turned to China for her first jet equipment and in 1973 received 12 Shenyang F-4s (Chinese-built MiG-17F) and two Shenyang F-2s (MiG-15UTI) at Mikumi air base near Dar-es-Salaam. These were joined in 1974 by 16 Xian F-7s (MiG-21MF) and 16 Shenyang F-6s (MiG-19SF). Attrition has reduced the MiG aircraft down to F-2 (2), F-4 (3), F-6 (5) and F-7 (11). Six Piper PA-28-140 Cherokees were delivered in 1971 (4) and 1973 (2) and these have replaced the Piaggio P.149Ds in the training role. The communications unit has at least seven Cessna 310Qs, two Cessna 404s and a Piper Cherokee Six. In addition to the Buffalo, transport capability is provided by three BAe HS 748 series 2Bs and a Pinkiang C-5 (Chinese-built An-2).

Helicopters in service include two Agusta-Bell 206B Jet Rangers, four Agusta-Bell 205As but confirmation is still awaited of delivery of two Meridionali-Vertol CH-47C Chi-nooks ordered in 1980. The Police air wing uses two Agusta-Bell 206As, two Bell 47G-3B-2s, two Bell 206L-1s and a single Cessna U.206. The government has a Fokker F.28 Fellowship, one BAe HS.748, one Piper PA-23 Apache, two Beech King Air A100s, four Piper PA-23 Aztecs two Aero Commander 690As, a Cessna 182 and a Beagle Husky.

Swaziland

Umbufto Swaziland Defence Force

Despite being granted independence in 1968, it was only in 1979 that two IAI Arava transports were ordered; one was lost on its delivery flight and replaced by a new aircraft in 1980. These are still in use. Flying training is provided by the Swaziland School of Flying in a Piper Cherokee 140.

Togo

Force Aérienne Togolaise

Another former French colony, Togo achieved independence during 1960 and received four Broussards (1963-65) and two C-47 Dakotas (1968-71). All have been replaced in the transport and utility role by a Cessna 337D, a Reims-Cessna F 337E, two Dornier Do 27A-4s and two DHC-5D Buffaloes delivered in 1976. The transport aircraft are operated by the Escadrille de Transport which also has a helicopter flight as one of its component units. This flight operated two Alouette IIs, both lost in accidents prior to 1979, and received an SA.330 Puma (1978) and an SA.315B Lama (1980) to replace them.

The combat element is the Escadrille de Chasse which was formed during 1976 with three CM.170R Magisters on loan from the Armée de l'Air. They were returned upon the arrival of five ex-Luftwaffe Magisters refurbished by Aérospatiale in 1976, and in which year three EMBRAER EMB.326 Xavantes were delivered followed by three more in 1978. The Escadrille was further strengthened with the arrival of six Alpha Jets during 1981. On order are three Aérospatiale TB.30 Epsilons for training purposes.

A Fokker F.28-1000 Fellowship is in use for VIP operations and the government operates a Boeing 720-047B and a DC-8-55F in civil markings. All the FAT and government aircraft are based at Lome-Tokoin.

Uganda Uganda Army Air Force

Since the Tanzanian overthrow of the Amin regime in 1979, the UAAF has been nearly non-operational. It was formed in 1964 with initial aid provided by Israel in the shape of 16 Piper Super Cubs, 12 Fouga CM.170 Magisters, six Douglas C-47 Dakotas and one Nord N.2501 Noratlas. The surviving Magisters and Dakotas were returned to Israel in 1972. Uganda fell into the Soviet sphere of influence during the mid-1960s, with four MiG-17Fs and two MiG-15UTIs being delivered in 1966, UAAF pilots then being trained in the Soviet Union. Further Soviet aid arrived in the shape of at least nine but possibly 12 Aero L-29 Delfins, 14 MiG-17Fs and 12 MiG-21MFs but the Israeli raid on Entebbe resulted in the loss of four MiG-17s and seven MiG-21s, though some of the latter may have been repaired as ten are reported to be still in service. Six MiG-17PFUs were delivered to replace those lost in accidents and in the Entebbe raid and about 12 should still survive.

Helicopter strength was provided initially by two Mil Mi-8s but subsequent deliveries have all been Bell types. These consisted of four Agusta-Bell 205s, four Bell 206As, four Bell 212s and one Bell 205A (since sold) in 1971, followed by three Bell 206Bs, three Bell 214s and a Westland-Bell 47G in 1982. These are shared jointly with the Police Air Wing which also operates a DHC-2 Turbo-Beaver and a DHC-6 Twin Otter, and at least the two Agusta-Bell 205s, four Bell 206s and a Bell 212 have been sold; confirmation of delivery of the Bell 214s is required. Initial aircraft in the training role were 10 Piaggio P.149Ds, but these were supplemented by six FFA AS.202 Bravos in 1978. Two Piper PA-23 Aztecs are in use for liaison along with the survivors of the ex-Israeli Super Cubs.

Venda Venda Defence Force

Another tribal homeland recently granted a form of independence by South Africa, Venda formed an air unit with the arrival of a new Aérospatiale SA.316B Alouette III.

Zaïre Force Aérienne Zaïroise

The Force Aérienne Congo was formed in 1961 following the grant of independence to the Belgian Congo in 1960. Independence brought civil war between Katanga (now known as Shaba Province) and Congo-Kinshasa (renamed Zaïre in 1971). Following the end of civil war in 1964, Belgian, Italian and US aid has since been given erratically and the air force is currently organised as detailed below.

2ème Groupement Aérien Tactique consists of two wings: 21e Groupe de Chasse et Assaut at Kamina with two component squadrons, 211e Escadrille flying the Dassault Mirage 5M and Mirage 5DM. All three 5DMs have been lost and of the 14 Mirage 5Ms at least two have been written off. 212e Escadrille operates Aermacchi MB.326s and North American AT-6G Texans; 17 MB.326GBs and eight MB.326GKs were delivered, but five were lost in the attack on Kolwezi in 1978 and financial problems delayed supply of he MB.326GKs; some may still be waiting delivery. Only a handful of AT-6G Texans are believed to be in service. The second wing at Kamina is the 22e Groupe de Transport Tactique with only one component squadron, 221e Escadrille operating the DHC-5D Buffalo. Of three delivered in 1976 only one remains in service.

1er Groupement Aérien at Kinshasa consists of three wings, first being the 12e Groupe de Liaison with 122e Escadrille operating the Aérospatiale SE.3160/SA.316B Alouette III of which 10 were delivered and half still survive, SA.330 Puma of which 14 were acquired and again about half remain in use, one SA.321J Super Frelon for Presidential use, one AS.332L Super Puma delivered in 1984 for Presidential use, and three Mitsubishi MU-2Js. The second wing is the 13e Groupe de Entrainement with two training squadrons, 131e Escadrille d'École-age Elémentaire operating the sur-vivors of 15 Cessna 150Ms delivered in 1976 and 12 SIAI SF.260Cs delivered in 1970-71. The nine surviving SF.260Cs are reportedly being returned in exchange for 12 new aircraft. The 132e Escadrille d'Écolage Avance has a few T-6G Texans and MB.326GBs detached from the 21e Groupe at Kamina. The last of the Kinshasa wings is the 19e Groupe d'Appui Logistique with two component squadrons. 191e Escadrille has the five surviving Lockheed C-130H Hercules (of seven), two Douglas DC-6s and two Douglas DC-4/C-54s. 192e Escadrille operates two more DC-4/C-54s and possibly a few surviving Douglas C-47 Dakotas and Curtiss C-46 Commandos, though these may by now be withdrawn from use.

Also in use are 16 Cessna 310Rs and one Britten-Norman BN.2A Islander for twin conversion training and liaison duties. The government operates a BAe HS 125-400B and a Dassault Falcon 20C.

Zambia Zambian Air Force and Air Defence Command

Formerly Northern Rhodesia, before independence in 1964, Zambia was initially assisted by the UK in forming an air force but more recently has turned to Italy and Eastern Europe for assistance. The interceptor role is fulfilled by 16 MiG-21MFs delivered in 1981-82. Fighter-bombers in service consist of 12 Shenyang F-6 (Chinese-built MiG-19s delivered 1977-78) which are totally ineffective in countering Rhodesian air strikes against ZAPU guerrilla bases in the country. The light strike and armed advanced trainer roles are served by six Soko Jastrebs and six Soko Galebs delivered in 1969 and based at Mbala, and the survivors from 24 Aermacchi MB.326GBs delivered in 1971 (6) and 1978 (18).

Initial training is carried out on a handful of Shenyang BT-6s (delivered 1977-78), nine SIAI-Marchetti SF.260MZs and 20 Saab MFI-17 Supporters (1977-78). Helicopters in use consist of 13 Agusta-Bell 205As and 13 Agusta-Bell 47G-4As (delivered 1977-78), seven Mil Mi-8s (mid-1970s) and two Agusta-Bell 212s (1977). The VIP flight uses a BAe HS 748-265, and three Yak-40s delivered in 1978. Transport and liaison aircraft consist of the survivors from 11 C-47 Dakotas, nine DHC-2 Beavers, five DHC-4 Caribous, eight Dornier Do 28D-1s, seven DHC-5 Buffaloes, and possibly two DC-6Bs may still be in service. Unconfirmed reports suggest a few Alouette IIs and Alouette IIIs may be in use.

Zimbabwe Air Force of Zimbabwe

This once highly effective and efficient air force has altered greatly since independence was gained and the introduction of majority rule in 1980. The light bomber role is fulfilled by No. 5 Sqn at New Sarum with 10 Canberra B.2 and three Canberra T.4s. In 1984 there was only one qualified Canberra pilot in Zimbabwe and none of the Canberras have flown for several years, but attempts were being made to get them airworthy again. Fighter-bomber capability is provided by No. 1 Sqn at Thornhill, flying Hunter FGA.9s. The most recent additions were five ex-RAF FGA.90s received in 1984. Two Hunters were written off prior to the sabotage attack at Thornhill in 1982, when three aircraft were destroyed and five damaged. Only 11 aircraft are still in service.

Another Thornhill based unit is No. 2 Sqn which received eight BAe Hawk Mk.60s during 1982. Four had been delivered at the time of the Thornhill sabotage incident on 25 July 1982; one was totally destroyed, one repaired locally and the other two, badly damaged, had to be returned to BAe for rebuild; they were returned to the AFZ during October 1984. No. 2 Sqn is also still operating a few Vampires from the 19 FB.9s and 16 T.11s delivered in 1955 and supplemented by at least 10 two-seat and eight single-seat Vampires during UDI. For obvious reasons, the COIN capability of the AFZ has reduced since independence. No. 4 Sqn at New Sarum only flies three of the 21 Reims-Cessna FTB 337G Lynx received. Two were shot down during anti-guerrilla operations, one destroyed in the Thornhill sabotage incident and the remainder are in storage. No. 6 Sqn at Thornhill operates all the remaining SIAI SF.260 Genets from 31 delivered during UDI. They are now used in the basic training role and have suffered heavily at the hands of indigenous pilots. Only 14 are confirmed as still being in service but a few turboprop powered SIAI SF.260TPs have been delivered recently to supplement them.

No. 8 Sqn at New Sarum with Agusta-Bell/Bell 205 Cheetahs has been disbanded. Of the 11 aircraft acquired, three were written off, five are currently unserviceable and the three remaining aircraft have been transferred to No. 7 Sqn. At the present time, there are no pilots qualified to fly the Cheetah but one has recently started training. No. 7 Sqn flies the Aérospatiale Alouette III;

as many as 95 were acquired or loaned from numerous sources, but 40 loaned South African Air Force aircraft were returned prior to independence and less than 20 remain in use. Two Bell 412s were delivered in 1983 and are still current. Transport facilities are provided by No. 3 Sqn at New Sarum. The AFZ received six CASA C-212s in 1983-84 but a chronic spares shortage means only two are airworthy at any one time. No. 3 Sqn also has the 11 survivors from 15 Douglas C-47 Dakotas received. Only one aircraft is kept airworthy

No. 2 Sqn at Thornhill operates the BAe Hawk for strike duties. Seven are still in service, one being totally written off during a 1982 sabotage incident. Three aircraft were out of service while being repaired.

because all the current pilots are reservists, but new pilots are in training. No. 3 Sqn also operates seven Britten-Norman BN-2A Islanders in the liaison role. Reports of the AFZ receiving Shenyang J-6 or J-7 are incorrect and somewhat ambitious.

Asia and the Far East

This area, particularly in Southeast Asia, has since the end of the war, been a melting pot of world politics. The emergence of China as a world power means that this nation now dominates the area. Strongly Communist countries such as North Korea and Vietnam, now equipped with MiG-23 'Flogger' fighters, are opposed by large, mainly US-supplied forces in Thailand and South Korea. India has a large and well-equipped air force which draws heavily on both Soviet and European sources. Its arch-enemy Pakistan is now receiving much US aid, including F-16s and Bell AH-1s. The two other military giants in the region are Japan and Australia. The former maintains a large defence force, spearheaded by the F-15 Eagle, while Australia is currently re-equipping with the McDonnell Douglas F/A-18 Hornet.

Afghanistan Afghan Republic Air Force

It is difficult to ascertain which aircraft are operated by the Afghan air force, as most Soviet aircraft adopt Afghan markings while operating there. The nearest of these Mi-24s is a 'Hind-A' and the furthest is a 'Hind-D'.

Ever since the Soviet invasion of Afghanistan in December 1979, it has been impossible to draw any meaningful distinction between Soviet and indigenous air power. ARAF markings are worn by most of the Mil Mi-8 and Mil-24 transport and gunship helicopters, the heavy use of which has been a feature of Kabul's attempt to subdue the Mujahideen guerrillas. Over 100 of each type are thought to be operating. The Mi-6 and giant Mi-26 types have also been seen. A number of new helicopter bases have been established, including Farah, Herat,

Jalalabad and Jurm.

The main fighter base is at Bagram, near Kabul. Here the original ARAF force of three Mikoyan MiG-21F squadrons was augmented by more Soviet MiG-21s and an estimated 30-45 MiG-23s, some of which wear ARAF markings. The ARAF force of MiG-17Fs (originally two squadrons at Mazar-I-Sharif and Kandahar) remains active, although the status of the 24 MiG-19s that were received is unknown. Sukhoi Su-7s, Su-17s and Su-22s are present in large numbers, and in Afghanistan during 1982 the newly developed Su-25 'Frogfoot' close support aircraft made its first appearance outside the Soviet Union. Some of these are based at Shindand. The ARAF once had three light bomber squadrons with Ilyushin Il-28s, but these have seemingly not played any part in the current hostilities. However, Soviet Tupolev Tu-16 medium bombers extended the scope of the conflict significantly in 1984 by flying round-trip sorties from bases inside the Soviet Union to attack guerrillas in the troublesome Panjsher Valley.

ARAF transports to be seen at Kabul are mainly Antonov An-26s, with a few Il-18s and An-24s for passenger flights. Soviet An-12 and An-32s are always in evidence and more than one of them has been shot down. An ARAF training school at Sherpur had Yakovlev Yak-18s, and Czech-supplied L-29 Delfin and L-39 Albatros jets. There was also an OCU flying MiG-15Us, but it may be that Afghan pilots now go to the Soviet Union for their training.

Australia Royal Australian Air Force

The RAAF is now receiving its long-awaited Mirage replacement in the form of 75 McDonnell Douglas F/A-18 Hornets, including 18 TF-18A two-seat conversion trainers. All but the first two will be assembled at the Government Aircraft Factory, Avalon. No. 2 OCU at Williamstown is the first unit to convert, followed by No. 75 Sqn in mid-1986, No. 3 Sqn in 1987 and No. 77 in 1988. At present these squadrons all operate the surviving 70-odd Mirage IIIO and IIIDO two-seat trainers. No. 77 is now at Williamstown and No. 75 is at Darwin, although a new base will be built for the Hornets at Tindal. No. 3 is at Butterworth, Malaysia, but will remain at Williamstown following conversion to the F/A-18; thereafter, Hornets and General Dynamics F-111s will be deployed to Butterworth periodically. The F-111C force totals 20 aircraft with Nos 1 and 6 Sqns at Amberley. Four of the latter's aircraft have been modified for reconnaissance and the Pave Tack laser designator pod has been introduced. Two maritime reconnaissance units at Edinburgh fly Lockheed Orions; No. 11 Sqn traded in its P-3Bs to join No. 10 Sqn, each now operating 10 P-3C Update II versions.

The RAAF has a substantial transport force. Richmond houses No. 36 Sqn with 12 Lockheed C-130Hs, No. 37 Sqn with 12 C-130Es, and No. 38 Sqn has 18 de Havilland Canada DHC-4 Caribou. At Fairbairn are two VIP and passenger transport squadrons, No. 33 with four Boeing 707-320C and No. 34 with two BAe One-Elevens, two BAe 748s and three Dassault Falcons. Twelve CH-47C Chinook heavylift helicopters operate in No. 12 Sqn at Amberley, backed up by Bell Iroquois medium-lift helicopters at Fairbairn, where No. 9 Sqn has 30 UH-1Hs and No. 5 Sqn has 14 UH-1Hs plus 18 Aérospatiale AS.350B Ecureuils. This latter unit has training and search and rescue commitments.

All the above units are part of Operational Command. Support Command controls training and maintenance activities. Flying training begins on the Aerospace CT-4A Airtrainer with No. 1 FTS at Point Cook, and continues on the Aermacchi MB.326H with No. 2 FTS at Pearce. Each school has 40 aircraft on strength. Six further CT-4As fly in a FAC role with No.4 Flt at Point Cook. Ten further MB.326Hs are used for instructor training with the CFS at East Sale. This base also houses the School of Air Navigation and its eight BAe 748s. The CT-4s will be replaced by 69 indigenous AAC A 10 Wamira turbo-trainers from 1987, but the MB.326s are undergoing a life extension programme which should keep them in service until 1995.

Royal Australian Navy

The RAN Fleet Air Arm has been an all-helicopter force since 1984, when the aircraft carrier HMAS *Melbourne* and its associated McDonnell Douglas A-4G Skyhawks and Grumman S-2E Trackers were axed. The eight Westland Sea King Mk 50s of HS-817 are now permanently shore-based at Nowra, alongside HU-816 with 15 Westland Wessex HAS 31Bs. This squadron will receive the Sikorsky S-70B Seahawk in late 1987, of which eight have been ordered, with the prospect of 24 more for operation on frigates. Also at Nowra is HC-723 with six AS.350B Ecureuils and three Bell 206B JetRangers.

Australian Army Aviation Corps

Limited to reconnaissance and support flying, the AAAC nevertheless received 75 Bell 206B light helicopters, which are organized into No. 161 Sqn and No. 171 Sqn at Holsworthy, and No. 162 Sqn at Townsville. No. 171 Sqn also operates some of the 13 GAF Nomads and 15 Pilatus PC-6 Turbo-Porters, the balance of these types serving with No. 173 at Oakey, which is the headquarters base.

The RAAF has two squadrons assigned to maritime patrol, Nos 10 and 11 which both fly from Edinburgh. Equipment is the Lockheed P-3C Update II Orion, of which 20 have been supplied to replace the earlier P-3B.

Bangladesh Bangladesh Defence Force (Air Wing)

The 36 Shenyang F-6 fighters supplied by China in the late seventies to equip two squadrons at Dacca and Jessore are not very active. In this impoverished country priority is given to maintaining the rotary-wing inventory, which comprises six Mil Mi-8s and 10 Bell 212s. Transport aircraft include a Yakovlev Yak-40 and a few Antonov An-24/-26s. A training school at Jessore received about 12 Chujiao-6 piston-engine trainers and eight refurbished Fouga Magisters from West Germany.

Brunei Royal Brunei Armed Forces – Air Wing

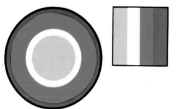

This primarily helicopter force is tasked with protecting the oilfields from which this small but wealthy country derives virtually all of its income; additional duties include fishery protection and search and rescue. No. 1 Sqn flies 11 Bell 212s, one of which is outfitted for VIP use, from Berakas Camp. At Brunei Airport No. 2 Sqn has six MBB BO 105 gunships and No. 3 Sqn is the training unit with two Bell 206Bs and two SIAI-Marchetti SF 260W Warriors. The Royal Flight is based at the Sultan's palace and has one Sikorsky S-76 and one BO 105.

Burma Union of Burma Air Force

This small force, which has its main bases at Meiktila and Rangoon, plays its part in the Rangoon government's continuing struggle to subdue regional and Communist insurgency. The UBAF bought 12 SIAI-Marchetti SF 260MBs in 1976 and 16 Pilatus PC-7 Turbo-Trainers three years later; both types have been used for armed reconnaissance in support of ground troops. The US supplied 15 Lockheed AT-33As and 12 Cessna T-37Cs in the late sixties, but the survivors are unlikely to be active. Karen insurgents accounted for two of the UBAF's 20-strong fleet of Bell UH-1H helicopters in mid-1983, and 14 Aérospatiale Alouette IIIs and 13 Kawasaki-built Bell 47Gs were purchased some years ago. For transport there are four Fairchild Hiller FH-227 Friendships with cargo doors, augmented by a Fokker F.27 and a Cessna Citation. Other assets include eight PC-6B Turbo-Porters, three PC-6 piston-engine Porters, 10 Cessna U-17s and nine de Havilland Canada DHC-3 Otters, but not all of these may remain in use.

China Air Force of the People's Liberation Army

The AFPLA is theoretically the world's third largest airforce, but most of its equipment is obsolescent and the new leadership has made it clear that military forces do not rate high on its list of priorities for modernizing the country. The Military Air Regions of the AFPLA correspond to the PLA's subdivision of the country into 11 areas (Beijing, Chengdu, Fuzhou, Guangzhou, Jinan, Kunming, Lanzhou, Nanjing, Shenyang, Urumqi and Wuhan). Within each area a number of functional air divisions (air defence, transport etc.) each control up to three regiments, which in turn comprise three or four brigades of about 15 aircraft each.

Most military aircraft in service are Chinese-produced copies of Soviet designs, and have generally been designated in the West by reference to the location of the factory where they were manufactured and to the role for which they are produced. For example, fighter aircraft built at the Shenyang factory have been known as Shenyang F-2, F-4, F-5 etc. However, the Chinese have only adopted such Western practice when dealing with the export market. Their own designation system uses the same numbers as those known in the West, but simply prefixes them with the assigned role, for example Fighter Type 2, 4, 5 etc. Using the now preferred Pinyin transliteration of Chinese characters, 'Fighter' becomes Jianjiji and hence the designations are J-2, J-4, J-5 etc. The other relevant transliterations are Jianjiji-Jiaolianji (JJ-Fighter Trainer e.g. two-seat conversion types), Jianjiji-Zhenehaji (JZ-Fighter Reconnaissance), Qiang (Q-Attack), Hong (H-Bomber), Yunshu (Y-Transport), Zhi (Z-Helicopter) and Chujiao (CJ-Basic Trainer). The suffix -Xin indicates a modernization or update to the basic design.

The Chinese resumed production of the J-7Xin (Mikoyan MiG-21F improved) some years ago at Xian, but have exported many to Egypt and Iraq. The number in AFPLA service (including earlier J-7s) may be as low as 200, compared with more than 3,000 of the fifties-vintage J-6 (MiG-19SF). Modifications include the camera-equipped JZ-6 and the J-6Xin with a short-range intercept radar. The second most numerous fighter/attack type is likely to become the Q-5 (NATO codename 'Fantan'), an indigenous development of the J-6 produced at Nanchang. The J-8 designation was long presumed in the West to belong to the new fighter under development at Shenyang, for which the Chinese made a licence agreement with Rolls-Royce for supply of the Spey turbofan. The Spey plan was apparently abandoned, although a few engines were supplied. Later a request to the US for the Pratt & Whitney PW1117 engine for the new fighter was refused. When pictures of the J-8 were eventually released in late 1984, it appeared to be a stretched twin-engine MiG-21 type, bearing great similarity to the Mikoyan E-152 design of the early sixties. It seemed that only about 50 of them were in AFPLA service, and that development of the long-awaited new fighter was continuing in a separate programme. An estimated 500 J-4 (MiG-17) and J-5 (MiG-17F) fighters are in service. The JJ-5 (two-seat MiG-17 derivative) seems to be the main initial jet trainer, with the JJ-6 (two-seat MiG-19 derivative) serving as a conversion type. Production of the H-6 (Tupolev Tu-16) medium bomber continues at Xian with more than 100 in service, supplemented by over 500 H-5 (Ilyushin Il-28) light bombers.

The AFPLA's transport capability is very small in relation to the overall size of the force, comprising around 16 BAe Tridents (CAAC has 20 more) and similar numbers of Soviet-supplied Antonov An-12s, An-26s and Il-18s. Ancient Lisunov Li-2s, Il-14s and Curtiss-Wright C-46 Commandos soldier on, but the Y-7 (An-24) and Y-8 (An-12) are both now in production in China at Xian and Harbin respectively. The few hundred Y-5s in service could be supplemented by the new Y-11/Y-12 light transports, the latter benefitting from an infusion of Western technology in the form of two Pratt & Whitney Canada PT6A engines. Basic trainers are the Yak-11/-18 and indigenous CJ-5/6 from Nanchang. New Western helicopters are joining a force of some 300 Z-5/6 (Mil Mi-4/8) machines. Fifty Aérospatiale SA.365N Dauphin IIs are being licence-produced at Harbin as the Z-9, and may be destined for the military. The 24 Sikorsky S-70C Black Hawks ordered in 1984 will be flown by the PLA in Tibet.

Aviation of the People's Navy

The Navy controls a separate force of

Most of the Chinese aviation industry is concerned with building Soviet types, but is now swinging to more modern Western and indigenous designs. The factory at Harbin produced the Z-5 copy of the Mil Mi-4 'Hound'.

four bomber and five fighter divisions equipped with torpedo-carrying H-5s and H-6s, J-5 and J-6 fighters which are integrated with the AFPLA alert system, Y-5 transports, and Z-5 helicopters. France delivered 13 Aérospatiale Super Frelons to China in the mid-seventies, and these may be in Navy ASW use.

Production is in full swing at Nanchang of the Q-5 'Fantan'. This aircraft is developed from the Shenyang J-6 (MiG-19) and is optimized for ground attack. Several have been supplied to Pakistan.

Hong Kong Royal Hong Kong Auxiliary Air Force

The RHKAAF provides surveillance patrols, medevac, firefighting, survey, SAR and transport services to the colony. Based at Kai Tak Airport, it comprises a Cessna 404 Titan, a Britten-Norman BN-2A Islander, two BAe Bulldog 128s and three Aérospatiale SA.365 Dauphin 2s.

Hong Kong operates a small fleet for general duties around the territory. As well as four fixed-wing aircraft, three Aérospatiale Dauphin 2 helicopters are also on strength. The speed and power of this twin-engined helicopter are particularly useful for the wide array of duties it is asked to perform.

India Indian Air Force

The IAF continues to strike a balance between the Soviet Union and Europe in its aircraft acquisition policy, which is motivated by both political and economic considerations.

For the moment India has decided to halt its deal with Dassault for Mirage 2000s, at the stage of acquiring 40 examples from the French production line (36 single-seat 2000H versions and four 2000TH two-seaters). Although the French aircraft reportedly beat the Mikoyan MiG-23 in a 1981 air-superiority evaluation, enough MiG-23MLs to equip two IAF squadrons were flown into India two years later. They carry the R23 (Soviet AA-7 'Apex') medium-range and R60 (Soviet AA-8 'Aphid') short-range AAMs. The Mirages will have the Matra Super 530 and Magic AAMs; deliveries to India are scheduled for mid-1985 to end-1986. In a surprising move, India was offered the MiG-29 'Fulcrum' advanced air superiority fighter in 1984, almost before it had entered service with the Soviet air force. It seems that 45 aircraft will be delivered to the IAF from Soviet production lines, followed with licence-production by Hindustan Aeronautics Ltd (HAL) at Nasik. The timescales involved have not been authoritatively revealed, and it is not clear whether the previous plan to assemble the MiG-27 'Flogger-J' at Nasik from 1986 is affected by the MiG-29 acquisition.

At the moment HAL Nasik is concluding its production of the MiG-21, which began in 1967 with the MiG-21FL. About 200 were completed before giving way in turn to the MiG-21MF in 1974 (about 150 produced), and the MiG-21bis in 1981 (a total of 200 scheduled to be built). Indian production of all three sub-types was preceded by deliveries of aircraft from the Soviet Union in flyaway or disassembled condition; they comprised 42 MiG-21Fs in the early sixties, 50 MiG-21MFs (along with a few MiG-21U two-seaters) in the early seventies and at least 30 MiG-21bis in 1977. Nineteen IAF Squadrons fly the MiG-21: Nos 1, 3, 4, 7, 8, 15, 17, 21, 23, 24, 26, 28, 29, 30, 37, 45, 47, 101 and 108. Most of the early MiG-21F models have been retired. The indigenous development of the Folland Gnat, the HAL Ajeet of which 100 were built, serves with Nos 2, 9, 18 and 22 Sqns.

In the deep penetration strike role the IAF will have five SEPECAT Jaguar squadrons with a total of 116 aircraft. The first 40 of these were built in the UK, including five two-seaters, and equipped Nos 5 and 14 Sqns. (An additional 18 were loaned to the IAF from RAF stocks from 1979 to 1984.) HAL is assembling 76 Jaguars at Bangalore, and these are replacing the remaining BAC Canberra B(I).58s, B.74s and B(I).12s with Nos 6, 16 and 35 Sqns. For shorter range air-ground missions the IAF has three squadrons of MiG-23BNs (Nos 10, 220 and 221) which were delivered in knock-down form and assembled at HAL Nasik in 1981/2. The survivors of 150 Sukhoi Su-7BM fighter-bombers still serve with Nos 32 and 222 Sqns, and the HAL HF-24 Marut survives in No. 31 Sqn. India wants to produce its own Light Combat Aircraft (LCA) in the nineties and is evaluating industrial partners from the European countries. Some Canberra PR.57s may be retained in No. 106 Sqn for reconnaissance, and a separate reconnaissance flight operates eight MiG-25Rs.

A total of 95 Antonov An-32 specialized 'hot-and-high' medium-range transports are replacing the geriatric Fairchild C-119G jet-boosted Flying Packets in Nos 12, 19 and 48 Sqns; they will also replace the de Havilland Canada DHC-4 Caribou in No. 33 Sqn, and the Douglas C-47s of Nos 43 and 49 Sqns. Deliveries of the Ilyushin Il-76 to replace some 30 Antonov An-12s in Nos 25 and 44 Sqns were due to begin in 1985. Since the Soviets also produce an AWACS version of the Il-76 (NATO codename 'Mainstay'), it may be this aircraft which will satisfy a long-standing IAF requirement for such a machine. The BAe 748 was licence-produced for many years at HAL Kanpur; military examples serve with No. 11 Sqn, No. 106 Sqn for photo-mapping, and with the Air HQ Communications Squadron for VIP transport (alongside three Boeing 737s). The Kanpur line switched to the Dornier 228 light transport in 1985, and about 40 of these will go to the IAF, replacing DHC-3 Otters which serve with Nos 41 and 59 Sqns.

Licence-produced Aérospatiale Alouette IIIs and the Soviet-supplied Mil Mi-8/-17 make up most of the IAF helicopter inventory. Over 260 Alouette IIIs were built at HAL Bangalore as the Chetak and serve with Nos 104, 107, 111, 113, 115 and 116 Sqns for transport and anti-armour missions. The Mi-8s are in Nos 105, 109, 110, 118, 119 and 121 Sqns. An initial 12 Mi-24/-25 'Hind' gunship helicopters were supplied by the Soviet Union in 1984. The HAL Cheetah (licence-built Alouette II) serves the army in four liaison squadrons, where it supplemented the fixed-wing HAL Krishak.

IAF pilots begin their flying training at the Elementary Flying School, Bidar, where approximately 50 HAL HT-2 trainers eventually began to be replaced by a new piston trainer, the HPT-32, in 1983.

However, in 1984 HAL flew a turboprop version (HTT-34) which could supplement the HPT-32. Jet training is conducted at the Air Force Academy, Dundigal, on the HAL Kiran 1/1A and the Polish PZL-Mielec TS-11 Iskra of which 50 were supplied in 1975-76. About 130 Kirans were delivered (they also serve at Tambaram in the Flying Instructor's School), and a Mk 2 version is supposed to augment them. A Fighter OCU at Kalaikunda operates the remaining HAL/Hawker Siddeley Hunters (F.56/T.66) and Gnats (F.1); the Transport Training Wing at Yelahanka has around 20 HAL/BAe 748s; the Helicopter Training School uses Chetaks and Cheetahs at Hakinpet; the Parachute Training School at Agra has some C-119s; and the Navigation and Signals School at Begumpet has seven 748s. Conversions to the Russian fighter types are conducted at squadron level on MiG21Us, MiG-23UMs and Su-7Us.

Indian Naval Aviation

The carrier *Vikrant* emerged from a 1984 refit with a ski-jump, ready to embark the new 300 Sqn with its six Sea Harrier FRS.51s; two T.60 trainers were also supplied, and a second squadron will be formed when 10 more FRS.51s and two T.60s are delivered. The ski-jump put an end to carrier operations by 310 Sqn's eight Breguet Alizés. However, 315 Sqn has five Ilyushin Il-38s for long-range maritime patrol and 321 Sqn has some HAL Chetaks for search and rescue, including shipborne detachments. All these units are based at Goa, which also houses an OCU (No. 551 Sqn) with Kirans.

The other INA base is at Cochin, where 12 Westland Sea King Mk 42/42As serve with Nos 330 and 336 Sqns. They fly ASW missions from the two newest 'Leander' class frigates and the 'Godavari' class missile frigates; 12 Sea King Mk 42Bs with anti-ship capability have been ordered for 1986 delivery. A few torpedo-carrying HAL Chetaks serve on the three older 'Leander' class frigates (331 Sqn), and five Kamov Ka-25 'Hormone' helicopters were supplied for use with the three Soviet supplied 'Krivak' class heavy destroyers (333 Sqn); 18 Ka-27 'Helix' are scheduled to join INA. A helicopter training squadron at Cochin (No. 562) has a few Hughes 300s and HAL Chetaks. Coastal patrol duties are undertaken by No. 550 Sqn from Cochin with 16 Pilatus Britten-Norman BN-2B Maritime Defenders, although a new Coast Guard Air Wing is being created. Its first base is at Daman, where a squadron of six Chetaks has been established.

The sizeable transport force is under modernization with Soviet Antonov An-32s replacing C-119s, Caribous and Buffaloes, while Ilyushin Il-76s are replacing the Antonov An-12s. This An-12 is typical of the 30 supplied which have been used extensively on runs to the Soviet Union.

Indonesia Tentara Nasional Indonesia-Angatan Udara

This huge country has taken some steps towards building a force capable of providing airborne protection over an area which stretches nearly 5,000 km (3,107 miles) from the western tip of Sumatra to the eastern border of Irian Jaya with Papua, New Guinea.

Although very small in relation to the vast size of the country, the jet fighter-attack element of the TNI-AU is looking more capable than in the seventies when a mere handful of fifties-vintage Sabres were in use. Based at Madian, but with frequent detachments elsewhere, it now comprises No. 14 Sqn with 12 Northrop F-5Es and four F-5Fs, and two squadrons (one of which is No. 11) flying 28 McDonnell Douglas A-4Es and four TA-4Hs, which were supplied in two batches from Israel with US military assistance funding. The A-4s are equipped for inflight-refuelling, and the TNI-AU can now make use of the tanker capability that was built into two of its C-130Bs some years ago. The Lockheed design provides the backbone of the transport force, with 10 C-130Bs, two C-130Hs, a C-130H-MP and nine C-130H-30s equipping No. 31 Sqn at Jakarta-Halim and No. 32 Sqn at Malang. Three Boeing 737-200 Maritime Patrol aircraft were delivered in 1983-84 and a single C-130H-MP is also used in this role. Sixteen Rockwell OV-10F Broncos were delivered in the mid-1970s for No. 3 Sqn and a few Lockheed AT-33s may still survive from 10 delivered in 1973 from the US.

Pilot training takes place at the Jogjakarta training school where students progress from the Swiss FFA AS.202 Bravo (20 received and with No. 101 Sqn) via the Beech T-34C (25 equipping No. 102 Sqn) to the BAe Hawk Mk 53 (of which 16 are now in use with No. 103 Sqn). A couple of Beech King Air A100s provide navigator training.

The expanding state aircraft production facility of Nurtanio at Bandung can be expected to provide most of the future helicopter and transport needs of the Indonesian armed forces. With this in mind the CASA-Nurtanio CN-235 airliner was designed with a rear ramp/cargo door, and 32 have now been ordered for the TNI-AU. The present medium transport force comprises eight F.27Ms with No. 2 Sqn at Jakarta-Halim, supplemented by some Douglas C-47s and 10 CASA-Nurtanio NC-212A Aviocars from the Bandung line. No. 4 Sqn at Jakarta-Kemayoran flies an assortment of Cessnas (180Bs, T207s, 401s and 402s).

Jakarta-based No. 6 Sqn has about 12 Aérospatiale SA.330J Pumas (later deliveries of NSA-330s were from the Bandung line) and seven NAS-332 Super Pumas are being received from the first batch to be produced at Bandung. Surprisingly, the TNI-AU is continuing to convert its elderly Sikorsky UH-34Ds to S-58T turbine configuration and purchased kits for this purpose in late 1983. Search and rescue missions are the responsibility of No. 5 Sqn at Semarang flying a few Grumman HU-16A Albatross, Agusta-Bell AB 204Bs and Aérospatiale Alouette IIIs. Twelve Hughes 500Cs were transferred from the state oil company airline Pelita to the TNI-AU for oil rig security patrols in 1984. The TNI-AU operates an agricultural air unit from Jakarta equipped with a few Pilatus Turbo-Porters, Cessna 188 Agwagons, Nurtanio-MBB BO 105s and a rather larger number of Gelatiks (an Indonesian-assembled Polish PZL Wilga).

Tentara Nasional Indonesia-Angatan Laut

The Surubaya-based TNI-AL (naval aviation) has also expanded and now flies 18 GAF Searchmasters. The first 12 delivered were 'B' models with Bendix radar, and the last six 'L' models with a larger Litton 360° radar. The Searchmasters were supplied under a defence aid pact with Australia, which also helped with the construction of new airfield facilities for them at Tanjung Pinang in the Riau Islands and Manado in North Sulawesi. A total of 26 NAS-332F Super Pumas are being delivered at a slow rate from Bandung

for the anti-ship role. Ten ex-Royal Navy Westland Wasps were acquired in 1983, and will presumably be operated from the three ex-British 'Tribal' class frigates which the Indonesians bought in 1984. Four NC-212 Aviocars are used for transport.

Tentara Nasional Indonesia-Angatan Darat

The Army's desire to expand its trooplift capability is behind the Nurtanio plan to build the Bell 412 under licence, with the TNI-AD wanting to operate as many as 80. It currently uses the Bell 205A-1 for this purpose (16 delivered in 1978) and also flies the MBB BO 105 (16

The Indonesian air force flies three Boeing 737s for maritime patrol duties. These are equipped with Motorola SLAMMR (side-looking airborne multi-mission radar).

delivered from Nurtanio). The Australian Army transferred 12 Bell 47Gs to the TNI-AD in 1978 for training, and the survivors are being updated with Soloy conversions. Nine Hughes 300Cs were ordered in late 1982 for assembly in Indonesia. The Army has four NC-212 Aviocars for transport, together with one or two Aero Commanders and Cessna 310Ps, as well as a Britten-Norman BN-2A Islander, a Beech H18 and a number of Gelatiks.

Japan Japan Air Self-Defence Force

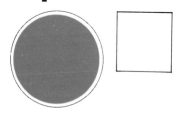

Japanese military spending has increased somewhat in response to US prompting, and the JASDF is building a new air defence force around the McDonnell Douglas F-15 Eagle and Grumman E-2C Hawkeye, allied to the long-established Base Air Defence Ground Environment (BADGE) system.

Japan originally ordered 88 single-seat F-15Js, all bar the first two to be assembled by Mitsubishi at Komaki; 12 F-15DJ two-seaters were to be supplied from the US line. In 1985, an initial commitment to 55 additional aircraft was made. The F-15s serve in Air Defence Command (ADC) alongside the McDonnell Douglas F-4EJ, of which 138 were assembled by Mitsubishi after two initial machines were delivered from the US. No. 2 Wing at Chitose has No. 203 Sqn (F-15) and No. 302 Sqn (F-4); No. 5 Wing at Nyutabaru is the OCU with No. 202 Sqn (F-15) and No. 301 Sqn (F-4); No. 6 Wing at Komatsu has No. 303 and No. 306 Sqn (both F-4); No. 7 Wing at Hyakuri has No. 204 Sqn (F-15) and No. 305 Sqn (F-4). The latter unit will become the fifth F-15 squadron in 1986/7, before which No. 302 Sqn is scheduled to swap places with No. 207 Sqn at Naha, Okinawa (currently flying the last F-104Js) so that No. 207 can convert to the F-15 and thus make the most northerly Wing at Chitose an all F-15 force. The eight E-2Cs of No. 601 Sqn are also deployed in the north, at Misawa. There is one further F-4 squadron, No. 304, which is part of No. 8 Wing at Tsuiki. The other squadron there,

No. 6, flies the indigenous Mitsubishi F-1 strike fighter, 77 of which were ordered. The others serve in No. 3 Sqn which, together with No. 6 forms part of No. 3 Wing at Misawa. The Hyakuri-based No. 501 Sqn has 14 RF-4EJ reconnaissance aircraft supplied from the US in 1974-5. An ECM Flight at Iruma has a Kawasaki C-1, three NAMC YS-11Es, and some Lockheed T-33s, and an aggressor squadron at Nyutabara has Mitsubishi T-2s and T-33s.

Japan's Maritime Self-Defence Force is currently re-equipping with the Lockheed P-3C Orion, which is being built by Kawasaki. These replace Kawasaki P-2J Neptunes.

Over 150 T-33As still serve the JASDF, many with the ADC squadrons and regional HQ flights for communications and proficiency flying purposes. They also make up one of the five wings in Flying Training Command, No. 1 at Hamamatsu

Japan (continued)

(Nos 33 and 35 Sqns). The T.2 advanced trainer is still in limited production, 92 having so far been ordered for No. 4 Wing (Nos 21 and 22 Sqns) at Matsushima, and for the F-1 squadrons. The veteran Fuji T-1A/B jet trainer still equips No. 13 Wing at Ashiya. Base flying training is conducted on 500 Fuji T-3s (a development of the Beech T-34 Mentor) in No. 11 Wing at Shizuhama and No. 12 Wing at Hofu. The new Kawasaki T-4 turbofan trainer will replace the T-1 and T-33.

The Air Transport Wing has six Lockheed C-130Hs in No. 401 Sqn at Komaki, with more to come. This squadron also flies the C-1, 28 of which are in service there, and at Iruma (No. 402 Sqn) and Miho (No. 403 Sqn). These latter two squadrons also have eight YS-11 transports between them. The Air Rescue Wing maintains detachments at all major bases from the 25 Mitsubishi MU-2S light transports and 36 Kawasaki-Vertol KV-107s in its inventory. A Flight Check Group at Iruma has a YS-11, four MU-2Js and some T-33s. The Air Proving Wing at Gifu carries out test flying for research and development.

Japan Maritime Self-Defence Force

Like the JASDF, the JMSDF is in the process of upgrading the major element in its line-up, the airborne ASW and anti-ship force. There are 45 P-3C Orions currently on order; the first three were built by Lockheed in 1981, after which production switched to the Kawasaki plant at Gifu. The two squadrons of Fleet Air Wing 4 at Atsugi (Nos 3 and 6) have now been converted from the Lockheed P-2J Neptune to the P-3C, and Fleet Air Wing 2 at Hachinobe will

follow suit in 1986-87 (Nos 2 and 4 Sqns). The other two P-2J squadrons (No. 1 in Fleet Air Wing 1 at Kanoya and No. 5 in Fleet Air Wing 5 at Naha) will also convert to the P-3C in the late eighties, and in 1989 the sole squadron flying the Shin-Meiwa PS-1 ASW flying-boat (No. 31 Sqn Fleet Air Wing 31, Iwukuni), will also convert to the P-3C. The JMSDF has 79 P-2Js and 15 PS-1s on strength.

Licence production of the Sikorsky SH-3 Sea King at Mitsubishi has continued for 20 years now. Known as the HSS-2/2A/2B in JMSDF service, about 70 are in use with six ASW helicopter squadrons. Three squadrons of Fleet Air Wing 21 at Tateyama (Nos 101 and 121, plus No 122 detached at Omura) make shipboard deployments, unlike the other three un-numbered units at Komatsushima, Omura and Ominato. The Sikorsky SH-60 Seahawk was selected in 1984 to replace the Sea Kings, starting in 1990, and Mitsubishi will integrate Japanese systems into the airframe. Search and rescue duties are performed by six flights at Atsugi, Hachinobe, Kanoya, Ozuki, Tokushima and Iwo-Jima, which are equipped with two helicopters each. Mitsubishi-built Sikorsky S-61As are replacing the smaller Sikorsky S-62Js. Long-range SAR is conducted with 10 Shin-Meiwa US-1 flying-boats, which fly alongside their PS-1 cousins in No. 71 Sqn, Fleet Air Wing 31 at Iwakuni. This wing also controls No. 81 Sqn with four UP-2Js for target-towing, drone-launching and ECM, and is due to receive six Gates Learjet 36As. No. 61 transport squadron at Atsugi has four YS-11Ms, and the JMSDF's test squadron is No. 51 at the same base. No. 111 Sqn is a mine-sweeping unit

at Shimofusa with the KV-107, which the JMSDF wants to augment with a squadron of Sikorsky MH-53Es.

There are four Air Training Groups, and that at Ozuki conducts primary flying training on 34 Fuji KM-2s (a development of the T-34A) in No. 201 Air Training Squadron. No. 202 ATS at Tokushima is responsible for multi-engine and instrument training on the Beech B-65 Queen Air; of the 22 supplied, some are used elsewhere for communications duties, and the B-65s have been supplemented by 21 Beech TC-90 King Airs. The Kanoya-based group has No. 203 Sqn with P-2Js and No. 211 Sqn with HSS-2s and eight Hughes OH-6D/J primary helicopter trainers. No. 205 Sqn is the main element of th Shimofusa group, flying the P-2J, B-65 and six YS-11T crew trainers.

Japan Ground Self-Defence Force

The JGSDF is developing an anti-tank helicopter force with the Bell AH-1S HueyCobra. Two were delivered from the US for evaluation in 1979-80, and the first licence-built

Fourteen RF-4EJ Phantom reconnaissance aircraft were supplied direct from McDonnell Douglas to equip No. 501 Sqn based at Hyakuri. A number of colour schemes have been seen.

example from Fuji flew in 1984; a total of 58 are planned for three squadrons. The two squadrons flying 54 KV-107 trooplifters at Kisarazu will be upgraded with about 40 Boeing Vertol CH-47 Chinooks, which will be produced by Kawasaki from 1986. The two mainstays of the JGSDF's helicopter force are the Fuji-built Bell HU-1 Iroquois (70 HU-1Bs and 80-odd HU-1Hs, the latter still in production) and the Kawasaki-built Hughes OH-6 (110 OH-6J and nearly 60 OH-6D, which are still in production). They are organized into 13 squadrons, one attached to each army division with a further 10 squadrons attached in pairs to the HQ of each of the five regional armies. Seventeen LR-1s (Mitsubishi M-2Cs) are in service in liaison and photo-reconnaissance roles. An aviation school operates all the types in the operational inventory, plus 36 Hughes TH-55Js for basic flying training.

Kampuchea

It is not thought that any of the US aircraft supplied to the Lon Nol regime in the early 1970s are still active. Vietnamese armed forces have occupied the country since over-

throwing the Pol Pot regime in early 1979, and have not made any attempt to revive a separate Kampuchean air arm (see Vietnam).

North Korea Korean People's Army Air Force

The KPAAF's front-line fighter is the Mikoyan MiG-21F/PF/PFM, about 100 of which equip two regiments. Many of them were assembled at the Lyongaksan aircraft plant in Pyongyang. The Soviet Union also supplied around 30 Sukhoi Su-7BM ground attack aircraft, and have been asked to supply MiG-23s. Earlier model MiGs were mainly sup-

plied from China; intelligence sources provide totals of around 300 Shenyang J-5s and 50 J-6s which equip 12 squadrons. The Chinese also supplied about 60 Harbin H-5 light bombers and over 200 Shijiazhuang Y-5 light transports. Transports include three Tupolev Tu-154s and an Ilyushin Il-62M for VIP use, a small number of Il-18s

and a rather larger number of Antonov An-24s. Mil Mi-8 and Mi-4 helicopters are operated. Pilot training starts on the CJ-6 and proceeds to the Yakovlev Yak-18 and CJ-5/6.

South Korea Republic of Korea Air Force

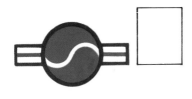

The ROKAF is scheduled to receive a squadron (16 aircraft) of General Dynamics F-16s in 1986. Until then, its front-line fighters are the McDonnell Douglas F-4 and the Northrop F-5. The 1st Fighter Wing

received a top-up of six former USAF F-4Ds in late 1982, adding to the 36 F-4Ds acquired earlier to equip two squadrons. A further two squadrons of this wing operate 37 F-4Es received in 1976-78 and four more F-4Es were transferred from the US in 1984. Eight AN/AVQ-26 Pave Tack target designator pods have been ordered for the F-4Es. The 10th and 11th Fighter Wings fly most of the near-200 strong force of Northrop F-5E/Fs. Northrop delivered 126 F-

South Korea's main fighter at present is the McDonnell Douglas F-4 Phantom, pending delivery of F-16s.

South Korea (continued)

5Es and 20 F-5Fs to replace the earlier F-5A/B force, before licensed production of another 36 F-5E and 32 F-5F began in September 1982. The 12th Fighter Wing received 27 former South Vietnamese Cessna A-37Bs via the US in 1976. The ROKAF also received 24 OV-10G Broncos in the late 1970s, and later began negotiating with Rockwell to resume production of the Bronco, possibly in Korea on a licensed basis. It would want an updated version,

equivalent to the US Marine Corps' OV-10D conversions with FLIR.

The Air Transport Wing flies a mix of about 30 Curtiss-Wright C-46s, Douglas C-54s and Fairchild C-123B/J/Ks, which obviously need replacing. The ROKAF has been looking at the Shorts 330 and CASA C-212 proposing a sufficiently large purchase to justify licensed production. A helicopter squadron had four Bell 212s and three 412s. Some Cessna O-2s and U-17s are also in the

inventory. The ROKAF has used five training types in recent years, comprising the Cessna T-37/T-41, Lockheed T-33, North American T-28 and Northrop F-5Bs, which may be replaced by current F-5F production.

The army helicopter force is being considerably expanded, thanks to licensed assembly of the Hughes 500 Defender. An anti-armour capability was acquired when Hughes delivered 34 TOW-armed Defenders

in 1981-82; since then orders for 70 more have been placed for delivery in kit form, along with 89 Defenders in the 'scout' configuration. For trooplift, 25 Bell UH-1Bs were acquired from surplus stocks in 1977 and refurbished and some UH-1Hs have since been supplied. A few Cessna O-1s may remain in service. The navy has had an ASW squadron of Grumman S-2s for some years. Six US surplus S-2Es were added in 1981.

Laos Air Force of the Liberation Army

The AFLA has a single Mikoyan MiG-21F squadron at Vientiane with about 20 aircraft. A few North American T-28s, Douglas AC-47s and Cessna O-1s supplied to the former RLAF may still be flying. The Soviets have supplied about 10 Mil Mi-8 helicopters and some Antonov An-2Ms. Most of the military transports supplied by the Soviet Union

in recent years have been put on the civil register and are formally operated by Lao Air. They include five An-24s, three An-26s and a Yakovlev Yak-40.

Malaysia Tentara Udara Diraja Malaysia

The TUDM finally took delivery of 40 reworked McDonnell Douglas A-4 Skyhawks in 1985. Although Malaysia purchased 88 A-4C/L aircraft from US storage in mid-1979, budgetry cutbacks led to the cancellation of an ambitious modernization programme, and the TUDM has had to settle for a smaller force of Skyhawks which have only been moderately upgraded during refur-

bishment by Grumman in the US. They have entered service with Nos 6 and 9 Sqns at Kuantan, allowing 10 Canadair CL-416 Tebuans to be retired to reserve status. A second fighter-interceptor squadron will probably be needed when the RAAF withdraws its Mirage IIIs from Butterworth in 1988. For the moment, No. 12 Sqn there has 13 Northrop F-5Es, three F-5Fs and a couple of RF-5E Tigereyes.

Three Lockheed C-130H-MP maritime patrol aircraft entered service with No. 5 Sqn at Kuantan in 1980; the TUDM had earlier purchased six C-130Hs for transport duties with No. 14 Sqn at Kuala Lumpur (Simpang). About 20 de Havilland Canada DHC-4 Caribous soldier on with No. 1 Sqn there and

No. 8 Sqn at Labuan in East Malaysia. The survivors of 40 Sikorsky S-61As and about the same number of Aérospatiale Alouette IIIs serve in four mixed helicopter squadrons, No. 3 at Butterworth, No. 5 at Labuan, No. 7 at Kuching and No. 10 at Kuantan. No. 2 VIP transport squadron at Kuala Lumpur has two Canadair CL-600 Challengers, two Fokker F.28s and 12 Cessna 402Bs. Basic flying training is at No. 1 Flying Training School, Alor Setar, on 40 Pilatus PC-7s (preceded by grading on 10 BAe Bulldogs now based at Kuala Lumpur). No. 2 FTS is the helicopter school at Keluang with the S-61A, Alouette III and eight Bell 47Gs. Advanced jet training is at No. 3.FTS, Kuantan, on 12 Aermacchi MB.339s.

Nepal

Royal Nepalese Air Force

The RNAF is a small force providing transport (including a Royal Flight) and army support. It has one BAe 748, a de Havilland Canada Twin Otter, four Short Skyvans, two HAL Chetaks and two Aérospatiale Pumas.

New Zealand Royal New Zealand Air Force

The RNZAF purchased Australia's unwanted McDonnell Douglas A-4G fleet in 1984, enabling it to form a second Skyhawk squadron at Ohakea. No. 75 Sqn has been joined by the new No. 2 Sqn; between them they have nine A-4K, eight A-4G, two TA-4G and three TA-4K Skyhawks, and all 22 aircraft are to be

upgraded with a new weapons package. No. 14 Sqn at Ohakea has 16 BAe Strikemasters for advanced flying training, with a secondary strike role. At Whenuapai No. 5 Sqn has five Lockheed P-3B Orions and No. 40 Sqn five C-130H Hercules; No. 42 Sqn has five Hawker Siddeley Andover C.1s and another five Andovers are stored. A transport flight of three Cessna 421Cs operates from Woodbourne. Helicopters are concentrated in the tri-service No. 3 Sqn at Hobsonville. Five Bell UH-1Ds, six UH-1Hs and five 47Gs fly mainly for the army; four Westland Wasp HAS.1s deploy to Navy frigates for ASW and four UH-1Hs are deployed to Singapore in support of the New Zealand battalion sta-

tioned there. The Flying Training Wing at Wigram has about 20 Aerospace CT4 Airtrainers, four 47Gs and three Fokker F.27-100s.

The RNZAF's combat assets rest largely on the McDonnell Douglas A-4 Skyhawk, 22 of which are on strength. The BAe Strikemaster assists in the strike role.

Pakistan Pakistan Air Force

The PAF has now received all 32 General Dynamics F-16As and 8 F-16Bs that were ordered in 1981. They replaced Chinese-supplied Shenyang F-6s in three squadrons. Well over 150 F-6s were acquired from 1966 and modified in-country to take AIM-9B/J Sidewinder missiles and Martin-Baker ejection seats. They served in eight squadrons: No. 11 at Rafiqui, No. 14 at Mianuhali, Nos 17 and 18 at Mas-

roor, Nos 19 and 26 at Peshawar and Nos 23 and 25 at Sargodha. They are being replaced by China's Nanchang A-5 attack fighter (NATO codename 'Fantan'), of which the first of 42 to equip three squadrons arrived in 1983. Another 100 are likely to be acquired as the PAF seeks to build its combat strength in a cost-effective manner. Dassault supplied 32 Mirage 5s in 1981-82, comprising 5PA2 air-interception, 5PA3 mari-

time strike and 5DPA2 conversion versions, and bringing the grand total of PAF Mirages to 107. Earlier versions, supplied in 1967-70, still equip No. 5 Sqn at Sargodha (24 IIIEPs and three IIIDPs originally), No. 9 Sqn at Rafiqui and No. 33 Sqn at Sargodha (33 5PAs and two 5DPs), and No. 20 Sqn also at Rafiqui (13 IIIRPs for reconnaissance). Of these earlier aircraft, 30 were updated in the early eighties.

A transport wing at Chaklala has No. 6 Sqn with ten assorted Lockheed Hercules which survive from the 17 acquired at various times (five C-130Bs, four C-130Es and an L-100), and No. 12 Sqn which is a VIP unit operating a Fokker F.27, a Dassault Falcon and a Piper Seneca. A few Beech Twin Bonanzas, Cessna 172s and Aero Commanders are in use for liaison flights. A SAR squadron has some Aérospatiale Alouette IIIs. The Primary Training Wing at Risalpur has about 20 Saab MFI-17s out of the 87 which were assembled in-country at Risalpur; the rest are with the army, which is receiving additional examples that are now being fabricated at Kamra. The Basic Flying School, also at Risalpur, operates the survivors from about 60 Cessna T-37B/Cs which have been supplied over the years. A Fighter Conversion Unit at Mianuhali has operated about 20 Chinese

FT-5s to prepare those pilots converting to the F-6.

Army Aviation Corps

Apart from the MFI-17s (known as the Mushshak in Pakistan), the Army has 30 Aérospatiale SA.330J Pumas in service as assault helicopters, along with about 20 Mil Mi-8s. A squadron of Agusta-Bell AB.205/ UH-1H Iroquois (16 received from the US and Iran) is only part operational. Some 20 Alouette IIIs were assembled at the main Army aviation base, Dhamial, and at least 12 Bell 47Gs and 206Bs were received. The MFI-17s replaced most of the Cessna O-1 Bird Dogs in roles such as artillery spotting and reconnaissance, but some survive for communications alongside a Cessna 421, two Rockwell Turbo Commanders and a few Beech Queen Airs. Twenty AH-1S HueyCobras are in

the process of delivery, complete with TOW anti-armour missiles.

Pakistan Navy Air Arm

Five of the Navy's six Westland Sea King Mk 45s were modified to carry Exocet anti-ship missiles and four

Pakistan is a large user of the Shenyang J-6, including the two-seat JJ-6 version. The J-6 is being slowly replaced by the Nanchang Q-5.

Alouette IIIs carry torpedoes for ASW. Three Breguet Atlantics provide maritime patrol.

Papua New Guinea

Papua New Guinea Defence Force

An air transport squadron based at Port Moresby has six Douglas C-47s and four GAF Nomads, all provided under Australian defence aid agreements. In 1985 three IAI Aravas were acquired from Israel for surveillance tasks.

Philippines Philippine Air Force

The PHILAF has been unable to upgrade much of its equipment in recent years because of the country's economic problems. The 5th Fighter Wing at Basa is operating midsixties vintage aircraft in the form of the Northrop F-5A (10 with the 6th TFS, plus two F-5Bs) and the Vought F-8H Crusader (25 reworked examples supplied in 1977-78) for the 7th TFS. Also at Basa is the 105th CCTS with about 10 Lockheed T-33As and a few Fuji-built T-34s. At Sangley Point the surviving North American T-28s in the 15th

Strike Wing were grounded in 1983, leaving the 17th Attack Sqn with about 12 SIAI-Marchetti SF-260WPs and the 27th SAR Sqn with four Grumman HU-16 Albatross. The 205th Airlift Wing at Villaonar AB (the military side of Manila Airport) has the 208th Air Transport Sqn with seven F.27 Friendship transports and three F.27 Maritime Patrol Aircraft acquired in 1981. A few Douglas C-47s may remain airworthy with the 207th ATS. The 505th Air Rescue Sqn has a few Bell UH-1Hs and MBB BO 105s.

The helicopter force has been expanded considerably, however. Repeat acquisitions of the UH-1H Iroquois have brought the total in service to over 60. In addition, 15 Bell 205As and eight UH-1Bs were also acquired. In 1984 Sikorsky delivered 12 utility versions of the S-76, plus two SAR and three VIP versions, as well as two S-70A Black Hawks. At Mactan the 220th Heavy

Airlift Wing has the 222nd HAS with four Lockheed C-130Hs and two L-100s and the 223rd TAS with 12 GAF Nomads. The 240th Composite Wing at Sangley Point has the 291st Sqn with 20 Britten-Norman BN-2A Islanders, which were locally assembled, and the 601st Sqn with about the same number of Cessna U-17A/Bs. The 100th Training Wing at Fernando has 101 Sqn with about 20 Cessna T-41Ds and 102 Sqn with about 30 SF-260MPs. The Presidential aircraft are a PHILAF responsibility, for which a BAe One-Eleven, an NAMC YS-11, a Fokker F.27 and two Aérospatiale SA.330 Pumas are based at Villaonar.

Philippine Navy

The Navy has a few Islanders and BO 105s for coastal patrol, and the Philippine Constabulary also have a few examples of these two types.

Singapore Republic of Singapore Air Force

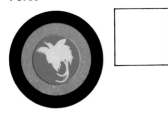

The RSAF has expanded rapidly to become a potent force of over 100 combat aircraft. Due to the small size of Singapore, permanent training detachments of the front-line fighters are made to Clark AB, Philippines and Williamtown, Australia.

The premier air defence squadron is No. 144 at Tengah with 24 Northrop F-5Es and six F-5Fs. Also at Tengah is a large fleet of McDonnell Douglas A-4 Skyhawks in Nos 142, 143 and 145 Sqns. The first 40 of these were introduced in 1973-74 being ex-US Navy A-4B models reworked by Lockheed as the A-4S with new avionics and engines; seven TA-4S conversion trainers were also supplied. Later, the government-owned aircraft industry purchased a further 70 US-surplus A-4Cs and 16 TA-4Bs for rework in Singapore and these have been steadily appearing in RSAF service as the A-4S1 and TA-4S1. The survi-

vors of 47 Hawker Hunter F.74/ T.75s are based at Paya Lebar with No. 140 Sqn. This new RSAF base (formerly the island's international airport) also houses the Flying Training School, comprising No. 130 Sqn with about 15 BAe Strikemasters and No. 131 Sqn with 20 Lockheed T-33As No. 150 Sqn is detached to Seletar where it flies 23 SIAI-Marchetti SF.260M/Ws in the basic course. Also at Paya Lebar is No. 122 Sqn with four Lockheed C-130Hs, and four C-130Bs which were converted as flying tankers in 1984. No. 121 Sqn operates four Short Skyvans from Changi.

The RSAF helicopter base is at

Sembawang where No. 123 Sqn has six Aérospatiale AS.350B Ecureuil and 30 UH-1Bs, and No. 120 Sqn has 20 UH-1H/Bell 205s plus three Bell 212s for rescue duties. RSAF re-equipment plans include the SIAI-Marchetti S.211 jet trainer, 10 of which have been ordered so far; the Aérospatiale AS.332B Super Puma, 22 on order; more air defence fighters for a second squadron; and two Grumman E-2C Hawkeye AEW aircraft for delivery in 1986. The S.211s and AS.332s will mostly be assembled in Singapore.

Sri Lanka Sri Lanka Air Force

The conflict that erupted in 1983 between the island's two ethnic groups led to expansion for the almost dormant SLAF, whose main activity hitherto had been the operation of a tourist flying service.

About 12 Bell 206B JetRangers were added to the seven already operated by No. 4 Sqn at Colombo, and two Bell 212s were also acquired; No. 4 Sqn also has two Aérospatiale SA.365 Dauphins. No. 2 Sqn

uses a mixed bag of communications types from Colombo, comprising two Douglas C-47s, three Cessna 337s, a 206 and a 421, and a Hawker Siddeley HS.748 and four Herons. The other SLAF base is at China Bay, where No. 3 Sqn has three Hawker Siddeley Doves and No. 1 FTS four Cessna 150s and seven de Havilland Canada DHC-1 Chipmunks.

Taiwan Chinese Nationalist Air Force

Northrop has continued to assist the ROCAF in maintaining a viable force since the ending of formal diplomatic relations between the US and Taiwan. They provided over 300 F-5E/F models, mainly in knock-down form for assembly by the country's Aero Industry Development Centre (AIDC). They have also helped AIDC to develop an indigenous jet trainer, the AT-TC-3 Tse Than.

The F-5E/F serves in 11 squadrons and four out of the five front-line wings. These are No. 1 Fighter Wing, Tainan (1FS, 3FS, 9FS); No. 2 Fighter Wing, Hsin-Chu (41FS, 44FS, 48FS); No. 455 Fighter Wing, Chit-Ye (21FS, 22FS, 23FS); and No. 5 Composite Wing, Taoyuan (26FS, 27FS). AIDC is now finishing assembly of the final batch of 30 F-5Es and 30 F-5Fs purchased in 1982. Earlier deliveries to the ROCAF totalled 226 F-5Es and 22 F-5Fs. The other front-line Fighter Wing is No. 30 at Ching Chuan Kang (7FS, 8FS and

28FS) flying Lockheed F-104 Starfighters. Here the survivors of 74 F-104A/G fighters and 19 F-104B/D/TF-104G conversion trainers were joined in 1984-85 by virtually all of the 66 TF/F-104G Starfighters operated formerly by Lockheed for pilot training in the US under contract to the West German Luftwaffe. Over the years 21 reconnaissance RF-104Gs have been supplied, and the survivors serve in 12 Sqn, part of 5CW at Taoyuan. The ROCFF has three reserve fighter bases operating older types: Makung, with about 50 North American F-100A/D/F; Hua Lien, about 50 F-5A, and Chi-Hong about 30 North American F-86Fs. An OCU is attached to 5CW with Lockheed T-33As and F-5A/Bs. Some of these types also serve alongside F-104s in 28FS, 30FW.

The ROCAF transport fleet is receiving much-needed modernization with the introduction of 12 Lockheed C-130Hs. Prior to their arrival, 20 Troop Carrier Wing at Taipei-Sung Shan was an all-piston entity, with two squadrons of Fairchild C-119Gs and one of C-123Ks, and one of Curtiss-Wright C-46s, and one of Douglas C-47s for multi-engine training. There was also a Douglas DC-6B and a few C-54s. A VIP squadron at the same base received four Boeing 727-100s from China Air Lines in 1982 (two with large cargo door), adding to a single Boeing 720B. Search and rescue is the responsibility of No. 4

Sqn attached to 455FW with about eight Grumman HU-16Bs and 10 Bell UH-1Hs.

The AT-TC-3 jet trainer, first flown by AIDC in March 1984, has provisions for air-to-air missiles, but its first assignment will be to replace Lockheed T-33s and Northrop T-38s in the advanced training role; about 30 of each type were transferred from the US in the seventies. Before starting design of the AT-TC-3, AIDC had developed the T-28 Trojan into the Lycoming-powered T-CH-1B trainer, but an order for the Beech T-34C in 1981 suggested that the project had been abandoned. The Beech aircraft might also replace most of the 40-strong fleet of PL-1B Chien Shou primary trainers (AIDC's copy of the Pazmany lightplane). The main training base is Kang Shan.

Many F-104 Starfighters still serve with the ROCAF. Since US aid has been stopped there is no replacement on the horizon, and F-104s will serve for a few years yet.

Taiwan Army

The army operates most of the 110 Bell UH-1Hs that were assembled by AIDC. It ordered 24 Sikorsky S-70s in 1984, and was also shopping for a light anti-armour helicopter at that time. Bell 47Gs and Hughes 500s are used for training.

Taiwan Navy

The navy has a squadron of Grumman S-2 Trackers for ASW, the original S-2As having been supplemented in 1979 by the transfer of 18 S-2E from US surplus stocks.

Thailand Royal Thai Air Force

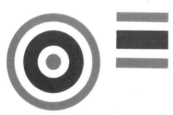

The RTAF was attempting to buy a squadron of General Dynamics F-16s in 1984-85 but was encountering difficulties in gaining the necessary finance. In the meantime, there is a fighter force of three squadrons of Northrop F-5s. At Korat with No. 1 Wing is No. 103 Sqn which has received a total of 15 F-5As, four RF-5As and five F-5Bs; and there are 32 F-5Es and six F-5Fs split between No. 102 Sqn here and No. 401 Sqn, No. 4 Wing at Takhli. Advanced training for the fighters is by Lockheed T-33s in No. 101 Sqn at Korat. Also in No. 4 Wing is No. 402 Sqn which had about six Douglas AC-47D gunships remaining; these were presumably replaced by the 20 GAF Nomads which were supplied from Australia in 1982-83 and armed with gun and rocket pods. The fighting between the Kampuchean resistance movement and the Vietnamese occupation force in that country has inevitably drawn the Thais into the conflict from time to time. Four of the 16 Cessna A-37B Dragonfly aircraft of No. 211 Sqn, No. 21 Wing, Ubon, have been shot down along the border since 1983. However, there have been no reports of losses to the RTAF inventory of other light attack types. These are North American T-28Ds (about 40 remaining with No. 231 Sqn, No. 23 Wing at Udorn but now grounded); Rockwell OV-10Cs (32 supplied to two units: No. 411 Sqn, No. 41 Wing at Chiang Mai and No. 531 Sqn, No. 53 Wing at Prachuap Khiri Khan);

and Fairchild AU-23As (28 supplied to two units: No. 202 Sqn, No. 2 Wing at Lopburi and No. 713 Sqn, No. 71 Wing at U-Taphao).

No. 2 Wing at Lopburi also has 18 Sikorsky S-58T Twin-Pac conversions in No. 201 Sqn, and some Bell UH-1Hs with No. 203 Sqn. Over 30 Helio U-10Bs were supplied, and about 12 Cessna O-1 Bird Dogs are flown by No. 711 Sqn at U-Taphao, No. 531 Sqn at Prachuap and No. 604 Sqn at Bangkok. In the air observation role they have been supplemented by three IAI Aravas.

Bangkok houses the transport force in No. 6 Wing. No. 601 Sqn now has three Lockheed C-130Hs and two stretched C-130H-30s and will acquire more. It also has a Douglas DC-8-62F. No. 602 Sqn operates the surviving Fairchild C-123B/Ks; over 40 were once in use, but they are being replaced by ex-Thai Airways Hawker Siddeley HS 748s (four transferred in 1983-84). No. 603 Sqn has about 16 C-47s and two Swearingen Merlin IVs with photo-reconnaissance provisions. The Royal Flight at Bangkok has a Boeing 737-200, a Merlin and two Bell 412s. An Aerial Mapping Unit has two Beech King Airs and a Rockwell Commander, and an Agricultural Flight has four CASA Aviocars, four Enstrom F-28F helicopters and a number of Transavia PL-12 Air-truks. The RTAF training syllabus will be drastically overhauled now that the unique RFB Fantrainer is entering service; 47 have been ordered, with all but the first two being assembled in Thailand. Up till now primary training has been on the Aerospace CT4 Airtrainer (24 supplied), progressing to the SIAI-Marchetti SF.260 (28) and the Cessna T-37 (about 22).

Royal Thai Army

There are a substantial number of

UH-1Hs in service, as well as six Bell 212s, three 214Bs and four 214STs; these may have superseded three Vertol KV-107s acquired from Kawasaki. Bell 206 JetRangers are also in service, and 25 Hughes TH-55As were acquired from US surplus for training. Two Shorts 330/UTTs have supplemented a Beech 99 and a King Air as fixed-wing transports.

Royal Thai Navy

The RTN air arm has expanded significantly in recent years. Two Canadair CL-215s were delivered in 1978 for coastal patrol. In 1983 four more armed Cessna 337s were added to the six supplied in 1980 by a US refurbisher. Three Fokker F.27 maritime patrol aircraft and five GAF Searchmaster Bs were added in

1984. Eight Bell 212s have been acquired for ASW, supplementing 10 Grumman S-2 Trackers, which have been overhauled. The main base is at U-Tapao.

Thai Border Police

The Border Police have taken delivery of a Shorts 330/UTT to supplement its fixed-wing force of three de Havilland Canada Caribous, three Short Skyvans, three Do 28Ds, four Pilatus Turbo-Porters and five AU-23As. It also flies 13 Bell 212s, about 30 UH-1H/205As and 20 206B/Ls.

The Northrop F-5E forms the bulk of Thailand's fighter defences, although General Dynamics F-16s are sought.

Vietnam Vietnam People's Air Force

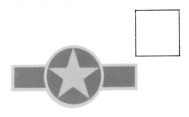

Fears amongst the neighbouring countries that the VPAF would receive a modern fighter had not been realised by the end of 1984. The front-line interceptor was still the Mikoyan MiG-21 in about 12 squadrons, mainly MiG-21M/bis which had replaced earlier models. However, Thai intelligence reports claimed that 14 Soviet AF MiG-23s were now based at Cam Ranh Bay. What is beyond dispute is that the Soviet Union has been allowed to build up a substantial medium/long-range bomber force at Cam Ranh Bay, where about four Tupolev Tu-95 'Bear-Ds and -Fs' are now permanently based along with 16 Tu-16 'Badger-As, -Cs and -Es'.

The VPAF has been active in Kampuchea, with aircraft based at Pnom Penh, Battembang, Pochentong and Siem Reap. The Soviet Union supplied Mil Mi-24 'Hind-D' helicopter gunships for the first time in 1984, and six of the reported total of 30 were deployed temporarily in Kampuchea from Ho Chi Minh City. There are as many as 200 Mi-6/-8/ -17 helicopters in service for trooplift.

VPAF ground attack types are the Sukhoi Su-20/-22 'Fitter' and the MiG-17 (about 70 of each). Some 50 MiG-19s were reportedly no longer in service from 1983. Since the US and South Vietnam Air Forces left behind 73 Northrop F-5s, 364 Bell UH-1s, 160 Cessna A-37s and 119 O-1s, and 23 Lockheed C-130s in 1975, it was not surprising that the VPAF managed to keep a proportion of them flying through cannibalization. However, the delivery of large numbers of Mi-8 helicopters and Antonov An-24/-26 and Yakovlev Yak-40 transports since 1979 indicates that the Western types are no longer needed. About 100 transports

are now in use. Basic trainers are the Nanchang CJ-6 and Yak-11/-18. Also reported in service are 12 Beriev Be-12 Chaika amphibians and 17 Kamov Ka-25s.

Following the unification of Vietnam in 1975, many US aircraft have been used. These have been kept in the air by cannibalization. This is a Northrop F-5B.

North America

Canada's air force is undergoing modernisation with the McDonnell Douglas F-18 Hornet replacing Lockheed F-104s and McDonnell F-101 Voodoos both at home and in Germany. The air defence of North America integrates the interceptor forces of both countries, F-15 Eagles now providing the bulk of the United States' contribution. The USAF has been systematically replacing Convair F-106s and McDonnell F-4s with F-15s and F-16s, and these types are now even on the strength of National Guard units. The major change to the USAF has been the introduction into service of the long-awaited Rockwell B-1 supersonic bomber, which supplements Boeing B-52s in the strategic bomber role. The USAF, USN and US Army still have massive resources around the world, particularly in the Far East and Europe, and there is no sign of these being rundown.

Canada Canadian Armed Forces – Air Command

The CAF (or Forces Armées Canadiennes to Francophile citizens) was formed in 1968 by amalgamation of the three armed services, but when the anticipated efficiencies failed to materialize, partial devolution followed in 1975 with creation of Air Command. All military flying is currently performed by Canadian Air Command (CAC), including that on behalf of Mobile Command (the army) and Maritime Command (the navy), but the new government elected late in 1984 has pledged a full return to the three-service system and re-introduction of distinctive uniforms. The most recent CAC restructure was completed in 1982 to give the Command an increased total of seven component Groups: Fighter, Air Transport, 10th Tactical, Maritime Air, No. 14 Air Training, Air Reserve and No. 1 Canadian Air.

No. 1 Canadian Air Group is CAC's direct contribution to NATO's 4th ATAF and is controlled from Lahr, West Germany, by the Commander, Canadian Forces Europe. The No. 1 CAG combat airfield is nearby Baden-Söllingen, where Nos 421, 439 and 441 Squadrons are in the process of retiring the last of Canada's 200 Canadair/Lockheed CF-104 and 38 TF-104D Starfighters. Replacement type is

the McDonnell Douglas CF-18 Hornet, 54 of which are assigned from 137 (including 25 two-seat CF-18B trainers) on order. After slight programme slippage, No. 439 was the first to convert, in July 1985, and No. 441, due to be the last, in January 1987. In addition to the CF-104's ground attack role with conventional weapons, the Hornets will have an air combat commitment with AIM-7M Sparrow and AIM-9L Sidewinder AAMs. Also in No. 1 CAG are 13 Bell CH-136 Kiowas of No. 444 Squadron at Lahr for support of 4 Mechanised Brigade Group; and a detachment of No. 412 Squadron with the CAF's two de Havilland Canada (DHC) CC-132 Dash 7s (due to be joined by two of the six Dash 8Ms ordered in 1984.)

Befitting an army-assigned formation, No. 10 Tactical Air Group comprises six helicopter units for battlefield transport and reconnaissance. Four squadrons have a mixed complement of six or seven Bell CH-135 TwinPacs and eight to 10 CH-136 Kiowas, these being No. 403 (Operational Training) Squadron at Gagetown; No. 408 at Namao; No. 427 at Petawawa; and 430e Escadrille at Valcartier. Some 43 CH-135s and 64 CH-136s remain from 50/74 delivered. Six Boeing Vertol CH-147 Chinooks are shared by No. 450 Squadron at Ottawa and No. 447 at Namao.

Principal equipment of Maritime Air Group is the Lockheed CP-140 Aurora, 18 of which were delivered in 1980-81, including four to VP 407 at Comox, on the West Coast. The remainder are pooled at Greenwood for Atlantic patrols by VP 404, VP 405 and VP 415, and trials by the Maritime Proving & Evaluation

Unit. Inshore fishery patrols are performed by most of the remaining 15 DHC-built Grumman CP-121 Trackers of MR 880 from Summerside, but fleet requirements unit VU 33 at Comox has three, plus three Canadair/Lockheed CT-133s. The East Coast FRU is VU 32 with six CT-133s and two CH-135s at Shearwater. MAG also operates the remaining 33 Sikorsky CH-124A Sea Kings from Shearwater with HT 406, HS 423 and HS 443, the OCU (HT 406) also having two CH-124Us with sonar removed. A pair of Sea Kings is operated from each of four DDH280-class destroyers, and one from six St Laurent and two Annapolis frigates.

Fighter Group collaborates with the USAF in defence of North America via the NAADC HQ inside Cheyenne Mountain. With retirement of its last McDonnell CF-101F Voodoos in December 1984, FG virtually completed modernization of home defences with the assigned 48 CF-18 Hornets. The OCU, No. 410 Squadron, formed at Cold Lake in

Re-equipment of Canadian combat units with the McDonnell Douglas CF-18 Hornet is now well under way. Many will serve in Germany.

June 1982 and received its first aircraft in October, and was followed at the same base by Nos 409 and 425 in June 1984 and January 1985, although a detachment will operate at Comox. ECM training is provided from North Bay by 414 Squadron with five Dassault EW-117 Falcon 20s (the other two CC-117s are also being converted) and 16 CT-133s. Three squadrons operate 55 Canadair-Northrop CF-5As and 30 CF-5D trainers, comprising No. 434 and 433e Escadrille at Bagotville and fighter school No. 419 at Cold Lake. Bagotville's units are assigned to reinforcement of Norway in wartime and will receive 36 replacement CF-18s, No. 434 converting in July 1987 and 433e disbanding to be replaced by 416 Squadron in January 1988. The remaining CF-5s will operate until the 1990s from Cold Lake, in whose remote surroundings the

Canada (continued)

'Maple Flag' training exercises are flown by CAF and allied combat aircraft. Cold Lake also houses the sole unit reporting direct to CAC HQ: the Aerospace Engineering & Test Establishment with one or two CF-18s.

An extensive internal route network (as well as overseas support flights) is flown by Air Transport Group with the five Boeing CC-137s of 437 Squadron at Trenton (two of them rapidly convertible to aerial tankers); and 21 Lockheed CC-130Es and seven CC-130H Hercules of Nos 429, 435 and 436 Squadrons at Winnipeg, Namao and Trenton respectively. Lighter transports include seven Convair CC-109 Cosmopolitans and two Canadair CC-144 Challenger 601s of No. 412 Squadron at Uplands, but there are other units with STOL transports and helicopters for general duties and SAR, their equipment including 11 DHC CC-115 Buffaloes, eight DHC CC-138 Twin Otters, nine Bell CH-118 Iroquois and 12 Boeing Vertol CH-113A Voyageurs. These are Nos 413 (at Summerside) and 442 (Com-

ox) Transport & Rescue Squadrons each with CC-115s and CH-113s; No. 424 T&R Squadron (Trenton), CC-115 and CH-135 TwinPac; No. 440 T&R Squadron (Namao) with CC-138; No. 103 Rescue Unit (Gander) with CH-113As; and Cold Lake, Moose Jaw, Bagotville and Chatham Base Flights operating CH-118s. No. 426 Squadron at Trenton handles multi-engine and fixed-wing type conversion, borrowing aircraft from other units as required.

Pilot instruction with No. 14 Air Training Group begins with grading on 20 Beech CT-134A Musketeer IIIs of No. 3 Flying Training School at Portage la Prairie, followed by 200 hours at No. 2 FTS, Moose Jaw, on some of the 130 remaining Canadair CT-114 Tutors (several of which are in storage). CT-114s are also operated by No. 431 Air Demonstration Squadron, otherwise known as the 'Snowbirds'. Helicopter pilots return to No. 3 FTS for 70 hours on 14 Bell CH-139 Kiowa IIIs. Seven Douglas CT-129 Dakotas serve the Air Navigation School at Winnipeg,

but will be replaced by four DHC-8Ms, the same base also housing the CT-114s and CT-133s of the Central Flying School. Some 55 venerable CT-133s remain with various units, including fighter squadrons, for communications and instrument rating.

Air Reserve Group has some flying units, including Nos 401 and 438 Squadrons at Montreal (forming No. 1 AR Wing); and Nos 400 and

Canadair-built Northrop CF-5s still serve in numbers, although some are scheduled for replacement by CF-18s. Ground attack is the prime duty.

411 at Toronto (No. 2 ARW), sharing 16 CH-136 Kiowas. Three more squadrons partner full-time units: No. 402 Reserve Squadron with the ANS Dakotas; No. 418 RS with No. 440's Twin Otters; and No. 420 RS sharing MR 880's Trackers.

United States of America   United States Air Force

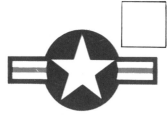

Responsibility for the US Air Force contribution to the strategic nuclear deterrent is entrusted to Strategic Air Command (SAC) which, from its headquarters at Offutt AFB, Nebraska, exercises control over two numbered Air Forces: the 8th at Barksdale, Louisiana, and the 15th at March, California, plus several smaller subordinate echelons.

Although withdrawal of the older Boeing B-52D has resulted in some reduction in force levels, the Stratofortress is still the mainstay of SAC's manned bomber fleet, about 260 B-52Gs and B-52Hs at present serving with 14 wings in the USA and one at Andersen AFB, Guam. Continual updating of the Stratofortress has enabled it to remain a viable system despite the fact that some of the aircraft now in service are of the order of 25 years old. The most recent development has been deployment of the AGM-86B ALCM (Air-Launched Cruise Missile), three B-52 wings having now received this new weapon. Eventual procurement of the ALCM is set at no less than 3,000 missiles, which will permit the B-52 to fulfil the stand-off role rather than the 'shoot-and-penetrate' philosophy which now prevails. Long-overdue force modernization has begun during 1985 when the first of 100 Rockwell B-1Bs enters service with the 96th Bomb Wing at Dyess AFB, Texas, and all of these should be in service by early 1987, some months earlier than originally planned. Looking even further into the future, 1992 should witness the introduction of the Northrop ATB (Advanced Technology Bomber), a prototype of which is now being built. Incorporating the much vaunted 'stealth' technology, this is reportedly a 'flying wing' design and it seems likely that 100 will be acquired for service with

SAC. The only other bomber asset currently available to SAC is the General Dynamics FB-111A, about 60 of these SRAM-armed aircraft equipping two wings on the eastern seaboard of the USA.

SAC also has responsibility for managing Intercontinental Ballistic Missiles (ICBMs), 1,000 examples of the Boeing Minuteman in six Strategic Missile Wings forming the mainstay of this formidable force. At the time of writing 450 of these are LGM-30F Minuteman IIs with a single warhead each, while there are also 550 LGM-30G Minuteman IIIs, each of which has three warheads. In addition, about 40 examples of the Titan II are still operational, although numbers are dwindling as these weapons are removed from their silos. Complete phase-out of the Titan system is targeted for 1987, but short-term updating of 50 Minuteman IIs to Minuteman III standard will more than compensate for the reduction in warheads. As far as new missiles are concerned the controversial MEM-118 Peacekeeper ICBM continues to draw heavy flak, particularly in the form of Congressional opposition which has curtailed funding in the recent past. Anticipated procurement of this 10-warhead weapon remains at 100, but there is still considerable debate over the question of the definitive basing mode to be employed. In the near-term, some Peacekeepers are

to be emplaced in existing Minuteman silos.

SAC also provides inflight refuelling support to all major USAF commands, several hundred Boeing KC-135 Stratotankers and an expanding number of McDonnell Douglas KC-10A Extenders fulfilling this mission; additional refuelling resources, comprising 128 Air Force Reserve and Air National Guard KC-135s, would be SAC-gained in the event of mobilization. Despite its age, the KC-135 will continue to furnish fuel for many more years to come; major updating of this aircraft centres around re-engining with the more fuel-efficient CFM-56 turbofan, producing a variant known as the KC-135R which began to enter service with the 384th ARW in 1984.

Other SAC units include several engaged in strategic reconnaissance, these being spearheaded by the 9th SRW at Beale AFB, California with the remarkable Lockheed SR-71A as well as the U-2R and TR-1. SR-71A detachments are also active at Kadena, Okinawa, and Mildenhall, England, while the TR-1 is now in process of joining the 17th RW at Alconbury. Less glamorous, but no less important, are the RC-135s and EC-135s which undertake a variety of functions ranging from the acquisition of electronic intelligence to the exercising of command and control of SAC forces; most resources of this nature are assigned to

The Lockheed SR-71A and McDonnell Douglas F-15 Eagle represent the costly yet highly technological approach to air power adopted by the United States.

the 55th SRW at Offutt, from where they regularly deploy overseas.

In terms of pure numbers of aircraft operated, Tactical Air Command is by far the largest element of the USAF and this has been the object of major modernization initiatives in recent times, vast numbers of new fighter and attack aircraft having entered service within the past ten years. Headquartered at Langley AFB, Virginia, TAC has two subordinate numbered air forces, these being the 9th at Shaw, South Carolina, and the 12th at Bergstrom, Texas.

After a period of almost unchallenged supremacy as the backbone of TAC, the McDonnell Douglas F-4 Phantom is gradually being ousted from the front line by newer and more capable types such as the F-15 Eagle and F-16 Fighting Falcon. Nevertheless, substantial numbers of the highly versatile Phantom are still to be found in TAC service, this type operating with five wings and among the variants in use at the time of writing are the F-4E, RF-4C and F-4G 'Wild Weasel'.

Procurement of both the F-15 and F-16 is continuing towards the objective of standardizing on these two types and newer and more capable versions are now in the pipeline, these being epitomized by the F-16C which recently began to enter service, and by the upcoming F-15E multi-role version of the Eagle. At present, five F-16 wings and four F-15 wings are assigned to TAC, these being ably supported by three wings of A-10A Thunderbolt IIs, the last of 713 examples of this potent tank-buster having been delivered in April 1984. From an earlier generation, but still a most capable aircraft, the F-111 serves with two wings, the original F-111A equipping the 366th TFW at Mountain Home AFB, Idaho, while the F-111D is assigned to the 27th TFW at Cannon AFB, New Mexico. Another version of the swing-wing General Dynamics aircraft is the EF-111A Raven, and most of the 42 planned conversions have now been delivered, the majority going to the 366th TFW at Mountain Home, although some are also operational with USAFE.

In addition to the more glamorous fighter units, TAC also possesses a number of ancillary elements engaged in such tasks as forward air control, airborne early warning, fighter lead-in training and other highly specialized operations. Types assigned to these functions include the OV-10A Bronco, OA-37B Dragonfly, Boeing E-3A Sentry, CH-3E 'Jolly Green Giant' and AT-38 Talon.

Like other US-based elements of the Air Force, TAC is a major 'gaining' command and would be augmented significantly in the event of mobilization. AFRES and ANG elements engaged in tactical missions have benefitted from the acquisition of much new equipment in recent years, such as the F-16 and A-10A, and this process is continuing with more F-16s and F-15s due to be assigned in the fairly near future. Older combat types active with these second-line echelons include the F-4 and A-7.

Responsibility for air defence of the continental USA now rests with an organization known as ADTAC (Air Defense Tactical Air Command), which has a modest number of interceptor squadrons. A stalwart of the US air defence network for many years, Convair's F-106A Delta Dart is being gradually retired from the front line and now equips just three squadrons, two former units having been re-equipped recently with the F-15A Eagle. One other ADTAC flying unit is the 57th FIS, which is unique in being the only non-US based component of this command, as well as the only one equipped with the F-4E Phantom. In addition to regular units, there are a number of Air National Guard squadrons assigned to the air defence task which operate either F-4 Phantoms or F-106A Delta Darts.

Turning to tactical elements based outside the contiguous USA, United States Air Forces Europe (USAFE) is easily the largest command and this has benefited from modernization in the recent past. A major portion of the NATO alliance, USAFE headquarters are located at Ramstein with subordinate forces being organized into three numbered Air Forces, each of which is responsible for operations in a specific area. In the UK, the 3rd AF is engaged mainly in long-range strike duties, possessing two wings, at Upper Heyford and at Lakenheath, equipped with the General Dynamics F-111E and F-111F respectively; the unit at Heyford has recently

enjoyed modest expansion through the addition of an Electronic Combat Squadron operating EF-111A Ravens. Tank-killing capability is also concentrated in Great Britain, the 81st TFW operating over 100 examples of the A-10A Thunderbolt II from bases at Bentwaters and Woodbridge. However, about a third of this force is maintained on permanent rotation over four bases in West Germany, much closer to the terrain over which they could reasonably expect to engage in combat.

Other UK-based units include the 10th TRW at Alconbury, this wing made up of an 'aggressor' squadron with the F-5E Tiger II and a reconnaissance squadron with RF-4C Phantoms; the same base is also home to SAC's 17th RW and its TR-1As, which are assigned to battlefield surveillance tasks. Airlift support is centred on Mildenhall, which acts as the UK-terminal for MAC flights as well as serving as a temporary home for C-130s deployed from the USA on a rotational basis. Mildenhall also hosts rotational SAC KC-135s, as does Fairford, some 25-30 tankers normally being attached to the ETTF (European Tanker Task Force) at any given time. Although not supporting permanently-based flying units, three other 3rd AF bases provide valuable support and other services. Of these, Greenham Common is easily the best known, the resident 501st Tactical Missile Wing now in process of receiving the highly controversial BGM-109G GLCM (ground-launched cruise missile), which has also begun to arrive at Comiso, Sicily. Sculthorpe and Wethersfield are the other two stations and these periodically host aircraft engaged in major exercises, or those displaced from their normal bases while repairs are being effected.

On the southern flank of the NATO alliance the USAFE contribution is managed by the 16th AF from its headquarters at Torrejon, Spain. This in fact has few combat aircraft assigned, controlling just a single F-16 wing at the same base. However, its area of jurisdiction is large, encompassing most of the Mediterranean and including support bases in Turkey, Italy, Greece and Spain.

The remaining USAFE sub-command is the 17th AF at Ramstein, which controls units situated in West Germany and Holland. Largely equipped with the Phantom just a few years ago, the 17th AF has been virtually transformed and today includes examples of the F-15, F-16 and F-4 in its line-up. Operational assets comprise one 72-aircraft wing of F-15Cs at Bitburg, an F-15C squadron at Soesterberg, an F-16 wing at Hahn and F-4 wings at Ramstein and Spangdahlem; the latter includes the equivalent of one squadron of F-4Gs for SAM-suppression tasks, one RF-4C squadron at Zweibrucken and a tactical control wing at Sembach. In addition, a permanently based C-130 squadron is located at Rhein-Main, while delivery of the EDSA (European Distribution System Aircraft) C-23A Sherpas to the 10th Military Airlift squadron at Zweibrucken got under way shortly before the end of 1984; both of these units are nominally a part of Military Airlift Command, although they are fully re-

Plattsburgh AFB and Pease AFB on the East Coast house the two wings of FB-111A bombers. These carry SRAM missiles and report to SAC.

sponsive to USAFE requirements. Finally, a number of smaller support units are equipped with communications aircraft such as the Beech C-12F Super King Air and the Gates C-21A Learjet.

On the other side of the world, Pacific Air Forces (PACAF) has also benefited from an infusion of new equipment, and has expanded still further in size in 1985 with the activation of a 48-aircraft F-16 wing at Misawa, Japan. Command headquarters are located at Hickam AFB, Hawaii, subordinate numbered air forces comprising the 5th in Japan and the 13th in the Philippines. The principal element under 5th AF control is the 18th TFW at Kadena, Okinawa, with three squadrons of F-15C Eagles and one of RF-4C Phantoms. Other units are dispersed amongst four air bases in South Korea, namely Kunsan where the 8th TFW has two F-16 squadrons; Osan where the 51st TFW is headquartered alongside one squadron of OA-37Bs and one squadron of F-4Es; Suwon where the 51st TFW has one A-10A squadron, and Taegu where the 51st TFW's second F-4E squadron is situated. The 13th AF is rather smaller, controlling just one fighter wing at Clark AB in the Philippines, this being made up of single squadrons operating the F-4E, F-4G and F-5E. Rather closer to PACAF headquarters, the 22nd TASS operates OV-10A Broncos from Wheeler Field, Hawaii.

Headquartered at Elmendorf AFB, Alaskan Air Command is one of the smallest USAF sub-commands but this too has been modernized more recently, with the F-4Es operated previously giving way to F-15s and A-10s. Operations are centred on two main bases, that at Elmendorf being home for the

The General Dynamics F-16 is now serving in large numbers throughout the USAF. It has a dual role in most cases.

21st TFW which has one squadron of F-15A Eagles and one squadron of Lockheed T-33As. F-15s from the 21st TFW are kept on air defence alert at their home base, detachments being maintained also at Eielson AFB, at King Salmon airport and at Galena AFS. The second major Alaskan base, at Eielson, hosts the 343rd Composite Wing with single squadrons of the A-10A and the O-2A for close air support. In addition, both MAC and SAC have a permanent presence in Alaska, with airlift, rescue and reconnaissance elements from these commands operating routinely from the two major bases.

Military Airlift Command (MAC) is the principal cargo-hauling agency, managing a large fleet of airlift dedicated aircraft from headquarters at Scott AFB, Illinois, as well as overseeing weather reconnaissance, rescue and aeromedical evacuation activities. Two numbered air forces, the 21st at McGuire AFB, New Jersey, and the 22nd at Travis AFB, California, are responsible for airlift, these serving the eastern and western hemispheres respectively. Three basic types of aircraft are operated, enabling MAC to satisfactorily fulfil all aspects of the demanding airlift mission. At the top end of the weight scale is the Lockheed Galaxy, and approximately 75 C-5As now on charge are due to be augmented by 50 C-5Bs in the near future. MAC's workhorse might fairly be regarded as the StarLifter, and all of the 270 or so aircraft assigned have now been stretched to C-141B configuration, a programme which has greatly enhanced their

value; at the other end of the payload scale is the C-130, which is generally employed on tactical rather than strategic airlift duties. Communications and VIP-type missions are performed by a motley assortment of aircraft including the VC-9C, VC-137C, VC-137B, VC-140, C-135B, C-12 Super King Air and C-21A Learjet. MAC is also a major 'gaining' command and would assume responsibility for ANG and AFRES units assigned to the airlift mission, these universally equipped with the C-130 at the present time.

MAC's newest addition to the chain of command is the 23rd Air Force, which is responsible for managing the large fleet of rescue-dedicated aircraft and the smaller number engaged in weather reconnaissance duties. Types in use include variants of the C-130 Hercules, as well as helicopters such as the HH-3E, HH-1N Iroquois, HH-53C/H and UH-60A Black Hawk; some of the last have been acquired on loan to assist with training, pending delivery of the command's own HH-60D/Es.

From its headquarters at Randolph AFB, Texas, Air Training Command controls seven pilot training wings equipped with the Cessna T-37B and Northrop T-38A Talon, although one of these is engaged solely in the production of suitably qualified instructors. There is also a navigator training wing at Mather AFB, California, with T-37s and Boeing T-43As. A long overdue new basic jet trainer is at last in the pipeline for introduction in 1987, this being the Fairchild T-46A Eaglet, and current plans call for the purchase of no less than 650 to replace the long-serving 'Tweety Bird'. Primary training, on the Cessna T-41 Mescalero, is conducted by the civilian-operated Officer Training School at Hondo, Texas. Some Bell TH-1Fs are also operated from Randolph to turn out helicopter instructors.

The most recent addition to the USAF is Space Command, which is located alongside NORAD headquarters in the Cheyenne Mountain complex in Colorado. Engaged mainly in surveillance activities at present, this manages missile warning radars situated in the continental USA, Alaska and Greenland. It is still very much in the formative stages, but when fully operational will have about 7,000 military and civilian personnel. Lacking 'teeth' at the time of writing, the development of ASAT (anti-satellite) weapons may eventually permit Space Command to take more positive action against potential enemy satellites.

United States Navy

The USN presently operates a fleet of 14 aircraft carriers, with the nuclear-powered USS *Carl Vinson* (CVN-70) being the most recent 'Nimitz' class vessel to attain operational status. Three additional nuclear-powered carriers are to be acquired, the *Theodore Roosevelt* (CVN-71), *Abraham Lincoln* (CVN-72) and *George Washington* (CVN-73), and the first of these was launched recently. Once they enter service the last of two surviving smaller 'Midway' class carriers will be retired.

Each aircraft carrier normally deploys with approximately 85-95 aircraft and helicopters, organized into a Carrier Air Wing (CVW). A typical CVW is made up of the following units: two interceptor squadrons with F-4S Phantoms or F-14A Tom-

cats; two light attack squadrons with A-7E Corsairs; one medium attack squadron with A-6E and KA-6D Intruders; one tactical electronic warfare squadron with EA-6B Prowlers; one airborne early warning squadron with E-2B or E-2C Hawkeyes; one fixed-wing anti-submarine warfare squadron with S-3A Vikings and one helicopter ASW squadron with SH-3H Sea Kings. The two older 'Midway' class carriers, which are somewhat smaller, are organized slightly differently; lacking the S-3 element, they sometimes carry a detachment of USMC RF-4B Phantoms for reconnaissance duties, this function being performed by TARPS (Tactical Air Reconnaissance Pod System) configured F-14s aboard the larger carriers. Ultimately this role will be undertaken by the RF-18 Hornet, which made its maiden flight in October 1984.

Despite recent problems with the aircraft's vertical tail surfaces, procurement of the McDonnell Douglas F-18 is continuing towards the eventual target of 1,366, some 300 of which will be operated by the Marine Corps. As far as the Navy is concerned, this type has made its operational debut in 1985 when it deploys for the first time, and it is expected eventually to replace the A-7 Corsair which currently equips some 22 light attack squadrons. Some Hornets are also earmarked for service with a handful of fighter units. With regard to the F-14A Tomcat, this is now active with 22 front-line squadrons, half of which normally deploy with three examples of the TARPS-configured aircraft in addition to pure interceptors. Plans to update the Tomcat are to go ahead, although production of the present F-14A variant will continue until 1988, at which time the F-14D will begin to join Fleet units. Re-engining with the hopefully more reliable General Electric F110 is the most significant feature of the F-14D, but this aircraft will also incorporate a digital weapons aiming system, improved radar and an advanced version of the AIM-54 Phoenix long-range air-to-air missile. Eventually, it is also intended to bring existing F-14As to F-14D standard by means of a major retrofit programme.

Another type which is to be the subject of major modernization initiatives is the long-serving Grumman Intruder, which is to acquire new avionics and instrumentation, and this too is a candidate for re-engining although no firm decision has yet been made as to the choice of powerplant. Production of the A-6E is due to terminate with the delivery

of six aircraft funded in FY85, but the EA-6B Prowler will continue to be built at a rate of six units per annum until at least FY89. A considerably more capable and sophisticated aircraft from that which entered service in the early 1970s, the Prowler is also earmarked for updating, this effort centres around mission-related avionics. Ten squadrons of EA-6Bs are now active.

On the airborne early warning front, continuing production of the E-2C Hawkeye has permitted the older and less capable E-2B to be virtually phased out of the operational inventory, and the E-2C is now being delivered also to Reserve units. Some 13 Fleet squadrons presently operate the Hawkeye, most equipped with the 'C' model, some aircraft incorporating the APS-138 Trac-A system, which offers greater resistance to jamming, are now in service.

Navy ASW assets will receive a significant boost in the not too distant future, 166 examples of the S-3A Viking being earmarked for modernization to S-3B standard with greater radar range, improved processing capability and compatibility with the Harpoon anti-shipping missile. As far as helicopter ASW units are concerned, although it has long been recognized that a replacement for the Sikorsky Sea King is well overdue, it looks as though these veteran helicopters will continue to appear in the active inventory for a good few years yet. Updating to SH-3H standard has increased their capability, however, and it would seem that age is the greatest enemy, many Sea Kings now being more than 20 years old.

Shore-based patrol forces, universally equipped with the Lockheed Orion, consist of some 24 fully-operational squadrons; the Navy is now approaching its goal of out-

The USAF's enormous transport fleet relies heavily on the Lockheed C-141 StarLifter, with 270 on strength. These have now all been stretched to C-141B standard.

fitting all these with the P-3C version, the few remaining P-3B squadrons stationed at Barbers Point, Hawaii, having begun to relinquish the 'Bravo' model during 1984. More or less continual refinement of the P-3C does, however, mean that several different configurations exist, and it will probably be some years before the fleet is fully standardized. The latest model to make its debut has been the P-3C Update-III, which began to enter service in May 1984. In addition to front-line units, 13 Reserve squadrons also fly the Orion, these having older P-3A and P-3B versions.

Following its success in the Navy's LAMPS competition, the Sikorsky SH-60B Sea Hawk has now begun to enter service at North Island, where HSL-41 is currently engaged on training in anticipation of large-scale deployment aboard Navy surface combatants in the near future. Total anticipated procurement stands at 204 SH-60Bs, while the interim LAMPS (Kaman's diminutive SH-2F Seasprite) remains in widespread service with some eight light helicopter ASW squadrons. Indeed, it was reinstated in production fairly recently, in order to ensure that sufficient helicopters of this type are available to meet requirements for the foreseeable future. Other helicopters used by the Navy include examples of the UH-1N Iroquois, the CH-53E Super Stallion and the CH/UH/HH-46 Sea Knight, while approximately 30

Two squadrons of F-14s are assigned to most carriers, with four E-2 Hawkeyes for airborne control.

RH-53D Sea Stallions are employed by three mine countermeasures squadrons; these last are scheduled to be replaced by the specialist MH-53E Super Stallion.

Navy airlift elements comprise just four squadrons, all of which are concerned primarily with Fleet support and are accordingly distributed widely. Equipment operated includes a handful of C-130Fs, backed up by modest quantities of Rockwell CT-39 Sabreliners, Grumman C-1A Traders and C-2A Greyhounds, with the last type also reinstated in production recently to overcome an anticipated shortfall in COD (Carrier Onboard Delivery) capability. Additional airlift aircraft, primarily McDonnell Douglas C-9B Skytrain IIs, are operated by the Reserves and are deployed periodically to augment the regular force on long-haul missions.

Other elements are engaged in what is loosely described as 'reconnaissance', two squadrons equipped with Lockheed EP-3E Orions and EA-3B Skywarriors are employed in the acquisition of electronic intelligence, and two further squadrons undertake the Tacamo function, which involves VLF/ELF communications with the large fleet of nuclear-armed submarines. At the present time the Lockheed EC-130Q is used for the latter mission, but these will be replaced eventually by the Boeing E-6A (essentially an E-3A airframe with CFM56 engines), it being intended to acquire 15 examples of this aircraft.

Navy training is handled by some 20 squadrons organized into six wings and this aspect of operations has also been the subject of modernization in recent years, this being very much an on-going process. Older piston-engine types, such as the North American T-28B/C Trojan and Grumman TS-2A Tracker, have been entirely replaced by the T-34C Turbo-Mentor and T-44A respectively, and plans are in hand to update advanced training elements through the acquisition of 300 T-45 Hawks from 1988 onwards. In the meantime, Rockwell's T-2C Buckeye and the McDonnell Douglas TA-4J continue to bear the burden of Navy fast-jet pilot training, successful candidates then progressing to a Fleet Replacement squadron or 'RAG', where they become acquainted with the type that they will fly when assigned to an operational unit. Helicopter training is now undertaken solely by the TH-57 Sea Ranger, acquisition of new helicopters of this type having permitted the TH-1Ls and UH-1Ls to be reassigned to other duties.

A rather more advanced form of training is that furnished by the small number of 'aggressor' units which concentrate on honing air combat manoeuvring skills throughout the Fleet. Now mainly equipped with such types as the F-5E Tiger II and A-4 Skyhawk, the Navy is taking part in joint studies with the USAF concerning the possibility of purchasing a common type to fulfil this task. In the short term, however, Navy aggressor units are to be bolstered by a dozen IAI Kfirs, obtained on lease for a period of three years.

In addition to the active units, the US Navy has a substantial Reserve force which is largely organized along similar lines to the regular Fleet. Two Reserve Carrier Air Wings exist as do two Reserve Patrol Wings, a Helicopter Wing Reserve and a Reserve Tactical Support

Wing, these controlling the activities of some 40 squadrons. Equipment operated is generally older than that of their front-line counterparts, although efforts at modernization are now being implemented with the introduction of such types as the F-14A, F-18A and E-2C, thus enhancing mobilization potential should it be necessary to call upon the Reserves at some future date. Reserve units now active comprise one fighter squadron with F-14s and three with F-4s, one light attack squadron with F-18s and five with A-7s, one light photographic reconnaissance squadron with RF-8Gs, two tanker squadrons with KA-3Bs, two ECM squadrons with EA-6As, two AEW squadrons with E-2B/Cs, three helicopter ASW squadrons with SH-3s, one light helicopter ASW squadron with SH-2Fs, two light attack helicopter squadrons with HH-1Ks, one helicopter combat search and rescue squadron with HH-3As, 13 patrol squadrons with P-3A/Bs and several transport squadrons with either C-9Bs or C-131Hs.

United States Marine Corps

Dedicated to providing aerial support to Marine ground troops wherever they operate, Marine air power is essentially subordinate to the US Navy but is, nevertheless, a very powerful force in its own right. As far as organization is concerned, aviation elements are divided amongst three Marine Air Wings, these being the 1st MAW at Iwakuni, Japan, the 2nd MAW at Cherry Point, North Carolina and the 3rd MAW at El Toro, California. Each MAW operates in conjunction with an associated Marine Division, and since the MAW is expected to undertake all aspects of the Marine air mission they are basically similar from the viewpoint of equipment. Thus, a typical MAW will consist of four fighter-attack squadrons with F-4 Phantoms or F-18 Hornets; two to three light-attack squadrons with A-4 Skyhawks or AV-8 Harriers; one or two medium attack squadrons with the A-6 Intruder; one tanker/transport squadron with the KC-130 Hercules; one observation squadron with the OV-10 Bronco; one attack helicopter squadron with the AH-1 Sea Cobra; six or seven transport/utility helicopter squadrons with CH-53 Sea Stallions, CH-46 Sea Knights and UH-1 Iroquois; and elements engaged in electronic warfare and reconnaissance with the EA-6B Prowler and RF-4B Phantom.

The principal fighter-attack type is still the McDonnell Douglas Phantom; nine squadrons currently operate the F-4S variant but replacement of this veteran has begun, three former F-4N squadrons at El Toro having already converted to the F-18 Hornet, of which the Marines will eventually receive just over 300 examples.

Light-attack assets comprise four squadrons with the McDonnell Douglas A-4M Skyhawk, a fifth unit now being in process of conversion to the AV-8B Harrier which will eventually entirely supplant all A-4s, as well as the AV-8A/C Harriers which now serve with three operational squadrons. Current planning calls for the procurement of 328 AV-8Bs, but there is as yet no Intruder replacement in view for the five squadrons which utilize this type in the all-weather/night attack role. However, as with the Navy, modernization of the A-6E will enable this combat veteran to continue to play an effective part in Marine operations. Electronic warfare capability is provided by the Grumman EA-6B Prowler, some 15 aircraft being assigned to a single squadron at Cherry Point, and this provides detachments to other Marine commands as and when required. Reconnaissance assets consist of about 25 RF-4B Phantoms with VMFP-3 at El Toro and, once again, this meets all USMC demands on a detachment basis.

USMC airlift capability is vested in three squadrons equipped with either the KC-130F or KC-130R tanker/transport versions of the Lockheed Hercules, with a modest number of Rockwell CT-39 Sabreliners, Beech UC-12B Super King Airs and McDonnell Douglas C-9B Skytrain IIs also being employed on liaison and communications tasks by higher command echelons.

Fixed-wing assets are completed by a couple of observation squadrons equipped with a mixture of Rockwell OV-10A and OV-10D Broncos, while the large helicopter force numbers some 28 squadrons at present, although more are due to form during the coming year. Some 13 medium helicopter squadrons (HMM) now exist, these being universally equipped with the Boeing-Vertol CH-46E Sea Knight, and this number is expected to rise to 15 fairly shortly. Eventual replacement will be the tilt-rotor JVX which is expected to enter Marine Corps service early in the next decade. Heavy-lift capability has been given a boost recently through the formation of three CH-53E Super Stallion squadrons, and procurement is continuing towards the eventual objective of standardizing on this type. Meanwhile, six more squadrons continue to operate older CH-53A/Ds Sea Stallions and these are likely to remain in the inventory for some time

Weapons training for the Marine Corps is carried out by the McDonnell Douglas TA-4. A further two-seat Skyhawk is used for FAC duties.

to come. Additional helicopter units comprise three squadrons with the Bell UH-1N Iroquois for light duties, while there are also three attack squadrons with an assortment of Bell AH-1J and AH-1T Sea Cobras, the latter version being compatible with TOW anti-tank missiles and earmarked eventually for re-arming with the Hellfire weapon.

Although fixed-wing elements of the Marine Corps do occasionally embark aboard Navy carriers for operational deployments, it is the helicopter units which go to sea most frequently, these regularly serving aboard amphibious vessels of the LHA or LPH type. In such instances, it is usual to assemble a composite squadron with examples of the four principal types of helicopter now in service (CH-46E, CH-53D/E, AH-1J/T, UH-1N). On occasions, small numbers of AV-8A/C Harriers may also be embarked to fulfil the CAS (close air support) mission.

Aircrew destined for service with the Marine Corps undergo initial flying training with the Navy, but the USMC does have extensive training resources of its own, these utilizing examples of most types presently to be found in the operational inventory and dedicated to polishing those skills unique to Marine tactical air.

Like other branches of the US armed services, the Marine Corps has substantial Reserve elements available for mobilization in the event of national emergency or war. These are organized along similar lines to the regular front-line forces, control being exercised by the 4th MAW which has its headquarters at New Orleans, Louisiana. Some 20 squadrons are now active, these comprising two with F-4N Phantoms, five with A-4E/F/M Skyhawks, one with EA-6A Intruders, one with KC-130F/T Hercules, one with OV-10 Broncos, one with AH-1J Sea Cobras, three with CH-53A/D Sea Stallions, four with CH-46 Sea Knights and two with the UH-1.

United States Army

With over 8,000 rotary-wing and some 600 fixed-wing aircraft on strength, the US Army is clearly one of the world's foremost exponents of military helicopter expertise. The backbone of this vast fleet is still the 'Huey', with close to 4,000 UH-1s in service, although the Sikorsky UH-60A Black Hawk is becoming available in ever increasing numbers, approximately 500 of the planned fleet of 1,107 having been delivered to date.

Acquisition of the Hughes AH-64A Apache anti-tank helicopter has also begun, the initial units now being in the process of working-up at Fort Eustis, Virginia, and Fort Rucker, Alabama, and this type is expected to reach European-based combat elements during the course of 1986. Funding has been released for 267 of the anticipated total of 446 Apaches and production is now gradually building towards the eventual peak of 12 per month. In the meantime, Bell's AH-1S HueyCobra continues to act as the primary anti-tank helicopter, close to 1,000 examples of this TOW-armed machine being active at present.

The principal liaison helicopter is still the Bell OH-58 Kiowa, some 2,000 or so now serving with the Army throughout the world, and substantial numbers are employed as scouts for the anti-tank force. Indeed, this mission has resulted in appearance of the OH-58D variant which incorporates a mast-mounted sight and is capable of undertaking the scouting function by both day and night. Rather than purchase new Kiowas, the Army is converting

some 720 existing machines to OH-58D configuration. Additional observation resources include a fairly large number of the Hughes OH-6A Cayuse, although this type is not assigned to European-based Army units.

Medium- and heavy-lift capability is provided by the Boeing Vertol Chinook, over 400 examples of the CH-47A/B/C variants currently being on charge; these too are being modernized, the resulting CH-47D being a far more versatile machine. Attaining an initial operational capability in early 1984, with elements of the 101st Airborne Division at Fort Campbell, Kentucky, some 436 conversions are expected to join the Army inventory during the next few years.

Turning to fixed-wing aircraft, the Army operates a fairly substantial quantity of OV/RV-1 Mohawks on what can loosely be described as reconnaissance duties, while much modified examples of the U-21 Ute and C-12 Huron are employed in the acquisition of communications and electronics intelligence. The rather more mundane liaison duties are performed by unmodified U-21s and C-12s and there are also a couple of Pilatus UV-20A Chincahuas on

charge, these being based in Berlin where their STOL performance is of particular value. A few examples of the Beech U-8 Seminole are also still to be found in the inventory.

With regard to training, primary helicopter tuition is given on the Hughes TH-55A Osage, while those earmarked to fly fixed-wing types utilize the Cessna T-41 for initial training, thereafter progressing to the Beech T-42 for the multi-engine phase.

In addition to the front-line elements, Army Reserve aviation resources are quite extensive, and the

The US Army's new assault helicopter is the Sikorsky UH-60 Black Hawk, 1,107 of which are on order. Despite this, the Bell UH-1 will serve in enormous numbers for some time, giving the US Army a total of around 5,000 assault helicopters.

Army National Guard also has a large number of flying units in the continental USA. For the most part, these formations are equipped with variants of the UH-1 and OH-58, although some CH-47s and CH-54A Tarhes are assigned to second-line units.

South America

Military events in this part of the world have been making headlines for the last couple of years. In 1982 the Argentine capture and subsequent loss of the Falkland Islands occurred, in 1983 the United States invaded Grenada to expel the Communist regime and in 1984 Nicaraguan support for left-wing guerrillas operating in Honduras and Salvador has caused the United States to send aid to those countries and warn Nicaragua not to get involved; though Nicaragua is receiving Soviet, Cuban and Libyan aid this has yet to include any frontline strike or fighter aircraft.

Argentina is in the process of rebuilding her air arms following the conflict with the UK over the Falkland Islands, but rampant inflation and existing massive overseas debts has limited her re-equipment programme. Cuba still receives equipment from the Soviet Union but none of it is likely to pose a threat to the United

States. Other than Nicaragua and Cuba, only Peru has been a large purchaser of Soviet equipment. Peru has been disappointed that many of the aircraft received have been downgraded to a basic export standard without sophisticated avionics.

The rest of South America operates aircraft that originate largely from the United States, France, Italy or Britain. Many countries are Mirage operators because the United States refused to sell combat aircraft above Northrop F-5E standard, but this has changed with the recent sale to Venezuela of General Dynamics F-16A/B Fighting Falcons.

Brazil and Argentina each have their own aircraft industries, but neither has yet designed and produced combat aircraft other than for COIN roles. Peru has started to licence-build the Aermacchi MB.339.

Argentina Fuerza Aérea Argentina

The FAA has five main commands but Comando de Operaciones Aéreas (Air Operations Command) controls all flying activities. In turn, Air Operations Command has several subsidiary commands, like Air Transport Command, Air Defence Command and Strategic Air Command. At base level a Brigada Aérea controls flying operations from the Base Aérea Militar (BAM).

I Brigada Aérea at BAM El Palomar controls Grupo 1 de Transporte which has five subordinate Escuadrones de Transporte, each named after the type of aircraft they fly, plus the Escuadrilla Presidenciale. Escuadron Hercules operates nine Lockheed Hercules comprising one C-130E, five C-130Hs, two KC-130Hs and an L-100-30. Escuadron Fellowship operates seven F.28-100s. Escuadron Guarani operates about 12 FMA IA.50 Guarani IIs which are usually dispersed around

the Brigada Aéreas on liaison and communications duties. Escuadron Friendship uses two F.27-400, seven F.27-400M, two F.27-600 and two F.27-500. Escuadron Boeing has six Boeing 707s of various models. Escuadrilla Presidenciale has an Aero Commander 500S and two Sikorsky S-58Ts but can call on other equipment as necessary. Friendships and Fellowships are loaned to the state airline LADE as required. II Brigada Aérea at BAM Parana has Grupo 1 Aérofotografico operating five FMA IA.50 Guarani IIs and four Gates Learjet 35As. Also with IIBA is Grupo 2 de Bombardeo flying the Canberra B.62/T.64 of which 10 Canberra B.62s and two Canberra T.64s were ordered from BAC; all but four Canberra B.62s are thought to be in service. A further order for two more has been embargoed. III Brigada Aérea at BAM Reconquista has Grupo 3 de Exploracion y Ataque operating the FMA IA.58A Pucará. Nearly 100 Pucarás have been built to date, but these may include export orders for Uruguay (six) and Venezuela (18). About 30 aircraft have been lost in accidents and during the Falklands conflict, and the remainder are used by Escuadrones II and III of Grupo 3 and Grupo 9. IV Brigada Aérea at BAM Mendoza has Grupo 4 de Caza

Bombardeo attached. Grupo 4 has two training units, Escuela de Caza Bombardeo 1 and Escuela de Caza Bombardeo 2. They operate the North American F-86F Sabre and MS.760 Paris respectively. A total of 28 F-86F Sabres and 48 MS.760 Paris were received but only 12 of each type is thought to survive. Also with Grupo 4 is III Escuadron de Caza Bombardeo, operating the McDonnell Douglas A-4C Skyhawk, of which 25 were received from the US in 1976; accidents and combat losses have reduced the unit to less than 10 aircraft. V Brigada Aérea at BAM Villa Reynolds controls Grupo 5 de Caza Bombardeo with its two

Gaining fame during the Falklands war, Argentina's Mirages have come from a variety of sources, including Israel, Peru and France.

subordinate units IV and V Escuadrones de Caza Bombardeo flying the McDonnell Douglas A-4B Skyhawk. 50 surplus USN Skyhawks were delivered to the FAA between 1966 and 1970. A very high attrition rate (believed to be approaching 50 per cent and approximately 10 losses during the Falklands conflict) casts some doubt as to whether both escuadrones are still operational. VI Brigada Aérea at BAM Tandil controls Grupo 6 de Caza with its two component units II and III Escuadrones de Caza flying the IAI Dagger. Between 30 and 40 IAI Dagger A (single-seat) and Dagger B (trainers) were bought from Israel and accidents and combat losses have reduced the number available by about 15. VII Brigada Aérea at BAM Moron is the main FAA helicopter unit and Grupo 7 de Helicopteros has the following squadrons: I Escuadron de Exploracion y Ataque operates three Bell UH-1Ds, three Bell UH-1Hs, two Hughes 395HEs, 12 Hughes 369HMs and six Bell 212s. I Escuadron Busqueda y Salvamento fulfils the search and rescue role with two Sikorsky S-61NRs, six Aérospatiale SA.315B Lamas, two CH-47C Chinooks, and a few Aero Commander 500s from 15 received. Escuadron Sanitario operates two Swearingen SA.226AT Merlin IVs. VIII Brigada Aérea at BAM Dr. Mariano Moreno controls Grupo 8 de Caza with I Escuadron de Caza flying the Dassault Mirage IIIEA/DA; 10 Mirage IIIEA (single-seat) and two Mirage IIIDA (trainers) were delivered in 1972-73. Seven more Mirage IIIEAs followed in 1980 and two Mirage IIIDAs in 1983. Four have been lost to date. IX Brigada Aérea at BAM Comodoro Rivadavia controls Grupo 9 de Transporte. The main squadron is IV Escuadron de Exploracion y Ataque flying the IA.58A Pucará, but Grupo 9 de Transporte also operates aircraft on behalf of LADE (the State airline) and several communications aircraft. These include five DHC-6 Twin Otters. X Brigada Aérea at BAM Rio Gallegos is a newly formed BA and is believed to comprise Grupo 10 with a squadron of Dassault Mirage 5Ps and one of Dassault Mirage IIIBJ/CJs. 10 Mirage 5Ps were received from Peru in 1982 along with about 20 ex-Israeli Mirage IIIBJs/CJs as replacements for combat losses.

The Escuela Aviacion Militar is located at Cordoba, in use are the survivors from 90 Beech/FMA T-34A Mentors, along with a few MS.760 Paris. Aircraft used for liaison and communications include 10 MH.1521M Broussards, 40 FMA-built Cessna A.182J/K/L/N, one Piper PA-31 Navajo, several C-47 Dakotas, Aero Commanders and four Hughes 500Ds. These aircraft are detached to various Brigada Aéreas. The Commander in Chief of the FAA uses a Rockwell Sabre 75A. A reservist training unit operates a few of the Mentors from Jose C. Paz Airport and a sport and gliding unit at Cordoba has several light aircraft for use as tugs or tourers. Escuadron Antartico borrows helicopters from VII BA and Twin Otters from IX BA. Escuadron Verificacion Radio use two Learjet 35As and two FMA IA.50 Guarani IIs for navaid calibration duties and are based at Moron.

INAC (National Civil Aviation Institute) carries out training for civil pilots and technicians using FAA provided aircraft, which include three Hughes 500Ds, three Chincul Piper PA-A-28RT-201 Arrow IVs, three Chincul Piper PA-A-34-220T Seneca IIIs and four Douglas C-47 Dakotas.

Comando de Aviación Naval Argentina (CANA)

The command structure of the CANA consists of six escuadras (wings) each with several escuadrillas (squadrons). These are all shore-based and deploy to the aircraft carrier *25 de Mayo* or frigates and

destroyers when required.

1 Escuadra Aeronaval based at BA Punta Indio is also known as the Escuela de Aviacion Naval. The two component escuadrilla are 4 Escuadrilla Aeronaval de Ataque operating the Beech T-34C-1 Turbo Mentor and the Escuadrilla Aérofotografico with a Beech Queen Air 80. Of the 15 T-34C-1s received, four were destroyed in the British raid on Pebble Island. 2 Escuadra Aeronaval at BA Comandante Espora, Bahia Blanca, has three component squadrons. The Escuadrilla Aeronaval Antisubmarine operates the Grumman S-2E Tracker; six were delivered in 1977-78 and all are thought to be in use. The Escuadrilla Aeronaval de Exploracion operates the Lockheed SP-2H Neptune; four were received from surplus USN stocks in 1977-78 and replaced older SP-2Es and P2V-5s. The 2 Escuadrilla Aeronaval de Helicopteros operates the Sikorsky S-61D-4; four were delivered in 1972, a fifth in 1975, and at least two Agusta-Sikorsky SH-3Ds were received in 1984. 3 Escuadra Aeronaval at BA Comandante Espora, Bahia Blanca, has three component escuadra. 1 Escuadrilla Aeronaval de Helicopteros operates the Alouette III. One of the two Westland Lynx HAS.23s was lost in an accident and the second has been withdrawn through lack of spares; the order for a further eight Lynx HAS.23s was embargoed by the UK government. Three Aérospatiale SE.3160 Alouette IIIs were delivered in 1969 followed by two more and two SA.316A versions in 1970. Two more SA.316Bs followed in 1971, one in 1975 and four SA.319Bs in 1978. At least six have been lost in accidents. 2 Escuadrilla Aeronaval de Caza y Ataque operates the Dassault Super Etendard, of which 14 were delivered between 1981-82. 3 Escuadrilla Aeronaval de Caza y Ataque operates the McDonnell Douglas A-4Q Skyhawk, but of the 16 received accidents and combat losses have reduced the available number of aircraft to about six. 4 Escuadra Aeronaval at BA Punto Indio has one component escuadrilla, 1 Escuadrilla Aeronaval de Ataque operating Aermacchi MB.326GBs, EMBRAER EMB.326GB Xavantes and the Aermacchi MB.339AA. Of eight Aermacchi MB.326GBs received two were lost in accidents and about half

of 10 Aermacchi MB.399AAs were lost during the Falklands conflict; 12 EMB.326GB Xavantes were received from Brazil in 1983 as replacements. 5 Escuadra Aeronaval at BA Ezeiza has one subordinate unit, 1 Escuadrilla Aeronaval de Sosten Logistico Movil, which is the transport unit of the CANA. It operates three Fokker F.28-3000 Fellowships, three Lockheed L-188A Electras and one Hawker Siddeley HS.125-400B. Finally, 6 Escuadra Aeronaval at BA Vice Almirante ZAR Trelew has the Escuadrilla Aeronaval de Propositos Generales as its sole unit. This operates four Grumman S-2A Trackers, three Pilatus PC-6B Turbo Porters (which are also deployed to Antarctica with the Escuadrilla Aeronaval Antartico), four Beech Queen Air 80s and four Lockheed L.188PEs.

Eight Beech Super King Air 200s are also in use for liaison duties and are believed to be with 2 and 4 Escuadra.

Comando de Aviación del Ejercito

The CAE is divided into two major groups, the Batallon de Aviación de Combate 601 which operates all the helicopters and the Compania de Aviación de Apoyo General 601 with all the fixed-wing aircraft. Both units are based at Campo de Mayo, Buenos Aires, but many aircraft are with the aviation sections of the five corps Ejercitos and spend protracted periods away from base.

The BAC 601 has at least five Agusta A.109As from nine delivered, and six Aérospatiale SA.315B Lamas are in service. The CAE received 25 Bell UH-1Hs but at least two have been lost in accidents and about nine captured on the Falkland Islands. Two Bell 212s are in use for VIP duties. Nine Aérospatiale SA.330 Pumas were delivered to the CAE in 1978-79 but most of them were lost on the Falkland Islands; between 12 and 24 Aérospatiale AS.332 Super Pumas are on order to replace them. Two Boeing CH-47C Chinooks were delivered to the CAE in the late 1970s but both were lost or captured in the Falklands.

The CAAG 601 operates the survivors from 16 Cessna U-17As, six Cessna 207s, two Beech Queen Air 80s and four DHC-6 Twin Otters.

The FMA IA.58 Pucará saw service in the Falklands, where they were feared by British troops. Counter-insurgency is the major role for this light attack aircraft.

Also in use are a Rockwell Sabre 75A, a Beech King Air 100, three Aeritalia G.222s, four Swearingen Merlin IIIAs, one Swearingen Merlin IV, one Cessna 500 Citation, one Cessna 550 Citation, a few Cessna T-41D Mescaleros and four Piper PA-31 Navajos.

Prefectura Naval Argentina

This small air arm of the coastguard service also saw action during the Falklands conflict and lost one of its three Aérospatiale SA.330L Pumas and two of its five Short Skyvan 3Ms. Also in service are six Hughes 500Cs. The Skyvans are normally based at Jorge Newbery Aeroparque, Buenos Aires, and the helicopters at the Puerto Nuevo Heliport, Buenos Aires.

Gendarmeria Nacional

This paramilitary unit operates two Cessna 310s, five Cessna 182Hs, one Cessna 185, one Cessna 337, two Cessna 206s, one Cessna 402, a Piper PA-31 Navajo and three Pilatus PC-6B Turbo Porters. The GN has its headquarters at Campo de Mayo but many aircraft are deployed around the country.

Belize

Belize Defence Force

Formerly British Honduras, Belize gained independence in 1981 and her first two aircraft, a pair of Pilatus Britten-Norman BN-2B Defenders, were delivered in 1983. A British presence is still maintained in Belize with flights of Harriers, Gazelles and Pumas based there and a frigate on patrol offshore.

Bolivia Fuerza Aérea Boliviana

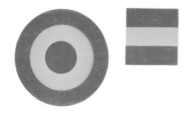

The FAB is in need of new combat aircraft, but the current economic situation within Bolivia precludes any major re-equipment programme.

The combat element is provided by Grupo Aéreo de Caza 32 based at El Trompillo, Santa Cruz, with the North American F-86F Sabre. Nine ex-Venezuelan F-86Fs were received but at least half are thought to be non-operational. The other jet equipped unit is Grupo Aéreo de Caza 41 at El Alto, La Paz, with the Canadair T-33AN, of which 21 were delivered during 1973-81. At Tarija are the Grupo de Perfeccionamiento

Tactico 41 and the Grupo Aéreo de Cobertura 42, both flying the Pilatus PC-7 Turbo Trainer of which 24 were delivered during 1979-81. Most are still in use and a few North American AT-6Gs remain with GAC 42.

For transport duties the Grupo Aéreo de Transporte has four component units. The Escuadrilla Ejecutiva carries out VIP duties with a Rockwell Sabre 60, three Beech Super King Air 200s, one Cessna 402B, one Cessna 421B, one Cessna 404, one Lockheed Electra and one Fairchild PC-6/C Turbo Porter. The SAN-Servicio Nacional de Aérofotogrametria has two Gates Learjet 25Bs and a Cessna 402B. The Transporte Aéreos Militares (TAM) has about 15 Douglas C-47 Dakotas, two Curtiss C-46A Commndos, four Convair 440s, four IAI Aravas and six Fokker F.27-400Ms. The final component of the GAT is Transporte Aéreo Boliviano which operates a Lockheed C-130H Hercules and a L.100-30 Hercules in civil markings

on behalf of the FAB.

The Grupo Aéreo Mixto carries out SAR, liaison and communications duties at Cochabamba and operates the Aérospatiale SA.315B Lama and Helibras Gavilanes (Brazilian-built Lamas). Five of each type were acquired to replace the Bell UH-1H and at least one has been lost. The Grupo de Operaciones Aéreas Especiales at Robore operates Hughes 500Ms. These armed helicopters were acquired in 1978 and most are thought to remain in use. Training is the responsibility of the Colegio Militar de Aviación at Santa Cruz. Basic training is carried out by the Escuadron Basico flying the Aerotec T-23 Uirapuru of which 19 were delivered. The Escuadron Primario operates six SIAI-Marchetti SF.260CBs, three Cessna 172Ks and six ex-USAF Cessna T-41D Mescaleros. Fixed-wing communications and liaison duties are carried out by the survivors of a variety of Cessna aircraft that includes seven A.185Es, eight A.185Fs, two

U.206Cs, one U.206F, seven U.206E/Gs along with five 210s.

A reported order for 12 Dassault Mirage 5s has yet to be confirmed.

Army

The Ejercito Boliviano only operates a few aircraft for VIP duties, those currently in use including a Cessna 421B, a Piper PA-31T Cheyenne II and a Beech Super King Air 200. The Carabinero (Paramilitary Police Unit) operates a Cessna 421B.

Navy

Bolivia being landlocked, the navy has only a very modest air arm for river patrol duties, comprising one Cessna U.206G and a Cessna 402C.

The Bolivian air force's two Lockheed Hercules transport aircraft are operated by Transporte Aereo Boliviano in civil markings. Both have a cargo transport role rather than assault transport.

Brazil Fôrça Aérea Brasileira

The largest air arm in Latin America has four major commands. Comando de Defesa Aérea has one subordinate Grupo de Defesa Aérea, 1 GpDA, based at Santa Cruz with two component squadrons, 1 and 2 Esquadroes, flying the Dassault Mirage III. Four Mirage IIIDBR trainers and 12 Mirage IIIEBRs were ordered in 1972 followed by a further five Mirage IIIEBRs in 1977 and two Mirage IIIDBRs in 1983; at least six have been lost.

Comando Aérotáctico has eight subordinate Grupos which are as follows: 1 Grupo de Aviação de Caca (1 GAvCa) with 1 and 2 Esquadroes at Santa Cruz and 14 GAvCa with 1 Esquadrao at Canoas operate the Northrop F-5; 36 F-5Es and six F-5Bs were delivered (1975-76), but at least six have been written off. 4 GAvCa at Fortaleza with 1 and 2 Esquadroes, 10 GAvCA with 1 Esquadrao at Santa Maria and 2 Esquadrao at Campo Grande, operate the EMBRAER EMB.326GB

Xavante in the attack role. Of the 168 Xavantes (including two Italian built pattern aircraft) that were delivered, 20 have been lost. The remaining four grupos all have liaison and helicopter support roles: they consist of 8 Grupo de Aviação 11 GAv, 13 GAv and 15 GAv with a total seven component squadrons, 1 Esq/8 GAv at Manaus with Bell UH-1H and EMBRAER EMB.810C Seneca; 2 Esq/8 GAv at Recife with UH-1H and Seneca; 3 Esq/8 GAv at

Campo dos Afonsos with UH-1H and SA.330 Puma; 5 Esq/8 GAv at Santa Maria with UH-1H and Seneca; 1 Esq/11 GAv at Sao Paulo with Bell UH-1H; 1 Esq/13 GAv at Santa Cruz with Neiva U-42 and Seneca; and 1 Esq/15 GAv at Campo Grande with EMB.110 Bandeirante. Six Bell SH-1Ds, eight Bell UH-1Ds and about 30 UH-1Hs were delivered, though several have been lost, and eight ex-IDF/AF UH-1Ds were delivered in 1983. Of the 32 EMB.810C Senecas

Brazil's advanced jet trainer is the EMBRAER AT-26 Xavante, a license-built version of the popular Aermacchi M.B.326.

delivered 1977-81, the first 12 were designated U-7 and the final 20 U-7A.

Comando de Transporte Aéreo has seven subordinate grupos with a total of 12 component esquadroes and six autonomous esquadroes. 1 Grupo de Transporte at Campo dos Alfonsos has 1 Esquadrao which operates five Lockheed C-130E and three Lockheed C-130H Hercules. 2 Esquadrao operates two Lockheed KC-130H Hercules in the tanker role. 2 Grupo de Transport (2 GT) at Galea has 1 Esquadrao operating the Bandeirante and HS 748-205, with 2 Esquadrao using the Bandeirante and HS 748-281; the six HS 748-205s and six HS 748-281s were delivered 1962-75. 1 Grupo de Transporte de Tropas has 1 and 2 Esquadroes operating the DHC-5A Buffalo; about 18 are in service with 1 GTT and 2 Esq/2 GTT at Campo Grande and 1 Esq/9 GAv at Manaus. 1 Esq/2 GTT at Galeao also has a few Lockheed C-130H Hercules.

The Grupo de Transporte Especial at Brasilia has 1 Esquadrao using two Boeing 737s and three Bell 206 JetRangers and 2 Esquadrao with five HS.125-3BRAs, four HS.125-400Bs and six EMB.121 Xingus. The six autonomous Esquadroes de Transporte Aéreo consist of 1 ETA at Belem, 2 ETA at Recife, 3 ETA at Galeao, 4 ETA at Cumbica, 5 ETA at Porto Alegre and 6 ETA at Brasilia; all operate a mix of EMB.810C Seneca and EMB. 110 Bandeirantes. EMB.110 Bandeirante production for the FAB consisted of two YC-95 prototypes (EMB.100), 56 C-95s (EMB.110), 201 C-95As (EMB.110K1), 20 C-95Bs (EMB.110P1K), four EC-95s (EMB.110A) navaids calibration aircraft, six R-95 (EMB.110B) for photo-survey, and eight SC-95Bs (EMB.110P1K) for SAR duties.

Comando Costeiro has four component Grupos which are as follows: 6 GAv at Recife with 1 Esquadrao flying three Lockheed RC-130E Hercules and six R-95 Bandeirantes; 7 GAv has 1 Esquadrao at Salvador

and 2 Esquadrao at Florianopolis, operating between them 16 EMB.111M Bandeiruhla patrol aircraft; and 10 GAv has two component esquadroes, 2 Esq. at Campo Grande with eight SC-95 Bandeirantes and 3 Esq. at Florianopolis with SH-1Ds and a few Bell 47s. The Grupo de Aviação Embarcada's sole component squadron, 2 Esquadrao, is shore-based at Santa Cruz but normally provides ASW cover on the aircraft carrier *Minas Gerais* with eight Grumman S-2E Trackers.

The Comando de Treino has several component units. Pilot training starts at the Centro de Formação de Pilotas Militares at Natal on the Aerotec T-23 Uirapuru, followed by the Neiva T-25 Universal before graduating onto jets in the shape of the EMB.326GB Xavante. An alternative training route is by starting on T-23 Uirapurus and Bandeirantes at the Escola Practica de Cadetes do Ar followed by transfer to the Academia de Fôrça Aérea at Pirassununga with T-27 Tucano, EMB.326 Xavantes and Bandeirantes for twin conversion. There is also an air force Clube de Voo a Vela at Pirassununga with Blanik, Libelle and Quero-Quero sailplanes and three EMB.201 Ipanema tugs. The FAB received 68 Aerotec T-23 Uirapurus and these will be replaced by around 100 Aerotec T-28 Tangaras. The T-25 Universal and Cessna T-37C are being replaced by 118 EMB.312 Tucanos (FAB designation T-27). The first T-27s were delivered to the Academia de Fôrça Aérea and the Esquadrilha da Fumaça aerobatic team in 1983.

Operational conversion is undertaken by the Centro de Aplicação Tatica e Recomplementação de Equipagens at Natal with three Es-

quadroes de Instrução Aérea operating Universals, Xavantes and Bandeirantes. Helicopter instruction is carried out by the Centro de Instrução de Helicopteros at Santos with Bell UH-1, OH-6A and 206A JetRangers. The Grupo Especial de Inspeção e Vigilancia operates four EC-95 Bandeirantes and two EC-93 HS.125s for navaids checking and calibration duties.

The Neiva C-42/U-42 Regente (more than 100 acquired) is still used in large numbers for communications and liaison duties, and Bell 47s also remain in use. The first two prototypes of the joint Italian/Brazilian AMX project flew in 1984; this light strike fighter is designed to replace the MB.326 in FAB service. An order has been placed for about 20 EMB.120 Brasilias.

Fôrca Aéronaval de Marinha do Brasil

This was re-formed in 1965 after 25

Under the military designation P-95, 2 Esquadrao, 7GAv operates 16 EMB-111 Bandeirantes for maritime patrol duties. These have search radar and underwing hardpoints for light anti-shipping strikes.

years as part of the FAB. All Marinha helicopter units are based at São Pedro de Aldeia and comprise 1 Esquadrao de Helicopteros Antisubmarine (HS-1) with three Sikorsky S.61D-3 Sea Kings, two SH-3D Sea Kings, four Agusta-Sikorsky SH-3D Sea Kings and nine Westland Lynx HAS.21s. The Sea Kings are often carrier-based and the Lynx are frigate-based. 1 Esquadrao de Helicopteros de Emprego Geral (HU-1) operates seven Westland Wasp HAS.1s. At least nine Helibras AS.350B Esquiloes are in use (licence built Aérospatiale AS.350 Ecureuils) and eight Bell 206B JetRangers are also in use. The 12 delivered in 1974 are used by 1 Esquadrao de Helicopteros de Instrução in the training role.

Chile Fuerza Aérea de Chile

The FACh has recently altered the command structure. All flying wings are allocated to the Comando de Combate with the exception of training wings which come under Comando de Personal. Below these commands are regional Brigada Aéreas which are set up as follows:

I Brigada Aérea which controls Ala (Wing) No. 4 at Los Condores, Iquique, with two components, Grupo 1 and Grupo 4. The first is in the process of equipping with CASA C.101CC and C.101BB Aviojets. The FACh initially ordered 12 T-36 Halcones (C.101BB – trainer version) and four A-36 Halcones (C.101CC – attack version). These are being assembled at El Bosque and the order was recently increased by 21 T-36/A-36s. Once Grupo 1 is fully converted on to the A-36, the T-36s will be transferred to the Escuela de Aviación at El Bosque to replace the Cessna T-37B. Grupo 4 has the Dassault Mirage 50; eight refurbished ex-Armée de l'Air Mirage 5Fs were delivered as Mirage 50FCH, with six new Mirage 50CHs and two Mirage 50DCH two-seat trainers from 1980 onwards. Also with I Brigada Aérea/Ala No. 4 is an Escuadrilla de Enlace

Aéreo with a few Bell UH-1 and Aérospatiale SA.315B Lama helicopters.

Ia Brigada Aérea which controls Ala No. 1 at BA Cerro Moreno, Antofagasta, has three component grupos and an Escuadrilla de Enlace Aéreo. The last flies Douglas C-47 Dakotas, SA.315B Lamas and Sikorsky S-55s. The three grupos are Grupo 8 and Grupo 9 flying the Hawker Hunter and Grupo 7 flying the Northrop F-5E (15) and F-5F (3). Hunter deliveries have totalled 51, comprising 30 F.71s, three F.71As, four T.72s, two FR.71s and finally, 11 FGA.9s and one F.6A delivered in 1982 from RAF stocks in return for Chilean help during the Falklands conflict.

II Brigada Aérea which controls Ala No. 2 at BA Los Cerrillos, Santiago, has three component units. Grupo 2 operates two Canberra PR.9s. Grupo 10 is the main FACh transport unit and has two Lockheed C-130H Hercules, one Boeing 727-22C and one 707-351C, one Beech King Air A.90 and one Super King Air 200, four Douglas C-47 Dakotas, a few Aérospatiale SA.315B Lamas, 11 Bell UH-1Hs and six Helitec-Sikorsky S-55Ts. Grupo 11 operates nine Beech 99s for Elint and maritime surveillance duties in addition to navigation training. Also at Los Cerrillos is the third FACh command, the SAN-Servicio Aéreo Aérofotogrametrico with two Learjet 35s, a few DHC-6 Twin Otters and a King Air A100.

III Brigada Aérea which controls Ala No. 5 at BA El Tepual, Puerto Montt, has Grupo 5 operating the

DHC-6 Twin Otter. Deliveries have totalled 20 comprising eight DHC-6-100s, six DHC-6-300s and six ex-LAN-Chile aircraft; at least four have been written off and one sold. Also at Puerto Montt is a SAR Escuadrilla with Sikorsky S-55s.

IV Brigada Aérea at BA Chabunco, Punta Arenas, has two component Grupos. Grupo 6 operates the Bell UH-1 and DHC-6 Twin Otter. Grupo 12 operates the Cessna A-37B Dragonfly.

The Escuela de Aviación at El Bosque is gradually replacing the survivors of 68 Beech T-34A Mentors and 29 Cessna T-37B/Cs with 80 Industria Aeronautica (IndAer) T-35A/B Pillans and at least 20 T-36 Halcones. The Escuela de Especialides at El Bosque has eight Cessna T-41D Mescaleros, one Cessna U.206 and nine Neiva T-25 Universals. The *Los Halcones* aerobatic team based at El Bosque has at least four Pitts S-2 Specials. The Escuela de Vilelo sin Motor at BA Los Con-

This Boeing 707-351C is on the strength of Grupo 10 for long-range transport duties. It was seen several times in Britain, presumably collecting equipment for the Hunters and Canberras operated by the Chilean air force.

des, Santiago, has a collection of Blanik and Cirrus gliders with Cessna O-1 Bird Dogs in use as tugs. Sundry types in liaison use include four Hiller UH-12Es, six UH-12SL-4s, 10 Cessna 182s and a few Beech Twin Bonanzas. Two EMB.120 Brasilias are on order for Grupo 10.

Servicio de Aviación de la Armada de Chile

The main base is Quilpue-El Belloto, near Valparaiso. The Armada currently has four active squadrons: VC-1 operates three EMBRAER EMB.110C(N) Bandeirantes and four CASA 212A Aviocars; HS-2

Chile (continued)

operates 10 Alouette IIIs and about seven Bell 206 JetRangers; VP-3 has six EMB.111A(N) Bandeirantes for maritime patrol duties; and VT-4 uses 10 Pilatus PC-7 Turbo-Trainers in a training role. A couple of Bell 47G/Js and a Piper PA-31 Navajo may also still be in use.

Comando de Aviación del Ejercito de Chile

Formed in 1970, the CAEC is still relatively well equipped. Helicopters in use include two Bell 206B JetRangers, and two UH-1Hs, 14 Aérospatiale SA.330 Pumas, at least one AS.332B Super Puma and the survivors of 16 SA.315B Lamas. Fixed-wing aircraft consist of three Cessna 337G Super Skymasters, six 212A Aviocars and four PA-31 Navajos. The training unit, the Escuela de Aviación at the main base of Tobalaba has 18 Cessna R.172K Hawk XPs and 16 IndAer-assembled Piper PA-28 Dakota 236s. The Carabineros, a para-military police force, has 10 MBB BO 105s and a Fairchild-Hiller FH.1100.

Colombia Fuerza Aérea Colombiana

The FAC is divided into several commands, four of which operate aircraft; they are the Comando Aéreo de Combate which has three component Grupos as follows: Grupo Aéreo de Combate 1 (GAC 1) based at BAM German Olando, Palenquero, operating the Dassault Mirage of which 18 were delivered comprising 14 5COAs, two 5COD trainers and two 5CORs for reconnaissance; at least one and probably more have been lost. GAC 1 is currently forming detachments at Barranquilla and San Andres. Grupo Aéreo de Combate 2 (GAC 2) based at BAM de Apiay now operates the 12 Cessna A-37B Dragonflies delivered in 1983. Grupo Aéreo de Combate 3 (GAC 3) is not officially formed but has 12 A-37B Dragonflies based at BAM Luis F. Gomez-Nino, Barranquilla with detachments at Palanquero, de Apiay and San Andres.

Comando Aéreo de Apoyo Tactico has only one component grupo. This is the Grupo Aéreo de Helicopteros based at BAM Luis F. Pinto, Melgar, operating nearly all the FAC's rotary-wing aircraft. They consist of the survivors from nine Bell UH-1Bs, 13 205A-1s and 18 UH-1Hs, 10 Hughes 500Ds and 12 OH-6A Cayuse helicopters delivered 1963-81. Reports of the FAC receiving Aérospatiale SA.315 Lamas appear to be unfounded.

Comando Aéreo de Transporte Militar has headquarters at BAM Techo, Bogota/El Dorado, but its aircraft can be found operating out of any of the small airstrips and airfields across the country. CATM has five component units comprising the Escuadron de Transporte operating one Lockheed C-130B and two C-130H-30 Hercules, two or three Douglas DC-7, one DC-7, two Douglas C-54s and six C-47 Dakotas, two Curtiss C-46A Commandos and three IAI Aravas. The Escuadron de Enlace operates one Aero Commander 560A, one Gulfstream 980, two Cessna 310s and three 404s, one Beech King Air, two Piper PA-31 Navajos, a few float-equipped Cessna A.185E Skywagons and the survivors from 20 DHC U-6 Beavers. Many other nearly new or modern light aircraft are available because of being impounded for drug running, but due to a lack of spares are no longer flown. The Escuadron de Helicopteros operates only a few Bell 206B JetRangers. The Escuadrilla Presidencial has a Boeing 707-373C, a Fokker F.28 Fellowship and a Bell 212 for Presidential and VIP use. Finally, SATENA, a para-military airline, operates three HS.748s, a few C-47 Dakotas, six Pilatus PC-6B Turbo Porters (with a further six reportedly on order), two F.28-3000 Fellowships and five CASA 212 Aviocars.

Comando Aéreo de Entrenmiento has two component training units. The Escuela Militar de Aviación at BAM Marco Fidel Suarez, Cali, operating the survivors from 30 Cessna T-41D Mescaleros and 42 Beech T-34A Mentors; six ex-US Navy T-34B Mentors were sent as attrition replacements in 1978. The Escuela de Helicopteros at BAM Luis F. Pinto, Melgar, operates 12 Bell 47D/G/Js and six Hughes TH-55. Jet training takes place in the United States.

A Naval air arm was formed in 1983 with the arrival of two MBB BO 105s for the Armada Colombiana.

Costa Rica

Seguridad Publica Seccion Aérea

Armed forces have been banned in Costa Rica since the 1948 Civil War, but a few aircraft are operated in support of the paramilitary force called the Guardia Civil. Currently operated are a Fairchild-Hiller FH.1100, two Hughes 269Cs, two Piper PA-23-250 Aztecs, three PA-32-300 Cherokee Six, one PA-23-235 Apache 160 and a PA-34-200T Seneca.

Cuba Fuerza Aérea Revolucioniaria

Following the overthrow of the Batista regime in 1959 and the conversion of Cuba to Communism under Fidel Castro, all the current FAR equipment is of Soviet origin. As well as providing her own air defence, many Cuban aircrew have seen action overseas, particularly in Africa. Cuban civil airlines are frequently used to airlift troops into Africa and South America, and Cuban bases are used frequently by Soviet long range reconnaissance aircraft, due to Cuba's close proximity to the USA.

Air defence is carried out by one squadron of MiG-23 'Flogger-Es' delivered from 1977 onwards and based at San Julian. Several squadrons operate the MiG-21 with 30 MiG-21Fs, 34 MiG-21PFMs, 20 MiG-21PFMAs and more recently at least 17 MiG-21bis. The MiG-19SF is believed to have been withdrawn from the day fighter role and the early MiG-21F cannot have many years service left. In the ground-attack role one squadron with MiG-17Fs at Guines is believed to soldier on but recent deliveries of between 18 and 36 MiG-23BNs will presumably soon replace them. Reports of a unit flying MiG-27s from Guines are still unconfirmed. Operational conversion is carried out on about 10 MiG-21Us, and a few MiG-23Us. Jet conversion is carried out on the Aero L-39C Albatros, which is believed to have replaced the old MiG-15UTI.

Basic training is carried out on Zlin Z526s with students then progressing to either the Ilyushin Il-14 for twin rating or the L-39C. Transport equipment includes at least 30 Antonov An-2s, 20 An-24s, 20 An-26s and 20 Il-14s. Helicopters in service include 60 ageing Mil Mi-4s, 40 Mi-8s (of which half are probably armed), and at least 12 Mi-24 attack helicopters.

The Cuban air force (Fuerza Aerea Revolucionara) is equipped entirely with Warsaw Pact aircraft. Security is high, and it is not known exactly what the FAR operates. The MiG-17 is still thought to be used in the ground-attack role, but may have been replaced or supplemented by MiG-23BN aircraft or even the MiG-27.

Dominican Republic Fuerza Aérea Dominicana

Most of the aircraft equipping the

Escuadron de Caza and the Escuadron de Caza-Bombardeo have been withdrawn from use and the nation's limited budget prevents their replacement. It is believed that the latter squadron may still operate about six North American T-28Ds.

The Escuadron de Transporte Aéreo operates at least seven Douglas C-47 Dakotas, two Curtiss C-46A Commandos, two Aero Commander 590As, a Beech Queen Air, a Cessna 310 and a Mitsubishi MU-2 for communications and transport duties. An Aérospatiale SA.365N is in service for Presidential use.

The Escuela de Aviación Militar at Ciboa operates eight Cessna T-41D Mescaleros in the basic training role, as well as 12 ex-US Navy Beech T-34B Mentors.

The FAD has operated numerous helicopters and at least eight Bell 205A-1s are in service along with seven Hughes OH-6A Cayuses, three Aérospatiale SE.3130/SA.318C Alouette II/Astazous and an SE.3160 Alouette III. Two Hiller UH-12A and two Sikorsky S-55C helicopters are believed to be no longer in service. Two Consolidated PBY-5A Catalinas were operated until the late 1970s and may still remain in use, and a Morane-Saulnier MS.893A Commodore is in service in a liaison role.

Ecuador Fuerza Aérea Ecuatoriana

The FAE has been expanding for some time now to meet the large Peruvian threat. A small border war broke out with Peru in 1981 but no major incidents have occurred since then.

Ala de Combate 21 based at La Taura has at least three component escuadrones. Escuadron 2111 'Aguilas' operates the SEPECAT Jaguar International; 10 single-seat strike aircraft and two trainer versions were acquired. Escuadron 2112 'Dragones' operates the Cessna A-37B Dragonfly of which 12 were delivered but seven have reportedly been lost. Escuadron 2113 operates the BAe Strikemaster Mk.89/89A of which eight Mk. 89 and eight Mk.89A were delivered; at least six have been written off. Also at La Taura and probably part of Ala de Combate 21 is an escuadron flying the survivors from 16 Dassault Mirage F.1JE and two F.1JB trainers.

The Grupo de Transportes Aéreos Militares based at Quito comprises two paramilitary airlines, TAME and Ecuatoriana, along with a small

liaison and communications flight. Transportes Aéreos Militares Ecuatorianos (TAME) operates a Boeing 727-2T3, two DHC-5D Buffaloes, two DHC-6 Twin Otters, three Lockheed L.188 Electras, three HS.748s, two Lockheed C-130H Hercules, and a few Douglas DC-6Bs and C-47 Dakotas continue to soldier on in service. Two second-hand Boeing 727-134s were acquired in 1984. Ecuatoriana carries out the long-haul traffic with three Boeing 720-021s and four 707-321s. Other aircraft in use for communications duties are a Beech King Air E.90, a Piper PA-31 Navajo and a Cessna 337D. Helicopters in service include the survivors from nine Aérospatiale SA.316 Alouette IIIs, one SA.330 Puma, four Bell UH-1Hs, one 212 and two 214Bs.

The Escuela de Aviación Militar at Guayaquil operates 24 Cessna 150L Aerobats, eight 172Fs, 12 T-41D Mescaleros and 16 Beech T-34C Mentors. 12 SIAI-Marchetti SF.260Ws were reportedly delivered in 1977 and 14 EMBRAER EMB.326GB Xavantes and 12 IAI Kfir C2s are on order.

The Aviación Naval Ecuatoriana has four flights of aircraft based at Guayaquil which are 1 Escuadrilla with one Cessna 320E and a T-41D for communications duties; 2 Escuadrilla with a Cessna 177, two T-41Ds, a Citation I, four 337s, three Beech T-34C-1s, a Super King Air 200 and an IAI Arava which are all used for training. 3 Escuadrilla

operates 2 Aérospatiale Alouette III and 4 Escuadrilla at present has no aircraft.

The Aviación del Ejercito Ecuatoriana has for transport use five IAI Aravas and one DHC-5D Caribou. A Beech Super King Air 200 is for VIP duties, and two Cessna T-41Ds, one 172G, two 185Ds and three Pilatus PC-6B Turbo Porters are in use for liaison and training. Helicopters on order include five Aérospatiale SA.330 Pumas, 36 SA.342L Gazelles and 10 AS.332B

It is thought that only five Cessna A-37Bs remain in service, these being used for light strike and COIN.

Super Pumas, but by late 1984 only two Super Pumas, one Puma and four Gazelles were reported. Two Bell 214Bs are also in service.

The Instituto Geografico Militar operates two Beech Queen Air A80s and one King Air, one Learjet 24D, one Cessna TU.206, two Aérospatiale SA.315B Lamas and one SA.316B Alouette III.

Guatemala Fuerza Aérea Guatemalteca

Following loss of US military aid during 1977, for human rights violations, it was not until 1982 that US

aid for Guatemala was reinstated after a more moderate military Junta had been established.

The Cessna A-37B Dragonfly is the only attack aircraft available to the FAG; the 16 received are operated by the Escuadron de Caza-Bombardeo at San Jose, which also maintains a detachment at Flores close to the Belize-Guatemalan border. Two Lockheed T-33As are also with this unit and three refurbished Aérospatiale CM.170R Magisters are in use for advanced training.

The Escuadron de Transporte at La Aurora operates about 10 Douglas C-47 Dakotas, one civil registered DC-6B and seven IAI Aravas. Four Fokker F.27 Friendships undergoing overhaul in Holland during 1984 were reportedly destined for the FAG despite having civil registrations. A Beech Super King Air 200 is in use as a Presidential transport.

Basic training is carried out by the Escuela de Aviación Militar with at least seven Cessna 172 Hawk XPs

and 12 Pilatus PC-7 Turbo Trainers. For liaison and communications duties six Cessna 170Bs, two 172Ks, three 180Hs, one 182F, one 185 and two U.206Cs are in use.

The FAG received nine Bell UH-1Ds but only two were serviceable until renewed US aid was received; also in use are at least six 206B JetRangers, five 206L LongRangers, three 212s and six 412s. The Bell 212/412s delivered in 1980-81 for civil use have since been armed for anti-guerrilla operations.

Guyana Guyana Defence Force/Air Command

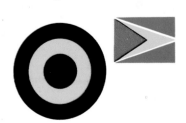

After Guyana gained independence from Britain in 1966, a small defence force was formed which operates a few light aircraft and helicopters for patrol and SAR use. These include six Britten-Norman BN-2A Islanders delivered from 1971-76 which are still in use; other fixed-wing aircraft comprise a Cessna U206F, a Beech Super King Air 200 and two Short Skyvan 3Ms. The helicopters in service are three Bell 212s, two 206B JetRangers and a 412.

Haiti Corps d'Aviation d'Haiti

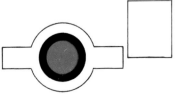

The HAC combat element now consists of six Cessna O-2s which were former USAF aircraft overhauled by Summit Aviation to 'O-2-337' standard. For liaison the HAC

uses one Cessna 310, one Beech Baron, one Debonair, one Twin Bonanza, one Britten-Norman BN-2A Islander and three DHC-2 Beavers. At least three Douglas C-47 Dakotas are still in use in the transport role, plus a DHC-6 Twin Otter. Six to eight SIAI-Marchetti SF.260TPs in service for pilot training are operated by Air Haiti. Two Hughes 269Cs and two 369Cs are the only helicopters in use. Delivery of three SIAI S.211 jet trainers was imminent at the end of 1984.

Honduras Fuerza Aérea Hondurena

Following the threat of infiltration from guerrillas operating in Marxist Nicaragua, US aid to Honduras has greatly increased recently.

The Escuadrilla de Caza at Base Aérea Coronel Hector Caracciolo, Moncada, La Ceiba operates the Dassault Super Mystère B.2, of which about 21 were received from Israel. The Escuadrilla de Caza-Bombardeo operates a few surviving North American F-86 Sabres from Base Aérea Coronel Armando Escalon Espinal, San Pedro Sula. The Sabres are in the process of being withdrawn as more Cessna A-37B Dragonflies are received to equip the Escuadrilla de Ataque. The first six A-37Bs were delivered in 1975 and six in 1982 and at least four more were seen in the US awaiting delivery in late 1984. One or two Lockheed T-33As may also be current.

The Escuadrilla de Transporte at Base Aérea Teniente-Coronel Herman Acosta Mejia at Tegucigalpa has two IAI Aravas, two IAI Westwinds about 12 Douglas C-47 Dakotas, one C-118A and one Lockheed Electra. The Escuadrilla de Comunicaciones, also at Tegucigalpa, operates two Rockwell Commander 114s and four Cessna A.180/185s. Helicopters in use include 20 Bell UH-1Hs delivered in two batches (a possibility exists that the first 10 aircraft were UH-1B), one Sikorsky S.76 and possibly three UH-19s. Orders are outstanding for four CASA C.101BB Aviojets and eight EMBRAER EMB.312 Tucanos.

Jamaica Air Wing, Jamaica Defence Force

After gaining independence from the UK in 1962, the Air Wing of the JDF was formed in 1963. The main base is at Up Park Camp, Kingston, with some maintenance being carried out at Tinson Pen. The principal duties of the Air Wing are anti-drug smuggling patrols, SAR, liaison duties and coastal surveillance.

Initial equipment was four Cessna A.185 Skywagons delivered in 1963 followed by two Bell 47G-3B-1 in 1963 and 1964. A flight of fixed-wing aircraft now operates from Norman Manley airfield with equipment that includes two Britten-Norman BN-2A Islanders, a DHC-6 Twin Otter, one Cessna U206G Centurion and one 337G Super Skymaster, a Beech King Air A100 and a Duke B60. The helicopters, which operate from Up Park Camp, include four Bell 206Bs and three 212s.

Mexico Fuerza Aérea Mexicana

The FAM is organised into nine Grupos Aeréos each of which theoretically consists of two squadrons. Expansion and modernization of the FAM began in the late 1970s funded by Mexican oil revenues, but subsequent devaluation of the peso has resulted in a drastic cut in defence spending.

Fighter strength is contained with 7 Grupo Aéreo (GA) at Base Aérea Militar (BAM) Santa Lucia with Escuadron Jet de Pélea 202 operating seven Lockheed T-33As. Also with 7 GA is Escuadron de Defensa 401 with 10 Northrop F-5Es and two Northrop F-5Fs. The FAM's 100 North American T-28A/B/D Trojans were replaced from 1980 by 55 Pilatus PC-7 Turbo-Trainers, delivered to 2 Grupo Aéreo with Escuadron Aéreo de Pélea (EAP) 206 at BAM Puebla, EAP 207 at BAM Ixtepec; 3 Grupo Aéreo with EAP 203 at BAM La Paz and EAP 204 at BAM El Cipres; 4 Grupo Aéreo with EAP 205 at BAM Merida and EAP 201 at BAM Cozumel. Each EAP retains a few T-28s for training use. Additional counter insurgency roles are fulfilled by 1 Grupo Aéreo, BAM Santa Lucia, with three component units. Escuadron Aéreo 208 operates 10 IAI Aravas, and also with 1 GA is Escuadron Aéreo 209 which operates the majority of the FAM helicopters. These include the survivors of nine Aérospatiale Alouette IIIs, plus five Bell 205A-1s, five 206Bs and three 212s. Finally, Escuadron Mixto de Entrenamiento Tactico with T-28A at Zapopan is also part of 1 GA.

5 Grupo Aéreo is at BAM Santa Lucia with component units Escuadron 101 operating Aero Commander 500s and Escuadron Aéreo de Reconocimiento Fotografico also with Aero Commander 500s; the FAM initially received 20 Aero Commanders. 6 Grupo Aéreo is also at BAM Santa Lucia with two component units, the first is Escuadron de Transporte Pesado 301 operating Douglas DC-6/7s and 302 with DC-4s. Of the five DC-4s, 20 DC-6s (mostly -6Bs) and four DC-7s received, four DC-4s, a few DC-6s and one DC-7 are believed to remain in service. 8 Grupo Aéreo has two component units, both based at Mexico City International Airport. The first is Escuadron Transporte Ejecutivo which operates and maintains the Government-owned civil registered Douglas C-47 Dakotas and Escuadron Aereo de Transporte Presidencial. The latter operates two Boeing 727-51s, five 727-14s, one 737-247, one Boeing 737-112, one Lockheed L.188AF Electra, three Fairchild FH-227s, one Swearingen Metro and one Merlin, one Beech King Air E90 and one King Air 200, five Short Skyvans, two DHC-5D Buffaloes, six Rockwell Sabres, six Britten-Norman Islanders, one Bell 212 and about 10 Aérospatiale SA.330J Pumas. Finally, 9 Grupo Aéreo at Santa Lucia has two component units, Escuadron Aéreo de Transporte Mediano 311 and 312, which fly the remaining 12 Douglas C-47 Dakotas.

Training is carried out by the Escuela Militar de Aviación at Zapopan using 20 Beechcraft Musketeer Sport IIIs and 20 F.33C Bonanzas for primary and basic training respectively. Advanced training is carried out on the PC-7 and 20 Mudry CAP.10Bs are used as aerobatic proficiency trainers.

One Gates Learjet Model 24D works on general and VIP transport duties for the Mexican navy.

Aviación de la Armada de Mexico

Shipborne helicopters consist of six MBB BO 105Cs, and other helicopters in use include four Aérospatiale SA.319B Alouette IIIs, two Bell UH-1Hs and the survivors from five Bell 47G/Js. For SAR at least nine Grumman HU-16D Albatross are still in use from a procurement of 16. In use in the liaison and transport roles are a Learjet 24D, three or four Fairchild F-27s, two Beech Barons five T-34A Mentors and four F33C Bonanzas, three Cessna 150Js, two 180s, three 310s, two or three 337Gs and two 402s, plus one DHC-5D Buffalo.

Nicaragua Fuerza Aérea Sandinista

Following the overthrow of the Samoza regime by Sandinista guerrillas in 1979, the Fuerza Aérea Guardia de Nicaragua became the FAS. Immediate cessation of US aid had left the FAS with few aircraft until Soviet aid started to arrive.

At least one, possibly a few more, Lockheed AT-33As were in use during 1984. Original FAGN equipment had included 10 T-33s, six North American T-28Ds and 10 armed Cessna 337s for COIN operations, but few are likely to be serviceable. Approximately six Let L-39Z Albatros and six Libyan-assembled SIAI-Marchetti SF.260Ws are now available for COIN use while more Let L-39Cs are available for advanced training.

Helicopters in service are believed to consist of seven Mil Mi-8s; eight are thought to have been delivered but at least one has been shot down. Transport duties are fulfilled by two CASA 212 Aviocars, one IAI Arava and the survivors of 10 Douglas C-47 Dakotas. A few Antonov An-26s, have reportedly been received and a Falcon 20 was donated to the Sandinista Government by Libya during 1984. Reports of the FAS receiving MiG-21s appear to be unfounded.

Panama Fuerza Aérea Panamena

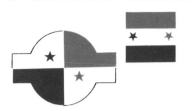

Panama has been dependent for protection on the US force based in the Canal Zone for many years; this is now started to change following an agreement with the United States to allow Panamanian armed forces to be in total control of the CZ by 1999, when all remaining US forces will be withdrawn.

As yet no jet combat aircraft have

officially been received but there have been reports of Cessna A-37B Dragonflies in national markings of Panama. For transport use there are three CASA 212A Aviocars, four Douglas C-47 Dakotas, four DHC-6 Twin Otters, one Short Skyvan, one Britten-Norman BN-2A Islander and two Piper PA-31/31T Navajos. For liaison duties two or three DHC-

3 Otters are still active along with two Cessna T-41D Mescaleros, four 180s and a few U-17s. Helicopters in service include eight Bell UH-1Bs, ten UH-1Hs, four UH-1Ns and a 412. For VIP and Presidential use two second-hand Boeing 727s are in service and a civil registered Dassault Falcon 20 is also in use.

Paraguay Fuerza Aérea Paraguaya

The FAP received its first jet combat equipment in 1980, in the shape of nine EMBRAER EMB.326GB Xavantes, which are operated by the Grupo Aérotactico at Asuncion; a

tenth was received subsequently as an attrition replacement.

Transport duties are carried out by TAM-Transporte Aéreo Militar. TAM operates from Aéropuerto Presidente Stroessner, Asuncion, and uses about 15 Douglas C-47 Dakotas, three ex-Brazilian air force DC-6Bs, one Convair C-131D, and four CASA 212 Aviocars delivered in 1984; for liaison duties TAM has a Cessna 337 and a 402. Training is carried out by the Academia de la Fuerza Aérea at Nhu-Guazu using 10 Cessna T-41D Mescaleros for

primary instruction and at least eight Aerotec A.122A Uirapurus for basic training. Five ex-Chilean air force Neiva T-25 Universals were acquired and unconfirmed reports suggest that 12 Cessna T-37Cs may have been received from Brazil. About 10 armed North American T-6 Texans remain in use with the Grupo de Entrenamiento Tactico, and the only helicopters in service are two Hiller UH-12SL-4s received from Chile in 1980.

Servicio de Aéronautica de la Marina

Though land-locked, Paraguay uses a few aircraft for river patrol; these consist of four Cessna 150Ms, two U.206As, two U.206Cs and one 210, one Douglas C-47 Dakota and two North American T-6G Texans.

Arma Aérea del Ejercito Paraguayo

The only aircraft in use by the army are three ex-Chilean Hiller UH-12Es delivered in 1980.

Peru Fuerza Aérea Peruana

The FAP operates a mix of Western and Soviet types, but the current economic plight of Peru has forced her to become more reliant on the purchase of Soviet aircraft because of more favourable repayment rates offered.

For air defence three grupos are equipped with either the Dassault Mirage 5 or Sukhoi Su-22 or, alternatively, a mix of both. Grupo 11 at BA La Joya, Mariano Melgar, has two component escuadrones operating Mirage 5Ps, Su-22s and a few Cessna A-37B Dragonflies. Grupo 12 at BA Capitan Montes, Talara, has Escuadrones 11 and 13 both flying Su-22/22Us. Grupo 13 at BA Capitan Jose Abelado Quinones Gonzales, Chiclayo, has Escuadrones 12 and 14 flying the Mirage 5P. Deliveries have included 48 Su-22s and four Su-22U trainers, but at least four and probably more have been written off. Mirage deliveries totalled 20 5Ps and three 5DP trainers by 1976; other orders have been placed but none have yet materialized so that the sale of 10 Mirage 5Ps to Argentina in 1982 suggests that fewer than 12 remain in FAP service. Grupo 7 at BA Capitan Concha, Piura, has two escuadrones operating the A-37B Dragonfly; 36 were procured in 1975-78 and most are still in service. Grupo 9 with Canberras at BA Renan Elias Olivera, Pisco, operates the survivors of 10

Canberra B(I).8s, three T.54s, six B.52s, six B(I).56s and 11 B(I).58s. Of the 36 acquired, about 24 are believed to be in service.

The transport unit is Grupo 8 with component Escuadrones 841, 842 and 843 based at BA Jorge Chavez International Air Port Lima-Callao, Las Palmas. It is equipped with 14 Antonov An-26s, two Douglas DC-8-62CFs, one Fokker F.28-100, seven Lockheed L-100-20 Hercules, one Fairchild-Hiller FH-227, 15 Beech Queen Air A80s and eight King Air C90s, a Dassault Falcon 20F and 15 DHC-5D Buffaloes. Grupo 42 at BA Coronel FAP Francisco Secada Vignetta, Iquitos, is also known as TANS-Transportes Aereos Nacionales de la Selva with 12 Pilatus PC-6B Turbo Porters, nine DHC-6 Twin Otters and a few surviving Douglas C-47 Dakotas. SAN-Servicio Aérofotografico Nacional carries out aerial survey duties with two Learjet 25Bs and two Learjet 36As.

The FAP helicopters are concentrated within Grupo 3 at BA Jorge Chavez IAP. The three component escuadrones are Escuadron 307 which flies a few Bell 47Gs, eight 206Bs and 5 Aérospatiale Alouette IIIs; Escuadron 332 which operates 10 MBB BO 105CBs and two or three Mil Mi-8s; and Escuadron 341 with six Mi-6s, two or three Mi-8s and about 20 Bell 212s. In the training role Grupo 51 at BA Las Palmas, Lima, has three component escuadrones. Escuadron 511, which is the basic training squadron, operates the survivors of 40 Cessna T-41Ds and also has over 20 T-41A/172Fs. Pilots then go to 512 Escuadron which is equipped with some 20 Cessna T-37Bs and about 12 T-37Cs.

Finally, Escuadron 513 flies the Aermacchi MB.339AP of which 16 were delivered during 1981-83.

Peru has started licence-construction of the MB.339; these are being built by Indaer SA at BA Collique, Lima, and will consist of about 20 trainers and 40 MB.339K single-seat attack aircraft. On order are about 20 Dassault Mirage 5Ps, 22 Mirage 2000Ps and four Mirage 2000DP trainers, but the nation's current economic situation makes their delivery uncertain. The intention to acquire 12 Mil Mi-24 'Hind' gunships has been announced. In late 1984 delivery was made of six of 12 Bell 214STs.

Aviación del Ejercito Peruano

The main Army base is at BA Jorge Chavez IAP. Here are based the survivors from eight Aérospatiale SA.315B Lamas, one SA.318C Astazou and eight Bell 47G helicopters, along with a few Cessna 185 Skywagons and Helio H.395 Super Couriers for liaison duties. The AEP had 42 Mi-8s, but many were written off and the remainder are believed to have been withdrawn from

Around two dozen BAC Canberras still survive in the Peruvian air force in various marks. The original bombing role is still maintained.

use due to their unsuitability for 'hot and high' conditions.

Servicio Aéronaval de la Marina Peruana

The naval air arm was merged with the army in 1929 and only reformed as a separate organization in the late 1950s. Current helicopter equipment includes about two Aérospatiale Alouette IIIs, five Bell UH-1Ds and four 206B JetRangers. For anti-submarine duties there are some five Agusta-Bell AB.212ASWs and of the five Agusta-Sikorsky AS.61Ds two delivered in 1980 were configured for an ECM/ESM role. Four more AS.61Ds were ordered in 1984. Fixed-wing complement of the Marina consists of 13 Grumman S-2E/-2G Trackers, and also in use are two Fokker F.27-400MPAs, about five Douglas C-47 Dakotas, six Beech T-34C Mentors and six Super King Air 200C/CT/Ts.

Salvador Fuerza Aérea Salvadorena

The current activities of left-wing guerrillas within the country has resulted in large quantities of US arms being delivered.

The FAS first began to expand in 1974-75 with the receipt of refurbished ex-Israeli equipment, comprising 18 Dassault Ouragans and six CM.170R Magisters. At least

three and possibly six more Magisters arrived in 1978 after overhaul by Aérospatiale and at least four Super Mystère B.2s followed, though it is still unclear if they came from a French or Israeli source. The surviving Ouragans are operated by the Escuadrilla de Caza and the Magisters by the Escuadrilla de Ataque,

both based at Ilopango. The Escuadrille de Transporte, also at Ilopango, operates four IAI 201 Aravas, two Douglas DC-6Bs and seven C-47 Dakotas. First helicopters for the FAS were six Bell UH-1Hs supplied by the US in 1980, followed by four more in 1981.

A guerrilla raid on Ilopango dur-

Salvador (continued)

ing January 1982 resulted in five Ouragans, six Bell UH-1Hs and three Dakotas being destroyed. The US immediately promised aid and 12 Bell UH-1Hs, eight Cessna A-37B Dragonflies, four O-2s and three Fairchild C-123Ks were delivered during 1982; six more A-37Bs followed in 1983. Numerous accidents and losses by guerrilla action had reduced the number of available Bell UH-1Hs to eight by early March 1984, when 14 more ex-US Army UH-1Hs were received; at that time more A-37Bs were scheduled for delivery. The Escuela de Aviación Militar operates a mix of three Beech T-34A Mentors, seven North American T-6 Texans and seven Cessna T-41D Mescaleros in the training role. For liaison seven Cessna 180s, one 182 and a 185 are used. Further helicopters include two Hughes 500s and two Aérospatiale SA.316 Alouette IIIs.

Surinam Surinam Air Force

Following independence from the Netherlands in 1975, the SAF was formed in 1981 with help from Dutch military advisers.

The first aircraft was a Hughes 500D (which crashed in 1982); in the same year delivery followed of four Pilatus Britten-Norman BN-2B Defenders and a Cessna U206, all of them based at Paramaribo. The December 1982/January 1983 massacre of opponents to the Marxist regime resulted in the Netherlands withdrawing her advisers.

Trinidad and Tobago

Trinidad and Tobago Defence Force

This small air arm currently operates one Cessna 401 which replaced an earlier Cessna 337A. The National Security Force operates one of the former TTDF Aérospatiale SA.341M Gazelles and two Sikorsky S-76As.

Uruguay Fuerza Aérea Uruguaya

The FAU is divided into two commands, the Comando Aéreo Tactico which has two component Brigada Aéreas, and the Comando Aéreo de Entrenamiento which controls the Escuela Militar de Aéronautica.

Brigada Aérea 1 of the Comando Aéreo Tactico is at Base Aérea 1, Capitan Boiso Lanza, Carrasco, with four component grupos. Grupo de Aviación 3 flies five CASA 212s, eight Beech Queen Air A65s and a few Douglas C-47 Dakotas. Grupo de Aviación 4 operates two Fokker F.27-100s. Grupo de Aviación 5 is the helicopter unit with five Bell UH-1Bs, three or five Bell UH-1Hs and two Bell 212s. Grupo de Aviación 6 has four EMBRAER EMB.110C Bandeirantes and one EMB.110B-1. The three transport units of BA 1 together form TAMU, the Transporte Aéreo Militar Uruguayo. Brigada Aérea 2 is at Base Aérea 2, Teniente Segundo Mario Walter Parallada, Santa Bernardina Airport, Durazno. Here are based Grupo de Aviación 1 with six FMA IA.58 Pucaras and a few Beech T-34Bs. Grupo de Aviación 2 operates the survivors of eight Cessna A-37B Dragonflies and a few remaining Lockheed AT-33As.

The second command is Comando Aéreo de Entrenamiento which controls the Escuela Militar de Aéronautica at Aéropuerto Militar General Artigas, Pando. In service are eight Cessna T-41D Mescaleros and the survivors from 25 Beech T-34B Mentors. An Escuadrilla de Enlace carries out communications duties with a Cessna 182A, two 182Ds and eight U-17As, and two Piper PA-18 Super Cubs.

Aviación Naval Uruguaya

With its main base at Base Aéronaval 2, Capitan Curbelo, Laguna Del Sauce, the fixed-wing element of this small naval air wing comprises three Grumman S-2A and three S-2G Trackers and a Beech Super King Air 200T; for training there are the survivors from six North American SNJ-4s and two SNJ-6s, three ex-Argentine Beech TC-45Hs, three T-34C-1s and nine North American T-28 Fennecs.

Helicopters in service include a Bell 47G-2 and a 222, and two Sikorsky SH-34Js.

Venezuela Fuerza Aérea Venezolana

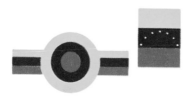

The FAV is formed into four commands, three of which have aircraft attached to them.

Comando Aéreo de Combate has five component grupos which are as follows: Grupo de Caza 11 at Base Aérea Teniente Vicente Landaeta, Barquisimeto, has Escuadron de Caza 36 as its sole squadron operating nine Dassault Mirage IIIEV interceptors, four Mirage 5V strike aircraft and two Mirage 5DV two-seat trainers. Grupo de Caza 12 at Barquisimeto has Escuadrones 34 and 35 sharing 16 ex-Canadian Canadair CF-5As and four CF-5Ds. Grupo de Bombardeo 13 at Base Aérea Teniente Luis des Valle Garcia, Barcelona, has two components, Escuadrones 38 and 39, which operate the BAC Canberra. A total of 30 were acquired, comprising six Canberra B.2s, two T.4s, eight Canberra B(I).8s, two PR.53s, eight B.52s and four B(I).52s; of this total, 22 were re-repaired in the UK between 1974 and 1980. A new Grupo de Caza 16 had received most of an order for 18 General Dynamics F-16As and six F-16Bs by the end of 1984. The Grupo is based at BA El Libertador but the component escuadrones are currently unknown. An additional 15 Canadair CF-5As and four Canadair CF-5Ds were acquired in 1982, but it is uncertain whether they formed a new unit or reinforced Grupo de Caza 12. The final Grupo is Grupo de Operaciones Especiales 2, based at Barcelona, with Escuadron de Ataque 40 operating 14 Rockwell OV-10E Broncos. In 1983 18 single-seat and six two-seat IA.58 Pucaras were ordered from Argentina.

Comando Aéreo Transporte has two component grupos, the first being Grupo de Transporte 6 at Base Aérea Francisco de Miranda, La Carlotta, with three component squadrons and a Presidential flight. Escuadron de Transporte 1 operates five Lockheed C-130H Hercules. Escuadron de Transporte 2 has the survivors of 18 Fairchild C-123B Providers, along with one Cessna 500 and one HS.748. Escuadron 42 is a helicopter unit flying two Bell 212s and two 412s. The Escuadrilla Presidencial operates a Gulfstream II, a Gulfstream III, a Boeing 737-2N1, a Douglas DC-9-15, two Bell 214STs and a Cessna 550. The last of the Comando's Douglas C-47 Dakotas are believed to have now been withdrawn. Six Aeritalia G.222s are on order and some of them had entered service in late 1984. Four Agusta-Sikorsky SH-3Ds are on order and should enter service in 1984. The remaining group of the Comando Aéreo Transporte is Grupo Mixto de Enlace y Reconocimiento 1 based at Barquisimeto in the liaison and reconnaissance role. The unit operates three Beech Queen Air 65s and eight Queen Air 80s, two Cessna 180s and 12 182Ns, several Bell UH-1Ds and possibly a few 47Gs, plus 12 Aérospatiale Alouette IIIs.

Grupo de Operaciones Especiales 1 operates the Bell UH-1H, and other transport and liaison aircraft in use include four Beech Super King Air 200 and a Piper PA-31 Navajo. The Grupo de Entrenamiento Aéreo at Base Aérea Mariscal Sucre, Boca del Rio-Maracay, is responsible for flying training. The subordinate unit, the Escuela de Aviación Militar, operates about 25 Beech T-34As, 21 Rockwell T-2D Buckeyes and a few Jet Provost T.52s at BA Mariscal Sucre. Plans to form an Escuela de Combate at Barquisimeto appear to have been shelved and an order for BAe Hawks cancelled.

Servicio de Aviación Naval Venezolana

The naval air arm has one ASW unit, the Escuadron Aéronaval Anti-submarino 11 based at Puerto Cabello AB with six Agusta-Bell 212ASs and eight ex-US Navy Grumman S-2E Trackers. For transport, patrol and liaison duties the Escuadron Aéronavale de Transporte at General Francisco de Miranda AB, Caracas, operates a Beech King Air E90, two Cessna 310Rs and one 402C, a DHC-7 Dash 7, an HS.748-215, and four CASA 212A Aviocars.

Regimiento Aéreo del Ejercito Venezolana

The army air arm had a modest beginning with two Beech Queen Airs, the first received in 1970 and the second in 1977. From 1977 there was rapid expansion with the acquisition of six Bell UH-1Hs, three 205A-1s, one 206B JetRanger and one 206L LongRanger, one Beech King Air E.90 and one King Air 200, one Britten-Norman Islander, six IAI Aravas, two Aeritalia G.222s, two Cessna U206 Stationair 6-IIs and four U206 Stationair 8-IIs, and six Agusta A.109A Hirundos.

Guardia Nacional

This paramilitary organization was formed at some time during the late 1940s/early 1950s. It operates some 10 Agusta A.109A Hirundos, a Beech King Air 200C, 15 Bell 206 JetRangers and two 214STs, two IAI Aravas and a few other light aircraft.

The World's Military Aircraft

Aeritalia G91Y

Whereas the original Fiat G91R was a single-engined fighter-bomber and tactical reconnaissance aircraft produced to a NATO specification, the G91Y (or 'Yankee' as it is unofficially known) was developed for the Italian Air Force. Compared with the G91R – which was declared non-operational in Italy during 1984 – the 'Yankee' differs fundamentally in having two J85 afterburning turbojets side-by-side in a revised fuselage, these providing 60 per cent more thrust for only a small increase in weight. In order to increase combat radius one engine can be shut down during the cruise phase of a mission. Fuel capacity of the fuselage and inner wing tanks is 3200 litres (704 Imp gallons) but drop tanks may be carried on the four underwing pylons, giving a ferry range of 3500 km (2,175 miles). Retaining its predecessor's three-camera nose and two internal cannon, the G91Y matches the ability to carry an increased warload with more sophisticated avionics, including a position and homing indicator, twin-axis gyro platform, doppler radar ranger, air-data computer, radar altimeter and head-up display. With the accent on STOL performance, the aircraft has provision for JATO rocket bottles which can halve the required take-off run, and an airfield arrester hook for use with SATS (Short Airfield for Tactical Support) installations. Operating from a semi-prepared surface, the aircraft will unstick in 914 m (3,000 ft) and land (unassisted) from 15 m (50 ft) in 600 m (1,970 ft). With the obvious exception of the engine compartment, the G91Y's airframe is based on the earlier G91T two-seat trainer, although accommodation is provided for a pilot only in an armoured, air-conditioned, pressurized cockpit fitted with a Martin-Baker 'zero-zero' ejection seat. The wing has 38° sweepback, full-span slats and electrically-actuated slotted flaps. Two airbrakes are hinged beneath the centre fuselage, and engine replacement requires removal of the rear fuselage, with the variable-incidence trimming tailplane. The first of two prototypes was flown on 27 December 1966. An order for 20 was followed by contracts for a further 53 but, in the event, production ended at No. 67. A projected G91YT two-seat trainer was not developed, and the sole G91YS was demonstrated to the Swiss air arm without success. Deliveries to the 8° Stormo at Cervia began in May 1970

Still going strong with Esc. 301 and 303 of the Portuguese air force is the Aeritalia G91, a handful of which are G91Ts operating in the training role. Single-seat variants carry Sidewinder AAMs for interception duties in addition to tactical strike and reconnaissance duties.

for a single over-size squadron (101° Gruppo), whilst 32° Stormo (13° Gruppo) at Brindisi re-equipped from August 1973, assigning some of its aircraft to anti-shipping strike and decorating them with 'sharkmouth' markings. The G91Y will remain in use for close support and day reconnaissance until replaced by the AMX from 1987.

Specification
Aeritalia G91Y
Type: single-seat fighter-bomber and tactical reconnaissance aircraft
Powerplant: two General Electric J85-13A turbojets each rated at 1851 kg (4,080 lb) thrust with afterburning
Performance: maximum speed 1110 km/h (690 mph) at sea level, or Mach 0.95 at 9150 m (30,020 ft); service ceiling 12500 m (41,010 ft); initial rate of climb 5180 m (17,000 ft)/minute; typical combat radius 750 km (466 miles)

Aeritalia G91Y

Weights: empty 3900 kg (8,598 lb); maximum take-off 7000 kg (15,432 lb) on semi-prepared surface, or 8700 kg (19,180 lb) on hard runway
Dimensions: span 9.01 m (29 ft 6.7 in); length 11.67 m (38 ft 3.4 in); height 4.43 m (14 ft 6.4 in); wing area 18.13 m^2 (195.16 sq ft)
Armament: two DEFA 30-mm (1.18-in) cannon, plus up to four 454 kg (1,000 lb) bombs, or 340 kg (750 lb) napalm tanks, or rocket launchers
Operator: Italy, Portugal

Aeritalia G222

The Fiat G222 proposal was drawn up to meet the outlines of NATO's Basic Military Requirement Four (NBMR4) of 1962, which sought to develop a practical V/STOL transport for service with NATO air forces. Although a number of advanced proposals came from several manufacturers, none seemed sufficiently practical and attractive to gain even a prototype contract. However, the Aeronautica Militare Italiana believed that Fiat's proposal could prove a useful transport, if finalized as a more conventional design in terms of powerplant and aerodynamics, and in 1968 signed a contract for two G222TCM prototypes and a static test airframe. Their manufacture was delayed by Fiat's restructuring which resulted in the name Aeritalia, and it was not until 18 July 1970 that the first prototype (MM582) was flown, the second (MM583) following on 22 July 1971. These began operational evaluation with the AMI on 21 December 1971, highly successful tests resulting in a contract for 44 production G222s, the first of them flown on 23 December 1975.

From the outset other major Italian manufacturers were involved in the programme, with Aermacchi responsible for the outer wings, CIRSEA for the landing gear, Piaggio for the wing centre-section, SIAI-Marchetti for the tail unit, and Aeritalia building the fuselage and being responsible for final assembly and testing. The G222 continued in production in 1985 and has been built in several versions. These include the G222 standard military transport which serves with the armed forces of Argentina, Dubai, Italy, Nigeria, Somalia and Venezuela; the G222R/M (Radio Misure) for radio/radar calibration; the G222SAA (Sistema Aeronautico Antincendio) fire-fighter with equipment to disperse water or fire retardants; the G222T (Rolls-Royce Tyne-powered version) for the Libyan Arab air force which designates it G222L; and the electronic warfare G222VS (Versione Speciale). Just over 70 of all versions had been built by the end of 1984.

Projected versions of the G222 which are actively being studied by Aeritalia include aircraft for airborne early warning (AEW), an inflight-refuelling tanker, a drone aircraft launcher, an Earth resources reconnaissance platform and a maritime patrol and ASW/ASV mission version.

Specification
Aeritalia G222
Type: general-purpose transport
Powerplant: two General Electric T64-GE-P4D turbojet engines (licence-built by Fiat) each flat-rated at 2535 kW (3,400 shp)
Performance: maximum speed 292 kts (540 km/h; 336 mph) at 4575 m (15010 ft); long-range cruising speed 237 kts (439 km/h; 273 mph) at 6000 m (19,685 ft); initial rate of climb 520 m (1,706 ft) per minute; service ceiling 7620 m (25,000 ft); range with maximum payload 1370 km (851 miles); ferry range with maximum fuel 4633 km (2,879 miles)
Weights: empty 14590 kg (32,165 lb); maximum take-off 28000 kg (61,729 lb)
Dimensions: span 28.70 m (94 ft 1.9 in); length 22.70 m (74 ft 5.7 in); height 9.80 m (32 ft 1.8 in); wing area 82.00 m^2 (882.67 sq ft)
Armament: none
Operators: Argentina, Italy, Libya, Nigeria, Somalia, UAE, Venezuela

Aeritalia G222 of 98° Gruppo, 46ª Aerobrigata Trasporto Medi, Italian air force, based at Pisa.

Aeritalia/Aermacchi/EMBRAER AMX

Looking a decade ahead to the anticipated rundown of its Fiat G91R/Ys and licence-built Lockheed F-104Gs by 1987, the Aeronautica Militare Italiana began to formulate its specification for a small tactical fighter-bomber. This was also required to be suitable for a reconnaissance role, of providing support to ground or naval forces, and to be able to complement the activities of the Aeritalia F-104S and Panavia Tornado. At about the same time the Fôrça Aérea Brasileira had identified its A-X requirement for a somewhat similar aircraft to complement its EMBRAER AT-26 Xavantes, and this led to discussions between the Brazilian company EMBRAER and Italy's Aermacchi to collaborate on an adaptation of an Aermacchi M.B.340 design to meet this need. Finalization of the AMI's specification brought a co-operative agreement between Aeritalia and Aermacchi, and following selection of the Rolls-Royce Spey turbofan for the aircraft, by then identified as the AMX, EMBRAER became a member of the design/production team.

A compact shoulder-wing monoplane with swept wings and tail surfaces, the AMX accommodates the pilot on a Martin-Baker Mk 10L zero-zero seat in a pressurized and air-conditioned cockpit. Manufacture is shared with major assemblies coming from Aeritalia (responsible for the fuselage centre section, fin, rudder and elevators), Aermacchi (forward and rear fuselage sections) and EMBRAER (wings, tailplane and engine inlets). The Rolls-Royce Spey Mk 807 non-afterburning turbofan engine is being licence-built in Italy by Alfa Romeo, Fiat and Piaggio. Current procurement plans cover the manufacture of 187 aircraft for the AMI and 79 for the FAB, with initial entry into service during late 1987.

The first prototype was rolled out by Aeritalia at Turin in February 1984, and this made the type's maiden flight on 15 May 1984; it was unfortunately lost in a flying accident during its fifth flight, on 1 June 1984, inevitably causing some programme delay. The second prototype, completed at Aermacchi's Venegono factory, completed a successful 70-minute first flight at Turin-Caselle on 19 November 1984.

Aeritalia/Aermacchi/EMBRAER AMX as it is anticipated to appear in Brazilian air force markings.

Specification
Aeritalia/Aermacchi/EMBRAER AMX
Type: single-seat multi-purpose combat aircraft
Powerplant: one 5003-kg (11,030-lb) thrust licence-built version of the Rolls-Royce Spey Mk 807 non-afterburning turbofan engine
Performance: (estimated) maximum speed Mach 0.95 or 626 kts (1160 km/h/721 mph) at 305 m (1,001 ft); cruising speed Mach 0.77 or 512 kts (950 km/h/590 mph) at 610 m (2,000 ft); attack radius with 907 kg (2,000 lb) of external stores (including 5 minutes combat and 10 per cent reserves) hi-lo-hi 890 km (533 miles) or lo-lo-lo 555 km (345 miles); ferry range with two 1000-litre (220-Imp gal) drop tanks 2965 km (1,842 miles)
Weights: empty 6000 kg (13,228 lb); typical mission take-off 10750 kg (23,700 lb); maximum take-off 11500 kg (25,353 lb)
Dimensions: span 8.874 m (29 ft 1.4 in); length 13.575 m (44 ft 6.4 in); height 4.576 m (15 ft 0.2 in); wing area 21.00 m² (226.05 sq ft)
Armament: one 20-mm M61A1 cannon (Brazilian version two 30-mm DEFA cannon) plus external stores carried on a single twin-pylon attachment on the fuselage centreline, four underwing hardpoints and wingtip rails for two AAMs, for a total external weapon load of 3500 kg (7,716 lb); this can include air-to-surface missiles, cluster bombs, electro-optical guided precision weapons, free-fall or retarded bombs and rocket-launchers; for reconnaissance three alternative pallet-mounted photographic systems can be mounted in a lower-fuselage camera bay

Aermacchi AL.60 Trojan

Manufacturing rights for the Lockheed 60 light utility transport were obtained by Aermacchi in 1960, and the first Varese-built machine appeared in April 1961. It is a strut-braced high-wing cabin monoplane of all-metal construction. Pilot and co-pilot sit side-by-side, and behind them is provision for two benches each seating three passengers; the space can be used alternatively for cargo, or for two stretchers plus one sitting patient and an attendant. A sliding door to the rear on the left side can be used for dropping parachutists or supplies.

Lockheed-Azacarte SA based in Mexico has sold a considerable number, especially of civil variants. The Mexican LASA 60 Santa Maria is used for search and rescue; the Aermacchi AL.60C5 has been purchased by customers in Africa.

Specification
Aermacchi AL.60C5 Conestoga/Trojan
Type: utility aircraft
Powerplant: one 298-kW (400-hp) Lycoming IO-720-A1A flat-eight piston engine
Performance: maximum speed at sea level 251 km/h (156 mph); economic cruising speed at 1525 m (5,005 ft) 174 km/h (108 mph); initial rate of climb 330 m (1,083 ft)/minute; service ceiling 4150 m (13,615 ft); range with maximum fuel 1037 km (644 miles)
Weights: empty equipped 1086 kg (2,394 lb); maximum take-off 2040 kg (4,497 lb)
Dimensions: span 11.99 m (39 ft 4 in); length 8.80 m (28 ft 10.5 in); height 3.30 m (10 ft 9.9 in); wing area 19.55 m² (210.4 sq ft)
Armament: none

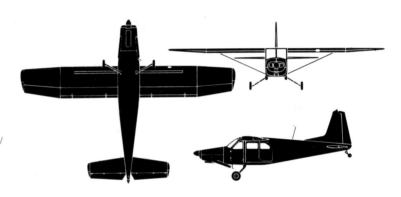

Operators: Central African Republic (Conestoga); Mauretania, South Africa, Zimbabwe (Trojan)

A simple design, the Aermacchi AL.60 has a roomy cabin enabling a number of roles to be performed including paradrops, casevac and cargo transport.

Aermacchi AM.3

Development of the AM.3 was begun during the 1960s to meet an Italian military requirement. Initial work was undertaken by Macchi in conjunction with Aerfer, as the MB.335, using a wing based on that of the Aermacchi-Lockheed AL.60. When Aerfer was incorporated in Aeritalia the aircraft was redesignated AM.3, and the first of three prototypes was flown at Aermacchi's Varese airfield on 12 May 1967; an Aerfer-built aircraft followed on 22 August 1968. Produced for the roles of forward air control, observation, liaison, light transport with cargo or up to three passengers, casualty evacuation and close air support, the aircraft was offered to the Italian Army as a replacement for the Cessna O-1E. Both flying prototypes were redesignated AM.3C in 1969, following the installation of 254-kW (340-hp) Lycoming GSO-480 engines prior to evaluation, but testing against the SIAI-Marchetti SM.1019 resulted in the latter being chosen during 1972. The first customer proved to be Rwanda, which took delivery of three in mid-1973, but subsequently returned them. The second and only other purchaser was the South African Air Force, which received its first AM.3CM in February 1974. Of 40 ordered and allocated the name Bosbok, the last 32 were assembled in South Africa by Atlas Aircraft. With its easy access for personnel and cargo, via an outward-opening full-length door on the right, the Bosbok is used for all manner of 'bush' operations by three units of Western Air Command: No. 41 Sqn at Lanseria and No. 42 Sqn and No. 84 Advanced Flying School, both at Potchefstroom. At a normal take-off weight of 1500 kg (3,301 lb) the AM.3CM will leave the ground in only 85 m (279 ft) and land in 66 m (217 ft). A dual-control aircraft, its rear cabin can accommodate two passengers, or two superimposed stretchers. Weapon options on its two underwing pylons include 7.62-mm (0.3-in) machine-gun pods, AS.11 or AS.12 anti-tank missiles, rocket pods, light bombs, supply containers or a photo-reconnaissance pod.

Specification
Aeritalia AM.3C
Type: four-seat utility aircraft
Powerplant: one 254-kW (340-hp) Piaggio-built Lycoming GSO-480-B1B6 flat-six piston engine
Performance: maximum cruising speed 246 km/h (153 mph) at 2440 m (8,005 ft); initial rate of climb 420 m (1,378 ft)/minute; service ceiling 8400 m (27,560 ft); endurance (no reserves) 6 hours 15 minutes
Weights: empty 1080 kg (2,381 lb); maximum take-off 1700 kg (3,748 lb)
Dimensions: span 12.64 m (41 ft 5.6 in); length 8.93 m (29 ft 3.6 in); height 2.72 m (8 ft 11 in); wing area 20.36 m² (219.16 sq ft)
Armament: up to 170 kg (375 lb) on each of two underwing pylons
Operator: South Africa

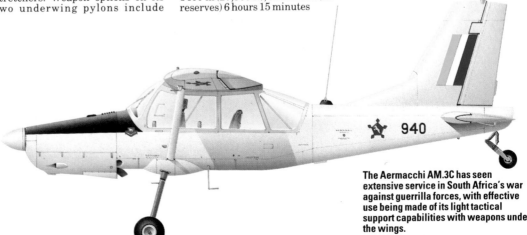

The Aermacchi AM.3C has seen extensive service in South Africa's war against guerrilla forces, with effective use being made of its light tactical support capabilities with weapons under the wings.

Aermacchi MB.326

One of five MB.326KDs operated by Dubai as part of the United Arab Emirates air force.

A popular and widely-used basic trainer and light attack aircraft, the MB.326 first flew on 10 December 1957 and has been built in four continents. Powered by a Rolls-Royce Viper turbojet, the MB.326 is stressed for flight load factors of +8 and −4g. Although it lost to the Magister in its bid to be adopted as the standard NATO basic trainer, it was ordered into production by the Italian Air Force (AMI), the initial 15 pre-series aircraft being made up to a total of 100. The first aircraft for the AMI was flown in October 1960, and student training at the Basic FTS (SVBIA) at Lecce began in March 1962. The AMI's total eventually increased to some 130, including six new and six converted MB.326Es with strengthened wings for weapons training, and three of the four MB.326Ds originally supplied to Alitalia for airline pilot training. Replaced at the SVBIA in October 1981, when they had flown about 400,000 hours, MB.326s continued in the training role for two more years. Surplus MB.326s have been issued to combat wings as instrument-rating and communications aircraft, and two were tested as pilotless drones. The MB.326H was adopted by the Australian armed forces, and produced under licence by Commonwealth Aircraft Corporation as the CA.30. Deliveries to the RAAF for the Central Flying School and No. 2 FTS totalled 87, whilst the Navy flew 10 more with VC-721 until the eight survivors were transferred to the RAAF in 1983. Addition of six underwing stores pylons and two machine-guns in the fuselage produced the MB.326B for training and light attack. Tunisia ordered eight, whilst nine MB.326Fs went to Ghana. South Africa followed initial Italian deliveries by production at Atlas Aircraft. In all, 151 MB.326M Impala Mk 1s were acquired, these continuing to serve the Central Flying School, Flying Training School and Nos 6 and 7 Sqns. The 1134-kg

(2,500-lb) thrust Viper Mk 22 of earlier models was replaced in early 1967 by the 1547-kg (3,410-lb) Viper Mk 540 to produce the MB.326G and MB.326GB trainer/COIN aircraft. Eight were supplied to the Argentine navy, and continue to serve 4° Escuadra; Zambia bought 24 and Zaire 17. In Brazil, EMBRAER built 182 MB.326GCs under licence as the AT-26 Xavante, described separately. A further boosting of power to 1814 kg (4,000 lb) in the Viper Mk 632-43 brought about the MB.326L and single-seat MB.326K. Equipped for attack, the MB.326K flew on 22 August 1970 and has additional provision for armament and armour plating, plus further fuel tankage in the fuselage. Ghana obtained eight Ks; Tunisia, eight, plus four Ls; and Dubai, three KDs and one L. Zambian orders covered six MB.326Ks, delivery of which has been postponed. In South Africa the single-seat aircraft entered production in 1974, as the MB.326KC Impala Mk 2, and at least 85 appear to have been built so far, almost entirely with indigenous components. Macchi's last MB.326 was supplied in 1979, but production by EMBRAER and Atlas continued to 1983.

Specification
Aermacchi MB.326GB
Type: two-seat basic trainer and light attack aircraft
Powerplant: one 1547 kg (3,410 lb) thrust Piaggio-built Rolls-Royce Viper Mk 540 turbojet engine
Performance (attack configuration): maximum permissible speed 871 km/h (541 mph); initial rate of climb 1080 m (3,543 ft)/minute; service ceiling 11900 m (39,040 ft); combat radius 648 km (403 miles)

Weights (attack): empty 2640 kg (5,820 lb); maximum take-off 5216 kg (11,500 lb)
Dimensions: span (over tanks) 10.15 m (33 ft 3.6 in); length 10.65 m (34 ft 11.3 in); height 3.72 m (12 ft 2.5 in); wing area 19.0 m² (204.52 sq ft)
Armament: two internal 7.7-mm (0.303-in) machine-guns, plus up to 1814 kg (4,000 lb) of ordnance on six strongpoints
Operators: Argentina, Australia, Brazil, Dubai, Ecuador, Ghana, Italy, Paraguay, South Africa, Togo, Tunisia, Zaire, Zambia.

An ideal basic jet trainer, the MB.326 has trained thousands of pilots, full use being made of its excellent handling characteristics and tandem seating configuration by air forces around the world.

Aermacchi MB.339

Realizing that a decade ahead it would need a second-generation jet trainer to supersede the MB.326 and Aeritalia (Fiat) G91T in service, the Aeronautica Militare Italiana awarded a study contract to Aermacchi. No fewer than nine different design studies were originated and considered under the designations MB.338 and MB.339 (respectively seven and two variants of each). Selection of one of the MB.338 proposals would probably have provided a greater increase in performance and capability, but would equally have proved more costly to finalize as hardware; with no vast sums available the AMI opted for close examination of the two MB.339 proposals, the MB.339L with a Larzac turbofan, the MB.339V retaining the Rolls-Royce Viper turbojet of the MB.326 series.

It was the latter which was finally selected for development, this being based upon the airframe of the MB.326K. Principal revision was to the forward fuselage to provide a new cockpit with a raised rear (instructor's) seat, an extended canopy for a better all-round view, and advanced avionics. Revised aerodynamics and enlarged vertical tail surfaces give performance improvements and better handling on the same powerplant as that of the MB.326K. The first (MM588) of

three MB.339X prototypes (one for static test) was flown on 12 August 1976. The second (MM589), which flew on 20 May 1977, was completed to production standard, and incorporated improved air-conditioning, a steerable nosewheel and an antiskid braking system.

The first of 100 production MB.339A trainers for the AMI was flown on 20 July 1978 and began service trials on 8 August 1979; the type entered service initially with the Scuola di Volo Basico Iniziale Aviogetti. The Frecce Tricolori, the AMI's aerobatic team, received 15 M.B.339PAN aircraft in 1982. These differ by installation of a smoke-generating system and deletion of the wingtip tanks. A number of

export orders were also gained, for the air arms of Argentina (10), Dubai (four), Malaysia (12) and Peru

Three squadrons of the Peruvian air force fly the Aermacchi MB.339AP in the training role. They are built locally by Indaer-Peru, with single-seaters on order for light attack duties.

(16), most of which have been delivered. Nigeria also ordered 12, but it is believed that these have been cancelled by the new government.

Specification
Origin: Italy
Type: basic and advanced trainer and close-support aircraft
Powerplant: one 1814-kg (4,000-lb) thrust Piaggio-built Rolls-Royce Viper 632-43 turbojet engine
Performance: limiting Mach number 0.85 or 499 kts (925 km/h/575 mph); maximum speed 485 kts (900 km/h/559 mph) at sea level; initial rate of climb 2010 m (6,594 ft) per minute; service ceiling 14630 m (48,000 ft); combat radius with maximum weapon load hi-lo-hi 539 km (368 miles), lo-lo-lo 371 km (231 miles); ferry range with external fuel 2110 km (1,311 miles)

Weights: empty 3125 kg (6,889 lb); maximum take-off, clean 4400 kg (9,700 lb) or with maximum weapon load 5895 kg (12,996 lb)
Dimensions: span over tiptanks 10.86 m (35 ft 7.6 in); length 10.97 m (36 ft 0 in); height 3.99 m (13 ft 1.1 in); wing area 19.30 m^2 (207.75 sq ft)
Armament: six underwing hardpoints with maximum load of 1814 kg (4,000 lb); the two inboard

points can carry 30-mm cannon or multi-barrel 7.62-mm (0.3-in) Miniguns in a Macchi pod, and the two centre points are 'wet' for the carriage of drop tanks; a wide variety of weapon loads includes bombs, napalm, AS.11/AS.12 or magic missiles, rocket-launchers, and a single four-camera reconnaissance pod
Operators: Argentina, Dubai, Italy, Malaysia, Nigeria, Peru

Aermacchi MB.339A of 1ª Escuadrilla, 4ª Escuadra, Argentinian navy. The aircraft perform the dual roles of continuation training and ground support.

Aero L-29 Delfin

An Aero L-29 Delfin of the Soviet air force, one of over 2,000 supplied to this air arm.

The first studies leading to the Aero L-29 were made in 1955, and the prototype flew for the first time on 5 April 1959, powered by a Bristol Viper turbojet; the second prototype, of July 1960, had as powerplant the nationally-designed Motorlet M 701. A year later the Delfin (dolphin) was subjected to competitive evaluation against the Yakovlev Yak-30 and PZL-Mielec TS-11 Iskra. As a result, all Warsaw Pact countries except Poland decided to adopt the Delfin as their standard basic and advanced jet trainer. The first production Delfin was completed in April 1963, and approximately 3,500 had been built before the run ended some 12 years later. More than 2,000 were supplied to the Soviet air force, whose L-29s were assigned the NATO reporting name 'Maya'.

The L-29 is simple to fly and uncomplicated. Flight controls are manual, with generous flaps and a perforated airbrake on each side of the rear fuselage. Both occupants are provided with ejection seats though, unlike modern trainers, the

instructor's (rear) seat is no higher than the pupil's. Runway requirements are modest, and it can operate from grass, sand or waterlogged airstrips.

Aero also built a small batch of the single-seat L-29A Delfin Akrobat, for aerobatic displays, but this did not go into large-scale production. Neither did an attack version, the L-29R, but the standard L-29 was supplied to a number of countries (including Egypt) equipped for this role, with a modest weapon load on two underwing pylons.

Specification
Aero L-29 Delfin

Type: tandem two-seat basic and advanced jet trainer
Powerplant: one 890-kg (1,962-lb) thrust Motorlet M 701 VC-150 or S-50 turbojet engine
Performance: maximum speed at 5000 m (16,405 ft) 655 km/h (407 mph); maximum speed at sea level 610 km/h (379 mph); initial rate of climb 840 m (3,756 ft)/minute; service ceiling 11000 m (36,090 ft); range on internal fuel 640 km (398 miles); maximum range with two underwing drop-tanks 895 km (556 miles)
Weights: empty 2280 kg (5,027 lb); maximum take-off 3280 kg (7,231 lb)

Dimensions: span 10.29 m (33 ft 9.1 in); length 10.81 m (35 ft 5.6 in); height 3.13 m (10 ft 3.2 in); wing area 19.80 m^2 (213.1 sq ft)
Armament provision for two 7.62-mm (0.3-in) gun pods, two 100-kg (220-lb) bombs, eight air-to-air rockets or two drop tanks on two underwing pylons
Operators: Afghanistan Bulgaria, China, Czechoslovakia, Egypt, East Germany, Guinea, Hungary, Indonesia, Iraq, Mali, Nigeria, Romania, Soviet Union, Syria, Uganda, Vietnam

Aero L-39

Designed before the Soviet intervention in Czechoslovakia in 1968, the L-39 is now well on the road to emulating its predecessor, the L-29 Delfin, as a trainer for Warsaw Pact and other air forces. Aero began with three prototypes, the second of these making the type's maiden flight on 4 November 1968, but development was somewhat protracted. By late 1970, at which time five flying prototypes had been completed, modified intakes of greater length and increased entry area had been introduced, but almost two years elapsed before series production began in late 1972. By 1985 more than 2,500 had been ordered, almost all of which have been completed. The basic L-39, for elementary and advanced jet training, has been supplied in quantity to all of the Warsaw Pact air forces, except Poland, as a successor to the L-29; it began to enter service in 1974. When equipped for weapons training, the two-seater is known as the L-39Z. A single-seat armed variant, for use in the light close-support and ground-attack roles, is designated L-39D; Iraq operates this version.

The L-39 offers a significant improvement in performance over its predecessor. Tandem seating on zero-height ejection seats is arranged with the rear (instructor's) seat elevated to improve his view forward. Construction is modular, the airframe being broken down into only three major sub-assemblies (one-piece wing, fuselage and rear fuselage/tail unit) to make for rapid maintenance and overhaul. A first-class all-round view is available from both pressurized cockpits, and in the L-39D the rear seat is removed and replaced by avionics or additional fuel. A small auxiliary power unit (APU) makes the aircraft independent of ground power.

Specification
Type: tandem two-seat basic and advanced jet trainer (L-39), weapons

trainer (L-39Z), and single-seat light ground-attack aircraft (L-39D)
Powerplant: one 1720-kg (3,792-lb) thrust Walter Titan turbofan engine (Ivchenko AI-25-TL licence-built by Motorlet)
Performance: maximum speed at sea level 700 km/h (435 mph); trainer, clean at 6000 m (19,685 ft) 780 km/h (485 mph), L-39D at same altitude with four rocket pods 630 km/h (391 mph); initial rate of climb, trainer 1320 m (4,331 ft)/minute, L-39D 960 m (3,150 ft)/minute; service ceiling, trainer 11500 m (37,730 ft), L-39D 9000 m (29,530 ft); range on internal fuel, trainer 850 km (528 miles), L-39D with rocket pods 780 km (485 miles); maximum range with two drop tanks and no weapons 1600 km (994 miles)

Aero L-39 Albatros

East Germany is one of several Warsaw Pact nations operating the Aero L-39 Albatros trainer.

Weights: empty 3330 kg (7,341 lb); maximum take-off, trainer with wingtip tanks empty 4570 kg (10,075 lb), L-39D with four rocket pods 5270 kg (11,618 lb)
Dimensions: span 9.46 m (31 ft 0.4 in); length 12.32 m (40 ft 5 in);

height 4.72 m (15 ft 5.8 in); wing area 18.80 m^2 (202.4 sq ft)
Armament: (L-39D): up to 1100 kg (2,425 lb) of weapons on four underwing points, including bombs of up to 500 kg (1,102 lb), pods of 57-mm (2.24-in) or 130-mm (5.12-in)

rockets, gun pods, a single five-camera reconnaissance pack, or two drop-tanks; underfuselage centreline point for podded gun, normally a Soviet GSh-23L of 23-mm (0.91-in) calibre with 180 rounds

Operators: Afghanistan, Bulgaria, Cuba, Czechoslovakia, East Germany, Hungary, Iraq, Libya, Guinea-Bissau, Nicaragua, Romania, Soviet Union, Syria, Vietnam

Aerospace Airtrainer CT/4

The Airtrainer stemmed from Australian Henry Millicer's winning design in a Royal Aero Club Competition of 1953, which was built in Australia by Victa as a civil tourer/trainer. In 1971 the New Zealand company Aero Engine Services Ltd (AESL), later amalgamated with Air Parts, bought the rights and decided to convert the design to the military training role. One of the first priorities was to strengthen the airframe; the prototype, re-stressed to +6 and −3g and featuring a hinged canopy, side-by-side seating for two (with an optional third seat aft) and stick-type controls, flew on 23 February 1972.

Some 80 Airtrainers now equip three air forces. They are used for training, although versions with underwing hardpoints have been evaluated. Australia operates 31 with No. 1 Flight Training School at

Point Cook, and has a further six at the Central Flying School, East Sale. Six, with underwing hardpoints, are used to train forward air controllers. New Zealand bought 13, and a further six airframes for spares, all based at Wigram. The Royal Thai air force trains its pilots on 24 Airtrainers based at Korat AB.

Specification
Type: two/three-seat aerobatic trainer
Powerplant: one 157-kW (210-hp) Teledyne Continental IO-360-H flat-six piston engine
Performance: maximum speed at sea level 286 km/h (178 mph); initial rate of climb 411 m (1,350 ft)/minute; service ceiling 5455 m (17,900 ft); range, internal fuel, 10 per cent reserves, 75 per cent power at 1525 m (5,000 ft) 1271 km (790

miles)
Weights: empty equipped 675 kg (1,488 lb); maximum take-off 1089 kg (2,400 lb)
Dimensions: span 7.92 m (26 ft 0 in); length 7.06 m (23 ft 2 in); height 2.59 m (8 ft 6 in); wing area 11.98 m^2 (129.0 sq ft)
Armament: two underwing hardpoints for a variety of light stores

The Royal New Zealand Air Force currently operates a small number of Airtrainer CT/4 two-seat trainers from Wigram. Export operators are Australia and Thailand, the former including six aircraft fitted with underwing hardpoints for the training of forward air controllers, while the Royal Thai air force has 24.

Operators: Australia, New Zealand, Thailand

Aérospatiale Alouette II and Lama/SE.313/SA.315/SA.318

As the first successful turboshaft-powered helicopter, the Alouette (lark) II achieved a long production run, being adopted by military and civilian agencies throughout the world. Produced by Sud-Est from the original SE 3120 Alouette I prototype of 1951, the SE 3130 Alouette II first flew on 12 March 1955, becoming the SE 313B two years later on formation of Sud-Aviation. This version, powered by a Turboméca Artouste of 268 kW (360 shp), entered large-scale production in 1956, with initial deliveries going to the French air force, army, navy, Gendarmerie and the Protection Civile. Germany adopted the Alouette II for its army and for Luftwaffe pilot training, and an eventual total of 17 was acquired by the British Army, which designated it Alouette AH.Mk 2 (a few surviving in service on Cyprus). In January 1961 the SA 318C flew with a Turboméca Astazou IIA derated from 395 kW (530 shp) to 268 kW (360 shp). This Alouette II Astazou was bought by several earlier customers, France and Germany included. In all, 419 Alouette II and Alouette II Astazou helicopters were built to French government order, and another 25 were the subject of miscellaneous acquisitions. Today, about 180 of the 230 supplied to the French army are in service (many for training); the Armée de l'Air still has 35 of its 139 for communications duties; and the Aéronavale has 15 of 26 left for second-line work. From 301 supplied

to West Germany, several have recently been transferred to Turkey or assigned to the Bundesgrenzschutz paramilitary border guard, whilst others will remain in service as liaison aircraft for a further decade. French production of SE 313s and SA 318s totalled 1,305, of which 963 were military. In response to an Indian request for improved 'hot and high' performance, the SA 315B Lama was flown on 17 March 1969, this combining a reinforced SE 313 airframe with the Artouste engine and rotor system of the parallel Alouette III line. French manufacture was mainly for the civil market (291 of the 374 built), but military users included the armed forces of Argentina, Chile and Ecuador, whose areas of responsibility include the Andes. Demonstrations to India in 1969 included the highest take-off at operational weight to be recorded by a helicopter at 7500 m

(24,605 ft). This was sufficient to secure a contract from the IAF for equipment of four army co-operation squadrons, production being undertaken by Hindustan Aeronautics at Bangalore as the Cheetah after supply of the first 20 kits by France. About 150 have been built, half of them for the IAF. A third production centre was established in Brazil during 1979 by Helibras – a joint company in which Aérospatiale has a 45 per cent holding – and assembly begun of an initial batch of 30. These are designated HB 315B Gavao (Lama), and six have been supplied to the Bolivian Air Force.

Specification
Aérospatiale SA 315B Lama
Type: five-seat utility helicopter
Powerplant: one 649-kW (870-shp) Turboméca Artouste IIIB turboshaft engine
Performance: maximum cruising

Aérospatiale SA.318C Alouette II of the French army.

speed 192 km/h (119 mph); initial rate of climb 330 m (1,083 ft)/minute; hovering ceiling (in ground effect) 5050 m (16,570 ft)
Weights: empty 1021 kg (2,251 lb); maximum take-off with slung load 2300 kg (5,071 lb)
Dimensions: main rotor diameter 11.02 m (36 ft 1.9 in); fuselage length 10.24 m (33 ft 7.1 in); height 3.09 m (10 ft 1.7 in); main rotor disc area 95.38 m^2 (1,026.69 sq ft)
Armament: none
Operators: Angola*, Argentina*, Belgium, Benin, Bolivia*, Brazil, Cameroun, Central African Republic, Chile*, Colombia*, Congo Republic, Djibouti, Dominican Republic, Ecuador*, France, West Germany, Guinea-Bissau, India*, Indonesia, Ivory Coast, Lebanon, Malaysia*, Mexico, Peru*, Senegambia, Sweden, Switzerland, Togo*, Tunisia, Turkey UK, Zimbabwe (*Lama)

Aérospatiale Alouette III/SA.316B/SA.319B

The reliability and sales success of the Alouette II prompted Sud-Aviation to initiate development of an advanced version; the incorporation of a more powerful turboshaft engine and improved aerodynamics was considered essential to give greater payload capability and enhanced performance and, at the

same time, the opportunity was taken to introduce new equipment. Initially designated the SE.3160, the prototype Alouette III incorporated a larger and more enclosed cabin than that of its predecessor, able to carry a pilot and six passengers with baggage holds for luggage and parcels, or a pilot and six equip-

ped troops. In a casevac role two stretchers and two sitting casualties or medical attendants could be accommodated behind the pilot, or alternatively the six seats could easily be removed for the carriage of cargo; there was also provision for an external sling for loads of up to 750 kg (1,653 lb).

A prototype was flown for the first time on 28 February 1959 and early production examples followed in 1961. The initial production SA.316A helicopter, built for home and export markets, became the subject of a licence agreement with Hindustan Aeronautics Ltd in India. Subsequent development produced the main-production SA.316B, first flown on 27 June 1968, which introduced the Turboméca Artouste IIIB turboshaft with uprated main and tail rotor transmissions, and was able to carry more payload. Last of the Artouste-powered Alouette IIIs was the SA.316C, built in only small numbers with an Artouste IIID engine. The SA.316B was also the subject of licence agreements with the Swiss Federal Aircraft Factory, and ICA-Brasov in Romania where the type continued in production in 1984. The capability of the SA.316B soon led to two-seat military versions deployed in a variety of roles, with a range of weapon options that made them suitable for light attack and ASW. As with the Alouette II, a version was introduced with the Turboméca Astazou turboshaft, this being the SA.319B Alouette III with a 649-kW (870-shp) Astazou XIV derated to 447 kW (600 shp). Both the SA.316B and SA.319B are included in the production total of 1,453, and a considerable number of these were for military service.

The Alouette III has been widely used by South African forces throughout their operations against guerrillas in Angola.

Specification
Aérospatiale SA.316B Alouette III (standard version)
Type: general-purpose civil/military helicopter
Powerplant: one 649-kW (870-shp) Turboméca Artouste IIIB turboshaft engine derated to 425 kW (570 shp)
Performance: maximum cruising speed 100 kts (185 km/h; 115 mph) at sea level; initial rate of climb 260 m (853 ft) per minute; service ceiling 3200 m (10,499 ft); range with maximum fuel at optimum altitude 540 km (336 miles)
Weights: empty 1143 kg (2,520 lb); maximum take-off 2200 kg (4,850 lb)
Dimensions: main-rotor diameter 11.02 m (36 ft 1.9 in); length, rotors

turning, 12.84 m (42 ft 1.5 in); height 3.00 m (9 ft 10.1 in); main-rotor disc area 95.38 m² (1,026.7 sq ft)
Armament: (military versions) one 7.62-mm (0.3-in) AA52 machine-gun with 1,000 rounds, tripod-mounted in cabin to fire to starboard (maximum four crew); or one 20-mm MG 151/20 or GIAT cannon; or four AS.11 or two AS.12 missiles on external mounts (maximum two

crew); or two Mk 44 homing torpedoes, or one Mk 44 torpedo and MAD gear (ASW role); or two AS.12 missiles (ASV role)
Operators: Abu Dhabi, Angola, Argentina, Austria, Belgium, Burkina-Faso, Chile, Congo, Denmark, Dominican Republic, Ecuador, Eire, El Salvador, Ethiopia, France, Gabon, Ghana, Guinea-Bissau, Guinea Republic,

The Alouette III has seen widespread service in a variety of roles, proving the type's versatility.

India, Indonesia, Iraq, Ivory Coast, South Korea, Libya, Malawi, Malaysia, Mexico, Morocco, Mozambique, Netherlands, Pakistan, Peru, Portugal, Rwanda, South Africa, Spain, Switzerland, Tunisia, Venezuela, Zaïre, Zimbabwe

Aérospatiale Caravelle/SE.210

One of two Aérospatiale Caravelle IIIs flown on Elint duties by Sweden's National Defence Research Institute on behalf of the Swedish military.

Unique in its day as the first rear-engined jet transport, the Caravelle has been used by air forces for transport and trials. Designed by Sud-Est, the first SE 210 Caravelle flew on 27 May 1955, and the last of 280 production aircraft was delivered in March 1973. France is the major military operator, with nine in use at various times. First in service was the Caravelle III, with Avon Mk 527s of 5171 kg (11,400 lb) thrust. Two were supplied to the VIP transport squadron (GLAM) in 1962-63, of which one was retired and the other passed to the CEV experimental establishment. A third served as test bed for the SNECMA M53 turbofan (Mirage 2000 engine). In the Caravelle VIR, thrust-reversing Avon 533Rs of 5715 kg (12,600 lb) were fitted. The GLAM and CEV obtained one each, the latter remaining in service as a flying avionics laboratory. Pratt & Whitney JT8D-1 turbofans of 6350 kg (14,000 lb) thrust power the single Caravelle 10R operated by GLAM.

This version was developed into the Caravelle 11R passenger/freight model with a 0.93 m (3 ft 0.6 in) fuselage exension and 'hot and high' JT8D-7 engines. Three 11Rs are based in Tahiti. Among export customer, only Jugoslavia bought a new aircraft (a VIN in 1969 – since sold), but in 1973-75 the Argentine air force acquired three VINs from Aerolineas. Two Caravelle IIIs supplied to Sweden in 1971 are used by the National Defence Research Institute as ECM and Elint aircraft, and another III serves the President of the Central African Republic.

Civil-registered aircraft are operated by the governments of Chad (VIR), Mauretania (VIR), Rwanda (III) and Senegambia (III).

Specification
Aérospatiale Caravelle 11R
Type: short-range passenger transport
Powerplant: two 6350-kg (14,000-lb) thrust Pratt & Whitney JT8D-7 turbofan engines
Performance: maximum cruising speed 800 km/h (497 mph); range 2300 km (1,429 miles) with full payload

Weights: empty 28841 kg (63,584 lb); maximum take-off 54000 kg (119,050 lb); maximum payload 9095 kg (20,051 lb)
Dimensions: span 34.30 m (112 ft 6.4 in); length 32.71 m (107 ft 3.8 in); height 8.72 m (28 ft 7.3 in); wing area 146.7 m² (1,579.12 sq ft)
Armament: none
Operators: Central African Republic, France, Sweden

Aérospatiale Dauphin/Dauphin 2/SA.360/SA.365

In the early 1970s Aérospatiale began development of a helicopter to supersede the Alouette III. The initial version, the Aérospatiale SA.360 Dauphin, had a four-blade main rotor, a 13-blade 'fenestron' (ducted-fan anti-torque rotor), tail-wheel landing gear and standard accommodation for a pilot and nine passengers. Initial powerplant was a 731-kW (980-shp) Turboméca Astazou XVI turboshaft, which

powered the first prototype (F-WSQL) flown on 2 June 1972. After 180 flights F-WSQL was re-engined by the 783-kW (1,050-shp) Astazou XVIIIA adopted for production aircraft. The roomy cabin gave scope for optional layouts seating pilot, co-pilot and 12 passengers, a two-crew ambulance with four stretchers and medical attendant, a mixed traffic version with six passengers and 2.50 m³ (88.3 cu ft) of cargo space,

and executive interior for four to six passengers. A military SA.361H for assault transport and anti-tank operations was developed as a private venture, powered by a 1044-kW (1,400-shp) Astazou XXB, but it was soon clear that for both civil and military use the future lay with a twin-turbine version, designated SA.365C Dauphin 2.
The prototype SA.365C (F-WVKE), with two 485-kW (650-shp)

Turboméca Arriel turboshafts, flew on 24 January 1975; production aircraft had 492-kW (660-shp) Arriel 1As. Variants include SA.365N, similar to the SA.365C except for retractable tricyle landing gear, 529-kW (710-shp) Arriel 1C engines and a high proportion of composite material in its structure. The SA.365F, ordered first by Saudi Arabia (24) is intended primarily for an anti-shipping role, but four are for SAR and have Omera ORB 32 search radar; for the anti-shipping role they carry Thomson-CSF

Aérospatiale Dauphin/Dauphin 2/SA.360/SA.365 (continued)

Agrion 15 radar, a Crouzet MAD 'bird', and two or four AS.15TT air-to-surface missiles. Ireland has ordered five SA.365Fs for SAR and fishery surveillance. Currently under development is a dedicated military version, the SA.365M with the new 634-kW (850-shp) Turboméca TM 333-1M turboshafts and crashworthy fuel tanks to provide a high-speed assault transport for 8 to 10 troops. Armament on the outriggers for anti-armour or fire support can include SNEB 68-mm (2.68-in) rockets, HOT anti-tank missiles and GIAT 20-mm cannon pods.

Specification
Aérospatiale SA.365F Dauphin 2
Type: anti-ship and SAR helicopter
Powerplant: two 522-kW (700-shp) Turboméca Arriel 520M turboshaft engines
Performance: maximum cruising

Licence-built examples of the SA.365 Dauphin II are known as the Harbin Z-9 Haitun, in operation with the Air Force of the People's Liberation Army of China.

speed at sea level 136 kts (252 km/h; 157 mph); economic cruising speed at optimum altitude 140 kts (259 km/h; 161 mph); hovering ceiling in ground effect 2600 m (8,530 ft); range with maximum fuel at sea level 898 km (558 miles); endurance with maximum fuel and

four missiles 2 hour 45 minutes
Weights: empty 2166 kg (4,775 lb); maximum take-off 4000 kg (8,818 lb)
Dimensions: main-rotor diameter 11.93 m (39 ft 1.7 in); length, rotors turning, 13.74 m (45 ft 0.9 in); height 3.51 m (11 ft 6.2 in); main-rotor disc area 111.78 m² (1,203.25 sq ft)

Armament: two or four AA.15TT air-to-surface all-weather anti-ship missiles
Operators: Cameroon, China, Dominican Republic, Eire, France, Hong Kong, Ivory Coast, Malawi, Morocco, Saudi Arabia, Sri Lanka, USA

Aérospatiale Ecureuil/AS.350

Produced as a successor to the Alouette II by Aérospatiale, the Ecureuil (squirrel) has not achieved the level of military sales attained by its predecessor, though civil sales are tremendous. A few air arms have adopted the Ecureuil for utility duties, and the number is likely to increase now that armed versions have been announced. Following the first flight of a prototype on 27 June 1974, both single- and twin-engined Ecureuils are being produced in parallel. The former are the AS 350B Ecureuil powered by a Turboméca Arriel of 478 kW (641 shp), and the AS.350C and AS.355D AStar, available only in North America, with the Lycoming LTS 101. A pair of Allison engines are installed in AS.355 models, known as TwinStars in North America and Ecureuil 2s elsewhere. Progressive increases in take-off weight have been attained through rotor blade and other modifications in the AS.355E, 355F and 355F1. Fitted with outrigger pylons for light armament, the designation becomes AS.350L or AS.355M according to powerplant. A version car-

rying Hughes TOW anti-tank missiles is offered, and a naval model for operation from ships' platforms is also in prospect, equipped with 360° radar. The Armée de l'Air took delivery of the first of 50 AS.355Fs in 1984, and from the eighth onwards (to be delivered in 1986), these will have two of the new 330-kW (443-shp) Turboméca TM.319 engines. AS.350Bs are being supplied to the Gendarmerie, which requires 30, and an order is also expected from the Aéronavale for 20. Australia ordered 24 AS.350Bs in 1982-83 for 1984 delivery to the RAAF (18) and Navy (six) to be used for pilot training. Singapore received six Ecureuils in 1983 for naval use. Small numbers of Ecureuils have been sold in Africa to the Tunisian Gendarmerie (AS.350s and 355s), Central African Republic (AS.350) Bophuthatswana Defence Force (AS.355) and Djibouti (two AS.355s). In Brazil, Helibras has been assembling the HB 350B Esquilo (squirrel) since 1979, and delivered six to the Brazilian navy for operation from the platforms of sur-

vey ships, followed by three more in early 1984.

Specification
Aérospatiale AS.355M Ecureuil 2
Type: six-seat utility helicopter
Powerplant: two 317-kW (425-shp) Allison 250-C20F turboshaft engines
Performance: maximum cruising speed 230 km/h (143 mph); initial rate of climb 456 m (1,496 ft)/ minute; hovering ceiling (in ground effect) 2350 m (7,710 ft)
Weights: (AS.355F1): empty 1288 kg (2,840 lb); maximum take-off 2500 kg (5,512 lb)

Dimensions: main rotor diameter 10.69 m (35 ft 0.9 in); fuselage length 10.91 m (35 ft 9.5 in); height 3.15 m (10 ft 4 in); main rotor disc area 89.75 m² (966.12 sq ft)
Armament: anti-tank missiles, 20-mm (0.79-in) cannon or 68-mm (2.68-in) rocket pods
Operators: Australia, Bophuthatswana, Brazil, Central African Republic, Djibouti, France, Gabon, Malawi, Singapore, Tunisia

Six AS.350E Ecureuil light general-purpose helicopters were delivered to No. 123 Squadron, Royal Singapore air force, for training duties.

Aérospatiale Epsilon/TB.30

The Epsilon military basic trainer was originated by the SOCATA light-aircraft subsidiary of Aérospatiale in 1977. It was based on the civil TB.10 Tobago, but was later redesigned to an Armée de l'Air specification as an entirely new tandem-seater under the designations TB.30A 194 kW (260 hp) and TB.30B 224 kW (300 hp). The TB.30B variant was awarded a French Government development contract in June 1979. The first prototype was flown on 22 December 1979 at Tarbes, an all-metal aircraft with a large aft-sliding canopy and retractable tricycle landing gear. It was designed to provide the student with a cockpit layout and flying characteristics which would prepare him for the Alpha Jet and Mirage. The Epsilon is fully aerobatic with an inverted-flight fuel system. It has a Hartzell two-blade constant-speed propeller. The second prototype, flown on 12 July 1980, was to production standard with increased wingspan and rounded tips, a redesigned rear fuselage, low-set tailplane and narrow-chord fin and rudder. The Armée de l'Air ordered 150, the first of which flew on 29 June

1983. It was delivered to CEAM at Mont de Marsan for establishment of the training syllabus. The type went into service with GE 315 at Cognac in spring 1984. Some 55 Epsilons will be operated at Cognac to weed out unsuitable pupils and supplant the Magister in the basic role, taking the student through the first 70 hours of training. This stage includes 23 hours up to solo standard, blind flying, IFR and night navigation, and 12 hours of formation flying. The Epsilon was being delivered to Salon de Provence during 1985 and to Aulnat in 1986. A tour of former colonies in Africa led to an order from Togo for three of the armed version, for export only, with four underwing hardpoints. In late

1985 a TB.30C development aircraft was to be flown with a 261-kW (350-shp) Turboméca turboprop based on the TM 319 helicopter powerplant.

Specification
Aérospatiale TB.30B Epsilon
Type: two seat primary trainer
Powerplant: one 224-kW (300-hp) Lycoming AEIO-540-L1-B5D flat-six piston-engine
Performance: maximum speed 520 km/h (323 mph); maximum cruising speed 370 km/h (230 mph); initial rate of climb 516 m (1,693 ft)/minute; service ceiling 6700 m (21,980 ft); maximum range 1350 km (839 miles)
Weights: empty 878 kg (1,936 lb); maximum take-off 1250 kg (2,756 lb)
Dimensions: span 7.92 m (25 ft 11.8 in); length 7.59 m (24 ft

Aérospatiale TB.30 Epsilon

10.8 in); height 2.66 m (8 ft 8.7 in); wing area 9.60 m² (103.34 sq ft)
Armament (export version): 262 kg (578 lb) of bombs, fuel or rockets on four underwing hardpoints
Operators: France, Togo

Aérospatiale TB.30B Epsilon trainer of Groupement Ecole 315, French air force.

Aérospatiale Gazelle/SA.340

Aérospatiale's project X.300 was originated to meet a requirement of the French Aviation Légère de l'Armée de Terre (ALAT) for a light observation helicopter. Soon designated Aérospatiale SA.340, it had kinship with the SA.318C Alouette II, using its Astazou II powerplant and transmission system, but introduced a new enclosed fuselage seating two pilots side-by-side, a rigid main rotor of the type developed by Bölkow in Germany, and the patented 'fenestron' tail rotor. Early interest in the UK for this helicopter to equip its three armed forces brought about the Anglo-French helicopter manufacturing/development programme, finalized on 22 February 1967, under which Gazelles were produced jointly with Westland Helicopters of Yeovil, Somerset.

The first SA.340 prototype (with conventional Alouette III tail rotor) was flown initially on 7 April 1967. The second (with fenestron) flew on 12 April 1968, these being followed by four pre-production SA.341 Gazelle helicopters. The first production SA.341, flown on 6 August 1971, introduced the uprated Turboméca Astazou IIIA engine, a lengthened cabin and enlarged tail surfaces. Initial versions included the SA.341B for the British Army (Gazelle AH.Mk 1) with Astazou IIIN engine, the similar SA.341C for the Royal Navy (Gazelle HT.Mk 2), the SA.341D trainer for the Royal Air Force (Gazelle HT.Mk 3), RAF communications SA.341E (Gazelle HCC.Mk 4), and with the Astazou IIIC engine the original SA.341F for the French ALAT. A military version known as the SA.341H was built under licence by SOKO in Jugoslavia, and these can operate in the anti-tank role with an armament of four Soviet AT-3 'Sagger' missiles.

Early utilization of the SA.341 Gazelle confirmed that this new helicopter was a worthy successor to the Alouette II, with satisfactory reliability and a minimum of teething troubles as a result of adoption of the powerplant and transmission of the Alouette and the rigid main rotor which Bölkow had developed for the MBB BO 105 five-seat helicopter. Aérospatiale realized very quickly that a version with uprated powerplant would offer greater scope for military use, and the changed designation Aérospatiale SA.342 emphasized the difference. For hot and dry conditions the SA.342K Gazelle introduced the 649-kW (870-shp) Turboméca Astazou XIVH with momentum-separation shrouds over the engine inlets. First flown on 11 May 1973, the type secured initial sales to Kuwait for a total 20 aircraft for use in attack and AOP roles.

Continuing development resulted in civil and military versions designated SA.342J and SA.342L respectively, these having the Astazou XIV turboshaft engine and an improved 'fenestron', both being approved at a higher gross weight. The military SA.342L was able to carry a wide range of weapons, and from this version has been developed for ALAT the more capable SA.342M especially for the anti-tank role. It has the Astazou XIVM with automatic start-up, and an instrument panel specified by ALAT. Crew workload in the demanding anti-tank role has been reduced by provision of an autopilot, a self-contained navigation system plus Doppler radar, and an SFIM APX 397 gyrostabilized sight for guiding the four Euromissile HOT anti-tank missiles. Current ALAT procurement of the SA.342M covers a total of 128 aircraft, delivered since 9 June 1980. The SA.342L is being built under licence by SOKO in Jugoslavia and the Arab British helicopter Company at Helwan in Egypt.

Iraq has made effective use of its Aérospatiale SA.342L Gazelles in its war with Iran. Equipment includes stub-wing mounted HOT missiles.

Specification
Aérospatiale SA.341 Gazelle
Type: five-seat utility helicopter
Powerplant: one 440-kW (590-shp) Turboméca Astazou IIIA turboshaft engine
Performance: maximum cruising speed 142 kts (264 km/h; 164 mph) at sea level; initial rate of climb 540 m (1,772 ft) per minute, service ceiling 5000 m (16,404 ft); hovering ceiling in ground effect 2850 m (9,350 ft); range with pilot and payload of 500 kg (1,102 lb) 360 km (224 miles); range with maximum fuel 670 km (416 miles)
Weights: empty 917 kg (2,022 lb); maximum take-off 1800 kg (3,968 lb)
Dimensions: main-rotor diameter 10.50 m (34 ft 5.4 in); length, rotors turning, 11.97 m (39 ft 3.3 in); height 3.18 m (10 ft 5.2 in); main-rotor disc area 86.59 m² (932.08 sq ft)
Armament: usually four (sometimes two) anti-tank missiles
Operators: Abu Dhabi, Burundi, Cameroon, Chad, Ecuador, Egypt, Eire, France, Guinea Republic, Iran, Jordan, Jugoslavia, Kenya, Lebanon, Libya, Morocco, Qatar, Rwanda, Senegambia, Syria, Trinidad & Tobago, UK

The anti-armour version of the Gazelle carries four Euromissile HOT weapons.

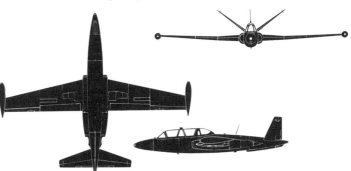

Aérospatiale (Fouga) CM.170 Magister/CM.175 Zéphyr

The Aérospatiale CM.170 Magister (originally produced by Air Fouga and then by Potez) was designed to meet an Armée de l'Air requirement for a jet trainer (the first in the world). The prototype flew on 23 July 1952 and the first production aircraft on 13 January 1954. Since then over 400 Magisters have been produced for the Armée de l'Air.

The Magister is all-metal and has long-span wings that incorporate single-slotted flaps and upper and lower airbrakes, and an unusual butterfly-type tail with surfaces separated by 110°. The tandem cockpits are pressurized, but ejection seats are not fitted. A carrier-equipped version was produced for the Aéronavale designated CM.175 Zéphyr. Two prototypes and 30 production aircraft were built to provide naval pilots with initial experience of carrier operations. Magisters were made under licence in West Germany by Flugzeug-Union-Sud for the Luftwaffe. Valmet OY in Finland built 62 under licence (in addition to 18 purchased from France) and Israel Aircraft Industries also acquired manufacturing rights for the type, building many for light tactical use as well as training. Total production was 916.

Specification
Aérospatiale CM.170 Magister
Type: two-seat jet trainer and light attack aircraft
Powerplant: two 400-kg (882-lb) thrust Turboméca Marboré IIA turbojet engines
Performance: maximum speed 715 km/h (444 mph) at 9150 m (30,020 ft); initial rate of climb 1020 m (3,345 ft)/minute; service ceiling 11000 m (36,090 ft); range 925 km (575 miles)
Weights: empty equipped 2150 kg (4,740 lb); maximum take-off 3200 kg (7,055 lb)
Dimensions: span over tip-tanks 12.15 m (39 ft 10.3 in); length 10.06 m (33 ft 0 in); height 2.80 m (9 ft 2.2 in); wing area 17.30 m² (186.22 sq ft)
Armament: two 7.5-mm (0.295-in) or 7.62-mm (0.3-in) machine guns in the nose, each with 200 rounds; underwing loads include two Matra Type 181 pods each with 18 37-mm (1.46-in) rockets, two launchers each mounting seven 68-mm (2.68-in) rockets, two 50-kg (110-lb) bombs, or two Nord AS.11 missiles
Operators: Algeria, Bangladesh, Belgium, Cameroon, El Salvador, Finland, France, Gabon, Guatemala, Ireland, Israel, Lebanon, Libya, Morocco, Rwanda, Senegambia, Togo, Uganda

Aérospatiale (Fouga) CM.170 Magister.

Still flying with the Algerian air force are 28 Magisters operating in the armed training role.

Aérospatiale/Westland SA.330 Puma

The Aérospatiale/Westland SA.330 Puma is serving with close to 40 air arms, plus many civil concerns. Conceived by Sud-Aviation in response to a French army (ALAT) requirement for an all-weather medium tactical helicopter capable of carrying 16 troops or equivalent freight, the SA.330 was also selected in 1967 for the Royal Air Force, becoming the subject of a joint production agreement between Aérospatiale and Westland. The first prototype flew in April 1968; the first production SA.330B for ALAT followed in September 1968, with deliveries beginning in March 1969. Deliveries to the RAF began in 1971, initially to No. 240 Operational Conversion Unit and later to No. 33 Sqn, RAF machines being part-built at the former Fairey works at Hayes and assembled at Yeovil.

The export SA.330C achieved considerable success but was supplanted by the SA.330H and SA.330L, with more power, composite blades and other improvements. Many operators have introduced large engine air filters and some have added de-icing systems. Licence production and assembly agreements were reached with ICA in Romania, Nurtanio in Indonesia and with Helibras in Brazil. The AS 332 Super Puma is described separately.

Specification
Aérospatiale/Westland SA.330L Puma
Type: medium transport/assault helicopter
Powerplant: two 1174.5-kW (1,575-hp) Turboméca Turmo IVC turboshaft engines
Performance: maximum cruising speed 271 km/h (168 mph); service ceiling 6000 m (19,685 ft); maximum range at normal cruising speed 570 km (354 miles)
Weights: empty 3615 kg (7,970 lb); maximum take-off 7400 kg (16,314 lb)
Dimensions: main rotor diameter 15.00 m (49 ft 2.6 in); length 18.15 m (59 ft 6.6 in); height 5.14 m (16 ft 10.4 in); main rotor disc area 176.72 m^2 (1,902.21 sq ft)
Armament: usually none, but various cannon, machine-guns, rockets and missiles can be carried
Operators: (military) Abu Dhabi, Algeria, Argentina, Belgium, Brazil, Cameroun, Chad, Chile, Congo, Ecuador, Ethiopia, France, Gabon, Guinea Republic, Indonesia, Iraq, Ivory Coast, Kenya, Kuwait, Lebanon, Malawi, Mexico, Morocco, Nepal, Nigeria, Pakistan, Portugal, Romania, Senegambia, South Africa, Spain, Sudan, Togo, Tunisia, UK, Zaire

Wearing the codes of Overseas Transport Squadron 58, this SA.330H is typical of Pumas operated by the French air force's Transport Aérien Militaire. This model is a marriage of the basic Puma airframe with the 1,575-shp Turmo IVC powerplant for 'hot and high' operations. Up to 20 troops can be carried in one of a multitude of operational roles. This machine operates in the SAR role from Pointe-à-Pitre in the French Caribbean.

Aérospatiale Super Frelon/SA.321

To meet requirements of the French armed services for a medium transport helicopter, on 10 June 1959 Sud-Aviation flew the prototype SA.3200 Frelon (hornet). Powered by three Turboméca Turmo IIIB turboshaft engines, the SA.3200 had large external fuel tanks that left the interior clear for a maximum 28 troops, and a swing-tail fuselage to simplify loading cargo. However, development was terminated in favour of a larger and more capable helicopter designed in conjunction with Sikorsky in the USA, and with Fiat in Italy producing the main gearbox and transmission. What was to become Europe's largest production helicopter clearly shows Sikorsky influence, the rotor system being of Sikorsky design, and in its watertight hull suitable for amphibious operation. Two military prototypes of the Super Frelon were built, the SA.3210.01 troop transport (F-ZWWE) flown on 7 December 1962, and the SA.3210.02 maritime version for the Aéronavale (F-ZWWF) on 28 May 1963.

The designation changed to SA.321 for the four pre-production aircraft, and the initial production version was the maritime SA.321G ASW aircraft for the Aéronavale. These were identifiable by having a small stabilizing float incorporating search radar mounted to the support structure of each main unit of the tricycle landing gear. This model was followed by the SA.321F commercial airliner for 34-37 passen-

gers, and the commercial SA.321J intended for use as a 27-seat passenger transport, as a cargo carrier with 4000-kg (8,818-lb) internal or 5000-kg (11,023-lb) external load, or for other utility purposes such as firefighting; it was later superseded by the SA.321Ja operating at a higher gross weight. Non-amphibious military export versions included 12 SA.321K transports for Israel, and the similar SA.321L supplied to the People's Republic of China, Libya (9) and South Africa (16). When production ended in 1983 a total of 99 Super Frelons had been built.

Specification
Aérospatiale SA.321G Super Frelon
Type: ASW helicopter
Powerplant: three 1,171-kW (1,570-shp) Turboméca Turmo IIIC$_6$ turboshaft engines
Performance: cruising speed 134 kts (248 km/h; 154 mph) at sea level; initial rate of climb 300 m (984 ft) per minute; service ceiling 3100 m (10,171 ft); hovering ceiling in ground effect 1950 m (6,398 ft); endurance in ASW role 4 hours
Weights: empty 6863 kg (15,130 lb); maximum take-off 13000 kg (28,660 lb)

The principal role for French navy Super Frelons is the shielding of the submarine force as they leave harbour, but other roles can be undertaken, including search-and-rescue.

Dimensions: main-rotor diameter 18.90 m (62 ft 0 in); length, rotors turning 23.03 m (75 ft 6.7 in); height 6.76 m (22 ft 2.1 in); main-rotor disc area 280.55 m^2 (3,019.94 sq ft)
Armament: four homing torpedoes or, in ASV role, two Exocet missiles
Operators: China, France, Iraq, Israel, Libya, South Africa, Zaire

Aérospatiale Super Puma/AS.332

Intended as a successor to the SA 330 Puma, Aérospatiale's AS.332 Super Puma is available in five models, two of which are essentially civil in nature (though suitable for military VIP missions). Eliminating many of its predecessor's shortcomings, the AS.332 discarded the Turboméca Turmo engine in favour of the Makila, and introduced several safety and damage-resistant features. Notable amongst these are glassfibre rotor blades, high-tensile steel gears and bearings (which will function for an hour without oil), self-sealing tanks, dual hydraulic and electrical systems, and optional armour for the crew. The basic military version is the AS.332B, carrying up to 20 troops in a cabin of the same size as the SA 330 Puma. This has a greater load-carrying capacity than the 'stretched' variants, which have larger internal volume provided by a 76 cm (29.9 in) increase in cabin length, namely the civil AS.332L and military AS.332M, with seating for 24. Another short-fuselage version is the AS.332F (for *Frégate*). Designed for shipboard operation, this has harpoon-type winch-down equipment, a folding tail, nose radar

and pylons for anti-ship missiles. The prototype Super Puma flew on 7 September 1978, and military deliveries began in 1982, one of the early recipients being Spain's air force which ordered ten winch-equipped SAR AS.332Bs with the local designation HD.21, and two VIP HT.21s. Largest of the known orders is 24 AS.332Bs for the Argentine army (and three for its air force), supplied from 1984 as replacements for Falklands war attrition. Singapore has 22 AS.332Bs on order, including 17 to be assembled locally. Customers for the naval version comprise Abu Dhabi, Kuwait and Qatar.

Specification
Aérospatiale AS.332B Super Puma
Type: transport helicopter
Powerplant: two 1327-kW (1,780 shp) Turboméca Makila 1A turboshaft engines
Performance: maximum cruising speed 280 km/h (174 mph); initial rate of climb 528 m (1,732 ft)/minute; service ceiling 4600 m (15,090 ft); range (standard tankage) 635 km (395 miles)

Weights: empty 4200 kg (9,259 lb); maximum take-off 8700 kg (19,180 lb)
Dimensions: main rotor diameter 15.60 m (51 ft 2.2 in); length 14.76 m (48 ft 5.1 in) excluding tail rotor; height 4.92 m (16 ft 1.7 in); main rotor disc area 191.13 m^2 (2,057.4 sq ft)
Armament: AS.332B, optional machine-guns, cannon or rocket

An AS.332B Super Puma disgorges its load of fully-equipped troops during assault exercises. Special conical inlet filters protect the engines.

pods; AS.332F, two Aérospatiale AM.39 Exocet or six AS.15TT anti-ship missiles
Operators: Abu Dhabi, Argentina, Chile, Ecuador, France, Kuwait, Oman, Qatar, Singapore, Spain

Aérospatiale/MBB HAP/PAH-2/ HAC-3G

In the late 1970s Aérospatiale and Messerschmitt-Bölkow-Blohm (MBB) began project definition for a new anti-tank helicopter, this following selection by their respective governments to make proposals for an aircraft in this class. After finalizing a single basic design that was acceptable to both nations, their defence ministers signed a memorandum of understanding on 29 May 1984 covering development of this aircraft, which is currently scheduled

to enter service in the 1990s. The manufacturers have established a joint company named Eurocopter to manage the programme, but while Aérospatiale and MBB will share the work equally, the latter has systems leadership. Eurocopter thus combines effectively the vast rotary-wing experience and expertise of Aérospatiale with the extensive research and development on four-blade rigid rotors which MBB has been conducting under its own and

Federal German defence ministry funding.

The basic Eurocopter combat helicopter is to be developed in three versions: the HAP (Hélicoptère d'Appui et de Protection), an escort and fire support version for the French army, and armed with a GIAT AM-3078 30-mm cannon in an undernose turret and four Matra Mistral IR AAMs, or two pods each carrying 22 68-mm (2.68-in) rockets at the tips of stub wings; the anti-

tank PAH-2 (Panzerabwehr Hubschrauber, 2nd generation) for the West German army, and able to carry up to eight HOT anti-tank missiles plus four Stinger 2 AAMs; and the anti-tank HAC-3G (Hélicoptère Anti-Char, 3rd generation) for the French army, and armed with up to four ATGW-3 advanced long-range IR-guided anti-tank missiles.

The four-blade rigid main rotor is expected to be some 10 per cent more efficient than those in current use, and other design features include a crash-resistant fuselage structure able to endure hits from weapaons of up to 23-mm calibre, flat-plate cockpit transparencies, self-sealing fuel tanks with explosion suppression, fixed tailwheel type landing gear, twin MTU/Turboméca MTM 385-R turboshaft engines, and the crew in tandem on impact-absorbing seats. Prototypes of the HAP and PAH-2 are expected to fly in late 1987, and that of the HAC-3G in early 1993. Initial anticipated production figures for the HAP, PAH-2 and HAC-3G are 75, 212 and 140 respectively.

Specification
Eurocopter HAP (provisional)
Type: two-seat escort/fire-support helicopter
Powerplant: two 895-kW (1,200-shp) MTU/Turboméca MTM 385-R turboshaft engines
Performance: maximum cruising speed 151 kts (290 km/h; 174 mph); endurance with 20-minute reserves 2 hours 50 minutes
Weights: mission take-off 4800 kg (10,528 lb); design maximum take-off 5000 kg (11,023 lb)
Dimensions: main rotor diameter 13.00 m (42 ft 7.8 in); length of fuselage 13.2 m (43 ft 3.7 in); height 3.75 m (12 ft 3.7 in); main rotor disc area 132.72 m² (1,428.76 sq ft)
Armament: as detailed in text

Aerotec A-122 Uirapuru/T-23

One of the first projects of the Brazilian manufacturer Aerotec S/A Industria Aeronautica, established during 1962 in São Paulo state, was the private-venture design and development of a two-seat side-by-side dual-control primary trainer which was designated Aerotec A-122 Uirapuru. A low-wing monoplane largely of light alloy construction, the A-122 has fixed tricycle landing gear with a steerable nosewheel. The first prototype (PP-ZTF), powered by an 81-kW (108-hp) Avco Lycoming O-235-C1 piston engine, was flown for the first time on 2 June 1965, followed by a second prototype (PP-ZTT) with a 112-kW (150-hp) Lycoming O-320-A. After extensive testing and evaluation of the type, the Fôrça Aérea Brasileira ordered an initial 30 examples under the designation T-23; the two pre-production T-23s (0940 and 0941, flown on 23 January and 11 April 1968 respectively) had Lycoming O-320-As, but standard powerplant for the production version was the O-320-B2B.

Production of the military A-122A totalled 100 for the FAB, plus 18 and eight respectively for the air forces of Bolivia and Paraguay, and many of these remain in service. In addition to the A-122A, Aerotec built 25 generally similar civil A-122B aircraft which differed primarily by having a revised cockpit canopy. When production of the type was suspended in 1977 a total of 155 had been built, this figure including prototypes, pre-production aircraft, A-122As and A-122Bs.

Under the designation A-132 Tangará, Aerotec developed in the late 1970s an improved primary trainer based on the Uirapuru. So far built only in prototype form, this model retains the powerplant of the A-122 but has revised wings, simplified fuselage construction, and a cockpit of improved layout with a canopy that provides a better field of view. First flown on 26 February 1981, this prototype has since been tested by the FAB under the designation YT-17, but no order has so far been announced.

Specification
Aerotec A-122 Uirapuru/T-23
Type: primary trainer
Powerplant: one 119-kW (160-hp) Lycoming O-320-B2B flat-four piston engine
Performance: maximum speed 122 kts (227 km/h; 141 mph); maximum cruising speed 100 kts (185 km/h; 115 mph); initial rate of climb 255 m (837 ft) per minute; service ceiling 14,765 ft (4500 m); maximum range 800 km (497 miles)
Weights: empty 540 kg (1,190 lb); maximum take-off 840 kg (1,852 lb)
Dimensions: span 8.50 m (27 ft 10.6 in); length 6.60 m (21 ft 7.8 in); height 2.70 m (8 ft 10.3 in); wing area 13.50 m² (145.32 sq ft)
Armament: none

A large number of Aerotec T-23s are in service with the Brazilian air force as primary trainers.

Agusta A 109A

Agusta, one of Italy's earliest aircraft manufacturers, became involved in helicopter construction in 1952, after acquiring a licence for the Bell Model 47, and still has licence agreements with Bell. Growing experience in helicopter design/construction led to the Agusta A 109, the company's first own-design helicopter to enter large-scale production, and by 1985 nearly 300 had been built. The initial A 109 was based on a single 515-kW (690-shp) Turboméca Astazou XII, but was revised in 1967 to use two 276-kW (370-shp) Allison 250-C14 turboshafts. The planned military utility A 109B was abandoned in 1969; instead, Agusta concentrated on the eight-seat civil A 109C Hirundo (swallow), the first of three prototypes (NC7101) flying on 4 August 1971. However, it was 1976 before deliveries of production aircraft, then redesignated A 109A, began. This model soon proved a commercial success, being used not only as a light passenger transport, but also as an air ambulance, for freight carriage and for search and rescue. From September 1981 the basic model was redesignated A 109A Mk II following introduction of detail improvements and an uprated transmission.

The type clearly had military potential, and testing in 1976-7 by the Aviazione Leggera dell'Esercito led to the development of versions for such roles as aerial scout, air ambulance, close support, command and control, ECM/ESM, light attack, and utility transport carrying up to seven troops. Primary naval applications include ASV, ASW, ECM/ESM, reconnaissance and stand-off missile guidance. Agusta also offers the A 109A Mk II for a variety of coastguard and police tasks.

With an eye on African and Middle East markets, more recent development has been concentrated on the multi-role 'hot-and-high' A 109K, with two 539-kW (723-shp) Turboméca Arriel IK turboshafts, uprated transmission, a lengthened nose to house increased avionics and detail improvements. With certification anticipated for mid-1985, initial deliveries of the A 109K were expected to follow later in the year.

Specification
Agusta A 109A Mk II (anti-tank role)
Type: light anti-tank helicopter
Powerplant: two 313-kW (420-shp) Allison 250-C20B turboshaft engines, each derated to 258 kW (346 shp) for twin-engine operation
Performance: maximum cruising speed 150 kts (278 km/h; 173 mph); initial rate of climb 503 m (1,650 ft) per minute; service ceiling 4450 m (14,600 ft); range with maximum fuel and no reserves 556 km (345 miles)
Weights: empty 1790 kg (3,946 lb); maximum take-off 2600 kg

Equipped for anti-tank duties with TOW missiles in fuselage-mounted pylons is an Agusta A 109A of the Italian army (ALE).

(5,732 lb)
Dimensions: main-rotor diameter 11.00 m (36 ft 1.1 in); length, rotors turning 13.05 m (42 ft 9.8 in); height 3.30 m (10 ft 9.9 in); main-rotor disc area 95.03 m² (1,022.96 sq ft)
Armament: eight Hughes BGM-71A TOW anti-tank missiles
Operators: Argentina, Iraq, Italy, Libya, UK, Venezuela

Agusta A 129 Mangusta

To meet a requirement of the Aviazione Leggera dell'Esercito for a dedicated light anti-tank helicopter, Agusta began work in 1978 on a development of the A 109A. It was soon realized this would not be good enough and a completely new design was initiated under the designation Agusta A 129 Mangusta (mongoose). Adopting the form that has become almost standard for anti-armour helicopters, this type has a narrow fuselage incorporating separate tandem cockpits for the co-pilot/gunner and pilot (in a raised rear cockpit) on energy-absorbing armoured seats, the cockpits enclosed by flat-plate low-glint canopies. The fuselage carries mid-mounted stub wings, each with two underwing pylons allowing the carriage of a 1000-kg (2,205-lb) weapon load. Other design features include robust impact-absorbing fixed tailwheel landing gear; twin Rolls-Royce Gem turboshaft engines which, in the case of production aircraft, will be partially built under licence in Italy by Piaggio; a crashworthy fuselage structure, with ballistic tolerance against 12.7-mm (0.5-in) AP ammunition; and a transmission, four-blade main and three-blade tail rotor all with similar ballistic tolerance. The first 'official' flight of the A 129 prototype (MM 590/E.I.901), following two 'unofficial' flights, was made on 15 September 1983.

An initial production batch of 66 has been approved by the Italian

Currently flying with the Italian army, the Agusta A 129 anti-tank helicopter is set to become a major part of the Italian military inventory.

government, six of them being allocated for training, with the remainder split between two operational squadrons; the Italian army hopes to procure an additional 30 aircraft, plus reserves, to equip a third squadron. Initial deliveries are expected to be made from the autumn of 1986, and the A 129 with full day and night operational capability is regarded by the Italian army as a very significant aircraft. This follows the adoption of an integrated multiplex system (IMS), managed by two computers, which controls and/or monitors aircraft performance, autopilot, caution/warning systems, communications, engine condition, flight director, fly-by-wire system, navigation, electronic warfare systems, rocket fire control

and the status of electrical, fuel and hydraulic systems. The A 129 Mk 2 may be developed in partnership with Westland for British and other markets.

Specification
Agusta A 129 Mangusta (anti-tank role)
Type: light anti-tank and multi-role military helicopter
Powerplant: two 608-kW (815-shp) Rolls-Royce Gem 2 Mk 1004D turboshaft engines, each with an emergency rating of 772 kW (1,035 shp)
Performance: (estimated) maximum speed 146 kts (270 km/h; 168 mph); initial rate of climb 637 m (2,090 ft) per minute; hovering ceiling in ground effect 3290 m

(10,795 ft); maximum endurance with no reserves 3 hours
Weights: (provisional) empty 2,529 kg (5,575 lb); maximum take-off 3700 kg (8,157 lb)
Dimensions: main-rotor diameter 11.90 m (39 ft 0.5 in); length, rotors turning 14.29 m (46 ft 10.6 in); height 3.35 m (11 ft 0 in); main-rotor disc area 111.22 m^2 (1,197.2 sq ft)
Armament: up to eight BGM-71A TOW missiles on the two outer stores pylons, plus a 7.62-mm (0.3-in), 12.7-mm (0.5-in) or 20-mm gun pod or a launcher for seven 70-mm (2.75-in) rockets on each of the two inner stores pylons
Operators: Abu Dhabi, Italy

Agusta-Bell AB 205

Basically similar to the earlier Bell Model 204, Bell Helicopter's Model 205 introduced a number of improvements. The most significant was a lengthened fuselage providing an enlarged cabin seating 11-14 troops or, when used in a medevac role, accommodating six litters, a seated casualty and a medical attendant. Several nations have built the Model 204/205 under licence, but the major licensee is Agusta which as early as 10 May 1961 flew its first AB 204.

The Agusta-Bell AB 205, which continues in production, corresponds to the UH-1D/UH-1H 'Huey' versions built by Bell in the USA. It is suitably equipped for one pilot operation by day (VFR or IFR) and night; in its AB 205 military version it proved a useful multi-role utility helicopter and has been built in considerable numbers for the Italian armed forces and for export. When stripped of all internal fittings the cabin has a clear volume of 6.2 m^3 (220 cu ft), making it easily adaptable as a cargo carrier and for search and rescue. Agusta has expanded the AB 205's capability by providing kits for the installation of auxiliary fuel tanks, rescue hoist or rotor brake; for the engine there is filter/separation protection and winter-

A nominal number of Agusta-Bell AB 205 transport helicopters are operated by the Greek army.

ization; and emergency flotation gear, regular flotation gear, and skis for operation from snow. There are also a number of weapon systems involving machine-guns, missiles and rockets (singly or in combination), plus armour protection and a gyrostabilized sight.

Since 1969 Agusta has also built as the AB 205A-1 a slightly modified version of the Bell civil Model 205A-1, and this has proved a useful utility helicopter. It has only minor modifications by comparison with the Bell 205A-1 and a similarly-powered sub-type of the T53 turboshaft engine which powers the military AB 205.

Specification
Agusta-Bell AB 205
Type: multi-role military helicopter
Powerplant: one 1044-kW (1,400-shp) Avco Lycoming T53-L-13B turboshaft engine, flat-rated at 820 kW (1,100 shp)
Performance: maximum speed 120 kts (222 km/h; 138 mph); cruising speed 110 kts (204 km/h; 127 mph); initial rate of climb 512 m (1,680 ft) per minute; service ceiling 4,575 m (15,010 ft); maximum range with standard fuel and no reserves 580 km (360 miles)
Weights: empty 2177 kg (4,800 lb); maximum take-off 4309 kg (9,500 lb)

Dimensions: main-rotor diameter 14.71 m (48 ft 3.2 in); length, rotors turning 17.39 m (57 ft 0.7 in); height 4.48 m (14 ft 8.4 in); main-rotor disc area 169.95 m^2 (1,829.36 sq ft)
Armament: can include, singly or in combination, 7.62-mm (0.3-in) and 12.7-mm (0.5-in) machine guns, 7.62-mm Minigun, 70-mm (2.75-in) rocket launchers, AS.12 and TOW missiles
Operators: Greece, Iran, Israel, Italy, Morocco, Oman, Saudi Arabia, Singapore, Spain, Tanzania, Tunisia, Turkey, Uganda, UAE, Yemen, Zambia, Zimbabwe

Agusta-Bell AB 212ASW

Under licence from Bell Helicopters, Agusta began production of the Agusta-Bell AB 212 utility transport helicopter which is essentially the same as the Bell Model 212 Twin Two-Twelve (UH-1N). It differs from the earlier Model 205 by introduction of a twin-turbine powerplant, a revised and uprated transmission, and generally improved dynamics, structure and systems. With the

same roomy cabin as its predecessor, the AB 212 is suitable for a wide range of both civil and military roles, and following initial deliveries in late 1971 soon found sales support in both markets.

The capability and reliability of this twin-turbine helicopter induced Agusta to develop a specialized version, primarily for ASW, but suitable also for deployment in ASV,

SAR, utility and Vertrep roles. Work began on this project in late 1971, the resulting AB 212ASW prototype being evaluated by the Aviazione per la Marina Militare during 1973. Externally it can be identified by its dorsal radome, but otherwise the structure differs little from the basic AB 212; changes include localized strengthening, the addition of deck mooring attachments and provision of increased resistance to salt-water corrosion. The major changes are internal, with a crew of three or

four for combat missions; two pilots, four litters and a medical attendant for casevac; and two pilots and seven passengers as a light transport.

Key to the AB 212ASW's mission capability is its onboard equipment. For ASW the automatic flight-control system combines inputs from the automatic stabilization system, radar altimeter, Doppler radar and other sensors to give hands-off flight from the cruise state to sonar hover under all weather conditions. Its automatic navigation system pinpoints the helicopter's position on the radar tactical display, together with target information from the AQS-18 low-frequency variable-depth sonar. For the ASV mission a high-performance long-range search radar is introduced, and optional avionics can equip the helicopter for mid-course passive guidance of the OTO Melara Otomat 2 anti-ship missile equipping the Italian navy. More than 100 AB 212ASW helicopters are in service worldwide.

Specification
Agusta-Bell AB 212ASW
Type: ASW and multi-role naval helicopter
Powerplant: one 1398-kW (1,875-shp) Pratt & Whitney Canada PT6T-6 Turbo Twin Pac twin-turboshaft engine
Performance: maximum speed 106 kts (196 km/h; 122 mph); cruising speed with weapons 100 kts (185 km/h; 115 mph); initial rate of climb 396 m (1,300 ft) per minute; hovering ceiling in ground effect 3200 m (10,500 ft); search

endurance 3 hours; maximum range with auxiliary fuel 667 km (414 miles)
Weights: empty 3420 kg (7,540 lb); maximum 5070 kg (11,175 lb)
Dimensions: main-rotor diameter 14.63 m (48 ft 0 in); length, rotors turning 17.40 m (57 ft 1 in); height 4.53 m (14 ft 10.25 in); main-rotor disc area 168.11 m² (1,809.6 sq ft)

Armament: two Mk 44/46 or two Moto Fides A 244/S homing torpedoes, or two Marte Mk 2 or two Sea Skua type air-to-surface missiles
Operators: Australia, Dubai, Greece, Iran, Iraq, Italy, Lebanon, Libya, Morocco, Peru, Saudi Arabia, Somalia, Spain, Sudan, Turkey, Yemen, Zambia

The prominent housing for the MM/APS-705 search radar above the cabin identifies this machine as an Agusta-Bell AB 212ASW, in use with the Italian navy for anti-submarine duties.

Agusta AB.412 Griffon

Derived from the Bell 412, the Griffon is a product of Italian licensee Agusta for armed operations. The Bell UH-1 was given a double engine unit to produce the Model 212, then added a four-blade rotor for smoother, quieter flight to become the Model 412. Agusta augmented this aircraft with a high-energy-absorbing landing gear, self-sealing fuel tanks, crash-resistant seats (armour-plated for the crew) and armament attachment points to produce the military AB.412 Griffon. Carrying up to 15 troops, or rapidly adapted for tactical support, logistic transport, SAR, patrol and casualty evacuation (six stretchers, plus two attendants), the Griffon may be used against ships, tanks and other hard targets when armed with appropriate missiles. Survivability can be enhanced by installation of radar/laser warning, missile detection, ECM, jamming and decoy systems. Ground handling and concealment is assisted by two folding blades, so that the Griffon occupies the same area as a 212. A prototype flew in August 1982. Spain's army ordered 24 to expand airborne forces supporting Mountain Brigades; a further four are unarmed VIP transports. An initial contract for nine has been placed by the Italian army, presaging replacement of its AB.205 Hueys, and two communications models were supplied to the Carabinieri (para-military police) in 1984. The Turkish navy has ordered an anti-ship version with BAe Sea Skua missiles. Delivery of at least six AB.412s ordered by Uganda has been deferred, but two have gone to the Zimbabwe air force.

Specification
Agusta AB.412 Griffon
Type: multi-role armed helicopter
Powerplant: one 1342-kW (1,800-shp) Pratt & Whitney Canada PT6T-3B Twin-Pac turboshaft unit (single-engine emergency rating 764 kW/1,025 shp)
Performance: maximum speed 226 km/h (140 mph); initial rate of climb 438 m (1,437 ft)/minute; service ceiling 2315 m (7,595 ft); range 498 km (309 miles)
Weights: empty 2858 kg (6,301 lb); maximum take-off 5261 kg (11,599 lb); useful load 2400 kg (5,291 lb)
Dimensions: main rotor diameter

Agusta-Bell AB 412 Griffon in the markings of the Lesotho Police Mobile Unit (Air Wing).

14.02 m (46 ft 0 in); fuselage length 12.92 m (42 ft 4.7 in); height to rotor head 3.29 m (10 ft 9.5 in); main rotor disc area 154.39 m² (1,661.9 sq ft)
Armament: (fire support) two 25-mm (0.98-in) cannon; (area suppression) two 19-tube 70-mm (2.75-in) rockets; (scout/recce) two 12.7-mm (0.5-in) machine-guns; (air defence) four air-to-air missiles; (anti-ship) four BAe Sea Skua missiles
Operators: Italy, Lesotho, Singapore, Spain, Zimbabwe

Agusta AS-61

Sikorsky having completed production of its S-61 series, Westland in Britain and Italy's Agusta remain as sources, but only the latter is building the HH-3 and civil versions. As a shipboard anti-submarine helicopter equipped with radar, sonar and anti-ship missiles, the ASH-3 Sea King has been produced in models equivalent to the US Navy's SH-3D and SH-3H. Differences are restricted to local strengthening, uprated engines and a revised horizontal tail. Deliveries to the Italian navy began in 1968, the short-range armament of Aérospatiale AS.12 missiles later being augmented by Marte 2s. A small top-up batch has increased the Italian navy's total to 36. The Iranian navy received 15 ASH-3Ds between 1976 and 1981 (most now being unserviceable); Peru's navy acquired six in 1978 and four more in 1984; Brazil took delivery of six ASH-3Hs in 1983; and Argentina, six in 1984. The two last-mentioned also operate US-built Sea Kings. For transport roles (invariably VIP) the ASH-3D/TS (Trasporto Speciale) is an unarmed model obtained by the air forces of Italy (two), Iran (two), Iraq (six) and Saudi Arabia (three), also being known as the AS-61A-4 or AS-61VIP. Similar contracts may be received for the new, Agusta-developed, short-fuselage AS-61N1 Silver which has increased fuel capacity and may later be re-engined with 1312-kW (1,760-shp) General Electric CT7s. Superficially different from the Sea King family, with its revised cabin and boom shape, the Sikorsky S-61R transport is built in Italy under the designation HH-3F Pelican as a parallel to the USAF's Jolly Green Giant. The Italian Air Force received 20 between 1977 and 1980 for SAR flights from four coastal bases, and ten are being built for export.

Specification
Agusta ASH-3H Sea King
Type: amphibious all-weather anti-submarine helicopter
Powerplant: two 1119-kW (1,500-shp) General Electric T58-GE-100 turboshaft engines
Performance: cruising speed 222 km/h (138 mph); initial rate of climb 672 m (2,205 ft)/minute; service ceiling 3720 m (12,205 ft); range 1166 km (725 miles)
Weights: maximum take-off 9525 kg (21,000 lb); cargo capacity 2722 kg (6,000 lb)
Dimensions: main rotor diameter 18.90 m (62 ft 0 in); fuselage length 16.69 m (54 ft 9 in); height to rotor head 4.74 m (15 ft 6.6 in); main rotor disc area 280.47 m² (3,019.1 sq ft)
Armament: four Aérospatiale AS.12, two Oto-Melara Marte 2, two Aérospatiale AM.39 Exocet or McDonnell Douglas AGM-84A Harpoon anti-ship missiles; four A 244/AS, Mk 44 or Mk 46 homing torpedoes; depth charges

A pair of Agusta-Sikorsky ASH-3H anti-submarine helicopters are operated by the Brazilian navy.

Operators: Argentina, Brazil, Iran, Iraq, Italy, Peru, Saudi Arabia

AIDC T-CH-1

The Chinese Nationalist air force has had an aircraft production facility in Taiwan since 1948. The AIDC (Aero Industry Development Center) was set up in March 1969, and with production know-how backed by design capability work began on the T-CH-1 in November 1970.

It is based on the North American T-28 Trojan and two prototypes (XT-CH-1A and XT-CH-1B) made their respective maiden flights on 23 November 1973 and 27 November 1974. The second was modified to weapons training and COIN (counter-insurgency) configuration, and this capability is retained in the 50 T-CH-1s of the Chinese Nationalist air force. Production of these began at Taichung in May 1976, and they replaced T-28s in service.

Specification
Type: tandem two-seat trainer and

light ground-attack aircraft
Powerplant: one 1081 ekW (1,450-eshp) Lycoming T53-L-701 turboprop engine
Performance: maximum speed 592 km/h (368 mph) at 4570 m (14,995 ft); maximum cruising speed 407 km/h (253 mph) at 4570 m (14,995 ft); initial rate of climb 1037 m (3,400 ft)/minute; service

ceiling 9755 m (32,005 ft); maximum range 2010 km (1,249 miles)
Weights: empty 2608 kg (5,750 lb); maximum take-off 5057 kg (11,150 lb)
Dimensions: span 12.19 m (39 ft 11.9 in); length 10.26 m (33 ft 7.9 in); height 3.66 m (12 ft 0 in); wing area 25.18 m² (271.0 sq ft)

The major basic trainer in the Chinese Nationalist air force is the T-CH-1, a derivative of the T-28 Trojan. Rockets and light bombs can be carried for COIN duties. Production has now ceased.

Armament: probably provision for light underwing weapons
Operator: Taiwan

AIDC AT-3

In July 1975, before production of the T-CH-1 had got under way, AIDC was awarded a contract to design and develop for the Chinese Nationalist air force a twin-turbofan military trainer. Its design, as the AIDC XAT-3, was finalized in collaboration with the Chungshan Institute of Science and Technology and the US manufacturer Northrop. The result is a cantilever low-wing monoplane with retractable tricycle landing gear; it seats the crew of two in tandem in an air-conditioned/pressurized cockpit on zero-zero ejection seats, beneath a long transparency incorporating individual side-opening canopies; unusually, the two members of the crew are separated by an internal windscreen. The powerplant comprises two Garrett TFE731 turbofan engines mounted in nacelles, one on each side of the fuselage; standard fuel is contained in two fuselage bladder tanks, but can be augmented by a 568-litre (150-US gal) drop tank on each inboard underwing paylon. Advanced avionics are provided for communication and navigation, and five external stores stations allow the carriage of a

AIDC AT-TC-3 advanced trainer in service with the Chinese Nationalist air force.

maximum external weapon load of 2268 kg (5,000 lb).

Following evaluation of the design proposal the CNAF ordered two prototypes and their construction began in January 1978; these flew respectively on 16 September 1980 and 30 October 1981. Following evaluation the CNAF has awarded AIDC a contract for more than 50 production aircraft under the designation AT-TC-3, and the first of these (0803) was rolled out and flew initially on 6 February 1984.

Specification
Origin: Taiwan

Type: military advanced trainer
Powerplant: two 1588-kW (3,500-lb) thrust Garrett TFE731-2-SL non-afterburning turbofan engines
Performance: maximum speed 485 kts (898 km/h; 558 mph) at sea level; initial rate of climb 2438 m (8,000 ft) per minute; service ceiling 14630 m (48,000 ft); maximum endurance on internal fuel 3 hours 12 minutes
Weights; empty 3855 kg (8,500 lb); maximum take-off 7485 kg (16,500 lb)
Dimensions: span 10.46 m (34 ft 3.75 in); length 12.90 m (42 ft 4 in); height 4.36 m (14 ft 3.75 in); wing

area 21.93 m² (236.05 sq ft)
Armament: one underfuselage and four underwing hardpoints for a maximum 2268 kg (5,000 lb) of external stores, which can include practice bombs, auxiliary tanks, and towed targets; a weapons bay beneath the rear cockpit is suitable for semi-recessed machine-gun packs and other stores; there are wingtip launch rails for two AIM-9 Sidewinder AAMs
Operator: Taiwan

Airtech CN-235

In 1979 Construcciones Aeronauticas SA (CASA) embarked on a new transport, larger than the C-212 but smaller than the Transall C-160 or Lockheed C-130 In this venture they joined with the Indonesian company PT Nurtanio which was already building the C-212 under licence, and thereby opened up sharing of the financial, design and production workloads and also the possiblity of valuable startup contracts from the Indonesian armed forces. Similar in configuration to the larger Transall, the CN-235s rear ramp allows loading of four LD-3 standard containers or equiivalent 224-cm (88-in) palletized loads. In passenger configuration the aircraft can be fitted with 11 rows of four seats. The cabin is pressurized at the low level of 0.25 kg/cm² (3.6 lb/sq in) to simplify the sealing of the ramp, and an unobstructed cabin cross section is achieved by placing the wing above the fuselage. Two prototypes, built by CASA and Nurtanio, were rolled out in September 1983, the first flying on 11 November 1983. The major requirement is from the Indonesian air force

(Tentara Nasional Indonesia Angkatan Udara) which will receive 32; the Indonesian navy (Angkatan Laut) has 18 on order. It is expected that the Spanish air force will also order the CN-235. Indonesian aircraft will be used as general transports with emphasis on paratroop operations, in which role they will have 41 slung seats facing inwards. The CN-235 is being marketed in the electronic warfare role with a nose-mounted radome, and as an ASW and maritime patrol type with 360° search radar and underwing installation of two Exocet missiles or Mk 46 torpedoes. A medical evacuation version can carry 24 stretchers together with a four-man medical team.

Specification
Airtech CN-235
Type: multi-role transport
Powerplant: two 1321-kW (1,772-shp) General Electric CT7-7 turboprop engines
Performance: maximum cruising speed 452 km/h (281 mph); initial rate of climb 542 m (1,778 ft)/

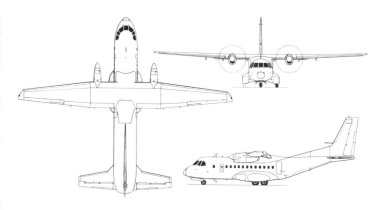

Airtech CN-235

minute; maximum range (cargo configuration) 4720 km (2,933 miles)
Weights: empty 7950 kg (17,527 lb); maximum take-off 14400 kg (31,747 lb)
Dimensions: span 25.81 m (84 ft

8.1 in); length 21.35 m (70 ft 0.6 in); height 8.17 m (26 ft 9.7 in); wing area 59.10 m² (636.17 sq ft)
Armament: none
Operator: Indonesia (on order), Saudi Arabia

Antonov An-2

Believed to have been built in larger numbers than any other aircraft designed since World War II, the Antonov An-2 'Colt' is still in production though the An-3 may replace it. As well as excelling as a light transport, with outstanding STOL performance, it is also used widely as a parachute trainer by associations such as DOSAAF, the Soviet Union's paramilitary training organization.

The first An-2 flew in August 1947. The biplane layout was chosen for its combination of good field performance with viceless low-speed handling, and the fuselage – like a DC-3 but shorter – filled the gap between the wings. The structure is all-metal, apart from the fabric-skinned wings and tailplane, and the wings incorporate slotted flaps and drooping ailerons.

More than 5,000 An-2s were built in the Soviet Union between 1948 and 1960, and production in that country ceased temporarily in 1962. In 1967 production started in China, as the Fong Chou No 2, several

thousand of which have been built. Since 1960, however, the Soviet bloc's main source of the aircraft has been WSK-Mielec in Poland, and production from this manufacturer exceeded 9,500 in 1984. Soviet production was resumed in 1964 with a small number of the An-2M type, with larger tail surfaces, a new variable-pitch propeller and other changes.

Specification
PZL-Mielec An-2
Type: 14-seat transport and general-purpose aircraft
Powerplant: one 746-kW (1,000-hp) PZL Kalisz (Shvetsov) ASz-62IR nine-cylinder radial piston engine
Performance: maximum speed 258 km/h (160 mph) at 1750 m (5,740 ft); economic cruising speed 185 km/h (115 mph); initial rate of climb 210 m (689 ft)/minute; service ceiling 4400 m (14,435 ft); range at optimum altitude with 500 kg (1,102 lb) payload 900 km (559 miles)

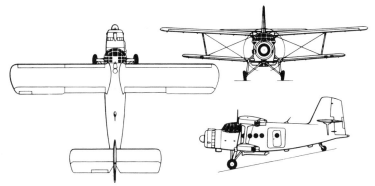

Antonov An-2 'Colt'

Weights: empty 3450 kg (7,606 lb); maximum take-off 5500 kg (12,125 lb)
Dimensions: span, upper 18.18 m (59 ft 7.7 in), lower 14.24 m (46 ft 8.6 in); length 12.74 m (41 ft 9.6 in); height 4.00 m (13 ft 1.5 in); wing area 71.60 m² (770.72 sq ft)
Armament: none

Operators (military): Afghanistan, Albania, Bulgaria, China, Cuba, Egypt, Ethiopia, East Germany, Hungary, Iraq, Mali, Mongolia, North Korea, Poland, Romania, Somalia, Sudan, Syria, Tanzania, Tunisia, USSR, Vietnam

---

Antonov An-12 'Cub-A'

Sometimes described as the Soviet equivalent version of the Lockheed C-130 Hercules, the Antonov An-12 civil and military cargo transport has not, however, been produced on such a large scale and in so many variants. Having said that, however, it must be realized that more than 900 An-12s were built before production ended in the USSR in 1973, and the type serves with the V-TA (Military Transport Aviation) of the Soviet air force, and with the air arms of Algeria, China, Ethiopia, India, Iraq, Madagascar, Poland, Syria and Yugoslavia. Civil versions are used by Aeroflot, as well as by Air Guinée, Balkan Air, CAAC in the People's Republic of China and the Polish airline LOT.

Flown in prototype form in 1958, the An-12 was built in parallel with the An-10 Ukraina and had the same basic airframe, but the rear fuselage was modified to incorporate two large longitudinal doors that hinge upwards into the cabin to allow direct loading of cargo from trucks. Behind the two doors is a full-width door that also hinges upwards into the cabin to improve headroom and access during loading/unloading operations, but it is not possible to load or unload vehicles without the use of a separate ramp.

In 1984, around 350 'Cub-A' transports were still in service with the Soviet air force, although these are being slowly replaced and augmented by Ilyushin Il-76s. The An-12s of the state airline Aeroflot are also available to military commanders for transport duties. The 'Cub-A' in V-TA service is designated An-12BP and at present is seeing much service in Afghanistan, where its rear-loading doors make it useful for paradropping supplies to outlying garrisons. Troops can also be dropped from the 'Cub-A'; it can carry up to 100 in full combat gear. Several of the Aeroflot aircraft have appeared alongside their V-TA counterparts on the airlift from the USSR to Afghanistan but these retain their gun turrets. A recent addition to the aircraft are attachments for flares which distract heat-seeking missiles from the aircraft, a necessity incorporated due to the successes of Afghan rebels firing SA-7 shoulder-launched surface-to-air missiles.

The good range and speed of the

Equipped for ECM duties, this An-12 'Cub-C' wears temporary Egyptian air force markings.

An-12, and above all its capacious fuselage made it an obvious choice for conversion for various electronic roles. The first to appear was the 'Cub-B' which has been seen over international waters, especially the Baltic Sea, sporting four additional blister fairings under the forward and centre fuselage and many small blade aerials. The role is electronic intelligence, and the 'Cub-B' has appeared over NATO exercises and snooping around Western coastlines. Ten aircraft were believed to have been modified, and these serve with the AV-MF (naval air arm). It seems they are involved primarily with RINT (radiation intelligence), producing electronic fingerprints of NATO ships and radar installations to gain information about radar wavelengths and strengths.

An important role for the An-12 is ECM, and there are two variants so far which operate with both the air force and the navy. The 'Cub-C' has a solid ogival tailcone which replaces the tail turret and the 'Cub-D' has two large blister fairings running

side by side longitudinally between the nose and main landing gear. This model retains the conventional tail turret. These are complex jamming platforms which house several tons of equipment, much of it palletized for ease of change. At least five wavebands can be jammed, and judging by the amount of extra aerials, foreign language specialists may be carried to transmit confusing messages on enemy frequencies. Physical ECM means are also carried in the form of chaff and flares. Several Soviet-flown aircraft were seen during the 1970s in Egyptian air force markings.

Specification
Antonov An-12BP 'Cub-A'
Type: military transport
Powerplant: four 2983-ekW (4,000-eshp) Ivchenko AI-20K turboprop engines
Performance: maximum speed 419 kts (777 km/h; 483 mph); cruising speed 361 kts (670 km/h; 416 mph); initial rate of climb 600 m (1,970 ft) per minute; service ceiling

10200 m (33,465 ft); range with maximum payload 3600 km (2,237 miles); range with maximum fuel 5700 km (3,542 miles)
Weights: empty 28000 kg (61,729 lb); maximum take-off 61000 kg (13,448 lb)
Dimensions: span 38.00 m (124 ft 8.1 in); length 33.10 m (108 ft 7.1 in); height 10.53 m (34 ft 6.6 in); wing area 121.70 m² (1,310.0 sq ft)
Armament: twin 23-mm NR-23 cannon in tail turret
Operators: Algeria, Bangladesh, China, Egypt, Ethiopia, Guinea Republic, India, Iraq, Jordan, Madagascar, Poland, Syria, USSR, Yemen, Yugoslavia

The primary heavy transport aircraft of the Indian air force, the Antonov An-12 is gradually being replaced by the more capable Ilyushin Il-76.

Antonov An-14 'Clod'

The An-14 was designed in response to a requirement for a light STOL transport for civil and military operations, and made its first flight in March 1958. Its development, however, was extremely slow, and production aircraft did not appear until 1964-65. Around 200 of the type were built.

Remaining An-14s serve in second-line transport duties with the Soviet Air Forces and the East German Luftstreitkräfte (LSK). The type is an indifferent design, even by Soviet standards, and is comparable in performance with the Canadian Otter, which preceded it into service by more than ten years. Most of the military roles for which the An-14 was designed are now performed by helicopters.

The An-14 shows clear design influence from the French Hurel-Dubois prototypes, which revived interest in the braced, high-aspect-ratio wing in the early 1950s, and the British Miles Aerovan. The design is quite complex, with powered leading-edge slats, full-span trail-

ing-edge flaps. It has influenced China's Harbin Y-11 and Y-11T utility aircraft.

Specification
Antonov An-14 'Clod'
Type: light STOL aircraft
Powerplant: two 224-kW (300-hp) Ivchenko AI-14RF radial piston engines

Performance: cruising speed 170-180 km/h (105-120 mph); maximum range with six passengers or 570 kg (1,200 lb) payload 650 km (400 miles); service ceiling 5000 m (16,400 ft); take-off run 100-110 m (330-360 ft); landing run 110 m (360 ft)
Weights: empty 2600 kg (5,700 lb); normal take-off 3450 kg (7,600 lb);

A simple design for various utility tasks, the An-14 'Clod' has rough-field capability and the ability to carry stretcher cases in a casevac role as well as standard passenger transport.

maximum take-off 3630 kg (8,000 lb)
Dimensions: span 22.00 m (72 ft 3 in); length 11.36 m (37 ft 3.5 in); height 4.63 m (15 ft 2.5 in); wing area 39.72 m² (422.8 sq ft)
Operators: East Germany, USSR

Antonov An-22 'Cock'

On 27 February 1965 Antonov flew the prototype of the Antonov An-22 Antei (Antheus) long-range heavy-lift transport, the appearance of the type at the Paris Air Show a few months later creating considerable surprise among Western nations. Subsequently given the NATO reporting name 'Cock', the An-22 is in overall configuration basically a scaled-up version of the An-12 'Cub', but differs in having a tail unit incorporating twin fins and rudders which extend above and below the tailplane.

With a high-aspect-ratio wing and a maximum take-off weight of 246 tons, it is understandable that the An-22 has a high wing loading; in fact, at 724.6 kg/m² (148.4 lb/sq ft) it is the highest of any military transport in service, comparing with 661.6 kg/m² (135.5 lb/sq ft) for the dimensionally larger Lockheed C-5 Galaxy. Under these conditions one might expect sluggish field performance, but four powerful turboprop engines, each driving a pair of four-blade contra-rotating propellers, provide a massive slipstream which is 'blown' over the double-slotted flaps that represent 60 per cent of

Soviet air force An-22 Antheus heavy transports are used primarily for military support duties.

the wing trailing edges. The combination proved good enough for a maximum weight take-off run of only 1300 m (4,265 ft), and to set and hold 27 FAI records for speed with payload and payload to height. The retractable tricycle landing gear is designed for off-runway operation, the nosewheel unit having twin wheels, and each main unit three twin-wheel levered-suspension units in tandem; tyre pressures are adjustable in flight or on the ground for optimum airfield performance. Pressurization is provided for the crew of five or six, and the 28-29 passengers seated in a forward cabin section, but the 33.0-m (108.3-ft) long main cargo hold is unpressurized. Crew/passenger access is via a door in each landing gear fairing,

cargo being loaded and unloaded by means of a ramp in the upswept aft fuselage. As with most Soviet aircraft, production is a matter for conjecture, but it is believed that Aeroflot and the Soviet air force's V-TA had each received about 50 when production ended in 1974. The An-22 is expected to be replaced by the Antonov An-124 'Condor' towards the end of the 1980s.

Specification
Antonov An-22 'Cock'
Type: long-range heavy transport
Powerplant: four 11186-kW (15,000-shp) Kuznetsov NK-12MA turboprop engines
Performance: maximum speed 399 kts (740 km/h; 460 mph); cruising speed 324 kts (600 km/h;

373 mph); range with maximum payload of 80000 kg (176,370 lb) 5000 km (3,107 miles); range with maximum fuel 10950 km (6,804 miles)
Weights: empty 114000 kg (251,327 lb); maximum take-off 250000 kg (551,156 lb)
Dimensions: span 64.40 m (211 ft 3.4 in); length 57.90 m (189 ft 11.5 in); height 12.53 m (41 ft 1.3 in); wing area 345.00 m² (3,713.67 sq ft)
Armament: none
Operator: Soviet Union

The 50 or so An-22s are split between the Soviet air force transport fleet (V-TA) and Aeroflot. The civil machines are often operated on military business, and have been seen regularly on the airlift to Afghanistan.

Antonov An-24 'Coke'

First flown in December 1959, the An-24 'Colt' was designed in the class of the F.27 but is heavier and more powerful, burning more fuel

for a similar job. The structure makes extensive use of welding and bonding, and the type entered service in October 1962. The major pro-

duction An-24V has 28-40 seats and, in some cases, a side freight door and convertible cabin. The AN-24V Series II of 1967 introduced the more powerful AI-24T engine to improve hot-and-high performance, and this seated up to 50.

All An-24Vs were delivered with a TG-16 gas-turbine APU (auxiliary power unit) in the right nacelle, but in the AN-24RV of 1967 this was replaced by a Tumansky turbojet APU to boost take-off performance. Take-off weight of the An-24RV is increased to 21800 kg (48,061 lb) and this can be maintained up to ISA +30°. More than 1,000 AN-24s were built, and some dozens are used by many air forces as VIP and government transports.

Specification
Antonov An-24V
Type: short-range and VIP transport
Powerplant: two 2148-KW (2,880-shp) Ivchenko AI-24 turboprops, plus (An-24RV) one 900-kg (1,984-lb) thrust Tumansky RU-19-300 turbojet engine
Performance: cruising speed 500 km/h (311 mph) at 6100 m (20,015 ft); maximum range with 30

passengers 2400 km (1,491 miles), range with 5520-kg (12,170-lb) payload 550 km (342 miles)
Weights: empty 13600 kg

(29,983 lb); maximum take-off 21000 kg (46,297 lb)
Dimensions: span 29.20 m (95 ft 9.6 in); length 23.53 m (77 ft 2.4 in);

height 8.32 m (27 ft 3.6 in); wing area 72.46 m² (780.0 sq ft)
Operators: Czechoslovakia, East Germany, Hungary, Poland

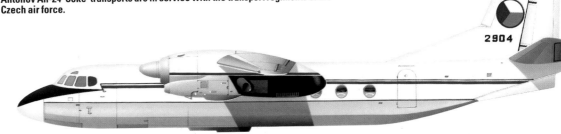
Antonov An-24 'Coke' transports are in service with the transport regiments of the Czech air force.

Antonov An-26 'Curl'

These two types were developed independently from the An-24 airliner, itself flown in December 1959 and designed as a replacement for the Lisunov Li-2 (the unlicensed Soviet copy of the C-47). The An-24 resembled earlier Antonov transports, with its high wing and drooped outer wing panels. Its engine and landing gear configuration was based on that of the Fokker F.27, with the main gear retracting rearwards into slender, aerodynamically efficient nacelles. The airframe used a great deal of welding and bonding, reducing manufacturing costs but incurring a considerable weight penalty. The final production version of the type was the An-24RV, with a small Tumansky turbojet in the starboard nacelle. The jet provided all systems power for hot-and-high take-offs, together with some additional thrust.

The size and configuration of the An-24 made it a logical basis for a light tactical and logistics transport. The first freighter version of the type, the An-24T, featured a modified rear fuselage and a loading hatch, but it was closely followed by the more developed An-26 'Curl', with a completely new rear fuselage. This incorporated a novel loading ramp, which could admit any load that would fit in the cabin and could be drawn forward beneath the fuselage for air-dropping. Like other modern Soviet freighters, the type was equipped with built-in powered winches and an overhead travelling crane to aid cargo handling. All An-26s have the booster engine installed, together with the higher-rated AI-24T turboprops.

The An-26 is now the standard light tactical transport of non-Soviet Warsaw Pact forces and other allies. (The Soviet Union's own airborne forces appear to use Il-76s and helicopters.) The An-26 can carry small vehicles or towed artillery, but would more commonly be used to carry palletized cargo. Most aircraft

of the type feature a side observation blister for navigation and air-dropping of troops or supplies. An improved version, the An-26B, was introduced in the early 1980s, and features a more efficient freight handling system. The only direct Western equivalent of the type was the BAe Andover.

Specification
Antonov An-26
Type: light tactical transport
Powerplant: two 2104-kW (2,820-eshp) Ivchenko AI-24VT turboprops and one 900-kg (1,985-lb) thrust Tumansky RU-19A-300 turbojet

The Antonov An-26 'Curl' has been supplied to many Soviet client states. Afghanistan uses the type for all manner of transport duties, including the airborne supply of outlying garrisons deep in guerrilla territory. Paradropping from the rear ramp door is widely used.

engine
Performance: maximum cruising speed 435 km/h (270 mph); range with 5500-kg (12,130-lb) payload 900 km (560 miles); take-off field length 1240 m (4200 ft); landing field length 1740 m (5,700 ft)
Weights: empty 15020 kg (33,120 lb); maximum take-off 24000 kg (53,000 lb)
Dimensions: span 29.2 m (95 ft 9 in); length 23.8 m (78 ft 1 in); height 8.58 m (28 ft 6 in); wing area 75 m² (807 sq ft)
Operators: Afghanistan, Algeria, Angola, Bangladesh, Benin, Bulgaria, Cape Verde, Congo, Cuba,

Czechoslovakia, East Germany, Ethiopia, Guinea Bissau, Hungary, Iraq, Laos, Libya, Madagascar, Mali, Mozambique, Mongolia, Nicaragua, Peru, Poland, Romania, Somalia, South Yemen, Syria, Tanzania, USSR, Vietnam, Yugoslavia, Zambia

Antonov An-26 'Curl' transports serve with Soviet Military Transport Aviation units.

Antonov An-32 'Cline'

Unveiled at the Paris air show in 1977, the An-32 is a highly modified development of the An-26, designed for much better hot-and-high, short-field performance. It retains the basic airframe of the An-26, but has some 80 per cent more turboprop power and many other changes.

The new engines drive much larger propellers; in order to provide adequate clearance between the propeller discs and the fuselage without increasing the span of the

centre-section, the designers installed the new engines above the wing, well above the greatest width of the fuselage. The powerplant installation is very similar to that of the Be-12 amphibian, and may have used common components to reduce the development effort.

Prototypes of the An-32 differed in configuration, but according to recent reports the standard production version of the aircraft features full-span leading-edge slats and triple-

slotted trailing-edge flaps. The powerful trim changes generated by these devices are countered by a redesigned tailplane which features a fixed, inverted slat. The type has bigger ventral fins than the An-26, to compensate for the added side area of the cowlings and the destabilizing effect of the propellers.

The An-32 is considerably heavier than the An-26, and carries a greater payload. It is designed to operate normally from airfields as much as 4600 m (15,000 ft) above sea level; however, a version is available with lower-rated AI-20M

engines, for more economical operation at lower elevations.

The An-32 appears to be unique in having been developed for export, in the absence of a firm Soviet need for the type. So far, the only known operator is the Indian Air Force, which ordered 95 of the type in 1979. These have now been delivered, and are equipped with Indian-developed navigation and electronic equipment. The type was also offered to Finland, but no firm order had resulted at the time of writing.

Specification
Antonov An-32
Type: light STOL tactical transport
Powerplant: to 3864-kW (5,180-eshp) Ivchenko AI-20M turboprops

Performance: maximum cruising speed 530 km/h (329 mph); range with 3000-kg (6,600-lb) payload 1100 km (685 miles); take-off field length 500 m (1,640 ft)

Weights: empty 16000 kg (35,275 lb); maximum take-off 2700 kg (59,500 lb)
Dimensions: span 29.2 m (9 ft 9 in); length 23.8 m (78 ft 1 i);

height 8.58 m (28 ft 6 in); wing area 75 m^2 (807 sq ft)
Operators: India, Tanzania

Antonov An-72 and An-74 'Coaler'

The requirement for a light STOL (short take-off and landing) transport to supersede the An-26 in service led to the design and development of the Antonov An-72, the first of two prototypes flying in 1977. This, the first jet-powered transport to emanate from the Antonov design bureau, is very similar in configuration to the Boeing YC-14 prototype which had flown a year earlier, differing primarily in design of the wing and by being dimensionally much lighter, smaller, and one-quarter as powerful. However, its two turbofan engines, mounted above and forward of the wing as on the YC-14, allow the turbofan efflux to be discharged over the wing surface to provide similar upper-surface blowing (USB), which relies on the Coanda effect upon the extended flaps to enhance lift. Other design similarities with the YC-14 include an upswept rear fuselage carrying a high T-tail clear of the engine efflux; multi-wheel landing gear, the main units retracting into external fuselage fairings; and rear cargo loading via the undersurface of the upswept rear fuselage. In the case of the An-72, the rear loading door/ramp is similar in design to that developed for the An-26.

Intended primarily as a cargo carrier, the An-72 (NATO reporting

Antonov An-72 'Coaler' STOL transports are set to replace An-24/26 fleets within the Soviet air force.

name 'Coaler') can also carry up to 32 passengers on folding seats mounted along the side walls of the cabin, or in an ambulance role accommodate 24 casualties and a medical attendant. The capability of the An-72 was demonstrated in late 1983 when 16 FAI accredited payload/height class records were established.

No production An-72 had been seen by mid-1985, although recent reports would indicate that production is under way. At least two examples have been used for evaluation by Aeroflot for some time and it is reasonable to assume that the type will have started to enter service both with Aeroflot and the V-TA of the Soviet air force. What

appears to be a developed version of the An-72 for use in Arctic and Antarctic regions has been reported under the designation An-74, this being equipped with wheel-ski landing gear, full de-icing equipment and advanced avionics to make the type suitable for all-weather operation.

Specification
Antonov An-72
Type: short-medium range STOL transport
Powerplant: two 6500-kg (14,330-lb) thrust Lotarev D-36 turbofan engines
Performance: maximum speed 410 kts (760 km/h; 472 mph); maximum cruising speed 388 kts

(720 km/h; 447 mph); service ceiling 11000 m (36,090 ft); range with maximum payload and 30-minute reserves 1000 km (621 miles); range with maximum fuel and 30-minute reserves 3800 km (2,361 miles)
Weights: maximum payload, STOL operation 3500 kg (7,716 lb); maximum payload normal operation 10000 kg (22,046 lb); take-off from 1000-m (3,280-ft) runway 26500 kg (58,422 lb); maximum take-off 33000 kg (72,753 lb)
Dimensions: span 25.83 m (84 ft 8.9 in); length 26.58 m (87 ft 2.5 in); height 8.24 m (27 ft 0.4 in)
Armament: none
Operator: Soviet Union

Antonov An-124 Ruslan

In summer 1977 it was reported that the Antonov bureau had started design work on a large transport, probably larger than the Lockheed C-5 Galaxy. It was required, in due course, to replace in service the Antonov An-22, but there was a requirement also for a large-capacity transport to accommodate in a single load a complete SS-20 intercontinental ballistic missile system. Initially the provisional designation An-40 was reported, subsequently 'confirmed' as being An-400 and later allocated the NATO reporting name 'Condor'. In early May 1985 came information that the aircraft's designation is Antonov An-124, and on 28 May one of three pre-production aircraft (SSSR-82002) flew in to Le Bourget to take part in the 1985 Paris Air Show.

As the An-124 has the same role as the C-5, it is not surprising that its configuration is similar, but the tailplane is low. Zero-fuel weight is much heavier, so payloads are much greater. Cabin width and height are larger than the C-5, and the supercritical wings (with gigantic one-piece skins) hold 220 tonnes of fuel, an all-time record. The high-bypass-ratio (5.7:1) engines incorporate reversers. The fuselage has a visor-type nose door/folding ramp and the upswept rear fuselage a four-section door/ramp. With both open there is a through cargo hold 36.0 m (118.1 ft) in length, 6.4 m (21 ft) wide and 4.4 m (14.4 ft) high. Two 10000 kg (22,046-lb) travelling gantries traverse the hold, front to rear, each mounting two 5000-kg (11,023-lb) hoists that travel the cabin width. Above the lightly pressurized cargo

hold is the fully pressurized flight deck for a crew of six, to its rear accommodation for a relief crew, and behind the wing a cabin for 88 passengers. Each main landing gear comprises five pairs of wheels in tandem, the front two pairs steerable, and there are two twin-wheel nosegear units. The An-124 is designed for operation from any 1200 m (3,800 ft) of rough field or packed snow. It has high ground manoeuvrability and a kneeling capability, to front or rear, to speed loading and unloading. Flight control is via a fly-by-wire system with quadruple redundancy, permitting

relaxed static stability, and the An-124's avionics includes triplicated inertial navigation.

Naturally intended for use by both Aeroflot and the Soviet air force V-TA, the An-124 was expected to be in service by mid-1986.

Specification
Antonov An-124
Type: heavy-lift strategic transport
Powerplant: four 23428-kg (51,650 lb) thrust Lotarev D-18T turbofan engines
Performance: maximum cruising speed 466 kts (865 km/h; 537 mph) at 12000 m (39,370 ft); cruising

Currently the largest aircraft in the world, Antonov's giant Ruslan will serve with both Aeroflot and the V-TA.

speed 432 kts (800 km/h; 497 mph); range with maximum payload 4500 km (2796 miles); ferry range with maximum fuel 16000 km (9942 miles)
Weights: maximum payload 150000 kg (330,693 lb); maximum take-off 405000 kg (892,872 lb)
Dimensions: span 73.30 m (240 ft 5.8 in); length 69.50 m (228 ft 0.2 in); height 22.00 m (72 ft 2.1 in)
Armament: none
Operator: Soviet Union

Atlas C4M Kudu

Atlas Aircraft developed the C4M Kudu entirely in South Africa, but there is a clear design connection with the Aeritalia/Aermacchi AM.3C. The Kudu is a general-purpose transport, accommodating a crew of two and four to six troops or passengers in the cabin. Freight (up to 560kg/1,235 lb) can be loaded via a double door on the left side, and there is a sliding door for parachute jumping on the right. Other military applications include supply dropping, casevac and aerial survey. The first prototype flew on 16 February 1974.

Specification
Type: six/eight-seat STOL utility light transport
Powerplant: one 254-kW (340-hp) Lycoming GSO-480-B1B3 flat-six piston engine
Performance: maximum speed at 2440 m (8,005 ft) 259 km/h (161 mph); maximum cruising speed at 3050 m (10,005 ft) 233 km/h (145 mph); initial rate of climb 244 m

(801 ft)/minute; service ceiling 4270 m (14,010 ft); range with 400 kg (882 lb) payload 740 km (460 miles); range with maximum fuel 1,297 km (806 miles)
Weights: empty 1230 kg (2,712 lb); maximum take-off 2040 kg

(4,497 lb)
Dimensions: span 13.08 m (42 ft 10.9 in); length 9.31 m (30 ft 6.5 in); height 3.66 m (12 ft 0 in); wing area 20.97 m² (225.7 sq ft)
Armament: none
Operator: South Africa

Based on the Aermacchi AL.60 airframe, the Atlas Kudu is an indigenous light observation aircraft that is used for a variety of duties, including spotting, liaison, medevac and supply dropping. The type's rugged structure is welcome in bush operations.

Atlas Impala

Advanced training and counter-insurgency duties within the South African Air Force are undertaken by a variant of the licence-produced MB.326M. Manufacture has been by Atlas Aircraft at its Kempton Park, Transvaal works, the aircraft being named for the Impala antelope. Contracts covered 16 Impala Mk 1 aircraft from Italy (the first of which flew in South Africa on 11 May 1966), a further 30 as kits of diminishing completeness, which were assembled from November 1966 onwards, and 105 made by Atlas. Initial deliveries were made to what is now the Flying Training School at Langebaanweg, where future jet pilots fly the aircraft for 110 hours in 32 weeks after a basic course on Harvards. They then transfer to No 7 Sqn at Ysterplaat for continuation training on Impala 1s, and finally to No 85 AFS at Pietersburg for Mirage conversion, aided by both Impala 1s and 2s. No 7 is an Active Citizen Force (ACF) Sqn as is No 6 at Port Elizabeth, the other operational user. South Africa's national aerobatic team, the Silver Falcons,

Atlas Impala Mk II of the South African Air Force.

flies Mk 1s based with the Central Flying School at Dunottar. On 13 February 1974, six months before the last Mk 1 was delivered, Atlas flew the first of seven single-seat, attack-optimised MB.326Ks received from Italy in kit form and known as the Impala Mk 2. At least 85 were produced before manufacture ended in 1983. This variant is flown by three ACF squadrons (No 4 at Lanseria, No 5 Durban and No 8 Bloemspruit) and has provided close air support for ground forces penetrating Angolan territory on several occasions. Impala 1s may carry armament for training purposes,

allowing them to participate in the regular 'Gemsbok' weapons competitions open to all Impala units.

Specification
Atlas Impala Mk 2
Type: single-seat light attack aircraft
Powerplant: one 1524-kg (3,360-lb) thrust Rolls-Royce Viper 540 turbojet engine
Performance: maximum speed 890 km/h (553 mph); initial rate of climb 1980 m (6,495 ft)/minute; radius of action 268 km (167 miles) with 1280 kg (2,822 lb) weapon load
Weights: empty 3123 kg (6,885 lb);

maximum take-off 5897 kg (13,000 lb)
Dimensions: span over tanks 10.85 m (35 ft 7.2 in); length 10.67 m (35 ft 0 in); height 3.72 m (12 ft 2.5 in); wing area 19.35 m² (208.29 sq ft)
Armament: twin internal 30-mm (1.18-in) DEFA cannon: up to 1814 kg (4,000 lb) of bombs, podded rockets, missiles or reconnaissance pod on six underwing pylons
Operators: South Africa

The Atlas Impala Mk I is used as a basic and advanced jet trainer of the SAAF. It is used for all forms of training

Beech 99

Beech Aircraft Corporation's Model 99 has been one of the most successful third-level commuter aircraft. Distantly related to the C-45, it is a stretched version of the Model 80 Queen Air which is powered by PT6 turboprops and was flown first on 25 October 1966. To facilitate cargo carrying there is a separate crew door, and a belly mounted cargo pod can be installed easily to contain 363 kg (800 lb) of freight. Large double-hinged rear doors admit stretchers, and light cargo can be carried in the nose. The main military user has been Chile, which since 1971 has operated eight for transport, and as a navigational and instrument flight trainer with the Grupo 11 at Valparaiso.

Chilean air force Beech Model 99s are assigned several roles including transport and navigation training.

Specification
Beech Model C99
Type: light passenger and freight transport
Powerplant: two 533-kW (715-shp) Pratt & Whitney Canada PT6A-36 turboprop engines
Performance: maximum speed 496 km/h (308 mph); maximum cruising speed 460 km/h (286 mph); initial rate of climb 677 m (2,220 ft)/minute; service ceiling 8559 m (28,080 ft); maximum range 1595 km (991 miles)
Weights: empty 2778 kg (6,124 lb); maximum take-off 5126 kg (11,300 lb)
Dimensions: span 13.98 m (45 ft 10.5 in); length 13.58 m (44 ft 6.75 in); height 4.38 m (14 ft 4.25 in); wing area 25.98 m² (279.7 sq ft)
Armament: none
Operators: Chile, Peru

Beech Model 24 Musketeer

On 23 October 1961 Beech flew the prototype Musketeer two/six-seater of which, ultimately, over 4,000 were built. Features of the design included electrically-actuated slotted flaps and fixed tricycle landing gear. In 1970 20 two-seaters were acquired by the Fuerza Aérea Mexicana for instrument training, and 25 were supplied to the Canadian Armed Forces as primary trainers, with the designation CT-134.

Specification
Type: two/six-seat light aircraft
Powerplant: one 112-kW (150-hp) Lycoming O-320-E3D flat-four piston engine
performance: maximum speed at sea level 225 km/h (140 mph); maximum cruising speed at 2135 m (7,000 ft) 211 km/h (131 mph); range with maximum fuel and allowances and reserves 1421 km (883 miles)
Weights: empty equipped 630 kg (1,390 lb); maximum take-off 1021 kg (2,250 lb)
Dimensions: span 9.98 m (32 ft 9 in); length 7.65 m (25 ft 1 in); height 2.51 m (8 ft 3 in), wing area 13.56 m² (146.0 sq ft)
Armament: none
Operators: Algeria, Canada, Hong Kong, Mexico

Beech Model 24 Musketeer

Beech Model A36 Bonanza/QU-22

The Bonanza prototype flew on 22 December 1945 and today well over 15,000 of all versions have been built. The Model A36, introduced in 1968, has a larger cabin with double cargo doors on the right side, but the most noticeable change is replacement of the distinctive V-tail by a conventional unit with swept vertical surfaces. Electrically-retractable landing gear is standard and powerplant consists of a Continental IO-520 engine. In Vietnam the A36 was used as a radio relay platform in the Igloo White programme in which air-dispensed sensors were distributed in areas where it was believed there was enemy activity. Some 40 A-36 aircraft were procured for conversion as relay aircraft. The intention was that they should be equipped with a microwave command guidance system, so that they could be flown to and operate in any area as RPV (drone) aircraft, and these entered service under the designation QU-22B. The normally sleek nose of the Bonanza was blunted by the installation of a 254-kW (340-hp) or 280-kW (375-hp) Continental GTSIO-520 engine which, through the medium of a large reduction gear, drove a large-diameter, slow-turning propeller. Special communications systems were installed. Despite their 'drone' capability, the QU-22Bs used operationally in Vietnam were flown conventionally by a pilot.

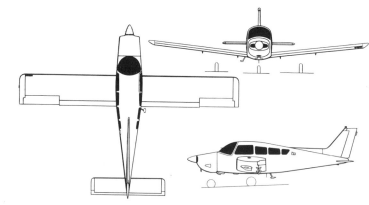

Dual controls and good handling characteristics make the Bonanza an excellent low-cost trainer. The Spanish air force operates 29 F-33As.

Specification
Beech A36 Bonanza
Type: six-seat lightweight utility aircraft
Powerplant: one 209-kW (280-hp) Continental IO-520 flat-six piston engine
Performance: maximum speed at sea level 328 km/h (204 mph); maximum cruising speed 314 km/h (195 mph); range with allowances and reserves 1577 km (980 miles)
Weights: empty equipped 916 kg (2,020 lb); maximum take-off 1633 kg (3,600 lb)
Dimensions: span 10.20 m (33 ft 5.5 in); length 8.13 m (26 ft 8 in); height 2.57 m (8 ft 5 in); wing area 16.81 m² (181.0 sq ft)
Armament: none
Operators: Haiti, Iran, Mexico, Spain

Beech T-34

Conceived as a tandem trainer derived from the Bonanza, the Beech Model 45 made its maiden flight on 2 December 1948, but it was not until March 1953 that it was ordered for the US Air Force as the T-34A Mentor. Eventually 450 were bought, plus 423 under the designation T-34B for the US Navy. The Model 45 was also produced under licence by Fuji for service in Japan and with the Philippine air force. Production was also undertaken in Argentina, and US-built examples were delivered to a number of recipients of aid from the Military Assistance Program. In 1973 the US Navy's hunt for a turboprop-powered primary trainer resulted in Beech being awarded a contract for the conversion of two T-34Bs to YT-34C configuration. Flown from 21 September 1973, the evaluation proved satisfactory and led to orders for more than 300 new-build T-34Cs for service with Naval Air Training Command. Delivery began in November 1977, and it is expected that the US Navy will procure a total of 450. In addition to the benefits of turbine power, the T-34C also has modern avionics and air-conditioning, the latter being particularly welcome in the southern United States where the navy has its main training bases. An armed T-34C-1 can undertake light attack and forward air control duties in addition to armament training. Although not acquired for service with the US armed forces, this has proved popular with seven overseas customers.

Specification
Beech T-34C
Type: two-seat primary trainer
Powerplant: one 533-kW (715-shp) Pratt & Whitney Canada PT6A-25 turboprop engine, torque-limited to 298 kW (400 shp)
Performance: maximum cruising speed at 5180 m (17,000 ft) 396 km/h (246 mph); initial rate of climb 451 m (1,480 ft)/minute; range at 333 km/h (207 mph), 1310 km (814 miles)
Weights: empty (T-34C) 1343 kg (2,960 lb), (T-34C-1) 1356 kg (2,990 lb); maximum take-off (T-34C) 1950 kg (4,300 lb), (T-34C-1)

2495 kg (5,500 lb)
Dimensions: span 10.16 m (33 ft 4 in); length 8.75 m (28 ft 8.5 in); height 2.92 m (9 ft 7 in); wing area 16.68 m² (179.6 sq ft)
Armament: (T-34C-1 only) four underwing hardpoints carrying a maximum 544 kg (1,200 lb);
Operators: Algeria, Argentina, Chile, Colombia, Dom. Rep., Ecuador, El Salvador, Gabon, Indonesia, Japan, Mexico, Morocco, Peru, Philippines, Spain, Taiwan, Turkey, United States, Uruguay, Venezuela

The T-34 has seen over 30 years' service with the US Navy as its principal basic trainer. Current aircraft are turboprop-powered T-34Cs, this aircraft serving with Training Wing 8 (TAW-8).

Beech Model 50 and 65/U-8 Twin Bonanza and Queen Air

In 1951 the US Army evaluated aircraft capable of meeting its requirement for communications and light transport. An 'off the shelf' design was desired and the choice was the Twin Bonanza, 184 early L-23 versions being delivered during 1952-56. Six L-23Es were similar to the D.50 Twin Bonanza, and by 1960 the RL-23D was developed with side-looking airborne radar (SLAR). In 1962 L-23D, RL-23D and L-23E aircraft became respectively the U-8D, RU-8D and U-8E, named Seminole.

In January 1959 Beech flew the L-23F derived from the Model 65 Queen Air. The more spacious fuselage seated up to 11 passengers, a payload increase of some 318 kg (700 lb). Evaluation at Fort Rucker led to procurement of 71 U-8F Seminoles. Standard seating is for a crew of two and six passengers; the U-8G

seats four VIP passengers, and has unsupercharged engines. Queen Airs serve as transports and navigation trainers with the JMSDF.

Specification
Beech U-8F Seminole
Type: six/eleven-seat light transport
Powerplant: two 254-kW (340-hp) Lycoming IGSO-480 flat-six piston engines
Performance: maximum speed at 3660 m (12,000 ft) 385 mph (239 km/h); maximum cruising speed at 4630 m (15,200 ft) 344 km/h (214 mph); range with allowances and reserves 1794 km (1,115 miles)
Weights: empty 2263 kg (4,990 lb); maximum take-off 3493 kg (7,700 lb)
Dimensions: span 13.98 m (45 ft 10.5 in); length 10.82 m (35 ft 6 in);

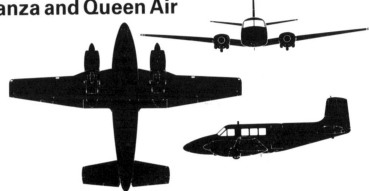

Beech U-8F

height 4.33 m (14 ft 2.5 in); wing area 25.74 m² (277.06 sq ft)
Armament: none
Operators: Argentina, Dom. Rep.,

Ecuador, Israel, Japan, Pakistan, Peru, Taiwan, Thailand, Uruguay, Venezuela, US Army

Beech C-12/T-44/U-21/King Air

Currently operated by all elements of the US armed forces, numerous variants of the Beech King Air and Super King Air undertake such diverse missions as multi-engine pilot training, utility transport, acquisition of signals and communications intelligence, battlefield surveillance and electronic warfare. Procurement began in October 1966 when the US Army placed a contract for 48 King Air 65-A90-1Cs, designating them U-21A Ute. Acquisition eventually totalled 141, these 12-seaters, being deployed mainly as utility transports, unlike the RU-21B (three), RU-21C (two) and RU-21D (16) all configured for Sigint (signals intelligence) and EW (electronic-warfare) duties. Subsequently, a number of U-21As were modified for EW missions; known conversions include four EU-21As for electronic reconnaissance, four RU-21As for Sigint/EW, and 18 RU-21Ds operating alongside the other RU-21Ds; all were later brought to RU-21E standard by installation of the more powerful T74-CP-700 engine. Subsequently, 12 RU-21Es were further updated to RU-21H configuration, this involving some structural strengthening for operation at higher gross weights. Some U-21As were also brought up to U-21G standard, while five examples of the U-21F utility model were acquired by the US Army in 1970, these being based

Battlefield surveillance and Elint gathering are the roles for US Army Beech RC-12Ds, including this West Germany-based example.

on the King Air 100 and capable of carrying up to 15. The US Navy received 61 King Air 90s during 1977-80 for use as multi-engine pilot trainers. Designated T-44A, these serve with Training Wing Four at Corpus Christi, Texas, having replaced the Grumman TS-2A Tracker.

The newest variant to enter US service has been the Super King Air 200. Following the delivery of three 'special mission' RU-21Js to the Army in 1974, production turned to the C-12A transport which entered service in July 1975, 60 for the Army and 30 for the Air Force. Some 66 UC-12Bs followed, 49 being allocated to the Navy, the remainder to the Marine Corps. Next came the C-12C for the Army, with PT6A-41 engines, 14 being built. The Army was again the major customer for the C-12D ordering 33, some of

which appeared as RC-12D EW surveillance aircraft while at least five were earmarked for overseas nations as part of the FMS programme. A similar utility version, the UC-12D, was delivered in 1984, six being for the USAF and six for the Army National Guard. The most recent major order was for a version of the Super King Air B200C, 40 being delivered to Military Airlift Command. Examples of the Super King Air have also been obtained for use as light transports by a small number of overseas air arms, some being fitted with search radar and observation windows which permit them to be employed in the maritime patrol role.

Specification
Beech C-12A
Type: utility transport/special missions and pilot trainer

Powerplant: two 559-kW (750-shp) Pratt & Whitney Canada PT6A-38 turboprop engines
Performance: maximum speed at 4265 m (14,000 ft) 481 km/h (299 mph); maximum cruising speed at 9145 m (30,000 ft) 438 km/h (272 mph); range at maximum cruising speed 2935 km (1,824 miles)
Weights: empty 3538 kg (7,800 lb); maximum take-off 5670 kg (12,500 lb)
Dimensions: span 16.61 m (54 ft 6 in); length 13.34 m (43 ft 9 in); height 4.57 m (15 ft 0 in); wing area 28.15 m² (303 sq ft)
Armament: none
Operators: Algeria, Argentina, Chile, Colombia, Ecuador, Eire, Greece, Indonesia, Israel, Jamaica, Japan, Mexico, Morocco, Peru, Spain, Thailand, Uruguay, US Air Force, US Army, US Marine Corps, US Navy, Venezuela

Beech T-42A Cochise/Baron

Operated by the US Army Aviation Flying School at Fort Rucker, Alabama, the Beech T-42A Cochise is primarily an instrument trainer. Evolved from the Beech Baron, it was a logical candidate for an 'off-the-shelf' order and following the assessment of several competing types a total of 65 was obtained, these being equipped with US Army radio, navigation aids and avionics. Designated T-42A these aircraft entered service in 1966; five were ordered in 1971 for supply to the Turkish army under the Military Assistance Program. A number of Barons have also found their way into military service around the world, the most notable customer being Spain which acquired 19 for use by the Ejercito del Aire in the training role.

Specification
Beech T-42A
Type: four/six-seat instrument

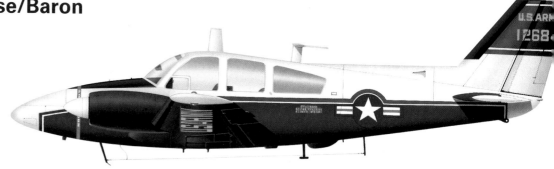

Sixty-five T-42As were supplied to the US Army for training and light transport. Many now serve with Army National Guard units.

trainer
Powerplant: two 194-kW (260-hp) Teledyne Continental IO-470-L flat-six piston-engines
Performance: maximum speed 372 km/h (231 mph); maximum cruising speed 348 km/h (216 mph); range at maximum cruising speed, including allowances and reserve,

1477 km (918 miles)
Weights: empty 1463 kg (3,226 lb); maximum take-off 2313 kg (5,100 lb)
Dimensions: span 11.53 m (37 ft 10 in); length 8.53 m (28 ft 0 in); height 2.92 m (9 ft 7 in); wing area 18.50 m² (199.1 sq ft)
Armament: none

Operators: (T-42A) Turkey, US Army; (Baron) Algeria, Brazil, Central African Republic, Haiti, Mexico, Pakistan, Spain, Venezuela

Bell Model 47

On 8 December 1945 Bell flew the prototype of a classic helicopter design, the Model 47; on 8 March 1946 this was awarded the first Approved Type Certificate to be issued for a civil helicopter. The type remained in production by Bell into 1973, and was also built widely under licence. The Model 47 has been used by armed forces all over the world, its simplicity and low cost more than outweighing its limited capabilities.

In 1947 the USAAF procured for service evaluation 28 of the improved Model 47A, powered by 157-hp (117-kW) Franklin O-335-1 piston engines; 15 were designated YR-13, three YR-13As were winterized for cold weather trials in Alaska, and the balance of 10 went to the US Navy for evaluation as HTL-1 trainers. The US Army's first order was issued in 1948, 65 being accepted under the designation H-13B; all US Army versions were later named Sioux. In 1952 15 were converted to carry external stretchers, with the designation H-13C (famed from TV's MASH series). Two-seat H-13Ds with skid landing gear, stretcher carriers and Franklin O-335-5 engines followed, and generally similar three-seat dual control H-13Es. The H-13G added a small movable elevator, the H-13H the 250-hp (186-kW) Lycoming VO-435 engine.

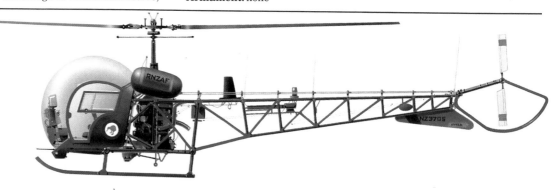

Bell Model 47G of the Royal Australian Air Force.

Some H-13Hs were used by the USAF, as were H-13Js with 240-hp (179-kW) Lycoming VO-435s acquired for the US President. Two H-13Hs converted for trial purposes, with a large rotor and 225-hp (168-kW) Franklin 6VS-335s, were designated H-13K. In 1962 US Army H-13E, -G, -H and -K Sioux were redesignated with prefix O, for observation; USAF H-13H and H-13J were given prefix U (utility). Later acquisitions were the three-seat OH-13S and TH-13T instrument trainer.

US Navy procurement began with 12 HTL-2s and nine HTL-3s, but the first major version was the HTL-4, followed by the HTL-5 with an O-335-5 engine; HTL-6 trainers incorporated the movable elevator.

Final USN versions were the HUL-1 acquired for ice-breaking ships and the HTL-7 all-weather instrument trainer. In 1962 the HTL-4, HTL-6, HTL-7 and HUL-1 were redesignated respectively TH-13L, TH-13M, TH-13N and UH-13P.

The Model 47 was built under licence during 1954-76 by Agusta in Italy, Kawasaki in Japan, and Westland Helicopters received a sub-licence from Agusta for the AB 47G-2 Sioux for the British Army.

Specification
Bell Model 47G-5A
Type: utility helicopter
Powerplant: one 265-hp (198-kW) Lycoming VO-435-B1A flat-six piston engine

Performance: maximum speed at sea level 169 km/h (105 mph); cruising speed 137 km/h (85 mph) at 1525 m (5,000 ft); service ceiling 3200 m (10,500 ft); range with maximum fuel 412 km (256 miles)
Weights: empty equipped 786 kg (1,732 lb); maximum take-off 1293 kg (2,850 lb)
Dimensions: main rotor diameter 11.32 m (37 ft 1.5 in); length, rotors turning 13.30 m (43 ft 7.5 in); height 2.84 m (9 ft 3.75 in); main rotor disc area 100.61 m² (1,083.0 sq ft)
Armament: currently, none
Operators: about 20 countries use the Model 47 in small numbers

Bell Model 204/UH-1 Iroquois

In the early 1950s the US Army notified its requirement for a helicopter with a primary casevac mission, but suitable also for utility use and as an instrument trainer; in 1955 the design submitted by Bell was announced the winner, three prototypes of the Bell Model 204 being ordered under the designation XH-40. The first of these (55-4459) was flown initially on 22 October 1956, its 615-kW (825-shp) Lycoming XT53-L-1 turboshaft engine, derated to 522-kW (700-shp), making it the first turbine-powered aircraft to be acquired by the US Army. The XH-40s were followed by six YH-40 service trials aircraft with small changes, the most important being a 30.5-cm (1.0-ft) fuselage 'stretch'. When ordered into production, the designation HU-1A was allocated, the HU prompting the 'Huey' nickname that survived the 1962 rede-

Austria flies the Bell Model 204 on general duties around the country. These are used for transport and rescue in the mountains.

signation to UH-1, and which became far better known than the official title of Iroquois. Initial production version was the HU-1A, with a crew of two, plus six passengers or two stretchers, and with the T53-L-1 engine. It was followed by the HU-1B with revised main rotor blades and an enlarged cabin seating two crew, plus seven passengers or three stretchers; early-production helicopters had the 716-kW (960-shp) Lycoming T53-L-5, late-production machines the 820-kW

(1,100-shp) T53-L-11. In 1962 the HU-1A and HU-1B were redesignated UH-1A and UH-1B respectively, and in 1965 the UH-1B was superseded in production by the UH-1C. This had a 'door-hinge' main rotor with wide-chord blades, giving improvements in performance and manoeuvrability. Other military versions include the US Marine Corps UH-1E (with rescue hoist, rotor brake and special avionics); the USAF UH-1F and similar TH-1F trainer (962-kW/1,290-shp General Electric T58-GE-3 and increased-diameter rotor); the US Navy search and rescue HH-1K (similar to UH-1E but with 1044-kW/1,400-shpT53-L-13), plus the TH-1L (training) and UH-1L (utility) versions of the UH-1E with T53-L-13 engine; and the US Army UH-1M with night sensor equip-

ment (three acquired for evaluation). In addition to production for the US armed forces Bell also built the Model 204B for military export, and this version was extensively licence-built for both civil and military use by Agusta in Italy (Agusta-Bell AB.204) and by Fuji in Japan, the latter also developing the Fuji-Bell 204B-2 with increased engine power and a tractor tail rotor.

Specification
Bell UH-1E Iroquois
Type: assault support helicopter
Powerplant: one 820-kW (1,100-

shp) Lycoming T53-L-11 turboshaft engine
Performance: maximum speed 120 kts (222 km/h; 138 mph); initial rate of climb 2,350 ft (716 m) per minute; service ceiling 5090 m (16,700 ft); range with maximum fuel 341 km (212 miles)
Weights: empty 2155 kg (4,750 lb); maximum take-off 3856 kg (8,500 lb)
Dimensions: main rotor diameter 13.41 m (44 ft 0 in); length of fuselage 12.98 m (42 ft 7 in); height 4.44 m (14 ft 7 in); main rotor disc area 141.26 m^2 (1,520.5 sq ft)

Armament: some UH-1A/-1Bs were operated in Vietnam with up to four side-mounted 7.62-mm (0.3-in) machine-guns, or two similarly mounted packs each containing 24 rockets
Operators: Costa Rica, Honduras, Indonesia, Japan, South Korea, Norway, Singapore, USA, Uruguay, Venezuela

Bell Model 205/UH-1D/H Iroquois/Agusta-Bell AB 205

In 1960 Bell proposed an improved version of the Model 204 with a longer cabin and relocated fuel cells, providing accommodation for a pilot and 14 troops, or six stretchers, or 1814 kg (4,000 lb) of freight. In July 1960 the US Army bought seven YUH-1D (Model 205) helicopters for service test. The first flew on 16 August 1961 and production UH-1Ds with the 820-kW (1,100-shp) Lycoming T53-L-11 and standard fuel of 833 litres (220 US gallons) supplemented by auxiliary tanks to a total of 1968 litres (520 US gallons), were delivered to the 11th Air Assault Division at Fort Benning, Georgia, from 9 August 1963. Dornier built 352 under licence for the Heeresfliegertruppen and Luftwaffe, and Agusta in Italy built the AB 205A-1 for many customers.

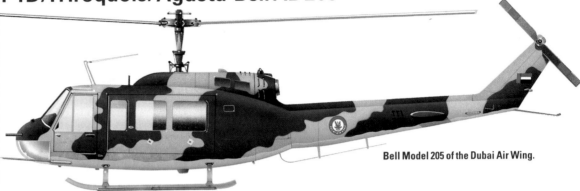

Bell Model 205 of the Dubai Air Wing.

The UH-1H introduced the 1044-kW (1,400-shp) Lycoming T53-L-13 and delivery began in September 1967. Taiwan licence-built 118. Variants include the CH-118 for the Canadian Armed Forces and the USAF's HH-1H for local base rescue. The UH-1D/H played the central role in special warfare operations in Laos, Cambodia and South Vietnam, and also evacuated nearly all battlefield casualties.

Specification
Agusta-Bell AB 205A-1
Type: utility helicopter
Powerplant: one 1044-kW (1,400-shp) Lycoming T5313B (T53) turboshaft engine, flat-rated at 932 kg (1,250 shp)
Performance: maximum speed 222 km/h (138 mph); initial rate of climb 619 m (2,030 ft)/minute; service ceiling 4480 m (17,700 ft);

range 532 km (331 miles)
Weights: empty 2356 kg (5,194 lb); maximum take-off 4763 kg (10,500 lb)
Dimensions: main rotor diameter 14.63 m (48 ft 0 in); length of fuselage 12.78 m (41 ft 11.1 in); height 4.48 m (14 ft 8.4 in); main rotor disc area 168.11 m^2 (1,809.6 sq ft)

Armament: usually none
Operators: Argentina, Australia, Bahrain, Bolivia, Brazil, Burma, Canada, Chile, Colombia, Dom. Rep., Dubai, Ecuador, West Germany, Greece, Guatemala, Honduras, Indonesia, Iran, Japan, South Korea, Libya, Mexico, New Zealand, Pakistan, Panama, Peru, Philippines, El Salvador, Singapore,

Large numbers of UH-1H Iroquois general-purpose helicopters continue to serve with the US Army, typified by this Arizona National Guard example.

Spain, Taiwan, Tanzania, Thailand, Tunisia, Turkey, Uganda, United States, Uruguay, Venezuela, Vietnam

Bell OH-58 Kiowa

Designed to meet a 1962 requirement for a new US Army LOH (Light Observation Helicopter), the Bell OH-4A was not selected, but the design was quickly developed into the Model 206A JetRanger, a type which went on to achieve considerable success in civil and export military markets. The contest was reopened in 1967 and this time the Bell OH-58 emerged as the winner in March 1968. The reward was a contract for 2,200, all of them delivered by 1973.

Differing from the 206A by virtue of a larger-diameter main rotor, military avionics and a few detail changes, the OH-58A Kiowa reached US Army units in May 1969 and was quickly deployed to South Vietnam. Since then it has seen service throughout the world, normally operating as a scout helicopter in conjunction with AH-1 HueyCobra

Pakistan army Bell 206 Jetrangers help perform liaison duties around the nation.

gunships, and is often armed with an XM27 Minigun. Bell rebuilt 275 to OH-58C standard, with the 313-kW (420-shp) C20B engine (with infrared suppressed exhaust nozzles), flat glass cockpit panels to cut down reflection, and wire cutters above and

below the cockpit. In 1981 the proposed Bell 406 development of the OH-58A won the AHIP (Army Helicopter Improvement Program) 'Near-Term Scout' contest, and 578 Kiowas are to be rebuilt as OH-58Ds by 1991. They have a ball-type mast-

mounted sight, 485-kW (650-shp) C30P engine, completely new avionics, inertial navigation, an airborne-target handoff system, night vision devices and twin FIM-92 Stinger missiles.

In addition to those operated by

Bell OH-58 Kiowa (continued)

the US Army, variants of the basic Bell 206 also serve with the US Navy and many other countries. Large numbers have been built (as the AB 206 family) by Agusta in Italy, some of these (eg Swedish) carrying anti-submarine torpedoes. The Australian Army JetRangers were assembled by Commonwealth Aircraft. Bell has also offered a military version of the seven-seat 206L LongRanger with TOW missiles and advanced attack avionics.

Specification
Bell OH-58A Kiowa
Type: light observation/scout helicopter
Powerplant: one 236-kW (317-shp) Allison T63-A-700 turboshaft engine
Performance: maximum speed 222 km/h (138 mph) at sea level; cruising speed 188 km/h (117 mph); maximum range at sea level with 10 per cent reserves 481 km (299 miles)
Weights: empty 664 kg (1,464 lb); maximum take-off 1361 kg (3,000 lb), (206L 1814 kg/4,000 lb)
Dimensions: main rotor diameter 10.77 m (35 ft 4 in); length, rotors turning 12.47 m (40 ft 11 in); height 2.91 m (9 ft 6.5 in); main rotor disc area 91.09 m² (980.53 sq ft)
Armament: usually none
Operators: (* = Agusta-built in

Italy) Argentina, Australia, Austria*, Brazil, Brunei, Canada, Chile, Colombia, Dubai*, Finland*, Guyana, Iran*, Israel, Italy*, Jamaica, Japan, Kuwait*, Liberia, Libya*, Malaysia, Malta*, Mexico, Morocco (some*), Oman*, Peru,

Saudi Arabia*, Somalia, Spain, Sri Lanka, Sweden*, Tanzania*, Thailand, Turkey*, Uganda*, United Arab Emirates*, USA (all services), Venezuela

Spotting duties by OH-58As are augmented by a six-barrel machine-gun for suppressive fire, a necessity when operating over hostile territory.

Bell Model 209/AH-1 HueyCobra

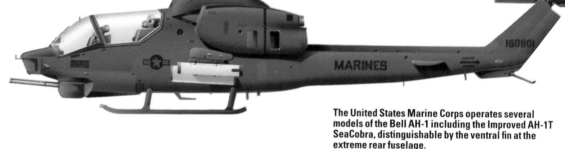

The US Army's early experience in Vietnam showed the need for an agile, fast and well-armed helicopter to escort and give fire support to transport helicopters. Bell had in 1963 started investigating the armed helicopter concept with the OH-13X Scout, derived from the Model 47. This gave an appreciation of the limitations and possibilities of such a type, and proved valuable when the US Army began to make urgent demands for an armed helicopter to serve in Vietnam. When details were issued of the Army's AAFSS (Advanced Aerial Fire Support System) requirement to replace the failed Lockheed Cheyenne, Bell initiated crash development of a company-funded prototype derived from the Model 204; it had the powerplant, transmission and wide-chord rotor of the UH-1C but introduced a new fuselage seating the gunner in the nose and the pilot higher in the rear. The fuselage was very narrow, only 0.97 m (3 ft 2 in) at its widest point and this, coupled with a low silhouette, made the aircraft easy to conceal on the ground and a more difficult target in the air. Designated Bell Model 209 Huey-Cobra, the prototype (N209J) was flown for the first time on 7 September 1965. It began service tests in December 1965, and then gained an order for two pre-production AH-1G helicopters on 4 April 1966 and an initial contract for 110 production AH-1Gs nine days later; such was the degree of urgency. Initial production deliveries reached the US Army in June 1967 and within weeks the type had become operational in Vietnam. AH-1G produc-

The United States Marine Corps operates several models of the Bell AH-1 including the Improved AH-1T SeaCobra, distinguishable by the ventral fin at the extreme rear fuselage.

tion totalled 1,119 for the US Army, of which 38 were transferred to the US Marine Corps for training; some US Army AH-1Gs were converted as TH-1G dual-control trainers. Later variants include 92 AH-1Q conversions from AH-1Gs to fire TOW missiles, and the AH-1R conversion with a 1342-kW (1,800-shp) T53-L-703 turboshaft and no TOW capability. The designation Modified AH-1S applies to 315 AH-1Gs retrofitted with a TOW system and the powerplant of the AH-1R, plus the 92 AH-1Qs brought up to this same

Twenty Modified AH-1S anti-tank helicopters have been supplied to Pakistan, these featuring TOW missile-capability and an uprated powerplant.

Israel's AH-1 fleet has varying marks, including this AH-1S. These have seen action against Syrian tanks during the fighting in the Bekaa valley.

standard. They were followed by 100 Production AH-1S which have improved avionics, uprated powerplant and transmission; 98 similar Up-gun AH-1S which introduce a universal 20/30-mm gun turret and other refinements; and the current Modernised AH-1S. This last incorporates the updates of the Production and Up-gun AH-1S, and improvements that include new air data, Doppler navigation and fire-control systems, and continuing programmes are under development to enhance the capability of the Huey-Cobra. In addition to the production of Modernised AH-1S Hueys for the US Army, others are in the process of manufacture and delivery for foreign air arms.

US Marine Corps interest in a well-armed close-support helicopter was heightened by the US Army's procurement of the Bell AH-1G HueyCobra. Following USMC evaluation of the AH-1G it was decided to acquire this for service, but with the extra reliability of twin-engine powerplant. In May 1968 the USMC ordered 49 AH-1J SeaCobra helicopters, and as an interim measure procured 38 AH-1G Huey-Cobras which were used for training and initial deployment until delivery of the SeaCobras during 1970-1. This initial AH-1J retained basically the same airframe as the AH-1G, with some detail changes to cater for the higher output of the Pratt & Whitney Canada T400-CP-

A total of 58 Bell AH-1S Cobras are planned for the JGSDF, most being licence-built by Fuji.

400 powerplant, a revolutionary engine with two powerplants which ran for the first time in July 1968. Flat-rated at 820 kW (1,100 shp) and with a take-off and emergency power rating of 932 kW (1,250 shp), the T400 is a militarized version of Pratt & Whitney Canada's PT6T-3 Turbo Twin-Pac, incorporating aluminium instead of magnesium in its construction, which is essential for a maritime or seaboard environment. Bell was also to build for Iran 202 similar AH-1Js, but these incorporated TOW-capability.

An additional 20 AH-1Js were delivered to the USMC in 1974-5, the last two of this batch being modified subsequently to serve as prototypes for the AH-1T Improved SeaCobra. Retaining many features of the AH-1J airframe, these have a slightly lengthened fuselage (to increase fuel capacity), a lengthened tail boom, improved main and tail rotors as developed for the Bell Model 214, and an uprated transmission to handle the full rated power of the upgraded T400-WV-402 power-

plant, developing 1469 kW (1,970 shp). The first of 57 AH-1Ts (59228) was flown on 20 May 1976 and delivered to the USMC on 15 October 1977. Of these, 51 have now been equipped to carry TOW missiles.

In 1980 Bell was loaned an AH-1T by the USMC, and demonstrated this machine with two General Electric T700-GE-700 turboshafts having a combined output of 2386 kW (3,200 shp). From this was planned an improved AH-1T which, in production form, will have T700-GE-401 turboshafts developing together 2424 kW (3,250 shp), a new combining gearbox and other improvements. Designated AH-1W SuperCobra, 44 have been ordered by the USMC with initial deliveries scheduled for March 1986.

Specification
Bell Modernised AH-1S
Type: anti-armour attack helicopter
Powerplant: one 1342-kW (1,800-shp) Lycoming T53-L-703 turboshaft engine, flat-rated at

820 kW (1,100 shp)
Performance: maximum speed with TOW missiles 122 kts (227 km/h; 141 mph); initial rate of climb 494 m (1,620 ft) per minute; service ceiling 3720 m (12,200 ft); range with maximum fuel and reserves 507 km (315 miles) at sea level
Weights: empty 2939 kg (6,479 lb); maximum take-off 4536 kg (10,000 lb)
Dimensions: main rotor diameter 13.41 m (44 ft 0 in); length, rotors turning 16.14 m (52 ft 11.5 in); height 4.12 m (13 ft 6.25 in); main rotor disc area 141.26 m^2 (1,520.53 sq ft)
Armament: eight TOW missiles, one General Electric universal turret for 20-mm or 30-mm cannon, plus launcher pods for 2.75-in (69.85-mm) folding-fin rockets
Operators: Greece, Iran, Israel, Japan, Jordan, South Korea, Pakistan, Turkey, United States

Bell Model 212 Twin Two-Twelve/UH-1N/Agusta-Bell AB 212

On 1 May 1968 Bell Helicopter announced that the Canadian government had approved development of the Bell 212 general-purpose helicopter. This was to be a UH-1 airframe combined with the Pratt & Whitney Canada PT6T Turbo Twin Pac powerplant. Canada ordered 50, with options on 20, under the designation CH-135. The US armed services also announced orders for the UH-1N, comprising 79 for the USAF, 40 for the Navy and 22 for the Marine Corps, with a follow-on order of 159 for the Navy and Marine Corps. Despite the Canadian origin, initial production deliveries went to the USAF in 1970.

The PT6T developed for the Model 212 provides 1342 kW (1,800 shp) and offers true engine-out capability, plus improved performance in hot day/high altitude operations. It has two PT6 engines driving into a combining gearbox; should one engine fail, the other continues to drive the helicopter's rotors.

Agusta in Italy builds the AB 212, first deliveries being made in late 1971, and has developed an ASW variant, the AB 212ASW. This introduces the PT6T-6 Twin Pac of 1398 kW (1,875 shp), local strengthening and deck hauldown equipment. Deliveries to the Italian navy's 5° Gruppo Elicotteri began in 1976.

Specification
Bell UH-1N

Type: general purpose/ASW/ASV helicopter
Powerplant: one 1342-kW (1,800-shp) Pratt & Whitney Canada PT6T-3 Turbo Twin-Pac, flat-rated to 962 kW (1,290 shp)
Performance: maximum cruising speed at sea level 229 km/h (142 mph); initial rate of climb 402 m (1,320 ft)/minute; service ceiling 4330 m (14,200 ft); maximum range with standard fuel, no reserves

Assigned to Marine Corps' squadron HMX-1, the Bell VH-1N is a VIP-configured derivative of the UH-1N.

420 km (261 miles)
Weights: empty 2787 kg (6,144 lb); maximum take-off 5080 kg (11,200 lb)
Dimensions: main rotor diameter 14.69 m (48 ft 2.25 in); length, rotors turning 17.46 m (57 ft 3.25 in); height 4.53 m (14 ft

10.25 in); main rotor disc area 173.90 m² (1,871.91 sq ft)
Armament (AB 212ASW): homing torpedoes, depth charges and air-to-surface missiles
Operators: Argentina, Bangladesh, Bahrain, Bangladesh, Bolivia, Brunei, Canada, Colombia,

The Canadian Armed Forces operates the twin engined UH-1N on SAR and utility tasks.

Ecuador, Ghana, Greece, Guyana, Israel, Jamaica, Japan, South Korea, Mexico, Norway, Peru, Philippines, Singapore, Sri Lanka, Thailand, Uganda, United States, Uruguay, Venezuela

---

Bell Model 214 Huey Plus

In October 1970 Bell announced that it had developed an improved Model 205, the Model 214 Huey Plus. The 1417-kW (1,900-shp) T53-L-702 Lycoming turboshaft engine drove a newly-developed main rotor of large diameter. Minor internal rearrangement provided accommodation for a crew of two and 12 passengers. Speed at maximum weight was 306 km/h (190 mph), but despite almost 50 per cent increase by comparison with the 205, the 214 failed to attract the US Army. However, on 22 December 1972, Bell received an order for 287 Model 214As for Iran; six more were ordered in March 1977, and 150 of this total still fly. These were powered by the 2185-kW (2,930-shp) LTC4B turboshaft. In 1975 Bell and the Iranian government agreed to establish a helicopter manufacturing industry in Iran.

The plan in March 1978 covered the construction in Iran of 50 Model 214As, and 350 of a completely new version designated 214ST with 19 passengers and power provided by two 1212-kW (1,625-shp) General Electric CT7-2A engines. Following the Iranian revolution Bell continued with the 214ST in the US as a company-funded project. Today it is in production at a weight of 7938 kg (17,500 lb), with the ability to carry a slung load of 3629 kg (8,000 lb), more than the loaded weight of the first 'Huey' helicopters.

Specification
Bell Model 214A

Type: utility helicopter
Powerplant: one 2185-kW (2,930-shp) Lycoming LTC4B-8D turboshaft engine
Performance: cruising speed

Bell Model 214A

259 km/h (161 mph); range with maximum payload 455 km (283 miles)
Weights: loaded 6260 kg (13,800 lb); maximum take-off with external load 6804 kg (15,000 lb)
Dimensions: main rotor diameter

15.24 m (50 ft 0 in); main rotor disc area 182.41 m² (1,963.5 sq ft)
Armament: none
Operators: (214 family, military): China, Iran, Thailand, Uganda, Venezuela

---

Bell Model 222

The decision to build five prototypes of the Bell Model 222 was announced in April 1974. Flown for the first time on 13 August 1976, the 222 was aimed mainly at the commercial market, first delivery being made in January 1980. Five variants have appeared, the initial 222A being superseded by an improved 222B incorporating a larger main rotor with blades of reduced chord, a taller

main-rotor mast, larger tail rotor and longer tail boom. The 222 Executive and 222 Offshore are self-explanatory. The 222UT (utility twin) differs visibly by having the standard retractable tricycle landing gear replaced by tubular skids. Two 222As were acquired by Urguay for service with the air force and navy.

Specification
Bell Model 222A

Type: utility helicopter
Powerplant: two 462-kW (620-shp) Lycoming LTS101-650C-3 turboshaft engines
Performance: economical cruising speed 246 km/h (153 mph); initial rate of climb 487 m (1,600 ft)/minute; service ceiling 3900 m (12,800 ft); range with maximum

fuel and reserves 523 km (325 miles)
Weights: empty 2204 kg (4,860 lb); maximum take-off, with external payload 3674 kg (8,100 lb)
Dimensions: main rotor diameter 12.12 m (39 ft 9 in); length of fuselage 12.50 m (41 ft 0 in); height 3.51 m (11 ft 6 in); main rotor disc area 115.29 m² (1241.0 sq ft)
Armament: none
Operator: Nigeria, Uruguay

Bell 412

A derivative of the Bell 212, incorporating a four-blade main rotor of advanced design, development of the Bell 412 was announced in September 1978. Two prototypes were produced by modifying 212s, the first of them flying in August 1979, and production deliveries began in January 1981. The 412 has since been adopted for service in Venezuela and by police/security forces in Bahrain and Nigeria.

In addition to production in the US, Bell's Italian licensee, Agusta, is also manufacturing the Model 412 for civil and military use, and has developed a multi-purpose military version which is described under the entry Agusta AB.412 Griffon.

Specification
Bell Model 412
Type: utility helicopter
Powerplant: one Pratt & Whitney Canada PT6T-3B-1 Turbo Twin-Pac flat-rated at 975-kW (1,308 shp)
Performance: maximum cruising speed 230 km/h (143 mph); initial rate of climb 442 m (1,450 ft)/minute; service ceiling 4330 m (14,200 ft); range, with maximum payload, standard fuel and 30 min reserves 370 km (230 miles)
Weights: empty 2854 kg (6,292 lb); maximum take-off 5262 kg (11,600 lb)
Dimensions: main rotor diameter 14.02 m (46 ft 0 in); fuselage length 12.92 m (42 ft 4.75 in); height 3.29 m (10 ft 9.5 in); main rotor disc area 154.40 m² (1,662.0 sq ft)
Armament: none
Operators: Bahrain, Indonesia, Nigeria, Venezuela

Bell/Boeing Vertol JVX

In 1982 Bell and Boeing Vertol teamed up to submit a proposal for the US government's JVX (Joint Services Advanced Vertical Lift Aircraft) project. The Bell/Boeing Model 901-X concept – broadly a scaled-up version of the Bell XV-15 tilt-rotor research aircraft – is an advanced design employing extensive use of composite material. It won a Naval Air Systems Command contract on 26 April 1983, covering two years of preliminary design work.

Full-scale development is expected to occupy a further two years and, if all goes well, the first of eight prototypes will fly in August 1987. Deliveries to the US Marine Corps are expected to begin in 1991, the type being employed as a tactical transport by this service which anticipates receiving 552 from the planned requirement for 1,086. Of this total 50 are earmarked for the Navy, 200 for the Air Force and 284 for the Army, the latter service planning to use JVX as an electronic warfare platform while those for the Navy and Air Force will undertake search and rescue.

The type's predecessor, the XV-15,

The US Marine Corps is expected to receive its first V-22 Ospreys in the early 1990s, with the Navy, Army and Air Force also likely recipients.

has already demonstrated successfully the validity of this concept; one of the two aircraft built spent much of 1983 engaged in a joint USN/USMC evaluation of projected JVX roles, these including search and rescue, slung cargo carrying and simulated weapons delivery.

Specification (provisional)
Type: multi-role tilt-rotor military aircraft
Powerplant: two 3620-kW (4,855-shp) General Electric T64-GE-717 turboshaft engines
Performance: maximum cruising speed 483 km/h (300 mph); mission radius with 24 troops at optimum altitude 370 km (230 miles); SAR mission radius at optimum altitude 853 km (530 miles)
Weights: maximum take-off weight, VTOL 19867 kg (43,800 lb), STOL 24948 kg (55,000 lb)
Dimensions: prop/rotor diameter (each) 11.58 m (38 ft 0 in); length 17.32 m (56 ft 10 in); height 6.15 m (20 ft 2 in)
Armament: not specified
Operators: procurement planned for US armed forces

Beriev Be-12 Tchaika

The Beriev design bureau, based at Taganrog on the Azov Sea, has been the main supplier of marine aircraft to the Soviet Navy since the 1930s, most of its aircraft going to the Northern and Black Sea fleets. The origins of the Be-12 go back to the LL-143 prototype of 1945. This led in 1949 to the Be-6, which served with success until 1967. Following testing of jet-powered flying-boats, the AV-MF selected a twin-turboprop amphibian bearing a family resemblance to the Be-6. The defensive armament of the Be-6 was deleted, being replaced by an MAD (magnetic anomaly detector) sting in the tail, while the search radar was carried in a long nose thimble instead of a ventral 'dustbin'.

The considerable weight-lifting capability of the Be-12 was demonstrated in a series of class records for amphibians set up in 1964, 1968 and 1970, suggesting a weapons load as high as 5000 kg (11,023 lb). The Be-12 can load on the water via side hatches in the rear fuselage and stores can be air-dropped through a watertight hatch in the hull aft of the step. The Be-12 could alight on the water in reasonably calm conditions to search with its own sonar equipment, rather than rely on sonobuoys. Since 1970 there has been a diminishing ASW role for the Be-12, but about 75 remain in service as a search-and-rescue (SAR) vehicle.

Specification (provisional)
Type: maritime patrol amphibian
Powerplant: two 3124-kW (4,190-shp) Ivchenko AI-20D turboprop

A substantial number of Be-12 'Mail' anti-submarine warfare aircraft continue to serve with Soviet Naval Aviation units.

engines
Performance: maximum speed 608 km/h (378 mph); patrol speed 320 km/h (199 mph); initial rate of climb 912 m (2,992 ft)/min; range (low level) 4000 km (2,485 miles)
Weights: estimated empty 20000 kg (44,092 lb); maximum take-off 29450 kg (64,925 lb)
Dimensions: span 29.71 m (97 ft 5.7 in); length 30.17 m (98 ft 11.8 in); height on land 7.00 m (22 ft 11.6 in); wing area 105.00 m² (1,130.25 sq ft)
Armament: bombs, rockets or other stores on underwing pylons; depth charges and sonobuoys in fuselage bays
Operators: Vietnam, Soviet Navy (AV-MF)

With its exaggerated gull-wing, lengthy engine nacelles and nose radar housing, the Be-12 has an unusual configuration, but its operational effectiveness cannot be doubted. A crew of five or six includes two ASW sensor operators charged with the handling of sonobuoy returns.

Boeing B-52 Stratofortress

Currently operational with 16 Strategic Air Command (SAC) wings, Boeing's B-52 Stratofortress entered service 30 years ago. It may well establish records for longevity as a front-line type, since present planning calls for it to remain active until well into the nineties.

Following the recent retirement of the B-52D, which was completely rebuilt for conventional bombing roles, only the B-52G and B-52H variants are now to be found in the SAC inventory, a combined total of 264 still being on charge. The subject of continuous updating throughout its life, the B-52 fleet is in the process of receiving the Phase VI avionics upgrade, this including installation of Motorola ALQ-122 SNOE (Smart Noise Operation Equipment), Northrop AN/ALQ-155(V) advanced electronic countermeasures, AFSATCOM (Air Force Satellite Communications) radio installations, Dalmo Victor ALR-46 digital radar warning receiver and ITT Avionics ALQ-172 jammers. To enhance low-level navigation and weapons delivery Boeing has also developed the OAS (Offensive Avionics System). Incorporating Tercom (Terrain Contour Matching) guidance, OAS also features a Teledyne Ryan doppler radar, Honeywell AN/ASN-131 inertial navigation system, IBM/Raytheon ASQ-38 analog bombing/navigation equipment, Sperry controls and displays and Norden Systems modernised radar. Since 1982 OAS has been installed in both variants, it being intended to modify 168 B-52Gs and 96 B-52Hs by Fiscal Year 1989. Both models are compatible with the Boeing AGM-69 SRAM (Short-Range Attack Missile), eight of these weapons being carried internally and 12 externally. The most significant recent development centres around the modifica-

tion of 99 B-52Gs and 96 B-52Hs to carry 12 examples of the AGM-86B ALCM (Air-Launched Cruise Missile) on the two underwing store stations in addition to internally-housed SRAMs and other weapons. A further development, planned for the late eighties, will entail the fitment of a lengthened rotary launcher to all 96 B-52Hs, permitting this variant to carry an additional eight ALCMs internally. ALCM entered service on the B-52Gs of the 416th Bomb Wing at Griffiss AFB, New York, during 1982, initial operational capability being attained in December of that year.

Approximately 70 B-52Gs which are not earmarked for the cruise missile have assumed responsibility for those conventional missions previously undertaken by the B-52D, including maritime surveillance and support and conventional bombing. To permit them to undertake the former mission, plans are in hand to fit Harpoon air-to-surface anti-shipping missiles to some aircraft, two wings of Harpoon-compatible B-52Gs being operational by the end of 1984. Looking to the future, conventional capability should be enhanced with the adoption of the AGM-109H Tomahawk MRASM (medium-range air-to-surface missile), trials of a B-52G fitted with a four-round MRASM pylon under each wing taking place in 1984.

Since 1982 Pratt & Whitney has been lobbying for a giant re-engining programme, the proposed powerplant being four PW2037 engines as used on some B.757s. So far this campaign has not been successful.

Specification
Boeing B-52G
Type: long-range strategic bomber
Powerplant: eight Pratt & Whitney J57-P-43WB turbojet engines each rated at 6237 kg

The B-52 soldiers on with the US Air Force in important roles, including cruise missile launch and maritime patrol/mine-laying. New avionics keep the 'Buff' in the forefront of strategic operations.

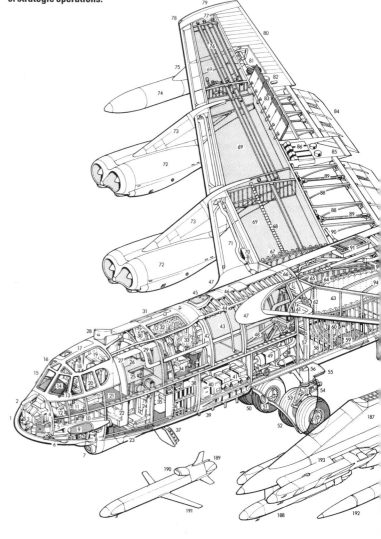

Boeing B-52 Stratofotress (continued)

(13,750 lb) thrust with water injection

Performance: maximum speed at altitude 958 km/h (595 mph); cruising speed at altitude 819 km/h (509 mph); low-altitude penetration speed 652-676 km/h (405-420 mph); service ceiling 16765 m (55,000 ft); unrefuelled range more than 12070 km (7,500 miles)

Weights: operating empty 72002 to 78048 kg (158,737 to 172,066 lb); maximum take-off 221353 kg (488,000 lb)

Dimensions: span 56.39 m (185 ft 0 in); length 49.05 m (160 ft 10.9 in); height 12.40 m (40 ft 8 in); wing area 371.60 m² (4,000 sq ft)

Armament: four 12.7-mm (0.5-in) machine-guns in tail barbette for defensive purposes plus up to 20 AGM-69A SRAM missiles (eight on rotary launcher plus six externally under each wing) plus free-fall nuclear bombs: aircraft being modified to carry up to 12 AGM-86B cruise missiles externally

Operator: US Air Force Strategic Air Command

Various lumps and bumps have been added to the B-52 nose area over the years, evidence of new equipment and sensors. The two large fairings on the undernose house a low-light television scanner and infra-red sensors with links to the cockpit.

Boeing B-52G Stratofortress cutaway drawing key

1 Nose radome
2 ALT-28 ECM antenna
3 Electronic countermeasurers (ECM) equipment bay
4 Front pressure bulkhead
5 Electronic cooling air intake
6 Bombing radar
7 Low-light television scanner turret (EVS system), infra-red on starboard side
8 Television camera unit
9 ALQ-117 radar warning antenna
10 Underfloor control runs
11 Control column
12 Rudder pedals
13 Windscreen wipers
14 Instrument panel shroud
15 Windscreen panels
16 Cockpit eyebrow windows
17 Cockpit roof escape/ejection hatches
18 Co-pilot's ejection seat
19 Drogue chute container
20 Pilot's ejection seat
21 Flight deck floor level
22 Navigator's instrument console
23 Ventral escape/ejection hatch, port and starboard
24 Radar navigator's downward ejection seat, navigator to starboard
25 Access ladder and hatch to flight deck
26 EWO instructor's folding seat
27 Electronics equipment rack
28 In-flight refuelling receptacle, open
29 Refuelling delivery line
30 Electronic warfare officer's (EWO) ejection seat
31 Rear crew members escape/ejection hatches
32 EWO's instrument panel
33 Gunner's remote control panel
34 Gunner's ejection seat
35 Navigation instructor's folding seat
36 Radio and electronics racks
37 Ventral entry hatch and ladder
38 Lower deck rear pressure bulkhead
39 ECM aerials
40 ECM equipment bay
41 Cooling air ducting
42 Upper deck rear pressure bulkhead
43 Water injection tank, capacity 1,200 US gal (4542 litres)
44 Fuselage upper longeron
45 Astro navigation antenna
46 Tank access hatches
47 Leading edge 'strakelets' fitted to identify cruise missile carriers
48 Forward fuselage fuel tank
49 Air conditioning plant
50 Forward starboard main undercarriage bogie
51 Landing lamp
52 Forward port main undercarriage bogie
53 Torque scissor links
54 Steering jacks
55 Main undercarriage door
56 Main undercarriage leg strut
57 Wing front spar/fuselage/ main undercarriage attachment frame
58 Main undercarriage wheel bay
59 Doppler aerial
60 Central electronic equipment bay
61 Air conditioning intake duct
62 Front spar attachment joint
63 Wing root rib
64 Wing panel bolted attachment joint
65 Centre section fuel tank bay
66 Wing centre section carry-through
67 Starboard wing attachment joint
68 Vortex generators
69 Starboard wing integral fuel tank bays; total fuel system capacity (includes external tanks), 48,030 US gal (181813 litres)
70 Engine ignition control unit
71 Bleed air ducting
72 Starboard engine nacelles
73 Nacelle pylons
74 Fixed external fuel tank, capacity 700 US gal (2650 litres)
75 Tank pylon
76 Fuel venting channels
77 Tip surge tank
78 Starboard navigation light
79 Wing tip fairing
80 Fixed portion of trailing edge
81 Starboard outrigger wheel, stowed position
82 Hydraulic equipment bay
83 Roll control spoiler panels, open
84 Outboard single-slotted, Fowler-type flap, down position
85 Inboard fixed trailing edge segment
86 Chaff dispensers and flare launchers
87 Inboard single slotted flap, down position
88 Flap guide rails
89 Flap screw jacks
90 Flap drive torque shaft
91 Life raft stowage
92 Wing centre section/ longeron ties
93 Central flap drive motor
94 Rear spar attachment joint
95 AGM-69 missile environmental control unit
96 Bomb bay rotary missile launcher
97 AGM-69 SRAM, air to ground missiles
98 Bomb bay rear bulkhead
99 Rear fuselage bag-type fuel tanks
100 Rear fuselage longeron
101 Fuel delivery and transfer piping
102 Fuselage skin panelling
103 Fuselage fuel system surge tank
104 Data link antenna
105 Rear fuselage frame construction
106 Rear equipment bay air conditioning plant
107 Ram air intake
108 Starboard tailplane
109 Vortex generators
110 Starboard elevator
111 Fin spar attachment joint: fin folds to starboard
112 Tailfin rib construction
113 VOR aerial
114 Lightning isolator
115 Fin tip aerial fairing
116 Rudder
117 Rudder tab
118 Hydraulic rudder control jack
119 Rudder aerodynamic balance
120 Rear ECM and fire control electronics pack
121 ECM aerial fairing
122 Brake parachute stowage
123 Parachute and door release mechanism
124 ALQ-117 retractable aerial fairing
125 AN/ASG-15 search radome
126 ALQ-117 and APR-25 ECM radome
127 Four 0.5-in (12.7-mm) machine-guns
128 AN/ASG-15 tracking radome
129 Remote control gun turret
130 Ammunition feed chutes
131 Ammunition tanks, 600 rounds per gun
132 Elevator tab
133 Port elevator
134 ALQ-153 tail warning radar
135 All-moving tailplane construction
136 Tailplane carry-through box section spar
137 Elevator aerodynamic balance
138 Centre section sealing plate
139 Tailplane trimming screw jack
140 Air conditioning ducting
141 Fuel system venting pipes
142 Ventral access hatch
143 Rear fuselage ECM equipment bay
144 ECM aerials
145 Strike camera compartment
146 Rear main undercarriage wheel bay
147 Bomb/wheel bay box section longeron
148 Main undercarriage mounting frame
149 Hydraulic retraction jack
150 Rear main undercarriage bogie units
151 Flap shroud ribs
152 ECM dispensers
153 Fixed portion of trailing edge
154 Port flaps, down position
155 Outboard single slotted flap
156 Port roll control spoiler panels
157 Hydraulic reservoir
158 Outrigger wheel bay
159 Fixed portion of trailing edge
160 Glass-fibre wing tip fairing
161 Port navigation light
162 Outer wing panel integral fuel tank
163 Port outrigger wheel
164 Fixed external fuel tank
165 Fuel tank pylon
166 Outrigger wheel retraction strut
167 Outer wing panel attachment joint
168 Engine pylon mounting rib
169 Pylon rear attachment strut
170 Engine pylon construction
171 Pratt & Whitney J57-P-43WB turbojet engine
172 Engine oil tank, capacity 8.5 US gal (32 litres)
173 Accessory equipment gearbox
174 Generator cooling air duct
175 Oil cooler ram air intakes
176 Engine air intakes
177 Detachable cowling panels
178 Leading edge rib construction
179 Front spar
180 Wing rib construction
181 Rear spar
182 Port wing integral fuel tank bays
183 Inboard pylon mounting rib
184 Leading edge bleed air and engine control runs
185 Weapons bay doors, open (loading) position
186 Bomb doors, open
187 Wing mounted cruise missile pylon
188 Boeing AGM-86B Air Launched Cruise Missiles (ALCM), six per wing pylon, stowed configuration
189 AGM-86B missile in flight configuration
190 Retractable engine air intake
191 Folding wings
192 AGM-69 SRAM, alternative load
193 Missile adaptors
194 Nacelle pylon
195 Port inboard engine nacelles
196 Central engine mounting bulkhead/firewall
197 Bleed air ducting
198 Generator cooling air ducting
199 Fuselage bomb mounting cradle
200 Free-fall 25-megaton nuclear weapons (four)

© Pilot Press Limited

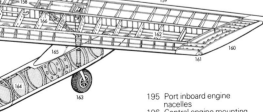

97

Boeing C-135/KC-135

On 15 July 1954 Boeing flew its private-venture Model 367-80 prototype for the first time. For the company it represented an enormous gamble, one that depended on winning orders from the US Air Force. In consequence the 'Dash-80' prototype, as it soon became known, was initially a military demonstrator equipped with a Boeing-developed inflight-refuelling boom. This boom, controlled by an operator, was thought to be more practical than the 'probe-and-drogue' system developed in the UK, and its effective demonstration in the 'Dash-80' convinced the USAF that not only was Boeing's rigid 'flying boom' an efficient refuelling system, but that the aircraft had important potential as a tanker/transport. In August 1954, a few weeks after the maiden flight of the 'Dash-80', the USAF placed its first contract for the Model 717 as the Boeing KC-135A Stratotanker; the first (55-3118) was flown on 31 August 1956 and initial deliveries, to Castle AFB, California, began on 28 June 1957; production totalled 724.

Generally similar to the 'Dash-80', but smaller than even the first Boeing Model 707 developed from it, the KC-135A was powered by four J57 turbojets. In the tanker role fuel system capacity was 118105 litres (31,200 US gal), but the cabin could be used alternatively to carry cargo, loaded via a large door in the port forward fuselage, or up to 80 passengers or a mixture of freight and passengers; with reduced fuel a maximum of 160 troops could be airlifted. Variants include a small number of KC-135As rebuilt for special duties as JC-135A and JKC-135A, subsequently becoming NC-135A and NKC-135A which was the designation of some 20 rebuilds for test and research programmes; 17 new-build KC-135B tankers had improved equipment, increased fuel capacity, TF33-5 turbofan engines and provisions for use as airborne command posts; and 56 conversions of KC-135As to transfer JP-7 fuel to Lockheed SR-71 'Blackbirds' were redesignated KC-135Q. The Model 717 airframe was adopted also as a cargo/troop transport for MATS (later MAC) as the C-135A Stratolifter accommodating 126 troops or 44 stretchers and 54 seated casualties. The C-135A (18 built) was followed by the C-135B (30 with TF33-5 engines), but the C-135F was a tanker/transport of which 12 were

Boeing RC-135V of the 55th Strategic Reconnaissance Wing based at Offutt AFB.

built for France. Seven VC-135A staff transports were conversions of C/KC-135A aircraft, two of them retaining tanker capability; five VC-135B staff transports were conversions of C-135Bs, as were 11 WC-135B weather reconnaissance aircraft, three of which later became reconverted as C-135C freighters.

When production of Boeing KC-135 tankers for the US Air Force ended a total of 724 had been built, of which about 650 remain in use. It was decided to ensure they would remain operational into the next century, the major requirement being replacement of the underwing skin. This task, started in 1975, has progressed steadily and by mid-1985 more than 500 KC-135s had benefited from this modification which should extend service life by some 27,000 hours. It was followed by a programme to re-engine Air National Guard and Air Force Reserve KC-135s with JT3D engines (civil equivalent of the TF33). These powerplants were removed and refurbished from ex-commercial Boeing 707s acquired by the USAF, and at the same time the KC-135s gain also tail units, engines pylons and cowlings from the Model 707s. Simultaneously new brakes and anti-skid units are installed and, upon completion of the work, the aircraft are redesignated KC-135E.

Far more comprehensive is the programme to update the main tanker fleet with the 9979-kg (22,000-lb) thrust General Electric/SNECMA F108-CF-100 turbofan (equivalent to the civil CFM56-2B-1), existing contracts covering 108 conversions. With this powerplant revision comes also an APU to give self-start capability; more advanced autopilot, avionics, controls and displays on the flight deck; strengthened main landing gear incorporating anti-skid units; revised hydraulic/pneumatic systems; and an enlarged tailplane. Redesignated KC-135R on completion of this update, the first example was rede-

Boeing KC-135A Stratotanker

livered to SAC's 384th Air Refueling Wing at McConnell AFB, Wichita, in July 1984.

The EC-135 family is a group of C-135 rebuilds which provide the USAF with airborne command posts and radio-relay stations. These are used for flying commanders in time of war. There are many different variants. The EC-135B and EC-135C have large steerable antennae in a bulbous nose for satellite and missile tracking.

The other main family of C-135 rebuilds are the RC-135s. These have been the main Elint gatherer for the United States for many years, serving in many marks. Currently 17 are in service around the world, consisting of RC-135S Telint specialists, RC-135U with large cheek SLARs, and RC-135V and W with smaller SLARs and 'thimble' noses. These have global responsibilities, collecting Elint and producing SLAR imagery.

Specification
Boeing KC-135A
Type: inflight-refuelling tanker/cargo/transport aircraft
Powerplant: four 6237-kg (13,750-lb) thrust Pratt & Whitney J57-59W turbojet engines
Performance: average cruising speed 460 kts (853 km/h; 530 mph) between 9399 and 12190 m (30,500 and 40,000 ft); initial rate of climb 393 m (1,290 ft) per minute; time to climb to 9300 m (30,500 ft) 27 minutes; transfer radius with 3040-kg (6,702-lb) reserve fuel 1850 km (1,150 miles)
Weights: empty 44663 kg (98,466 lb); maximum take-off 136078 kg (300,000 lb); maximum fuel load 86047 kg (189,702 lb)
Dimensions: span 39.88 m (130 ft 10 in); length 41.53 m (137 ft 3 in); height, short fin 11.68 m (38 ft 4 in); wing area 226.03 m² (4,433.0 sq ft)
Armament: none

A standard USAF KC-135A Stratotanker supplies fuel to the test aircraft for the F108-powered KC-135R conversion.

Boeing Model 707/C-137

Having decided in 1953 that it needed the Boeing KC-135A tanker, the US Air Force authorised Boeing to proceed with a commercial transport development. Boldly, Boeing designed this with a wider fuselage, and then built the Model 707 in small medium-range and large long-range versions, and later added the short-range 720. The first production version was the 707-120, powered by four 6123-kg (13,500-lb) thrust Pratt & Whitney JT3C-6 turbojet engines, and three of these were ordered by the USAF in May 1958 to serve as VIP transports, or for high priority cargo, with the designation VC-137A. When re-engined with TF33 turbofans they were redesignated VC-137B. A single VC-137C was added in 1962, similar to the large long-ranged 707-320B, and this became the Presidential 'Air Force One'.

Subsequently, the E-3A Sentry and E-6A Tacamo have been developed around the 707-320C airframe, as described separately. The appearance of large numbers of fairly low-time 707s on the secondhand market spurred Boeing to offer specially rebuilt tanker/transports. In 1984 Boeing Military Airplane Co was rebuilding 25 of these T/T aircraft for four air forces. In addition, in 1986-88 BMAC will deliver to Saudi Arabia eight tankers designated KE-3A with the same CFM56-2-A2 turbofans as that country's Sentries. Also in preparation for delivery to the USAF, as EC-18B ARIA (Ad-

vanced Range Instrumentation Aircraft), are six ex-airline 707-320s with masses of new gear, including in the nose the world's largest steerable airborne aerial.

The German Luftwaffe also acquired four 707-320Cs as VIP or special freight transports, the Portuguese air force four, the Iranian Imperial air force five, the Israeli air force 10, the Royal Moroccan air force three and the Canadian Armed Forces five, this last designating them CC-137. The last four countries use wingtip inflight refuelling hosereel pod installations.

Specification
Boeing VC-137C
Type: VIP or special freight transport
Powerplant: four 8165-kg (18,000-lb) thrust Pratt & Whitney JT3D-3 turbofan engines
Performance: maximum cruising speed 966 km/h (600 mph) at 7620 m (25,000 ft); range with maximum fuel, allowances for climb and descent, no reserves 12247 km

The Royal Australian Air Force operates its Boeing 707s on VIP transport and inflight-refuelling tasks.

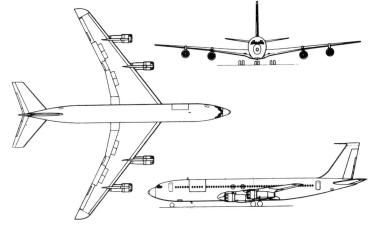

Boeing Model 707-320

(7,610 miles)
Weight: maximum take-off 148325 kg (327,000 lb)
Dimensions: span 44.42 m (145 ft 9 in); length 46.61 m (152 ft 11 in); height 12.93 m (42 ft 5 in); wing

area 279.63 m² (3,010.0 sq ft)
Armament: none
Operators: Argentina, Australia, Brazil, Canada, Chile, Colombia, Ecuador, West Germany, Indonesia, Iran, Israel, Morocco, Portugal, Saudi Arabia, US

Boeing E-3 Sentry

Although a requirement for a new Airborne Warning and Control System (AWACS) aircraft to replace the Lockheed EC-121 Warning Star variant of the Constellation existed in 1963, development of the E-3 did not get under way until Boeing Aerospace Company was awarded a contract in July 1970 for two rebuilt 707 testbeds designated EC-137D. Acting as prime contractor, the company's first task was to establish which of the two competing radars – one designed by Hughes and the other by Westinghouse – was best suited for this mission. On 5 October 1972 the Westinghouse AN/APY-1 was chosen, approval for full scale development being announced by the US Air Force on 26 January

1973. Plans to use eight GE TF34 engines for extra endurance were abandoned. Designated E-3 Sentry, it was at first anticipated that a total of 34 would be procured, but it now seems likely that production for the US Air Force will approach 50. In

addition, 18 Sentries were acquired by NATO, while Saudi Arabia is to operate five powered by CFM56 engines, although these will not be full-system aircraft, lacking certain sensitive items of equipment such as JTIDS (Joint Tactical Information

Eighteen E-3A Sentry AWACS aircraft are operated by the multi-national NATO AEWF, home-based at Geilenkirchen in West Germany.

Easily distinguished by the large rotodome atop the rear fuselage, the E-3 Sentry keeps a watchful eye over the world's 'hot-spots', helping to direct US air power in times of tension.

Boeing E-3 Sentry (continued)

Distribution System).

To date, four variants have been delivered. The initial 'Core E-3A' was acquired solely by the USAF, 24 being completed. These are equipped with pulse-doppler radar with the ability to detect high- and low-flying aircraft and also have the IBM CC-1 computer plus nine situation display consoles (SDCs) for use by crew members engaged in surveillance, direction and battle staff duties, and two auxiliary display units (ADUs) to support communications, maintenance and data-processing. Very extensive communications equipment is standard. All Core E-3As are being brought to E-3B configuration. This entails the installation of the JTIDS secure communications package, far more resistant to jamming, a new and faster CC-2 computer with expanded memory, and additional radios and SDCs. The E-3B also has an 'austere' maritime surveillance capability. The Standard E-3A, is the original configuration of aircraft 25-34 for the USAF, and the 18 for NATO. Delivery of these began in December 1981, and they differ from earlier variants by having full maritime surveillance capability. They also incorporate the CC-2 computer

and JTIDS, although plans are in hand to modify the 10 USAF Standards to E-3C configuration. This features five more SDCs, additional UHF radios and provision for Have Quick anti-jamming improvements. Any further new-build aircraft will probably be completed as E-3Cs.

Delivery of the Sentry to Tactical Air Command began on 24 March 1977, initial operational capability being attained during April 1978. Today, USAF E-3s routinely operate from Keflavik (Iceland), Elmendorf (Alaska) and Kadena (Okinawa) on a permanent forward-deployed basis, while the NATO fleet, normally based at Geilenkirchen with multinational crews, attained operation capability in 1983.

In 1985 delivery began of an even more astronomically costly version, the E-3A/Saudi, five of which formed a $4,660 million package including support. These have increased performance and greatly extended endurance because of their later engines.

Specification
Boeing E-3 Sentry
Type: airborne early-warning and control aircraft
Powerplant: four 9525-kg (21,000-

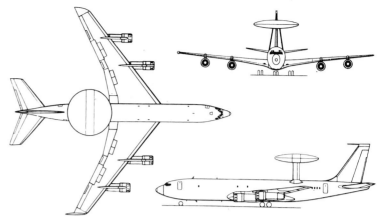

Boeing E-3A Sentry AWACS platform.

lb) thrust Pratt & Whitney TF33-PW-100A turbofan engines; (Saudi 9979-kg/22,000-lb CFM56-2-A2 turbofans)
Performance: maximum speed 853 km/h (530 mph); service ceiling 8850 m (29,000 ft); endurance on station 1609 km (1,000 miles) from base 6 hours; maximum unrefuelled endurance exceeds 11 hours
Weights: empty 73482 kg (162,000 lb); maximum take-off

147418 kg (325,000 lb)
Dimensions: span 44.42 m (145 ft 9 in); length 46.61 m (152 ft 11 in); height 12.73 m (41 ft 9 in); wing area 283.35 m² (3,050 sq ft)
Armament: none
Operators: US Air Force, NATO (registered in Luxembourg)

Boeing Model 747/E-4

To ensure that it retains a deterrent capability, even in the aftermath of a nuclear attack, the United States Department of Defense created a National Military Command System, through which National Command Authorities are able to issue orders and receive a feedback of information to show whether the orders were effective or not. Part of the system comprises the Airborne National Command Post (ABNCP), and these special versions of the Boeing C-135 family have been operational in this role for many years. Confidence in the system depends upon a survivable ABNCP, and DoD planners had augmented the EC-135 force by the Boeing E-4 Advanced Airborne Command Post (AABNCP) which it is believed would have much increased survivability. This aircraft, the most costly ever bought, is based upon the Boeing 747 airframe, and the initial contract for two 747Bs to be converted to serve as AABNCP aircraft was announced in February 1973. Follow-up contracts for two more aircraft were awarded in July and September 1973. It has been announced that the total planned force is six aircraft, but in 1985 the last two had yet to be funded.

Three of the initial four aircraft became operational as E-4As in an interim ABNCP role. They were built with F105 engines, and had avionics and equipment removed from EC-135 aircraft; the first was handed over in December 1974. The second and third followed in May and September 1975, all three delivered initially to Andrews AFB, Maryland. Since that time they have shown that, as a result of their ability to accommodate a larger battle staff of up to 60, they are more flexible than the EC-135s which they have replaced.

The fourth aircraft has much more advanced avionics and equipment, nuclear shielding, different engines, and was the first definitive E-4B. It was delivered to the USAF in August 1975; the installation of its advanced command, control and communications equipment took

four years, the aircraft being redelivered on 21 December 1979. The E-4B carried three times the payload of the EC-135, and of a totally different nature. The main deck comprises a work area for the battle staff, briefing room, conference room, communications control centre, National Command Authorities' area, and a rest area. Under the floor are extensive power stations, maintenance sections and large quantities of mission equipment including very special communications. The latter include an 8-km (5-mile) trailing wire VLF aerial, and an SHF aerial in a dorsal 'doghouse' fairing.

The three E-4As have been updated to E-4B standard, and all four aircraft are normally based at Offutt AFB, Nebraska. Boeing 747s are also used by Iran in transport, inflight refuelling and electronic roles.

Specification
Boeing E-4B
Type: special-purpose airborne command post
Powerplant: four 23814-kg (52,500-lb) thrust General Electric

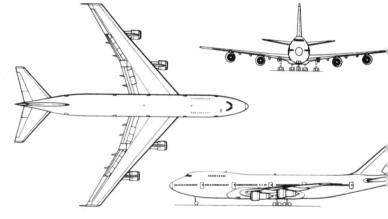

Boeing Model 747

F103-GE-100 (CF6-50E2) turbofan engines
Performance: unrefuelled endurance over 12 hours; mission endurance 72 hours
Weights: fuel 150395 kg (331,565 lb); maximum take-off 362874 kg (800,000 lb)
Dimensions: span 59.64 m (195 ft 8 in); length 70.51 m (231 ft 4 in);

height 19.33 m (63 ft 5 in); wing area 510.95 m² (5,500.0 sq ft)
Armament: none
Operators: Iran, Saudi Arabia, USAF

Iran received several Boeing Model 747s before the revolution. These have been used for transport and electronic duties.

Boeing E-6

Intended to fulfil the US Navy's TACAMO (Take Charge And Move Out) mission, the Boeing E-6 will serve as a survivable airborne communications system providing a link between the US National Command Authorities in Washington or aboard the NEACP (National Emergency Airborne Command Post) and the fleet of Trident-armed 'Ohio'-class nuclear submarines (SSBNs). This vital mission was previously undertaken by the Lockheed EC-130Q Hercules of VQ-3 at Agana, Guam, and VQ-4 at Patuxent River, Maryland.

The selection of Boeing to develop this system was revealed on 3 May 1983. Eventual cost is expected to be of the order of $1,600 million, this encompassing development and supply of the proposed fleet of 15 aircraft as well as crew instruction, provision of training aids and system support once the E-6A becomes operational in 1987-90. Employing an airframe virtually identical to that

Boeing E-6 TACAMO in the markings of the US Navy, its intended first customer.

of the E-3 Sentry, the E-6 is powered by four CFM56 engines and incorporates the same AVLF (airborne very low frequency) communications system fitted to the EC-130Q. Transmission of signals is effected by means of a trailing wire antenna (aerial) about 4 km (2.5 miles) long. The aircraft orbit with the wire almost vertical, submarines receiving messages via a towed buoyant wire antenna. Other noteworthy features of the E-6A include harden-

ing against electromagnetic pulse, thermal and blast effect, and gamma and neutron radiation. Assuming the programme continues to progress as planned, the 15 aircraft will be delivered between 1987 and 1991.

Specification
Type: airborne communications relay platform
Powerplant: four 9979-kg (22,000-lb) thrust CFM International CFM56-2-A2 turbofan engines

Performance: maximum speed 980 km/h (609 mph); operating ceiling 12190 m (40,000 ft); range 11748 km (7,300 miles)
Weights: empty 78393 kg (172,828 lb); maximum take-off 154805 kg (341,286 lb)
Dimensions: span 45.20 m (148 ft 3.5 in); length 46.61 m (152 ft 11 in); height 12.73 m (41 ft 9 in); wing area 279.63 m² (3,010 sq ft)
Armament: none
Operator: US Navy (from 1987)

Boeing Model 737/T-43/Surveiller

Experience in Vietnam showed that the USAF had inadequate facilities for the training of navigators, and a decision was made to procure an 'off the shelf' aircraft to replace the Convair T-29 in this role. In May 1971 the USAF announced that the Model 737 had been selected, and Boeing was awarded an $82.4 million contract for 19 T-43A aircraft. First flight was made on 10 April 1973, and all were delivered to Mather AFB, California, by the end of July 1974. Although the airframe is generally similar to that of the commercial 737-200, passenger doors and windows were reduced to one and nine respectively on each side of the fuselage, the cabin floor was strengthened to accommodate avionics consoles, and a 3028 litre (800 US gallon) auxiliary tank was installed in the aft cargo compartment. Each T-43A accommodates (in addition to the flight crew), 12 trainees, four advanced trainees and three instructors. They are operated in conjunction with ground-based simulators for high- and low-level

flight by day or night, high-speed flight, and the requirements of airways navigation. Onboard equipment is updated from time to time to ensure that it is the same as that used in USAF operational aircraft.

The Surveiller is a versatile transport which is also fully equipped for the maritime surveillance role. The main sensor comprises a Motorola SLAMMR (side-looking airborne multi-mission radar), its aerials being housed in two 4.88 m (16 ft) canoe blades above the rear fuselage extending on each side of the fin. This can see a small craft in a heavy sea from a distance of 185 km (115 miles). The Indonesian Surveillers

can also carry 102 passengers, 14 of them in a VIP forward section.

Specification
Boeing T-43A
Powerplant: two 6577-kg (14,500-lb) thrust Pratt & Whitney JT8D-9 turbofan engines
Performance: cruising speed Mach 0.7 or 747 km/h (464 mph) at 10670 m (35,000 ft); operational range, MIL-C-5011A reserves, 4820 m (2,995 miles); endurance 6 hours
Weights: empty 27442 kg (60,500 lb); maximum take-off 52390 kg (115,500 lb)
Dimensions: span 28.35 m (93 ft

0 in); length 30.48 m (100 ft 0 in); height 11.28 m (37 ft 0 in); wing area 91.04 m² (980.0 sq ft)
Armament: none
Operators: Brazil, India, Indonesia (Surveiller), Mexico, US (T-43A), Venezuela

The Indonesian air force (TNI-AU) operates a trio of Boeing 737-2X9 Surveiller aircraft on long-range maritime-patrol tasks.

Boeing Model 727

Boeing's 727 has the distinction of being the most extensively-built commercial transport in aviation history, with 1,832 built in 1963-84. A 'narrow-body' trijet, it began life at a gross weight of 68946 kg (152,000 lb) and was 'stretched' and given increased fuel capacity to allow for far greater range and payload, all-passenger versions seating up to 189. None were procured by the US armed forces, but a few were obtained by other countries for military service.

Specification
Boeing 727-200
Type: short/medium-range transport

Powerplant: three 6577-kg (14,500-lb) thrust Pratt & Whitney JT8D-9A turbofan engines
Performance: maximum cruising speed 964 km/h (599 mph) at 7530 m (24,700 ft); maximum range with 12474 kg (27,500 lb) payload, ATA domestic reserves, and 36832 litres (9,730 US gallons) fuel 4818 km (2,994 miles)
Weights: empty 44271 kg (97,600 lb); maximum take-off 95028 kg (209,500 lb)
Dimensions: span 32.92 m (108 ft 0 in); length 46.69 m (153 ft 2 in); height 10.36 m (34 ft 0 in); wing area 157.93 m² (1,700.0 sq ft)
Armament: none
Operators: Belgium, Chile, Ecuador, Mexico, Panama, New Zealand, Taiwan, Yugoslavia

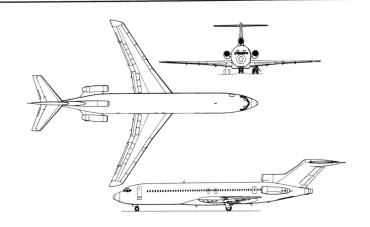

Boeing Model 727-200

Boeing Vertol CH-46 Sea Knight

General Electric's new T58 engine made possible this improved tandem-rotor transport helicopter, first flown as the civil Vertol Model 107 on 22 April 1958. The US Army ordered 10 for evaluation, but later

cut this to three after deciding that it required a larger machine. By the time the first of these YHC-1As was flown, on 27 August 1959, it had attracted US Marine Corps interest, and it won a USMC competition for a

utility transport. Designated initially HRB-1 (becoming CH-46A in 1962), production totalled 160, plus 14 UH-46As for US Navy combat support squadrons. Many CH/UH-46As have since been modified to

HH-46A configuration for base rescue. Re-engining with the T58-GE-10 resulted in the H-46D, produced for the US Navy (10 UH-46D) and the Marines (266 CH-46D). Manufacture then switched to the CH-

46F, with improved avionics and other equipment, 174 being delivered to the Marines.

Surviving CH-46Ds and CH-46Fs are being remanufactured as CH-46Es by the Naval Air Rework Facility at Cherry Point, North Carolina. Two prototype CH-46Es were subjected to intensive evaluation before the 1977 decision to go ahead with the conversion of 273 more for the Marine Corps. Noteworthy features included more powerful T58-GE-16 engines, enhanced crashworthiness, the ability to withstand a greater degree of battle damage, and the introduction of glassfibre rotor blades.

The Vertol 107 was produced in limited numbers by the parent company for export, and has been built in Japan (see Kawasaki KV-107). Swedish HKP-4s have Rolls-Royce Gnome engines.

Specification
Boeing Vertol CH-46D
Type: utility transport helicopter
Powerplant: two 1044-kW (1,400-shp) General Electric T58-GE-10 turboshaft engines
Performance: maximum speed 267 km/h (166 mph); range (at weight of 9435 kg/20,800 lb with 2064-kg/4,500-lb payload and 10 per cent reserves) 383 km (238 miles)
Weights: empty 5927 kg (13,067 lb); maximum take-off 10,433 kg (23,000 lb)
Dimensions: rotor diameter (each) 15.54 m (51 ft 0 in); length (rotors turning) 25.70 m (84 ft 4 in); height 5.09 m (16 ft 8.5 in); rotor disc area (total) 379.56 m² (4,085.6 sq ft)
Armament: none
Operators: Canada, Sweden, US Marine Corps, US Navy

Progressive updating of the existing CH-46 Sea Knight force will allow the US Marine Corps to operate this stalwart utility helicopter for many years to come in a variety of transport tasks.

Boeing Vertol CH-47 Chinook

Evolution of the Boeing Vertol CH-47 Chinook can be traced back to the late 1950s when the US Army began to search for a medium transport helicopter capable of all-weather operation in virtually all conditions. First flown on 21 September 1961, the Chinook began to enter service just over a year later. Capable of airlifting 40 troops, the CH-47 soon proved of great value in airlifting supplies to the numerous garrisons which lay dotted around South Vietnam, while it was also used frequently to recover downed aircraft and helicopters and for the evacuation of casualties and refugees, lifting up to 147 refugees and their belongings at once. The CH-47A was soon replaced in production by the more powerful CH-47B, but the CH-47C was the definitive Chinook, this entering service in 1968. Considerably more powerful engines were installed, this version also having a strengthened transmission system and additional fuel capacity. By 1972 over 550 had served in Vietnam.

Most US Army Chinooks are being rebuilt to CH-47D standard, with revised electrical system, glassfibre rotor blades, modular hydraulic systems, triple cargo hook, an advanced flight-control system, single-point pressure refuelling, an auxiliary power unit and revised avionics. The modernization programme began in 1979, and CH-47Ds began to join the US Army in 1983. Large numbers of several variants of the Chinook have been produced under licence by Elicotteri Meridionali in Italy, most of which have been exported.

Twenty Italian-built CH-47Cs were delivered to the Libyan Arab Republic air force in the late 1970s.

Argentina, Australia, Canada, Egypt*, Greece*, Iran*, Israel, Italy*, Japan, South Korea, Libya*, Morocco*, Nigeria*, Spain, Tanzania*, Thailand, UK, USA

Spanish operation of the Chinook is assigned to the Army with the designation Z.17. These are basically CH-47Cs with D-model flight-control systems.

Boeing Vertol CH-47 Chinook

Specification
Boeing Vertol CH-47C Chinook
Type: medium transport/assault helicopter
Powerplant: two 2796-kW (3,750-shp) Lycoming T55-L-11A turboshaft engines
Performance: maximum speed 304 km/h (189 mph); service ceiling 4570 m (15,000 ft); mission radius with 3294 kg (7,262-lb) payload 185 km (115 miles)
Weights: empty 9351 kg (20,616 lb); maximum take-off 20865 kg (46,000 lb)
Dimensions: rotor diameter, each 18.29 m (60 ft 0 in); length, rotors turning 30.18 m (99 ft 0 in); height 5.66 m (18 ft 7 in); rotor disc area, total 525.34 m² (5,654.88 sq ft)
Armament: none
Operators: (* = EM-built in Italy)

British Aerospace 748

The Avro 748 was designed in the late 1950s as a civil transport. First flown in prototype form on 24 June 1960, it remains in production, deliveries approaching 400 having included many to military operators. By far the largest of these is the Indian Air Force, following acquisitions of a licence by Hindustan Aeronautics Ltd (HAL) and assembly from British kits of an initial four Hawker Siddeley 748 Srs 1s from 1961. A further 68 'Avros' (as they are known in India) were supplied to the IAF with three diverted from a civil order, these being to 748 Srs 2 standard, except for the final 20 which are HAL 748Ms with wide freight doors. Others are used for VIP transport, navigation and signals training, twin-conversion and communications. The RAF ordered six Series 2 aircraft, powered by 1570-kW (2,105-shp) Rolls-Royce Dart 531 engines as the Andover CC.2, and these were delivered in 1964-65. Three are still used by The Queen's Flight, but scheduled for replacement by BAe 146s. In mid-1967 production turned to the HS.748 Srs 2A powered by 1700-kW (2,280-shp) Darts, while this gave way in early-1979 to the Series 2B with a 1.22-mm (4-ft) increase in wing span and optional hushed engines. Most military aircraft have been to Series 2A standard, some with the optional cargo door measuring 2.67 m × 1.72 m (8 ft 9 in × 5 ft 7.75 in). Principal military customers for British-built aircraft have been Australia, which flies 12 including two for ECM training, and Brazil, also operating 12 under the designation C-91. Other operators have generally bought the 748 in small numbers, some for VIP transport. Seeking to provide a competitor to the Fokker F.27MPA Maritime, BAe modified a 748 Srs 2A as the prototype Coastguarder in 1977 and offered production versions to Srs 2B standard. Recognisable by its belly radome for MEL MAEREC radar, the Coastguarder is equipped with additional navigation equipment and a tactical navigator's position in the cabin, plus wing and centreline strongpoints for armament. Despite a good showing during demonstrations, no orders have been placed for the Coastguarder. In 1983-84 versions of the 748 were considered with Searchwater radar in response to Indian requirements for an early-warning aircraft and a South African need to replace Shackleton maritime patrol aircraft.

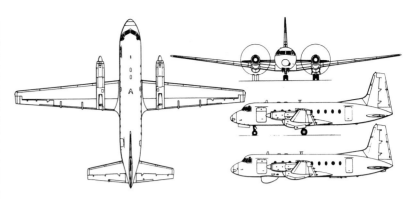

British Aerospace 748 (lower side view: Coastguarder)

Specification
BAe 748 Srs 2B
Type: military transport aircraft
Powerplant: two 1700-kW (2,280-shp) Rolls-Royce Dart 536-2 turboprop engines
Performance: cruising speed (at weight of 17237 kg/38,000 lb) 452 km/h (281 mph); initial rate of climb 433 m (1,420 ft)/minute; service ceiling 7620 m (25,000 ft); range (with maximum payload plus reserves) 1455 km (904 miles)
Weights: empty 12206 kg (26,910 lb); maximum take-off 23133 kg (51,000 lb)
Dimensions: span 31.23 m (102 ft 5.5 in); length 20.42 m (67 ft 0 in); height 7.57 m (24 ft 10 in); wing area 77.00 m² (828.87 sq ft)
Armament (748 Coastguarder only): homing torpedoes, anti-ship missiles or depth charges on underwing or centreline pylons
Operators: Australia, Belgium, Burkina-Faso, Brazil, Colombia, Ecuador, India, Malagasy, South Korea, Nepal, Sri Lanka, Tanzania, Thailand, UK, Venezuela, Zambia

British Aerospace Andover C.1

The Andover C.1 was developed by A.V. Roe (later Hawker Siddeley) from the civil HS.748 Series 2 to meet a Royal Air Force requirement for a rear-loading transport capable of operating from short unprepared strips. Work on the design began in 1962, and the original HS.748 prototype (G-APZV) acted as guinea pig for the new layout. Modifications to the aircraft included 'stretching' the fuselage, extending the centre-section but reducing the outer wings, and fitting a new tail layout which is raised to make room for the rear ramp door; the tailplanes have dihedral and root extensions but are of shorter span. The much more powerful engines drive larger propellers. Other distinguishing modifications include a fully-controllable kneeling landing gear, which permits vertical alignment of the cabin floor with the deck of a ground vehicle or conveyor. This complete controllability of sill height remains a unique feature.

The first of 31 C.1s (Avro 780 or 748MF) ordered by the Royal Air Force did not fly until 9 July 1965, and deliveries to No. 46 Sqn began in the following year. No. 52 Sqn of the Far East Air Force, based in Singapore, also received C.1s in 1966. Typical loads comprise 58 troops, 40 paratroops, 24 stretcher patients or up to 6963 kg (15,350 lb) of freight or vehicles. A typical air-supply mission would involve carrying a Ferret scout car and a ¼-ton Land-Rover or three Land-Rovers, all of which can be para-dropped. An inward-opening rear fuselage door is provided for paratroopers. The floor is equipped with lightweight roller tracks and guide rails for palletized cargo.

Ten C.1s were sold to the Royal New Zealand Air Force in 1976. Six serve with 1 Sqn at Whenuapai for trooping and freighting duties, while the remaining four equip 42 Sqn and are tasked with VIP flying and twin continuation training. Six of the RAF's remaining C.1s were converted in 1977 as Andover E.3s for flight checking and calibration duties, replacing Argosy E.1s. The first E.3s were delivered to No. 115 Sqn in 1978. Other RAF Andovers are described in the entry on the BAe 748.

Specification
BAe Andover C.1
Type: STOL transport
Powerplant: two 2420-ekW (3,245-eshp) Rolls-Royce Dart 201 turboprop engines
Performance: maximum cruising speed 426 km/h (265 mph) at 4570 m (15,000 ft); initial rate of climb 360 m (1,180 ft)/minute; range, with payload of 3869 kg (8,530 lb) and reserves 1864 km (1,158 miles)
Weights: empty 11574 kg (25,516 lb); maximum take-off 23133 kg (51,000 lb)
Dimensions: span 29.95 m (98 ft 3 in); length 23.77 m (78 ft 0 in); height 9.17 m (30 ft 1 in); wing area 77.24 m² (831.40 sq ft)
Armament: none
Operators: New Zealand, UK

The Royal Air Force has several BAe Andovers for radar and navaid calibration work around the military airfields of the UK. Illustrated is one of the brightly-coloured E.Mk 3s.

British Aerospace Buccaneer

Three squadrons – one of them abroad – and a training unit are all that remain operational to fly the survivors of 209 Buccaneers built at Brough by Blackburn Aircraft and its successor, Hawker Siddeley. The B.103 project, designated YB.3 by the SBAC, was a response to the exacting Royal Navy NR/A.39 requirement for a carrier-borne strike aircraft robust enough to penetrate enemy airspace in the turbulent air beneath defensive radar coverage. Twenty prototypes and a development aircraft, incorrectly known as NA.39s, were ordered initially and made their first flights beginning on 30 April 1958. These, and the 40 Buccaneer S.1s which followed in 1962-63, were powered by two de Havilland Gyron Junior engines of 3221 kg (7,100 lb) of thrust, but a considerable improvement in performance resulted from a change to Rolls-Royce Spey turbofans in the Buccaneer S.2 (otherwise YB.6), first flown on 17 May 1963. Production of this variant, which ended in 1977, totalled 133 for Britain, the later aircraft going to the RAF (46) and experimental units (three) in the aftermath of the TSR.2's cancellation. Last survivors of the naval S.2 batches joined them after the final RN carrier de-commissioned in December 1978. Aircraft to Navy standard and fitted for carriage of the Martel anti-radar or TV-guided missile were designated S.2D while their unmodified counterparts were

S.2C. RAF equivalents were the S.2B and S.2A respectively, although only one squadron (No 12) was fully committed to Martel operations in the maritime strike/attack role. Other units flying the

Maritime strike within the RAF is assigned to two squadrons of BAe Buccaneer S.Mk 2Bs based at RAF Lossiemouth in Scotland. Forthcoming equipment will include the BAe Sea Eagle anti-shipping missile.

aircraft included Nos 15 and 16 Sqns based at Laarbruch, West Germany, for nuclear strike missions (now equipped with Panavia Tornados). In the UK, No 216 Sqn was equipped for similar overland operations until it disbanded in 1980 when Buccaneer numbers were reduced as the result of serious fatigue being found in some aircraft. The two anti-shipping squadrons (Nos 12 and 208), plus No 237 OCU for crew training, would have disbanded in the mid-1980s but for the diversion of their intended Tornado replacements elsewhere. All three are based at Lossiemouth, No 12 with both Martel variants; No 208 with two Paveway laser-guided bombs on wing pylons and Pave Spike laser designators (plus four 'back-up' 454 kg/1,000 lb free-fall bombs in the rotary internal bay) or an alernative load of two anti-radar Martels; and No 237 OCU with a wartime role of providing in-flight refuelling with the Buddy system. Self-defence is assisted by AN/ALQ-101 jamming pods, a single AIM-9 Sidewinder air-to-air missile and a 454-kg (1,000-lb) bomb (for dissuading a low-level

tail-chaser). Enhanced stand-off capability will be available from mid-1985 with first deliveries of BAeD Sea Eagle anti-ship missiles, although an avionics improvement programme due to begin in 1986 and involving changes to the Blue Parrot radar is in doubt because of escalating costs. The surplus aircraft would make excellent EW jamming platforms.

In addition to RAF aircraft, six of the original 16 Buccaneer Mk 50s delivered to South Africa in 1965-66 remain with No 24 squadron at Waterkloof. Equivalent to S.2s except for a BS.605 rocket motor pack for 'hot-and-high' take-offs, they have been used for attack missions on neighbouring Angola and can carry AS.30 missiles.

Specification
BAe Buccaneer S.28
Type: two-seat strike/aircraft
Powerplant: two 5035-kg (11,100-lb) thrust Rolls-Royce Spey 101 turbofans plus (Mk 50 only) one 3,628-kg (8,000-lb) thrust Bristol-Siddeley/Rolls-Royce BS.605 rocket engine of 30 second burn time

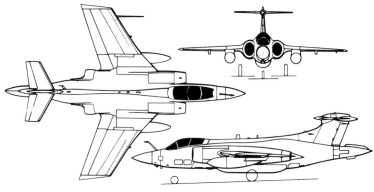

British Aerospace Buccaneer S. Mk 2B

Performance: maximum speed Mach 0.85 or 1038 km/h (645 mph) at 61 m (200 ft); typical range 3701 km (2,300 miles); endurance 9 hours with two inflight refuellings
Weights: typical landing 15876 kg (35,000 lb); maximum take-off 28123 kg (62,000 lb)
Dimensions: span 13.41 m (44 ft 0 in); length 19.33 m (63 ft 5 in); height 4.95 m (16 ft 3 in); wing area 47.82 m^2 (514.7 sq ft)

Armament: 20 kiloton nuclear weapon; BAeD Martel, or Sea Eagle anti-ship missiles; up to 16 454-kg (1,000-lb) bombs; self-defence AIM-9L Sidewinder air-to-air missile. Mk 50: four Aérospatiale AS.30 air-to-surface missiles, or four 68-mm (2.68-in) rocket pods, plus internal ordnance
Operators: South Africa, UK

British Aerospace Canberra

Long ago replaced by the RAF in its intended role as a light jet bomber, the Canberra possesses qualities which have kept it gainfully employed for over three decades, and will undoubtedly see it complete a fourth. Now assigned to peripheral tasks in its home country, the aircraft is front-line equipment elsewhere, as ironically demonstrated by Argentine use in the Falklands war of 1982. Flown in prototype form on 13 May 1949, the aircraft entered service in its Canberra B.2 form in May 1951. A few of this variant remain, although many survivors have been converted to other standards, notably the Canberra T.17, a bulbous-nosed electronic jamming training aircraft – and the Canberra TT.18 target tug. Included in this first generation are Canberra T.4 dual-control trainers. Replacement of 2948-kg (6,500-lb) thrust Rolls-Royce Avon 101s by 3402-kg (7,500-lb) Avon 109s, together with increased fuel capacity, brought about the Canberra B.6 (now used in Britain only by experimental establishments) and its reconnaissance derivative, the Canberra PR.7, the latter remaining in service for towing banner and sleeve targets and forming the basis of the Royal Navy's Canberra T.22 radar trainer. B.6 developments include the Canberra E.15 electronic warfare exercise aircraft. Following a substantial revision of the design with increased wing area and 5103-kg (11,250-lb) thrust Avon 206s, the Canberra PR.9 high-altitude reconnaissance model flew in 1958 featuring the offset pilot's canopy adopted by the B(I).8 interdictor (a pair of which are used for trials work). Five PR.9s are being completely rebuilt by the original manufacturer, Shorts, for multisensor reconnaissance until the end of the century. The PR.9 also remains in service for survey work, as does a a unique testbed conversion known as the Short SC.9. About 80 aircraft are still in British service, of which 54 were extensively refurbished by BAe during 1976-82.

Although the US and Pakistani air forces have withdrawn their licence-produced Martin B-57s, the aircraft remains in widespread use with several other overseas air

arms. India is a large-scale operator of Canberra 58s based on the RAF's B(I).8, Canberra 57s (PR.7 equivalents) and ex-New Zealand Canberra 12s. Aircraft of other air arms have almost invariably benefitted from refurbishment by BAe, including Venezuela's 22 updated Canberra B.82, B(I).82, B(I).88, PR.83 and T.84s, the last of which was overhauled in 1980. BAe holds a stock of aircraft against further sales prospects, while in 1984 a PR.9 version with Thorn-EMI radar was proposed for the British Army's CASTOR (Corps Airborne Stand-Off Radar) requirement.

Specification
BAe Canberra 58
Type: two-seat light bomber/interdictor

Powerplant: two 3402-kg (7,500-lb) thrust Rolls-Royce Avon 109 turbojet engines
Performance: maximum speed at sea level 832 km/h (517 mph); initial rate of climb 1036 m (3,400 ft)/minute; service ceiling 14630 m (48,000 ft); range (maximum load at 610 m/2,000 ft) 1296 km (805 miles)
Weights: empty 10511 kg (23,173 lb); maximum take-off 25515 kg (56,250 lb)
Dimensions: span 19.51 m (64 ft 0 in); length 19.96 m (65 ft 6 in); height 4.75 m (15 ft 7 in); wing area 89.18 m^2 (960.0 sq ft)
Armament: four internal 20-mm (0.79-in) Hispano cannon; up to six 454 kg (1,000 lb) bombs internally; 907 kg (2,000 lb) of stores on wing pylons
Operators: Argentina, Chile,

British Aerospace (BAC) Canberra PR.Mk 9

Ecuador, West Germany, India, Peru, South Africa, UK, Venezuela, Zimbabwe

The Canberra continues to serve the RAF in several important roles, though a replacement is being sought. No. 100 Squadron carries out target-towing and calibration duties, one of the unit's PR.Mk 7s being illustrated.

A dozen Canberras were supplied to Argentina, with operations including ineffective sorties against British forces during the Falklands conflict.

South African air force Canberra B.(I).Mk 12s are used for strike duties and high-altitude reconnaissance flights.

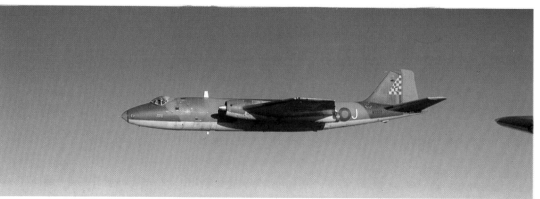

British Aerospace 125/Dominie

The first de Havilland 125 flew on 13 August 1962 and was followed by the first production model on 12 February 1963. Impressed by the performance, roominess and low cost of this business jet, the Royal Air Force ordered 20 in September 1962, and the first Dominie T.1 flew on 30 December 1964. It normally carries one pilot, a supernumerary flight crew member, two students and an instructor. The students sit in rearward-facing seats with a navigation panel and worktop. Avionics include Collins HF, Marconi VHF, Plessey UHF, Marconi VOR/ILS, ADF, intercom, Decca Navigator Mk 1, Sperry compass, Ekco weather radar, Decca doppler, two ground-position indicators and a periscopic sextant in the centre fuselage. A belly fairing houses the Decca and doppler aerials.

The RAF also uses two communications variants: the CC.1 based on the civil Series 400, and the CC.2 based on the 'stretched' and improved-performance 125-600. Nineteen Dominies remain in service, and two communications squadrons, Nos 32 and 207, fly four -400s (CC.1s) and two -600s (CC.2s).

A mixed fleet of VIP-configured BAe 125s fly with No. 32 Squadron, RAF, this being one of the turbofan-powered CC.Mk 3s.

The Argentine navy operates a single BAe 125-400 for navaid calibration, and the Brazilian air force's Grupo de Transporte Especial operates eight. The Royal Malaysian Air Force's No. 2 Sqn continues to fly two on VIP/government missions. No. 21 Sqn of the South African air force flies four (known as Mercurius) on VIP duties. Production is now centred on the much more capable and quieter 125 Series 700 and the refined Series 800, all with Garrett engines and enhanced systems.

Specification
BAe 125-600
Type: crew trainer and communications aircraft
Powerplant: two 1701-kg (3,750-lb) thrust Rolls-Royce Viper 601-22 turbojet engines
Performance: maximum speed 917 km/h (570 mph); cruising speed 840 km/h (552 mph) at 8535 m (28,000 ft); initial rate of climb 1494 m (4,900 ft)/minute; range with maximum payload of 907 kg (2,000 lb); maximum fuel and reserves 2890 km (1,796 miles)
Weights: empty 5684 kg (12,530 lb); maximum take-off 11340 kg (25,000 lb)
Dimensions: span 14.33 m (47 ft 0 in); length 15.39 m (50 ft 6 in); height 5.26 m (17 ft 3 in); wing area 32.79 m² (353.0 sq ft)
Armament: none
Operators: Argentina, Australia, Brazil, Malawi, Malaysia, South Africa, UK

British Aerospace (DH/HS) Dominie T.Mk 1

British Aerospace EAP

Since the mid-1970s British Aerospace at Warton has been studying possible successors to the Tornado programme. By late 1981 these were hardening into a twin-RB199 canard delta, with conventional engine nozzles and no pretensions at V/STOL performance. The idea was that a high ratio of thrust-to-weight would result in a very fast and quite short take-off, the landing problem being much more difficult to solve. Despite the obvious need for aircraft genuinely able to disperse away from known geographical locations the same basic design was in parallel accepted by France, West Germany and Italy, whose industries collaborated in the design and research for a prototype; Aeritalia, for example, built one complete carbon-fibre composite wing for static tests.

In 1982 the British government announced it would contribute towards the cost of an EAP (Experimental Aircraft Programme), and on 26 May 1983 a contract was signed between BAe and MoD(PE) for a single flight prototype. This has been built at Warton and was rolled out for weighing and fuel-system tests on 27 October 1985. This aircraft is to be completed by January 1986, flying by mid-year. French insistence on a lighter aircraft configured mainly for ground attack, with a carrier-based variant, led to a go-ahead on a tri-national aircraft by Britain, West Germany and Italy on 9 August 1985, Spain joining later.

Features of the EAP, which will form the basis of the collaborative design, include conventional configuration and aerodynamics, with variable-camber wing and powered anhedralled canard, advanced cockpit, active fly-by-wire flight controls and unprecedented agility. Much thought has been devoted to electronic-warfare provisions (though it is clearly not a stealth design) and low-drag carriage of stores.

The EAP remains a one-aircraft programme, in many respects slightly compromised by shortage of funds and the need to take swift decisions. The colossal workload needed to define the eventual production aircraft really requires no fewer than 20 aircraft, on which no decision had been announced in late 1985. To start by building a single prototype is unique, and underlines the need within NATO for planning that, instead of producing paper, produces properly structured weapon programmes.

Specification
British Aerospace EAP
Type: multi-role combat aircraft demonstrator
Powerplant: two 7938 kg (17,500 lb) Turbo-Union RB199 Mk 104 augmented turbofan engines
Performance: maximum speed over Mach 2
Weights: basic equipped about 9750 kg (21,500 lb)
Dimensions: span 11.17 m (36 ft 7.8 in); length 14.70 m (48 ft 2.7 in); wing area 52.02 m² (560.0 sq ft)
Armament: not disclosed, but will probably include two 25-mm Aden or 27-mm Mauser guns, four Sky Flash medium-range AAMs and four AIM-9L Sidewinder close-range AAMs, plus various attack options

Seen on its first public airing, BAe's EAP combines most of the features of new technology fighters, such as canards, FBW controls and LERXs.

British Aerospace Harrier

Developed from the Hawker P.1127 vertical-take-off technology demonstrator, whose intended supersonic development (the P.1154) was cancelled in a defence 'economy drive', the BAe Harrier remains the only STOVL (short take-off, vertical landing) combat aircraft in the world. Originally regarded as an impractical novelty, it confounded its critics by giving outstanding service during the Falklands war of 1982, combining technical reliability with accuracy in attack and resistance to heavy defensive fire,

RAF Harriers are cleared to carry the Paveway LGB, which were used on two occasions in the Falklands. This aircraft is a GR.Mk 3 from No. 3 Sqn.

while its naval cousin achieved undisputed air superiority over the islands. Though the P.1127 flew in October 1960, it was not until almost nine years later that the RAF's first squadron was formed with Harrier GR.1 aircraft, the designation changing subsequently to Harrier GR.1A and then Harrier GR.3 as the Pegasus progressed from the 8709-kg (19,200-lb) thrust Mk 101 through Mk 102 to the present Mk 103. The two-seat trainer, with longer fuselage and taller fin, was similarly designated Harrier T.2, Harrier T.2A and Harrier T.4. Despite great power increases, the Harrier is unable to take off vertically with a maximum fuel and weapon load, but can take off from a short length of road or semi-prepared strip in the STOVL mode. In an all-out war context its priceless asset is tactical concealment, and thus survivability. Equipped from the outset with a Ferranti FE.541 inertial navigation system with head-up display, the RAF aircraft were fitted from 1976 with a Marconi LRMTS (laser ranger and marked-target seeker) contained in a 'thimble' added to the nose. A Marconi ARI 18223 E-J band radar warning receiver was added to the fin and extreme rear fuselage at the same time. The Harrier carries a single oblique camera in the port side of the nose, but two RAF squadrons are equipped with a multi-sensor pod beneath the fuselage for more extensive reconnaissance. Following the construction of six pre-production aircraft, the RAF received 114 single-seat Harriers, and added four more for replacement of Falklands losses (three, all pilots being recovered). US Marine Corps contracts covered 102 AV-8A aircraft (survivors now converted to AV-8C standard) and the Spanish navy acquired 11 VA.1 Matadors similar to the AV-8A. Two-seat trainer orders covered 23 for the RAF (those newly-built with with Mk 103 engine and LRMTS being Harrier T.4A), eight TAV-8A, two VAE. 1

In wartime, Harriers would spend most of their time away from fixed airfields, flying from specially prepared airstrips and hides. These hides are usually cut into treelines so as not to break the general countryside. This makes the Harrier much less vulnerable to attacks from enemy aircraft than its non-STOVL cousins, and enables it to operate closer to the front.

Matador, four navalized Harrier T.4M aircraft for the Royal Navy, two T.60s with Blue Fox radar for the Indian navy, plus a Mk 52 company demonstrator. The Harrier GR.5 is described under the AV-8B Harrier II.

Specification
BAe Harrier GR.3
Type: single-seat close-support and reconnaissance fighter
Powerplant: one 9752-kg (21,500-lb) Rolls-Royce Pegasus 103 vectored-thrust turbofan engine
Performance: maximum speed, clean Mach 0.95 or 1159 km/h (720 mph) at 305 m (1,000 ft); tactical radius hi-lo-hi mission 418 km (260 miles)
Weights: empty 5425 kg (11,960 lb); maximum vertical take-off 8165 kg (18,000 lb); maximum short take-off 11431 kg (25,200 lb)
Dimensions: span 7.09 m (23 ft 3 in); length 14.27 m (46 ft 10 in); height 3.45 m (11 ft 4 in); wing area 18.67 m^2 (201.0 sq ft)
Armament: two 30-mm (1.18-in) Aden cannon (with up to 130 rpg) on fuselage strongpoints; four wing pylons carrying up to 2268 kg (5,000 lb) of ordnance (or 455-litre/100-Imp gal tanks, inboard only), including 454-kg (1,000-lb) free-fall or retarded bombs, 68-mm (2.68-in) SNEB rocket pods, BL.755 cluster bombs, AIM-9L Sidewinder AAMs (single round each outer pylon), or optionally, Pave Way laser guided bombs and, Royal Navy, 51-mm (2-in) rocket pods
Operators: India (T.60 only), Spain, UK, USA

British Aerospace (HS) Harrier GR.3 cutaway drawing key

1 Pitot tube
2 Laser protective 'eyelids'
3 Ferranti Laser Ranger and Marked Target Seeker unit (LRMTS)
4 Cooling air duct
5 Oblique camera
6 Camera port
7 Windshield washer reservoir
8 Inertial platform
9 Nose pitch reaction control air duct
10 Pitch feel and trim actuator
11 IFF aerial
12 Cockpit ram air intake
13 Yaw vane
14 Cockpit air discharge valve
15 Front pressure bulkhead
16 Rudder pedals
17 Nav/attack 'head-down' display unit
18 Underfloor control linkages
19 Canopy external handle
20 Control column
21 Instrument panel shroud
22 Windscreen wiper
23 Birdproof windscreen panels
24 Head-up display
25 Starboard side console panel
26 Nozzle angle control lever
27 Engine throttle lever
28 Ejection seat rocket pack
29 Fuel cock
30 Cockpit pressurization relief valve
31 Canopy emergency release
32 Pilot's Martin-Baker Type 9D, zero-zero ejection seat
33 Sliding canopy rail
34 Miniature detonating cord (MDC) canopy breaker
35 Starboard air intake
36 Ejection seat headrest
37 Cockpit rear pressure bulkhead
38 Nose undercarriage wheel well
39 Boundary layer bleed air duct
40 Port air intake
41 Pre-closing nosewheel door
42 Landing/taxiing lamp
43 Nosewheel forks
44 Nosewheel
45 Supplementary air intake doors (fully floating)
46 Intake ducting
47 Hydraulic accumulator
48 Nosewheel retraction jack
49 Intake centre-body
50 Ram air discharge to engine intake
51 Cockpit air conditioning plant
52 Air conditioning system ram air intakes
53 Boundary layer bleed air discharge ducts
54 Starboard supplementary air intake doors
55 UHF aerial
56 Engine intake compressor face
57 Air refuelling probe connection
58 Forward fuselage integral fuel tank, port and starboard
59 Engine bay venting air scoop
60 Hydraulic ground connections
61 Engine monitoring and recording equipment
62 Forward nozzle fairing
63 Fan air (cold stream) swivelling nozzle
64 Nozzle bearing
65 Venting air intake
66 Alternator cooling air ducts
67 Twin alternators
68 Engine accessory gearbox
69 Alternator cooling air exhausts
70 Engine bay access doors
71 Gas turbine starter/Auxiliary power unit, GTS/APU
72 APU exhaust duct
73 Aileron control rods
74 Wing front spar carry-through
75 Nozzle bearing cooling air duct
76 Engine turbine section
77 Rolls-Royce Pegasus Mk 103 vectored thrust turbofan engine
78 Wing panel centreline joint rib
79 APU intake
80 Wing centre section fairing panels
81 Starboard wing integral fuel tank, total internal fuel capacity 630 Imp gal (2865 litres)
82 Fuel system piping
83 Pylon attachment hardpoint
84 Aileron control rod
85 Reaction control air duct
86 Leading-edge dog-tooth
87 Starboard inner stores pylon
88 100-Imp gal (455-litre) jettisonable combat fuel tank
89 1,000-lb (454-kg) HE bomb
90 BL.755, 600-lb (272-kg) cluster bomb
91 Starboard outer stores pylon
92 Wing fences
93 Outer pylon hardpoint
94 Aileron hydraulic power control unit
95 Roll control reaction air valve
96 Starboard navigation light
97 Wing tip fairing
98 Profile of extended span ferry tip
99 Starboard outrigger fairing
100 Outrigger wheel retracted position
101 Starboard aileron
102 Fuel jettison pipe
103 Starboard plain flap
104 Trailing edge root fairing
105 Water-methanol filler cap
106 Anti-collision light
107 Water-methanol injection system tank
108 Fire extinguisher bottle
109 Flap hydraulic jack
110 Fuel contents transmitters
111 Rear fuselage integral fuel tank
112 Ram air turbine housing
113 Turbine doors
114 Emergency ram air turbine (extended position)
115 Rear fuselage frames
116 Ram air turbine jack
117 Cooling air ram air intake
118 HF tuner
119 HF notch aerial
120 Rudder control rod linkages
121 Starboard all-moving tailplane
122 Temperature sensor
123 Tailfin construction
124 Forward radar warning receiver
125 VHF aerial
126 Fin tip aerial fairing
127 Rudder upper hinge
128 Honeycomb rudder construction
129 Rudder trim jack
130 Rudder tab
131 Tail reaction control air ducting
132 Yaw control port
133 Aft radar warning receiver
134 Rear position light
135 Pitch reaction control valve
136 Tailplane honeycomb trailing edge
137 Extended tailplane tip
138 Tailplane construction
139 Tail bumper
140 IFF notch aerial
141 Tailplane sealing plate
142 Fin spar attachment
143 Tailplane centre section carry through
144 All-moving tailplane control jack
145 Ram air exhaust duct
146 UHF standby aerial
147 Equipment air conditioning plant
148 Ground power supply socket
149 Twin batteries

The US Marine Corps have procured the Harrier as the AV-8A/C, to be based on its assault carriers. These operate with Marines assault units covering beach landings alongside Bell AH-1 Cobra gunships. The Harrier has proved an excellent addition to the Marine armoury, being able to respond rapidly to any call for help.

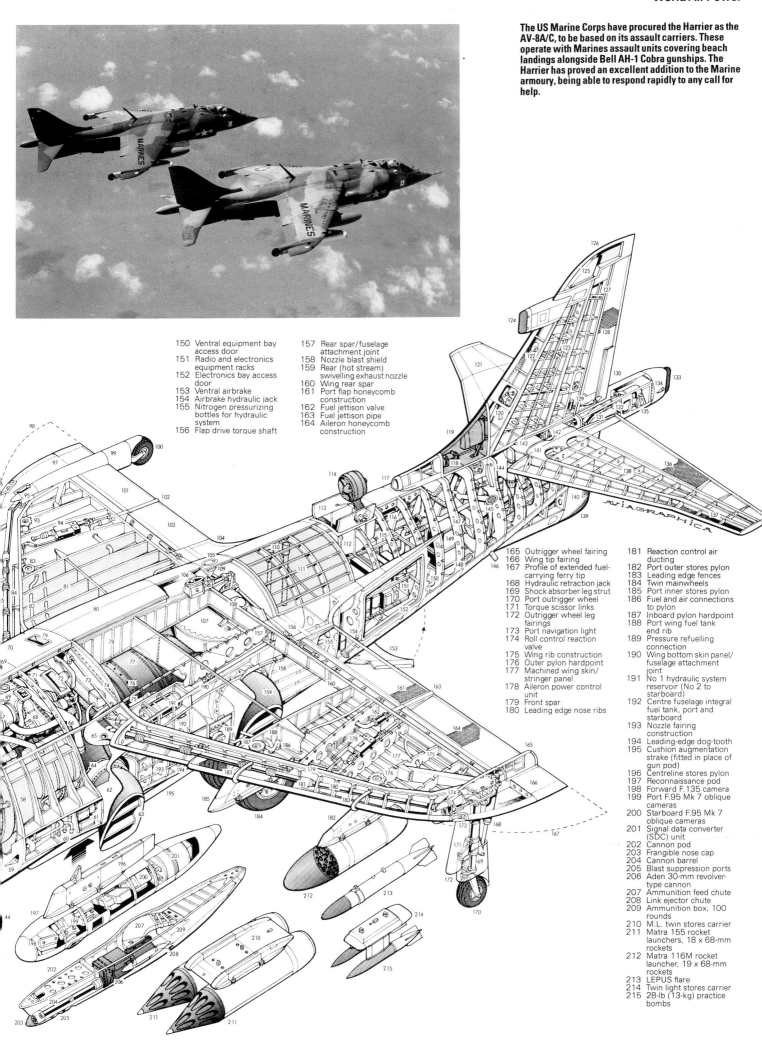

150 Ventral equipment bay access door
151 Radio and electronics equipment racks
152 Electronics bay access door
153 Ventral airbrake
154 Airbrake hydraulic jack
155 Nitrogen pressurizing bottles for hydraulic system
156 Flap drive torque shaft
157 Rear spar/fuselage attachment joint
158 Nozzle blast shield
159 Rear (hot stream) swivelling exhaust nozzle
160 Wing rear spar
161 Port flap honeycomb construction
162 Fuel jettison valve
163 Fuel jettison pipe
164 Aileron honeycomb construction

165 Outrigger wheel fairing
166 Wing tip fairing
167 Profile of extended fuel-carrying ferry tip
168 Hydraulic retraction jack
169 Shock absorber leg strut
170 Port outrigger wheel
171 Torque scissor links
172 Outrigger wheel leg fairings
173 Port navigation light
174 Roll control reaction valve
175 Wing rib construction
176 Outer pylon hardpoint
177 Machined wing skin/stringer panel
178 Aileron power control unit
179 Front spar
180 Leading edge nose ribs

181 Reaction control air ducting
182 Port outer stores pylon
183 Leading edge fences
184 Twin mainwheels
185 Port inner stores pylon
186 Fuel and air connections to pylon
187 Inboard pylon hardpoint
188 Port wing fuel tank end rib
189 Pressure refuelling connection
190 Wing bottom skin panel/fuselage attachment joint
191 No 1 hydraulic system reservoir (No 2 to starboard)
192 Centre fuselage integral fuel tank, port and starboard
193 Nozzle fairing construction
194 Leading-edge dog-tooth
195 Cushion augmentation strake (fitted in place of gun pod)
196 Centreline stores pylon
197 Reconnaissance pod
198 Forward F.135 camera
199 Port F.95 Mk 7 oblique cameras
200 Starboard F.95 Mk 7 oblique cameras
201 Signal data converter (SDC) unit
202 Cannon pod
203 Frangible nose cap
204 Cannon barrel
205 Blast suppression ports
206 Aden 30-mm revolver-type cannon
207 Ammunition feed chute
208 Link ejector chute
209 Ammunition box, 100 rounds
210 M.L. twin stores carrier
211 Matra 155 rocket launchers, 18 x 68-mm rockets
212 Matra 116M rocket launcher, 19 x 68-mm rockets
213 LEPUS flare
214 Twin light stores carrier
215 28-lb (13-kg) practice bombs

British Aerospace Hawk

The Hawk has had excellent overseas sales, proving the overall soundness of design and concept. Among the less well-fated exports were the aircraft for Zimbabwe, four of which were badly damaged in the attack on Thornhill base by terrorists.

The British Aerospace Hawk, known originally as the Hawker Siddeley HS.1182, replaced the Royal Air Force's Gnat and Hunter advanced trainers. It was selected in 1971, and in March 1972 an order was placed for 176 designated Hawk T.1. The Hawk is a transonic ground-attack and training aircraft, with a low-mounted wing and stepped tandem seats. Its primary structure is designed for 6,000 hours in the exacting conditions demanded by the RAF. Simplicity in design and manufacture were emphasized to ensure that it has a high utilization rate and is inexpensive to operate. One man can prepare a Hawk for its next flight in less than 20 minutes, and in the weapon-training role it can be rearmed by four men in less than 15 minutes. Costs are reduced by the fuel-efficient Adour turbofan, an unaugmented version of that employed in the SEPECAT Jaguar, any module of which can be changed without a need to rebalance the rotating assemblies, large doors beneath the engine bay allowing easy access and removal. The gas-turbine starter runs off the aircraft fuel supply.

The Hawk had trouble-free development, entering service 27 months after first flight. All performance objectives were easily met and most,

including level speed and Mach limit, were far exceeded. Hawk is cleared to Mach 1.2 in a dive, and has a level Mach of 0.88, allowing student pilots to experience transonic handling at low cost. The RAF uses the T.1 for weapon instruction as well as flying training, with three pylons available for stores. The centreline pylon is normally occupied by an Aden cannon pod, with Matra rocket launchers or practice bombs beneath the wings. Potential export customers often demand heavier armament, however, and British Aerospace has tested the Hawk with about 40 combinations of air-to-surface and air-to-air weaponry. Two stations can be provided beneath each wing, giving a total of five, and the use of multiple racks allows the aircraft to carry 3084 kg (6,800 lb) of stores, such as six 454-kg (1,000-lb) bombs plus the Aden gunpack. For airfield defence and day attack duties 72 RAF Hawks had been modified to T.1A standard with wing pylons carrying AIM-9L Sidewinder AAMs or BL.755 cluster bombs.

Large numbers of export Hawks began with the 50-series, with the Mk 851 engine, and continued with the 60-series with the Mk 861 en-

gine of further increased power. All have five pylons, large drop tanks, a special high-lift wing with leading edge devices and four-position flaps, adaptive anti-skid brakes, and greatly increased weight and weapons options.

In November 1981 a special carrier-equipped version of the Hawk was selected by the US Navy as the T-45 to replace the T-2 as standard pilot trainer. Over 300 will be supplied by BAe's American partner, McDonnell Douglas, in 1990-93. Features include new long-stroke main gears, twin-wheel nose-tow nose gear, twin rear airbrakes, a hook and a new cockpit. Many parts will be of carbon-fibre composite.

In June 1984 BAe went ahead with a long-planned single-seat Series 200, with the uprated Mk 861 (or even later Mk 871) engine. Each of four pylons will be rated at 907 kg (2,000 lb), two internal guns (of 20-, 25- or 27-mm/0.79-, 0.98- or 1.06-in) free the centreline for other stores, and there are large and varied avionics installations including Ferranti Blue Fox radar.

Armament: up to 2948 kg (6,500-lb) of stores, including 30-mm (1.18-in) Aden cannon pod and six 454-kg (1,000-lb) bombs; also Matra 155 rocket launchers, practice bombs, and a wide range of other stores
Operators: Abu Dhabi (Mk 63), Dubai (61), Finland (51), Indonesia (53), Kenya (52), Kuwait (64), USA (T-45) UK (T.1 and 1A), Zimbabwe (60)

Specification
BAe Hawk
Type: multi-role trainer and light attack aircraft
Powerplant: one Rolls-Royce/Turboméca Adour turbofan, rated at from 2422 to 2835 kg (5,340 to 6,250 lb) thrust
Performance: maximum speed at sea level 1006 km/h (625 mph); maximum speed 1065 km/h (662 mph); time to 9145 m (30,000 ft) 6 minutes 6 seconds; service ceiling 15240 m (50,000 ft); ferry range clean 2446 km (1,520 miles); ferry range with two 864-litre (190-Imp gallon) auxiliary fuel tanks 4072 km (2,530 miles)
Weights: empty, depending on sub-type 3221 to 3856 kg (7,100 to 8,500 lb); take-off, trainer 5035 kg (11,100 lb); maximum take-off, attack 8568 kg (18,890 lb)
Dimensions: span 9.39 m (30 ft 9.75 in); length 11.86 m (38 ft 11 in); height 3.99 m (13 ft 1.25 in); wing area 16.68 m^2 (179.60 sq ft)

The Hawk T.Mk 1A is the version flown by the RAF's Tactical Weapons Units based at Chivenor and Brawdy. These have a centreline Aden cannon and hardpoints for either rocket pods (shown) or Sidewinder missiles. The TWU Hawks will take part in the air defence of Britain in any war, being flown by the weapons instructors. This aircraft is from No. 234 Squadron ('shadow' number for No. 1 TWU) at Brawdy.

XX192

192

British Aerospace Hunter

A single-seat interceptor which transferred successfully to the attack role, Hunters continue in service with the air arms of 11 nations, the Hawker P.1067 having flown on 20 July 1951. Production totalled 1,971, including 269 by Armstrong Whitworth, and licence-manufacture by Avions Fairey of Belgium (249) and Holland's Fokker (183). Early marks were the Rolls-Royce Avon-engined F.1 and Armstrong-Siddeley Sapphire-powered F.2, these becoming F.4 and F.5 respectively with additional fuel tankage in the wing leading edge. Most were short-lived, though produced in considerable numbers, but relegation of F.4s to training (and removal of the four 30-mm/1.18-in Aden cannon) produced the Hunter GA.11 which serves the Royal Navy's Fleet Requirements and Air Direction Unit at Yeovilton. Three were fitted with camera noses (Hunter GA.11) and others have a Harley light in the nose for acquisition by naval gunners when acting the part of targets.

Most aircaft currently extant are based on the F.6, powered by the Avon 203. Forming the day interceptor backbone of the RAF during the 1950s, the Hunter found a new lease of life as a ground attack aircraft with RAF units in the Middle East. Under the designations Hunter FGA.9 it featured an Avon 207, a braking parachute and strengthened wings for 1046-litre (230-Imp gallon) drop-tanks and ordnance. All were conversions of the F.6s, as were similar tactical reconnaissance Hunter FR.10s with their three-camera nose. Until 1984, FGA.9s served in the armament training

role with No. 1 Tactical Weapons Unit at Brawdy, partnered by ground attack practice Hunter F.6A conversions and side-by-side Hunter T.7 dual trainers. The Hunter T.7A is fitted with some BAe Buccaneer avionics and used for continuation training by units operating that aircraft. Similarly, the Hunter T.8M has the Sea Harrier's Blue Fox radar, as a pilot conversion variant of the hooked (but not carrier operable) naval Hunter T.8A.

In addition to building aircraft for export, Hawker refurbished large numbers of surplus Hunters (including buy-backs from Belgium, Holland and Sweden) for re-sale, the last in 1975. The most advanced are Switzerland's Mk 58/58As which are in the process of local conversion to "Hunter 80" standard with a chaff/

flare dispenser, radar warning receiver and no less than ten wing strongpoints for 28 80-mm (3.15-in) rockets, or (on some aircraft) Hughes AGM-65A Maverick air-to-surface missiles.

Specification
Hunter F.6
Type: single-seat fighter
Powerplant: one 4559-kg (10,050-lb) thrust Rolls-Royce Avon 203 turbojet engine
Performance: maximum speed 1151 km/h (715 mph) at 10975 m (36,000 ft); initial rate of climb 5334 m (17,500 ft)/minute: service ceiling 16765 m (55,000 ft); range 2961 km (1,840 miles) with 4782 litres (1052 Imp gallons) fuel; operational radius 925 km (575 miles)

Britain's greatest fighter success, the Hunter is still in service around the world. The Royal Navy's FRADU operates the GA.Mk 11 as gunnery practice targets, but these are occasionally fitted with rocket pods for weapons training.

Weights: empty 6019 kg (13,270 lb); maximum take-off 10886 kg (24,000 lb)
Dimensions: span 10.26 m (33 ft 8 in); length 13.98 m (45 ft 10.5 in); height 4.01 m (13 ft 2 in); wing area 32.42 m² (349.0 sq ft)
Armament: military load of up to 3357 kg (7,400 lb) including four 30-mm (1.18-in) Aden cannon and ordnance on four wing pylons
Operators: Chile, India, Iraq, Lebanon, Oman, Qatar, Singapore, Somalia, Switzerland, UK, Zimbabwe

British Aerospace (Hunting/BAC) Jet Provost

Around 1950 it began to be appreciated in the RAF that pupil pilots might be trained on jets from the start. Percival (later Hunting Percival) Aircraft decided as a private venture to produce a jet conversion of the Provost piston-engined trainer, though this inevitably had to be a largely redesigned machine with a 745-kg (1,640-lb) Armstrong Siddeley Viper turbojet amidships, fed from inlets just ahead of and above the wing roots, a jetpipe leading to the tail, the side-by-side cockpit in the nose and retractable tricycle landing gear. Fuel capacity was also increased, and there were many minor changes. The Jet Provost flew on 26 June 1954 and looked rather crude. After much refinement (while 10 T.1s were evaluated in RAF service) the much neater T.2 was produced with short landing gear and a 794-kg (1,750-lb) Viper engine, and with provision for tip tanks, underwing pylons and two nose guns (none of which were used by the RAF).

Results were satisfactory, and from 1958 Hunting Aircraft (as it had become) delivered 201 T.3s with Martin-Baker seats, a clear-view canopy, tip-tanks and updated avionics. The first pupil course began in October 1959 (though RAF pupils had trained on the JP1 evaluation batch from 1955). BAC, which took over Hunting, continued development by fitting the much more powerful Viper 11 (Mk 200 series) of 1134-kg (2,500-lb) rating, and 198 of the resulting T.4 version were delivered to the RAF in 1961-4. By this time the emphasis on high-altitude

British Aerospace Jet Provost T.Mk 5

sorties, for example to increase the duration possible on the available fuel, had highlighted the need for cockpit pressurization. The result was the JP5, or Jet Provost T.5, first flown in February 1967 after transfer of the programme from Luton to BAC Warton. Distinguished by its smooth bulged canopy/windshield and longer nose, the JP5 also has increased internal fuel capacity and other updates, and is significantly heavier and less sprightly than the JP4. BAC delivered 110 of this final version. BAC and BAe have subsequently carried out many refurbishing programmes on JPs for example fitting VOR/DME to produce the T.5A.

Exports comprised 22 Mk 51 based on the JP3 but with weapons, for Sri Lanka, Kuwait and Sudan, 43 Mk 52 based on the JP4 for Iraq, South Yemen, Sudan and Venezuela and five Mk 55 based on the JP 5 for the Sudan.

Specification
BAe Jet Provost T.5
Type: basic trainer
Powerplant: one 1134-kg (2,500-lb) Rolls-Royce Viper 202 turbojet
Performance: maximum speed 382 kts (708 km/h; 440 mph) at

7620 m (25,000 ft); initial climb 1082 m (3,550 ft) per minute; service ceiling 1118 m (36,700 ft); range with tip-tanks 1448 km (900 miles)
Weights: 2271 kg (4,888 lb); maximum take-off 4173 kg (9,200 lb)
Dimensions: span 10.77 m (35 ft 4 in); length 10.36 m (34 ft 0 in); height 3.1 m (10 ft 2 in); wing area 19.85 m² (213.7 sq ft)

The Jet Provost has provided the RAF with a jet trainer since 1959, and is now due for replacement, which will come in the shape of a turboprop aircraft.

Armament: provision for two 7.62-mm FN or similar machine-guns in nose, and for variety of light loads under wings (not used by RAF)

British Aerospace (HP/Scottish Aviation) Jetstream

The original prototype of the Handley Page H.P.137 Jetstream flew on 18 August 1967. An attractive pressurized third-level and executive passenger transport, it featured a circular-section fuselage offering stand-up headroom, large elliptical passenger windows, manual flight controls, inwards-retracting main landing gears, twin steerable nose wheels and optional rubber-boot deicers. Powered by two 720-kW (965-shp) Turboméca Astazou XVI turboprops, it had a maximum weight of 5670 kg (12,500 lb) and could seat up to 18 and cruise at 448 km/h (278 mph). The factory was in full production when the firm went bankrupt in 1970, but many Jetstreams were completed by Scottish Aviation, among them 26 ordered by the RAF as Jetstream T.1 MEPTs (multi-engined pilot trainers). The first was delivered to 5 FTS in June 1973. Policy changed and the aircraft were stored. Later eight entered service with 3 FTS, while 14 went to the Fleet Air Arm as Jetstream T.2 observer trainers, equipped with MEL E.190 nose radar used in weather and mapping modes.

After Scottish Aviation was ab-

British Aerospace Jetstream T.Mk 1 of No. 6 FTS at Finningley. The aircraft is used for navigation training.

sorbed into BAe the decision was taken in 1978 to develop the new Jetstream 31. A fault of the original machine had been inadequate difference between empty and gross weight, but substitution of Garrett engines, driving new Dowty Rotol four-blade propellers, enabled gross weight to be greatly increased. The Mk 31 has proved a great success, and one of many customers is the Fleet Air Arm which purchased four T.3 observer trainers, operating alongside the T.2 at 750 Squadron at RNAS Culdrose. A totally updated aircraft, the T.3 has a Racal ASR.360 multi-mode radar with the scanner in a ventral blister. Other

equipment includes Doppler and a TANS (tactical air navigation system) computer. It is almost certain that further orders will be placed by various customers for other military variants. Numerous interior configurations are available, and the Jetstream is an ideal candidate for various paramilitary roles such as casevac, light cargo, prospecting and survey, and offshore patrol.

Specification
BAe Jetstream T.3
Type: radar observer trainer
Powerplant: two 701-kW (940-shp) Garrett TPE331-10UF turboprop
Performance: maximum cruising

speed 263 kts (488 km/h; 303 mph) at 4570 m (15,000 ft); initial rate of climb 635 m (2,080 ft) per minute; service ceiling 7620 m (25,000 ft); range about 1975 km (1,225 miles) with full reserves
Weights: empty about 4355 kg (9,600 lb); maximum take-off 6900 kg (15,212 lb)
Dimensions: span 15.85 m (52 ft 0 in); length 14.37 m (47 ft 1.8 in); height 5.32 m (17 ft 5.4 in); wing area 25.2 m² (271.3 sq ft)
Armament: none
Operator: UK

British Aerospace (EECo/BAC) Lightning

Unique in many respects, the English Electric Lightning was the first (and so far only) all-British type of supersonic military aircraft. Derived from the P.1 research aircraft of 1954, the Lightning F.Mk 1 flew in 1958 and entered service with No. 74 Squadron in 1960. Features included two afterburning engines staggered one above the other in a narrow slab-sided fuselage, untapered wings swept at 60° with ailerons at right-angles to the airstream joining the leading and trailing edges, tall main landing gear units hinged in the centre of each wing and retracting outwards so that the large but thin wheels lie near the tips, low-mounted slab tailplanes, a monopulse radar mounted inside the large conical centrebody in the nose air inlet, and a choice of gun, rocket or AAM armament. Shortcomings included modest fuel capacity, giving very short endurance, and extremely limited armament.

In the teeth of official opposition (because the fighters were said to be obsolete as a class) the type was developed through the Lightning F.Mk 1A, F.Mk 2, F.Mk 2A, F.Mk 3 and F.Mk 6 single-seaters, and the Lightning T.Mk 4 and T.Mk 5 dual trainers with a side-by-side cockpit with an upward-hinged canopy. Internal fuel was greatly increased by successively larger belly tankage which finally made room for guns as well. The vertical tail was enlarged, more powerful engines installed, and in the final export models (Lightning F.Mk 53) for Saudi Arabia and Kuwait provision was

British Aerospace (EECo/BAC) Lightning F.Mk 6 of No.11 Sqn based at Binbrook.

made for outboard wing pylons which in principle can carry tanks above the wing and bombs or rocket launchers.

Total production of all marks was 338, but only a handful remain in service. The LTF (Lightning Training Flight) and RAF Nos 5 and 11 Squadrons operate the T.Mk 5 and F.Mk 6, using the armament outlined below. Saudi and Kuwaiti F.Mk 53 and Lightning T.Mk 55 aircraft are withdrawn from the inventory but several remain operational and, such is the enduring popularity of this exciting and enjoyable fighter, several continue to be flown.

Specification
BAe Lightning F.Mk 6
Type: all-weather interceptor
Powerplant: two 7112-kg (15,680-lb) thrust Rolls-Royce Avon Mk 302 afterburning turbojet engines
Performance: maximum speed with AAMs Mach 2.25 or 1300 kts (2415 km/h; 1,500 mph) at 12190 m (40,000 ft); initial rate of climb 15240 m (50,000 ft) per minute;

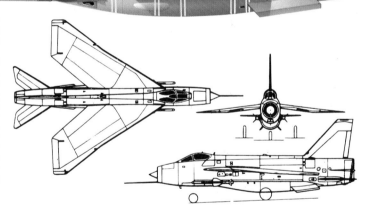
British Aerospace (EECo/BAC) Lightning F.Mk 6

service ceiling 18290 m (60,000 ft); range on internal fuel with high-altitude cruise 1290 km (800 miles)
Weights: empty 12701 kg (28,000 lb); maximum take-off 22680 kg (50,000 lb)
Dimensions: span 10.62 m (34 ft

10 in); length 16.25 m (53 ft 3 in); height 5.95 m (19 ft 7 in); wing area 35.31 m² (380.1 sq ft)
Armament: two Red Top infra-red homing AAMs and two 30-mm Aden cannon with 120 rounds per gun
Operators: Saudi Arabia, UK

British Aerospace Nimrod

The world's first jet-powered patrol aircraft, the Nimrod was developed from the Comet 4C airliner as a replacement for the Avro Shackleton. Under the designation Hawker Siddeley 801, work began in June 1964, and the design was officially adopted 12 months later. Differences immediately apparent from the Comet include the double-bubble fuselage,

British Aerospace Nimrod MR.Mk 2P of the Kinloss Wing.

British Aerospace Nimrod (continued)

the lower (unpressurized) component of which houses operational equipment and a giant weapons bay, 14.78 m (48 ft 6 in) long. In the Nimrod MR.1 an ASV.21 radar was located in the nose, ARAR/ARAX passive ESM equipment in the fin-tip pod, MAD gear in the tail 'stinger' and a 70 million candlepower searchlight in the right wing pod. The first prototype retained Avon engines, but the first to fly (at Woodford on 23 May 1967) was the second aircraft with the production engine, the Rolls-Royce Spey which is better suited to the mission profile, requiring rapid high-level transit to the operational area and then protracted search at low altitudes. Patrol times may be extended by shutting down one, two or even three engines as weight/altitude safety limits are achieved. The first of an initial batch of 38 MR.1s entered service in October 169 with what is now No. 236 Operational Conver-

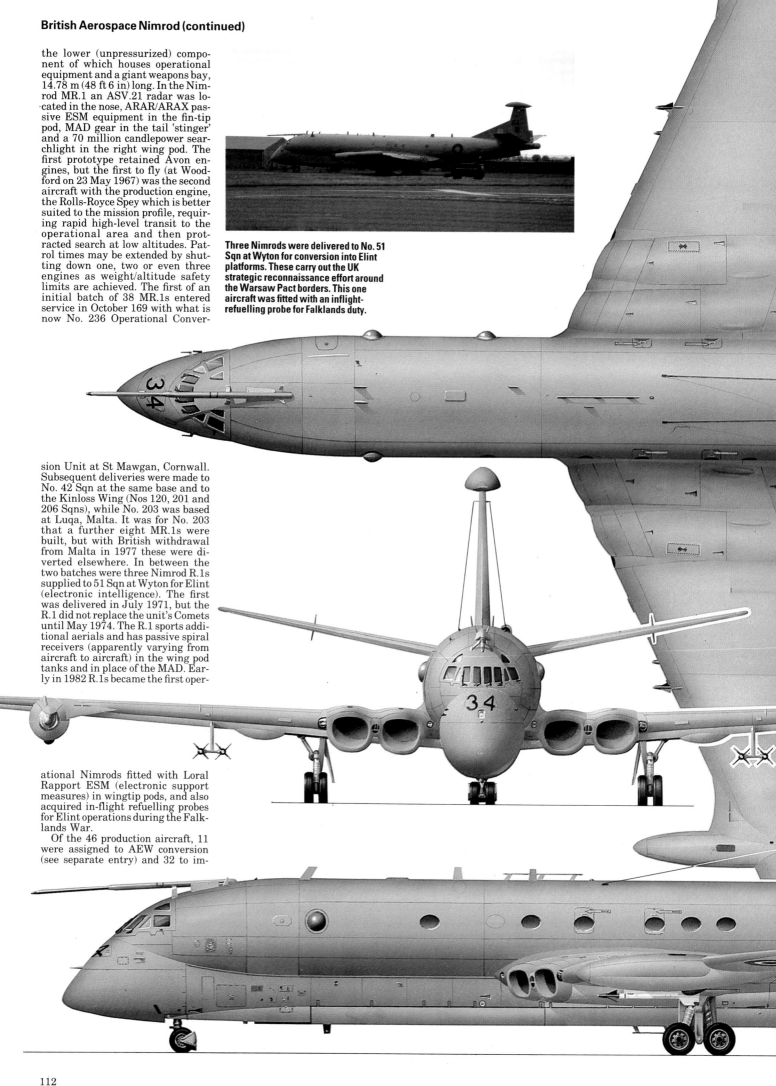

Three Nimrods were delivered to No. 51 Sqn at Wyton for conversion into Elint platforms. These carry out the UK strategic reconnaissance effort around the Warsaw Pact borders. This one aircraft was fitted with an inflight-refuelling probe for Falklands duty.

sion Unit at St Mawgan, Cornwall. Subsequent deliveries were made to No. 42 Sqn at the same base and to the Kinloss Wing (Nos 120, 201 and 206 Sqns), while No. 203 was based at Luqa, Malta. It was for No. 203 that a further eight MR.1s were built, but with British withdrawal from Malta in 1977 these were diverted elsewhere. In between the two batches were three Nimrod R.1s supplied to 51 Sqn at Wyton for Elint (electronic intelligence). The first was delivered in July 1971, but the R.1 did not replace the unit's Comets until May 1974. The R.1 sports additional aerials and has passive spiral receivers (apparently varying from aircraft to aircraft) in the wing pod tanks and in place of the MAD. Early in 1982 R.1s became the first oper-

ational Nimrods fitted with Loral Rapport ESM (electronic support measures) in wingtip pods, and also acquired in-flight refuelling probes for Elint operations during the Falklands War.

Of the 46 production aircraft, 11 were assigned to AEW conversion (see separate entry) and 32 to im-

112

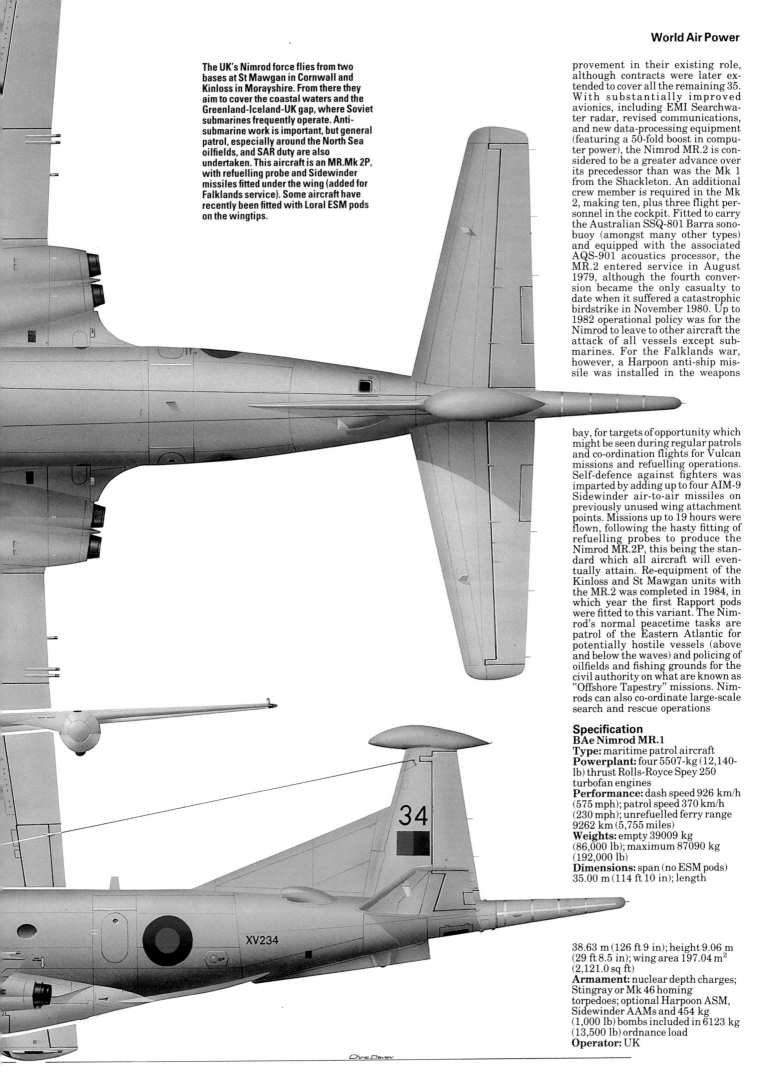

The UK's Nimrod force flies from two bases at St Mawgan in Cornwall and Kinloss in Morayshire. From there they aim to cover the coastal waters and the Greenland-Iceland-UK gap, where Soviet submarines frequently operate. Anti-submarine work is important, but general patrol, especially around the North Sea oilfields, and SAR duty are also undertaken. This aircraft is an MR.Mk 2P, with refuelling probe and Sidewinder missiles fitted under the wing (added for Falklands service). Some aircraft have recently been fitted with Loral ESM pods on the wingtips.

provement in their existing role, although contracts were later extended to cover all the remaining 35. With substantially improved avionics, including EMI Searchwater radar, revised communications, and new data-processing equipment (featuring a 50-fold boost in computer power), the Nimrod MR.2 is considered to be a greater advance over its precedessor than was the Mk 1 from the Shackleton. An additional crew member is required in the Mk 2, making ten, plus three flight personnel in the cockpit. Fitted to carry the Australian SSQ-801 Barra sonobuoy (amongst many other types) and equipped with the associated AQS-901 acoustics processor, the MR.2 entered service in August 1979, although the fourth conversion became the only casualty to date when it suffered a catastrophic birdstrike in November 1980. Up to 1982 operational policy was for the Nimrod to leave to other aircraft the attack of all vessels except submarines. For the Falklands war, however, a Harpoon anti-ship missile was installed in the weapons bay, for targets of opportunity which might be seen during regular patrols and co-ordination flights for Vulcan missions and refuelling operations. Self-defence against fighters was imparted by adding up to four AIM-9 Sidewinder air-to-air missiles on previously unused wing attachment points. Missions up to 19 hours were flown, following the hasty fitting of refuelling probes to produce the Nimrod MR.2P, this being the standard which all aircraft will eventually attain. Re-equipment of the Kinloss and St Mawgan units with the MR.2 was completed in 1984, in which year the first Rapport pods were fitted to this variant. The Nimrod's normal peacetime tasks are patrol of the Eastern Atlantic for potentially hostile vessels (above and below the waves) and policing of oilfields and fishing grounds for the civil authority on what are known as "Offshore Tapestry" missions. Nimrods can also co-ordinate large-scale search and rescue operations

Specification
BAe Nimrod MR.1
Type: maritime patrol aircraft
Powerplant: four 5507-kg (12,140-lb) thrust Rolls-Royce Spey 250 turbofan engines
Performance: dash speed 926 km/h (575 mph); patrol speed 370 km/h (230 mph); unrefuelled ferry range 9262 km (5,755 miles)
Weights: empty 39009 kg (86,000 lb); maximum 87090 kg (192,000 lb)
Dimensions: span (no ESM pods) 35.00 m (114 ft 10 in); length 38.63 m (126 ft 9 in); height 9.06 m (29 ft 8.5 in); wing area 197.04 m² (2,121.0 sq ft)
Armament: nuclear depth charges; Stingray or Mk 46 homing torpedoes; optional Harpoon ASM, Sidewinder AAMs and 454 kg (1,000 lb) bombs included in 6123 kg (13,500 lb) ordnance load
Operator: UK

British Aerospace Nimrod AEW.3

Detailed studies for an AEW (airborne early warning) version of the Nimrod began in 1973, to replace the RAF's 1950s-vintage Shackletons in a role vital to the defence of the United Kingdom. The capacious fuselage and ample reserve of power in the standard MR version of this aircraft (described separately) give it considerable potential for carrying alternative equipment and performing other duties. After much deliberation whether or not to buy the Boeing E-3A to meet the RAF's AEW requirement the British government decided in March 1977 to finance instead a developed version of the Hawker Siddeley (now British Aerospace) Nimrod. This enabled the design to be tailored more closely to British needs, than would otherwise have been the case.

In terms of outward appearance, the AEW.3 is distinguished by the appearance at each end of the fuselage of a grotesquely large radome. These contain the fore and aft scanners, each covering 180°, for a brand new Marconi radar system, designed specifically for the Nimrod 3 to fulfil the basic overwater mission, but compatible also with the air defence requirements of central Europe. Because of their fore-and-aft location, the scanners' efficiency is not obscured by other parts of the airframe, as is the case with the dorsally-mounted rotodomes; further advantages are stability of the radar in a banked turn and elimination of cyclic errors. An on-board digital computer controls the flow of data from the scanners (target range,

speed and height), also correlating this information with a control station on the ground. The scanners, which are interfaced with the Nimrod's IFF (identification, friend or foe) system, are also part of a pulse-Doppler radar installation capable of ship surveillance as well as aircraft detection, and highly resistant to electronic jamming. It interleaves high and low PRF (pulse recurrence frequency) to gain best results from all targets. Thus, despite its AEW designation, the Nimrod's function is accurately defined by the AWACS (airborne warning and control system) description applied by the Americans to the Boeing E-3A. The Nimrod's system can detect, track and classify aircraft, missiles or ships; control an interceptor fighter force; direct retaliatory strike aircraft; serve as an airborne air-traffic control centre; or carry out search and rescue duties.

The other outward sign of change in the Nimrod 3 is the presence at each wingtip of a pod containing electronic support measures (ESM) equipment. Internally, the sophisticated detection/tracking/classification gear goes hand in hand with

dual inertial navigation systems; tactical situation displays (six consoles); UHF, HF, U/VHF and LF radio sets; a flight director; plus many other electronic aids. An inflight refuelling probe is fitted.

The first testbed aircraft was a Comet 4C used for special communications trials, and this made its first flight on 28 June 1977 fitted with the forward radome only. The other three development aircraft, flown in 1980-82, were set aside from the supplementary batch of eight Nimrod MR.1s. The second and third had full mission avionics. The first 'production' aircraft flew on 9 March 1982, but the colossal magnitude of the development task resulted in the programme running late, and this has left a big gap in UK coverage. Aircraft began to reach No. 8 Sqn at Waddington for crew-training in 1984, but there were still major development hurdles to be cleared in early 1986. By this time most of the 11 production machines (including the three development aircraft, which are being brought up to operational standard) were fast becoming ready for delivery, once the complete system has been cleared of 'bugs'.

British Aerospace Nimrod AEW.Mk 3 as it is expected to appear with No. 8 Sqn, RAF.

British Aerospace Nimrod AEW.Mk 3

Specification
Type: airborne early warning and control aircraft
Powerplant: four 5507-kg (12,140-lb) thrust Rolls-Royce Spey Mk 250 turbofan engines
Performance: endurance more than 10 hours; no other figures released officially, but likely to differ marginally from MR versions (which see)
Weights: no details released
Dimensions: span over ESM pods 35.08 m (115 ft 1 in); length 41.97 m (137 ft 8.5 in); height 10.67 m (35 ft 0 in); wing area 197.05 m² (2,121.0 sq ft)
Armament: none
Operator: UK

British Aerospace/Rombac One-Eleven

Announced in May 1961, the BAC (British Aircraft Corporation) One-Eleven twin-jet airliner was ordered straight from the drawing board by British United Airways. The prototype made its maiden flight on 20 August 1963, followed by the first production aircraft, a Model 201, on 19 December 1963. Two Model 217s were supplied to the Royal Australian Air Force as executive transports and remain in service with No. 34 (VIP) Sqn at Fairbairn, Canberra, for government transport and liaison. In Mexico a One-Eleven serves with the Escuadron de Transport Presidencial at Mexico City. Following the 'stretched' 500-series came the 475 tailored to short unpaved airstrips. Three 475s were ordered by the Sultan of Oman's Air Force in June 1974. These replaced

Viscounts in No. 4 Sqn at Seeb, Airport, Muscat. All three incorporate a wide cargo door forward of the wing, and have quick-change pasenger/cargo interiors. Rombac (Romania/BAC) One-Eleven Srs. 495 and 560s are now built under licence by Grupul Aeronautic Bucuresti (Bucharest Aircraft Group) in Romania, the former with up to 89 seats and the latter with 109.

Specification
BAe One-Eleven Srs. 475
Type: short-range jet transport
Powerplant: two 5693-kg (12,550-lb) thrust Rolls-Royce Spey 512 DW turbofan engines
Performance: maximum cruising speed 871 km/h (541 mph) at

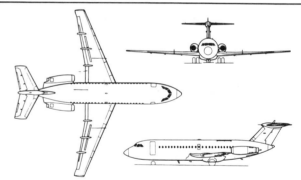

British Aerospace (BAC) One-Eleven Series 475

6400 m (21,000 ft); range with maximum fuel and reserves 3677 m (2,285 miles)
Weights: empty 23465 kg (51,731 lb); maximum take-off 44679 kg (98,500 lb)
Dimensions: span 28.50 m (93 ft

6 in); length 28.50 m (93 ft 6 in); height 7.47 m (24 ft 6 in); wing area 95.78 m² (1,031.0 sq ft)
Armament: none
Operators: Australia, Oman, Romania

British Aerospace Pembroke

First flown in 1948, the Percival Prince high-wing feederliner transport was a good basis for a military training and communications aircraft. The resulting Pembroke made its maiden flight in November 1952. Major differences included a 2.59 m (8 ft 6 in) increase in span, reinforced floor and strengthened twin-wheel main gears. The RAF received 44 Pembroke C.1 staff transports. The generous ground clearance afforded the propellers by the high wing, together with the sturdy tricycle landing gear, permitted operation from unprepared airstrips inaccessible even to its predecessor, the Avro Anson. Interior appointments were suitable for an air marshal in tall full-dress uniform and wearing a sword; other Pem-

brokes carried cargo. Six Pembroke C(PR).1s with fuselage-mounted cameras were supplied to the RAF for air survey and mapping, mostly overseas. Twelve similar aircraft were supplied to the Belgian air force, the first of several foreign customers. A few Pembrokes were diverted by the RAF to the Royal Rhodesian Air Force. The major overseas operator was the Luftwaffe, which had 33 in freighter, ambulance, survey and crew-training versions. The Royal Navy Sea Prince C.1 communications aircraft was similar to the civil Prince II. The four ordered in 1951 were followed by 42 larger T.1 radar trainers with consoles for three trainee operators. The Sea Prince C.2 was a passenger transport. By 1986 several Pem-

brokes remained on the strength of RAF Germany. Flying with 60 Sqn from Wildenrath, they provide regular communications to West Berlin and between RAF installations in West Germany. All have-

Several Pembrokes soldier on with No. 60 Squadron, based at Wildenrath with RAF Germany. These are employed for liaison duties between the German bases.

had to be re-sparred by British Aerospace, a costly task that terminated the lives of foreign Pembrokes.

Specification
BAe Pembroke C.1
Type: staff transport, light freighter and utility aircraft
Powerplant: two 403-kW (540-hp) Alvis Leonides 127 radial piston engines
Performance: maximum speed 354 km/h (220 mph) at 610 m (2,000 ft); cruising speed 241 km/h (150 mph) at 2440 m (8,000 ft); initial rate of climb 457 m (1,500 ft)/minute; service ceiling 6705 m (22,000 ft); range 1658 km (1,030 miles)
Weights: empty 4069 kg (8,970 lb); maximum take-off 5897 kg (13,000 lb)
Dimensions: span 19.66 m (64 ft 6 in); length 14.02 m (46 ft 0 in); height 4.90 m (16 ft 1 in); wing area 37.16 m² (400.0 sq ft)
Armament: none
Operator: UK

British Aerospace Sea Harrier

The exceptional versatility of the Harrier took an inordinate time to be translated into naval terms, but when Britain achieved operational capability with the Sea Harrier FRS.1 it was just in time for participation in combat. Three special vessels were built to carry the aircraft into the 21st century, and it will also be seen regularly in Indian waters as a result of what is currently the sole export contract. The Sea Harrier's *raison d'être* is to be found in a political decision to abandon the aircraft carrier, leaving the Royal Navy with the prospect of having to discontinue fixed-wing operations by Buccaneer Strike/attack aircraft and Phantom interceptors. The problem was overcome by combining these diverse qualities in an aircraft which could operate from a smaller and much cheaper V/STOL carrier, maximizing its weapon load by employing a rolling take-off over a ski ramp on the vessel's bows.

Development authorization for the Hawker Siddeley P.1184 and its Ferranti Blue Fox radar was announced in May 1975, and the first of 24 (plus a Harrier T.4A trainer to pay-back the RAF for crew training) was flown at Dunsfold on 28 August 1978. Compared with its forebear, the Sea Harrier has an airframe protected against salt and with no magnesium parts, a raised cockpit with far more panel space, avionics and a folding nose radome. (The small wings do not fold.) The Mk 104 engine has corrosion protection and generates more electrical power. RNAS Yeovilton was the designated Sea Harrier home base, deliveries beginning in June 1979 to what is now 899 Sqn, the HQ unit. Commissioning of the ASW carrier HMS *Invincible* took place in July 1980, followed by HMS *Illustrious* in 1982 and HMS *Ark Royal* in 1985. The helicopter carrier HMS *Hermes* was equipped with a 'ski jump' for Sea Harrier operations until it retired in 1983. First of the operational squadrons was 800, formed in March 1980 for *Invincible*, then moved to *Hermes*, and now *Illustrious*. The *Invincible* unit is 801, formed in January 1981. A follow-on batch of ten Sea Harriers was near completion when the Falklands war began in April 1982 and 20 aircraft were dispatched with the Task Force, soon to be joined by eight collected to form No. 809 Sqn which disbanded after the conflict. Employed as fighter-bombers and interceptors, Sea Harriers destroyed at least 20 and possibly 27 Argentine aircraft without air combat loss, demonstrating outstanding serviceability in very severe conditions; two fell to ground fire and four were lost in accidents. After the war, 14 further Sea Harriers were ordered, and in September 1984 another nine were added (making 57 in all) to allow the three squadrons to increase their complement from five to eight aircraft. This was the result of Falklands war experience which showed that eight aircraft were needed to maintain two on round-the-clock combat air patrol. Most air

British Aerospace Sea Harrier FRS.Mk 1 of No. 801 Squadron, Royal Navy.

Six Sea Harriers are currently operated from INS *Vikrant* by No. 300 Squadron, Indian Navy.

combat successes were achieved by AIM-9L Sidewinder air-to-air missiles, for which a twin-launcher – to give four per aircraft – was produced just too late for combat. Other wartime developments included fitment of Tracor ALE-40 chaff/flare dispensers in the rear fuselage and 864-litre (190-gal) drop tanks. From 1985 the very advanced BAe Sea Eagle anti-ship missile is carried. For training three dual-control Harrier T.4Ns were delivered to Yeovilton in 1983, these equipped with Blue Fox radar. A boost to operational capability may come in 1987 with the implementation of an update programme that will introduce the Ferranti Blue Vixen multi-mode pulse-Doppler radar in an enlarged radome, which will provide lookdown/shoot-down capability. Two self-guided Hughes AIM-120 air-to-air missiles will be carried on each outer wing pylon, displacing the Sidewinders to new wingtip rails.

LERX (leading-edge root extensions) will augment the Sea Harrier's already impressive manoeuvrablity, while new wing fences and vortex generators are under consideration. Other improvements will include redesigned engine nozzles and improved radar warning and ECM equipment.

The only other customer for the aircraft so far has been the Indian Navy with an order placed in November 1979 for six Sea Harrier Mk 51s supplemented in 1984 by a further ten, and two Sea Harrier Mk 60 trainers. Handed over from January 1983, they trained at Yeovilton until flying to India in December 1983. Operated from Goa/Dabolim by 300 Sqn, they will fly from INS *Vikrant*.

Specification
BAe Sea Harrier FRS.1
Type: single-seat fighter/reconnaissance/strike aircraft

Powerplant: one 9752 kg (21,500 lb) thrust Rolls-Royce Pegasus Mk.104 turbofan engine
Performance: maximum speed 1186 km/h (737 mph) at low altitude, or Mach 1.25 at high altitude; intercept radius 740 km (460 miles) at high altitude; strike radius 463 km (288 miles)
Weights: empty 5879 kg (12,960 lb); maximum take-off 11884 kg (26,200 lb)
Dimensions: span 7.70 m (25 ft 3 in); length 14.50 m (47 ft 7 in); height 3.71 m (12 ft 2 in); wing area 18.68 m² (201.1 sq ft)
Armament: two 30-mm (1.18-in) Aden cannon pods; four AIM-9L Sidewinder AAMs; two Sea Eagle (or Harpoon) ASMs; or rocket pods, bombs and flares
Operators: India, UK

Seen on Falklands duty, this Sea Harrier launches for a combat air patrol armed with two Sidewinders.

British Aerospace Sea Hawk

Prototypes of the Sea Hawk were completed to Specification N.7/46 and the first was flown on 2 September 1947. The 2268-kg (5,000-lb) thrust Nene 101 engine had twin nozzles, allowing fuel to be carried in the rear fuselage.

Sole remaining operator is the Indian navy, which in the autumn of 1959 ordered 24. Some were new-built (although the production line had closed three years earlier), the rest being refurbished ex-Royal Navy Mk 6s. They equipped No. 300 Sqn in the aircraft carrier INS *Vikrant* and were joined later by 12 more ex-RN Mk 4s and Mk 6s, plus 28 Mk 100/101s from Germany. A score or so remain in service, though relegated to training duties by arrival of the BAe Sea Harrier.

Specification
BAe Sea Hawk Mk 6
Type: single-seat shipboard attack/fighter aircraft
Powerplant: one 2449-kg (5,400-lb) Rolls-Royce Nene Mk 103 turbojet engine
Performance: maximum speed at sea level Mach 0.79 or 969 km/h (602 mph); initial rate of climb 1737 m (5,700 ft)/minute; service ceiling 13565 m (44,500 ft); combat radius, clean 370 km (230 miles); combat radius with two 227-kg (500-lb) bombs and two 409-litre (90-Imp gal) drop-tanks 463 km (288 miles)
Weights: empty 4409 kg (9,720 lb); maximum take-off 7348 kg (16,200 lb)
Dimensions: span 11.89 m (39 ft 0 in); length 12.09 m (39 ft 8 in); height 2.64 m (8 ft 8 in); wing area 25.83 m² (278.0 sq ft)
Armament: four 20-mm (0.79-in) Hispano cannon in nose with 200 rpg; underwing points for four 227 kg (500 lb) bombs; or two 227 kg (500 lb) bombs plus 20 76-mm (3-in)

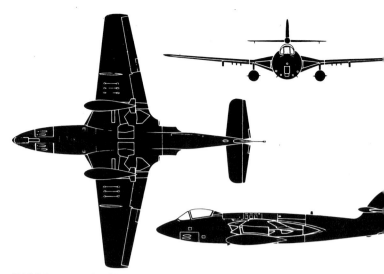

British Aerospace (AW) Sea Hawk

or 16 127-mm (5-in) rockets; or two AIM-9 Sidewinder AAMs, or four

409 litre (90 Imp gal) drop tanks
Operator: India

British Aerospace Shackleton

Currently employed by the RAF on a task different from that for which it was designed – and performing it 15 years past the anticipated retirement date – the Shackleton was designed as a maritime reconnaissance aircraft. Produced by Avro, the first of three Type 696 prototypes flew at Woodford on 9 March 1949, and the MR.1 entered service with RAF Coastal Command in February 1951. After 76 had been built, manufacture turned to the MR.2 for a further 70 before the MR.3 appeared, both marks being major redesigns. The RAF took 34 MR.3s and eight more went to No 35 Sqn of the South African Air Force, which retired the last survivors in 1984. Mainspar fatigue resulted in RAF MR.3Bs being withdrawn in 1970-71, but surviving MR.2s were rebuilt for AEW (airborne early warning). This came about because the Royal Navy needed a replacement for its Gannet AEW.3s after aircraft carriers were withdrawn. As such aircraft could also detect a low-level intruder attempting to slip under the UK's defensive radar coverage, the Shackleton was given the additional duty of co-ordinating low-level interceptor operations for Strike Command's No 11 Group (Lightnings and Phantoms). A prototype Shackleton AEW.2 conversion flew at Woodford on 30 September 1971, its APS-20B radar being taken straight from a redundant Gannet. A further 11 AEW.2s followed from the Bitteswell conversion line, these entering service with No. 8 Sqn at Lossiemouth. While the aircraft is little changed forward of the main spar, with positions for two pilots and two navigators, the centre fuselage houses three radar screens and their operators, the centre being the Tactical Co-ordinator. The APS-20B has no height-finding capability and is of 'stone age' technology compared with today's avionics, but experienced operators manage to coax results out of screens which other operators would classify as impossibly cluttered or completely devoid of useful information. Having demonstrated their ability in numerous fighter control exercises and in real encounters with Soviet aircraft transiting the UK Air Defence Region, the force was halved in 1981 as an economy measure prior to imminent service-entry of the far more sophisticated Nimrod AEW.3. The latter failed to appear on time, so the Shacketon is likely to soldier on into 1986 at least.

While Nimrod AEW undergoes seemingly interminable problems, the handful of Shackleton AEW.Mk 2s of No. 8 Sqn continue to provide AEW coverage for the UK's defence.

Specification
BAe Shackleton AEW.2
Type: airborne early warning aircraft
Powerplant: four 1827-kW (2,450-shp) Rolls-Royce Griffon 57A piston engines
Performance: maximum speed 438 km/h (272 mph) at 3050 m (10,000 ft); cruising speed 410 km/h (255 mph) at 3660 m (12,000 ft);
initial rate of climb 280 m (920 ft)/minute
Weight: maximum take-off 44452 kg (98,000 lb)
Dimensions: span 36.58 m (120 ft 0 in); length 28.19 m (92 ft 6 in); height 5.11 m (16 ft 9 in); wing area 132.01 m² (1,421.0 sq ft)
Armament: nil
Operators: South Africa (MR.3), UK

British Aerospace (BAC) Strikemaster

Based on the Jet Provost, the Strikemaster has found favour with a number of Middle East countries, including Saudi Arabia, which operates the Mk 80.

The obvious appeal of the Jet Provost as a highly developed and economical trainer prompted BAC to do what Hunting had lacked the funds to achieve: develop the type into a tactical multi-role aircraft able to fly both pilot training and weapon training sorties and also, should the occasion demand, go to war in attack and tactical reconnaissance roles. Via the BAC.145, virtually an armed version of the pressurized Jet Provost T.Mk 5, BAC developed the BAC.167 Strikemaster by fitting a more powerful version of the Viper engine and increasing the number of stores hardpoints to eight.

Features include side-by-side Martin-Baker Mk PB4 ejection seats, short landing gears suitable for operation from rough airstrips, fuel housed entirely in integral and bag tanks in the wings and in fixed tip tanks, hydraulic spoiler/airbrake surfaces above the wings, manual flight controls, a pressurized and air-conditioned cockpit, and comprehensive navigation and communications equipment which some customers have upgraded to include EW (electronic-warfare) installations.

The first Strikemaster flew in October 1967 and the Strikemaster Mk 80 series entered service a year later. Customers comprise Ecuador, Kenya, Kuwait, New Zealand, Oman, Saudi Arabia, Singapore, Sudan and South Yemen. The final batch of new Strikemaster Mk 90 aircraft were delivered to the Sudan in 1984, assembly of this batch having been relocated from Warton to Hurn. Sudan had previously been a customer for the less powerful BAC.145. Many Strikemasters have seen prolonged active service; for example, all 20 of the Sultan of Oman's Strikemaster Mk 82 and Mk 82A aircraft have sustained battle damage. The Strikemaster has a reputation for almost Russian toughness and longevity in austere circumstances, and most have a long career ahead of them.

Specification
Type: close-support and reconnaissance aircraft and weapons trainer
Powerplant: one 1547-kg (3,410-lb) thrust Rolls-Royce Viper Mk 535 turbojet engine
Performance: maximum speed, clean 418 kts (774 km/h; 481 mph) at 5485 m (18,000 ft); initial rate of climb 1600 m (5,250 ft) per minute; service ceiling 12190 m (40,000 ft); combat radius on a hi-lo-hi mission with 1361-kg (3,000-lb) weapons load and full reserves 397 km (247 miles)
Weights: empty 2810 kg (6,195 lb); maximum take-off 5216 kg (11,500 lb)

Dimensions: span 11.23 m (36 ft 10 in); length 10.27 m (33 ft 8.3 in); height 3.34 m (10 ft 11.5 in); wing area 19.85 m² (213.7 sq ft)
Armament: two forward-firing 7.62-mm (0.3-in) FN machine-guns each with 550 rounds, plus up to 1361 kg (3,000 lb) of other weapons on four underwing strongpoints, including bombs, rocket launchers, tanks, gun pods or a five-camera reconnaissance pod
Operators: New Zealand, Oman, Saudi Arabia, Singapore, Sudan

British Aerospace Strikemaster Mk 88 of the Royal New Zealand Air Force.

British Aerospace/Vickers VC10/Super VC10

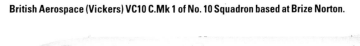

British Aerospace (Vickers) VC10 C.Mk 1 of No. 10 Squadron based at Brize Norton.

The VC10 airliner was designed and built by Vickers-Armstrongs (Aircraft) at Weybridge, Surrey, to satisfy a shortsighted BOAC specification for a civil transport to carry a payload of approximately 15876 kg (35,000 lb) over a range of up to 6437 km (4,000 miles), operating from 1950s-size hot-and-high airfields (which of course were lengthened to suit 707s and DC-8s). Thus, at a penalty in operating cost, excellent short-field performance and low-altitude manoeuvrability was conferred by the high-lift features incorporated in the wing design. These included leading edge slats, two-section ailerons and five-section Fowler flaps, spoilers to augment roll control and usable as airbrakes, and a boundary-layer fence on the upper surface. Accommodation of the basic VC10 was for 135-151 passengers; the Super VC10, which entered service in 1965, had a lengthened fuselage and structural strengthening to accommodate 163-174 passengers.

The prototype flew on 29 June 1962, and one of the orders was for 14 Model 1106s for service with RAF Air Support Command under the designation VC10 C.1. These combine the VC10-size airframe with various features of the Super, including high-power Mk 550 engines and a wet (integral tank) fin. They have large cargo doors forward and provisions for handling and lashing heavy vehicles or palletized freight, and during the Falklands campaign some served as casevac transports

between Brize Norton and Montevideo. Most have a flight-refuelling probe, special navaids and can operate independent of ground services. They are the usual vehicles for transport of the Royal Family or senior government VIPs on intercontinental journeys.

The RAF's VC10 K.2 and K.3 have transport capability but are dedicated tankers, being respectively rebuilds of ex-airline VC10s and Super VC10s. They have completely refurbished airframes, five fuselage fuel tanks for transfer fuel, a Mk 17B HDU (hose-drum unit) under the rear fuselage and a Mk 32 refuelling pod under each outer wing panel, closed-circuit TV refuelling surveillance, night floodlights, receiver probe, APU and military avionics. Whereas the Victor K.2 carries 50 long tons of transfer fuel, the VC10 K.2 carries 75 tons and the K.3 85. At the forward end of the cabin is an 18-seat passenger compartment, and underfloor holds can carry refuelling pods or essential spares. British Aerospace at Bristol delivered five K.2s and four K.3s, but is expected to rebuild a further six aircraft, all to be operated from

Brize Norton by No. 101 Sqn.

Specification
Vickers VC10 C.1
Type: long-range transport
Powerplant: four 9471-kg (20,880-lb) thrust Rolls-Royce Conway Mk 301 turbofan engines
Performance: cruising speed 885 km/h (550 mph) at 11580 m (38,000 ft); service ceiling 12800 m (42,000 ft); maximum range 8111 km (5,040 miles)
Weights: empty about 68039 kg (150,000 lb); maximum take-off

Ex-airline VC10s have been bought by the RAF for conversion into tankers to augment and replace the ageing Victor fleet. These are in the service of No. 101 Sqn at Brize Norton. This is a VC10 K.Mk 2.

146510 kg (323,000 lb)
Dimensions: span 44.55 m (146 ft 2 in); length 48.36 m (158 ft 8 in); height 12.04 m (39 ft 6 in); wing area 264.86 m² (2,851.0 sq ft)
Armament: none
Operators: Oman, UAE, UK (RAF)

British Aerospace Victor

Third and last of Britain's 'V-Bomber' trio, the Handley Page Victor was originally a strategic bomber that entered service in April 1958, but which was withdrawn from the deterrent force in 1968. A programme for conversion of early Victor B.1As into tankers, with an in-service date of 1967, had already been set in motion when the RAF's Valiant fleet had to be grounded nearly two years early with fatigue, and conversions became a matter of high priority. An initial batch of six was fitted with a Flight Refuelling FR.20B hose pod beneath each wing to become the Victor B.1A (K2P), while 24 remaining conversions were to the envisaged 'three-point' standard with two FR.20Bs plus an FR.17 hose drum unit in the fuselage, resulting in ten Victor K.1s and 14 K.1As. The last of these was withdrawn in 1977, by which time 24 later B.2s had been modified to Victor K.2s for Nos 55 and 57 Sqns, plus No. 232 Operational Conversion

British Aerospace (Handley Page) Victor K.Mk 2 of No. 57 Squadron, RAF Marham.

Unit, all based at Marham. K.2s were a major redesign job by BAe at Woodford and have a small reduction in span to reduce fatigue, and the usual three refuelling points, of which only two – the wing pods, each delivering 682 litres (150 Imp gal)/minute – are used simultaneously. Heavy aircraft may receive 2273 litres (500 Imp gal)/minute from the centre hose. Until recently, Victors were the RAF's sole tankers, responsible for refuelling interceptors policing local airspace and supporting overseas deployments of strike/attack aircraft. With the advent of the 1982 Falklands war, in which

Victor SR.2s flew important radar reconnaissance sorties, workload was considerably increased and the aircraft supported numerous epic flights. Remaining airframe hours are now rapidly diminishing, and aircraft are being withdrawn as the VC10 and TriStar tanker forces becomes operational.

Specification
BAe Victor K.2
Tpe: inflight refuelling tanker
Powerplant: four 9344-kg (20,600-lb) thrust Rolls-Royce Conway 201 turbofan engines
Performance: maximum speed

Mach 0.95 (982 km/h, 610 mph) at altitude; cruising speed 982 km/h (610 mph); range 7403 km (4,600 miles)
Weights: empty 50036 kg (110,310 lb); maximum take-off 107683 kg (237,400 lb); fuel load 55883 kg (123,200 lb)
Dimensions: span 35.66 m (117 ft 0 in); length 35.03 m (114 ft 11 in); height 8.57 m (2 ft 1.5 in); wing area 204.38 m² (2,200.0 sq ft)
Armament: nil
Operator: UK

Britten-Norman Islander/ Defender

The highly successful BN-2 Islander light STOL transport, which is also available with 298-kW (500-shp) Allison 250-B17C turboprops, has been supplied to a large number of military operators. Most have bought the armed version known as the Defender. Mauretanian Defenders have seen action against Polisario guerrillas, at least one being shot down in 1978 and an aircraft of the Botswana Defence Force made an abortive attack on three Rhodesian helicopters carrying a raiding party. Most Defenders are used for troop transport, logistic support, casevac, forward air control, internal security and long-range patrol. Seating capacity is nine plus pilot, or three stretchers and two attendants. The Maritime Defender has a 111-km (69-mile) range, 120° scanning Bendix RDR-1400 radar in an enlarged nose, with radar operator and two optional additional navigation equipment. Weapon options include the BAe Sea Skua anti-ship missile and Marconi Stingray lightweight torpedo. Maritime versions serve in India, the Philippines, Cyprus and the Seychelles, one of their principal characteristics being an increase in length to 11.07 m (36 ft 3.75 in). In 1984 an AEW Defender turboprop was revealed with Thorn-EMI 360° scanning radar (in a grotesquely enlarged nose radome) optimised for the detection and tracking of high-and low-level, fast-moving targets, supplemented by a data link, ESM/ECM and other equipment. The

The Britten-Norman Defender has found use with many nations around the world including the Belize Defence Force Air Wing which operates two of the type.

CASTOR Islander, evaluated by the British Army in 1984 for the Corps Airborne Stand-off Radar requirement, again has turboprop powerplants and is fitted with a 360° Ferranti radar capable of detecting battlefield targets from friendly territory. Information is automatically relayed to the ground, so only two crew are needed.

Specification
Britten-Norman Defender
Type: armed light transport
Powerplant: two 224-kW (300-shp) Lycoming IO-540-K1B5 flat-six piston engines
Performance: maximum speed, clean 280 km/h (174 mph), with external stores 267 km/h (166 mph); initial rate of climb 396 m (1,300 ft)/minute; service ceiling 5180 m (17,000 ft); range, clean 672 km (418 miles)
Weights: empty 1823 kg (4,020 lb); maximum take-off 2994 kg (6,600 lb)
Dimensions: span 14.94 m (49 ft 0 in); or 16.15 m (53 ft 0 in) with raked fuel-carrying tips; length 10.86 m (35 ft 7.6 in); height 4.18 m (13 ft 8.6 in); wing area 30.19 m² (325.0 sq ft) or 31.31 m² (337.0 sq ft)
Armament: four pylons for bombs, rockets, guided missiles (AIM-9 Sidewinders) or other stores, with 317.5 kg (700 lb) on each inboard

pylon and 204 kg (450 lb) on each outer
Operators: (including military Islanders) Abu Dhabi, Belgium, Belize, Botswana, Ciskei, Cyprus, Egypt, Ghana, Guyana, Haiti, Hong Kong, India, Iraq, Israel, Jamaica, Lesotho, Liberia, Malagasy, Malawi, Mauretania, Mauritius, Mexico, Oman, Panama, Philippines, Qatar, Rwanda, Seychelles, Somalia, Sudan, Surinam, Venezuela, Zaïre, Zambia, Zimbabwe

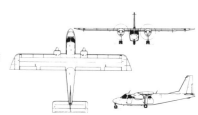

Britten-Norman Defender

Canadair CL-41 Tutor

Because of a lack of official interest, the CL-41 was funded privately by Canadair. Two prototypes were built, each powered by the 1088-kg (2,400-lb) thrust Pratt & Whitney JT12A-5 turbojet, the first flying on 13 January 1960. In September 1961 the Canadian government ordered 190 CL-41As for the RCAF (now CAF) with the designation CT-114 Tutor. These were powered by the 1293-kg (2,850-lb) thrust GE J85-CAN-40 turbojet, licence-built by Orenda. Features include side-by-side seats, upward opening canopy, lateral airbrakes, T-tail and steerable nosewheel.

Main user is Training Command's No. 2 Flying Training School at Moose Jaw, Saskatchewan. After primary training on the CT-134, pupils complete some 200 hours on the CT-114 to gain their 'wings'. Ten Tutors were modified for the Golden Hawks (renamed Golden Centennaires in 1967 for the nation's Centennial year and later snowbirds) aerobatic team, and the type also serves with the Flying Instruc-

tors' School, all at Moose Jaw. In 1976 the CAF began a 113-aircraft modification programme which includes provision of external tanks, upgrading of avionics, changes to the canopy, electrical system and relocation of the engine ice probe.

The CL-41G armament trainer and light attack aircraft has an uprated engine, six underwing hardpoints, landing gear modified for soft-field operation, and 'zero level' seats. In March 1966 the Royal Malaysian air force ordered 20, naming the type Tebuan (wasp). Deliveries began in 1967 to a new base at Kuantan on the east coast. Two squadrons fly Tebuans, No. 9 is a training unit, while No. 6 operates in the light strike role.

Specification
Canadair CL-41G
Type: training/light attack aircraft
Powerplant: one 1338-kg (2,950-lb) thrust General Electric J85-J4 turbojet engine
Performance: maximum speed at optimum altitude 756 km/h (470

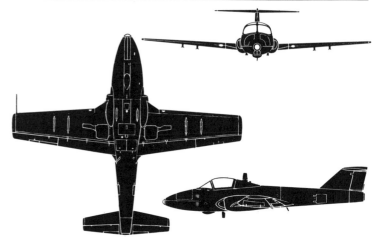

Canadair CL-41A Tutor

mph); service ceiling 12800 m (42,000 ft); range 2300 km (1,429 miles)
Weights: empty 2400 kg (5,291 lb); maximum take-off 5130 kg (11,310 lb)
Dimensions: span 11.13 m (36 ft 6.2 in); length 9.75 m (31 ft 11.9 in);

height 2.81 m (9 ft 2.6 in); wing area 20.44 m² (220.02 sq ft)
Armament: six wing hardpoints can carry up to 1814 kg (4,000 lb) of bombs, rocket launchers, gun pods or Sidewinder air-to-air missiles
Operators: Canada (CL-41A), Malaysia

Canadair CL-215

The CL-215 was the first purpose-designed firefighting amphibian used in the 'water bomber' role, the prototype flown on 23 October 1967. However, the robust and versatile amphibian has also been sold to military customers for search and rescue (SAR) and utility roles. Of substantial size, the CL-215 has a single-step hull, fixed stabilizing floats and retractable tricycle landing gear. For firefighting it can lift 5455 litres (1,200 Imp gallons) of water or retardant fluid in two fuselage tanks. The water is scooped from a

convenient lake or river through two retractable scoops under the hull while the CL-215 flies very low across the surface; the load can be jettisoned over a fire in under a second, and in most situations a load can be dropped at least every 10 minutes. Configured for the SAR role the CL-215 carries a crew of six: pilot and co-pilot, flight engineer, navigator, and two observers in the rear fuselage. The basic avionics (typically HF, VHF, and VHF/HM transceivers, ADF, marker beacon and VOR/ILS/glide slope) are aug-

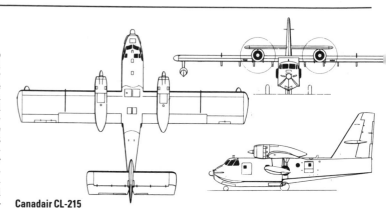

Canadair CL-215

mented by an AVQ-21 radar, radio altimeter, UHF/VHF homer and DME. Maximum endurance is 12 hours.

The Spanish air force was favourably impressed by the CL-215, acquiring 10 which equip Escuadron 404; based at Torrejon, this unit operates in the firefighting, SAR and casevac roles. Eight were acquired by the Greek air force, primarily for forest protection, but they are also used as troop transports. The Royal Thai navy operates two on patrol and SAR missions.

Specification
Type: firefighting and utility amphibian
Powerplant: two 1566-kW (2,100-hp) Pratt & Whitney R-2800-83 18-cylinder radial piston engines
Performance: cruising speed at 3050 m (10,000 ft) 290 km/h (180 mph); initial rate of climb 305 m (1,000 ft)/minute; range 1851 km (1,150 miles)
Weights: empty 12000 kg (26,455 lb); maximum take-off, land 19730 kg (43,497 lb), water 17100 kg (37,699 lb)
Dimensions: span 28.60 m (93 ft 10 in); length 19.82 m (65 ft 0.3 in); height, on wheels 8.92 m (29 ft 3.2 in); wing area 100.33 m^2 (1,080.0 sq ft)
Armament: none
Operators: France, Greece, Italy, Jugoslavia, Spain, Thailand

Canadair CL-601 Challenger

The Challenger was created by the late Bill Lear in a deal agreed by Canadair in October 1976. A business jet intended to compete with the Gulfstream III, the basic CL-600 version is powered by two ALF-502L turbofan engines. The circular-section fuselage allows 1.85 m (6 ft 1 in) of headroom in the cabin; typical interior layouts feature two separate cabins, together with a bar, galley and separate toilet. A 40-passenger arrangement is available but, to date, all Challengers have had a 15- or 17-passenger VIP layout.

The prototype CL-600 flew on 8 November 1978, and FAA Type Certification was granted on 7 November 1980. Extremely severe problems, both technical and financial, almost killed the entire programme. Initial deliveries have been to corporate customers except for two CL-600s for the Malaysian air force and two for the CAF's No. 412 Sqn at Ottawa. The Canadian Department of National Defense and the Department of Transport also use the CL-200 for communications.

First flown on 10 April 1981, the CL-601 has winglets, more powerful General Electric engines and fuel for range extended by 1280 km (795 miles) over the CL-600. The CL-601 won an exhaustive West German evaluation as a replacement for Jet-Stars and HFB 320 Hansa Jets in service with the Flugbereitschaftstaffel at Cologne, and the seven being delivered in 1985 have a variety of internal fits including a mixed six-pasenger and freight layout. Canadair has also proposed the CL-601 as an AEW aircraft.

Specification
Canadair CL-601
Type: intercontinental VIP transport
Powerplant: two 4146-kg (9,140-lb) thrust General Electric CF34-1A turbofan engines
Performance: high cruising speed 850 km/h (528 mph); normal cruising speed 800 km/h (497 mph); initial rate of climb 1372 m (4,500 ft)/minute; maximum range 6482 km (4,028 miles)

Weights: empty operating 11605 kg (25,585 lb); maximum take-off 19,550 kg (43,100 lb)
Dimensions: span 19.61 m (64 ft 4 in); length 20.85 m (68 ft 4.9 in); height 6.30 m (20 ft 8 in); wing area 48.31 m^2 (520.0 sq ft)
Armament: none
Operators: (600) Canada, Malaysia, (601) West Germany

No. 412 Squadron, Canadian Armed Forces, operates a pair of CL-144 Challengers on transport duties, with a further 12 machines on order for electronic support and training (EST), avionics testing and liaison duties.

CASA C-101 Aviojet

CASA flew the prototype C-101 trainer on 27 June 1977. Assistance in the design and test stage came from Messerschmitt-Bölkow-Blohm in West Germany and Northrop in the US. Imported components include Dowty landing gear, Garrett engine, Martin-Baker Mk 10E seats, Sperry STARS flight control and US air-conditioning and pressurization. Production began in early 1978, the Spanish air force using 88 as E-25 trainers. The C-101BB armed version is being assembled in Chile, which also uses the C-101CC with the 1950-kg (4,300-lb) thrust TFE731-5 engine. The C-101DD is planned to have Ferranti ISIS weapon-aiming avionics and a HUD-sight.

Specification
CASA C-101EB Aviojet
Type: trainer and light strike aircraft
Powerplant: one 1588-kg (3,500-lb) thrust Garrett TFE731-2-2J turbofan engine
Performance: maximum speed at sea level 676 km/h (420 mph); maximum speed at 7620 m (25,000 ft) Mach 0.69; initial rate of climb 1020 m (3,345 ft)/minute; service ceiling 12500 m (41,010 ft); ferry range 4000 km (2,485 miles)
Weights: basic operating 2957 kg (6,519 lb); maximum take-off, trainer 4700 kg (10,362 lb); ground-attack 5600 kg (12,346 lb)
Dimensions: span 10.60 m (34 ft 9.3 in); length 12.25 m (40 ft 2.3 in); height 4.25 m (13 ft 11.3 in); wing area 20.00 m^2 (215.29 sq ft)
Armament: underfuselage attachment for a 30-mm (1.18-in) cannon, a 12.7-mm (0.5-in) gun, reconnaissance camera or laser designator; three hardpoints under

each wing can carry up to 2000 kg (4,409 lb) of stores consisting of bombs, missiles, napalm canisters, or rocket pods
Operators: Chile, Honduras, Jordan, Spain

Locally designed and produced by CASA, the C-101 Aviojet is the Spanish air force's basic jet trainer, though a ground attack version is available. The first examples were delivered in March 1980.

CASA C-212 Aviocar

A Spanish air force requirement, to replace elderly transports such as the DC-3, Ju 52/3m and Azor, led to CASA drawing up a specification for a twin-turboprop general-purpose aircraft. The result is the C-212 Aviocar, designed for a crew of two and up to 16 troops or 19 passengers. The first prototype flew on 26 March 1971, and was demonstrated with verve at the 1971 Paris air show, although the main spar suffered damage when reverse thrust was applied with the aircraft well above the runway. An initial batch of eight flew between November 1972 and February 1974, the type being given the Spanish air force designation T.12. The C-212A (T.12B) is a utility transport for 18 troops or small vehicles, and the first of 61 was delivered on 20 May 1974; the first squadron to equip was No. 461 at Gando, Canary Islands. Five C-212AV were ordered as VIP transports, and other versions are the C-212B (TR.12A) for photo survey, and the C-212E navigation trainer.

This Spanish air force CASA Aviocar example serves with Escuadron 721 as a paratroop trainer at Alcantarilla.

CASA concluded a licence agreement with Nurtanio Aircraft Industries in Indonesia; production began in mid-1976 and 29 C-212-100s were built. In April 1978 CASA flew the prototype C-212-10 with a strengthened airframe and two 645-kW (865-shp) Garrett TPE331-10s. Today's C-212-200 has 746-kW (1,000 shp) TPE331-501Cs. The more powerful engines allow the maximum take-off weight to be increased to 7450 kg (16,424 lb) and the maximum payload to 3200 kg (7,055 lb). By 1985 over 360 of all

versions had been ordered, production being about four per month by CASA and Nurtanio combined, who are now also building the larger CN-235.

Specification
CASA C-212-100 Aviocar
Type: utility transport
Powerplant: two 533-kW (715-shp) Garrett TPE331-5-251C turboprop engines
Performance: maximum speed at 3660 m (12,000 ft) 359 km/h (223 mph); cruising speed at 3660 m

(12,000 ft) 275 km/h (171 mph); initial rate of climb 549 m (1,800 ft)/minute; range with maximum fuel and 1045-kg (2,304-lb) payload 1760 km (1,094 miles)
Weights: empty 3905 kg (8,609 lb); maximum take-off 6500 kg (14,330 lb)
Dimensions: span 19.00 m (62 ft 4 in); length 15.20 m (49 ft 10.4 in); height 6.30 m (20 ft 8 in); wing area 40.00 m² (430.57 sq ft)
Armament: none
Operators: Chile, Indonesia, Jordan, Panama, Portugal, Spain, Uruguay, Venezuela, Zimbabwe

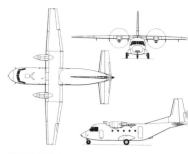
CASA 212 Aviocar

CASA/MBB 223 Flamingo

When aircraft manufacture was again permitted in Germany, in 1955, SIAT (Siebelwerke-ATG) produced first the SIAT 222 four-seat sporting monoplane. SIAT's second design, the Flamingo, won a competition for a fully-aerobatic aircraft and first flew on 1 March 1967. Two production versions emerged as the 223A1, a 'two plus two' utility model, and the 223K1 single-seat aerobatic variant. In 1970 SIAT became a member of the Messerschmitt-Bölkow-Blohm Group, production continuing as the MBB 223 until 1972 when MBB transferred the whole programme to Hispano Aviación. German production totalled 50, including 15 for the Turkish air force. The first Spanish-built Flamingo was flown on 14 February 1972, but later in the same year Hispano was taken over by CASA, which also built 50, of which 30 were reportedly for Syria.

Forty-eight CASA-built Flamingoes were supplied to the Syrian Arab air force, around 30 of which remain in service alongside several other types as part of the air force training school.

Specification
Type: one/four-seat trainer and utility aircraft
Powerplant: one 149-kW (200-hp) Lycoming IO-360-C1B flat-four piston engine
Performance: (two-seat)

maximum speed 243 km/h (151 mph); initial rate of climb 258 m (846 ft)/minute; service ceiling 3750 m (12305 ft); range with reserve fuel 88 km (547 miles)
Weights: (two-seat) empty equipped 685 kg (1,510 lb); maximum take-off

1050 kg (2,315 lb)
Dimensions: span 8.28 m (27 ft 2 in); length 7.43 m (24 ft 4.5 in); height 2.70 m (8 ft 10.3 in); wing area 11.50 m² (123.79 sq ft)
Armament: none
Operators: Spain, Syria, Turkey

Cessna 150/152

In 1959 Cessna Aircraft introduced a new side-by-side trainer, the Model 150, to fill a gap in their product range. The 150 had fixed tricycle landing gear and a 75-kW (100-hp) Continental O-200-A engine. The 150D of 1964 introduced a cut-down rear fuselage and 'omni-vision' rear cabin transparencies. Two years later, the 150F was given a swept fin and rudder. In 1970, the aerobatic A150K Aerobat with full harness, strengthened airframe and cabin-roof windows joined the standard trainer.

During the 1960s few 150s went to military users but some purchases have been made since introduction of the Aerobat. Two were delivered to the Armada de Mexico in July 1970 as primary trainers, and Reims Aviation (the French affiliate of Cessna) delivered a large batch of FRA150M basic trainers to Zaire

and smaller numbers to Burundi and the Ivory Coast. Progressive refinements have included tubular steel (instead of leaf-spring) landing gear legs for the 1971 Model 150L, but the Model 152 of 1978 introduced the O-235 engine and a taller vertical tail.

Specification
Cessna 152
Type: two-seat primary trainer
Powerplant: one 81-kW (108-hp) Lycoming O-235-N2C flat-four piston engine
Performance: maximum speed 202 km/h (126 mph); maximum cruising speed 96 km/h (122 mph); initial rate of climb 218 m (715 ft)/minute; service ceiling 4480 m (14,700 ft); range with allowances and fuel reserves 684 km (425 miles)
Weights: 501 kg (1,104 lb); maximum take-off 757 kg (1,670 lb)

Cessna 150 Aerobat

Dimensions: span 10.11 m (32 ft 2 in); length 7.34 m (24 ft 1 in); height 2.59 m (8 ft 6 in); wing area 14.82 m² (159.5 sq ft)
Armament: none

Operators: Botswana, Burundi, Ecuador, Gabon, Haiti, Ivory Coast, Lesotho, Liberia, Libya, Mexico, Paraguay, Somalia, Sri Lanka, Zaire

Cessna 170

The Cessna 170 was a four-seater, developed from the Models 120/140, and first flown in September 1947; 5,173 were produced, ending in 1956. A modest number reached military users, primarily in Latin America. Others have been purchased secondhand. The wing was fabric-covered on the initial models but metal-skinned on the 170A. There

are dual controls for the two front occupants and a rear bench seat. The large doors allow a stretcher to be loaded into a sling fitting on the left side.

Specification
Cessna 170B
Type: four-seat light aircraft
Powerplant: one 108-kW (145-hp)

Continental C-145-2 flat-six piston engine
Performance: maximum speed 225 km/h (140 mph); cruising speed 193 km/h (120 mph); initial rate of climb 210 m (690 ft)/minute; service ceiling 4725 m (15,500 ft); range 1107 km (688 miles)
Weights: empty 547 kg (1,205 lb); maximum take-off 998 kg (2,200 lb)

Dimensions: span 10.97 m (36 ft 0 in); length 7.61 m (24 ft 11.5 in); height 2.01 m (6 ft 7 in); wing area 16.26 m² (175.0 sq ft)
Armament: none
Operator: Dominica

Cessna 172 and T-41

The Cessna 172 four-seater has been produced in astronomic numbers for

30 years. It was the result of a need to modernize the Cessna 170, and

introduced tricycle landing gear, a squared-off vertical tail and normal-

ly has dual controls. The 1960 Model 172A was given a swept vertical tail

and in 1963 a cut-down rear fuselage added all-round vision. With the de luxe Skyhawk the design has remained similar up to the current 172P, built both at Wichita and at Reims in France. The largest French order was from the Royal Saudi air force, which during 1966-67 received 16 for the King Faisal Air Academy at Riyadh.

In 1958 the Model 175 added the higher powered, geared Continental GO-300-E engine. Cessna also proposed a virtually standard 172F to the USAF as an ab initio trainer and the T-41A Mescalero with a 108-kW (145-hp) Continental O-300-D commenced delivery in May 1965. They are operated by civil contractors for pre-acceptance screening of aircrew applicants and for initial tuition prior to the T-37 or T-46 jet course. Several T-41As were delivered to other air forces. The Model R.172 (USAF T-41B) offered substantial improvements in performance, with a constant-speed propeller, strengthened landing gear and separate rear seats. The T-41C of the US Air Force Academy has a fixed-pitch

propeller. Air Forces receiving aid under MAP get the T-41D, basically a T-41B with 28V electrics and simpler radio and avionics. From Reims a total of nine R.172 Rockets were sold to the Irish Air Corps; they can be fitted with light underwing stores and are used on patrols along the Northern Ireland border.

Specification
Cessna T-41B
Type: four-seat primary trainer
Powerplant: one 157-kW (210-hp) Continental IO-360-D flat-six piston engine
Performance: maximum speed 246 km/h (153 mph); maximum cruising speed 233 km/h (145 mph); initial rate of climb 189 m (620 ft)/minute; service ceiling 5335 m (17,500 ft); maximum range 1625 km (1,010 miles)
Weights: empty 637 kg (1,405 lb); maximum take-off 1157 kg (2,550 lb)
Dimensions: span 10.92 m (35 ft 10 in); length 8.20 m (26 ft 11 in);

height 2.68 m (8 ft 9.5 in); wing area 16.16 m² (174.0 sq ft)
Armament: none
Operators: Argentina, Bolivia, Burkina-Faso, Chile, Colombia, Dominican Republic, Ecuador, Ethiopia, Greece, Guatemala, Haiti, Honduras, Indonesia, Ireland, Korea, Laos, Liberia, Libya, Malagasy, Pakistan, Panama, Paraguay, Peru, Philippines, Salvador, Saudi Arabia, Singapore, Thailand, Turkey, Uganda, Uruguay, USA

A nominal number of Cessna T-41Ds were supplied to the Argentine army's aviation command under MAP agreements with the United States.

Cessna 177

Cessna's Model 177 Cardinal was designed as a replacement for the 172 Skyhawk (T-41). The 172J prototype flew on 15 July 1966 and became the 177 in 1967. Some 2,750 were built by 1978. It has a highly streamlined fuselage, 360° visibility and a new cantilever wing. The main landing gear units have tubular steel legs, and the 177RG intro-

duced in 1971 provides retractable gear. The Cardinal is in use with the Sri Lanka air force, where one task is to earn revenue from tourist flying.

Specification
Cessna Model 177
Type: four-seat light aircrft
Powerplant: one 112-kW (150-hp)

Lycoming O-320-E21D flat-four piston engine
Performance: maximum speed 232 km/h (144 mph); maximum cruising speed 216 km/h (134 mph); initial rate of climb 204 m (670 ft)/minute; service ceiling 3870 m (12,700 ft); maximum range 1328 km (825 miles)
Weights: empty 642 kg (1,415 lb);

maximum take-off 1066 kg (2,350 lb)
Dimensions: span 10.86 m (35 ft 7.5 in); length 8.22 m (26 ft 11.5 in); height 2.77 m (9 ft 1 in); wing area 16.13 m² (173.65 sq ft)
Armament: none
Operators: Ecuador, Sri Lanka

Cessna Model 180/185/U-17

Cessna's Model 170, from which the O-1 Bird Dog was evolved, developed in 1953 into the Model 180 with more power, constant-speed propeller, and larger tail. In July 1960 Cessna flew the Model 185, with even more power, a further increase in fin/rudder area, and increased gross weight. In 1962 it was picked by the USAF as the U-17 for export under the Military Assistance Program. Following 169 powered by a 194-kW (260-hp) Continental IO-470-F, came 136 U-17Bs with a 224-kW (300-hp) Continental IO-520D. Peru and South Africa bought U-17Bs direct from Cessna. Final production version was the U-17C with the O-470-L engine, with a carburettor instead of direct injection.

Specification
Cessna U-17B
Type: one/six-seat utility aircraft
Powerplant: one 224-kW (300-hp) Continental IO-520D flat-six piston engine
Performance: maximum speed 286 km/h (178 mph); maximum cruising speed 272 km/h (169 mph) at 2285 m (7,500 ft); range 966 km (600 miles), or with maximum fuel 1730 km (1,075 miles)
Weights: empty 717 kg (1,580 lb); maximum take-off 1497 kg (3,300 lb)
Dimensions: span 11.02 m (36 ft 2 in); length 7.77 m (25 ft 6 in); height 2.29 m (7 ft 6 in); wing area 16.16 m² (174.0 sq ft)
Armament: none

Operators: Bolivia, Costa Rica, Ecuador, El Salvador, Honduras, Iran, Israel, Laos, Liberia, Mexico, Nicaragua, Panama, Peru, Philippines, South Africa, Turkey, UAE, Uruguay, Venezuela,

The original militarized version of the Model 185, the U-17A features an enlarged fin and rudder and increased gross weight.

Vietnam

Cessna Model 206/207 Skywagon/Stationair

The Cessna 185 was developed into the six-seat Super Skywagon. Changes included tricycle gear, conical-camber wingtips, enlargement of the tailplane and flaps, addition of double cargo doors on the right side, and a more powerful engine. On 3 January 1969 Cessna flew the first production example of the lengthened seven-seat Model 207 Skywagon, later called Stationair. This introduced a second door on the right, a 54-kg (120-lb) baggage bay ahead of the cabin, and an even more powerful engine. The 206 and 207 can be flown with cargo doors removed to air-drop supplies, as well as for parachuting and photography. Both can carry a 136-kg (300-lb) glassfibre cargo pack beneath the fuselage, and ambulance kits

(stretcher, oxygen and attendant seat) are available.

Specification
Cessna 206/207
Type: utility aircraft
Powerplant: (206) one 213-kW (280-hp), (207) one 224-kW (300 hp) Continental IO-520 flat-six piston engine
Performance: maximum cruising speed (206) 262 km/h (163 mph) at 1830 m (6,000 ft), (207) 256 km/h (159 mph) at 1980 m (6,500 ft); range, maximum fuel no reserves (206) 1642 km (1,020 miles), (207) 1489 km (925 miles)
Weights: empty (206) 767 kg (1,690 lb), (207) 862 kg (1,900 lb); maximum take-off (206) 1633 kg (3,600 lb), (207) 1724 kg (3,800 lb)

Cessna Model 207 Stationair 7

Dimensions: span (206) 11.15 m (36 ft 7 in), (207) 10.92 m (35 ft 10 in); length (206) 8.46 m (27 ft 9 in), (207) 9.68 m (31 ft 9 in); height (206) 2.97 m (9 ft 9 in), (207) 2.91 m (9 ft 6.5 in); wing area (206)

16.30 m² (175.5 sq ft), (207) 16.16 m² (174.0 sq ft)
Armament: none
Operators: Bolivia, Guatemala, Guyana, Indonesia, Israel, Liberia, Mexico, Paraguay, Surinam

Cessna Model 305A/O-1 Bird Dog

In 1948 a US Army specification for a two-seat liaison and observation aircraft was circulated, and in June 1950 an initial contract was awarded for 418 Cessna Model 305As, designated L-19A. Derived from the Cessna 170, it had all-round view and provision to load a stretcher.

By October 1954 Cessna had supplied 2,486, 60 going to the US Marine Corps as OE-1s. An L-19A-IT instrument trainer version was developed in 1953, TL-19D trainers with constant-speed propellers in 1956, and the improved heavier L-19E in 1957 to bring the total to 3,431. In 1962 the L-19A, TL-19D and L-19E became O-1A, TO-1D and O-1E respectively. The Marines' OE-1 became O-1B, augmented by 25 higher-powered O-1Cs. Army trainer conversions were the TO-1A and TO-1E. The USAF used many for Forward Air Control in Vietnam,

former TO-1Ds and O-1As becoming O-1F and O-1G in this role. O-1s were supplied to many nations, and built under licence by Fuji in Japan.

Specification
Cessna O-1E
Type: liaison and observation aircraft
Powerplant: one 159-kW (213-hp) Continental O-470-11 flat-six piston engine
Performance: maximum speed 209 km/h (130 mph); range 853 km (530 miles)
Weights: empty 732 kg (1,614 lb); maximum take-off 1089 kg (2,400 lb)
Dimensions: span 10.97 m (36 ft 0 in); length 7.85 m (25 ft 9 in); height 2.22 m (7 ft 3.5 in); wing area 16.16 m² (174.0 sq ft)
Armament: none
Operators: Austria, Brazil,

Cessna O-1 Bird Dog

Cambodia, Canada, Chile, Colombia, Ecuador, France, Guatemala, Italy, Japan, South Korea, Laos, Lebanon, Norway, Pakistan, Thailand, Turkey, US (Army and Marine Corps), Vietnam

---

Cessna Model 318E Dragonfly/A-37

In 1962 two T-37s were evaluated by the USAF Special Air Warfare Center in the counter-insurgency (COIN) role. Re-engined with two 1089-kg (2,400-lb) thrust J85-GE-5s, these YAT-37Ds could be flown at weights up to 6350 kg (14,000 lb). Cessna was asked to convert 39 new T-37B trainers to a light strike configuration. These were given eight hardpoints, provided with wingtip tanks, and powered by derated J85 turbojets. Delivery began on 2 May 1967, and they saw much duty in South Vietnam. Cessna then built a purpose-designed strike aircraft, flown in September 1967; as the A-37B this was being delivered by May 1968.

The A-37B is stressed for 6g loading, fuel capacity is increased to 1514 litres (400 US gallons), and there is provision for inflight refuelling. A GAU-2B/A 7.62-mm (0.3-in) Minigun can be installed, and both gun and strike cameras. Layered nylon flak-curtains line the cockpit.

By 1977 577 A-37Bs had been built, for various nations. Many were transferred to the US Air National Guard, where the majority have been re-equipped as OA-37Bs.

for the FAC (Forward Air Control) role.

Specification
Cessna A-37B
Type: two-seat light strike aircraft
Powerplant: two 1293-kg (2,850-lb) thrust General Electric J85-GE-17A turbojet engines
Performance: maximum speed 843 km/h (524 mph) at 4875 m (16,000 ft); maximum cruising speed 787 km/h (489 mph) at 7620 m (25,000 ft); range, maximum fuel at 7620 m (25,000 ft) with reserves 1629 km (1,012 miles); range with maximum payload, including 1860 kg (4,100 lb) external weapons 740 km (460 miles)
Weights: empty 2817 kg (6,211 lb); maximum take-off 6350 kg (14,000 lb)
Dimensions: span 10.93 m (35 ft 10.5 in); length 8.62 m (28 ft 3.25 in); height 2.71 m (8 ft 10.5 in); wing area 17.08 m² (183.9 sq ft)
Armament: can include bombs, incendiary bombs, cluster bombs, rocket pods and gun pods
Operators: Brazil, Burma, Cambodia, Chile, Colombia,

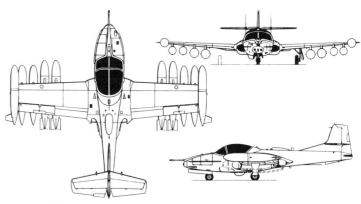

Cessna A-37B Dragonfly

Ecuador, El Salvador, West Germany, Greece, Guatemala, Honduras, Jordan, South Korea, Pakistan, Panama, Paraguay, Peru, Portugal, Thailand, Turkey, US (Air Force and Air National Guard), Vietnam

An excellent, low-cost platform for counter-insurgency (COIN) duties, the A-37 Dragonfly has given several smaller air arms a potent attack aircraft, as well as serving the US Air Force in a Forward Air Control (FAC) capacity.

Cessna 337/O-2

With the Model 336 Skymaster Cessna adopted the 'centreline twin' concept to sidestep the need for a twin licence. The fixed-landing gear 336 flew on 28 February 1961. In 1965 the Model 337 Super Skymaster introduced retractable gear. Cessna sold several 337s to military and governmental forces, and the French affiliate delivered more than 20 FTB.33 Milirole aircraft to Zimbabwe where they are known as the Lynx. These have turbocharged engines, STOL wing with two hardpoints under each wing and tail modifications. Roles include passenger and freight transport, casevac (two stretchers), SAR, navigation/IFR training and (Lynx) ground attack.

In 1966 the USAF selected the 337 as a replacement for the FAC (Forward Air Control) O-1. The O-2A (337M) had observation panels in the doors and cabin roof, military avionics and twin underwing hardpoints on each wing to carry 7.62-mm (0.3-in) Minigun pods, flare dispensers, rocket launchers and other light stores. A total of 501 was built,

and 31 Model 337D Super Skymasters were also purchased as O-2B psy-war platforms with loudspeakers to broadcast propaganda messages to Vietcong troops. Many O-2s were later civilianized and used by the US Bureau of Narcotics and Dangerous Drugs to patrol the Mexican border. The O-2 was also delivered from USAF stocks to various South Asia air arms, but 12 new aircraft were delivered to the Iranian air force in 1970 for ground attack. Cessna tested the O-2T with two Allison 250 B-15 turboprop engines, but with no USAF interest this variant was abandoned.

Specification
Cessna 337M (O-2A)
Type: four/six-seat utility, FAC and attack aircraft
Powerplant: two 157-kW (210-hp) Continental IO-360 flat-six piston engines
Performance: maximum speed 332 km (206 mph); maximum cruising speed 315 km/h (196 mph); initial rate of climb 366 m (1,200 ft)/minute; service ceiling 5945 m

(19,500 ft); maximum range with standard fuel 1255 km (780 miles)
Weights: empty 1227 kg (2,705 lb); maximum take-off 2100 kg (4,630 lb)
Dimensions: span 11.63 m (38 ft 2 in); length 9.07 m (29 ft 9 in); height 2.79 m (9 ft 2 in); wing area 18.81 m² (202.5 sq ft)
Armament: Minigun pods, flare dispensers, rocket launchers, etc
Operators: Argentina, Burkina-Faso

With its 'push-pull' powerplant and twin-boom configuration, the O-2 is a highly recognizable design which can operate in a variety of roles. USAF O-2s equip Tactical Air Support units.

Chad, Chile, Ecuador, Ethiopia, Gabon, Guinea Bissau, Iran, Ivory Coast, South Korea, Liberia, Malagasy, Mauretania, Mexico, Nicaragua, Paraguay, Portugal, Senegambia, Sri Lanka, Togo, Zaïre, Zimbabwe

Cessna Model 402/411/421

When it flew for the first time on 18 July 1962, Cessna's Model 411 then represented the largest business aircraft in the company's extensive range of light aircraft. In overall configuration it had a close relationship to the company's Model 310, which had served the USAF as a light twin-engine administrative liaison and cargo aircraft from December 1960 under the designations L-27A or U-3. It differed, however, by having a slightly increased wing span and area, a longer fuselage, and the introduction of more pwerful 340-hp (254-kW) Continental GTSIO-520-C flat-six turbocharged engines. There had, of course, been some detail refinements, and the permanently fixed wingtip fuel tanks had undergone some alteration in their shape to improve their aerodynamic characteristics and reduce drag. These aircraft have accommodation for a crew of two, and can seat from four to six passengers according to the interior layout, and each seat in the main cabin has a reading light, ventilator and oxygen outlet.

On 26 August 1965 Cessna flew the prototype of a generally similar aircraft which served for two new models, Models 401 and 402, and when FAA certification of the Model 401 prototype was awarded on 20 September 1966 it covered also the Model 402. These two aircraft represented lower-cost versions of the Model 411, differing primarily by having two 300-hp (224-kW) Continental TSIO-520-E flat-six engines and some reduction in basic installed equipment. The Model 401 accommodates a crew of two and four to six passengers, but the Model 402 has a cabin layout permitting a quick change from nine-seat commuter use to an all-cargo configuration.

Cessna's six-seat Model 421, first announced on 28 October 1965, is generally similar to these two latter aircraft, except for the provision of a new fuselage of fail-safe structureto permit pressurization, installation of 375 hp (280 KW) Continental GTISO-520-D turbocharged engines, and the introduction of AiResearch air-conditioning and pressurization system.

All three of these aircraft, namely the Models 402, 411 and 421, have proved useful to air forces for 'off the shelf' buys of communications aircraft.

Specification
Type: six/nine-seat light transport aircraft
Powerplant: (421) two 280-kW (375-hp) Continental GTSIO-520-L flat-six geared and turbocharged

piston engines
Performance: (421) maximum speed 478 km/h (297 mph) at 6095 m (20,000 ft); maximum cruising speed 449 km/h (279 mph) at 7620 m (25,000 ft); range (with allowances for start, taxi, take-off, climb, descent and 450-minutes reserves at 45 per cent power, at maximum cruising speed with maximum fuel at 7620 m (25,000 ft) 2317 km (1,440 miles)
Weights: (421) empty 2074 kg (4,572 lb); maximum take-off 3379 kg (7,450 lb)

A quartet of Cessna 402Bs serve the Royal Malaysian air force as VIP transports, with a further eight examples acting as crew trainers.

Dimensions: (421) span 12.53 m (41 ft 1½ in); length 11.09 m (36 ft 4½ in); height 3.49 m 911 ft 5½ in); wing area 19.97 m² (215 sq ft)
Armament: none
Operators: Bolivia, Comores, Finland, France, Haiti, Indonesia, Malaysia, Mexico, Saudi Arabia, Trinidad & Tobago, Turkey, Venezuela

Cessna T-37

In 1952 the USAF requested proposals for a jet primary trainer. The winner in early 1953 was Cessna's Model 318, with side-by-side ejection seats. The XT-37 flew on 12 October 1954, powered by Continental YJ69-T-9s (licence-built Turboméca Marborés). Introduced in 1957, the 534 T-37As were used initially as basic trainers, pupils completing primary training on Beech T-34 Mentors. In April 1961 all-through jet training was started, but fuel costs were high. In 1964 the USAF reverted to the use of light primary trainers, so that T-37 pupils were those remaining after the first weeding-out.

T-37Bs with more powerful engines and improved nac/com systems were introduced into service in November 1959, and surviving T-37As were converted. The T-37C has provision for armament and wingtip tanks. By 1977 1,268 T-37s

had been built.

Specification
Cessna T-37B
Type: jet-powered primary trainer
Powerplant: two 465-kg (1,025-lb) thrust Continental J69-T-25 turbojet engines
Performance: maximum speed 684 km/h (425 mph) at 6095 m (20,000 ft); cruising speed 612 km/h (380 mph) at 10670 m (35,000 ft); range, standard fuel 1400 km (870 miles)
Weight: maximum take-off 2982 kg (6,574 lb)
Dimensions: span 10.29 m (33 ft 9.3 in); length 8.92 m (29 ft 3 in); height 2.79 m (9 ft 2 in); wing area 17.08 m² (183.9 sq ft)
Armament: (T-37C) two 113-kg (250-lb) bombs, or four Sidewinder AAMs, or jettisonable pods which can contain an 12.7-mm (0.5-in) machine-gun with 200 rounds, two

70-mm (2.75-in) folding-fin rockets and four 1.4-kg (3-lb) practice bombs
Operators: Brazil, Burma, Cambodia, Chile, Colombia, West Germany, Greece, Guatemala, Jordan, Pakistan, Peru, Portugal, Thailand, Turkey, US

For years the basic trainer for the US Air Force, the T-37 'Tweety Bird' has provided excellent service within Air Training Command, though replacement of the existing fleet is now imminent.

Cessna T-47A Citation

Based on the Citation II executive jet, 15 examples of the T-47A reached the US Navy from summer 1984. Ordered at $159.4 million, they replace T-39Ds in Naval Air Training Command's VT-86 squadron at Pensacola, Florida, in the tuition of Naval Flight Officers in the use of airborne radars. The T-47A differs from the Citation II in having more powerful turbofans engines and a wing of shorter span to give increased climb and greater speed at altitude. All 15 have Emerson Electric radar. They are flown by civilian contract pilots, and each carries a Navy instructor and two students. Standard Citations have been purchased by other air arms, mainly

for VIP duties.

Specification
Cessna T-47A
Type: radar trainer
Powerplant: two 1315-kg (2,900-lb) thrust Pratt & Whitney Canada JT15D-5 turbofan engines

Performance: (Citation II) cruising speed at 7620 m (25,000 ft) 713 km/h (443 mph); initial rate of climb 1027 m (3,370 ft)/minute; maximum certificated altitude 13105 m (43,000 ft); range with six passengers, maximum fuel, allowances and fuel reserves 3069 km (1,907 miles)
Weights: (Citation II) empty equipped 3306 lb (7289 kg);

maximum take-off 6033 kg (13,300 lb)
Dimensions: span 14.17 m (46 ft 6 in); length 14.60 m (47 ft 10.8 in); height 4.51 m (14 ft 9.5 in); wing area 28.98 m² (312.0 sq ft)
Armament: none
Operators: (Citation I) Argentina, Ecuador, Turkey, Venezuela; (Citation II) Argentina, Spain, Venezuela; (T-47A) US Navy

Major military use of the Cessna Citation is by the US Navy in the form of the T-47A trainer, but small numbers serve with other air arms, including this Venezualan air force example.

Chujiao CJ-6

Though derived from the CJ-5 (Yakovlev Yak-18A) the CJ-6 (Chuji Jiaolianji 6, basic training aircraft No 6) is a Chinese design and was one of the first indigenous aircraft to enter full production in the People's Republic. The chief distinguishing feature is the larger straight-tapered vertical tail, but there are numerous detail differences compared with the CJ-5 and altogether this aircraft has proved most successful.

The engine, derived from the Vedeneev AI-14RF, drives a J9-G1 two-blade variable-pitch propeller. The compressed-air system is retained to operate the split flaps and tricycle landing gear. Provision is made for aerobatic and instrument training, and a glider tow hook is commonly fitted.

Production began in 1961 and it is estimated that the Nanchang factory had produced 2,500 examples by the end of 1985. Virtually all current Chinese military pilot training begins on the CJ-6, though many CJ-5s are still in use. Substantial numbers have been exported, recipients known to include Bangladesh, North Korea, Vietnam and Zambia.

Specification
Type: tandem-seat basic trainer
Powerplant: one 213-kW (285-hp) Quzhou Huosai 6A nine-cylinder air-cooled radial piston engine
Perrormance: maximum speed 286 km/h (178 mph); initial rate of climb 380 m (1,247 ft)/min; service ceiling 5080 m (16,665 ft); range 750 km (466 miles)
Weights: empty 1095 kg (2,414 lb); maximum take-off 1400 kg (3,086 lb)
Dimensions: span 10.70 m (35 ft 1.3 in); length 8.40 m (27 ft 6.7 in); height 3.30 m (10 ft 9.9 in); wing area 17.90 m² (192.68 sq ft)
Armament: normally none
Operators: Albania (?), Bangladesh, China, North Korea, Vietnam, Zambia

Convair CV-580

A conversion of the piston-engined CV-340 or CV-440, the turboprop powered CV-580 was certificated in 1960; a few found their way into military service. Four USAF C-131s were re-engined as VC-131Hs, and these are now operated by the US Navy Reserve. Eight of the 10 Napier Eland-powered Canadair CC-109 Cosmopolitans of the Royal Canadian Air Force were fitted with Allison engines following takeover of the Napier company, and seven remain active with the Canadian Armed Forces' No. 412 Sqn at Trenton. Other CV-580s are known to serve with TAM (Transportes Aéreos Militares) in Bolivia, which is a military airline operating scheduled passenger services.

Specification
Type: short-range transport
Powerplant: two 2796-ekW (3,750-eshp) Allison 501-D13D turboprop engines
Performance: cruising speed at 5485 m (18,000 ft) 560 km/h (348 mph); range with maximum payload 1038 km (645 miles)
Weight: maximum take-off 24131 kg (53,200 lb)
Dimensions: span 32.11 m (105 ft 4 in); length 24.13 m (79 ft 2 in); height 8.88 m (29 ft 1.5 in); wing area 85.47 m² (920.0 sq ft)
Armament: none
Operators: (CV-580) Bolivia; (CC-109) Canada; (VC-131H) US Navy Reserve

Convair F-106 Delta Dart

Now at last being replaced by the F-15 Eagle with front-line interceptor squadrons of the USAF's Air Defense Tactical Air Command, the Convair F-106 Delta Dart will remain active with a handful of Air National Guard units until 1986-87.

Conceived as the F-102B 'ultimate interceptor' in late 1951, and flown in late 1956, the F-106 was plagued initially by development problems, the complex Hughes MA-1 fire control system suffering more than the usual crop of bugs. Thus it was not until May 1959 that the first examples reached the 498th Fighter Interceptor Squadron at Geiger AFB, Washington, and the 539th FIS at McGuire AFB, New Jersey. A year later 14 Air Defense Squadrons had converted. Production of 277 F-106As ended in December 1960, 63 two-seat combat proficiency F-106Bs also being built. Between September 1960 and September 1961 all Delta Darts were brought up to an identical standard, but aircraft have since been repeatedly updated, with inflight refuelling receptacles, new canopies offering an enhanced view, an infra-red sensor,

Convair F-106A Delta Dart of the 87th Fighter Interceptor Squadron. A handful of these venerable fighters carry on in the defence of the United States.

a belly-mounted M61 20-mm (0.79-in) gun and improvements to the fire control system.

Specification
Type: all-weather interceptor
Powerplant: one 11 113-kg (24,500-lb) afterburning thrust Pratt & Whitney J75-P-17 turbojet engine
Performance: maximum speed 2445 km/h (1,519 mph) at 10975 m (36,000 ft); initial rate of climb 12130 m (39,800 ft)/minute; combat ceiling 15850 m (52,000 ft); combat radius with external tanks 1173 km (729 miles)
Weights: empty 10726 kg (23,646 lb); maximum take-off 17554 kg (38,700 lb)
Dimensions: span 11.67 m (38 ft 3.5 in); length 21.56 m (70 ft 8.75 in); height 6.18 m (20 ft 3.25 in); wing area 58.65 m² (631.3 sq ft)
Armament: one AIR-2A Genie or AIR-2B Super Genie nuclear-tipped unguided air-to-air missile and four AIM-4F semi-active radar homing, or AIM-4G infra-red homing Super Falcon air-to-air missiles housed in internal weapons bay; most aircraft also incorporate provision for belly-mounted Vulcan M61 20-mm gun, although this is seldom carried
Operators: USAF (Air Defense Tactical Air Command and Air National Guard)

Convair F-106A Delta Dart

Curtiss-Wright C-46 Commando

On 26 March 1940 Curtiss-Wright flew the prototype of a 36-seat commercial airliner which had the company designation CW-20. Its large-capacity fuselage aroused US Army interest for cargo/transport/casualty evacuation, and a militarized version with 1491-kW (2,000-hp) Pratt & Whitney R-2800-43 engines was ordered into production under the designation C-46 and named Commando. When these first entered service in July 1942 they were the largest and heaviest twin-engine aircraft to serve with the USAAF, and proved such a valuable transport in the Pacific theatre of operations that well over 3,000 were to be built before produciton ended.

Apart from differing engines and fewer cabin windows, the original C-46s were generally similar to the CW-20 prototype. The C-46As which followed had a large cargo door on the left side, a strengthened floor, and folding seats for 40 troops. Pratt & Whitney R-2800-51 engines provided better performance at altitude. This was to prove of great importance, for C-46As transporting vital supplies 'over the hump' of the eastern Himalayas to China from India were found to have better performance than the Douglas C-47 at the altitudes involved. They were to make a vital contribution to the success of this airlift.

In the Pacific the C-46 played a significant role in the island-hopping operations which culminated in Japanese surrender, and 160 R-5C-1s of the US Marine Corps made an important contribution. Later versions for the USAAF included C-46D/E/F and a single C-46G, and Commandos were to remain in service with both the USAAF/USAF and USMC after World War II had ended. Since 1945 the C-46 has been the leading aerial workhorse of Latin America, and some also remain in service with air forces in 1985.

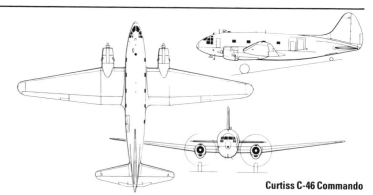

Curtiss C-46 Commando

Specification
Type: troop and cargo transport
Powerplant: (C-46A) two 1491-kW (2,000-hp) Pratt & Whitney R-2800-51 double Wasp radial piston engines
Performance: (C-46A) maximum speed 433 km/h (269 mph) at 4570 m (15,000 ft); cruising speed 295 km/h (183 mph); range 1931 km (1,200 miles)
Weights: (C-46A) empty 14696 kg (32,400 lb); maximum take-off 25401 kg (56,000 lb)
Dimensions: wing span 32.94 m (108 ft 1 in); length 23.27 m (76 ft 4 in); height 6.63 m (21 ft 9 in); wing area 126.34 m² (1,360 sq ft)
Armament: none
Operators: Brazil, Dominica, Honduras, Japan, South Korea, Peru, Taiwan, Uruguay, Zaire

Dassault-Breguet Alizé

Principal anti-submarine patrol air-craft of the French and Indian navies for more than two decades, the Breguet Br.1150 Alizé evolved from a late-1940s design for a carrier-based attack aircraft and the prototype was flown in 1956. Equipped with a radar in a retractable 'dustbin', the Alizé also carries sonobuoys in the large wing fairings aft of the main landing gear. Production amounted to 87 aircraft, of which 12 were ordered by the Indian navy for the carrier *Vikrant* and at shore bases, these seeing heavy utilization during the 1971 war with Pakistan. Part of the Alizé's duties with the Aéronavale have been assumed by the Super Frelon helicopter, but two squadrons continue to fly the type and deployments are undertaken regularly aboard the carriers *Foch* and *Clemenceau*. A refit programme is currently in progress to update 28 aircraft for service into the 1990s. Amongst new systems being installed are a Thomson-CSF Iguane radar, SERCEL-Crouzet VLF Omega inertial navigation system, electronic support measures (ESM) equipment, new communications systems, and other improved avionics. Under original

Dassault-Breguet Br.1150 Alize

plans the Alizé would now be extinct, but both its operators plan to keep the aircraft in service at least until their present carriers are withdrawn.

Specification
Type: three-seat carrier-based anti-submarine aircraft
Powerplant: one 1473-kW (1,975-shp) Rolls-Royce Dart Mk 21 turboprop engine
Performance: maximum speed 518 km/h (322 mph) at 3050 m (10,000 ft); patrol speed 240-370 km/h (149-230 mph); service ceiling 8000 m (26,245 ft); normal range 2500 km (1,553 miles); maximum endurance 7 hours 40 minutes

Weights: empty 5700 kg (12,566 lb); maximum take-off 8200 kg (18,078 lb)
Dimensions: span 15.60 m (51 ft 2.2 in); length 13.86 m (45 ft 5.7 in); height 5.00 m (16 ft 4.9 in); wing area 36.00 m² (387.51 sq ft)
Armament: underfuselage weapons bay for one AS torpedo or three 160-kg (353-lb) depth charges; racks under inner wings for two 160-

A French navy Alizé approaches the carrier *Foch*. Still retained for ASW, the Alizé has outlasted its planned service life.

kg (353-lb) or 175-kg (386-lb) depth charges, and racks beneath outer wings for six 127-mm (5-in) rockets, or two Nord AS.12 or other air-to-surface missiles
Operators: French and Indian navies

Dassault-Breguet Atlantic and Atlantique

The Italian air force operates this Atlantic for maritime patrol duties in the Mediterranean. They are operated by the 30° and 41° Stormi.

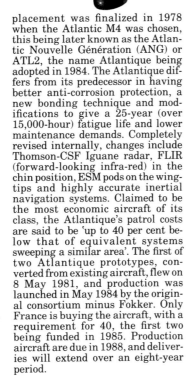

The Atlantic originated in a 1957 NATO requirement (NBMR-2) for a long-range maritime reconnaissance aircaft. The winner was the Breguet Br.1150 (this company being absorbed by Dassault in 1971), fabricated by the SECBAT (Société d'Etudes et de Construction de Breguet Atlantic) consortium, which then included Belgium's SABCA and SONACA, Holland's Fokker, Dornier and MBB of West Germany, Aérospatiale of France and Italy's Aeritalia. Assembly was by Breguet, the first of four prototypes being flown at Toulouse on 21 October 1961. Carrying 12 crew, the Atlantic is equipped with CSF search radar in a retractable bin and American ASW avionics similar to those of the Lockheed Neptune. Standard NATO stores can be carried in the 9.15 m (30 ft) unpressurized weapons bay in the lower section of the 'double bubble' fuselage, most of the external skin being light alloy sandwich. The first of 20 for the West German navy and 40 for France's Aéronavale entered service in December 1965. French aircraft serve with 21F and 22F at Nîmes-Garons and 23F and 24F at Lann-Bihoué. German Atlantics of MFG 5, Kiel-Holtenau (except for five equipped for Elint missions), have completed an update programme involving new Texas Instruments radar, Emerson Electric sonar and Loral ESM equipment in wingtip pods. In parallel, airframe improvements have doubled flying life to 10,000 hours. By contrast, the six survivors of nine delivered to 321 Sqn of the Royal Netherlands navy at Valkenburg in 1969-72 began phasing out in January 1984, replaced by P-3Cs. Italy's 18 were supplied between June 1972 and July 1974 to complete production, and are operated by 30° Stormo at Cagliari/Elmas and 41° Stormo at Catania/Sigonella on patrols of the Mediterranean. Pakistan obtained three from the French navy in 1975-76; wearing 'navy' titles, they are flown by 29 Sqn of the Pakistan Air Force from Sharea Faisal (Drigh Road).

A French requirement for a re-

placement was finalized in 1978 when the Atlantic M4 was chosen, this being later known as the Atlantic Nouvelle Génération (ANG) or ATL2, the name Atlantique being adopted in 1984. The Atlantique differs from its predecessor in having better anti-corrosion protection, a new bonding technique and modifications to give a 25-year (over 15,000-hour) fatigue life and lower maintenance demands. Completely revised internally, changes include Thomson-CSF Iguane radar, FLIR (forward-looking infra-red) in the chin position, ESM pods on the wingtips and highly accurate inertial navigation systems. Claimed to be the most economic aircraft of its class, the Atlantique's patrol costs are said to be 'up to 40 per cent below that of equivalent systems sweeping a similar area'. The first of two Atlantique prototypes, converted from existing aircraft, flew on 8 May 1981, and production was launched in May 1984 by the original consortium minus Fokker. Only France is buying the aircraft, with a requirement for 40, the first two being funded in 1985. Production aircraft are due in 1988, and deliveries will extend over an eight-year period.

Specification
Dassault-Breguet Atlantique 2
Type: maritime patrol aircraft
Powerplant: two 4638-ekW (6,220-eshp) SNECMA-built Rolls-Royce Tyne 21 turboprop engines
Performance: maximum speed

657 km/h (408 mph); patrol speed 315 km/h (196 mph) at 1525 m (5,000 ft); service ceiling 9145 m (30,000 ft); endurance 18 hours; typical mission, eight-hour patrol at 1110 km (690 miles) radius
Weights: empty 25700 kg (56,659 lb); maximum take-off 46200 kg (101,854 lb)
Dimensions: span 37.70 m (123 ft 8.25 in); length 32.62 m (107 ft 0.25 in); height 10.80 m (35 ft 5.2 in); wing area 120.0 m² (1,291.71 sq ft)
Armament: bombs, depth charges, torpedoes and up to two

The Atlantique is the revised and completely upgraded version of the original design. The structure is better adpted to the rigours of low-level over-water operation, while the avionics have been updated, including FLIR, ESM and INS. This example is seen carrying four AM.39 Exocet anti-ship missiles.

Aérospatiale AM 39 Exocet anti-ship missiles in bomb bay, plus 3500 kg (7,716 lb) of ordnance on four wing pylons
Operators: France; (Atlantic) France, West Germany, Italy, Pakistan

Dassault-Breguet Etendard

Although it is a carrier-based attack aircraft, the Dassault Breguet-Etendard (flag or standard) was conceived as a land-based ground-support aircraft for the French air force. It was developed in response to a NATO requirement for a light, high-subsonic strike fighter capable of operating from unpaved forward strips. Included in the NATO specification was a requirement that the aircraft should be powered by the Bristol Orpheus turbojet, which had a modest (for this class of aircraft) thrust of around 2177 kg (4,800 lb). The designation of this aircraft was Etendard VI. Another variant, the Etendard II with two 1097-kg (2,420-lb) Turboméca Gabizo turbojets, was the first to fly, on 23 July 1956. But Marcel Dassault, highly sceptical of such low-powered designs, privately financed a version powered by the much more powerful SNECMA Atar 08 and designated Etendard IV. This version flew on 24 July 1956, and was turned down by the French air force, which decided against the concept of such a light interceptor. Italy's Fiat G91 subsequently won the NATO competition. But far from going under, the Etendard IV was put through a protracted modification programme to meet the needs of the Aéronavale, which required carrier-based attack and reconnaissance aircraft. The Etendard IV was put into production in two forms to fulfil these roles. The first was designated Etendard IVM and deployed aboard the carriers *Foch* and *Clemenceau* (Flotilles 11F and 17F). The prototypes of this variant flew for the first time on 21 May 1958, and was followed by six pre-production aircraft. The first of 69 production Etendard IVMs for the French navy was delivered on 18 January 1962, and production was completed in 1964. The seventh Etendard was the prototype of the IVP, a reconnaissance/tanker version, of which 21 were ordered. First flight was made on 19 November 1960. The primary design changes include nose and ventral stations for three and two reconnaissance cameras, an independent navigation system, a fixed nose-probe for flight-refuelling, and 'buddy-pack' hose-reel unit designed by Douglas to allow Etendard-to-Etendard refuelling. Compared with the original land-based aircraft, both maritime versions are equipped with such standard naval features as long-stroke undercarriage, arrester hook, catapult attachments and associated strengthening, folding wing-tips and a high-lift system which combined leading-edge and trailing-edge flaps, as well as two perforated belly air-brakes. The Etendard IVM also carries Aïda all-weather fire-control radar. Both marks of Etendard are being replaced by the Super Etendard, but for the present Flotilles 11F and 17F (home bases Landivisiau and Hyères respectively when not deployed aboard the 27,300-ton carriers *Clemenceau* and *Foch*) fly 36 Etendard IVM aircraft. A reconnaissance squadron (Flotille 16F) based at Landivisiau operates 14 Etendard IVPs.

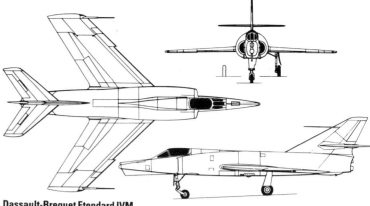

Dassault-Breguet Etendard IVM

Specification
Type: single-seat transonic carrier-based strike aircraft
Powerplant: one 4400-kg (9,700-lb) SNECMA Atar 8B single-shaft turbojet
Performance: maximum speed 1099 km/h (683 mph) at sea level; maximum speed 1083 km/h (673 mph) or Mach 1.02 at 11000 m (36,090 ft); maximum design speed Mach 1.3; combat range (low-level attack role) 600 km (370 miles); combat range (medium-altitude mission) 1600 km (1,000 miles); (ferry) range 3000 km (1,860 miles); rate of climb (sea level) 6000 m (19,685 ft) per minute
Weights: empty equipped 6123 kg (13,500 lb); normal take-off 8170 kg (18,010 lb); maximum take-off 10275 kg (22,650 lb)
Dimensions: span 9.6 m (31 ft 6 in); span (folded) 7.8 m (25 ft 7 in); length 14.4 m (47 ft 3 in); height 4.3 m (14 ft 2 in); wing area 29 m² (312 sq ft)
Armament: two 30-mm DEFA cannon; four underwing hardpoints for a maximum of 1360 kg (3,000 lb) of rockets, bombs, Nord 5103 air-to-surface or Sidewinder air-to-air missiles, or external fuel tanks
Operator: France (Aéronavale)

Etendards are still used for various secondary roles by the Aéronavale, including reconnaissance, tanking and conversion training.

Dassault-Breguet Falcon 10MER and 100

Dassault's Falcon 10 small business jet first flew on 1 December 1970 and deliveries began in November 1973. This aircraft seats up to seven passengers, with two pilots and provision for a third crew member on a jump seat. Manufacture is an international effort, the fuselages assembled by Potez (Aérospatiale) from parts provided by SOGERMA, SOCEA and SOCATA; wings from CASA (Spain); nose and tail from IAM (Italy); and other components from Latécoère. Over 200 have been built, recent models being to the improved Falcon 100 standard with a 225-kg (496-lb) increase in take-off weight, a fourth cabin window on the right side, opposite the door, provision for a Collins EFIS-85 instrument package with five-CRT display and a larger rear baggage compartment. The sole military customer is the French Aéronavale which ordered two, with option on three, before acquiring a further pair (one an attrition replacement) to increase procurement to seven by 1982. Designated Falcon 10MER, the type was declared operational on 1 July 1975 with the Section Réacteur Légèr at Landivisiau. The SRL became Escadrille 57s in September 1981, and now operates three 10MERs for instrument and night training, as well as acting as intruders for Super Etendard interception practice and calibration aircraft for ship radars. Three more operate with Escadrille 3S at Hyères, where additional duties include light transport and casevac. The second prototype was acquired in 1980 for trials work by the Centre d'Essais en Vol.

Specification
Dassault-Breguet Falcon 100
Type: executive transport
Powerplant: two 1465-kg (3,230-lb) thrust Garrett TFE731-2 turbofan engines
Performance: maximum cruising speed Mach 0.84 at 10670 m (35,005 ft), or 912 km/h (576 mph) at 7620 m (25,000 ft); range with four passengers and fuel reserves 3480 km (2,162 miles)

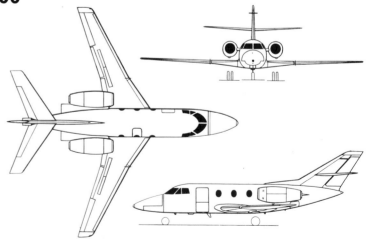

Dassault-Breguet Falcon 10

Weights: empty 5055 kg (11,144 lb); maximum take-off 8755 kg (19,301 lb)
Dimensions: span 13.08 m (42 ft 11 in); length 13.86 m (45 ft 5.7 in); height 4.61 m (15 ft 1.5 in); wing area 24.10 m² (259.42 sq ft)
Armament: none
Operator: France

Dassault–Breguet Falcon 20 Series G/HU-25A Guardian

The Dassault-Breguet Falcon 20 (originally known as the Mystère 20) was designed as a transport for the executive market. It has since found favour with military customers as a VIP transport and in more specialized roles. Developed in conjunction with Sud-Aviation (Aérospatiale), who produced the wings and tailplane, the prototype flew on 4 May 1963. The first production aircraft, flown on 1 January 1965, differed in being fitted with CF700 turbofans; Aérospatiale built the fuselage and tail, while Dassault was responsible for the wings.

The fuselage of the Falcon 20 is an all-metal, fail-safe structure of cir--cular cross-section. The low-mount ed wing has a sweepback of 30° at quarter chord and incorporates a leading-edge slats inboard and outboard of a fence. Hydraulic airbrakes are mounted forward of double-slotted flaps.

A total of 5200 litres (1,374 Imperial gallons) of fuel is carried in two fuselage and two wing tanks. Twin wheels are fitted on all three units of the tricycle undercarriage, and a braking parachute is standard. Controls are powered.

The French armed forces still use the designation Mystère 20. The 65e Escadre handles short-range liaison from Villacoublay, while five with VIP furnishing operate from the same airfield with the Groupe des Liaison Aériennes Ministerielles. A trainer, also operated by Libya, is fitted with Mirage IIIE radar and navigation equipment. The Aéro-navale's Falcon 10s provide proficiency, radar and continuation training.

The Canadian Armed Forces has four CC-117 VIP transports with No. 412 Squadron. There are ECM trainers with No. 414 Squadron at North Bay, Ontario. The Royal Norwegian Air Force's No. 333 Squadron operates two in the dual role of ECM and navaid calibration.

In January 1977 the US Coast Guard ordered 41 Falcon 20Gs, to be designated HU-25A Guardian for maritime surveillance. Japan has expressed interest in the Falcon 20G for similar duties.

The primary difference between the 20G and earlier Falcons is the fitting of the more-powerful ATF 3 turbofan, sufficient for greater weights with avionics for overwater search and rescue, maritime law and treaty enforcement and environmental protection. Communications

equipment includes dual HF, VHF-AM, IFF, single VHF-FM and UHF. Navaids include an inertial platform. Omega, dual VOR/ILS/MB, DME, ADF, radio altimeters, R-nav system and Tacan. Search and weather radar is standard, while optional sensors include SLAR, infra-red and ultra-violet scanners, reconnaissance camera and television. The 20G's fuselage is modified to incorporate two observation windows and a drop hatch, and provision is made for external pods to be carried. The US Coast Guard's HU-25As are being delivered one per month to replace the Grumman HU-16E Albatross amphibian at six Coast Guard stations.

Specification
Type: twin-turbofan transport
Powerplant: two 2040-kg (4,500-lb) General Electric CF-700-2D-2 turbofans; (20G) two 2400-kg (5,300-lb) Garrett AiResearch ATF 3-6 turbofans
Performance: maximum cruising speed at 7620 m (25,000 ft) 862 km/h (536 mph); (20G) maximum cruising speed at 12200 m (40,000 ft) Mach 0.8; absolute ceiling 12800 m (42,000 ft); range with maximum fuel (45-minute reserves) 3350 km (2,080 miles); (20G) range with crew of 5,

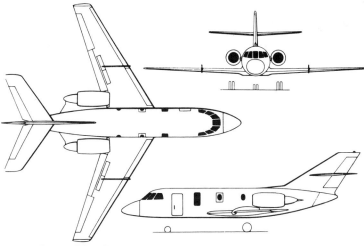

Dassault-Breguet Falcon 20

4185 km (2,600 miles)
Weights: empty equipped 7530 kg (16,600 lb); maximum take-off 13000 kg (28,660 lb); (20G) maximum take-off 14500 kg (32,000 lb)
Dimensions: span 16.3 m (53 ft 6 in); length 17.15 m (56 ft 3 in); height 5.32 m (17 ft 5 in); wing area 41m^2 (440 sq ft); (20G) wing area 41.8 m^2 (450 sq ft)

Armament: none
Operators: Australia, Belgium, Canada, Central African Empire, Egypt, France, Gabon, Iran, Iraq, Libya, Nicaragua, Norway, Oman, Pakistan, Spain, UK, US Coast Guard

Under the designation HU-25, the Falcon 20 has been procured for the US Coast Guard for search and rescue duties.

Dassault-Breguet Falcon 50

The Falcon 50 (Mystère 50 in France) meets the demand for a business aircraft capable of flying the Atlantic without refuelling. First flown on 7 November 1976, production aircraft were delivered from July 1979. The French government acquired the fifth Falcon 50 in January 1980 as a transport for the President. Named *Rambouillet*, it is operated from Villacoublay by Groupe des Liaisons Aériennes Ministerielles 1/60. The 12th aircaft was supplied to King Hassan of Morocco, with its call-sign presented as a registration. The government of Yugoslavia bought the 25th and 43rd; though both were supplied with military serial numbers, the second soon adopted civil guise. Spain acquired the 84th Falcon 50, designating it T-16, and this aircraft is operated by Escuadron 401 from Madrid/Barajas. After considering

four Falcon 20s to update its VIP fleet, the Italian Air Force opted in 1984 for a single Gulfstream III and two Falcon 50s. These replace Piaggio PD.808s in 306° Gruppo Reparto Volo Stato Maggiore of 31° Stormo at Rome/Ciampino.

Specification
Dassault-Breguet Falcon 50
Type: long-range executive transport
Powerplant: three 1678-kg (3,700-lb) thrust Garrett TFE731-3 turbofan engines
Performance: maximum cruising speed Mach 0.82 (880 km/h; 547 mph); service ceiling 13800 m (45,275 ft); range with 8 passengers and fuel reserves 6480 km (4,026 miles)
Weights: empty 9150 kg (20,172 lb); maximum take-off 17600 kg (38,801 lb)

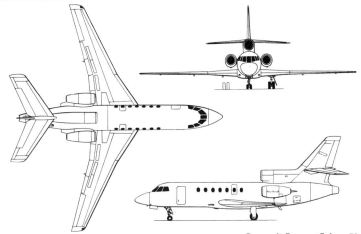

Dassault-Breguet Falcon 50

Dimensions: span 18.86 m (61 ft 10.5 in); length 18.50 m (60 ft 8.3 in); height 6.97 m (22 ft 10.4 in); wing area 46.83 m^2 (504.09 sq ft)

Armament: none
Operators: France, Italy, Jordan, Morocco, Spain, Yugoslavia

Dassault-Breguet Mirage III and IIING

Argentina has a large fleet of Mirages obtained from various sources. These were not used much during the Falklands war, having been diverted for home defence duties.

The Mirage III is by far the most successful combat aircraft produced in Western Europe. At one time forming the backbone of the French interceptor and strike/attack forces (and still prominent in the latter), the first-generation Mirages have found export success throughout the globe because of their combination of simplicity, reliability and performance. The aircraft takes its name from the diminutive Dassault 550 Mirage I which first flew on 25 June 1955, but the basic design was enlarged and refined before the Mirage III prototype was airborne on 17 November 1956. This was designed as a Mach 2 high-level interceptor, having a detachable booster rocket pack in the rear fuselage. Soon, however, its potential for ground attack and tactical reconnaissance was realised. Ten pre-production Mirage IIIAs paved the way for the Mirage IIIC interceptor, of which 95 entered local service with CAFDA (air defence command) from July 1961. The survivors fly with Escadron de Chasse 2/10 at Creil (shortly to disband) and with EC 3/10 detached to Djibouti. The Mirage IIIB dual trainer was ordered in several forms, the first 26 for pilot conversion, except for one modified to a IIIB-SV for variable-stability experiments. Next came six IIIB-1 testbeds for the Centre d'Essais en Vol trials establishment, and then ten dummy-probe-equipped IIIB-RVs for inflight refuelling training with the strategic force OCU, CIFAS 328. From February 1971, 20 IIIBEs were supplied as equivalents to the IIIE fighter-bomber, these lacking radar and being generally similar to the simplified Mirage 5 (described separately). The surviving IIIBs and IIIBEs are flown from Dijon by Escadron de Chasse et de Transformation 2/2, although the training task (and thus the aircraft) is expected to transfer to EC 13 at Colmar in 1986-87. The major production variant for France was the IIIE of the FATac (tactical air force), first flown on 5 April 1961. Room for extra avionics was made by lengthening the fuselage 30 cm (11.8 in) – moving the bottom edge of the cockpit canopy completely forward of the air intakes – and raking the main gear legs to accommodate an AN 52 tactical nuclear weapon on the centreline pylon. Two squadrons of EC 13 at Colmar received IIIEs from April 1965 for attack and tactical air defence roles, and one unit still flies the IIIE. Other IIIEs are flown by EC 3/2 at Dijon; two squadrons of EC 3 at Nancy in the defence-suppression role with AS 37 Martel ASMs; and two squadrons of EC 4 at Luxeuil, with AN 52s. The last of 183 Mirage IIIEs was delivered in 1973. Based on the IIIE, the tactical reconnais-

sance IIIR has the IIIC's navigation system and air-to-ground weapons capability, plus a five-camera nose, and so was able to enter service earlier, with the 33e Escadre de Reconnaissance (now at Strasbourg) in June 1963. Fifty were delivered to ER 1/3 and 2/3 (the latter converting to Mirage F1.CRs in 1983), and 20 IIIRDs (with the IIIE's doppler bulge below the forward fuselage) to ER 3/33. While many export customers preferred to buy Mirage 5s, the all-weather Mirage III with its early Cyrano radar has been acquired by ten, some of which have used it in combat. Israel's 72 IIICJs were supplied in 1961-64, followed by five IIIBJ trainers, and used with great effect in the 1968 Arab-Israeli war. In 1982 22 survivors, including three trainers, were sold to Argentina to equip Grupo 4 de Caza at El Plumerillo (IV Brigada Aérea). They thus augment the survivors of 17 IIIEAs and two IIIDA trainers in Grupo 8 de Caza at BAM Mariano Moreno (VIII Brigada Aérea) which lost two of its number to Sea Harriers during the 1982 Falklands conflict. Pakistan's first 18 IIIEPs were delivered in 1967-69 and fought against India in the December 1971 war, claiming eight air victories and two on the ground. Also received were three IIIRPs and three IIIDPs. South African operations have involved 58 aircraft: 16 IIICZ interceptors delivered in 1962-64; followed by three IIIBZs, four IIIRZs, 17 IIIEZs, three IIIDZ, 11 IIID2Zs and four IIIR2Zs. Those with '2' in their designation are powered by the uprated Atar 9K50 engine and are thus to Mirage 50 standard. In Australia the RAAF still flies some 70 Mirages, remaining from 116 that comprised 100 single-seat IIIOAs and 16 IIID trainers. These are flown by Nos 3, 75 and 77 Sqns and No. 2 OCU, all of which will have received F-18 Hornets by 1988. Switzerland's aircaft, however, are due for many more years of operation, to which end one has been experimentally fitted with canards (like the Kfir C2) to test improvements in manoeuvrability. This modification will be fitted to the survivors of 36 IIIS of 16 and 17 Staffeln, 18 IIIRS of 10 Staffel and six trainers. Spain has two squadrons of Ala de Caza 11 which received 24 IIIEEs and six IIIDEs under the local designations C.11 and CE.11 respectively. Mirage IIIs

The Mirage IIIs of South Africa have seen much action against guerrillas operating in Angola and Namibia. Unguided rockets are a favourite weapon.

Dassault Mirage IIIE cutaway drawing key

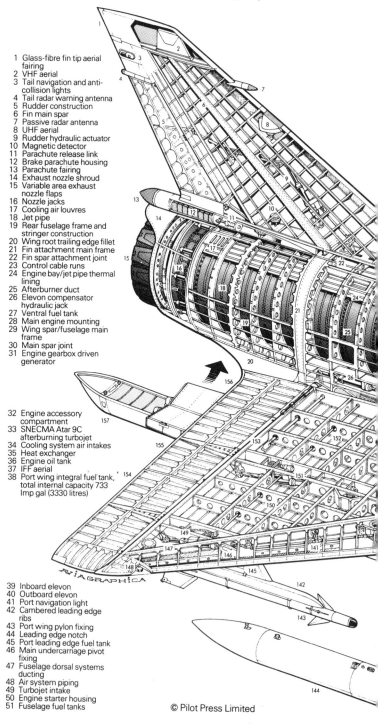

1 Glass-fibre fin tip aerial fairing
2 VHF aerial
3 Tail navigation and anti-collision lights
4 Tail radar warning antenna
5 Rudder construction
6 Fin main spar
7 Passive radar antenna
8 UHF aerial
9 Rudder hydraulic actuator
10 Magnetic detector
11 Parachute release link
12 Brake parachute housing
13 Parachute fairing
14 Exhaust nozzle shroud
15 Variable area exhaust nozzle flaps
16 Nozzle jacks
17 Cooling air louvres
18 Jet pipe
19 Rear fuselage frame and stringer construction
20 Wing root trailing edge fillet
21 Fin attachment main frame
22 Fin spar attachment joint
23 Control cable runs
24 Engine bay/jet pipe thermal lining
25 Afterburner duct
26 Elevon compensator hydraulic jack
27 Ventral fuel tank
28 Main engine mounting
29 Wing spar/fuselage main frame
30 Main spar joint
31 Engine gearbox driven generator

32 Engine accessory compartment
33 SNECMA Atar 9C afterburning turbojet
34 Cooling system air intakes
35 Heat exchanger
36 Engine oil tank
37 IFF aerial
38 Port wing integral fuel tank, total internal capacity 733 Imp gal (3330 litres)

39 Inboard elevon
40 Outboard elevon
41 Port navigation light
42 Cambered leading edge ribs
43 Port wing pylon fixing
44 Leading edge notch
45 Port leading edge fuel tank
46 Main undercarriage pivot fixing
47 Fuselage dorsal systems ducting
48 Air system piping
49 Turbojet intake
50 Engine starter housing
51 Fuselage fuel tanks

© Pilot Press Limited

The Mirage IIIC is the pure interceptor version of this aged fighter. This aircraft serves in Djibouti with the Armée de l'Air's EC 3/10, providing a measure of air defence for that country. A special sand and brown camouflage is applied for these operations.

also fly in Brazil (two squadrons of IIIEBRs), Lebanon (IIIEL) and Venezuela (IIIEV). The ultimate variant has yet to gain orders, this being the Mirage 3NG (Nouvelle Génération). Powered by a 9K50, it has canards and leading edge root extensions, fly-by-wire controls; and an advanced nav/attack system including Cyrano IV or Agave radar,

head-up display and inertial navigation.

Specification
Dassault-Breguet Mirage IIIE
Type: single-seat interceptor and ground-attack aircraft
Powerplant: one 6200-kg (13,669-lb) afterburning thrust SNECMA Atar 09C turbojet engine

Performance: maximum speed 1400 km/h (870 mph) clean at sea level, or Mach 2.2 at 12000 m (39,370 ft)
Weights: empty 7050 kg (15,543 lb); maximum take-off 13500 kg (29,762 lb)
Dimensions: span 8.22 m (26 ft 11.6 in); length 15.03 m (49 ft 3.7 in); height 4.50 m (14 ft 9.2 in);

wing area 34.85 m² (375.13 sq ft)
Armament: two 30-mm (1.18-in) DEFA 552 cannon; various combinations of bombs, rockets and guided missiles (AS 30 or 37) to limit of 1814 kg (4,000 lb)
Operators: Argentina, Australia, Brazil, France, Lebanon, Pakistan, South Africa, Spain, Switzerland, Venezuela

52 Equipment cooling system air filter
53 Computer system voltage regulator
54 Oxygen bottles
55 Inverted flight fuel system accumulator
56 Intake ducting
57 Matra 530 missile computor
58 VHF radio transmitter/receiver
59 Gyro platform multiplier
60 Doppler transceiver
61 Navigation system computor
62 Air data computer
63 Nord missile encoding supply
64 Radio altimeter transceiver
65 Heading and inertial correction computer
66 Armament junction box

67 Radar program controller
68 Canopy external release
69 Canopy hinge
70 Radio and electronics bay access fairing
71 Fuel tank stabilizing fins
72 286 Imp gal (1300 litres) auxiliary fuel tank (374 Imp gal/1700 Litre alternative)
73 132 Imp gal (600 litre) drop tank
74 Cockpit canopy cover
75 Canopy hydraulic jack
76 Ejection seat headrest
77 Face blind firing handle
78 Martin-Baker (Hispano licence) RM.4 ejection seat
79 Port side console panel
80 Canopy framing
81 Pilot's head-up display
82 Windscreen panels
83 Instrument panel shroud
84 Instrument pressure sensors

85 Thomson-CSF Cyrano II fire control radar
86 Radar scanner dish
87 Glass-fibre radome
88 Pitot tube
89 Matra 530 air-to-air missile
90 Doppler radar fairing
91 Thomson-CSF doppler navigation radar antenna
92 Cockpit front pressure bulkhead
93 Rudder pedals
94 Radar scope (head-down display)
95 Control column
96 Cockpit floor level
97 Starboard side console panel
98 Nosewheel leg doors
99 Nose undercarriage leg strut
100 Landing/taxying lamps
101 Levered suspension axle unit
102 Nosewheel
103 Shimmy damper
104 Hydraulic retraction strut

105 Cockpit rear pressure bulkhead
106 Air conditioning ram air intake
107 Moveable intake half-cone centre-body
108 Starboard air intake
109 Nosewheel well door (open position)
110 Intake centre-body screw jack
111 Air conditioning plant
112 Boundary layer bleed air duct
113 Centre fuselage bomb rack
114 882-lb (400-kg) HE bombs
115 Cannon barrels
116 30-mm DEFA cannon (2) 250-rounds per gun
117 Ventral gun pack
118 Auxiliary air intake door
119 Electrical system servicing panel
120 Starboard 30-mm DEFA cannon
121 Front spar attachment joint
122 Fuel system piping

123 Airbrake hydraulic jack
124 Starboard airbrake, upper and lower surfaces (open position)
125 Airbrake housing
126 Starboard leading edge fuel tank
127 AS.37 Martel, radar guided air-to-ground missile
128 Nord AS.30 air-to-air missile
129 Starboard mainwheel
130 Mainwheel leg door
131 Torque scissor links
132 Shock absorber leg strut
133 Starboard main undercarriage pivot fixing
134 Hydraulic retraction jack
135 Main undercarriage hydraulic accumulator
136 Wing main spar
137 Fuel system piping
138 Inboard pylon fixing
139 Leading edge notch
140 Starboard inner stores pylon
141 Control rod runs

142 Missile launch rail
143 AIM-9 Sidewinder air-to-air missile
144 JL-100 fuel and rocket pack, 55 Imp gal (250 litres) of fuel plus 18x68-mm unguided rockets
145 Outboard wing pylon
146 Outboard pylon fixing
147 Front spar
148 Starboard navigation light
149 Outboard elevon hydraulic jack
150 Starboard wing integral fuel tank
151 Inboard elevon hydraulic actuator
152 Wing multi-spar and rib construction
153 Rear spar
154 Outboard elevon construction
155 Inboard elevon construction
156 Elevon compensator
157 110 Imp gal (500-litre) auxiliary ventral fuel tank

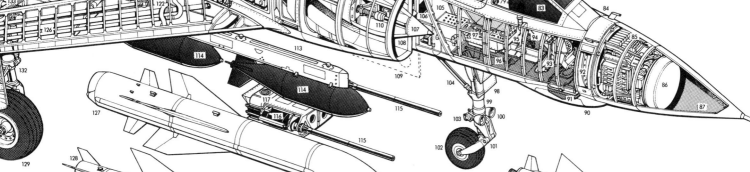

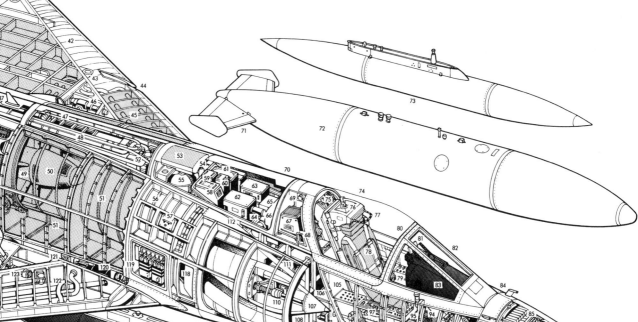

Dassault-Breguet Mirage IVA

Now entering probably the last phase of its career in the French nuclear deterrent forces, the Mirage IVA is being modified to carry the Aérospatiale ASMP (Air-Sol Moyenne Porté) nuclear stand-off bomb for more effective strikes against heavily-defended targets. Dassault received design responsibility for the French airborne nuclear delivery system in April 1957. When the first of four prototypes took to the air on 17 June 1959 it was seen to be some 50 per cent larger than the Mirage III with twice the wing area, thrust and weight. It also carried two crew, the navigator sitting almost totally enclosed behind the pilot. An initial contract for 50 in 1960 was later increased by 12, deliveries beginning in 1964 and being competed in November 1966. Nine squadrons (escadrons) in three wings (escadres) operated a front-line total of 36 aircraft, each carrying a 65 kT AN 22 free-fall bomb (later retarded, when the force turned to low-level attack), semi-recessed in the belly. Support was provided by three squadrons of Boeing C-135F tankers, while availability of French nuclear missiles prompted a reduction to six squadrons in 1976: Escadron de Bombardement 1/91 at Mont-de-Marsan, EB 2/91 at Cazaux, EB 3/91 at Orange (disbanded 1983), EB 1/94

at Avord, EB 2/94 at Dizier and EB 3/94 at Luxeuil. Training was provided by Centre d'Instruction des Forces Aériennes Stratégiques 328 at Bordeaux, whose four aircraft have provision for a 1000-kg (2,205-lb) CT 52 reconnaissance pod (visual only, or optionally including IR linescan) in the bomb recess. In 1979 plans were announced to convert 15 – later increased to 18 – aircraft to Mirage IVP (= Pénétration) standard, these having a 100-150 kT 100-km (62-mile) range ASMP on a new centreline pylon. In order to ensure accuracy in delivery the IVP is equipped with Antilope 5 ground mapping radar and a modern nav/attack system. After dummy separation trials the IVP fired its first ASMP in June 1983, prior to beginning an official acceptance trials programme at CEAM Mont-de-Marsan in the following month. Other conversions, by the air force's Atelier Industriel de L'Air at Aulnat/Clermont Ferrand, will be delivered in 1986-88 to EB 1/91 and EB 2/91, plus CIFAS 328 for training. Mirage IVAs are now slowly being withdrawn, but the IVP is destined to remain in the front line until 1996.

Specification
Dassault-Breguer Mirage IVP
Type: two-seat strategic bomber

The Dassault-Breguet Mirage IVA maintains France's airborne nuclear deterrent, armed with both free-fall weapons and the new ASMP stand-off missile (IVP version).

Powerplant: two 6700-kg (14,771-lb) thrust SNECMA Atar O9K afterburning turbojet engines, plus optional ATO rocket pack
Performance: maximum speed Mach 2.2 at 11000 m (36,090 ft); combat speed Mach 1.8 (1915 km/h; 1,190 mph) at high altitude; service ceiling 20000 m (65,615 ft); range 4000 km (2,485 miles) with one air refuelling
Weights: empty 14500 kg (31,967 lb); maximum take-off 31600 kg (69,666 lb)

Dassault-Breguet Mirage IVA

Dimensions: span 11.85 m (38 ft 10.5 in); length 23.50 m (77 ft 1.2 in); height 5.65 m (18 ft 6.4 in); wing area 78.0 m² (839.61 sq ft)
Armament: one Aérospatiale ASMP nuclear supersonic cruise missile
Operator: France

Dassault-Breguet Mirage 5-50

Responding to customer (Israeli) demand, Dassault-Breguet produced a simplified version of its highly successful Mirage III, the new variant, designated Mirage 5, making its first flight on 19 May 1967. Optimized for visual ground attack and interception, the aircraft lacked several features of its Mirage IIIE progenitor, notably the Cyrano II nose-mounted radar (replaced by a simple radar rangefinder) and other mission avionics. Retaining full flight performance, the Mirage 5 emerged with notable enhancements, such as greater range, easier maintenance and no less than seven weapon attachment points beneath its fuselage and wings. No sooner had this simplification been achieved, and the first of many orders booked, than the process began of developing a family of aircraft based on the Mirage 5 and incorporating various degrees of sophistication according to customer preference. For political reasons the Israeli aircraft, paid for, went instead to France and Libya. After the camera-nosed Mirage 5R and two-seat Mirage 5D trainer came versions equipped with SAGEM inertial navigation and nav/attack systems incorporating a head-up display and the choice of Aida II radar and an air-to-surface laser ranger or Agave multi-purpose radar. These option produced a plethora of sub-marks, some of which were almost indistinguishable from the Mirage III, such as the Egyptian Mirage 5E2 which incorporates the nav/attack system of the Alpha Jet MS2. The next evolutionary step was to replace the 6200-kg (13,669-lb) thrust Atar 9C afterburning turbojet by a more powerful Atar 9K50 to produce the Mirage 50, first flown on 15 April 1979 and exemplified by the non-radar Mirage 50FC and radar-equipped 50C delivered to Chile. The Mirage 50's usual radar fit is Agave or Cyrano IV, allowing the aircraft to operate in the attack or interceptor modes. Current models

Libya's Mirage 5 fleet is enormous, but appears to have been severely depleted due to maintenance problems and neglect.

Zaïre operates five Mirage 5M interceptors and two 5DM two-seat trainers. These form the main jet force for the country.

offered by Dassault are the Mirage 3-50 and 5-50 versions of the III and 5, both of which are powered by 9K50 engines, but no further sales appear likely.

Specification
Dassault-Breguet Mirage 50
Type: single-seat fighter-bomber
Powerplant: one 7200-kg (15,873-lb) thrust SNECMA Atar 9K50 afterburning turbojet engine
Performance: maximum speed Mach 2.2 or 2350 km/h (1,460 mph) at altitude; maximum speed Mach 1.13 or 1390 km/h (864 mph) at sea level; initial rate of climb 11100 m (36,415 ft)/minute' service ceiling 18000 m (59,055 ft); combat radius lo-lo-lo mission with 800 kg (1,764 lb) of weapons 630 km (391 miles)
Weights: empty 7150 kg (15,763 lb); loaded, clean 9900 kg (21,826 lb); maximum take-off 13700 kg (30,203 lb)
Dimensions: span 8.22 m (26 ft 11.6 in); length 15.56 m (51 ft 0.6 in); height 4.50 m (14 ft 9.2 in); wing area 35.00 m² (376.75 sq ft)
Armament: two 30-mm (1.18-in)

DEFA cannon in fuselage with 125 rpg; five weapon pylons for ASMs, rocket pods and bombs to theoretical limit of 4000 kg (8,818 lb)
Operators: (Mirage 5) Abu Dhabi, Belgium, Colombia, Egypt, France,

Belgium received Mirage 5s in the three main versions of single-seat fighter, two-seat trainer and reconnaissance Mirage 5BR (illustrated).

Gabon, Pakistan, Peru, Venezuela, Zaïre (Mirage 50); Libya

Dassault-Breguet Mirage 2000

After the Mirage F.1 was ordered, Dassault spent much effort on the large variable-sweep Mirage G series. This led to the ACF (Avion de Combat Futur) with a wing fixed at 55°, but in December 1975 this too was cancelled. In its place came another of the small single-Atar machines, and it marked a return to the tailless delta configuration. It was, however, a totally different aircraft, designed to CCV (control-configured vehicle) technology with variable camber wings having hinged leading and trailing edges, electrically signalled controls and artificial stability. Structure was entirely new, as was the engine whose extremely low bypass ratio was designed for Mach 2 at high altitudes, calling for small frontal area, rather than for subsonic fuel economy. Choice of a single-shaft engine also greatly increased weight: the basic engine weighing 1450 kg (3,195 lb). The prototype Dassault-Breguet Mirage 2000 flew on 10 March 1978 and following successful development, the first production fighter flew in December 1982, with 1983 production including tandem-seat Mirage 2000B trainers. A total of 127 was expected to be on order (48 by 1982), all in basic air-defence configuration. Initial production examples of the Mirage 2000 have the RDM multi-mode radar, while machines from no. 51 will have the considerably more capable RDI air interception equipment for delivery from 1986. The pulse-Doppler RDI radar is designed to pick up a target of 5 m² (54 sq ft) at a range of 100 km (62 miles). RDM radar is standard for all export sales of the Mirage 2000. Despite the extremely high price, said by Egypt to be US$50 million per aircraft, the same type is also on order for Egypt (20) and India (40), and plans were laid for the type to be built under licence in both countries if negotiations were successful. India has now cancelled its plan to build. Peru has also ordered 26. The Mirage 200N is a developed version with airframe strengthened to fly at 1110 km/h (690 mph) at sea level (this is very slow by modern standards) and is equipped with terrain-following radar and other modern attack systems for the delivery of the ASMP nuclear stand-off missile.

The Mirage 2000N is a two-seat version especially designed for low-level penetration carrying the ASMP stand-off nuclear missile. Matra Magic missiles provide self-defence.

Specification
Type: fighter
Powerplant: one 9000-kg (19,840-lb) afterburning thrust SNECMA M53-5 bypass turbojet
Performance: maximum speed, clean at high altituded 2350 km/h (1,460 mph) or Mach 2.2; service ceiling 20000 m (65,600 ft) range at high altitude with two tanks 1480 km (1,118 miles)
Weights: empty 7400 kg (16,315 lb); maximum take-off 16500 kg (36,375 lb)
Dimensions: span 9.0 m (29 ft 6 in); length 14.35 m (47 ft 1 in); height 5.30 m (17 ft 6 in); wing area 41.0 m² (441.3 sq ft)
Armament: two 30-mm cannon each with 125 rounds; normal missile load two Super 530 AAMs inboard under wings and two Magic AAMs outboard (Mirage 2000N attack version is planned to carry heavy and varied weapon loads)

Dassault-Breguet Mirage 2000

Operators: Egypt, France, Greece, India, Peru, UAE

EC 3/2 is the first French air force unit to receive the Dassault-Breguet Mirage 2000C. These aircraft are equipped to carry the Matra Magic and Super 530 AAMs for air defence duties.

Dassault-Breguet Mirage F1

The F1 was designed as a successor to the Mirage III/5. In 1964 Dassault was awarded a French Government contract to build a prototype of the Mirage F2 two-seat fighter powered by a SNECMA (Pratt & Whitney) TF306 turbofan. This made its maiden flight in June 1966, and in December of that year it was followed by the first F1, a smaller single-seat fighter matched to the existing Atar, which Dassault had designed as a private venture. Though this soon crashed, the F1 proved a more attractive proposition than its problematical brother, and in September 1967 the French Government ordered three pre-production F1s. The first flew in March 1969 and, despite being powered by the Atar 9K31, notched up a series of impressive performances including Mach 2.12 or 2260 km/h (1,404 mph) at 11000 m (36,090 ft), and 1300 km/h (808 mph) at low level.

The F1 quickly proved superior to the Mirage III, with 40 per cent more fuel, larger bombload, easy handling at low altitude and a high rate of climb. Compared with its predecessor, it has three times the endurance at high Mach numbers, three times the patrol time before and after an interception, twice the tactical range at sea level, a 30 per cent shorter take-off run at maximum weight, 25 per cent lower approach speed and improved manoeuvrability at subsonic and supersonic speeds, maximum speed being higher. The short take-off and landing performance results from the high-lift system, comprising leading edge droops and large flaps, fitted to the efficient wing. At average mission weight the F1 can take off and land within 500 to 800 m (1,640 to 2,625 ft). The ability to operate from short, rough strips is enhanced by the twin wheels with medium-pressure tyres, combined with a landing speed of only 230 km/h (143 mph).

The F1's weapon system is based on the Thomson-CSF Cyrano IV monopulse radar, which has 80 per cent greater range than the Cyrano II fitted in Mirage IIIs and allows intruders to be intercepted at all altitudes, even if they are flying low in ground clutter. Targets to be traced are selected manually by the pilot, who then transfers his attention to the electromechanical HUD (head-up display) while the radar continues to track its designated target automatically. The weapons can be fired automatically by the fire-control computer, or manually, with the computer supplying the pilot with clearance to engage the target.

Ground handling equipment is kept to a minimum and is air-transportable. A self-starter is used, and the high-pressure refuelling system allows all internal tanks to be filled in about six minutes; this contributes to a turn-round time of 15 minutes between two missions where identification of intruders rather than interception is required. An F1 can be scrambled within two minutes, thanks to the self-propelled GAMO truck which supplies electric power to preheat the navigation and weapon-system equipment, circulates fluid to cool the radar and controls cockpit air-conditioning, as well as carrying a sunshade on a telescopic arm so that the pilot can sit at readiness for extended periods in the heat of the day. When the pilot is ordered to scramble he starts the engine in the normal way. The sunshade is automatically withdrawn, the air-conditioning and radar-cool-

ing switched off, and as soon as the alternators supply sufficient electrical power GAMO is automatically disconnected and the pilot can taxi.

The first variant in the Armée de l'Air was the F1.C interceptor, the first production example of which flew in February 1973. The F1.A used by South Africa is a ground-attack and VFR (visual flight rules) fighter version, in which a ranging radar replaces the Cyrano IV and a more simple avionics fit is installed. This allows extra fuel to be carried. The F1.A is licensed to Atlas Aircraft in South Africa as well as being supplied directly by Dassault. The F1.E resembles the F1.C but has modifications to the fire-control system and updated avionics. The F1.B, flown in May 1976, is a two-seat trainer. No internal guns are carried, and the HUD and radar dis-

play are repeated in the rear cockpit.

Specification
Type: single-seat multi-role fighter and two-seat trainer
Powerplant: one 7215-kg (15,906-lb) thrust SNECMA Atar 9K50 afterburning turbojet engine
Performance: maximum speed Mach 2.2 or 2320 km/h (1,442 mph) at 12190 m (40,000 ft); maximum speed at sea level Mach 1.2 or 1470 km/h (913 mph); time to climb to Mach 2 at 12190 m (40,000 ft) 7 minutes 30 seconds; stabilized supersonic ceiling 18500 m (60,695 ft); maximum combat radius at low level with 1600 kg (3,527 lb) load 640 km (398 miles); endurance 3 hours 45 minutes
Weights: empty 7400 kg (16,314 lb); operational (pilot, guns and internal fuel) 10900 kg (24,030 lb); with AAMs 11500 kg (25,353 lb); maximum take-off (F1.E) 15200 kg (33,510 lb)

Dimensions: span 8.44 m (27 ft 8.3 in); length 15.25 m (50 ft 0.4 in); height 4.50 m (14 ft 9.2 in); wing area 25.00 m^2 (269.11 sq ft)
Armament: two 30-mm (1.18-in) DEFA 553 cannon with 125 rpg and (intercept mission) two R.550 Magic AAMs, or two Sidewinder AAMs, or two R.530 or Super 530 AAMs with option of two Magic; (attack mission) up to 4000 kg (8,818 lb) on seven hardpoints. Combinations can include eight 454-kg (1,000-lb) bombs, or four 36-round rocket launchers, or gun pods, or one AS.30 or AS.37 Martel air-to-surface missile, or other payloads such as napalm
Operators: Ecuador, France, Greece, Iraq, Jordan, Kuwait, Libya, Morocco, Qatar, South Africa, Spain

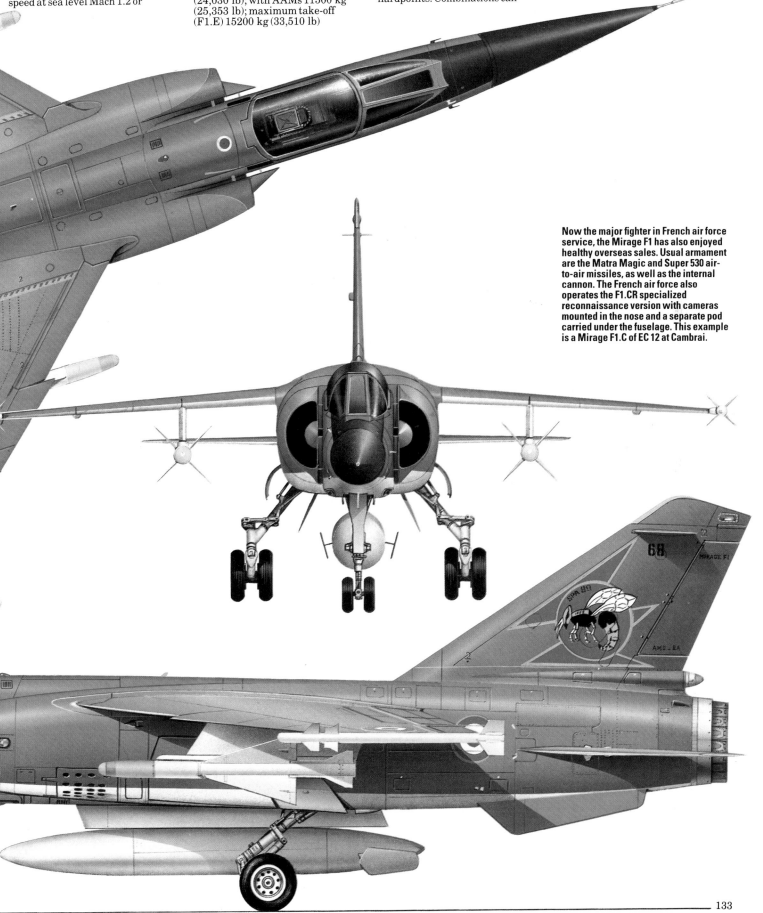

Now the major fighter in French air force service, the Mirage F1 has also enjoyed healthy overseas sales. Usual armament are the Matra Magic and Super 530 air-to-air missiles, as well as the internal cannon. The French air force also operates the F1.CR specialized reconnaissance version with cameras mounted in the nose and a separate pod carried under the fuselage. This example is a Mirage F1.C of EC 12 at Cambrai.

Dassault-Breguet Rafale

Though originally promoted forcefully as the basis for a European fighter, this aircraft failed to meet either the technical or the political requirements of other countries and from August 1985 has continued as a purely French programme. A single prototype is being built for first flight in mid-1986. From this may be developed two production versions, an airfield-based aircraft for the Armée de l'Air, tailored for the ground attack role with air combat as a secondary mission, and a carrier-based version for the Aéronavale.

The Rafale (squall) follows the fashionable configuration for conventional (non jet lift) combat aircraft with a variable-camber delta wing and canards. The cockpit has a Martin-Baker F10QA seat inclined sharply backwards, a side-stick controller and advanced displays, and it is expected to incorporate spoken voice control. Structure will be mainly carbonfibre composite, with many other parts of Kevlar composite or allithium type alloys. Other features include a retractable flight-refuelling probe, landing gear stressed for no-flare touchdown on runways, and the possibility of later replacing the fly-by-wire flight control system with an optical fibre 'fly-by-light' system.

The prototype will have imported US engines, but it is hoped to substitute subsequently the SNECMA M88 which is much later in design and all-French. The engine inlets, originally very like those of the Mirage 2000 but mounted on the flanks of the fuselage, were later simplified without variable centre-bodies, reflecting the diminished importance of Mach numbers beyond 2. Everything possible is being done to reduce weight, though this runs directly counter to the wish also to minimize cost. There are expected to be six stores attachments under the fuselage and two under each wing, with dogfight AAM rails on the wingtips. It is planned to instal a single gun on the left side of the fuselage under the wing, though some reports have stated that two guns will be fitted.

It goes without saying that the radar will have look-down shootdown capability, that there will be comprehensive electronic-warfare installations and excellent manoeuvrability with 'carefree' handling up to extreme angles of attack. The land-based version is to be able to operated (without ground attack stores) from runways 500 m (1,640 ft) long.

Specification
Type: tactical attack aircraft for use from airfields or carriers, with secondary air-combat role
Powerplant: two 7257 kg (16,000 lb) General Electric F404-400 augmented turbofan engines
Performance: (estimated) maximum speed over Mach 2; take-off run with offensive stores 700 m (2,300 ft); combat radius about 600 km (373 miles) with attack load of 3500 kg (7,716 lb)
Weights: empty 9250 kg (20,390 lb); internal fuel 4250 kg (9,370 lb); air combat mission 14000 kg (30,865 lb); attack (max) 20000 kg (44,090 lb)
Dimensions: span 11.0 m (36 ft 1 in); length 15.80 m (51 ft 10 in); wing area 47.0 m^2 (506.0 sq ft)
Armament: one gun (probably 30-mm GIAT 791B), and very wide range of AAMs, bombs, rockets, smart weapons and other stores

Dassault-Breguet Super Etendard

In 1957 the Dassault Etendard failed to win a NATO competition for a light attack aircraft, but it was later selected by France's Aéronavale for shipboard operation as the Etendard IV. Seven pre-series aircraft flew from 21 May 1958 onwards, and deliveries began in January 1962. Production comprised 69 Etendard IVMs and 21 photo-reconnaissance Etendard IVPs, the latter with three cameras in the nose and a twin vertical installation in the gun bay, replacing the two 30-mm (1.18-in) DEFA cannon. Etendards were issued to operational squadrons 11F, 14F, 15F and 17F for service aboard the aircraft carriers *Foch* and *Clemenceau*, the IVM acting as both interceptor, with two AIM-9 Sidewinder air-to-air missiles, and an attack aircraft carrying two Aérospatiale AS 30 anti-ship missiles or bombs. The reconnaissance aircraft are still flown by 16F from both vessels, but the IVMs are restricted to operational training with 59S at Hyères following their last carrier disembarkation in June 1980. From 1979 at least four IVMs were converted to IVPs.

The navalized SEPECAT Jaguar M was tested as a replacement, but Dassault instead offered a minimum-change version of the original aircraft, designated Super Etendard, and succeeded in keeping out the half-British machine. The Super now has a mere 10 per cent commonality with the IVM. It has an improved leading edge and flaps which, in conjunction with a better Atar engine, gives a considerable boost to maximum weight. The more modern nav/attack system includes an Agave radar. Three IVMs were converted to Super Etendard prototypes, the first flying on 28 October 1974, and deliveries of 71 (reduced from 100) began in June 1978. They are operated from Landivisiau by 11F and 14F and from Hyères by 17F; 50 are being equipped to carry the Aérospatiale ASMP nuclear cruise missile. Five of an order for 14 had been delivered to the Argentine navy by the start of the Falklands war in April 1982, and carrying an Aérospatiale AM 39 Exocet missile under the right wing, and a tank under the left, they destroyed two British ships. Similar successes with this missile have been achieved by five late production aircraft diverted from French navy stocks to Iraq in October 1983, for use in the war against Iran.

Specification
Dassault-Breguet Super Etendard
Type: single-seat carrier-based strike/attack aircraft
Powerplant: one 5000 kg (11,023-lb) thrust SNECMA Atar 8K50 turbojet engine
Performance: maximum speed (clean) nearly Mach 1 at altitude; 1204 km/h (748 mph) at low level; service ceiling 13700 m (44,950 ft); radius of action 650 km (404 miles) with Excocet
Weights: empty 6450 kg (14,220 lb); maximum take-off 11500 kg (25,353 lb)
Dimensions: span 9.60 m (31 ft 6 in); length 14.31 m (46 ft 11.4 in); height 3.86 m (12 ft 8 in); wing area 28.40 m^2 (305.71 sq ft)
Armament: two 30-mm DEFA cannon, one AN 52 nuclear store, one Aérospatiale ASMP nuclear missile, one Aérospatiale AM 39 Exocet anti-ship missile, or Matra Magic AAM for self-defence; conventional bombs to overall limit of 2100 kg (4,630 lb)
Operators: Argentina, France, Iraq

Dassault-Breguet Super Etendard of the Argentine navy.

The Super Etendard provides the Aéronavale with shipborne strike. Among the likely targets will be enemy shipping, and for this they can carry Exocet missiles.

Dassault-Breguet Super Mystère B2

First flown on 2 March 1955, the Super Mystère shortly afterwards exceeded the speed of sound in level flight, subsequently becoming the first European aircraft with a Mach 1-plus performance to enter full-scale service. Developed from Dassault's earlier Mystère IVA via an interim Rolls-Royce Avon-engined variant known as the Mystère IVB, the Super Mystère differed chiefly in having a new thinner-section wing with more marked sweepback, a flat oval air intake (similar to that of the North American F-100 Super Sabre), and a larger and more swept fin and rudder.

By the late 1970s only a few B-2s remained in French service, most having been replaced by Mirage F1s, but the type still served with the Israeli air force, and Israel has supplied Honduras with 12 examples of a version modified locally to take a 4218-kg (9,300-lb) thrust non-afterburning Pratt & Whitney J52-P-8A turbojet. Despite the absence of an afterburner, ths version is said to have a considerably longer rear

fuselage than the standard Super Mystère, and can carry a wider variety of external stores. This version appeared in the early 1970s and was used with some success in the Yom Kippur War of October 1973.

Specification
Type: single-seat fighter and fighter-bomber
Powerplant: one 4460-kg (9,833-lb) SNECMA Atar 101G-2 or G-3 afterburning turbojet
Performance: maximum speed at sea level 1040 km (646 mph); maximum speed at 12000 m (39,370 ft) 1195 km/h (743 mph); normal range 965 km (600 miles); maximum rate of climb at sea level at normal loaded weight 5340 m (17,500 ft) per minute; service ceiling 17000 m (55,775 ft)
Weights: empty equipped 6932 kg (15,282 lb); normal loaded 9000 kg (19,842 lb); maximum take-off 10000 kg (22,046 lb)
Dimensions: span 10.51 m (34 ft 5¾ in); length 14.04 m (46 ft 1 in); height 4.55 m (14 ft 11 in); wing

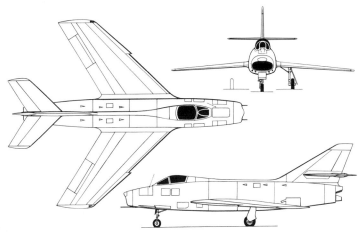

Dassault-Breguet Super Mystère B2

area 35.00 m^2 (376.74 sq ft)
Armament: two 30-mm DEFA 552 cannon in underside of forward fuselage; retractable internal pack of 35 68-mm SNEB rockets in lower fuselage; and up to 1000 kg

(2,205 lb) of bombs, rocket pods, drop-tanks or other stores on two underwing hardpoints
Operators: El Salvador, Honduras

Dassault-Breguet/Dornier Alpha Jet

On 22 July 1969 the governments of France and West Germany announced that they had a requirement for a new training aircraft, to replace T-33s and Magisters. On 24 July 1970 it was announced that the Alpha Jet had been selected, not only in the training role but also for battlefield-reconnaissance duties and light strike following a change in Luftwaffe requirements. Despite these differing specifications, the basic structure, engines and landing gear are virtually identical in all versions.

The whole programme suffered a slippage of approximately two years. The go-ahead came on 26 March 1975, though the prototype had flown on 26 October 1973. Luftwaffe aircraft No. 1 flew on 12 April 1978, and the first of 33 for Belgium (which had shared manufacture) on 20 June 1973. Luftwaffe Alpha Jet As can carry some 2500 kg (5,512 lb) of stores on one underfuselage and four underwing hardpoints. They have a full nav/attack system. Hi-lo-hi radius of action with a representative warload is around 555 km (345 miles), or for a lo-lo-lo sortie 370 km (230 miles). They have the Mauser 27-mm (1.06-in) gun, pointed pitot-nose, Stencel instead of M-B seats, BOZ-10 chaff pod and can carry a Super Cyclop reconnaissance pod. The only remaining production is centred in Helwan, Egypt, where both the MS1 trainer and MS2 close-support version (with inertial navigation, HUD, laser ranger and radar altimeter) are built. The even more advanced NGEA (new generation trainer/attack) had not found a buyer in early 1986.

Specification
Dassault-Breguet/Dornier Alpha Jet
Type: two-seat advanced trainer and light strike/reconnaissance aircraft
Powerplant: two 1350-kg (2,976-lb) thrust SNECMA/Turboméca Larzac C5 turbofan engines
Performance: maximum speed clean at sea level 927 km/h (576 mph); maximum speed at 9145 m (30,000 ft) Mach 0.85; time to 9145 m (30,000 ft) under 7 minutes; combat radius with maximum external load, and

Dassault-Breguet/Dornier Alpha Jet of the Ivory Coast air force.

Alpha Jets have been supplied in large numbers to Morocco, which uses them for both training and light attack.

including 5 minutes combat 410 km (255 miles); ferry range with maximum internal fuel 2780 km (1,727 miles)
Weights: empty, trainer 3345 kg (7,374 lb), strike 3,500 kg (7,716 lb); normal take-off, trainer clean 5000 kg (11,023 lb); maximum take-off, strike with external stores 7500 kg (16,535 lb)
Dimensions: span 9.10 m (29 ft 10.7 in); length 12.29 m (40 ft 3.9 in); height 4.19 m (13 ft 9 in); wing area 17.50 m^2 (188.37 sq ft)
Armament: underfuselage pod containing 30-mm (1.18-in) DEFA or 27-mm (1.06-in) Mauser cannon (150 rpg each) or 7.62-mm (0.3-in) machine-gun (250 rounds); four underwing hardpoints, each carrying up to 36 68-mm (2.68-in) rockets, or high explosive/retarded bombs up to 400 kg (882 lb), or cluster dispensers, or drop tanks; provision for air-to-air (Magic) or air-to-surface (Maverick) missiles
Operators: Belgium, Cameroon, Egypt, France, West Germany, Ivory Coast, Morocco, Nigeria, Qatar, Togo

France has adopted the Alpha Jet as its main trainer, and these also equip the national aerobatic team. This aircraft is from Groupement Ecole 314 at Tours.

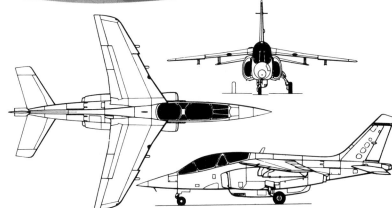

Dassault-Breguet/Dornier Alpha Jet.

de Havilland Sea Heron

As a communication aircraft and 'admiral's barge', the Heron serves in small numbers with the Royal Navy to this day, the basic design tracing its history back to the DH.114 prototype flown on 10 May 1950. With an exceptional short-field performance, the 14/17-seat aircraft was a 'stretched' development of the Dove/Devon. RAF Heron C.2, CC.3 and CC.4s are no longer in use. Naval light transport requirements were met by the Sea Heron C.1; the sole loss occurred in June 1972 when one ditched in the Irish Sea, the replacement being an ex-Queen's Flight aircraft now designated Heron C.4. Three Sea Herons and the Heron C.4 remain in service, all with the Station Flight at RNAS Yeovilton, Somerset.

Specification
de Havilland Sea Heron C.1
Type: executive transport
Powerplant: four 186-kW (250-hp) de Havilland (Rolls-Royce) Gipsy Queen 30 Mk. 2 piston engines
Performance: cruising speed 295 km/h (183 mph) at 2440 m (8,000 ft); initial rate of climb 328 m (1,075 ft)/minute: range 2494 km (1,550 miles)
Weights: empty 3848 kg (8,484 lb); maximum take-off 6123 kg (13,500 lb)
Dimensions: span 21.79 m (71 ft 6 in); length 14.78 m (48 ft 6 in); height 4.75 m (15 ft 7 in); wing area 46.36 m² (499.0 sq ft)
Armament: none
Operator: UK

de Havilland Sea Heron

de Havilland DH.104 Dove/Devon

Design of the DH.104 Dove began in 1944, and the prototype flew on 25 September 1945. It was of all-metal construction and scored two firsts, the first British transport with tricycle landing gear and the first with braking propellers. It rapidly went into production and soon attracted the attention of the military authorities. The 48th was modified to Air Ministry specification C.13/47 as a military communications version, and this variant went into production as the Devon C.1; the RAF received 41 from late 1947. The Devon incorporated several minor changes to meet RAF requirements, such as removal of the right-hand front passenger seat for installation of a dinghy, and addition of a jettisonable cabin door. RAF Devons were later re-engined with 298-kW (400-hp) Gipsy Queen 70-3 engines to bring them up to Dove 8 standard.

The Royal Navy acquired 13 Doves from 1955 under the designation Sea Dove C.20. India and New Zealand were the only other countries to use the designation Devon: the first of 22 for the Indian air force was delivered in March 1948, and the RNZAF received 30, deliveries beginning in mid-1948. The biggest

overseas military customer for the Dove was Argentina, which received 54, but these are out of service. A total of 544 Doves, including two prototypes, had been built when production ended in 1967, and of these about 240 were at one time in military use.

Specification
de Havilland Devon C.1
Type: twin-engine light transport
Powerplant: two 246-kW (330-hp) DH Gipsy Queen 71 or 298-kW (400-

hp) Gipsy Queen 70-3 inline piston engines
Performance: maximum speed 338 km/h (210 mph) at 2440 m (8,000 ft); cruising speed 288 km/h (179 mph) at 2440 m (8,00 ft);initial rate of climb 259 m (800 ft)/minute; service ceiling 6095 m (20,000 ft); range with 771-kg (1,700-lb) payload 1609 km (1,000 miles)
Weights: empty 2622 kg (5,780 lb); maximum take-off 3856 kg (8,500 lb)
Dimensions: span 17.37 m (57 ft

Doves have been retired recently from RAF service, but a few aircraft are still on strength as hacks. This aircraft wears the badge of No. 207 Sqn, the last RAF unit officially to operate the type.

0 in); length 11.99 m (39 ft 4 in); height 4.06 m (13 ft 4 in); wing area 31.12 m² (335.0 sq ft)
Armament: none
Operators: Ethiopia, Jordan, Sri Lanka, UK (Royal Navy)

de Havilland Vampire

Having made its first flight on 20 September 1943, the DH.100 Vampire has almost completed a remarkably long period of service. Extensively used by the RAF, the Vampire was exported to or built under licence in Australia, Ceylon (Sri Lanka), Egypt, Finland, France, India, Iraq, Italy, Jordan, Lebanon, New Zealand, Norway, Rhodesia (Zimbabwe), South Africa, Sweden, Switzerland and Venezuela. Until recently, the oldest still in service were a handful of ex-Swedish Vampire 1s and 50s in the air force of the Dominican Republic. The night fighter Vampire NF.10 and 54 has long disappeared, but a few examples remain of another side-by-side version, the Airspeed-developed DH.115 Vampire Trainer. Zimbabwe's aircraft are survivors of nine FB.9s and nine T.11 trainers, believed to have been augmented by some South African T.55s. No. 2 Sqn at Thornhill has re-equipped with BAe Hawks, but the Vampires are still flying. South Africa's T.55s have been withdrawn except for one or two with the Test Flight & Development Centre, while in the RAF a single T.11 of the 'Vintage Pair'

display team partners a Gloster Meteor from Scampton. Switzerland is thus the last major operator, with some 20 survivors of 178 FB.6s and 37 of 39 T.55s (including nine ex-RAF T.11s). Based at Emmen with Fliegerschule II-Teil, they provide advanced flying training, some T.55s having been fitted with Venom noses and one with a two-camera nose for recording other aircraft's firing-range sorties.

Specification
de Havilland Vampire T.55
Type: two-seat jet trainer
Powerplant: one 1588-kg (3,500-lb) thrust de Havilland (Rolls-Royce) Goblin 35 turbojet engine
Performance: maximum speed 866 km/h (538 mph) at sea level; initial rate of climb 1372 m (4,500 ft)/minute; service ceiling 12190 m (40,000 ft); range 1352 km (840 miles)
Weights: empty 3348 kg (7,380 lb); maximum take-off 5058 kg (11,150 lb)
Dimensions: span 11.58 m (38 ft 0 in); length 10.53 m (34 ft 6.5 in); height 1.88 m (6 ft 2 in); wing area 24.34 m² (262.0 sq ft)

Armament: optional up to four 20-mm (0.79-in) Hispano cannon
Operators: Dominican Republic, Switzerland, Zimbabwe

Switzerland is the last major operator of the Vampire, with around 50 still in service. This FB.Mk 6 serves as a target tug, hence the orange and black stripes. Others are used for weapons training and survey work.

de Havilland Canada DHC-1 Chipmunk

The de Havilland Canada DHC-1 Chipmunk was designed to succeed the Tiger Moth biplane trainer. Flying for the first time at Downsview, Toronto on 22 May 1946, the tandem-seat stressed-skin monoplane was the first indigenous design of de Havilland Aircraft of Canada Ltd. The prototype was powered by a 108-kW (145-hp) de Havilland Gipsy Major 1C. Chipmunks built to this specification were designated DHC-1B-1, while those with a Gipsy Major 10-3 were designated DHC-1B-2. Most Canadian-built Chipmunks had a bubble canopy; Downsview built 218, the last in 1951. The Chipmunk was ordered from Hatfield and Chester to Specification 8/48 as an ab initio trainer for the RAF, which received 735 out of 1,014 manufactured in the UK. They replaced Tigers with all 17 university air squadrons, as well as equipping many RAF Volunteer Reserve flying schools in the 1950s. A few of No. 114 Sqn were pressed into service in Cyprus on internal security flights during 1958. Under

de Havilland Canada DHC-1 Chipmunk of an RAF Air Experience Flight.

an agreement concluded between de Havilland and the General Aeronautical Material Workshops (OGMA) of Portugal, 60 Chipmunks were licence-manufactured from 1955 for the Portuguese air force.

Specification
Type: primary trainer
Powerplant: one 108-kW (145 hp) de Havilland Gipsy Major 8 inline piston engine
Performance: maximum speed 222 km/h (138 mph) at sea level; cruising speed 187 km/h (116 mph); service ceiling 4815 m (15,800 ft); range 451 km (280 miles)
Weights: empty 646 kg (1,425 lb); maximum take-off 914 kg (2,014 lb)
Dimensions: span 10.46 m (34 ft 4 in); length 7.75 m (25 ft 5 in); height 2.13 m (7 ft 0 in); wing area 15.98 m² (172.0 sq ft)
Armament: none
Operators: Portugal, Sri Lanka, Thailand, UK, Uruguay

de Havilland Canada DHC-2 Beaver

Design of the Beaver light utility transport was started in Toronto in late 1946, influenced by the Ontario Department of Lands and Forests. The prototype took to the air on 16 August 1947 with Russ Bannock at the controls. Type certification in Canada followed in March 1948. A ready market had been foreseen in the numerous 'bush' operators in the Canadian prairies, but it was as a military transport that the type was to make its mark.

The Beaver was entered in a design competition jointly sponsored by the US Air Force and US Army in 1951 for a new liaison aircraft. These services purchased six Beavers for evaluation purposes under the designation YL-20; the type's subsequent success made it only the second non-American type to be purchased in quantity since World War II. Once Congressional approval had been obtained, large-scale acquisition commenced. It is a tribute to the sturdy construction of the Beaver that very few changes were specified by the US authorities; these were mainly concerned with instrumentation and equipment, and were easily complied with.

Most Beavers were for the export market. Some 970 were delivered to the American armed forces, designated L-20 and, since 1962, U-6A. The Beaver's wide track and ease of

maintenance, features intended for operation in the snowy wastes of the Canadian north, proved equally attractive in Korea in the 1950s and, later, Vietnam. A pilot and up to seven passengers can be accommodated; alternatively, the tough floor with cargo attachments can carry 680 kg (1,500 lb). Wheels, skis, floats, or an amphibious float assembly can be fitted.

Over 20 countries ordered the Beaver for military use in addition to the United States and Canada. Some, such as the 24 supplied to the Royal Netherlands Air Force, were covered by America's postwar Mutual Assistance Defense Program. The British Army Air Corps took delivery of 46 Beavers. DH at Hatfield produced a Beaver 2, with Alvis Leonides engine and enlarged tail surfaces; no production resulted. Greater success was achieved by Toronto's Beaver 3 or Turbo-Beaver with a 431-kW (578-ehp) Pratt & Whitney PT6A-6 turboprop; 60 were built, the majority for civil use.

The US Army operated more Beavers than any other single type of fixed-wing aircraft. Several were operated on skis in Alaska during the early 1950s in a high-visibility red and white. USAF Strategic Air Command used 58 to support ICBM sites. The few dual-control Beavers of the US Navy were designated

de Havilland Canada DHC-2 Beaver

TU-6A.
Nearly 1,700 Beavers had been manufactured when de Havilland Canada made the decision to curtail production in the mid-1960s, in favour of larger and more powerful STOL aircraft.

Specification
Type: light utility transport
Powerplant: one 336-kW (450-hp) Pratt & Whitney R-985 Wasp Junior air-cooled radial engine
Performance: maximum speed at 1520 m (5,000 ft) 262 km/h (163 mph); cruising speed at 1520 m (5,000 ft) 230 km/h (143 mph);

cruising range on internal fuel 732 km (455 miles); maximum range 1180 km (733 miles); service ceiling 5485 m (18,000 ft); initial climb rate 311 m (1,020 ft) per minute
Weights: empty 1290 kg (2,850 lb); loaded 2310 kg (5,100 lb)
Dimensions: span 14.63 m (48 ft 0 in); length 9.22 m (30 ft 3 in); height 2.74 m (9 ft 0 in); wing area 23.22 m² (250 sq ft)
Armament: none
Operators: Canada, Colombia, Dominican Republic, Haiti, Turkey, UK, US Navy, Zambia

de Havilland Canada DHC-3 Otter

The all-metal de Havilland Canada DHC-3 Otter was designed to carry 1016 kg (2,240 lb) of freight or 14 passengers and was built in quantity in the early 1950s for the US Army, US Navy and Royal Canadian Air Force. The majority of the 466 examples built went to military customers although about 100 were delivered to civil operators.

Powered by a single 448-kW (600-hp) Pratt & Whitney R-1340 Twin Wasp radial, the Otter followed closely the layout and design philosophy of the early DHC-2 Beaver. It was, in fact, originally known as the King Beaver. Despite Canada's harsh climate and large areas with sparse populations, single-engined aircraft had been used successfully

to open up the country. The choice of a single engine for a new 14-seater was not therefore unusual. The parallel-chord wing is fitted with double-slotted flaps. The prototype made its maiden flight in December 1951 and the first of 66 Otters for the RCAF were employed on search and rescue operations in the Arctic, paratrooping, and aerial photography.

Quantity deliveries to the US Army began in 1955 under the designation U-1A, US Navy aircraft being designated UC-1 (changed to U-1B in 1962). The US armed forces expected to use the Otter for supplying troops in forward areas but in peacetime the aircraft found a number of roles reflecting its Canad-

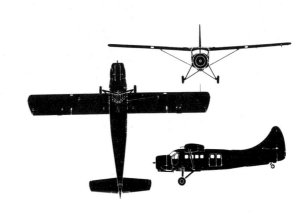

de Havilland Canada DHC-3 Otter

de Havilland Canada DHC-3 Otter (continued)

ian origins. The first six US aircraft were assigned to Alaska. Others were used on expeditions to the Antarctic and in 1957-58 an Otter provided the sole transport and reconnaissance support for the British component of the Commonwealth Trans-Antarctic Expedition. At one time or another 10 nations used the Otter for Arctic surveys. An RAF crew flew the Trans-Antarctic Expedition Otter on an 11-hour sortie

across the South Pole.

Like the Beaver and Twin Otter, the Otter is capable of operating from wheels, floats or skis, and an amphibian version was also offered. Production ceased in 1968.

Specification
Type: STOL light utility aircraft
Powerplant: one 448-kW (600-hp) Pratt & Whitney R-1340-S1H1-G air-cooled radial piston engine

Performance: (at maximum take-off weight, landplane at sea level) maximum speed 246 km/h (153 mph); maximum cruising speed 212 km/h (132 mph); economical cruising speed 195 km/h (121 mph); range with 953-kg (2,100-lb) payload and reserves 1410 km (875 miles)
Weights: (landplane) empty 2010 kg (4,431 lb); maximum take-off and landing 3629 kg (8,000 lb)

Dimensions: (landplane) span 17.69 m (58 ft 0 in); length 12.80 m (41 ft 10 in); height 3.83 m (12 ft 7 in); wing area 34.83 m² (375 sq ft)
Armament: none
Operators: Australia, Burma, Canada, Chile, Colombia, Costa Rica, Ghana, India, Indonesia, Norway, Paraguay, US Army, US Navy

de Havilland Canada DHC-4 Caribou

The first large STOL aircraft to be produced by the company, the de Havilland Canada DHC-4 Caribou came about in response to a 1956 conference between this manufacturer and representatives of the US and Canadian armies, both of which had identified a need for a genuine STOL type.

The prototype flew for the first time on 30 July 1958, and a batch of five early Caribous was ordered by the US Army for an exhaustive evaluation programme, these initially being given the designation YAC-1 although it was later changed to CV-2A. The aircraft's performance was such that the US Army subsequently ordered no fewer than 159 production specimens (56 to CV-2A and 103 to CV-2B standard) for delivery in 1961-4 although in the event this service was destined to operate the type for only a fairly short time, 134 of the surviving machines being transferred to the US Air Force on 1 January 1967.

Known in USAF parlance as the C-7A and C-7B, most of these machines were employed in South Vietnam, where the Caribou's startling short-field characteristics enabled it to fly into airstrips which

could not handle the two other major transport aircraft used in the theatre, namely the Fairchild C-23 Provider and Lockheed C-130 Hercules.

With the US withdrawal, a considerable number of Caribous was turned over to the South Vietnamese air force, but some also returned to the USA for service with the Air National Guard and the Air Force Reserve, the last examples of which have only fairly recently been retired.

Capable of carrying up to 32 fully-equipped troops, 24 paratroops or a 3048-kg (6,720-lb) cargo payload, the Caribou also proved attractive to a number of overseas air forces and by the time production terminated in 1973 examples of the DHC-4A major production type had been delivered to such countries as Abu Dhabi, Australia, Ghana, India, Malaysia, Spain, Tanzania and Zambia. In addition, it was also purchased by the then Royal Canadian Air Force, which obtained nine as the CC-108, these seeing extensive service with United Nations peacekeeping echelons in the Middle East and Africa.

Production of the Caribou totalled

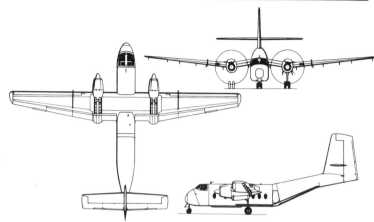

de Havilland Canada DHC-4 Caribou

approximately 300 aircraft, many of which are still active today.

Specification
de Havilland DHC-4A Caribou
Type: STOL tactical transport aircraft
Powerplant: two 1082-kW (1,450-hp) Pratt & Whitney R-2000-D5 Twin Wasp radial piston engines
Performance: cruising speed 293 km/h (182 mph) at 2285 m

(7,500 ft); service ceiling 7560 m (24,800 ft); range 390 km (242 miles) with maximum payload
Weights: empty 8283 kg (18,260 lb); maximum take-off 12298 kg (28,500 lb)
Dimensions: span 29.15 m (95 ft 7.5 in); length 22.12 m (72 ft 7 in); height 9.67 m (31 ft 9 in); wing area 84.7 m² (912 sq ft)
Operators: Australia, Cameroun, India, Malaysia, Mauretania, Spain, Thailand, US, Zambia

de Havilland Canada DHC-5 Buffalo

Fourth in the remarkable line of STOL aircraft to emerge from the de Havilland Canada stable, the de Havilland Canada DHC-5 Buffalo began life in response to a 1962 US Army request for proposals for a new 41-seat tactical transport. Originally known as the Caribou II, the DHC project was just one of the 25 submissions but emerged victorious and was subsequently rewarded with a contract for four evaluation specimens, the first of which made its maiden flight during April 1964. In the event, the CV-7A (as it was designated in US service) failed to secure any further US Army orders, most probably as a result of that service being forced to dispose of its relatively few large fixed-wing aircraft to the US Air Force at the beginning of 1967.

Despite this unfortunate beginning, the Buffalo has since enjoyed a measure of success, more than 100 being built and production continuing at a rate of around 12 per year at the time of writing. The first noteworthy order originated from the Canadian Armed Forces which acquired 15 DHC-5A aircraft for tactical transport tasks under the designation CC-115, although some have since been adapted for maritime functions. One major source of export orders has been Latin America, the Buffalo's almost unrivalled short-field performance being particularly attractive to both Brazil and Peru which between them obtained no less than 40 before

de Havilland Canada DHC-5 Buffalo of the Togo air force.

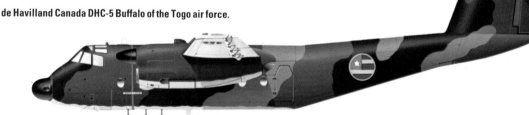

Canada operates the Buffalo for various transport duties. In CAF service it is designated CC-115.

production terminated on 1972.

In the event this hiatus proved to be only temporary, DHC endeavouring to secure an order from India in 1974 with a couple of proposals based on the use of different powerplants. These efforts proved unsuccessful, but several other potential customers had appeared on the scene and the line was therefore reopened in 1974, the variant now in production being the DHC-5D which possesses rather better performance. Since then, it has remained open, customers which have purchased the type in recent years including Egypt, Ecuador, Kenya, Mauritania, Mexico, Oman, Sudan, Tanzania, Zaire, and Zambia. This model can carry 8164 kg (18,000 lb)

of freight, or 41 troops, or 34 paratroops, or 24 litters plus six attendants.

Specification
de Havilland Canada DHC-5D Buffalo
Type: STOL tactical transport aircraft
Powerplant: two 2336-kW (3,133-shp) General Electric CT64-820-4 turboprops
Performance: cruising speed 420 km/h (261 mph) at 3050 m (10,000 ft); range 1112 km (691

miles) with maximum 8164-kg (18,000-lb) payload
Weights: empty 11412 kg (25,160 lb); maximum take-off 22316 kg (49,200 lb)
Dimensions: span 29.26 m (96 ft 0 in); length 24.08 m (79 ft 0 in); height 8.73 m (28 ft 8 in); wing area 87.8 m² (945 sq ft)
Operators: Brazil, Burma, Cameroun, Canada, Ecuador, Egypt, Ethiopia, Mexico, Peru, Sudan, Tanzania, Togo, UAE, US, Zaire, Zambia

de Havilland Canada DHC-6 Twin Otter

DHC had in the early 1960s fitted two PT6 turboprops to a basic DHC-3 Otter airframe. But the new 20-seat DHC-6 Twin Otter, while retaining the same fuselage cross-section, involved an almost complete redesign. The span was extended, struts moved inboard, nosewheel landing gear fitted, doors enlarged, cockpit revised and fin area increased. The first development aircraft flew on 20 May 1965. Orders built up slowly for the production Series 100, and there was a steady flow of military and paramilitary orders.

Among the current military users are the Canadian Armed Forces, with eight examples designated CC-138s. These are used for search and rescue, and utility duties. The US Army also has two examples, designated UV-18As, for service with the Alaska National Guard.

After 110 Series 100 aircraft had been delivered, DHC introduced the long-nosed Series 200. This aircraft has a much-improved baggage volume and 115 were built. The short nose remained a standard production fit on float-equipped aircraft.

The Series 300 introduced more power and more payload/range.

Twin Otters are available with a fire-bombing tank attached to the belly. A 3.05-m (10-ft) ventral pod is an optional extra for carrying up to 272 kg (600 lb) of cargo or baggage. Like other DHC designs, the Twin Otter can operate from wheels, skis or floats. When floats are fitted small fins are added to the horizontal tailplane to maintain directional stability. DHC expects the Twin Otter will remain in production for some years and in 1979 began negotiations with the Chinese government about possible licence production.

Specification
Type: light STOL utility transport
Powerplant: (Srs 100) two 430-kW (578-ehp) Pratt & Whitney Aircraft of Canada PT6A-20 turboprops; (Srs 300) two 486-kW (652-ehp) Pratt & Whitney Aircraft of Canada PT6A-27 turboprops
Performance: (Srs 100 at maximum take-off weight) maximum cruising speed 297 km/h (184 mph); range with 977-kg (2,150-lb) payload 1344 km (835 miles); (Srs 300 at maximum take-off weight) maximum cruising speed 338 km/h (210 mph); range with 1160-kg (2,550-lb) payload 1435 km (892 miles)
Weights: (landplane Srs 100) empty 2653 kg (5,850 lb); maximum take-off and landing 4763 kg (10,500 lb); maximum zero fuel 4603 kg (10,150 lb); (landplane Srs 300) empty 3363 kg (7,415 lb); maximum take-off 5700 kg (12,500 lb); maximum landing 5579 kg (12,300 lb); maximum zero fuel 5304 kg (11,695 lb)
Dimensions: (landplane) span 19.81 m (65 ft 0 in); (Srs 100) length 15.09 m (49 ft 6 in); (landplane Srs 300) length 15.77 m (51 ft 9 in); height 5.94 m (19 ft 6 in); wing

Two de Havilland Canada Twin Otters are flown for parachuting training by the US Air Force Academy. These are known as the UV-18B. Two UV-18As serve with the Alaskan Army National Guard.

area 39.02 m^2 (420 sq ft)
Armament: none
Operators: Argentina, Botswan, Canada, Chile, Ecuador, Ethiopia, France, Gabon, Jamaica, Norway, Panama, Peru, Senegambia, Sudan US

de Havilland Canada DHC-7R Ranger Dash-7

De Havilland Canada began project studies of a 40-50 seat STOL airliner in the late 1960s when enthusiam in inter-city STOL operations was at its height. The airlines remained sceptical, however, and worries about noise and pollution were compounded by the 1973 oil crisis. While other aircraft manufacturers rapidly lost interest, DHC pressed ahead and the 50-seat pressurized Dash-7 made its first flight on 27 March 1975. Development costs, paid for largely by the Canadian government, DHC and major equipment suppliers, were $120 million.

Airline orders for such a costly and specialized vehicle have been slow to materialize, but DHC remains confident that sales will build up in a similar manner to other DHC designs. In order to broaden the market, DHC has examined a number of developments including the DHC-7R Ranger maritime patrol aircraft. The Dash-7R has an increased fuel tankage to push endurance to 10-12 hours. A belly-mounted search radar provides 360° scan, and the package of electronics and reconnaissance equipment would be chosen to match the requirements of the customer. The Ranger can double as a transport with 26 seats in the rear of the cabin without any of the reconnaissance equipment having to be removed. Like the civil aircraft, the Ranger can be fitted with a freight door measuring 1.778 m (5 ft 10 in) by 2.31 m (7 ft 7 in).

The Dash-7 is unique amongst airliners in being able to operate with a full payload from grass runways as short as 701 m (2,300). The Dash-7 Ranger can use short, semi-prepared airstrips close to an area of military interest, and this is claimed to give it a distinct advantage over the more traditional maritime reconnaissance aircraft. Good take-off performance also means it can operate from hot-and-high airfields in remote areas.

Specification
Type: (Dash-7) STOL tactical transport; (Ranger) STOL maritime reconnaissance aircraft
Powerplant: four 835-kW (1,120-shp) Pratt & Whitney Aircraft of Canada PT6A-50 turboprops

Performance: (Dash-7) maximum cruising speed at 2440 m (8,000 ft) and with 18600 kg (41,000 lb) of payload 436 km/h (271 mph); maximum range with 50 passengers and full reserves 1303 km (810 miles); (Ranger) maximum cruising speed at sea level and at maximum take-off weight 432 km/h (268 mph); patrol endurance 9 hours 30 minutes
Weights: (Dash-7) empty 12178 kg (26,850 lb); maximum take-off 19731 kg (43,500 lb); maximum landing 18824 kg (41,500 lb); maximum zero fuel 17690 kg (39,000 lb); (Ranger) empty 12927 kg (28,500 lb); maximum take-off (20411 kg (45,000 lb); maximum landing 18597 kg (41,000 lb); maximum zero fuel 17690 kg (39,000 lb)

This de Havilland Canada CC-132 Dash Seven is on the strength of No. 412 Sqn CAF, based at Lahr in West Germany. These are used for liaison and transport around the CAF bases in Europe.

Dimensions: span 28.35 m (93 ft 0 in); length 24.58 m (80 ft 8 in); height 7.98 m (26 ft 2 in); wing area 79.9 m^2 (860 sq ft)
Armament: unarmed, but Ranger is equipped for flare and life-raft dropping
Operators: (Dash-7) Canadian Armed Forces, Venezuela

Dornier Do 27

The Dornier Do 27 was the first aircraft to enter production in Germany after World War II. Claudius Dornier recommenced activities in Spain in 1949, Oficinas Tecnicas Dornier working closely with the Spanish CASA. The initial fruits of this collaboration were evident with the first flight of the Do 25 in June 1954. Prepared to meet a Spanish air

ministry specification, the STOL transport was powered by a single 112-kW (150-hp) ENMA Tigre engine; 50 similar aircraft subsequently appeared under the designation CASA C-127.

Developed from this, the prototype Do 27 was flown on 8 April 1955. Production took place in Germany at Dornier-Werke, the first

example flying in October 1956. With a large 'wraparound' windscreen and generous five-seat layout, the Do 27A proved popular. Deliveries began at 20 aircraft per month.

The main military Do 27A and dual-control Do 27B differed little. The strutless, high wing provided ease of access for loading passengers

or freight. Large flaps gave an amazing STOL capability. By far the largest user was the Federal German Republic, with well over 400. Another early customer was the Swiss *Flugwaffe*, whose initial seven aircraft sported a wheel-and-ski undercarriage. A prototype float-plane, the Do 27S, was built and flown; another was re-engined with

Dornier Do 27 (continued)

the Turboméca Astazou turboprop. Production of the standard Do 27 exceeded 600 units in all its subversions before the line closed in 1965.

Specification
Type: STOL liaison and utility transport
Powerplant: one 205-kW (275-hp) Lycoming GO-480 air-cooled piston engine
Performance: maximum speed at 1000 m (3,280 ft) 250 km/h (155 mph); cruising speed at 1000 m (3,280 ft); 205 km/h (127 mph); cruising range 870 km (540 miles); service ceiling 5500 m (18,400 ft);

climb to 1000 m (3,280 ft) 2 minutes 36 seconds; take-off to clear 15-m (50-ft) obstacle (maximum fuel, no wind) 160 m (558 ft); landing from 15 m (50 ft) 170 m (525 ft)
Weights: empty 983 kg (2,167 lb); loaded 1570 kg (3,460 lb)
Dimensions: span 12.0 m (39 ft 4½ in); length 9.54 m (31 ft 4 in); height 3.28 m (8 ft 10¾/4⅓in); wing area 19.4 m² (208.8 sq ft)
Armament: none
Operators: 14 countries including Israel, Spain, Switzerland, West Germany

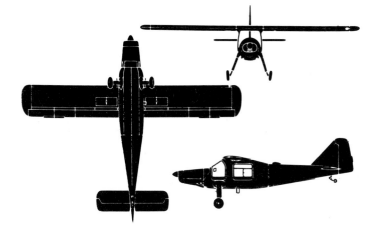

Dornier Do 27

Dornier Do 28D Skyservant

In the early 1950s Dipl-Ing Claudius Dornier established design offices in Madrid. Under the auspices of the Spanish aircraft manufacturer CASA, prototypes of the Do 25 and Do 27 STOL transports were flown in 1954 and 1955 respectively.

When the postwar embargo on aircraft manufacture in Germany was lifted, production of the Do 27 was transferred there, deliveries commencing in late 1956. Some 30 months later came the first flight of the twin-engined Do 28. To retain the STOL qualities of an aerodynamically clean high wing, the 190-kW (225-hp) Lycoming engines were mounted in stub-wings, flanking the six-seat cabin. Some 120 examples of the Do 28A and B were built.

The prototype Do 28D Skyservant flew on 23 February 1966. Its box-like fuselage seated 12 passengers, and 283-kW (380-hp) engines were installed.

The production Do 28D-1 was later fitted with wheel spats, wing fences and a large dorsal spine. These first four were the first to be delivered to the Luftwaffe, as VIP transports, in 1970.

By this time 101 had been allocated to the Luftwaffe as communications transports, a further 20 going to the Marineflieger. Skyservant roles include photographic survey, ambulance and para-dropping duties. Its robust construction and a take-off run of the order of 280 m (920 ft) has made its suitable for operation in Africa and other harsh environments. Most Skyservants exported are Do 28D-2s, lengthened by 15 cm (6 in), with larger fuel tanks in the engine nacelles and aerodynamic improvements to wing and tailplane. Dual controls became standard, and new landing lights were installed in the detachable fibreglass wingtips, which have

anhedral. Freight doors may be replaced by a sliding door for supply-dropping. Additional stores or 250-litre (55-gallon) fuel tanks may be attached to underwing hardpoints.

A Skyservant with turboprops flew in April 1978. Dubbed the TurboSky, the Do 28D-5 has two 298-kW (40-shp) Avco Lycoming LTP 101-600s. Turboprop power and reliability, combined with a near aerobatic manoeuverability and STOL capability, may attract orders from several air arms.

Specification
Type: light passenger, freight and liaison transport
Powerplant: two 283-kW (380-hp) Lycoming IGSO-540 piston engines

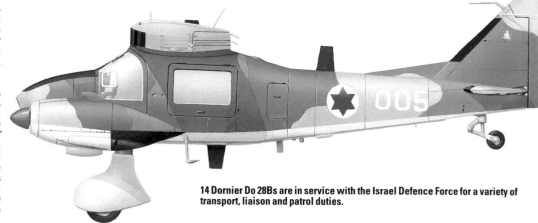

14 Dornier Do 28Bs are in service with the Israel Defence Force for a variety of transport, liaison and patrol duties.

Performance: maximum speed at 3050 m (10,000 ft) 323 km/h 201 mph); cruising speed 230 km/h (143 mph); range with 12 passengers 800 km (497 miles); range with 1360-kg (3,000 lb) payload 200 km (125 miles); service ceiling 7300 m (24,000 ft)
Weights: empty 2095 kg (4,615 lb); maximum loaded 3647 kg (8,040 lb)
Dimensions: span 15.0 m (49 ft 2½ in); length 11.4 m (37 ft 4¾ in); height 3.9 m (12 ft 9½ in); wing

area 28.06 m² (302 sq ft)
Armament: none
Operators: Cameroon, Ethiopia, Israel, Kenya, Malawi, Nigeria, Somalia, Thailand, West Germany, Zambia

Both the Luftwaffe and Marineflieger use the Dornier Do 28D for light transport, communications, paradropping and survey work. This example is from MFG 5 based at Kiel.

Douglas A-3 Skywarrior

The largest and heaviest aircraft designed for operation from an aircraft carrier when the Douglas El Segundo division's project design was completed in 1949, the A3D Skywarrior originated from a US Navy requirement of 1947. An attack-bomber with strategic strike capability was envisaged, tailored to the giant new aircraft-carriers that were ultimately (after prolonged opposition from the USAF) to materialize as the 'Forrestal' class of four ships, as it was believed that the moment had come to exploit the potential of the rapidly-developing gas turbine engine.

The Douglas design was that of a high-wing monoplane, with retractable tricycle landing gear, two podded turbojets beneath the wing, and a large internal weapons bay to accommodate up to 5443 kg (12,000 lb) of varied weapons. The wings were swept back 36° and had high aspect ratio for long range, all tail surfaces were swept, and the outer wing panels and vertical tail folded.

The first of two prototypes made its maiden flight on 28 October 1952, powered by 3175-kg (7,000-lb) Westinghouse XJ40-WE-3 engines, but the failure of this engine programme meant that the 4400-kg (9,700-lb)

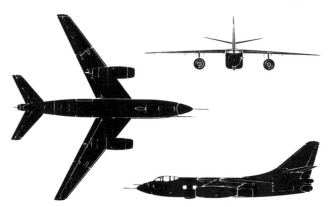

Douglas TA-3B Skywarrior

thrust Pratt & Whitney J57-P-6 powered the production A3D-1. The first of the A3D-1s flew on 16 September 1953, and deliveries to the US Navy's VAH-1 attack squadron began on 31 March 1956.

In 1962 the designation was changed to A-3, the initial three-seat production version becoming A-3A. Five of these were modified subsequently for ECM missions under the designation EA-3A. A-3Bs which entered service in 1957 had more powerful J57-P-10 engines and a flight-refuelling probe. A reconnaissance variant with cameras in the weapons bay was designated RA-3B, and EA-3B identified ECM aircraft with a four-man crew in the weapons bay. Other designations include 12 TA-3B trainers for radar operators, one VA-3B executive transport, and the final variants in US Navy service were KA-3B flight-refuelling tankers and 30 EKA-3B tanker/countermeasures/strike air-

craft. The EA-3 remains in service with VQ-1 and VQ-2 in 1985.

Specification
Type: carrier-based attack bomber
Powerplant: (A-3B) two 4763-kg (10,500-lb) thrust Pratt & Whitney J57-P-10 turbojets
Performance: (A-3B) maximum speed 982 km/h (610 mph) at 3050 m (10,000 ft); range 1690 km (1,050 miles)
Weights: (A-3B) empty 17876 kg (39,409 lb); maximum take-off 37195 kg (82,000 lb)
Dimensions: span 22.10 m (72 ft 6 in); length 23.27 m (76 ft 4 in); height 6.95 m (22 ft 9½ in); wing area 75.43 m² (812 sq ft)
Armament: two 20-mm guns in radar-controlled rear turret, and up to 5443 kg (12,000 lb) of assorted weapons in internal weapons bay; (EKA-3B) none
Operator: US Navy

The only variant of the Skywarrior in current front-line service is the EA-3B, a specialist Elint platform which, although land-based (at Rota and Guam), can operate from carriers as and when required.

Douglas DC-3/C-47 Skytrain/Skytrooper

Probably the most famous civil transport in aviation history, the Douglas DC-3 has achieved a reputation in military usage which is if anything even greater. During World War II these aircraft were to be found wherever there were Allied military services, and General 'Ike' Eisenhower was to comment that, in his opinion, the C-47 was one of the four major tools of war which had contributed to Allied victory.

In 1935 Douglas had flown the prototype of a Douglas Sleeper Transport which led to a 21-seat 'day-plan' designated DC-3. The latter was soon in wide use with US and world airlines. It was the most advanced airliner then in service, offering new standards of comfort, safety and reliability. The last factor is significant, but the DC-3 was not just reliable: partly by chance it has proved also to be enduring and fatigue-free, and 50 years after that first flight, many remain in both civil and military use.

The US Army's interest in the new generation of civil transports evolved by the Douglas Aircraft Company began with the purchase of a single DC-2 from FY 1936 funds. This was designated C-32 when it entered service, and was followed by C-33s with a cargo-loading door; C-34s with a passenger door and different interior; a single C-38 with DC-3

type tail unit; C-39s with more powerful engines; and single examples of the C-41 and C-42 with more powerful engines, C-41A with a *de luxe* interior for 23 passengers, and two extra C-42s converted from C-39s.

The potential of these aircraft had convinced the US Army of their excellence of design and construction, and a study of the DC-3 enabled the US Army to outline to Douglas the modifications required for its use as a military transport. These included more powerful engines, strengthening of the rear fuselage and cabin floor, and the provision of large loading doors. The airline-type interior disappeared, replaced by utility seats lining the cabin walls. Powerplant of the initial production version, and of most subsequent production, comprised two 895-kW (1,200-hp) Pratt & Whitney R-1830-92 Twin Wasp radial engines. Ordered in large numbers in 1940, these aircraft became designated C-47 and acquired the name Skytrain.

After a distinguished war record, many were dispersed around the world, where many still serve.

Specification
Type: basically troop, personnel and cargo transport
Powerplant: (C-47) two 895-kW (1,200-hp) Pratt & Whitney R-1830-

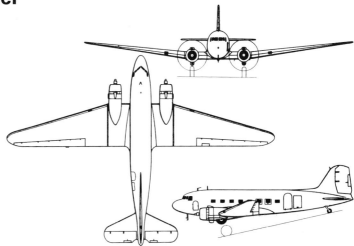

Douglas C-47 Skytrain

92 Twin Wasp 14-cylinder two-row radial piston engines
Performance: (C-47) maximum speed 370 km/h (230 mph); time to 3050 m (10,000 ft) 9 minutes 36 seconds; range 2414 km (1,500 miles)
Weights: (C-47) empty 8255 kg (18,200 lb); maximum take-off 11793 kg (26,000 lb)
Dimensions: span 29.11 m (95 ft 6 in); length 19.43 m (63 ft 9 in); height 5.18 m (17 ft 0 in); wing area

91.69 m² (987 sq ft)
Armament: AC-47D gunships were equipped with General Electric 7.62-mm (0.3-in) Miniguns, firing through the fuselage door aperture and right-hand windows
Operators: about 50 air forces are still equipped with military transports of the C-47 type

Douglas DC-6/C-118 Liftmaster

The DC-6 was little more than a larger pressurized DC-4 with more powerful engines. It flew on 15 February 1946. Civil DC-6s were built alongside 166 aircraft for the USAF and US Navy. The C-118A could accommodate 74 passengers, or 12247 kg (27,000 lb) of cargo, or 60 stretcher cases. Other military services have acquired DC-6s, the majority of them ex-civil, and a few continued in service in 1985.

Specification
Douglas C-118A
Type: transport aircraft
Powerplant: four 1864-kW (2,500-hp) Pratt & Whitney R-2800-52W Double Wasp radial piston engines
Performance: maximum speed 573 km/h (356 mph) at 5975 m (19,600 ft); cruising speed 504 km/h (313 mph) at 6220 m (20,400 ft);

normal range 6148 km (3,820 miles); ferry range 7419 km (4,610 miles)
Weights: empty 23358 kg (51,495 lb); maximum take-off 44089 kg (97,200 lb)
Dimensions: span 35.81 m (117 ft 6 in); length 30.66 m (100 ft 7 in); height 8.66 m (28 ft 5 in); wing area 135.36 m² (1,457.0 sq ft)
Armament: none
Operators: Chile, Colombia, Ecuador, Mexico, Paraguay, Zaïre

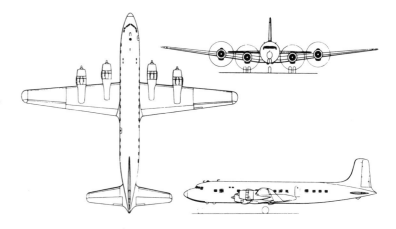

Douglas C-118 Liftmaster

141

Douglas DC-8

Only three air arms are known to have operated DC-8s in a military capacity. Of these France is the most significant, having used seven different aircraft, of which four are still on strength; three DC-8-72Fs and a DC-8-33. The Series 72s serve as long-range staff transports with the 60th transport wing from Paris-Roissy. The other aircraft is an Elint platform assigned to Escadrille Electronique 51 at Evreux.

Peru has two DC-8-62CFs assigned to the Escuadrilla Presidencial, and Spain uses two DC-8-52s for long-range airlift.

France's ET 60 uses the DC-8 for VIP transport, three of the aircraft having been re-engined. One other aircraft serves with EE 51 in the Elint role.

Specification
McDonnell Douglas DC-8 Series 72
Type: VIP/staff transport
Powerplant: four 10886-kg (24,000-lb) thrust CFM International CFM56-2-C1 turbofan engines

Performance: economical cruising speed 855 km/h (531 mph); range in still air with maximum payload 11619 km (7,220 miles)
Weights: empty 69218 kg (152,600 lb); maximum take-off 151953 kg (335,000 lb)

Dimensions: span 45.24 m (148 ft 5 in); length 47.98 m (157 ft 5 in); height 12.93 m (42 ft 5 in); wing area 271.92 m² (2,927.0 sq ft)
Armament: none
Operators: France, (other series) Peru, Spain

EH Industries EH 101

The most powerful helicopter yet designed in Western Europe, the EH 101 has its genesis in an SKR (Sea King Replacement) study by the British MoD (Navy) in 1977. Westland responded with a proposal designated WG.34, but meanwhile the Italian navy had come up with a similar requirement, though one with the accent on shore basing rather than operations from warships. Westland and Agusta decided to collaborate, formed EHI (Elicotteri Helicopter Industries Ltd) and began design of the EH 101 as the next-generation helicopter for the British and Italian navies. The same basic machine is also being developed as a civil and military passenger and utility cargo transport. In configuration the EH 101 is like a scaled-down CH-53E, though the three engines are arranged symmetrically around the main rotor. The main cabin is 6.50 m (21 ft 3.9 in) long, 2.50 m (8 ft 2.4 in) wide and, allowing for interior soundproofing, 1.82 m (5 ft 11.7 in) high, all dimensions exactly tailored to the missions, as are the folded dimensions for shipboard stowage.

The maritime roles for which the EH 101 has been designed are ASW, anti-ship surveillance, anti-surface vessel strike, amphibious operations, SAR, AEW and vertrep (vertical replenishment). An unusual feature is that, despite the striking differences between the missions, for example between cargo vertrep, ASW and AEW with a giant radar, all role changes can be carried out on board ship. The main five-blade rotor has power-folding composite blades with extended-chord tips of the BERP type as fitted to the latest Lynx and Westland 30. The tail carries a four-blade rotor and can also be power-folded. Use of three engines confers outstanding safety for flight at maximum weight in all weathers, and General Electric expect the same basic engine to mature at powers some 50 per cent greater than that given in the specification. The Rolls-Royce Turboméca RTM 322 is an alternative. First flight has been set for 1985, with production machines following in 1988. The EH 101 has been specified as equipment for the Royal Navy Type 23 frigates, and it is expected to replace the Sea

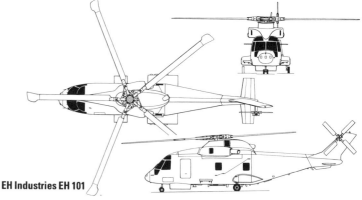

EH Industries EH 101

King all over the world.

Specification
EH 101 Naval
Type: multi-role maritime helicopter
Powerplant: three 1289-kW (1,729-shp) General Electric T700-401 turboshaft engines
Performance: (estimated) maximum cruising speed 278 km/h (173 mph); endurance at distant station with full weapon load (dunking ASW role) 5 hours
Weights: (design) maximum take-

off 13000 kg (28,660 lb); transport version 14290 kg (31,504 lb)
Dimensions: main rotor diameter 18.59 m (61 ft 0 in); length overall 22.90 m (75 ft 1.6 in), and folded 15.85 m (52 ft 0 in); height 6.50 m (21 ft 3.9 in); main rotor disc area 271.42 m² (2,921.69 sq ft)
Armament: enclosed weapons bay able to accommodate a wide range of torpedoes (Marconi Stingray for RN version) and other stores; total disposable load 6085 kg (13,415 lb)
Operators: none in 1985

EMBRAER EMB-110 and EMB-111

This best-seller was designed to replace the Beech 18 family in the Força Aérea Brasileira. Directed by French designer Max Holste, the prototype was flown on 26 October 1968. Entry to the cabin is through an airstair door ahead of the wing, with a cargo door in the rear fuselage. The EMB-110 entered production in 1972. It became an important commuter aircraft, but the FAB launched the military C-95 (EMB-110P1K). This has a 1.8 m (5 ft 10.9 in)-wide hydraulically powered cargo door incorporating an inward-opening paratroop door, hard interior trim and military avionics. The FAB also purchased the maritime surveillance P-95 (EMB-111) with uprated PT6A-34 engines, wingtip tanks providing 9-hour endurance, nose radar, wing-mounted searchlight and four outer wing hardpoints for rocket launchers or sonobuoy dispensers. The 12 for the FAB entered service with Brazil's coastal command in April 1978. Six similar EMB-111(N) had been sold to the Chilean navy. For SAR Brazil uses the SC-95B (EMB-110P1K). The photographic R-95 (EMB-110B) has a ventral hatch and extended pannier with a Zeiss RMK A8.5/23 camera. Four EC-95 (EMB-110A) check airfield navigation aids. The cargo C-95A has an 0.85-m (2 ft 9.5-in) fuselage plug and ventral fin.

In Brazilian air force service, the transport version of the Bandeirante is designated C-95.

Specification
EMB-110
Type: 16-passenger transport, utility and surveillance aircraft
Powerplant: two 507-kW (680-shp) Pratt & Whitney Canada PT6A-27 turboprop engines
Performance: maximum speed 452 km/h (281 mph); maximum

cruising speed 341 km/h (212 mph); initial rate of climb 442 m (1,450 ft)/minute; range with reserves 2038 km (1,266 miles)
Weights: empty 3200 kg (7,055 lb); maximum take-off 5300 kg (11,684 lb)
Dimensions: span 15.33 m (50 ft

3.5 in); length 14.22 m (46 ft 7.8 in); height 4.73 m (15 ft 6.2 in); wing area 29.00 m² (312.16 sq ft)
Armament: (EMB-111) four underwing rocket launchers
Operators: Brazil, Chile, Gabon, Uruguay

The Brazilian navy uses the P-95A for maritime patrol duties.

EMBRAER EMB-121 Xingu

EMBRAER's EMB-121 Xingu owes much to the EMB-110 but is, in fact, the first of the company's -12 series of light pressurized twins. The EMB-120 Argauiai and EMB-123 Tapajos are pressurized commuter airliners, and the EMB-121 Xingu has a short fuselage for nine passengers and air-stair door. Wings are similar to those of the Bandeirante but are reduced in span; the large fin is swept and has the tailplane on top. Full de-icing is provided.

The prototype flew on 10 October 1976. Initial demand came from the Brazilian air force which ordered five as VIP transports. Designated VU-9, these were delivered to the Grupo de Transporte Especial in 1978. The VU-9 seats five; the two-crew cockpit is screened off and re-sembles that of the C-95 Bandei-rante. standard avionics include dual RCA VHF, Collins ADF and Sunair HF, Sperry SPZ-200 auto-pilot and Bendix Weathervision

radar. The chief customer has been France, which has bought 25 for the Armée de l'Air and 16 for the Aéro-navale as MEPTs (multi-engine pilot trainers) and liaison aircraft.

Specification
Type: five/nine-seat light transport
Powerplant: two 507-kW (680-shp) Pratt & Whitney Canada PT6A-28 turboprop engines
Performance: maximum cruising speed 450 km/h (280 mph) at 3350 m (10,990 ft); economic cruising speed 365 km/h (227 mph) at 6100 m (20,015 ft); initial rate of climb 580 m (1,905 ft)/minute; service ceiling 7925 m (26,000 ft); range with 780 kg (1,720 lb) payload and 45 minutes reserve 2270 km (1,411 km)
Weights: empty 3620 kg (7,981 lb); maximum take-off 5670 kg (12,500 lb)
Dimensions: span 14.05 m (46 ft 1.1 in); length 12.25 m (40 ft 2.3 in);

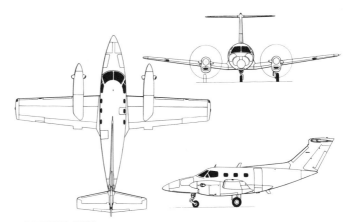

EMBRAER EMB-121 Xingu

height 4.84 m (15 ft 10.6 in); wing area 27.50 m^2 (296.02 sq ft)
Armament: none
Operators: Brazil, France

The 16 Xingus of the Aéronavale have dual multi-engined trainer and liaison roles.

EMBRAER AT-26 Xavante

EMBRAER AT-26 Xavante of the Brazilian air force. Brazilian-built MB.326s have been exported to a number of South American countries.

Under a licence agreement of May 1970, the Brazilian manufacturer EMBRAER has produced 182 Aer-macchi MB.326GC trainer and light attack aircraft. The dual role capa-bility is reflected in the Brazilian air force designation AT-26 Xavante (an Amazonian tribe). Home orders were placed for 168, the first two Ita-lian-built, the last being handed over in February 1983. By July 1983 41 had been lost in accidents. The 4° Grupo de Aviacão de Caca at For-teleza has two squadrons, of which the first (1° Esquadrão) is an OCU; 10° Grupo de Aviacão has three com-ponent squadrons: 1° Esq at Santa Maria for reconnaissance, 2° Esq at Campo Grande for COIN, and 3° Esq at an undisclosed base. Schools oper-ating the Xavante are the Academia da Força Aérea at Pirassununga; Centre de Aplicacão Tática e Recom-plementacão de Equipagens (Tactic-al Conversion and Crew Training Centre) at Natal; and Centro de For-maçao de Pilotos Militares (Military

Pilots' Finishing Centre), also at Natal.

The African state of Togo received three Xavantes in December 1976 and three in 1978, and Paraguay took delivery of nine in May 1980, plus an attrition replacement in January 1982. In 1983 12 surplus Brazilian Xavantes were supplied to the Argentine navy's 1a Escuadrilla Aeronavale de Ataque of 4a Escua-dra at Punta del Indo, and 14 were promised to Ecuador.

Specification
EMBRAER EMB-326/AT-26 Xavante
Type: two-seat trainer and light attack aircraft
Powerplant: one 1547-kg (3,410-lb) thrust Rolls-Royce Viper 540 turbojet engine
Performance: as for MB.326
Weights: (attack configuration) empty 2558 kg (5,640 lb); maximum take-off 5216 kg (11,500 lb)
Dimensions: span over tip-tank

10.85 m (35 ft 7.2 in); length 10.64 m (34 ft 10.9 in); height 3.72 m (12 ft 2.5 in); wing area 19.35 m^2 (208.3 sq ft)
Armament: combinations of light bombs, 7.62-mm (0.3-in) gun pods; LM-70/7, LM-70/19 or LM-37/36 rocket pods; or camera pods on six underwing strongpoints
Operators: Argentina, Brazil, Ecuador, Paraguay, Togo

EMBRAER EMB-312 Tucano

A turboprop basic trainer which set a trend in having ejection seats (Martin-Baker BR8LC), the Tucano (toucan) was produced to replace the Cessna T-37 and has the Força Aérea Brasileira designation T-27. Design began in January 1978, to FAR Pt 23 standard, and the aircraft first flew on 16 August 1980. In late

1980 the FAB ordered 118, plus an option on 50. T-27s entered service on 29 September 1983. Six have a scarlet, black and white colour scheme for the national aerobatic team, the Esquadrilha da Fumaca. The 1st air instruction squadron of the Academia da Força at Pirassu-nunga began student training in

February 1984.

Honduras received eight Tucanos in spring 1984, with COIN provi-sions. In October 1983 Egypt signed for 40, plus an option on 80, the first 10 supplied by Brazil in December 1984. Assembly from decreasingly complete kits is at the AOI factory at Kadar, with test flying at Helwan

from 1985 onwards. The second batch of 80, divided equally between Egypt and Iraq, will see the AOI manufacturing 70 per cent of the air-

As well as providing basic training in a type which is close in format to military jets, the Tucano can be used for attack and COIN, armed with rocket pods.

frame; deliveries will be complete in 1987. In 1984 the Tucano was offered to the RAF for Air Staff Target 412, arrangements being made for Shorts to build all except the first 25 of the 130 (plus 15 options) needed. The RAF specification required a ventral airbrake, provision for increased manoeuvre load factors, an extended fatigue life, landing gear for a 3.96 m (13 ft)/second sink rate, controls compati-ble with the BAe Hawk and a fixed windscreen with separate low-distortion canopy for birdstrike resistance. In early April 1985 it was announced that this submission by EMBRAER/Shorts had been selected for the RAF's requirement; the first of the initial 130 are due for delivery in 1986. Named Shorts Tucano, this model will be powered by the 820-kW (1,100-shp) Garrett TPE331-12B turboprop engine part-manufactured and assembled in the UK by Rolls-Royce.

Specification
EMBRAER T-27 Tucano
Type: trainer and COIN aircraft
Powerplant: one 559-kW (750-shp) Pratt & Whitney Canada PT6A-25C turboprop engine flat rated at 436 kW (585 shp)
Performance: maximum speed at 3050 m (10,000 ft) 433 km/h (269 mph); initial rate of climb 664 m (2,180 ft)/minute; service ceiling 7315 ft (24,000 ft); range at 6095 m (20,000 ft) with fuel reserves 1844 km (1,146 miles)
Weights: empty 1810 kg (3,990 lb); maximum take-off, external load 3175 kg (7,000 lb)
Dimensions: span 11.14 m (36 ft 6.6 in); length 9.86 m (32 ft 4.2 in); height 3.40 m (11 ft 1.9 in); wing area 19.40 m^2 (208.8 sq ft)

ENAER T-35 Pillan

To replace Beech T-34 Mentors the Fuerza Aérea de Chile co-operated with Piper, who designed a new trainer derived from the Saratoga SP and Arrow. The two Piper prototypes were transferred to Santiago in January 1982 and three T-35 aircraft were built by the FAC maintenance wing at El Bosque AB. A new company at El Bosque was named Indaer (Industrial Aéronautica de Chile), but in March 1984 this was absorbed into ENAER (Empresa Naçional de Aéronautica) to undertake production of the T-35 Pillan and the CASA C-101. An initial FAC production order for 80 T-35s was placed and the Spanish air force then ordered a further 40. Initial deliveries were made in January 1985, with CASA assembling the Spanish machines. It is expected that the T-35 will be developed for armament training and possibly light ground attack, with rocket pods, 113-kg (250-lb) bombs or 12.7-mm (0.5-in) guns.

Specification
Type: two-seat basic trainer
Powerplant: one 224-kW (300-hp) Lycoming AEIO-540-H1K5 flat-six piston engine
Performance: maximum speed 311 km/h (193 mph); maximum cruising speed 298 km/h (185 mph); initial rate of climb 465 m (1,525 ft)/minute; service ceiling 5820 m (19,095 ft); maximum range 1334 km (829 miles)
Weights: empty 833 kg (1,836 lb); maximum take-off 1315 kg

(2,899 lb)
Dimensions: span 8.81 m (28 ft 10.9 in); length 7.97 m (26 ft 1.8 in); height 2.34 m (7 ft 8.1 in); wing area 13.64 m^2 (146.82 sq ft)
Armament: none
Operators: Chile, Spain (on order)

The Pillan is basically a Piper PA-28 with tandem cockpit. The type has so far been adopted by Chile and Spain, and an armed version may follow.

Fairchild A-10A Thunderbolt II

Produced in reply to an exacting specification of 1967 which called for a hard-hitting close-support aircraft, the Fairchild A-10A Thunderbolt II first flew on 10 May 1972 and was selected by the USAF in preference to the Northrop A-9 after a competitive fly-off. The Thunderbolt's unusual appearance derives from the care taken to enhance its survival prospects over the battlefield and incorporate maximum fire-power. Absorbing much of the centre fuselage is an enormous GAU-8 Avenger seven-barrel 30-mm cannon, the muzzle protruding slightly beneath the nose, which can be fired at rates of 2,100 or 4,200 rounds per minute (much higher than other weapons of this calibre). Although unconventional, the engine location is considered optimum for minimizing hits by ground-fire and has the additional advantage that the wing and tailplane masks the infra-red emissions of exhaust gases and therefore affords some protection against heat-seeking SAMs. An extremely strong airframe, which includes a titanium armour 'bathtub' surrounding the pilot, it has numerous constructional features resistant to battle-damage or conducive to swift repair, such as interchangeable (left or right) flaps, fuselage components, rudders, elevators and main landing gear units. There are two primary hydraulic systems, each with manual back-up, and the landing gear can be extended under gravity if necessary. Well protected electronically by AN/ALQ-119 jamming pods, plus an AN/ALE-40 chaff and flare dispenser, the Thunderbolt carries a Pave Penny laser designation pod on a pylon starboard of the forward fuselage for accurate marking. In prospect is the far more advanced LANTIRN (Low-Altitude Navigation Targeting Infra-Red for Night) pod. In 1976-84 Fairchild delivered 713 production aircraft which are expected to be progressively updated. A private-venture two-seat Thunderbolt N/AW (Night Adverse Weather) has been offered, as yet without success.

Specification
Type: single-seat close-support aircraft
Powerplant: two 4112-kg (9,065-lb) thrust General Electric TF34-GE-100 high bypass ratio turbofan

Fairchild A-10A Thunderbolt II cutaway drawing key

1 Cannon muzzles
2 Nose cap
3 ILS aerial
4 Air-to-air refuelling receptacle (open)
5 Nosewheel bay (offset to starboard)
6 Cannon barrels
7 Rotary cannon barrel bearing
8 Gun compartment ventilating intake
9 L-band radar warning aerial
10 Electrical system relay switches
11 Windscreen rain dispersal air duct
12 Pave Penny laser receiver and tracking pod
13 Windscreen panel
14 Head-up display symbol generator
15 Pilot's head-up display screen
16 Instrument panel shroud
17 Air-to-air refuelling pipe
18 Titanium armour cockpit enclosure
19 Rudder pedals
20 Battery
21 General Electric GAU-8/A 30-mm seven-barrelled rotary cannon
22 Ammunition feed ducts
23 Steering cylinder
24 Nose undercarriage leg strut
25 Nosewheel
26 Nosewheel scissor links
27 Retractable boarding ladder
28 Ventilating air outlets
29 Ladder stowage box
30 Pilot's side console panel
31 Engine throttles
32 Control column
33 McDonnell Douglas ACES 2 ejection seat
34 Headrest canopy breakers
35 Cockpit canopy cover
36 Canopy hinge mechanism
37 Space provision for additional avionics
38 Angle-of-attack probe
39 Emergency canopy release handle
40 Ventral access panels to gun compartment
41 Ammunition drum (1,174 rounds)
42 Ammunition drum armour plating
43 Electrical system servicing panel
44 Ventral fin
45 Spent cartridge-case return chute
46 Control cable runs
47 Avionics compartments
48 Forward/centre fuselage joint bulkhead
49 Aerial selector switches
50 IFF aerial
51 Anti-collision light
52 UHF/TACAN aerial
53 Starboard wing integral fuel tank
54 Wing skin plating
55 Outer wing panel attachment joint strap
56 Starboard fixed wing pylons
57 ALE-37A chaff dispenser pod
58 ALQ-119 electronic countermeasures pod
59 Pitot tube
60 Starboard drooped wing tip fairing
61 Split aileron/deceleron mass balance
62 Deceleron open position
63 Starboard aileron/deceleron
64 Deceleron hydraulic jack
65 Aileron hydraulic jack
66 Control linkages
67 Aileron tab
68 Tab balance weight
69 Slotted trailing edge flaps
70 Outboard flap jack
71 Flap synchronizing shafts
72 Fuselage self-sealing fuel cells (maximum internal fuel capacity 10,700 lb/4853 kg)
73 Fuselage main longeron
74 Longitudinal control and services duct
75 Air conditioning supply duct
76 Wing attachment fuselage main frames
77 Gravity fuel filler caps
78 Engine pylon fairing
79 Pylon attachment joint
80 Starboard intake
81 Intake centre cone
82 Engine fan blades
83 Night/adverse weather two-seater variant
84 Radar pod (forward looking infra-red in starboard pod)
85 Engine mounting struts
86 Nacelle construction
87 Oil tank
88 General Electric TF34-GE-100 turbofan
89 Rear engine mounting
90 Pylon trailing edge fillet
91 Engine exhaust duct
92 Fan air duct
93 Rudder hydraulic jack
94 Starboard tail fin
95 X-band aerial
96 Rudder mass balance weight
97 Starboard rudder
98 Elevator tab
99 Tab control rod
100 Starboard elevator
101 Starboard tailplane
102 Tailplane attachment frames
103 Elevator hydraulic jacks
104 Tailcone
105 Tail navigation light
106 Rear radar warning receiver aerial
107 Honeycomb elevator construction
108 Port vertical tailfin construction
109 Honeycomb rudder panel
110 Rudder hydraulic jack
111 Formation light
112 Vertical fin ventral fairing
113 Tailplane construction
114 Tailplane control links
115 Port engine exhaust duct
116 Tailboom frame construction
117 VHF/AM aerial
118 Fuel jettison
119 VHF/FM aerial
120 Fuel jettison duct
121 Hydraulic reservoir
122 Port engine nacelle attachment joint
123 Cooling system intake and exhaust duct
124 Engine bleed air ducting
125 Auxiliary power unit
126 APU exhaust
127 Engine nacelle access door
128 Air conditioning plant
129 Port engine intake
130 Trailing edge wing root fillet
131 Fuselage bomb rack
132 Inboard slotted flap
133 Flap guide rails
134 Rear spar
135 Flap shroud structure
136 Honeycomb trailing edge panel
137 Outboard slotted flap
138 Port deceleron open position
139 Aileron tab
140 Aileron hinges
141 Port split aileron/deceleron
142 Drooped wing tip fairing construction
143 Port navigation light
144 Honeycomb leading edge panels
145 Wing rib construction
146 Centre spar
147 Leading edge spar
148 Two outer fixed pylons (1,000-lb/453.6-kg capacity)
149 ALQ-119 electronic countermeasures pod
150 ALE-37A chaff dispenser
151 Port mainwheel
152 2,500-lb (1134-kg) capacity stores pylon
153 Main undercarriage leg strut
154 Undercarriage leg doors
155 Main undercarriage leg pivot fixing
156 Port mainwheel semi-recessed housing
157 Pressure refuelling point
158 Undercarriage pod fairing

engines
Performance: maximum speed, clean 706 km/h (439 mph) at sea level; combat speed 704 km/h (438 mph) at 1525 m (5,000 ft) with six 227-kg (500-lb) bombs; initial rate of climb 1829 m (6,000 ft)/minute; operational radius with 1.7 hours over target

area 460 km (285 miles)
Weights: empty 11321 kg (24,959 lb); maximum take-off 22680 kg (50,000 lb)
Dimensions: span 17.53 m (57 ft 6.2 in); length 16.26 m (53 ft 4.2 in); height 4.47 m (14 ft 8 in); wing area 47.01 m² (506.0 sq ft)
Armament: one internal General

Electric GAU-8/A Avenger 30-mm (1.18-in) seven-barrel cannon with 1,174 rounds; eight wing and three fuselage pylons for up to 7257 kg (16,000 lb) of ordnance, including laser-guided bombs and AGM-65 Maverick ASMs
Operator: USA (AF and ANG)

One of the least attractive warplanes, the A-10 'Warthog' is one of the most effective, possessing great agility and strength for operation at low altitude. The gun is the most powerful fitted to an aircraft, and can throw out over 4,000 depleted-uranium rounds per minute. Mavericks, LGBs and cluster bombs make up the A-10's fearsome arsenal.

159 Outer wing panel attachment joint
160 Port wing integral fuel tank
161 Inboard leading edge slat
162 Slat hydraulic jacks
163 Slat endplate
164 2,500-lb (1134-kg) stores pylon

165 3,500-lb (1588-kg) capacity fuselage pylon
166 Bomb ejector rack
167 Mk 82 500-lb (226.8-kg) bombs
168 Rockeye anti-armour cluster bomb
169 600-US Gal (2271-litre) long range ferry tank
170 Mk 84 2,000-lb (907-kg) bomb
171 Maverick air-to-ground missile
172 Paveway 3,000-lb (1360-kg) laser guided bomb

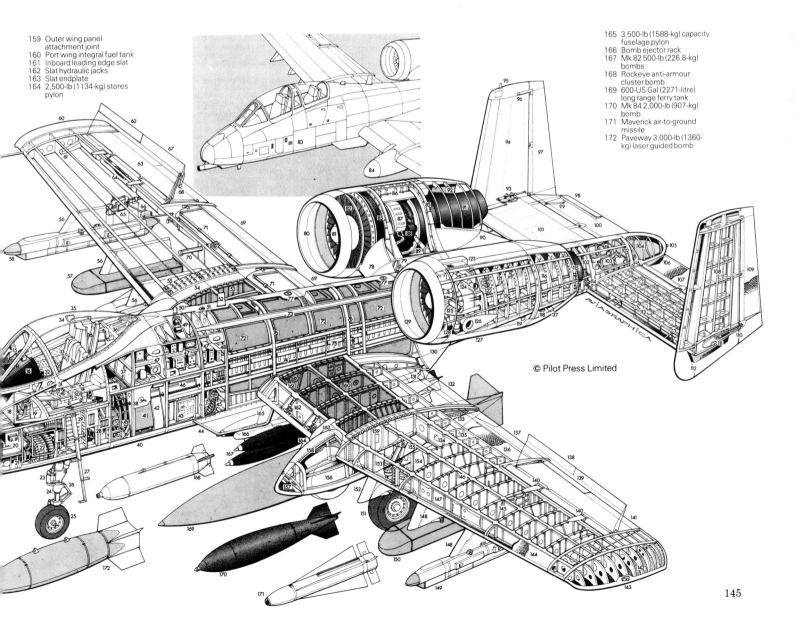

© Pilot Press Limited

Fairchild C-123 Provider

In 1943 the Chase Aircraft Company was founded to undertake the design, development and production of a heavy assault cargo glider for the US Army. Following the successful demonstration of the XCG-18A cargo glider, five YG-18A pre-production examples followed, and one of these was converted to a light assault transport aircraft by the addition of two wing-mounted radial engines, under the designation YC-122. YC-122A/B/C aircraft followed for service trials, leading to the construction of two prototypes of an even larger troop/cargo transport. These were designated XCG-20 in glider form and XC-123 in a powered configuration, the latter being known to Chase as the MS-8 Avitruc. It was powered by two Pratt & Whitney R-2800 Double Wasp engines and flew for the first time on 14 October 1949. Chase received a contract in 1952 for five pre-production aircraft under the designation C-123B, and these were built and flown in 1953. In that same year the Kaiser-Frazer Corporation acquired a majority holding in the Chase Aircraft Company, and was awarded a production contract by the USAF for

300 C-123Bs. Subsequently, the USAF had reason to cancel this contract in mid-1953, renewing it with the Fairchild Company which assumed responsibility for continued development and production of the C-123B.

Fairchild's interest in the development of the C-123 resulted in the provision of a large dorsal fin to improve directional stability, and the first of the Fairchild-built C-123Bs flew for the first time on 1 September 1954. Production of 302 for the USAF followed, this total including one airframe for static testing, and 24 for delivery to Saudi Arabia (6) and Venezuela (18) under the Military Assistance Program. At a later date four were supplied to Thailand.

In 1962 Fairchild produced a prototype YC-123H, which had a wide-track landing gear to overcome problems which had been experienced with C-123Bs being taxied in strong crosswinds. In addition, a General Electric CJ610 (similar to the J85) turbojet was pod-mounted beneath each wing to evaluate the performance improvement offered by this auxiliary power. This flew for the

first time on 30 July 1962, and was subsequently tested in South Vietnam in a counter-insurgency role, which resulted in a modification programme to convert 183 aircraft to this configuration. These were powered by the military version of the General Electric J85, and the first modified aircraft, designated C-123K, made its initial flight on 27 May 1966.

Some C-123s in southeast Asia were transferred for service with the Cambodian air force, and all but one squadron of C-123s operating in Vietnam were transferred to the South Vietnam air force. Nothing is known of the fate of these aircraft after the fall of South Vietnam.

Specification
Type: tactical transport aircraft
Powerplant: (C-123K) two 1865-kW (2,500-hp) Pratt & Whitney R-2800-99W Double Wasp radial piston engines, and two 1293-kg (2,850-lb) thrust General Electric J85-GE-17 turbojets
Performance: (C-123B) maximum speed 394 km/h (245 mph); cruising speed 330 km/h (205 mph); range 2366 km (1,470 miles)

Fairchild C-123K Provider

Weights: (C-123B) empty 13562 kg (29,900 lb); maximum take-off 27216 kg (60,000 lb)
Dimensions: span 33.53 m (110 ft 0 in); length 23.92 m (76 ft 3 in); height 10.39 m (34 ft 1 in); wing area 113.62 m² (1,223 sq ft)
Armament: (AC-123K) 7.62-mm (0.3-in) Minigun installations
Operator: Thailand

Fairchild-Hiller FH-1100

The US Army's Light Observation Helicopter (LOH) competition resulted in competitive prototypes being built by Bell, Hiller and Hughes. Hiller's first HO-5 prototype flew on 21 January 1963. Following a takeover Hiller became Fairchild-Hiller, and in 1965 it was announced that a developed version of this utility helicopter would enter production as the FH-1100; the first was completed in June 1966.

Standard accommodation is for a pilot and passenger in front, with three passengers, cargo, or two stretchers and an attendant behind. Following the initiation of production several armed forces acquired these helicopters for a range of utility roles but few remain in

service.

Specification
Type: four/five-seat helicopter
Powerplant: one 236-kW (317-shp) Allison 250-C18 turboshaft engine derated to 204 kW (274 shp) flat rating
Performance: maximum crusing speed 204 km/h (127 mph); range, maximum payload standard fuel 506 km (348 miles)
Weights: empty 633 kg (1,396 lb); maximum take-off 1247 kg (2,750 lb)
Dimensions: diameter of main rotor 10.79 m (35 ft 4.75 in); diameter of tail rotor 1.83 m (6 ft 0 in); length, rotor turning 12.13 m (39 ft 9.5 in); height 2.83 m (9 ft

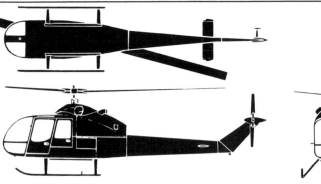
Fairchild-Hiller FH-1100

3.5 in); main rotor disc area 91.41 m² (984.0 sq ft)
Armament: can include 7.62-mm (0.3-in) machine-guns or two grenade launchers in armament

packs, or light anti-submarine weapons.
Operators: Argentina, Costa Rica, El Salvador, Panama, Philippines, Thailand

Fairchild T-46/AT-46

First flown in October 1985, the Fairchild Republic T-46 won the US Air Force's NGT (New Generation Trainer) award to replace the Cessna T-37 as its basic jet trainer. On 2 July 1982 Fairchild received a $104 million contract to cover two prototypes and two test specimens, plus an option on the first 54 aircraft. Garrett Turbine Engine Company received $121.2 million for 29 specially designed TFE76-4A (F109-GA-100) engines, with an option on 119. Metal was cut in April 1983, and in February 1984 $6 million was released to cover long-lead items for the first production aircraft, to fly in May 1986. A further 649 aircraft are required by the USAF.

Fairchild Republic revealed in 1983 that it intended to develop an export derivative, known within the

company as the FRC-225 Full Spectrum Trainer or AT-46A. This features a head-up display (HUD) and stores management system, as well as four underwing hardpoints capable of carrying gun pods, bombs, rockets or tanks.

Specification
Fairchild T-46A
Type: side-by-side basic trainer
Powerplant: two 603-kg (1,330-lb) thrust Garrett F109-GA-100 turbofan engines
Performance: (estimated) maximum speed at 10670 m (35,000 ft) 740 km/h (460 mph); cruising speed at 13715 m (45,000 ft) 616 km/h (383 mph); initial rate of climb 1362 m (4,470 ft)/minute; service ceiling 14175 m (46,500 ft); ferry range 2205 km (1,370 miles)

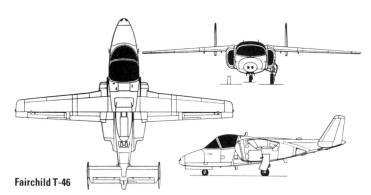
Fairchild T-46

Weights: (estimated) empty 2351 kg (5,184 lb); maximum take-off 3092 kg (6,817 lb)
Dimensions: span 11.78 m (38 ft 7.75 in); length 8.99 m (29 ft 6 in);

height 3.04 m (9 ft 11.75 in); wing area 14.95 m² (160.9 sq ft)
Armament: none
Operator: US Air Force (on order)

FFA C.3605

Origins of this type go back to the C.3601 fighter bomber produced by EKW (Eidgenössisches Konstruktions Werkstätte), flown on 15 March 1939. With a more powerful engine and retractable landing gear 160 were built as the EKW.3603.

Survivors were relegated to target towing in the 1960s. When the question of their replacement was considered, the most effective solution was judged to be the installation of a turboprop engine in the original fabric-winged airframes. In 1967 the

aircraft were assessed as having ten years of life remaining, though 18 years have now passed with no sign of retirement. As the T5307 engine was lighter than the old Hispano-Suiza, it had to be installed 1.19 m (3 ft 10.9 in) further forward, a third

fin also being needed. The C.3605 flew on 19 August 1968, following conversion by the Swiss Federal Aircraft Factory. In all 24, including two trainers, were produced. Almost all remain in service with the Zeilfliegerkorps, deploying their targets from beneath the rear cockpit with a hydraulic winch.

Specification
FFA C.3605
Type: two-seat target tug

Powerplant: one 820-kW (1,100-shp) Lycoming T5307A turboprop engine
Performance: maximum speed at 3050 m (10,000 ft) 432 km/h (268 mph); initial rate of climb 753 m

(2,470 ft)/minute; service ceiling 10000 m (32,810 ft); range 980 km (609 miles)
Weights: empty 2634 kg (5,807 lb); maximum take-off 3716 lb (8,192 lb)
Dimensions: span 13.74 m (45 ft

0.9 in); length 12.40 m (40 ft 8.2 in); height 4.05 m (13 ft 3.4 in); wing area 28.70 m² (308.93 sq ft)
Armament: none
Operator: Switzerland

FFA AS.202 Bravo

Conceived originally as a joint Swiss/Italian project, the Flug- und Fahrzeugwerke AG Altenrhein (FFA) Bravo was in the event produced exclusively in Switzerland. The first prototype flew on 7 March 1969 and a production line was established at Altenrhein, the former Swiss branch of the Dornier company. The first production Bravo flew on 22 December 1971 and two versions were initially available, the AS.202/15 and AS.202/18A, powered respectively by 112-kW and 134-kW (150-hp and 180-hp) Lycoming engines respectively, the latter with an inverted-flight oil system. Total sales by 1985 were about 170 aircraft.

An AS.202/26A powered by a 194-kW (260-hp) Lycoming O-540 engine was flown in 1978, but despite greatly improved performance only one was built. In 1984 a standard Bravo was flying with a 313-kW (420-shp) Allison B17C turbo-

prop (flat-rated at 239-kW/320-shp) to support the planned AS.32T Turbo Trainer, construction of which awaits an industrial partner.

Specification
FFA AS.202/18A
Type: two/three-seat light touring/training aircraft
Powerplant: one 134-kW (180-hp) Lycoming AEIO-360-B1F flat-four piston engine
Performance: maximum speed 241 km/h (150 mph) at sea level; cruising speed 226 km/h (140 mph); initial rate of climb 281 m (922 ft)/minute; service ceiling 5490 m (18,000 ft); range, no reserves 1133 km (704 miles)
Weights: empty 665 kg (1,466 lb); maximum take-off 1050 kg (2,315 lb)
Dimensions: span 9.75 m (31 ft 11.9 in); length 7.50 m (24 ft 7.3 in); height 2.81 m (9 ft 2.6 in); wing area 13.86 m² (149.19 sq ft)

Armament: none
Operators: Indonesia, Iraq, Jordan, Morocco, Oman, Switzerland, Uganda

Seen in civil guise, the FFA Bravo has been ordered for several air forces as an ab initio trainer. It is fully aerobatic and features an inverted fuel system in the 18A version.

FFA AS.202/18A Bravo

FMA IA 50 Guarani

Remaining in small-scale service with the Fuerza Aérea Argentina, the Guarani is a twin-engine light transport which has additional photo-survey and navaid-calibration roles. Design began in the 1950s, and a prototype was flown on 6 February 1962. With a crew of two and seating for 10, 12 or 15 passengers, the Guarani I was a twin-fin aircraft derived from the IA Huanquero. The definitive Guarani II however, has a shorter tail section and a single fin swept back at 52° 30′, an extra 60 kW (80 hp) from each of the Bastan turboprops, together with pneumatic de-icing equipment. Production began in 1967 with 18 for the I Escuadrón de Transporte of I Brigada Aérea at Buenos Aires/BAM El Palomar. One tested skis for antarctic use; another was a VIP machine for the General Staff; a third had civil markings for official duties. Most remain in ser-

vice, as does the first prototype with the Central de Ensayos de Vuelo (flight test centre). Further new aircraft comprised five photo-survey models for the I Escuadrón Fotográfico of II Brigada Aérea at Paraná and two calibration aircraft operated by the Instituto Nacional de Aviación Civil in paramilitary markings. All these have recently had many of their tasks taken over by Learjets.

Specification
FMA IA 50 Guarani II
Type: light transport
Powerplant: two 694-kW (930-shp) Turboméca Bastan IVA turboprop engines
Performance: maximum speed 500 km/h (311 mph); initial rate of climb 805 m (2,640 ft)/minute: service ceiling 12500 m (41,010 ft); range with maximum payload 1995 km (1,240 miles)

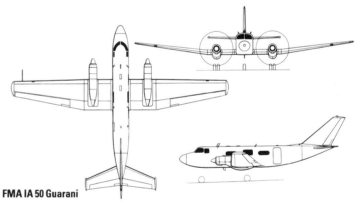

FMA IA 50 Guarani

Weights: empty 3924 kg (8,651 lb); maximum take-off 7350 kg (16,204 lb)
Dimensions: span 19.53 m (64 ft 0.9 in); length 14.86 m (48 ft 9 in); height 5.81 m (19 ft 0.7 in); wing

area 41.81 m² (450.05 sq ft)
Armament: none
Operator: Argentina

FMA IA 58 Pucará

The Pucará twin-turboprop COIN (counter-insurgency) aircraft was developed by FMA (Fabrica Militar de Aviones) which is part of the FAA (Argentine air force) Area de Matérial Cordoba division. The first powered Pucará made its maiden flight in August 1969 and the first production machine flew in November 1974.

The Pucará is named after stone forts built on mountain tops. Its main mission is armed reconnaissance over land and sea, together with fire support. The two-man crew sit in tandem, with the co-pilot's seats 25 cm (10 in) higher than that of the pilot in front. The windscreen and cabin floor are strengthened to stop impact of 7.62-mm (0.3-in) machine-gun bullets fired from close range.

The IA 58 is of all-metal construction and designed for good man-

FMA IA 58 Pucará of the Argentine air force.

oeuvrability at low altitude. It can operate from small grass fields, and the take-off run can be reduced to 80 m (262 ft) by fitting three JATO (Jet Assisted Take-Off) rockets. Large propeller-tip ground clearance ensures no damage will result from landings on uneven surfaces. The Matra 83A-3 sight with adjustable depression angle gives great flexibility in weapon-aiming. A programmer allows external stores to

be released in any quantity from two to 40, with two firing modes available: step (single, pairs or salvo) and ripple (single, pairs and salvo).

In the Falklands war the Pucará played a major role, and shot down a Westland Scout AH.1, but all 20 were destroyed or captured. Production of a planned 100 was completed in 1984, but it is hoped to build further batches for the FAA or for export, possibly with a single seat, Garrett engines and other options.

Specification
Type: two-seat armed reconnaissance/close-support COIN aircraft
Powerplant: two 737-kW (988-shp) Turboméca Astazou XVIG turboprop engines
Performance: maximum speed 500 km/h (311 mph) at 3000 m (9,845 ft); initial rate of climb 1080 m (3,543 ft)/minute; service ceiling 10000 m (32,810 ft); attack radius (maximum weapon load, hi-lo-hi, 10 per cent reserves) 250 km

(155 miles)
Weights: empty 4037 kg (8,900 lb); maximum take-off 6800 kg (14,991 lb)
Dimensions: span 14.50 m (47 ft 6.9 in); length 14.25 m (46 ft 9 in);

height 5.36 m (17 ft 7 in); wing area 30.30 m² (326.16 sq ft)
Armament: two Hispano-Suiza HS-804 20-mm (0.79-in) cannon each with 270 rounds and four Browning 7.62-mm (0.3-in) machine-guns each

with 900 rounds, plus up to 1620 kg (3,571 lb) of external ordnance on five stations; typical weapons include general-purpose, fragmentation, napalm and incendiary bombs; seven 19×70-mm

(2.75-in) rocket launchers, or a cannon pod and two 330-litre (72.5 Imp gallon) drop tanks
Operators: Argentina, Central African Republic, Uruguay, Venezuela

FMA IA 63 Pampa

Argentine requirements for a new trainer are to be met by the IA 63 design, produced at Córdoba by Fabrica Militar De Aviones (FMA), a department of the air force. The IA-62 turboprop under consideration in the late 1970s was abandoned in favour of a light jet which could be armed for COIN missions. This has

led to the IA-63 Pampa, its resemblance to the Alpha Jet not surprising in view of the fact that Dornier was heavily involved in the development phase, during which some Argentine personnel worked in West Germany. One difference is that the Pampa has unswept wings, these being of an advanced Dornier trans-

onic section. Of six prototypes, two are for static testing. The first flight was made in October 1984, 10 months late, and it may be presumed that the December 1985 flight date for the first production aircraft will also be delayed. The Pampa is expected to start work in March 1988 with the Escuela de Aviación Mili-

tar at Córdoba, where it will replace the Argentine-built MS.733 Paris. An order for 64 has been placed, with delivery schedule at three per month. Peru and Venezuela have expressed interest.

Fokker F.27 Friendship

Small numbers of the early production F.27 Friendship Mk 100, Mk 200 and Mk 300 models are still active with several air forces, but the most prolific derivative with regard to military contracts has been the F.27 Friendship Mk 400 combiplane, about 70 F.27 Friendship Mk 400M aircraft being built specifically for military customers such as Algeria, Argentina, Iran, Nigeria and Sudan. Incorporating a freight door on the port side of the forward fuselage and flown for the first time during 1961, the F.27 Mk 400 can carry up to 45 paratroops or 6025 kg (13,283 lb) of cargo. Other variants which have been adapted for military transport duties with some success are the F.27 Friendship Mk 500, which features a 1.5-m (5-ft) fuselage stretch, and the current F.27 Friendship Mk 600, both of these also possessing a freight door to facilitate the loading of large items of cargo. Combined military sales number about 15, most being F.27 Mk 600s.

The most recent addition to the Friendship family has been the F.27MPA Maritime. Now operational with the Netherlands, Peru, the Philippines and Spain, this features Litton search radar with its aerial in a prominent ventral radome, as well as inertial equipment, a radar altimeter, a new autopilot and a more modern and comprehensive array of communications gear. Auxiliary fuel tanks mounted under the wings help to increase

endurance, and the F.27MPA can remain aloft for up to 10 hours at a time. Lacking the facility to carry armament, the Maritime derivative is mainly intended for peacetime duties such as fishery patrol, oil platform surveillance, search and rescue and environmental control, but Fokker has recently been conducting studies of a version with more warlike applications. This is the Maritime Enforcer for anti-submarine warfare, optional equipment including new radar, forward-looking infra-red sensors, sonobuoy processing equipment, and a magnetic anomaly detector, whilst armament could include either anti-ship missiles or torpedoes.

Specification
Fokker Friendship Mk 400M
Type: medium range tactical transport aircraft
Powerplant: two 2,140-shp (1596-kW) Rolls-Royce Dart Mk 532-7R turboprops
Performance: cruising speed 480 km/h (298 mph) at medium altitude; range 2210 km (1,373 miles) with maximum 6420-kg (14,153-lb) payload
Weights: empty 10600 kg

(23,369 lb); maximum take-off 20410 kg (44,996 lb)
Dimensions: span 29.00 m (95 ft 2 in); length 23.56 m (77 ft 3.5 in); height 8.50 m (27 ft 11 in); wing area 70.00 m² (753.5 sq ft)
Operators: Algeria, Angola, Benin, Finland, Ghana, Indonesia, Iran, Ivory Coast, Netherlands, New Zealand, Nigeria, Pakistan, Philippines, Senegambia, Spain, Thailand, Uruguay

Senegambia is one of the countries that has bought the F.27. This transport is economic to operate and easy to maintain, and does not require a long runway.

Fokker F.27M Troopship of the Netherlands air force (No. 334 Sqn).

Fokker F.28 Fellowship

Fokker announced in 1962 plans for a 65-seat jet successor to the F.27 Friendship. Like the F.27, the F.28 Friendship is a collaborative venture. Fokker builds the cockpit, centre-section and wing root fairing, and carries out assembly at Schiphol. MBB supplies most of the fuselage aft of the wing, including engine nacelles and support stubs, the rear fuselage and tail unit, and Shorts builds the outer wings and landing gear doors. Equipment includes Dowty-Rotol landing gear and Smiths autopilot. A unique airbrake, forming the rear of the fuselage, is employed in place of thrust reversers.

Initial production version was the Mk 1000. In 1971 the Mk 2000 introduced a 2.21-m (7 ft 3-in) fuselage 'stretch' to accommodate 79 passengers. A new wing, with slats and extended tips, was applied to the Mk

1000 to produce the Mk 5000, and to the Mk 2000 to give the Mk 6000. Fokker then deleted the slats and continues to sell the short body and new wing as the Mk 3000, and the long body and the new wing as the Mk 4000. A large cargo door is an optional extra. Fokker still hopes the F.28 may be chosen for various USAF and US Navy requirements.

Specification
Fokker F.28 Mk 3000
Type: medium-range transport
Powerplant: two 4491-kg (9,900-lb) thrust flat-rated Rolls-Royce RB.183-2 Mk 555-15P turbofan engines
Performance: maximum cruising speed 842 km/h (523 mph) at 7000 m (22,965 ft); economic cruising speed 678 km/h (421 mph) at 9150 m (30,020 ft); maximum cruising height 10675 m (35,025 ft);

maximum range with maximum fuel and reserves 3169 km (1,969 miles)
Weights: empty operating 16965 kg (37,401 lb); maximum take-off 33110 kg (72,995 lb)
Dimensions: span 25.07 m (82 ft 3 in); length 27.40 m (89 ft 10.7 in); height 8.47 m (27 ft 9.5 in); wing area 79.00 m² (850.38 sq ft)

The Colombian air force uses this F.28 Fellowship for VIP duties. Most operators have fitted VIP interiors.

Armament: none
Operators: Argentina, Benin, Colombia, Congo, Ghana, Ivory Coast, Malaysia, Netherlands, Nigeria, Peru, Philippines, Tanzania, Togo

Fuji T-1

To replace the T-6 trainer, the Japanese Defence Agency issued in 1955 a design requirement that Fuji Heavy Industries won, three T-1s being ordered on 11 July 1956. The first prototype flew on 8 January 1958. It had been intended to use the Ishikawajima-Harima J3 turbojet, but delays led to use of the Orpheus in the first two batches of 20 aircraft; these were designated by the company T1F2 and deliveries were completed in July 1962; they received the JASDF designation T-1A. A prototype was flown on 17 May 1960 as a T1F1 with the J3-IHI-3 engine developing 1200 kg (2,646 lb) thrust. The JASDF ordered 20, with designation T-1B, and delivery took place between September 1962 and June 1963. The developed T1F3 flew in April 1965 with a J3-IHI-7 of 1400 kg (3,086 lb) thrust, and the JASDF re-engined all T-1Bs with this powerplant, redesignating

them T-1C. Most of the 50-plus arte with No. 13 Wing at Ashiya. Because of the unit's proximity to the sea, the T-1s are painted in anti-corrosion white.

Specification

Fuji T-1A
Type: intermediate jet trainer
Powerplant: one 1814-kg (4,000-lb) thrust Rolls-Royce Orpheus 805 turbojet engine
Performance: maximum speed 925 km/h (575 mph) at 6100 m (20,015 ft); cruising speed 620 km/h (385 mph) at 9150 m (30,020 ft); initial rate of climb 1980 m (6,496 ft)/minute; service ceiling 15850 m (52,000 ft); range, standard fuel 1300 km (808 miles), with underwing tanks 1950 km (1,212 miles)
Weights: empty 2420 kg (5,335 lb); maximum take-off 5000 kg (11,023 lb)

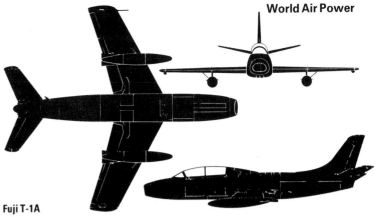

Fuji T-1A

Dimensions: span 10.50 m (34 ft 5.4 in); length 12.12 m (39 ft 9.2 in); height 4.08 m (13 ft 4.6 in); wing area 22.22 m^2 (239.18 sq ft)
Armament: one Colt Browning 12.7-mm (0.5-in) machine-gun in nose; if underwing tanks are not

carried each rack can be used for a gun pod, an AIM-9 Sidewinder AAM, a 340-kg (750-lb) bomb, a napalm bomb, or a cluster of 70-mm (2.75-in) air-to-air rockets
Operator: Japan

GAF (Government Aircraft Factories) Nomad

The GAF Nomad is a twin-turboprop STOL utility transport developed for a variety of roles, including maritime patrol. The first of two Model N2 prototypes made its maiden flight in July 1971, and the standard N22B short-fuselage production version is known as the Mission Master in military service. A fishery-production and anti-smuggling variant, the Search Master, has since been developed.

The basic Nomad is a high-wing aircraft with full-span double-slotted flaps for STOL operations, allowing it to take off in 183 m (600 ft). The aircraft is designed for single-pilot operation and can accommodate 12 passengers or a typical disposable load of 1931 kg (4,250 lb). Double doors on the left give access to the cabin and a dropping hatch, the doors of which can be operated from the cockpit, has a capacity of 227 kg (500 lb). The Mission Master can also carry up to 909 kg (2,000 lb) of stores on four underwing pylons, which can be fitted with gun or rocket pods. Surveillance and night-vision aids may be fitted in a nose bay, and removable seat armour and self-sealing fuel tanks can be incorporated.

The aircraft's powerplant, a pair of Allison 250-B17B turboprops, is the first application of this engine in a fixed-wing type. Its more normal use is as a turboshaft in helicopters such as the Bell JetRanger and Hughes 500, but the Model 250 has proved to be very suitable for driving a propeller instead of a rotor, with excellent power and reliability combined with low noise.

The maritime-patrol Search Master is based on the standard N22B Mission Master and is available in two versions: the Search Master B, with a nose-mounted Bendix RDR-1400 search radar; and Search Master L, incorporating the Litton LASR-2 radar in a radome beneath the belly. The RDR-1400 can detect a wooden boat 12 m (40 ft) long at a range of 36 km (22.5 miles) and can pick up a 46 m (150 ft) object at twice this distance. The radar can also detect weather out to ranges of

432 km (270 miles) and can interrogate transponders within 296 km (185 miles). The LASR-2, a derivative of the APS-503 which equips the Canadian Armed Forces' Sikorsky Sea King helicopters, is similar to the radar installed in the Fokker-VFW F27M maritime-patrol aircraft and is mounted under the Search Master's fuselage, giving it a 360° scan.

The L version has a slightly slower cruising speed than the B, but the superior radar performance more than compensates for this reduction. Against a target with a cross-section of 1000 m^2 (10,750 sq ft) in sea state three and with the aircraft cruising at 1524 m (5,000 ft), the Search Master L has a radar range of 184 km (115 miles) – twice that of the B version – and on an eight-hour patrol it surveys more than 3.2 times the ocean area. Even greater improvements are achieved against small targets.

The Search Master is also equipped with a Litton LTN-72 inertial navigator which supplies all the information needed to carry out a search over large tracts of sea and which can feed directly into the autopilot to fly an efficient pattern. Larger fuel tanks are fitted to increase the aircraft's range, and bubble windows are fitted in the fuselage sides to aid visual observation. The Search Master's crew comprises pilot, co-pilot, radar operator and navigator. Stores such as reconnaissance pods can be attached to underwing pylons, and an RC-9 camera may be fitted in a hatch in the cabin floor, which is also suitable for delivering flares or other droppable items.

Specification
Type: utility transport and maritime-patrol aircraft
Powerplant: two 298 kW (400-shp) Allison 250-B17B turboprops
Performance: maximum speed at sea level 309 km/h (193 mph); economical cruising speed 258 km/h

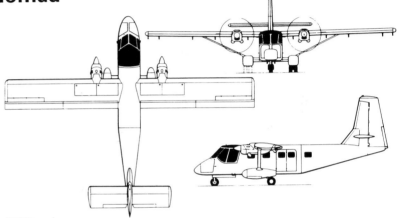

GAF Nomad

(161 mph); sea-level rate of climb 429 m (1,410 ft) per minute; ceiling 7164 m (23,500 ft); maximum range at 258 km/h (161 mph) with reserves and auxiliary fuel 2080 km (1,300 miles)
Weights: typical operating 2159 kg (4,750 lb); maximum take-off 4090 kg (9,000 lb)
Dimensions: span 16.46 m (54 ft); length 12.56 m (41 ft 2½ in); height 5.53 m (18 ft 1½ in); wing area 13.50 m^2 (154.3 sq ft)
Operators: (Mission Master) Australia, Indonesia, Papua New Guinea, Philippines; (Search Master) Indonesia

The Nomad is ideally suited to operations in the outback of Australia, where its ruggedness and short-field capability are put to good use.

Gates Learjet

The US Air Force's decision to acquire 80 Learjet 35As on lease was welcome to the Gates Learjet Corporation which had, in August 1982, elected to suspend production of this model. Under the designation C-21A, these aircraft are the subject of a $175.4 million contract covering a five-year lease, contractor support and crew training. They are replacing CT-39A Sabreliners, deliveries having begun in March 1984. Under the terms of the agreement, the USAF may purchase the aircraft outright in 1987 or extend the lease for a further three years before buying. Options exist for an additional batch of 20. The C-21As are operated by Military Airlift Command, detachments of two or three being dotted around the USA while a few are based at Ramstein and Stuttgart. The 80th aircraft was handed over in 1985.

Learjet variants have found favour elsewhere in the world, ex-amples of the Model 24D, 25B, and 35A operating with some eight air arms on aerial mapping, target towing, conventional and VIP transport, and air ambulance duties The parent company has also developed a variant for sea patrol missions, with a belly radome scanning through 360°. Three aircraft in this configuration are now in service with Finland, mainly employed as target tugs.

Specification
Learjet Model 35A/C-21A
Type: light communications aircraft
Powerplant: two 1588-kg (3,500-lb) thrust Garrett TFE731-2 turbofan engines
Performance: maximum cruising speed 859 km/h (534 mph); normal cruising speed 818 km/h (508 mph); initial rate of climb 1494 m (4,900 ft)/minute; service ceiling 12955 m (42,500 ft); range with four

passengers and fuel reserves 4482 km (2,785 miles)
Weights: empty 4341 kg (9,571 lb); maximum take-off 7711 kg (17,000 lb)
Dimensions: span 12.04 m (39 ft 6 in); length 14.83 m (48 ft 8 in); height 3.73 m (12 ft 3 in); wing area 23.53 m² (253.3 sq ft)
Armament: none
Operators: (Model 24) Ecuador,

Mexico; (Model 25) Bolivia, Jugoslavia, Oman (Police), Peru, Saudi Arabia; (Model 35) Argentina, Finland, Oman (Police), USAF (C-21A)

Learjets have found favour with a number of air forces for VIP duties and patrol, owing to the excellent range performance of the type.

General Dynamics F-16 Fighting Falcon

Small, lightweight, agile, hard to see and hard to hit, the General Dynamics F-16 is one of the most promising fighter designs to emerge in recent years. Its origins go back to February 1972, when General Dynamics, Boeing, LTV, Lockheed and Northrop all submitted proposals to the US Air Force for a new lightweight fighter (LWF) with exceptional manoeuvrability. Two months later General Dynamics and Northrop were each given contracts for two prototypes, to be flown against one another (and other contemporary USAF fighters) in a competitive fly-off to decide the winner. The choice was not an easy one, for both General Dynamics' YF-16 and Northrop's twin-engined YF-17 gave equally excellent performances, neither really deserving to lose. As subsequent events have shown, neither did actually lose, for although the YF-17 was unsuccessful in the US Air Force competition, its direct derivative, the F-18 Hornet, now seems assured of large orders from the US Navy and possibly other customers.

Originally there was no intention of building an LWF in quantity, but this was overturned by the emergence of a large export market, initially in Europe. Selection of the F-16 for the USAF was announced in January 1975, and five months later came the news that four European air forces (those of Belgium, the Netherlands, Norway and Denmark) had chosen the F-16 to replace Lockheed Starfighters and other types in their respective modernization programmes. Another substantial contract followed in October 1976, when the Imperial Iranian air force ordered 160, to carry total orders beyond the 1,000 mark. The USAF now plans to have nearly 1,400 eventually, and the four NATO countries have ordered 348. The latter is divided 116 to Belgium, 102 to the Netherlands, 72 to Norway and 58 to Denmark; there are assembly lines in Belgium and the Netherlands in addition to that in the USA. In all cases the totals include a proportion of tandem two-seat F-16B fighter/trainers.

The first YF-16 prototype made its maiden flight on 20 January 1974. An exhaustive fly-off against the YF-17 occupied almost the whole of that year, and after the F-16's acceptance a further eight modified

NATO forces in Europe have adopted the F-16A as their main fighter, and these are used in dual roles. Denmark has 53 single-seaters and 16 F-16B two-seaters.

development aircraft were built: six single-seat F-16As and two two-seat F-16Bs. The first of the these flew on 8 December 1976 and the last in June 1978. The production go-ahead was announced in the spring of 1978, when General Dynamics was authorized to start building the first 105 aircraft for the USAF and the first 192 for Europe. August 1978 saw the initial flight of a series-build F-16A, and in the winter of 1978-79 the 388th Tactical Fighter Wing at Hill AFB, Utah, became the first USAF unit to receive the new fighter; European deliveries began shortly afterwards.

The engine is almost identical with that used in the twin-engined McDonnell Douglas F-15, and although the USAF is deeply concerned about the engine's continuing troubles, the type's maturity must be an advantage. The inlet is a simple fixed-geometry hole on the underside of the fuselage, though this reduces the maximum speed.

Some of the latest technology can be seen in the aerodynamic structure, and in the avionics and fire control systems of the F-16. For example, the way that the wings are blended into the body, instead of being 'stuck on', not only helps to save weight but increases the overall lift at high angles of attack and reduces drag in the transonic speed range. Moveable flaps on the wing

F-16 cutaway drawing key

1 Pitot tube
2 Radome
3 Planar radar scanner
4 Scanner drive motors
5 ADF antenna
6 Front electronics equipment bay
7 Westinghouse radar electronics
8 Forward radar warning antenna
9 Cockpit front bulkhead
10 Instrument panel shroud Missile control electronics
12 Fuselage forebody strake fairing
13 Marconi-Elliot-head-up display
14 Side stick controller (fly-by-wire control system)
15 Cockpit floor
16 Frameless bubble canopy
17 Canopy fairing
18 Ejection seat (30° tilt-back)
19 Pilot's safety harness
20 Throttle
21 Side control panel
22 Cockpit frame construction
23 Ejection seat headrest
24 Cockpit canopy seat
25 Canopy hinge
26 Rear avionics bay (growth area)
27 Cockpit rear bulkhead
28 Boundary-layer splitter-plate
29 Fixed-geometry air intake
30 Antenna
31 Air retracting nosewheel
32 Shock absorber scissor link
33 Retraction strut
34 Nosewheel door
35 Intake trunking
36 Cooling louvres
37 Gun gas suppression nozzle
38 Air conditioning system pipes
39 Forward fuselage fuel tanks
40 Canopy aft glazing
41 Drop tank, capacity 370 US gal (1400 litres)
42 Forebody blended wing root
43 TACAN aerial
44 Fuel tank access panel
45 Cannon barrels
46 Forebody frame construction
47 M61 rotary cannon

48 Ammunition feed and link return chutes
49 Ammunition drum (500 x 20-mm rounds)
50 Ammunition drum flexible drive shaft
51 Hydraulic gun drive motor
52 Leading edge control shaft
53 Hydraulic service bay
54 Hydraulic reservoir
55 Leading edge manoeuvre flap drive motor
56 Antenna
57 No. 2 hydraulic system reservoir
58 Leading edge control shaft
59 Inboard pylon
60 Wing centre pylon
61 Mk 82 500-lb (227-kg) bombs
62 Outboard wing pylon
63 Missile launcher shoe
64 AIM-9 Sidewinder missile
65 Starboard navigation light
66 Aluminium honeycomb leading edge construction
67 Static dischargers
68 Fixed trailing edge section
69 Multi-spar wing construction
70 Integral wing fuel tank
71 Starboard flaperon
72 Fuel system piping
73 Access panels
74 Centre fuel tank panels
75 Centre fuselage fuel tank
76 Intake duct
77 Wing mounting bulkheads
78 Flight refuelling receptacle
79 Pratt & Whitney F100-PW-100(3) turbofan

80 Engine gearbox, airframe mounted
81 Gearbox drive shaft
82 Ground pressure refuelling receptacle
83 Flaperon servo-actuator
84 Rear fuselage frame construction
85 Integral fuel tank
86 Front engine mounting
87 Antenna
88 Fin root fairing
89 Flight control system hydraulic accumulators
90 Anti-collision light power supply
91 Starboard tailplane
92 Graphite-epoxy fin skins
93 Fin construction
94 Aluminium honeycomb leading edge construction
95 Steel leading edge strip
96 Antenna
97 Anti-collision light
98 Tail radar warning antenna

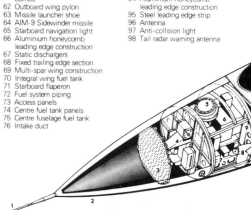

99 Aluminium honeycomb rudder construction
100 Rudder servo-actuator
101 Radar warning power supply
102 Tail navigation light
103 Fully variable exhaust nozzle
104 Split trailing edge airbrakes (upper and lower surfaces)

General Dynamics F-16 Fighting Falcon (continued)

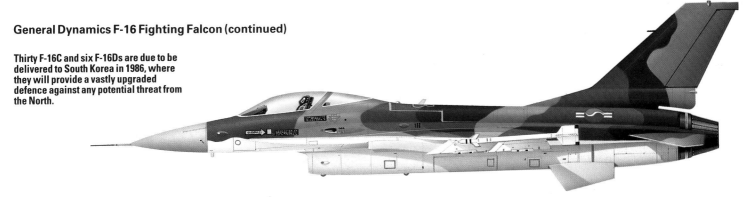

Thirty F-16C and six F-16Ds are due to be delivered to South Korea in 1986, where they will provide a vastly upgraded defence against any potential threat from the North.

leading- and trailing-edges, controlled automatically by the aircraft's speed and altitude, enable the wing to assume an optimum configuration for lift under all conditions of flight. The highly-swept strakes that lead forward alongside the nose provide further lift; they also prevent wing-root stall, reduce buffeting, and improve directional stability and roll control. A lot of thought has gone into cockpit design to get the canopy shape, seat angle and instrument layout just right, so that the pilot has the maximum field of view and maximum efficiency with a minimum of fatigue: a most important factor in an aircraft liable to pull up to 9g in an air-to-air combat. All flying controls are electrically operated through a 'fly-by-wire' system that replaces the old-fashioned mechanical linkages, enabling the aircraft to respond faster and more accurately to pilot commands, whilst also simplifying maintenance. A head-up display, side-stick controller and zero-zero ejection seat are also included in the cockpit.

The 8th TFW 'Wolf Pack' is based at Kunsan in Korea to augment the local defence. Other PACAF F-16s are based at Misawa in Japan.

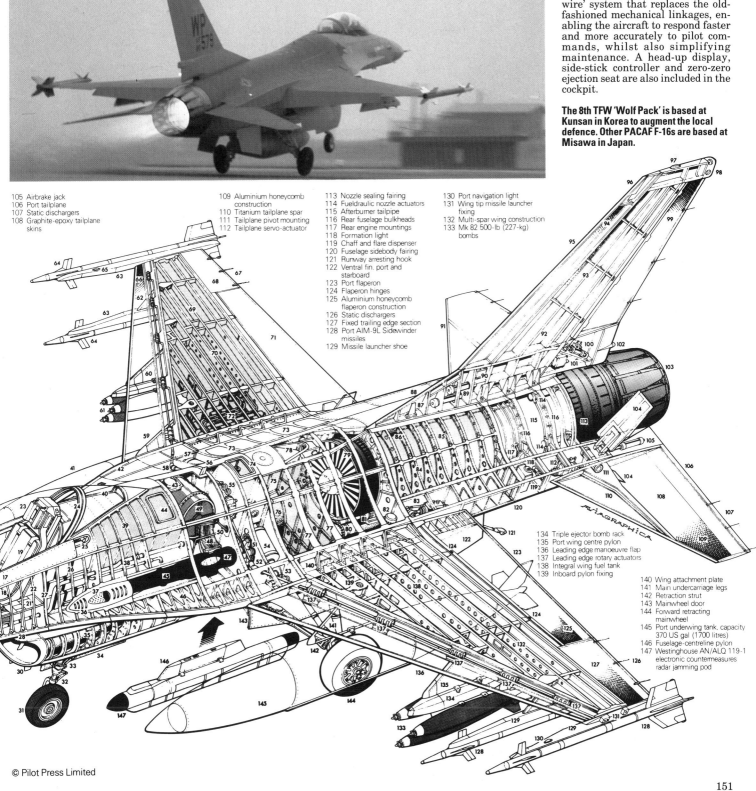

105 Airbrake jack
106 Port tailplane
107 Static dischargers
108 Graphite-epoxy tailplane skins
109 Aluminium honeycomb construction
110 Titanium tailplane spar
111 Tailplane pivot mounting
112 Tailplane servo-actuator
113 Nozzle sealing fairing
114 Fueldraulic nozzle actuators
115 Afterburner tailpipe
116 Rear fuselage bulkheads
117 Rear engine mountings
118 Formation light
119 Chaff and flare dispenser
120 Fuselage sidebody fairing
121 Runway arresting hook
122 Ventral fin, port and starboard
123 Port flaperon
124 Flaperon hinges
125 Aluminium honeycomb flaperon construction
126 Static dischargers
127 Fixed trailing edge section
128 Port AIM-9L Sidewinder missiles
129 Missile launcher shoe
130 Port navigation light
131 Wing tip missile launcher
132 Multi-spar wing construction
133 Mk 82 500-lb (227-kg) bombs
134 Triple ejector bomb rack
135 Port wing centre pylon fixing
136 Leading edge manoeuvre flap
137 Leading edge rotary actuators
138 Integral wing fuel tank
139 Inboard pylon fixing
140 Wing attachment plate
141 Main undercarriage legs
142 Retraction strut
143 Mainwheel door
144 Forward retracting mainwheel
145 Port underwing tank, capacity 370 US gal (1700 litres)
146 Fuselage-centreline pylon
147 Westinghouse AN/ALQ 119-1 electronic countermeasures radar jamming pod

Since the original LWF specification was drawn up in the early 1970s the role envisaged for the F-16 has predictably changed from that of a purely defensive air-superiority demonstrator to that of a multi-role tactical fighter able to carry out air-to-ground attack. To that extent, the performance originally to be expected when armed only with a built-in gun plus a pair of wingtip-mounted AIM-9L air-to-air missiles has been compromised by the 'Christmas tree' effect of hanging all kinds of external weapons or other equipment under the wings and fuselage; but the F-16 remains a most impressive performer even under full-load conditions, naturally regaining its full measure of agility once the weapons have been delivered. Though limited in all-weather attack capability, the F-16A carries a good range of modern avionics. Based on a digital computer, the Westinghouse multi-mode pulse-Doppler radar has a look-down range, eliminating ground 'clutter', of 37-56 km (23-35 miles), and a look-up range of 46-74 km (29-46 miles). In air-to-air fighting, the pilot has a choice of one missile-firing mode plus two gunnery modes ('snap-shoot' and optical lead-computing) available on the cockpit stores control panel. Air-to-ground attacks can be made under

visual, blind or electro-optical delivery conditions, by day or night and to some degree in adverse weather. Combined with a long-life structure, low radar signature, active and passive countermeasures, plus first-class manoeuvrability, the F-16 is clearly destined to make an outstanding contribution to the tactical defence of the United States and Western Europe for many years to come. By 1986 the USAF had equipped eight front-line wings and two reserve squadrons with the F-16, with two wings in the Far East and three in Europe.

The F-16 has seen combat with Israel, whose aircraft have downed 46 Syrian MiGs, as well as carrying out a daring raid on a nuclear reactor in Iraq.

Specification
Type: single-seat tactical fighter (F-16A) and two-seat combat trainer (F-16B)
Powerplant: one 10800-kg (23,810-lb) Pratt & Whitney F100-PW-100(3) afterburning turbofan
Performance: maximum speed (YF-16) at 10970 m (36,000 ft) with two sidewinders Mach 1.95 or 2074 km/h (1,289 mph); combat radius (YF-16) with two Mk 82 bombs 547 km (340 miles), (F-16A) about 925 km (575 miles); ferry range (F-16A) with drop-tanks 3705 km (2,303 miles); service ceiling (F-16A) about 18290 m (60,000 ft); maximum rate of climb (YF-16) with two Mk 82 bombs 12802 m (42,000 ft) per minute
Weights: (F-16A) operational

empty 6607 kg (14,567 lb); internal fuel 3162 kg (6,972 lb); maximum external load 6894 kg (15,200 lb); design take-off gross, clean 10205 kg (22,500 lb); maximum take-off without external tanks 10335 kg (22,785 lb), with external load 14968 kg (33,000 lb)
Dimensions: span (over missiles) 10.01 m (32 ft 10 in); length 14.52 m (47 ft 7¾ in); height 5.01 m (16 ft 5¼ in); wing area 27.87 m² (300 sq ft)
Armament: one 20-mm General Electric M61A-1 multi-barrel cannon in left wing/body flaring, with 500 rounds; one AIM-9J/L Sidewinder infra-red homing missile at each wingtip (radar-homing Sparrow or AMRAAM later) for air-to-air interception; six underwing hardpoints and one under fuselage for up to 6894 kg (15,200 lb) of attack weapons or drop-tanks (4763 kg/10,500 lb if full internal fuel is carried). Stores under wings/fuselage can include four more Sidewinders or Sparrows, Pave Penny laser tracking pod, single or cluster bombs, flare pods, air-to-surface missiles, laser-guided and electro-optical weapons
Operators: Belgium, Denmark, Egypt, Israel, Netherlands, Norway, Pakistan, Thailand, Turkey, Singapore, South Korea, USAF, Venezuela

Typical of the F-16s in USAF service is this Hahn-based machine of the 50th TFW. F-16s are slowly replacing the F-4 Phantom in the USAF, and in Germany the 86th TFW at Ramstein will soon join the 50th as an F-16 user. This aircraft is configured for a dual-role mission, with slick iron bombs and four AIM-9P Sidewinders for air defence.

Keith Fretwell.

153

General Dynamics F-111/FB-111

It needs no more than a glance at the current inventory of the Soviet air force to see how one country, at least, has moved heavily in favour of variable-geometry or 'swing-wing' aircraft in the past 10 to 15 years. For the nation that first put this principle into practice in a production aircraft, however, the progress from prototype to successful service warplane made a far from happy story. The major advantages offered by a variable-geometry aircraft are a high supersonic performance with the wings swept back; economical subsonic cruising speed with them fully spread; a long operational or ferry range; and relatively short take-off and landing runs at very high weights. So, when the US Air Force's Tactical Air Command was seeking a strike aircraft to replace the Republic F-105 Thunderchief, as outlined in its SOR (Specific Operational Requirement) 183 of 14 June 1960, it was very interested in the results of experiments with variable-geometry wing configurations that had recently been conducted by NASA's Langley Research Center at Hampton, Virginia. The US Navy, at the same time, was looking for a new fleet air defence fighter to succeed the McDonnell Douglas F-4 Phantom, and eventually the Department of Defense decreed that the two requirements should be combined in a single programme known as TFX, or Tactical Fighter, Experimental.

The Defense Secretary, Robert McNamara, stuck to this decision despite strong objections from both services, and his department rejected all six designs originally submitted in late 1961. However, a design from Boeing, and a joint offering by General Dynamics and Grumman, were considered worthy of study contracts. At three subsequent 'paper' evaluation conferences, after successive refinements of the two designs, the Boeing contender appeared to be a clear favourite and was almost universally recommended for adoption. To McNamara, however, it was technologically too advanced and lacked the commonality between the air force and navy versions that he believed was essential. He therefore overruled his advisers, and on 24 November 1962 a development contract for 23 aircraft was awarded to General Dynamics. Of these, 18 were to be basic tactical F-111As for the USAF and five were F-111Bs, developed primarily by Grumman for the US Navy.

The F-111B began to run into trouble almost immediately; despite a long and intensive flight development programme the type was eventually cancelled in July 1968. The aircraft had consistently proved overweight, quite unable to meet the performance required of it, and only seven examples were completed: the five development machines, plus two of the 231 production F-111Bs which the US Navy had planned to order.

The F-111A, on which all subsequent models were based, had an almost equally unhappy early history after its first flight on 21 December 1964, but eventually it was cleared for service and deliveries of 141 production examples began in October 1967, to the 474th Tactical Fighter Wing at Nellis AFB, Nevada. In spring 1968 the 428th Tactical Fighter Squadron took six F-111As to Thailand for operational trials over Vietnam – and lost three of them in four weeks. Groundings and modifications fol-

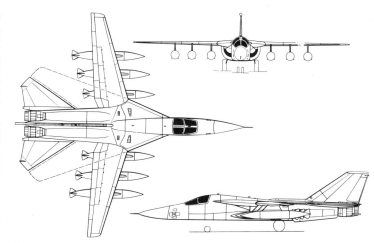
Australia is the only foreign operator of the F-111, flying 24. The 15 F-111Cs are being updated to carry Pave Tack laser-guided bomb and missile system.

lowed, and when 48 more F-111As were sent to Vietnam in 1972-73, they flew over 4,000 combat sorties in seven months for the loss of only six aircraft. One of the modfications was to the engine air inlet geometry; the next 94 aircraft were built with an enlarged inlet (to suit more powerful engines which were not fitted) and designated F-111E.

Meanwhile, the designation F-111C had been applied to 24 aircraft ordered by the Royal Australian Air Force in 1963, but as a result of extensive modifications and escalating costs the delivery of these did not begin until 1973. They have the increased-span wings of the FB-111A and a strenthened landing gear. Another export order was placed in 1966 when the Royal Air Force ordered 50 F-111Ks, but these were cancelled two years later and the two that were almost complete became YF-111As for the USAF, the rest becoming FB-111As.

The third production tactical version for the USAF (96 were built) was the F-111D, which combined a slightly more powerful engine with the modified inlets of the E model. It also introduced 'Mk II' avionics, which included an AN/APQ-30 attack radar, a digital (instead of analogue) computer, AN/APN-189 Doppler navigation equipment, and head-up displays for both crew members; several other installations were improved versions of the 'Mk I' systems in the F-111A and E. These avionics had great potential but proved extremely costly and troublesome.

General Dynamics FB-111A

The fourth and last tactical version, the F-111F, has been described as 'the aircraft that the F-111 should have been from the beginning'. It has a much more powerful TF30 engine, with which it first flew in May 1973, and 'Mk IIB' avionics which, while more advanced than those in the A and E, are less complex than those in the F-111D. In this form the F-111 finally emerged as the superb combat aeroplane that it was planned to be, with excellent range, efficiency and reliability in the worst possible weather. Unfortunately, by the time that it was ready for production, costs had risen so much that the USAF could only buy 94 aircraft. The F-111E wing is

based at RAF Upper Heyford, and the F-111F wing at RAF Lakenheath, both in England.

However, the combined force of Ds and Fs make the total look rather more respectable, and the Tactical Air Command has improved its overall effectiveness still further with the introduction of the EF-111A.

The other major basic version, serving with the Strategic Air Command, is the FB-111A. This has, in effect, the F-111D fuselage and in-

The F-111 has been in TAC service since 1967, and two wings are still equipped with the type. This is an early F-111A, some of which are being converted to EF standard.

154

takes, the larger wings of the F-111B/C, strengthened landing gear, and yet another variant of the TF30 engine. For the strategic role, the avionics are related to the Mk IIB fit. As an alternative to bombs the FB-111A can be equipped with six Boeing AGM-69A SRAMs (short-range attack missiles), two carried internally. Seventy-six FB-111As were built (instead of the 210 planned), and these equip two 30-aircraft wings.

Specification

Type: two-seat all-weather attack aircraft (F-111), electronic warfare aircraft (EF-111) and strategic bomber (FB-111)

Powerplant: two Pratt & Whitney TF30 afterburning turbofans: TF30-P-3s of 8390-kg (18,500-lb) static thrust in A and C; TF30-P-9s of 8890-kg (19,600-lb) static thrust in D and E; TF30-P-100s of 11385-kg (25,100-lb) static thrust in F; TF30-P-7s of 9230-kg (20,350-lb) static thrust in FB-111

Performance: maximum speed (clean) at 10670 m (35,000 ft) and above, Mach 2.2 2335 km/h (1,450 mph); maximum speed (clean) at low level Mach 1.2 (1287 km/h; 800 mph); range with internal and external fuel (A and C) 5093 km (3,165 miles), (F) more than 4707 km (2,925 miles), (EF) 3889 km (2,416 miles); service ceiling (clean) (A) 15550 m (51,000 ft), (F) 18300 m (60,000 ft), (EF) 15250 m (50,000 ft)

Weights: empty (A) 20943 kg (46,172 lb), (C) 21455 kg (47,300 lb), (D and E) about 22226 kg (49,000 lb), (F) 21398 kg (47,175 lb), (FB) about 22680 kg (50,000 lb), (EF) 24313 kg (53,600 lb); maximum take-off (A) 41504 kg (91,500 lb), (C) 51846 kg (114,300 lb), (D and E) 44906 kg

(99,000 lb), (F) 45359 kg (100,000 lb), (FB) 53977 kg (119,000 lb), (EF) 39825 kg (87,800 lb)

Dimensions: span fully spread (A, D, E and F) 19.20 m (63 ft 0 in), (C and FB) 21.34 m (70 ft 0 in); span fully swept (A, D, E and F) 9.74 m (31 ft 11½ in), (C and FB) 10.34 m (33 ft 11 in); length 22.40 m (73 ft 6 in), (EF) 23.47 m (77 ft 0 in); height 5.22 m (17 ft 1½ in), (EF) 6.10 m (20 ft 0 in); wing area fully spread (A, D, E and F) 48.77 m² (525

sq ft), (C and FB) 51.10 m² (550 sq ft); wing area fully swept (A, D E and F) 61.07 m² (657.3 sq ft)

Armament: (F) two 341-kg (750-lb) B-43 bombs, or one 20-mm M61 multi-barrel cannon and one B-43 bomb, in internal weapons bay; three underwing hardpoints on each outer wing panel, the inner four pivoting to keep stores aligned as wings sweep, the outer two non-pivoting and jettisonable. All six wing points 'wet', for carriage of drop-tanks instead of weapons;

The FB-111A carries the Boeing SRAM for nuclear strike. Two wings serve on the east coast of America as part of SAC's nuclear bomber force.

maximum ordnance load (E) 13154 kg (29,000 lb), (FB) 17010 kg (37,500 lb) as fifty 341-kg (750-lb) bombs, two in internal bay and 48 on wing pylons

Operators: Australia, USAF

General Dynamics (Grumman) EF-111A Raven

Development of this ECM variant of the F.111 can be traced back to 1972. In January 1975 Grumman received an $89.5 million contract covering the conversion of two F-111As into EF prototypes. Flight testing of an F-111A with mock up canoe fairing began in December 1975, followed on 10 March 1977 by a fully representative aerodynamic prototype, and on 17 May 1977 the first full-system EF-111A took to the air.

Testing of this aircraft took several years, involving deployment to the 366th Tactical Fighter Wing at Mountain Home AFB, Idaho.

For many years USAF planning has envisaged the conversion of 42 aircraft (including the two prototypes) for service with Tactical Air Command and USAFE. The initial production conversions were allocated to the 366th TFW at Mountain Home, which is the prime US base

for the Raven. Deliveries to USAFE began in February 1984 with the assignment of the first of 12 to the 42nd Electronic Combat Squadron at RAF Upper Heyford, Oxon.

Recognisable from the fin cap fairing which contains the receivers for the tactical jamming system, the EF-111A fulfils three basic missions: stand-off, penetration and close air support. In the first task, jamming aircraft operate in sup-

posedly secure airspace close to the battle area but outside the range of enemy ground-based weaponry, providing jamming to screen the approach of friendly strike aircraft In penetration missions EF-111As

Under a Grumman contract, early F-111As are being converted to EF-111A for tactical jamming. These escort other strike aircraft on the mission.

'ride shotgun', accompanying strike aircraft to the target area. In the close air support role they will neutralize enemy radars to permit friendly attack forces to engage enemy armour more or less unhindered.

The heart of the EF-111A is the ALQ-99E tactical jamming system, this being a derivative of the ALQ-99 in the Grumman EA-6B Prowler. A much improved system, the ALQ-99E can be handled by just one EWO (electronic warfare officer). Jamming transmitters are located in what was the weapons bay of the F-111A, antennae (aerials) for these being covered by a 4.88 m (16 ft) canoe-radome along the belly. Specialized mission equipment weighs about 2722 kg (6,000 lb).

Grumman EF-111A Raven, used for equipment tests.

Specification

Type: tactical jamming platform
Powerplant: two 8391-kg (18,500-lb) afterburning thrust Pratt & Whitney TF30-P-3 turbofan engines
Performance: maximum speed at optimum altitude 1867 km/h (1,160 mph); penetration speed at combat weight 919 km/h (571 mph); combat radius with reserves in stand-off role

370 km (230 miles); combat radius with reserves in penetration role 1495 km (929 miles); ferry range 3706 km (2,303 miles)
Weights: empty 25072 kg (55,275 lb); combat take-off 31751 kg (70,000 lb); maximum take-off 40370 kg (89,000 lb)
Dimensions: span, spread 19.20 m (63 ft 0 in), swept 9.74 m

(31 ft 11.4 in); length 23.16 m (76 ft 0 in); height 6.10 m (20 ft 0 in); wing area 48.77 m^2 (525.0 sq ft)
Armament: none
Operator: USAF (Tactical Air Command and United States Air Forces in Europe)

Grumman A-6 Intruder

During the Korean War the US services flew more attack missions than any other, in the case of the US Navy and US Marine Corps mostly with elderly piston-engined aircraft. What they learned during this conflict convinced them of the need for a specially-designed jet attack aircraft that could operate effectively in the worst weather. In 1957 eight companies submitted 11 designs in a US Navy competition for a new long-range, low-level tactical strike aircraft. Grumman's G-128, selected on the last day of the year, was to fulfil that requirement admirably, becoming a major combat type in the later war in South-East Asia, and leading to a family of later versions.

Eight development A-6As (originally designated A2F-1) were ordered in March 1959, a full-scale mockup was completed and accepted some six months later, and the first flight was made on 19 April 1960. The jetpipes of its two 3856-kg (8,500-lb) static thrust Pratt & Whitney J52-P-6 engines were designed to swivel downwards, to provide an additional component lift during take-off, but this feature was omitted from production aircraft, which instead have jet-pipes with a permanent slight

The Marine Corps flies the A-6 Intruder as its heavy attack component. This aircraft is from VMA(AW)-121.

downward deflection. The first production A-6As were delivered to US Navy Attack Squadron VA-42 in February 1963, and by the end of the following year deliveries had reached 83, to VA-65, VA-75 and VA-85 of the US Navy and VMA(AW)-242 of the US Marine Corps. First unit to fly on combat duties in Vietnam was VA-75, whose A-6As began operating from the USS *Independence* in March 1965, and from then onwards Intruders of various models became heavily involved in fighting in South-East Asia. Their DIANE (Digital Integrated Attack Navigation Equipment) gave them a first-

Marine Corps planning envisages no replacement for the A-6, preferring instead to update existing aircraft.

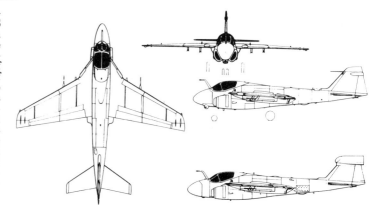
Grumman A-6E Intruder
Below: Grumman EA-6A

class operating ability and efficiency in the worst of the humid, stormy weather offered by the local climate, and with a maximum ordnance load of more than 7711 kg (17,000 lb) they were a potent addition to the US arsenal in South-East Asia.

Production of the basic A-6A ran until December 1969 and totalled 482 aircraft, plus another 21 built as EA-6As, retaining a partial strike capability but developed primarily to provide ECM (electronic countermeasures) support for the A-6As in Vietnam and to act as Elint (electronic intelligence) gatherers. The first EA-6A was flown in 1963, and six A-6As were also converted to EA-6A configuration. A more sophisticated electronic warfare version, the EZ-6B, is described separately.

The next variants of the Intruder were also produced by the conversion of existing A-6As. First of these (19 converted) was the A-6B, issued to one USN squadron and differing from the initial model primarily in its ability to carry the US Navy's AGM-78 Standard ARM (anti-radiation missile) instead of the AGM-12B Bullpup. For identifying and acquiring targets not discernible by the aircraft's standard radar, Grumman then modified 12 other A-6As into A-6Cs, giving them an improved capability for night attack by installing FLIR (forward-looking infra-red) and low light level TV equipment in a turret under the fuselage. A prototype conversion of an A-6A to KA-6D in-flight refuelling tankers flown on 23 May 1966, and production contracts for the tanker version were placed, These were subsequently cancelled, but 62 A-6As were instead converted to KA-6D configuration, equipped with Tacan (tactical air navigation) instrumentation and mounting a hose-reel unit in the rear fuselage to refuel other A-6s under the 'buddy' system. The KA-6D is also to operate as a day bomber, or as an air/sea rescue control aircraft, and since the withdrawal of the EKA-3B from seagoing duty has been the standard carrier-based tanker.

On 27 February 1970, Grumman flew the first example of the A-6E, an advanced, upgraded development of the A-6A, which the A-6E succeeded in production. Procurement of nearly 350 of this version is planned for the USN and USMC squadrons, of which some 120 are newly-built and about 230 are converted from A-6As. The basis of the A-6E, which retains upgraded forms of the airframe and powerplant of the earlier models, is a new avionics fit, founded on the addition of a Norden AN/APQ-148 multi-mode navigation/attack radar, and IBM/Fairchild AN/ASQ-133 computerized navigation/attack system, Conrac armament control unit, and an RCA video-tape recorder for assessing the damage caused during a strike mission. The Norden radar provides ground mapping, terrain avoidance/clearance, and target identification/tracking/range-finding modes, with cockpit displays for both the pilot and navigator/bombardier, who sit side by side in the well-forward cockpit. It replaces the two older radars of the A-6A.

Following the first flight of a test aircraft on 22 March 1974, all US Navy and US Marine Corps Intruders are to be progressively updated still further under a programme known as TRAM (Target Recognition Attack Multisensor). To the A-6E-standard Intruder, this adds a Hughes turreted electro-optical package of FLIR and laser detection equipment, integrated with the Norden radar; adds CAINS (Carrier Airborne Inertial Navigation System) to the existing navigation equipment; provides the capability for automatic landings on carrier decks; and incorporates provision for the carriage and delivery of automatic-homing and laser-guided air-to-surface weapons. The first US Navy squadron to be equipped with the A-6E/TRAM version was VA-165, which was deployed aboard the USS *Constellation* in 1977.

The title of 'miniature B-52' (bestowed by the North Vietnamese and Viet Cong) is well earned, for the Intruder's maximum weapon

load, all carried externally represents about 30 per cent of its maximum land take-off weight, and can be made up of a greater variety of weapons, nuclear or conventional, than any previous US naval attack aircraft. With its truly all-weather operating ability, plus a highly sophisticated set of avionics for navigation and pin-point precision bombing by day or night, it is certain to maintain a highly important contribution to US naval air power on land or at sea for many years to come.

Specification
Type: two-seat carrier or shore-based attack aircraft
Powerplant: two 4218-kg (9,300-lb) static thrust Pratt & Whitney J52-P-8A or -8B turbojets
Performance: maximum speed at sea level (A-6A, clean) 1102 km (685 mph), (A-6E, clean) 1043 km/h (648 mph); maximum speed at high altitude (A-6A, clean) 1006 km/h (625 mph); range with full weapon load (A-6E) 3096 km (1,924 miles); ferry range with maximum internal and external fuel (A-6E) 4382 km (2,723 miles); maximum rate of climb at sea level (A-6E, clean) 2804 m (9,200 ft) per minute; service ceiling (A-6A) 12700 m (41,660 ft), (A-6E, clean) 14480 m (47,500 ft)
Weights: empty (A-6A) 11650 kg

One squadron of A-6s is based on each US Navy carrier, complete with extra KA-6D tanker aircraft. The A-6s are being constantly updated.

(25,684 lb), (EA-6A) 12596 kg (27,769 lb), (A-6A) 11675 kg (25,740 lb); maximum take-off (A-6E, catapult) 26580 kg (58,600 lb), (A-6E, field) 27397 kg (60,400 lb)
Dimensions: span 16.15 m (53 ft 0 in); span folded 7.72 m (25 ft 4 in); length 16.69 m (54 ft 9 in); height 4.93 m (16 ft 2 in); wing area 49.1 m² (528.9 sq ft)
Armament: one underfuselage and four underwing attachments for maximum external load of 6804 kg (15,000 lb) in A-6A, or 8165 kg (18,000 lb) in A-6E; wide variety of nuclear or conventional weapons, typical loads ranging from 30 227-kg (500-lb) bombs, in clusters of six, to three 907-kg (2,000-lb) bombs plus two 1135-litre (250-gallon) drop-tanks; air-launched missiles can include Bullpup (A-6A), standard ARM (A-6B) or Harpoon (A-6E)
Operators: US Marine Corps, US Navy

Recent addition to the A-6E is the TRAM turret under the nose. This contains LLTV and FLIR to improve adverse weather and night accuracy.

Grumman EA-6 Prowler

Derived from the A-6 Intruder, the Grumman EA-6B Prowler qualified for a new name by reason of its highly specialized role: electronic warfare (EW). Having identified the need for such an aircraft in the early 1960s, the US Navy and US Marine Corps took the interim step of converting a dozen A-6As to EA-6A Intruder standard before the prototype Prowler, with much increased capability, made its first flight in 1968.

Intended to precede or accompany naval aircraft penetrating hostile airspace and to provide them with protection against radars associated with enemy SAMs, interceptors and surveillance units, the Prowler is equipped with the powerful Cutler-Hammer ALQ-99 active jamming system. Recognition features include the fin-tip pod (containing the ALQ-99's receiver aerials) and the extended cockpit for two EW operators positioned in Martin-Baker GRU.7 ejection seats immediately behind the flight crew. The active jammers are carried in up to five external pods, each powered by a windmill generator and each handling one (or two) waveband(s), which may be replaced by fuel tanks when the aircraft is engaged on passive intelligence-gathering.

Sole shipboard fixed-wing EW aircraft in the world, the Prowler will remain in production until 1990 against a requirement for 132 aircraft to equip 12 squadrons, plus 18 for the US Marines. Intended to remain in service for a further 20 years or so, the Prowler has already been subject to three modification

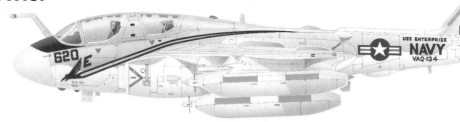

Grumman EA-6B Prowler of VAW-134. The Prowler provides both ECM protection for strike aircraft and for the carrier in the event of an attack. As well as internal gear, the EA-6B carries four ALQ-99 pods under the wings.

programmes to extend its jamming repertoire and improve receivers, displays and software, the most recent being known as ICAP (Improved Capabilty) II. The Prowler can provide active jamming in Bands 1-2 and 4-9, covering the full range of Soviet radars, including those used in conjunction with surface-to-surface missiles likely to be encountered.

Specification
Grumman EA-6B Prowler
Type: four-seat carrier- or land-based ECM jammer aircraft
Powerplant: two 5080-kg (11,200-lb) thrust Pratt & Whitney J52-P-408 turbojet engines
Performance: (with five external pods) maximum speed 1003 km/h (623 mph); cruising speed 774 km/h (481 mph) at optimum altitude; service ceiling 11580 m (38,000 ft); ferry range 3254 km (2,022 miles)
Weights: empty 14588 kg (32,161 lb); loaded, stand-off jamming 24703 kg (54,461 lb); maximum take-off 29484 kg

(65,000 lb)
Dimensions: span 16.15 m (53 ft 0 in); length 18.24 m (59 ft 10 in); height 4.95 m (16 ft 3 in); wing area 49.13 m^2 (528.90 sq ft)
Armament: none
Operator: USA (US Navy, US Marine Corps)

In common with other US Navy aircraft, the EA-6Bs have received a low visibility grey scheme, with toned-down markings. This aircraft is from VAQ-135, part of Air Wing One and usually deployed on USS *America*. The inflight-refuelling probe is prominent.

Grumman C-2 Greyhound

COD (carrier on-board delivery) is the prime function of the Grumman C-2A Greyhound, and only the US Navy possesses a specially-built aircraft for this role. Typical COD missions between distant carriers, or from shore to ship, involve transfer of personnel and urgently-needed equipment and supplies, ensuring maximum operability of embarked combat forces. With the Greyhound the cargo can be up to 28 passengers, 12 stretchers and attendants or 4536 kg (10,000 lb) of catapulted freight, increasing to 6804 kg (15,000 lb) for a land take-off. As such, the aircraft represents a faster and more efficient method of transport than the commando-type of helicopter which most navies have to use for this work.

Derived from the Hawkeye, the Greyhound entered service in 1966 and differed in having a bigger fuselage with rear-loading doors. The first batch totalled 19 aircraft based on the E-2A, of which most remain in use with land-based squadrons at locations as far apart as Japan and Italy. The C-2A has all-weather capability and, of course, provision for catapult take-offs and arrested landings.

Considerable expansion of the COD force, involving replacement of the veteran Grumman C-1 Trader, began in January 1985 with delivery of the first of a further 39 Greyhounds from re-launched production. Though still known as C-2As, the new aircraft parallel the current E-2C through installation of uprated engines and avionics, plus better

The Grumman C-2 Greyhound is used for carrier onboard delivery (COD), and is usually based on land, although it gets assigned to a carrier when it is needed.

corrosion protection. Passenger comfort has also received attention, and an APU (auxiliary power unit) gives a measure of autonomy when operating into poorly-equipped airfields.

Specification
Type: carrier on-board delivery aircraft

Powerplant: two 3661-ekW (4.910-eshp) Allison T56-A-425 turboprop engines
Performance: maximum speed 574 km/h (357 mph) at optimum altitude; cruising speed 482 km/h (300 mph); service ceiling 10210 m (33,500 ft); range, with 4356 kg (10,000 lb) of cargo 1931 km (1,200 miles); ferrying range 2890 km

(1,796 miles)
Weights: empty 14131 kg (31,154 lb); maximum take-off 24655 kg (54,354 lb)
Dimensions: span 24.56 m (80 ft 7 in); length 17.32 m (56 ft 10 in); height 4.83 m (15 ft 10 in); wing area 65.03 m^2 (700.0 sq ft)
Armament: none
Operator: US Navy

The C-2 has much in common with the E-2 Hawkeye, but features a far deeper fuselage to house cargo.

Grumman E-2 Hawkeye

First line of defence from air attack for the US Navy, the Grumman E-2 Hawkeye carrier-based airborne early warning system has been in service since 1964, and now flies with a dozen seagoing squadrons. Patrolling at 9145 m (30,000 ft) in all weathers with a flight crew of two and three systems operators, the Hawkeye's principal sensor is the APA-171 aerial (antenna) which is housed in a distinctive rotodome. In conjunction with the APS-125 ARPS (Advanced Radar Processing System) and an ALR-59 passive system, the discus shaped scanner, turning six times each minute, can locate aircraft up to 483 km (300 miles) distant, its main concern being those attempting to infiltrate beneath the coverage of ships' radars.

Highly automated systems aboard the Hawkeye enable the aircraft to track more than 250 targets and simultaneously to control at least 30 interceptions by friendly fighters. It can identify an object the size of a cruise missile 185 km (115 miles) distant, and is also capable of monitoring shipping and movements of vehicles on land.

Production of the original E-2A totalled 56, survivors of which became E-2B aircraft with the addition of enhanced computing capability. The present E-2C, which has a more reliable and easily maintained radar, is due to be manufactured until the last of 103 for the US Navy is delivered in 1991. Improvements now being incorporated in this version include a TRAC-A (total radiation aperture control antenna) for increased resistance to electronic jamming from Soviet aircraft such as 'Backfire', and even greater radar range.

No other navies fly the Hawkeye, but small numbers have been supplied to Israel, which has used the aircraft to direct attack missions as well as co-ordinate defence, notably over the Bekaa Valley in 1983.

Four Hawkeyes are flown by Israel for AEW duties, and have more than proved their worth in recent fighting with Syria.

France and Spain may become customers, acquiring aircraft for land-based operation.

Specification
Grumman E-2C Hawkeye
Type: airborne early warning radar and control aircraft
Powerplant: two 3661-kW (4,910-shp) Allison T56-A-425 turboprop engines
Performance: maximum speed 598 km/h (372 mph); long-range cruising speed 499 km/h (310 mph); service ceiling 9390 m (30,800 ft); endurance with maximum fuel 6 hours 6 minutes
Weights: empty 17265 kg (38,063 lb); maximum take-off 23556 kg (51,933 lb)
Dimensions: span 24.56 m (80 ft 7 in); length 17.55 m (57 ft 6.75 in); height 5.58 m (18 ft 3.75 in); wing area 65.03 m^2 (700.0 sq ft)

VAW-125 flies the E-2C from USS *America* providing the carrier with AEW and control coverage.

Armament: none
Operators: Egypt, Israel, Japan, Singapore, USA (Navy)

Japan has eight Hawkeyes for land-based operations. These are used in conjunction with the F-15 fighter force, a combination which Israel uses.

Grumman F-14 Tomcat

Unquestionably one of the finest warplanes in the world today, the Grumman F-14 Tomcat is fulfilling the role that, in the mid-1960s, it was hoped would be undertaken by the naval version of the General Dynamics F-111 variable-geometry strike aircraft. By the time that the F-111B programme was eventually cancelled in the summer of 1968, Grumman (also responsible for developing the F-111B version) had already reached an advanced stage in designing a new swing-wing carrier fighter, following a US Navy competition in which four other designs were in contention. From these, the US Navy selected Grumman's G-303 proposal in January 1969 to fill this major gap in its front-line inventory. In the following May a detailed mock-up was completed for US Navy evaluation, and in the same year an initial contract was placed with Grumman for six development aircraft (later increased to 12). The first of these made its maiden flight on 21 December 1970, but nine days later, on only its second flight, this prototype was coming in to land when the complete hydraulic system failed, resulting in the loss of the aircraft, although both crew members were able to eject to safety. Despite this setback, however, the development programme proceeded without further serious mishap, the second aircraft making its first flight on 24 May 1971.

Designed to later technology than the pioneering F-111, the F-14A was designed from the outset for operation from USN fleet carriers, and is unique among variable-geometry aircraft so far designed in having, in addition to variable-sweep outer wings, a smaller moveable foreplane (Grumman calls it a glove vane) inside the leading-edge root of the fixed inboard portion of each wing (the glove box). The outer wings pivot to give a leading-edge sweep of 68° when in the fully-aft position, reducing to 20° when the wings are in the fully-forward position. As the main wings pivot backwards, the glove vanes can be extended forward into the airstream to regulate any alterations in the centre of pressure and prevent the aircraft from pitching. By deploying its variable-sweep wings to the best advantage, the Tomcat is thus able to vary its flying configuration to the different aerodynamic and performance requirements needed when taking off from, or landing on, a carrier, taking part in an air-to-air dogfight, or carrying out a low-level attack mission against a surface target. In air-to-air fighting, the variation of wing sweep

can be undertaken by the in-built automatic flight control system, relieving the pilot of this task and enabling him to concentrate on out-manouvring and shooting down his opponent. With additional control surfaces which include full-span trailing-edge flaps, spoilers, leading-edge slats and all-moving horizontal tail surfaces, the Tomcat is a superbly manoeuvrable warplane; longitudinal stability is assured by the use of twin outward-canted fins and rudders. It is also very strong structurally, many of the airframe components being manufactured of boron-epoxy or other composites, or titanium.

The primary role of the Tomcat is to provide long-range air defence of the US fleet, and the two-man crew (pilot and naval flight officer) are seated in tandem on zero-zero ejection seats under a single long, upward-opening canopy, which affords a fine all-round view. An excellent weapons platform, the Tomcat's armament for the air defence role includes air-to-air missiles such as the medium-range AIM-7 Sparrow and close-range AIm-9 Sidewinder (the latter being mounted on launchers attached beneath the glove box on each side), and for unexpected dogfights a Gatling-type multi-barrel cannon. Primary interception armament, however, consists of six Hughes Phoenix air-to-air missiles, four of which are mounted on pallets which fit into the semi-recessed Sparrow positions under the aircraft's belly, with an additional Phoenix underneath each wing glove box alongside the underwing Sidewinders. The Phoenix is currently the longest-range air-to-air missile in use anywhere in the world (more than 200 km/124 miles), and the Tomcat is the only combat aircraft equipped to carry it. In conjunction with the extremely powerful Hughes AWG-9 radar mounted in the nose, it provides the Tomcat with the unique ability to detect and

Although primarily a long-range interceptor, the F-14 is a formidable dogfight opponent. The wing sweep is controlled automatically by computer to extract the best from the aircraft during a fight.

attack an airborne target while it is still 160 km (100 miles) away. The F-14A also has a secondary capability in the lowe-level attack role, in which event the air-to-air missiles can be replaced by up to 6577 kg (14,500 lb) of externally-mounted bombs or other weapons.

The initial F-14A model of the Tomcat has been in service with the US Navy since October 1972, when the first deliveries were made to squadron VF-1 and VF-2. It was an aircraft of the latter unit which, in March 1974, flew the first operational Tomcat sortie from the carrier USS *Enterprise*. Subsequent acceptance of the fighter into regular USN service was both enthusiastic and without major incident until 1975, when a series of powerplant, structural and systems problems were encountered. However, these have been largely resolved, one resultant modification being an increase in available thrust with the after-burners on. (Shortage of power and other propulsion problems have threatened replacement with different engines, however.)

The Naval Air Training Center at Patuxent River, Maryland, developed a tactical air reconnaissance pod system (TARPS) to extend further the versatility of the F-14A, seen as

the interim replacement for the Rockwell RA-5C Vigilante.

Eighty F-14As were exported to the Imperial Iranian Air Force in the mid-1970s, but these have suffered greatly with maintenance problems since the revolution.

Specification
Type: tandem two-seater carrier-borne multi-role fighter
Powerplant: two 9480-kg (20,900-lb) static thrust Pratt & Whitney TF30-P-412A after-burning turbofans
Performance: maximum speed at altitude Mach 2.34 or 2517 km/h (1,564 mph); maximum speed at sea level Mach 1.2 or 1470 km/h (910 mph); range (interceptor, with external fuel) about 3200 km (2,000 miles); service ceiling over 17070 m (56,000 ft); maximum rate of climb at sea level (normal gross weight) over 9145 m (30,000 ft) per minute
Weights: empty 17830 kg (39,310 lb); normal take-off 26553 kg (58,539 lb); take-off with four Sparrows 26718 kg (58,904 lb); take-off with six Phoenix 31656 kg (69,790 lb); maximum take-off 33724 kg (74,348 lb)
Dimensions: span unswept 19.45 m (64 ft 1½ in); span swept 11.65 m (38 ft 2½ in); length 18.89 m (61 ft

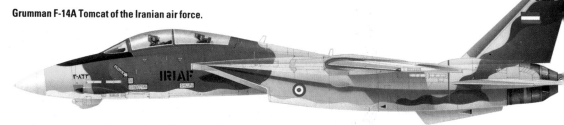
Grumman F-14A Tomcat of the Iranian air force.

Grumman F-14A Tomcat cutaway drawing key

1 Pitot tube
2 Radar target horn
3 Glass-fibre radome
4 IFF aerial array
5 Hughes AWG-9 flat plate radar scanner
6 Scanner tracking mechanism
7 Ventral ALQ-126 deception jamming antenna
8 Gun muzzle blast trough
9 Radar electronics equipment bay
10 AN/ASN-92 inertial navigation unit
11 Radome hinge
12 Inflight refuelling probe (extended)
13 ADF aerial
14 Windscreen rain removal air duct
15 Temperature probe
16 Cockpit front pressure bulkhead
17 Angle of attack transmitter
18 Formation lighting strip
19 Cannon barrels
20 Nosewheel doors
21 Gun gas vents
22 Rudder pedals
23 Cockpit pressurization valve
24 Navigation radar display
25 Control column
26 Instrument panel shroud
27 Kaiser AN/ANG-12 head-up display
28 Windscreen panels
29 Cockpit canopy cover
30 Face blind seat firing handle
31 Ejection seat headrest
32 Pilot's Martin-Baker GRU-7A ejection seat
33 Starboard side console
34 Engine throttle levers
35 Port side console panel
36 Pitot static head
37 Canopy emergency release handle
38 Fold out step
39 M61A1 Vulcan 20-mm six-barrel rotary cannon
40 Nose undercarriage leg strut
41 Catapult strop link
42 Catapult strop, launch position
43 Twin nosewheels
44 Folding boarding ladder
45 Hughes AIM-54A Phoenix air-to-air missile (six)

46 Fuselage missile pallet
47 Cannon ammunition drum (675 rounds)
48 Rear boarding step
49 Ammunition feed chute
50 Armament control panels
51 Kick-in step
52 Tactical information display hand controller
53 Naval Flight Officer's instrument console
54 NFO's ejection seat
55 Starboard intake lip
56 Ejection seat launch rails
57 Cockpit aft decking
58 Electrical system controller
59 Rear radio and electronics equipment bay
60 Boundary layer bleed air duct
61 Port engine intake lip
62 Electrical system relay controls
63 Glove vane pivot
64 Port air intake
65 Glove vane housing
66 Navigation light
67 Variable area intake ramp doors
68 Cooling system boundary layer duct ram air intake
69 Intake ramp door hydraulic jacks
70 Air system piping
71 Air data computer
72 Heat exchanger
73 Heat exchanger exhaust duct

74 Forward fuselage fuel tanks
75 Canopy hinge point
76 Electrical and control system ducting
77 Control rod runs
78 UHF/TACAN aerial

79 Glove vane hydraulic jack
80 Starboard glove vane, extended
81 Honeycomb panel construction
82 Navigation light
83 Main undercarriage wheel bay
84 Starboard intake duct spill door
85 Wing slat/flap flexible drive shaft
86 Dorsal spine fairing
87 Fuselage top longeron
88 Central flap/slat drive
89 Emergency hydraulic generator
90 Bypass door hydraulic jack
91 Intake bypass door
92 Port intake ducting
93 Wing glove sealing horn
94 Flap/slat telescopic drive shaft
95 Port wing pivot bearing
96 Wing pivot carry-through (electron beam welded titanium box construction)
97 Wing pivot box integral fuel tank
98 Fuselage longeron/pivot box attachment joint
99 UHF data link/IFF aerial
100 Honeycomb skin panelling
101 Wing glove stiffeners/dorsal fences
102 Starboard wing pivot bearing

2 in); height 4.88 m (16 ft 0 in); wing area 52.49 m^2 (565.0 sq ft)
Armament: one General Electric M61A-1 20-mm multi-barrel Vulcan cannon in forward fuselage with 675 rounds; four AIM-7 Sparrow or AIM-54 Phoenix air-to-air missiles under fuselage; two AIM-9 Sidewinder air-to-air missiles, or one Sidewinder plus one Phoenix or Sparrow, under each wing glove box; tactical reconnaissance pod containing cameras and electro-optical sensors; or up to 6577 kg (14,500 lb) of Mk 82/83/84 bombs or other weapons
Operators: Iran, US Navy

The Tomcat's main weapon is the Hughes AIM-54 Phoenix; a maximum of six may be carried. These have enormous range, can hit targets at all heights and can be launched against six targets.

103 Slat/flap drive shaft gearbox
104 Starboard wing integral fuel tank (total internal fuel capacity 2,364 US gal/8949 litres)
105 Leading edge slat drive shaft
106 Slat guide rails
107 Starboard leading edge slat segments (open)
108 Starboard navigation light
109 Low-voltage formation lighting
110 Wing tip fairing
111 Outboard manoeuvre flap segments (down position)
112 Port roll control spoilers
113 Spoiler hydraulic jacks
114 Inboard, high lift flap (down position)
115 Inboard flap hydraulic jack
116 Manoeuvre flap drive shaft
117 Variable wing sweep screw jack
118 Starboard main undercarriage pivot fixing
119 Starboard engine compressor face
120 Wing glove sealing plates
121 Pratt & Whitney TF30-P-412 afterburning turbofan
122 Rear fuselage fuel tanks
123 Fuselage longeron joint
124 Control system artificial feel units
125 Tailplane control rods
126 Starboard engine bay
127 Wing glove pneumatic seal
128 Fin root fairing
129 Fin spar attachment joints
130 Starboard fin leading edge
131 Starboard all-moving tailplane
132 Starboard wing (fully swept position)
133 AN/ALR-45 tail warning radar antenna
134 Fin aluminium honeycomb skin panel construction
135 Fin-tip aerial fairing
136 Tail navigation light
137 Electronic counter-measures antenna (ECM)
138 Rudder honeycomb construction
139 Rudder hydraulic jack
140 Afterburner ducting
141 Variable area nozzle control jack
142 Airbrake (upper and lower surfaces)
143 Airbrake hydraulic jack
144 Starboard engine exhaust nozzle
145 Anti-collision light
146 Tail formation light
147 ECM aerial
148 Port rudder
149 Beaver tail fairing
150 Fuel jettison pipe
151 ECM antenna
152 Deck arrester hook (stowed position)
153 AN/ALE-29A chaff and flare dispensers
154 Nozzle shroud sealing flaps
155 Port convergent/divergent afterburner exhaust nozzle
156 Tailplane honeycomb construction
157 AN/ALR-45(V) tail warning radar antenna
158 Tailplane boron fibre skin panels
159 Port wing (fully swept position)
160 All-moving tailplane construction
161 Tailplane pivot fixing
162 Jet pipe mounting
163 Fin/tailplane attachment mainframe
164 Cooling air louvres
165 Tailplane hydraulic jack
166 Hydraulic system equipment pack
167 Formation lighting strip
168 Oil cooler air intake
169 Port ventral fin
170 Engine accessory compartment
171 Ventral engine access doors
172 Hydraulic reservoir
173 Bleed air ducting
174 Port engine bay
175 Intake compressor face
176 Wing variable sweep screw jack
177 Main undercarriage leg strut
178 Hydraulic retraction jack
179 Wing skin panel
180 Fuel system piping
181 Rear spar
182 Flap hinge brackets
183 Port roll control spoilers
184 Flap leading edge eyebrow seal fairing
185 Port manoeuvre flap honeycomb construction
186 Wing tip fairing construction
187 Low-voltage formation lighting
188 Port navigation light
189 Wing rib construction
190 Port wing integral fuel tank
191 Front spar
192 Leading edge rib construction
193 Slat guide rails
194 Port leading edge slat segments, open
195 Slat honeycomb construction
196 Port mainwheel
197 Torque scissor links
198 Main undercarriage front bracing strut
199 Mainwheel well door
200 Ventral pylon attachment
201 External fuel tank (capacity 265 US gal/1003 litres)
202 Sparrow missile launch adaptor
203 AIM-7F Sparrow air-to-air missile
204 Wing glove pylon attachment
205 Cranked wing glove pylon
206 Sidewinder missile launch rail
207 AIM-9C Sidewinder air-to-air missile
208 Phoenix launch pallet
209 AIM-54A Phoenix air-to-air missile

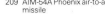

Pilot Press Limited

Typical of the US Navy's Tomcats, this aircraft is from VF-143 'Pukin' Dogs', based on USS *Dwight D. Eisenhower*. 'Dogs' Tomcats saw action covering the failed UN peace initiative in Beirut, where TARPS pod-equipped aircraft carried out many recce missions over the war-torn city. This aircraft is carrying a typical air defence mix of Phoenix, Sparrow and Sidewinder missiles. Allied to the internal cannon, this gives the Tomcat all-round capability for almost every type of air-to-air fighting.

Grumman S-2 Tracker/C-1 Trader

Grumman's S-2 Tracker is still used by several nations, some of which have only recently updated their fleets through acquisition of surplus US Navy S-2s taken from storage at Davis-Monthan AFB, Arizona.

One of the most successful ASW (Anti-Submarine Warfare) aircraft yet conceived, the Tracker first flew in December 1952 and entered service with the US Navy in 1954. It was continuously updated, while many of the earlier machines found work as utility and training aircraft, as the US-2A/B and TS-2A respectively. Some TS-2As remained active as multi-engine pilot trainers until replaced by the Beech T-44A in 1980. Most Trackers currently active are late-production S-2Es and S-2Gs, the latter being an S-2E with updated electronics.

The US Navy still uses the C-1A Trader COD (carrier on-board delivery) variant, roughly 30 remaining airworthy. With a revised fuselage configured for the carriage of cargo and personnel between ship and shore, a single example is assigned to each of the 15 active carriers as a back up to C-2As, while small numbers serve with transport squadrons in Europe, the Far East and the USA.

Specification
Grumman S-2 Tracker
Type: carrier ASW aircraft
Powerplant: two 1137-kW (1,525-hp) Wright R-1820-82WA Cyclone piston engines
Performance: maximum speed at sea level 426 km/h (265 mph); patrol speed 241 km/h (150 mph); ferry range 2092 km (1,300 miles); endurance with maximum fuel and reserves 9 hours
Weights: empty 8505 kg (18,750 lb); maximum take-off 13222 kg (29,150 lb)
Dimensions: span 22.12 m (72 ft 7 in); length 13.26 m (43 ft 6 in);

height 5.05 m (16 ft 7 in); wing area 46.08 m² (496.0 sq ft)
Armament: one Mk 47 or Mk 101 nuclear depth bomb or similar store in internal weapons bay plus a variety of bombs, rockets and torpedoes on six external stores stations. Up to 60 echo-sounding depth charges accommodated internally, with 32 sonobuoys ejected from the engine nacelles.
Operators: (S-2) Argentina, Brazil, Canada, Peru, Taiwan

Ten elderly Trackers still maintain Thailand's ASW patrols. As well as dropping sonobuoys, the S-2 has a retractable MAD sting in the tail.

Grumman S-2G Tracker

Thailand, Turkey, Uruguay, Venezuela; (C-1A Trader) US Navy

Grumman HU-16 Albatross

The prototype ordered by the US Navy for service as a utility aircraft had the designation XJR2F-1, and this flew for the first time on 23 October 1947. In service these became designated UF-1, and a modified version introduced in 1955 was UF-2. This latter aircraft had increased wing span, a cambered wing leading-edge, ailerons and tail surfaces of increased area, and more effective de-icing boots for all aerofoil leading edges. In the tri-service rationalisation of designations in 1962, these aircraft became UH-16C and HU-16D respectively. Winterized aircraft for Antarctic service were UF-1Ls, later LU-16Cs, and five UF-1T dual-control trainers became TU-16Cs.

The USAF found the G-64 attractive for rescue operations, the majority of the 305 ordered serving with the MATS Air Rescue Service under the designation SA-16A. An improved version, equivalent to the US Navy's UF-2, entered service in 1957 as the SA-16B: in 1962 these became HU-16A and HU-16B respectively. HU-16E was the designation of Albatross aircraft operated

by the US Coast Guard, and 10 supplied to Canada were CSR-110s. An anti-submarine version with nose radome, retractable MAD gear, ECM radome and searchlight was introduced in 1961, and this could carry a small number of depth charges. The versatile Albatross continues in service with Greece in the SAR role.

Specification
Type: utility and rescue amphibian
Powerplant: two 1063-kW (1,425-hp) Wright R-1820-76A/76-B

Cyclone radial piston engines
Performance: maximum speed 380 km/h (236 mph) at sea level; cruising speed 241 km/h (150 mph); range with 4826 litres (1,275 US gallons) of fuel 4345 km (2,700 miles); maximum endurance 22 hours 54 minutes

Twelve ancient HU-16Bs are still flown by Greece for maritime patrol duties around its myriad islands.

Dimensions: (HU-16D) span 29.46 m (96 ft 8 in); length 18.67 m (61 ft 3 in); height 7.87 m (25 ft 10 in); wing area 96.15 m² (1,035 sq ft)
Armament: (ASW) depth charges
Operators: Greece, Indonesia, Mexico, Philippines, Taiwan

Grumman OV-1 Mohawk

Intended primarily for battlefield reconnaissance duties, the bug-eyed Grumman OV-1 Mohawk dates back to the late 1950s, but despite its vulnerability it remains in widespread service with the US Army. Initial variants were unsophisticated, but progressive modification over the years has brought a significant improvement in sensor systems. The first model to see service was the OV-1A photographic reconnaissance aircraft with night flares and advanced navigation equipment. It was succeeded by the OV-1B equipped with APS-94 SLAR (Side-Looking Airborne Radar), in a pod under the fuselage. The next derivative was the OV-1C, which used the AAS-24 infra-red sensor in place of SLAR. The final new-build

model was the OV-1D, basically a quick-change aircraft capable of operating with either infra-red or SLAR. Deliveries terminated in 1970, these bringing total production to 375. Additional OV-1Ds were made available by converting most of the 100-plus surviving OV-1Bs and OV-1Cs.

More recently a number of other derivatives have appeared, including the RV-1D and the EV-1E. The former is a conversion of the OV-1B for Elint (electronic intelligence) duty, the dozen or so known to exist being fitted with a multiplicity of

The Mohawk is still operated in large numbers by the US Army for battlefield reconnaissance duties. Chief sensor is the large SLAR mounted in the canoe pod.

Israel flies an unknown number of Mohawks, perhaps including electronic intelligence RV-1D and EV-1E versions.

passive receivers, analysers and recorders with which to gather unknown or 'hostile' signals. The EV-1E, again a rebuilt OV-1B, is fitted with ALQ-133 'Quick Look II' surveillance radar, additional Elint equipment and electronic warfare pods. At least 16 conversions have been produced, and continued updating seems likely to result in further conversions.

Specification
Grumman OV-1D Mohawk, infra-red version
Type: two-seat STOL battlefield surveillance aircraft
Powerplant: two 1044-kW (1,400-shp) Lycoming T53-L-701 turboprop engines
Performance: maximum speed 491 km/h (305 mph); cruising speed 389 km/h (242 mph); service ceiling 7620 m (25,000 ft); range with two 568-litre (150 US gallon) drop tanks 1738 km (1,080 miles)
Weights: empty 5328 kg (11,747 lb); maximum take-off 8086 kg (17,826 lb)
Dimensions: span 14.63 m (48 ft 0 in); length 12.50 m (41 ft 0 in); height 3.86 m (12 ft 8 in); wing area 33.44 m^2 (360.0 sq ft)
Armament: normally none, but bombs, Minigun pods and rockets have been carried on underwing pylons
Operator: USA (Army)

Grumman Gulfstream II

Grumman's satisfaction with the Rolls-Royce turboprops that powered the Gulfstream I led to the selection of two Spey turbofans for the Gulfstream II, first flown on 2 October 1966. Accommodation is more spacious than that of the GI, with provision for two or three on the flight deck and up to 19 passengers in the main cabin. The US Coast Guard acquired a single VC-11A VIP transport, and a number of armed forces have procured these aircraft for similar duties.

Specification
Grumman Gulfstream II
Type: long-range executive jet
Powerplant: two 5171-kg (11,400-lb) thrust Rolls-Royce Spey 511-8 turbofan engines
Performance: maximum cruising speed 935 km/h (581 mph) at 7620 m (25,000 ft); range (NBAA VFR with 907-kg/2,000-lb payload and 30 minutes reserve) 6635 km (4,123 miles)
Weights: empty, no tip tanks 16576 kg (36,544 lb), with tip-tanks 16867 kg (37,185 lb); maximum take-off 29710 kg (65,500 lb)
Dimensions: span 20.98 m (68 ft 10 in), over tip-tanks 21.87 m (71 ft 9 in); length 24.36 m (79 ft 11 in); height 7.47 m (24 ft 6 in); wing area 75.21 m^2 (809.60 sq ft)
Armament: none
Operators: Cameroon, Gabon, Ivory Coast, Nigeria, USA (Coast Guard)

Grumman Gulfstream II

Gulfstream Aerospace Gulfstream III

First flown on 2 December 1979, the 'GIII' is an updated Gulfstream II executive jet, having a redesigned wing with winglets, longer fuselage and more fuel. Aerodynamic refinement gives an 18 per cent increase in fuel economy. The first military customer was Denmark, whose three aircraft are intended primarily for fishery patrol, although they can also undertake airdrop, medevac, SAR and transport tasks. Serving with Esk.721 at Vaerlose AB, one aircraft is usually detached to Sondrestrom, Greenland. On patrol a crew of seven is carried, including an observer and photographer. Equipment includes a Texas Instruments APS-127 sea surveillance radar, Litton 72R INS and flare launcher; the Danish 'GIII's have the ability to airdrop emergency supplies or survival equipment via a baggage door.

USAF procurement was initiated in June 1983 with the lease for two years of three C-20As. The first was delivered to the 89th Military Airlift Wing at Andrews AFB, Maryland, on 20 September 1983. Replacing the VIP C-140, contract support is being furnished by the manufacturer. Before this book appears the USAF should have purchased the aircraft and taken options for another eight to be delivered in 1988.

The Gulfstream III has been procured for the USAF under the designation C-20A, and will serve on light staff transport duties around the country. These aircraft are slowly replacing the C-140 JetStar.

Gulfstream Aerospace Gulfstream III

Specification
Type: executive jet
Powerplant: two 5171-kg (11,400-lb) thrust Rolls-Royce Spey 511-8 turbofan engines
Performance: maximum cruising speed 927 km/h (576 mph); long-range cruising speed 818 km/h (508 mph); maximum operating altitude 13715 m (45,000 ft); range with eight passengers, baggage and reserves 7593 km (4,718 miles)
Weights: empty 17237 kg (38,000 lb); maximum take-off 31615 kg (69,700 lb)
Dimensions: span 23.72 m (77 ft 10 in); length 25.32 m (83 ft 1 in); height 7.43 m (24 ft 4.5 in); wing area 86.82 m^2 (934.6 sq ft)
Armament: none
Operators: Denmark, Italy, Ivory Coast, Nigeria, USA

Gulfstream Aerospace Gulfstream IV

Continued refinement of the 'GIII' has resulted in this new variant which introduced a new engine, structurally redesigned wing, longer fuselage and carbon-fibre rudder. The flight deck has CRT displays and digital avionics. Design began in April 1982 and first flight was due in December 1985. In 1984 firm orders and options exceeded 70 aircraft.

Specification
Type: executive transport
Powerplant: two 5647-kg (12,450-lb) thrust Rolls-Royce RB183-03 Tay turbofan engines
Performance: (estimated) maximum cruising speed at 10670 m (35,000 ft) 908 km/h (564 mph); economical cruising speed at 13715 ft (45,000 ft) 850 km/h (528 mph); range with maximum payload and fuel reserves 7411 km (4,605 miles)
Weights: (estimated) empty 18098 kg (39,900 lb); maximum take-off 31615 kg (69,700 lb)
Dimensions: span 23.72 m (77 ft 10 in); length 25.93 m (85 ft 1 in); height 7.42 m (24 ft 4 in); wing area 88.29 m^2 (950.39 sq ft)
Armament: none
Operators: none yet confirmed

Harbin H-5

First flown in August 1948, the Ilyushin Il-28 proved a most successful Soviet light attack bomber, having the advantages of simplicity, low cost, reliabiality and outstanding handling at all airspeeds. Over 400 were supplied to the People's Republic of China from 1958, and after the political break with the Soviet Union the Chinese decided to build the type for themselves, together with its Klimov VK-1A engine derived from the Rolls-Royce Nene. The Chinese Hongzhaji (bomber aircraft) No. 5 has always been produced only at the Harbin complex of factories, another of which builds the engine and provides service support.

The Hongzhaji 5 differs only in details from the original Soviet aircraft, and until the late 1970s the Ilyushin aircraft remained in service alongside it. There are three main variants: the H-5 light bomber, HZ-5 reconnaissance aircraft and HJ-5 trainer. All have the same all-metal airframe, with hydraulically driven slotted flaps, sweepback on the tail surfaces only, manual flight controls, steerable nosewheel, and anti-skid brakes on the main gears, all retracted pneumatically with main wheels rotating to lie flat under the rear of each engine.

About 1,000 of the three versions are built in China, the last being completed as recently as 1981-82.

About 500 equip 12 regiments of the People's Liberation Army air force, and about 100 serve in the torpedo-bomber role with the PLA navy. The only known export customer was Albania, but there is doubt that this country's H-5s are operational.

Specification
Harbin H-5
Type: light bomber
Powerplant: two 2700 kg (5,952 lb) Harbin WP-5 turbojet engines
Performance: maximum speed 800 km/h (497 mph) at sea level, 902 km/h (560 mph) at 4500 m (14,760 ft); service ceiling 12300 m (40,350 ft); range with maximum fuel at high altitude 2180 km (1,355 miles)
Weights: empty 12890 kg (28,417 lb); normal loaded 18400 kg (40,565 lb); maximum take-off 21200 kg (46,738 lb)
Dimensions) span (no tip-tanks) 21.45 m (70 ft 4.5 in); length 17.65 m (57 ft 10.9 in); height 6.70 m (21 ft 11.8 in); wing area 60.80 m² (654.47 sq ft)
Armament: four 23-mm NR-23 guns, two in tail turret and two fixed in nose firing ahead; various loads of bombs (believed to include nuclear) or torpedoes, mines or other stores to maximum limit of 3000 kg (6,614 lb)
Operator: China

Harbin Y-11 and Y-12

One of the first aircraft designed wholly within China, the Yunshuji (transport aircraft) No. 11 is believed to have made its first flight in 1975. An outstanding STOL machine in the class of the Scottish Aviation Twin Pioneer, it is a neat and simple utility transport with an all-metal structure, apart from fabric covering on the tail control surfaces and on the full-span double-slotted flaps and drooping ailerons. Powered by two 213-kW (285-hp) Quzhou Huosai 6A (Vedeneev AI-14RF) radial engines, the Y-11 seats a pilot and up to nine other persons, the rear cabin normally seating seven. STOL field length is remarkable, take-off and landing runs typically being 140 m (459 ft), but with a payload of only 870 kg (1,918 lb) the Y-

11 was considered rather limited and it is believed only 15 were built.

The design team at Harbin studied ways of 'stretching' the Y-11 and powering it with turboprop engines, and on 14 July 1982 flew the prototype of the Y-12. This is a much more capable aircraft, and though it has almost the same wing it has a fuselage that is longer, wider and deeper, giving considerably more than twice the internal volume for twice the payload. Metal bonding is used in much of the structure, there are single (instead of twin) main wheels and comprehensive avionics for all weather operation.

Production of the Y-12 is now in hand for military as well as civil duties. The basic aircraft is versatile, and could be used for such

special purposes as parachute training, photo-surveying and mapping, firefighting and agricultural duties. Most are simple utility transports, seating two in the cockpit and up to 17 in the cabin. The main double door at the rear on the left is 1.45 m (4 ft 9 in) wide, and maximum cargo load is 1700 kg (3,748 lb). The Y-12 appears to be a most successful and useful machine which may be built in large numbers.

So far no Y-12 has been identified in service with the PLA army or navy air units, but large numbers are expected to serve with both. This aircraft is also being marketed outside China (with the name Turbo-Panda) and is again expected to find both military and civil customers.

Specification
Harbin Y-12
Type: multi-role STOL utility

transport
Powerplant: two 373-kW (500-shp) Pratt & Whiney Canada PT6A-11 or 462-kW (620-shp) PT6A-27 turboprops driving reverse-pitch Hartzell propellers
Performance: (PT6A-27 engines), maximum speed 302 km/h (188 mph); cruising speed 240 km/h (149 mph); take-off run 180 m (590 ft); service ceiling 7000 m (22,965 ft); range with full payload of 17 passengers, baggage and 45 minutes reserve 410 km (255 miles)
Weights: empty 2840 kg (6,261 lb); maximum 5500 kg (12,125 l)
Dimensions: span 17.235 m (56 ft 6.5 in); length 14.86 m (48 ft 9 in); height 5.275 m (17 ft 3.7 in); wing area 34.27 m² (368.9 sq ft)
Armament: none
Operator: China

Heliopolis Gomhouria 2

The Gomhouria, an Egyptian-built Bücker Bü 181 Bestmann, was the first military aircraft to be produced in quantity in Egypt. The Heliopolis factory was founded in 1950, and its initial production consisted of a batch of 60 Gomhouria 1s for the Egyptian air force, powered by a 78-kW (105-hp) Walter Minor. These were followed by the Continental-powered Mk 2. More than 300 Gomhourias were built. The Egyptian air force academy at Bilbeis uses these for initial student training, and a few are in use elsewhere.

Specification
Heliopolis Gomhouria 2
Type: primary trainer

Powerplant: one 108-kW (145-hp) Continental C-145 flat-six piston engine
Performance: maximum speed 220 km/h (137 mph); cruising speed 200 km/h (124 mph) at sea level; initial rate of climb 246 m (807 ft)/minute; service ceiling 4800 m (15,750 ft); range 790 m (491 miles)
Weights: empty 515 kg (1,135 lb); maximum take-off 800 kg (1,764 lb)
Dimensions: span 10.60 m (34 ft 9.3 in); length 7.90 m (25 ft 11 in); height 2.05 m (6 ft 8.7 in); wing area 13.50 m² (145.32 sq ft)
Armament: none
Operators: Algeria, Egypt, Sudan

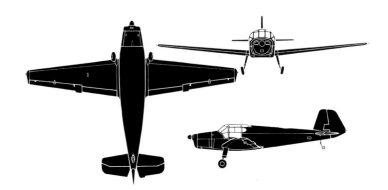

Heliopolis Gomhouria

HFB 320 Hansa

West Germany entered the business-jet field when the first prototype HFB 320 Hansa flew on 21 April 1964. It was Hamburger Flugzeugbau's first design. The first 15 had General Electric CJ610-1 engines, the following 20 the more powerful CJ610-5, and subsequent production the CJ610-9. A unique feature was the forward-swept wing, to enable the wing spars to pass behind the cabin. The biggest, and only military, user was the West German Luftwaffe, which accepted 16 of approximately 40 Hansas built. The FBS (*Flugbereitschafsstaffel*) operated six from Köln/Bonn on VIP and support flights (replaced by CL-601s), while at Lechfeld the FVSt (*Flugvormeesungsstaffel*) uses two Hansas on calibration work and four others for ECM training.

Specification
Type: twin-jet executive transport
Powerplant: two 1406-kg (3,100-lb) thrust General Electric CJ610-9 turbojet engines
Performance: maximum cruising speed 825 km/h (513 mph) at 7620 m (25,000 ft); initial rate of climb 1295 m (4,249 ft)/minute; service ceiling 12200 m (40,025 ft); range with 545 kg (1,202 lb) payload and 45 minutes reserve 2370 km (1,473 miles)
Weights: empty 5425 kg (11,960 lb); maximum take-off 9200 kg (20,283 lb)
Dimensions: span 14.45 m (47 ft 4.9 in); length 16.60 m (54 ft 5.5 in); height 4.75 m (15 ft 7 in); wing area 30.14 m² (324.43 sq ft)
Armament: none
Operator: West Germany

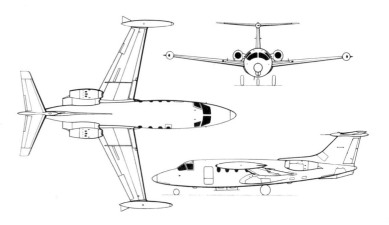

HFB 320 Hansa

Hindustan Ajeet

An Indian aircraft with a Hindu name (meaning invincible), the Ajeet is an improved version of the Folland Fo.141 Gnat. A low-powered Gnat prototype flew on 11 August 1954, and although Britain, Jugoslavia, India and Finland assessed it as a lightweight fighter, only the two last-mentioned placed production orders, and the RAF acquired a trainer version. India made a major commitment, the first step being the supply of 25 aircraft in 1958-60, plus 20 kits for assembly by Hindustan Aeronautics Ltd (HAL). This company built further batches of 80, 43 and 50. The first was flown on 21 May 1962, and the last delivered on 31 January 1974 as the 213th Gnat assembled in India. Gnats were issued to Nos 2, 9, 15, 18, 21, 22, 23 and 24 Squadrons of the IAF, and had their combat debut in the 1965 war with Pakistan. Although dubbed the 'Sabre Slayer', there is no evidence that a single F-86 was shot down. By 1981 all had been retired, some of them replaced by the Indian-developed Ajeet. Optimized for attack, the Ajeet has a wet (i.e. integral tank) wing to allow weapons to be carried on the drop-tank pylons. It also has improved tail unit controls and additional communications and navigation equipment. What would have been the 214th and 215th Gnats were completed as Ajeet prototypes, and began flying on 5 March 1975, being followed in September 1976 by production aircraft. No. 9 Sqn was the first to con-

vert from Gnats and was followed by Nos 2, 18 and 22 before the 79th and last series Ajeet was delivered in February 1982. (The other four Gnat squadrons now fly MiG-21s.) In addition, ten Gnats have been converted to Ajeets.

Interest has been shown in a trainer version, known as the Ajeet Mk 2, or Ajeet Trainer. This has two additional fuselage sections, each 0.7 m (2 ft 3.5 in) long, forward and aft of the wings to allow the extra seat to be inserted. Two fuselage tanks are deleted, though the loss may be made good by new tanks in place of the cannon. The twin-canopy prototype Ajeet Trainer flew on 20 September 1982, followed 12 months later by a second. HAL expects to build 12 for the IAF as replacements for two-seat Hunters, and eight for the Indian navy.

Specification
Hindustan Ajeet Trainer
Type: operational trainer
Powerplant: one 2200-kg (4,850-lb) thrust Rolls-Royce Orpheus 701-01AT turbojet engine
Performance: maximum speed

HAL-built Folland Gnat of the Indian Air Force.

The HAL Ajeet Trainer has a lengthened fuselage to accommodate the extra seat. These are to replace Hunter trainers in Indian Air Force service.

1070 km/h (665 mph); initial rate of climb 3240 m (10,625 ft)/minute; service ceiling 14000 m (45,925 ft); range 900 km (559 miles)
Weights: empty 2940 kg (6,482 lb); maximum take-off 4536 kg (10,000 lb)
Dimensions: span 6.73 m (22 ft 1 in); length 10.45 m (34 ft 3.4 in); height 2.58 m (8 ft 5.6 in); wing area 12.65 m² (136.17 sq ft)
Armament: two 30-mm (1.18-in) Aden cannon inside air inlets and up to 850 kg (1,874 lb) of bombs, rockets or gun pods on four underwing pylons
Operator: India

Hindustan HJT-16 Kiran

HAL decided on side-by-side Martin Baker seats for the pressurized Kiran, which bears a passing resemblance to the Jet Provost. Design began in 1959, and first flight was on 4 September 1964. Delivery of six Mk I aircraft took place in March 1968, the 50th was completed in late 1973, the 125th by the beginning of 1978, and the last of 190 Mk I and armed Mk IAs delivered in 1982.

Specification
HAL HJT-16 Kiran Mk II
Type: jet basic trainer
Powerplant: one 1873-kg (4,130-lb) thrust HAL-built Rolls-Royce Orpheus 701-01 turbojet engine
Performance: maximum speed 704 km/h (437 mph) at sea level; initial rate of climb 1600 m

(5,249 ft)/minute; service ceiling 12000 m (39370 ft); range with maximum internal fuel 615 km (382 miles)
Weights: empty 2966 kg (6,539 lb); maximum take-off 4950 kg (10,913 lb)
Dimensions: span 10.70 m (35 ft 1.3 in); length 10.25 m (33 ft 7.5 in); height 3.64 m (11 ft 11.3 in); wing area 19.00 m² (204.52 sq ft)
Armament: the Mk I has no hardpoints, the IA has two and the Mk II has four; each hardpoint can carry one 250-kg (551-lb) bomb, or a 227-litre (50-Imp gallon) drop tank, or a pod with two 7.72-mm (0.3-in) machine-guns, or seven 68-mm (2.68-in) SNEB rockets
Operator: India

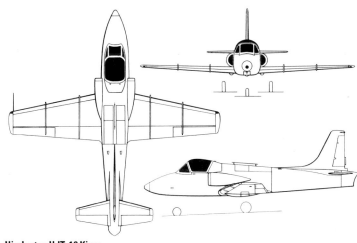
Hindustan HJT-16 Kiran

Hindustan HPT-32

First flown on 6 January 1977, the HPT-32 is intended for a wide variety of instructional roles including *ab initio*, instrument, navigation, night flying, formation flying and weapon training; it can be used also for glider or target towing. With four underwing pylons available for weapons or other stores, it can also perform light attack or supply dropping missions. With a third (and possibly fourth) seat behind the side-by-side front seats, other roles include search and rescue, liaison, reconnaissance or civil touring. The batch of 40 to replace the HT-2 in the Indian air force is being delivered during 1985-86.

Specification
Type: two/three-seat basic trainer

or four-seat liaison aircraft
Powerplant: one 194-kW (260-hp) Lycoming AEIO-540-D4B5 flat-six piston engine
Performance: maximum cruising speed 252 km/h (157 mph) at 3050 m (10005 ft); initial rate of climb 336 m (1,102 ft)/minute; service ceiling 6000 m (19,685 ft); maximum range 745 km (463 miles)
Weights: empty 890 kg (1,962 lb); maximum take-off 1250 kg (2,756 lb)
Dimensions: span 9.50 m (31 ft 2 in); length 7.72 m (25 ft 3.9 in); height 2.88 m (9 ft 5.4 in); wing area 15.00 m² (161.46 sq ft)
Armament: up to 255 kg (562 lb) of weapons or other stores on four underwing hardpoints
Operator: India

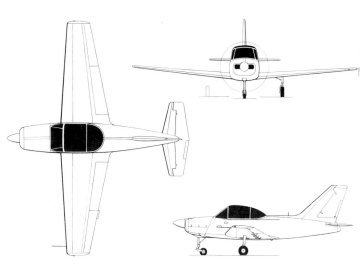
Hindustan HPT-32

Hindustan HT-2

Broadly similar to the contemporary Chipmunk, the HT-2 is an all-metal trainer first flown on 13 August 1951. Production of 169 continued until 1958. Some 50 remained in service in 1984, and these are due for replacement by the HAL HPT-32 primary trainer.

Specification
Type: primary trainer
Powerplant: one 116-kW (155-hp) Blackburn Cirrus Major III inverted 4-inline piston engine
Performance: maximum speed 210 km/h (130 mph); initial rate of climb 245 m (804 ft)/minute; service ceiling 4420 m (14,500 ft); range 565 km (351 miles)
Weights: empty 699 kg (1,541 lb); maximum take-off 1015 kg (2,238 lb)
Dimensions: span 10.72 m (35 ft 2 in); length 7.53 m (24 ft 8.5 in); height 2.72 m (8 ft 11.1 in); wing area 16.11 m² (173.41 sq ft)
Armament: none
Operator: India

India's basic trainer for some years has been the HT-2, but this is due for replacement by the HPT-32. The type is similar to the DHC-1 Chipmunk.

Hindustan HF-24 Marut

Producton of the Marut ended in 1977. Over the years a number of attempts have been made to push the performance up to Mach 2, one of which involved the conversion of a pre-production Marut to Mk IA standard by fitting afterburners to the existing Orpheus 703 engines. Entirely new powerplant studied have varied from the German-designed, Egyptian-built Brandner E300 turbojet in the latter half of the 1960s to the more recent study of the Turbo-Union RB.199 turbofan which powers the Panavia Tornado. None of these has yielded satisfactory results, and the idea of a Mach 2 Marut now seems to have been abandoned. The Marut has been relegated to training since the SEPECAT Jaguar has been bought by India. Only a handful remain in service in 1985.

Specification
Type: single-seat ground-attack fighter (Mk I); two-seat operational trainer (Mk IT)
Powerplant: two 2200-kg (4,850-lb) static thrust HAL-built Rolls-Royce Orpheus 703 turbojets
Performance: maximum speed at 1200 m (39,375 ft) (Mk I) Mach 1.02 or 1086 km/h (675 mph), (Mk IT) Mach 1.00 or 1064 km/h (661 mph); maximum indicated airspeed at sea level (Mk I) 1112 km/h (691 mph); combat radius (Mk IT) at low level 238 km (148 miles), at 1200 m (39,375 ft) 396 km (246 miles); ferry range (Mk IT) at 9150 m (30,000 ft) 1445 km (898 miles); time to 12200 m (40,000 ft), aircraft clean 9 minutes 20 seconds
Weights: empty (Mk I with ventral drop-tank) 6195 kg (13,658 lb), (Mk IT) 6250 kg (13,778 lb); maximum take-off (Mk I) 10908 kg (24,048 lb),

The HAL Marut is still just about in service with the Indian Air Force, being used as a weapons trainer.

(Mk IT) 10812 kg (23,836 lb)
Dimensions: span 9.00 m (29 ft 6¼ in); length 15.87 m (52 ft 0¾ in); height 3.60 m (11 ft ¾ in); wing area 28.00 m² (301 sq ft)
Armament: four 30-mm Aden Mk 2 cannon with 120 rounds per gun, two on each side of forward fuselage;

retractable Matra Type 103 pack of fifty 68-mm SNEB air-to-air rocket in belly (not in Mk IT); four underwing points for 454-kg (1,000-lb) bombs, Type 116 SNEB rocket pods, T10 air-to-surface rocket clusters, napalm, or drop-tanks
Operator: India

Hispano HA-200

Following the example set with Britain's Jet Provost, Hispano used piston-engined HA-100 components in the HA-200 flown on 12 August 1955. The 30 production HA-200A Saeta (arrow) trainers became the E-14 in Spanish air force service, the first flying on 11 October 1962. Hispano (later CASA) supplied 110 HA-200/220s to the Spanish air force, of which over 40 are still in service. Escuadron 214 has about 20 HA-220 light attack aircraft at Moron as part of 21 Wing, Tactical Air Command. A few HA-200s remain with Escuadron 793, though the standard trainer is now the CASA Aviojet.

Specification
Hispano/CASA HA-200D
Type: twin jet advanced trainer
Powerplant: two 400-kg (882-lb) thrust Turboméca Marboré IIA turbojets
Performance: maximum speed 650 km/h (404 mph); cruising speed 530 km/h (329 mph); initial rate of climb 840 m (2,756 ft)/minute; service ceiling 12,000 m (39,370 ft); maximum range 1,500 km (932 miles)
Weights: empty 1830 kg (4,034 lb); maximum take-off 3350 kg (7,385 lb)
Dimensions: span 10.42 m (34 ft 2.2 in); length 8.97 m (29 ft 5.1 in); height 2.85 m (9 ft 4.2 in); wing area 17.40 m² (187.30 sq ft)

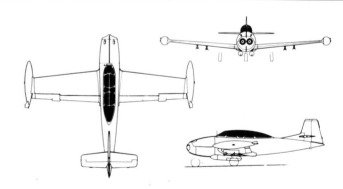

Hispano HA.200 Saeta

Armament: two underwing hardpoints for rocket launchers, bomb racks and photo-reconnaissance equipment; there is provision to mount a 20-mm Hispano-Suiza cannon in the fuselage
Operators: Spain

Hughes Model 269/300/300C/ TH-55A Osage

The diminutive but versatile Hughes TH-55A Osage is a military version of Hughes' original Model 269A, a re-engineered and simplified follow-on from the two-seat Model 269. Five 269As bought by the US Army under the designation YHO-2HU in 1958 completed a highly successful evaluation programme in the command and observation roles. In 1964 the type was chosen as the force's standard two-seat training helicopter, and by 1969 a total of 792 had been delivered. The conflict in Vietnam demanded the use of helicopters on a scale never seen before, and TH-55As are still operated to train pilots to fly the 9,000 rotary-wing aircraft in the US Army inventory.

The basic design has undergone many changes over the years. The first three-seat variant, the Model 300, appeared in 1963. Five years later, in July 1968, construction began of the Model 300C. Powered by a 142-kW (190-hp) Lycoming piston engine, this three-seat aircraft first flew in August 1969 and offered a 45 per cent increase in payload. The 300CQ is 75 per cent quieter than earlier models. Adoption of a larger engine called for a number of structural changes, including enlarging the tail rotor and fin, and lengthening the tailboom and rotor mast to accommodate the longer and heavier main rotor blades. Hughes has also experimented with alternative powerplants. A TH-55A has flown with a 138-kW (185-hp) Wankel RC2-60 rotary engine, and another has been fitted with a 236-kW (317-shp) Allison 250-C18 turboshaft. Kawasaki built 48 Osages for the Japan Ground Self-Defence Force (JGSDF), and the Model 300C is also built under licence in Italy by Breda Nardi.

Algeria uses six 269s for training; the Brazilian navy's 10 269/300s are now being phased out of the training role; Guyana has the type primarily for police work; Italy has trained many pilots on the type; Japan's GSDF is replacing much of its fixed-wing fleet with helicopters, including 38 TH-55Js; Nicaragua has one for training; the Sierra Leone Defence Force, formed in 1973 and with

Despite its age, the Hughes TH-55 Osage is still in widespread service with the US Army, mainly as a trainer. Others around the world are still used for light spotting and liaison duties.

an inventory of six aircraft, operates one; and the largest user, the US Army, still operates some 200.

Specification
Type: two/three-seat light training helicopter
Powerplant: one 134-kW (180-hp)

Avco Lycoming HIO-360-A1A four-cylinder, horizontally opposed, air-cooled piston engine
Performance: maximum design speed 138 km/h (86 mph); maximum cruising speed 121 km/h (75 mph); range (maximum fuel, no reserve) 328 km (204 miles); maximum rate

of climb at sea level 347 m (1,140 ft) per minute
Weights: empty equipped 457 kg (1,008 lb); maximum take-off and landing 757 kg (1,670 lb)
Dimensions: rotor diameter 7.71 m (25 ft 3½ in); length overall (rotors turning) 8.8 m (28 ft 10¾ in); height

(tail rotor turning) 2.5 m (8 ft 2¾ in); main rotor disc area 46.73 m^2 (503 sq ft)
Armament: none
Operators: Algeria, Brazil, Guyana, Haiti, Indonesia, Italy, Japan, Nicaragua, North Korea, Sierra Leone, Spain, Sweden, Turkey, US Army

Hughes Model 369/OH-6 Cayuse

The Hughes 369 was designed to meet the US Army Light Observation Helicopter (LOH) specification of 1960. It was given the designation OH-6 Cayuse before it flew on 27 February 1963. On 26 May 1965 it was announced that it had won the design competition with an initial order for 714. Hughes received further contracts to bring the total to 1,434, but Bell built the balance of 4,000 LOH after the US Army became dissatisfied with falling OH-6 production rates and rising costs.

Yet the OH-6 is a superb light helicopter with sparkling performance, which set a series of international records including a distance of 3561 km (2,213 miles), sustained altitude of 8061 km (26,448 ft), speed of 227.74 km/h (141.51 mph) over a 2000 km circuit, and a maximum speed of 277.47 km/h (172.41 mph).

Accommodation is for a crew of two, with two folding seats in the rear which, when folded, provide room on the floor for four fully equipped troops or cargo. In Vietnam the Cayuse was employed for many duties, including offensive operations with armament kits.

Hughes OH-6 Cayuse of the US Army with experimental mast-mounted sight and TOW launcher tubes.

Specification
Hughes OH-6A Cayuse
Type: light observation helicopter
Powerplant: one 236-kW (317-shp) Allison T63-A-5A turboshaft engine derated to 188 kW (252 shp)
Performance: maximum cruising

speed 230 km/h (143 mph) at sea level; normal range 665 km (413 miles); ferry range 2511 km (1,560 miles)
Weights: empty 524 kg (1,156 lb); maximum take-off 1225 kg (2,700 lb)

Dimensions: main rotor diameter 8.03 m (26 ft 4 in); length, rotors turning 9.24 m (30 ft 3.75 in); height 2.48 m (8 ft 1.5 in); main rotor disc area 50.60 m^2 (544.63 sq ft)
Armament: XM-27 7.62-mm (0.3-in) machine-gun, or XM-75 grenade launcher
Operator: US Army

Hughes Model 500 Defender

In 1965 the US Army held an LOH (Light Observation Helicopter) contest, with potential production for a four-figure total. When the Hughes OH-6A Cayuse won there was a storm of protest, it being claimed the company was selling below cost. Despite this, 1,415 OH-6s gave splendid service in Vietnam, and as its tadpole shape was extremely compact, and performance on a 236-kW (317-shp) Allison engine the highest in its class, the OH-6 was most popular. From it the company developed the Hughes Model 500 family, the company astutely seeing the considerable market for a versatile, high-performance military helicopter of low cost and proven reliability. The basic Model 500M with improved 236-KW (317-shp) engine was sold to nine countries and licence-made in Argentina and Japan. Operated by Spain as a light anti-submarine platform, the Model 500M has AN/ASQ-81 magnetic anomaly detection (MAD) gear with a towed 'bird', and provision for two Mk 44 torpedoes. The Model 500MD Defender has the more powerful

Allison 250-C20B engine and can have self-sealing tanks, inlet particle filter. IR-suppressing exhausts, and many role fits including seven seats, or two stretchers and two attendants, or various weapons, including the TOW anti-tank missile and nose-mounted sight. Licensed production proceeds at BredaNardi (Italy) and KAL (South Korea), and the type is in worldwide service for training, command and control,

Kenya operates eight Hughes 500ME armed helicopters.

The Defender is a useful light anti-armour helicopter, especially when fitted with mast-mounted sight. This example with nose sight fires a TOW.

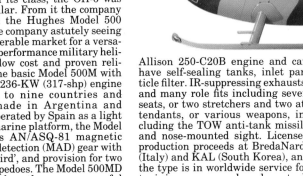

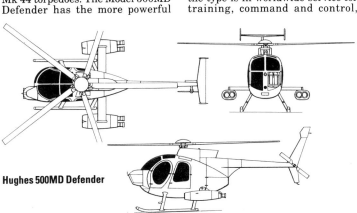

Hughes 500MD Defender

With mast-mounted sight, the Defender can keep behind trees while firing its weapons, thereby keeping itself hidden from gunfire.

light attack, observation, logistic support, troop transport and ASW. The Model 500MD Scout Defender is the basic armed version (with a baseline fit of 14 70-mm/2.75-in rockets plus one 7.62-mm/0.3-in Minigun with 2,000 rounds, or one 7.62-mm/0.3-in EX-34 Chain Gun with 2,000 rounds, or one 30-mm Chain Gun with 600 rounds, or one 40-mm grenade-launcher) and a sub-type (Model 500MD Quiet Advanced Scout Defender) has the MMS (mast-mounted sight) for 'hull-down' surveillance or missile guidance, and quiet-running features. The Model 500MD/TOW Defender has four TOW missiles, original deliveries having a stabilized nose sight. The Model 500MD Defender II is an update multi-role model now being delivered with quiet rotors (including a five- rather than four-blade main unit), MMS, IR suppression, FLIR (forward-looking IR) night vision and many other devices including APR-39 passive radar warning.

Specification
Hughes Model 500MD Defender
Type: multi-role combat helicopter
Accommodation: two
Powerplant: one 313-kW (420-shp) Allison 250-C20B turboshaft
Performance: maximum speed 217 km/h (152 mph); range 509 m (366 miles)
Weights: empty typically 572 kg (1,260 lb); maximum take-off 1361 kg (3,000 lb)
Dimensions: main rotor diameter 8.05 m (26 ft 4¾ in); fuselage length 7.01 m (23 ft 0 in); height 2.71 m (8 ft 10¾ in); main rotor disc area 50.7 m² (456.0 sq ft)
Armament: options include Hughes 30-mm Hughes Chain Gun (firing rate reduced to 350 rounds per minute), four TOW missiles and two Stinger MLMS AAMs
Operators: Bahrain, Denmark, El Salvador, Finland, Haiti, Iran, Iraq, Israel, Jordan, Kenya, Morocco, North Korea, South Kotea, Taiwan

Hughes AH-64 Apache

Designed in 1972-3 to meet the US Army's need for an AAH (Advanced Attack Helicopter), the Hughes AH-64A beat a Bell competitor which had reversed the traditional Cobra arrangement of seating the pilot above and behind the co-pilot/gunner, an arrangement maintained by Hughes. Features include two T700 engines flat-rated to provide high emergency power and with large IR-suppressing exhaust systems to reduce the chances of a hit by an IR-homing missile, a large flat plate canopy with boron armour, multi-spar stainless steel and glass-fibre rotor blades designed to withstand 23-mm hits, extremely comprehensive avionics and weapon fits, and numerous crash-resistant features to protect the crew. All these features are vital in a combat helicopter designed to undertake the most arduous of battlefield roles by day and night, and even under the most adverse of weather conditions. Development was unfortunately prolonged, the first prototype flying on 30 September 1975 and the programme being hard hit by modification, so that inflation has multiplied the price and not all the planned 536 Apaches may be funded. Appearance changed dramatically during development, especially at the nose and tail, and the nose carries the Martin Marietta TADS/PNVS (Target Acquisition and Designation Sight/Pilot's Night Vision System). New missiles have become available, and as well as laser designation and ranging an IHADSS (Integrated Helmet And Display Sighting System) is fitted, both crew members being able to acquire targets by head movement. The type entered service late in 1983, and procurement has already been sent back to 446 units. There can be little doubt as to the technical excellency and ingenuity represented by the AH-64A Apache, but even the most enthusiastic supporter of the type must have grave reservations about the location of the sensor package in the nose rather than above the rotor, as this means the type has to leave cover to acquire targets, and is thus very vulnerable.

The sharp of end of the Apache: armed with Hellfire anti-tank missiles, chain gun and rocket pods, the AH-64 presents an evil sight to its intended victims.

Hughes AH-64 Apache cutaway drawing key

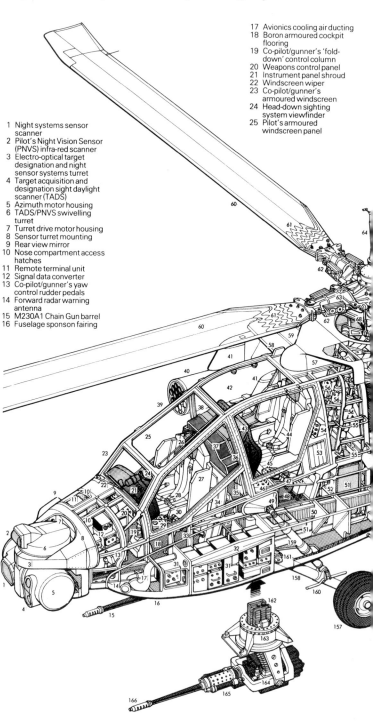

1 Night systems sensor scanner
2 Pilot's Night Vision Sensor (PNVS) infra-red scanner
3 Electro-optical target designation and night sensor systems turret
4 Target acquisition and designation sight daylight scanner (TADS)
5 Azimuth motor housing
6 TADS/PNVS swivelling turret
7 Turret drive motor housing
8 Sensor turret mounting
9 Rear view mirror
10 Nose compartment access hatches
11 Remote terminal unit
12 Signal data converter
13 Co-pilot/gunner's yaw control rudder pedals
14 Forward radar warning antenna
15 M230A1 Chain Gun barrel
16 Fuselage sponson fairing
17 Avionics cooling air ducting
18 Boron armoured cockpit flooring
19 Co-pilot/gunner's 'fold-down' control column
20 Weapons control panel
21 Instrument panel shroud
22 Windscreen wiper
23 Co-pilot/gunner's armoured windscreen
24 Head-down sighting system viewfinder
25 Pilot's armoured windscreen panel

Hughes AH-64 Apache (continued)

26 Windscreen wiper
27 Co-pilot/gunner's Kevlar armoured seat
28 Safety harness
29 Side console panel
30 Engine power levers
31 Avionics equipment bays, port and starboard
32 Avionics bay access door
33 Collective pitch control lever
34 Adjustable crash-resistant seat mountings
35 Pilot's rudder pedals
36 Cockpit side window panel
37 Pilot's instrument console
38 Inter-cockpit acrylic blast shield
39 Starboard side window entry hatches
40 Rocket launcher pack
41 Starboard wing stores pylons
42 Cockpit roof glazing
43 Instrument panel shroud
44 Pilot's Kevlar armoured seat
45 Collective pitch control lever
46 Side console panel
47 Engine power levers
48 Rear cockpit floor level
49 Main undercarriage shock absorber mounting
50 Linkless ammunition feed chute
51 Forward fuel tank; total fuel capacity 375 US gal (1419 litres)
52 Control rod linkages
53 Cockpit ventilating air louvres
54 Display adjustment panel
55 Grab handles/maintenance steps
56 Control system hydraulic actuators (three)
57 Ventilating air intake
58 UHF aerial
59 Starboard stub wing
60 Main rotor blades
61 Laminated blade-root attachment joints
62 Vibration absorbers
63 Blade pitch bearing housing
64 Air data sensor mast
65 Rotor hub unit

66 Offset flapping hinges
67 Elastomeric lead/lag dampers
68 Blade pitch control rod
69 Pitch control swashplate
70 Main rotor mast
71 Air turbine starter/auxiliary powered unit (APU) input shaft
72 Rotor head control mixing linkages
73 Gearbox mounting plate
74 Transmission oil coolers, port and starboard
75 Rotor brake
76 Main gearbox
77 Gearbox mounting struts
78 Generator
79 Input shaft from port engine
80 Gearbox mounting deck
81 Tailrotor control rod linkage
82 Ammunition magazine, 1,200 rounds
83 Stub wing attachment joints
84 Engine transmission gearbox
85 Air intake
86 Engine integral oil tank
87 General Electric T700-GE-701 turboshaft
88 Intake particle separator
89 Engine accessory equipment gearbox
90 Oil cooler plenum
91 Gas turbine starter/auxiliary power unit
92 Starboard engine cowling panels/fold-down maintenance platform
93 Starboard engine exhaust ducts
94 APU exhaust
95 Pneumatic system and environmental control equipment
96 Cooling air exhaust louvres
97 Particles separator exhaust duct/mixer

Specification
Hughes AH-64A Apache
Type: armed battlefield helicopter
Accommodation: pilot and co-pilot/gunner
Powerplant: two 1146-kW (1,536-shp) General Electric T700-700 turboshafts
Performance: maximum speed (at 6316 kg/13,925 lb) 309 km/h (192 mph); range (internal fuel) 611 km (380 miles), and (ferry) 1804 km (1,121 miles)
Weights: empty 4657 kg (10,268 lb); maximum take-off 8006 kg (17,650 lb)

Dimensions: main rotor diameter 14.63 m (48 ft 0 in); fuselage length 14.97 m (49 ft 1½ in); height 4.22 m (13 ft 10 in); main rotor disc area 168.11 m² (1,809.5 sq ft)
Armament: one 30-mm Hughes Chain Gun with 1,200 rounds and remote aiming; four stub-wing hardpoints for normal anti-armour load of 16 Hellfire missiles (initially with laser guidance); other loads can include four 18-round pods of 70-mm (2.75-in) rockets
Operator: US Air Force

Apache pilots will try to present the least size of target to their enemies, but with so large a machine there is still plenty to hit. Perhaps the greatest single accomplishment in the AH-64A structural and systems design is that, as far as possible, the entire helicopter is proof against hits up to 23-mm calibre.

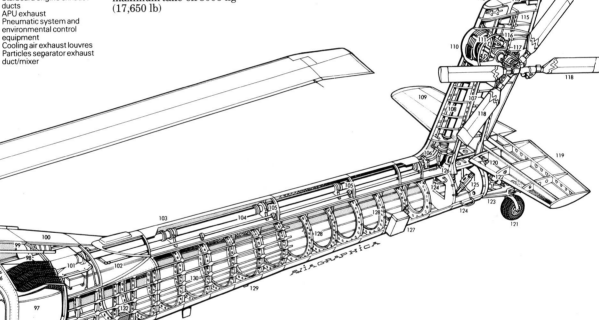

98 'Black Hole' infra-red suppression engine exhaust ducts
99 Hydraulic reservoir
100 Gearbox/engine bay tail fairings
101 Internal maintenance platform
102 Tail rotor control rod
103 Spine shaft housing
104 Tail rotor transmission shaft
105 Shaft bearings and couplings
106 Bevel drive intermediate gearbox
107 Fin/rotor pylon construction
108 Tail rotor drive shafts
109 All moving tailplane
110 Tail rotor gearbox housing
111 Right-angle final drive gearbox
112 Fin tip aerial fairing
113 Rear radar warning antennae
114 Tail navigation light
115 Cambered trailing edge section (directional stability)
116 Tail rotor pitch actuator
117 Tail rotor hub mechanism
118 Asymmetric (noise attenuation) tail rotor blades

119 Tailplane construction
120 Tailplane pivot bearing
121 Castoring tailwheel
122 Tailwheel shock absorber
123 Tailwheel yoke attachment
124 Handgrips/maintenance steps
125 Tailplane control hydraulic jack
126 Fin/rotor pylon attachment joint
127 Chaff and flare dispenser
128 Tailboom ring frames
129 Ventral radar warning aerial
130 Tailcone frame and stringer construction
131 UHF aerial
132 ADF loop aerial
133 ADF sense aerial
134 Access hatch
135 Handgrips/maintenance steps
136 Radio and electronics equipment bay
137 Rear fuel tank
138 Reticulated foam fire suppressant tank bay linings
139 VHF aerial
140 Main rotor blade stainless steel spars (five)
141 Glass-fibre spar linings
142 Honeycomb trailing edge panel
143 Glass-fibre blade skins

144 Trailing edge fixed tab
145 Swept blade tip fairing
146 Static discharger
147 Stub wing trailing-edge flap
148 Stub wing rib construction
149 Twin spar booms
150 Port navigation and strobe lights
151 Port wing stores pylons
152 Rocket pack: 19 2.75-in (7-cm) FFAR rockets
153 Rockwell Hellfire AGM-114A anti-tank missiles
154 Missile launch rails
155 Fuselage sponson aft fairing
156 Boarding step
157 Port main wheel
158 Main undercarriage leg strut
159 Shock absorber strut
160 Boarding steps
161 Main undercarriage leg pivot fixing
162 Ammunition feed and cartridge case return chutes
163 Gun swivelling mounting
164 Azimuth control mounting frame
165 Hughes M230A-1 Chain Gun 30-mm automatic cannon
166 Blast suppression cannon muzzle

One of the most complete battlefield weapons systems designed to date, the AH-64 is built around survivability. The crew are housed in a crash-proof cockpit while the engine exhausts have extremely effective IR suppressors. Surprisingly no mast-mounted sight has been fitted, which means the Apache has to show itself in order to aim the weapons.

Keith Fretwell

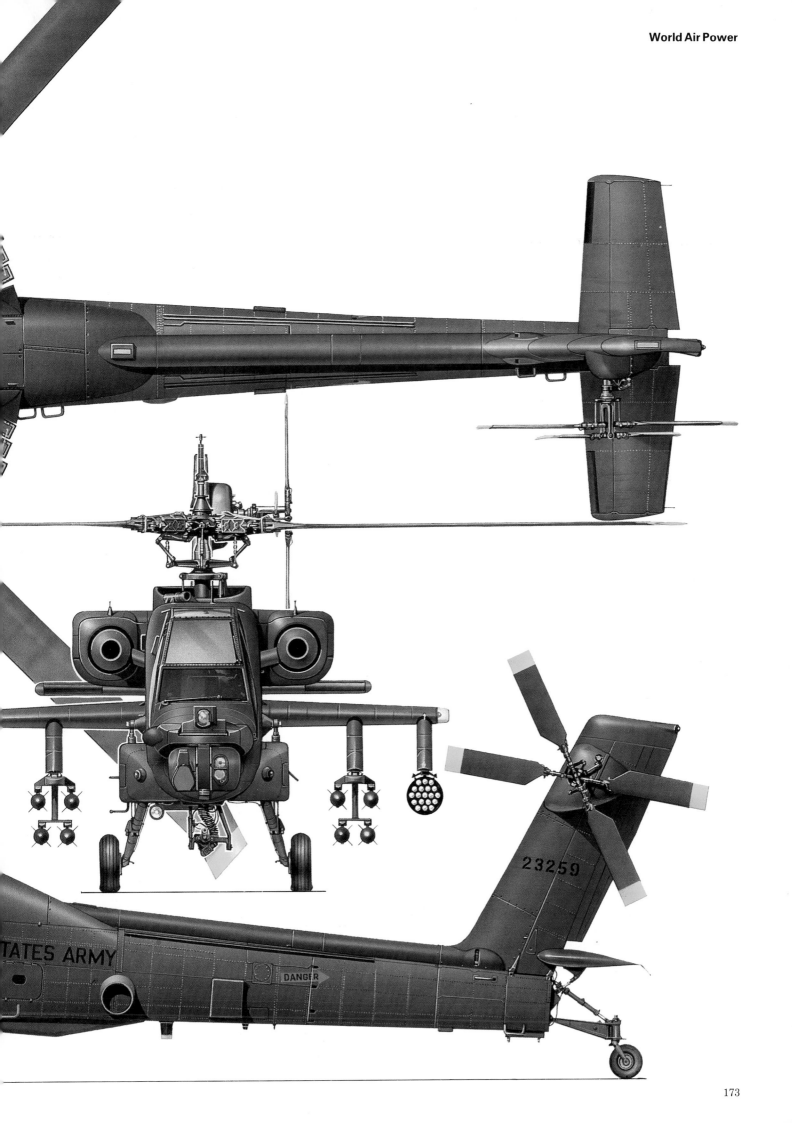

DANGER

TATES ARMY

23259

Hunting Firecracker

Britain's only turboprop trainer, the Hunting Firecracker was offered to the RAF as a replacement for the Jet Provost. The design has its origin in the NDN.1 Firecracker, powered by a 194-kW (260-hp) Lycoming AEIO-540-B4D5 piston engine, and first flown on 26 May 1977. Lack of sales partly stemmed from the fact that NDN was able to offer the design only for overseas manufacture, and had no facilities for production. The burgeoning turboprop trainer market prompted the NDN.1T Turbo Firecracker, flown on 1 September 1983. Production is by a new company, Firecracker Aircraft (UK) Ltd, which has the British Hovercraft Corporation as one of its sub-contractors. In September 1982 three

NDN.1Ts were ordered, with an option on four, by Specialist Flying Training Ltd, training mainly foreign students. Although BAe regarded the aircraft as a logical partner to its Hawk when offering training packages abroad, Firecracker was left to its own devices in tendering to the RAF's AST.412 requirement. After brief liaison with Westland, Hunting Firecracker Aircraft was formed in 1984 with participation from Hunting Associated Industries and Guinness Mahon & Co (merchant bankers). With its PT6A increased in power from 410 kW (550 shp) flat rating to 559 kW (750 shp), and a revised cockpit with American Stencel Ranger rocket seats, the Hunting Firecracker is

generally similar to its civil companion. Common features include stressing to +6.7/−3.5 g, a flush Centrisep air intake on each side of the nose as protection against debris ingestion, a raised instructor seat, reversible-pitch propeller and a belly airbrake.

Specification
Hunting Firecracker
Type: turboprop trainer
Powerplant: one 559-kW (750-shp) Pratt & Whitney Canada PT6A-25D turboprop engine
Performance: (estimated) maximum speed at sea level 414 km/h (257 mph); time to 4507 m (15,000 ft) altitude 6.25 minutes
Weights: (estimated) empty

1210 kg (2,668 lb); maximum take-off 1928 kg (4,250 lb)
Dimensions: span 7.92 m (26 ft 0 in); length 8.33 m (27 ft 4 in); height 3.25 m (10 ft 8 in); wing area 11.89 m² (128.0 sq ft)
Armament: provision for four wing pylons carrying light bombs, rocket and gun pods
Operators: none

IAR-823

Built by Industriei Aeronautice Romane (IAR), the all-metal IAR-823 first flew in July 1973. Roof transparencies for the front seats improve visibility in the training role, and it is stressed for limited aerobatics. The two-seat version for the Romanian air force (FA-RSR) has dual stick controls, a full night-flying panel and two underwing hardpoints for tanks or weapons. Approximately 50 have been delivered in camouflage finish. So far as

is known no examples of the IAR-823 are in service outside Romania.

Specification
Type: two-seat training or five-seat communications aircraft
Powerplant: one 216-kW (290-hp) Lycoming IO-540-G1D5 flat-six piston engine
Performance: maximum speed 300 km/h (186 mph); initial rate of climb 420 m (1,378 ft)/minute; service ceiling 5600 m (18,370 ft);

maximum range 1800 km (1,118 miles)
Weights: (normal category) empty 950 kg (2,094 lb); maximum take-off 1500 kg (3,307 lb)
Dimensions: span 10.00 m (32 ft 9.7 in); length 8.32 m (27 ft 3.6 in); height 2.86 m (9 ft 4.6 in); wing area 15.00 m² (161.46 sq ft)
Armament: two underwing pylons for practice bombs, rocket launchers or machine-gun pods
Operator: Romania

IAR-823

IAR-825TP Triumf

The Romanian national aircraft factory at Brasov has produced its IAR-825TP Triumf as an alternative to jet primary training for the Romanian air force (FA-RSR). The Triumf uses much of the wing and landing gear of the IAR-823, and has been developed in parallel with the IAR-831 piston-engine trainer. The tandem cockpits closely resemble the IAR-93 Orao. The prototype has a 507-kW (680-shp) Pratt & Whitney

Canada PT6A-15AG turboprop, but the lower-rated PT6A-25 or Czech M-601B will power production aircraft. The prototype has supplementary wingtip tanks and first flew on 12 June 1982. An initial batch of 15 is in production for the FA-RSR, and the type may see service with other Comecon countries.

Specification
Type: primary trainer

Powerplant: one 410-kW (550-shp) Pratt & Whitney Canada PT6A-25 turboprop engine
Performance: maximum speed 470 km/h (292 mph); initial rate of climb 720 m (2,362 ft)/minute; maximum range without tip tanks 1170 km (727 miles)
Weights: empty 1200 kg (2,646 lb); maximum take-off 2200 kg (4,850 lb)
Dimensions: span 10.30 m (33 ft

9.5 in); length 8.90 m (29 ft 2.4 in); height 2.38 m (7 ft 9.7 in); wing area 14.95 m² (160.93 sq ft)
Armament: two underwing hardpoints for gun pods, rocket launchers or bombs
Operator: Romania (on order)

IAR-831 Pelican

The IAR-831 is the piston-engine counterpart of the IAR-825TP Triumf. The wing, derived from that of the IAR-823, has fuel tanks in the outerpanels. The Pelican, of which the prototype was flown in 1982, is expected to provide characteristics similar to those of jet aircraft while reducing the capital cost of a basic-training fleet. Two underwing hard-

points permit the aircraft to carry a variety of stores including drop tanks. No order has yet been placed, but the IAR-831 is expected to replace the IAR-823 for initial FA-RSR training.

Specification
Type: primary trainer
Powerplant: one 216-kW (290-hp)

Lycoming IO-540-G1D5 flat-six piston engine
Performance: maximum speed 320 km/h (199 mph); initial rate of climb 420 m (1,378 ft)/minute; maximum range 1300 km (808 miles)
Weights: empty 700 kg (1,543 lb); maximum take-off 1200 kg (2,646 lb)

Dimensions: span 10.00 m (32 ft 9.7 in); length 8.90 m (29 ft 2.4 in); height 2.37 m (7 ft 9.3 in); wing area 15.00 m² (161.46 sq ft)
Armament: two underwing hardpoints for gun pods, rocket launchers, bombs or tanks
Operator: not yet ordered

Ilyushin Il-14 'Crate'

Obsolescent as a transport, the Ilyushin Il-14 remains in service for secondary duties, and in 1976 was first seen with Frontal Aviation electronic intelligence (Elint) units operating in East Germany. The type is also used as a navigation trainer.

Specification
Ilyushin Il-14M
Type: transport, navigation trainer and electronic intelligence (Elint)

aircraft
Powerplant: two 1417-kW (1900-hp) Shvetsov ASh-82T 14-cylinder radial piston engines
Performance: cruising speed 320-350 km/h (199-217 mph); range with maximum 3330-kg (7,275-lb) payload 400 km (249 miles)
Weights: empty 12700 kg (27,999 lb); maximum take-off 17500 kg (38,581 lb)
Dimensions: span 31.70 m (104 ft 0 in); length 22.31 m (73 ft 2.3 in);

height 7.95 m (26 ft 1 in); wing area 100.00 m² (1,076.43 sq ft)
Armament: none
Operators: Albania, Afghanistan, Bulgaria, Congo, Cuba, Czechoslovakia, East Germany, Guinea Republic, Egypt, Iraq, Mongolia, Nigeria, Poland, Romania, Soviet Union, Vietnam, North Yemen, South Yemen

Ilyushin Il-14M 'Crate'

Ilyushin Il-18/20 'Coot'

The first Soviet aircraft to be designed from the outset as a turbine-powered commercial transport, the

Ilyushin Il-18 made its first flight in July 1957 and entered service with Aeroflot in the following year. Early

aircraft used the Kuznetsov NK-4 engine, but by 1959 the improved AI-20 engine was standard. Weight,

payload and fuel capacity were steadily increased between 1959 and 1965, culminating in the Il-18D version.

The design was straightforward and sound, although the aircraft was marginally heavier and less efficient than its closest Western contemporary, the Vickers Vanguard. The Il-18 proved reliable in service, and some were converted for freight transport after being replaced on passenger routes by the Tu-154 trijet. About 800 of the type are thought to have been built before production ended in 1970. Some Il-18s remain in service for government use or for special transport missions.

In the mid-1970s, a number of Il-18s were converted to fill an important and specialized role with the Soviet air arm. The various activities which in the West are described as electronic countermeasures (ECM), electronic intelligence (Elint), communications intelligence (Comint) and defence suppression are covered by a single Soviet term which translates as 'radio-electronic combat' (REC). An REC platform based on the Il-18, known to NATO as 'Coot-A' and believed to be designated Il-20, was first observed in early 1978.

In peacetime, the Il-20 functions mainly as an Elint/Comint aircraft, gathering information on the fringes of NATO airspace. (For this reason, the aircraft are photographed frequently by NATO fighters.) In wartime, the IL-20 is designed to stand off behind the line of battle, or, in an overwater engagement, outside the main engagement zone, and support the strike force with information on radar locations and, possibly, powerful active jamming.

The most prominent piece of equipment on the Il-20 is a massive ventral 'canoe' radome. While this is often described as a side-looking airborne radar (Slar) installation, it would be odd to find an active surveillance sensor on an Elint/jamming aircraft. It is more likely to be

a highly sensitive directional receiver, designed to provide an accurate bearing on a hostile radar. (The canoe installation resembles that fitted to the US Navy's EP-3E, which fills a similar role.) On the sides of the forward fuselage are twin housings, the forward halves of which contain some type of optical or electro-optical installation; these may be intended to aid in the classification of targets detected and located by the main ventral sensor. More electronic sensors or, possibly, jamming transmitters occupy the rear of these housings. A number of antennae, including two large dorsal blades and an array of blisters and stings, are either dedicated to the Comint role or are used for communicating with friendly units. There is presumably a crew of 12 or more systems operators in the main cabin.

The number of Il-20s in service is not known, but is probably not much more than 20-30 at most. It is not certain to which arm of the Soviet forces the aircraft are assigned, but it is possible that they are attached to the 'reconnaissance directorates' of the five TVDs (theatres of military operations) which control the operation of Warsaw Pact forces. The aircraft have not been deployed outside Warsaw Pact territory.

Ilyushin Il-18s have served both civil airlines and air forces well for many years. The Soviet Union uses Il-18s for Antarctic support.

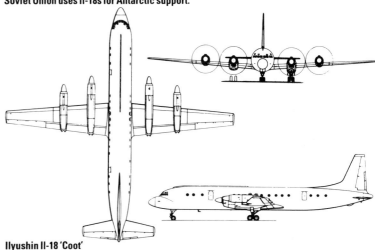

Ilyushin Il-18 'Coot'

Specification
Ilyushin Il-20
Type: heavy electronic warfare platform
Powerplant: four 3170 kW (4,250 eshp) Ivchenko AI-20M turboprops
Performance: maximum cruising speed 625 km/h (390 mph); range with maximum fuel 6,500 km (4,000 miles); endurance 11-13 hours
Weights: empty 37500 kg (82,700 lb); maximum take-off 64000 kg (141,000 lb)
Dimensions: span 37.4 m (122 ft 8 in); length 35.9 m (117 ft 9 in); height 10.17 m (33 ft 4 in); wing area 140 m² (1,507 sq ft)
Operator: Soviet air force

Ilyushin Il-28 'Beagle'

The first jet bomber of Soviet design to become known in the West, the Ilyushin Il-28 'Beagle' was an outstandingly straightforward machine which, though smaller and in many ways less capable than a Tupolev rival, was unanimously preferred by the bomber crews assigned to evaluate the two types competitively. Accordingly something like 3,000 were built at a rapid rate, followed 20 years later by another 1,500 (designated H-5 or Hong-5) in China. A small number were Il-28U trainers, but most are regular light bombers with the pilot on the centreline in a fighter-type cockpit, the navigator/bombardier in the glazed nose and a rear gunner in a separate pressurized turret. Main landing gears house the large wheels lying flat directly under the jetpipes. Unlike the wings, the tail surfaces are swept back to avoid control problems at high speed in dives. The Il-28 proved to be very easy to maintain and delightful to fly, and even today large numbers are still flying in many countries. Except for those in China, most have been modified as targets, meteoroligical aircraft, and for other second-line duties, but in the period prior to 1970 the Il-28 saw a great deal of action in such places as Nigeria, Egypt, Vietnam, Syria, Yemen, North Korea and Iraq. On most occasions the aircraft was used as a level bomber from medium altitude, typically dropping 250-kg (551-lb) bombs on such targets as airfields and suspected troop concentrations. The Il-28T torpedo-bomber was used by the Soviet AV-MF, even though that service also bought a torpedo-bomber based on the Il-28's original Tupolev rival.

Specification
Type: three-seat light bomber
Powerplant: two 2700-kg (5,952-lb) thrust Klimov VK-1 turbojets
Performance: maximum speed 900 km/h (559 mph); service ceiling 12300 m (40,355 ft); range at high altitude with maximum fuel 2180 km (1,355 miles)
Weights: empty 12890 kg (28,417 lb); maximum take-off in original bomber role 21000 kg (46,296 lb)
Dimensions: span excluding tip tanks 21.45 m (70 ft 4½ in); length (typical) 17.65 m (57 ft 10¾ in); height 6.70 m (22 ft 0 in); wing area 60.80 m² (654.5 sq ft)
Armament: tail turret with two 23-mm cannon, and in most versions two similar 23-mm guns fixed firing

ahead; internal bay for up to 3000 kg (6,614 lb) comprising six FAB-500 bombs each of 500 kg (1,102 llb) or two AV-45-36 torpedoes
Operators: Bulgaria, China, Egypt, Finland, North Korea, Poland, Romania, Somalia, North Yemen, South Yemen, Vietnam

Ilyushin Il-28 'Beagle'

Ilyushin Il-28U 'Mascot' of the Indonesian navy, now in storage.

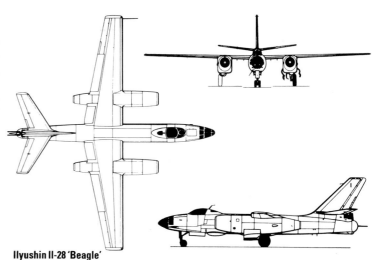

Ilyushin Il-38 'May'

Like the contemporary Lockheed Electra, the Il-18 airliner formed the basis for a long-range anti-submarine warfare (ASW) aircraft. The reasons were the same: the turboprop-powered airliner offered the best compromise between the propeller's low fuel consumption and the turbine's high-speed efficiency and smoothness. The large airliner cabins provided space for the large crew and extensive display console demanded by new and more sophisticated submarine-detection equipment.

The Soviet navy was a latecomer to the field of deep-water airborne ASW, which the West pioneered in the 1940s and 1950s, launching its first programme in the early 1960s. This decision was partly a response to the deployment of Polaris submarines by the USA, but was probably more influenced by the growing size and lethality of the US attack submarine fleet, and the Soviet Navy's own plans to expand its operations worldwide. The resulting aircraft was the Ilyushin Il-38, which flew in 1967 and entered service in 1970.

While the early P-3A and P-3B Orion could be built with improved versions of the P-2 Neptune's systems, the Ilyushin Il-38 represented entirely new technology for the Soviet Union. Its effectiveness depends mainly on the efficiency of its acoustic equipment: expendable sonobuoys, and the onboard proces-

sor that converts their data into a tactical picture. The sonobuoys are believed to be copied from salvaged Western units, and electronic computing is one area in which the Soviet Union trails behind the West. It is unlikely, therefore, that the Il-38 excels modern Western systems, and the odds are that it is inferior.

The design presents some unusual features, particularly the massive forward shift of the wing relative to the Il-18. This clearly indicates a concentration of heavy fixed equipment in the forward fuselage. Also noteworthy are what appear to be air inlets/outlets on either side of the fuselage. These features could both be due to the presence of a very large electronic processor in the forward fuselage, more akin to a shipboard computer or a 'ruggedized' fixed system than to Western airborne equipment. Such systems are heavy and dense, and require a dedicated cooling system.

In addition to its acoustic system, the Il-38 carries a J-band Wet Eye radar in a fixed underfuselage radome, providing a simple system with 360 deg coverage at a considerable price in drag. A magnetic anomaly detector (MAD) is installed in the forward fuselage, and other antennae are presumably connected with an electronic surveillance measures (ESM) system. Both sonobuoys and weapons – homing torpedoe and nuclear depth charges – are carried in a ventral weapons

bay. There are no wing stations visible and the type is not reported to carry external stores.

About 50 Il-38s are in service with the Soviet navy, sharing the ASW role with Tu-142s. In addition, three refurbished Il-38s were delivered to the Indian Navy in 1977, replacing Lockheed Constellations. Although no significant external changes have been noted since the type entered service, there is no sign of a direct replacement.

Specification
Ilyushin Il-38
Type: anti-submarine warfare aircraft
Powerplant: four 3170 kW (4,250 eshp) Ivchenko AI-20M turboprops
Performance: maximum cruising

The Il-18 airframe forms the basis for the Il-38 anti-submarine aircraft; the radar under the nose has necessitated moving the wings forward to counterbalance the extra weight. A MAD sting projects from the tail.

speed 625 km/h (390 mph); range with maximum fuel 6500 km (4,000 miles); endurance 13-16 hours
Weights: empty 40000 kg (88,700 lb); maximum take-off 68000 kg (150,000 lb)
Dimensions: span 37.4 m (122 ft 8 in); length 39.6 m (129 ft 10 in); height 10.17 m (33 ft 4 in); wing area 140 m² (1,507 sq ft)
Armament: internal weapons bay for homing torpedoes, nuclear and conventional depth charges, and sonobuoys
Operators: India, Soviet navy

Ilyushin Il-76 'Candid'

Throughout the 1960s the Ilyushin design bureau, increasingly led by Genrikh Novozhilov, planned a strategic airlift transport to replace the Antonov An-12. First flown on 25 March 1971 the resulting Il-76 proved an outstanding aircraft in all respects, fully meeting severe requirements and providing a basis for various derived versions.

The high wing has modest sweep, but very powerful high-lift slats and flaps, and 16 spoiler segments used as airbrakes, roll augmenters and lift dumpers. The landing gear is unique, with 20 wheels to spread the load on snow, mud or other soft surfaces. The engines, much more powerful than those of the Lockheed C-141, all have effective reversers. The cargo hold measures more than 3.40 m (11 ft 2 in) wide and high, appreciably less constricted than the Lockheed C-130 and C-141, and is pressurized and equipped with comprehensive computer-controlled loading and positioning systems.

As well as being able to operate from any rough airstrip, for which purpose tyre pressure can be reduced in flight to only 2.53 kg/cm² (36 lb/sq in), the Il-76 has to be independent of all ground services and also have avionics for precise navigation and blind automatic landing. Most have two nose radars (one weather, one mapping) and a navigator's glazed compartment. Military Il-76m and longer-range Il-76MD versions have armament. About 300 M and MD versions serve with the Soviet V-TA, a few being equipped as air refuelling tankers with three hose-drum units. NATO codename is 'Candid'.

According to US reports, a further version (codename 'Mainstay') is entering service as an AWACS type platform, with a large pylon-mounted radar rotodome above the

Iraq has a fleet of Il-76s which it uses for tactical transport. These are often seen in quasi-civil markings, complete with rear gun turret.

rear fuselage. Intended to replace the interim Tupolev Tu-126 'Moss', 'Mainstay' is said to be able to track small cruise missiles at low level, and to have endurance extended by a flight-refuelling probe. On paper its altitude performance ought to surpass that of any Western surveillance aircraft except the Lockheed TR-1. Production is said to be at the rate of five per year.

Specification
Ilyushin Il-76 'Candid'
Type: strategic airlift transport an air refuelling tanker
Powerplant: four 12000 kg (26,455 lb) thrust Soloviev D-30KP turbofan engines
Performance: maximum speed 850 km/h (528 mph); maximum cruising speed 800 km/h (497 mph); take-off run 850 m (2,790 ft); ceiling 15500 m (50,850 ft); range with 40000 kg (88,185 lb) payload and

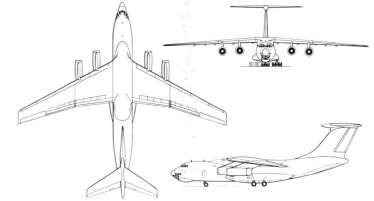

Ilyushin Il-76 'Candid'

reserves 5000 km (3,107 miles)
Weights: empty (typical) 75000 kg (165,347 lb); maximum 170000 kg (374,786 lb)
Dimensions: span 50.50 m (165 ft 8.2 in); length 46.59 m (152 ft 10.3 in); height 14.76 m (48 ft 5.1 in); wing area 300 m² (3,229.29 sq ft)

Armament: (M and MD) K-series rear turret with gunner and twin radar-directed NR-23 23-mm guns
Operators: (military) Czechoslovakia, El Salvador, India, Iraq, Papua New Guinea, Poland, Soviet Union, Swaziland, Syria

Israel Aircraft Industries 201 Arava

The civil prototype IAI-101 twin-turboprop STOL passenger/freight transport made its initial flight on 27 November 1969. The military Arava 201 (prototype 04) flew on 7 March 1972. It was demonstrated in gunship configuration, but is used mainly as an assault transport for 24 troops, or 17 paratroops plus dispatcher, or to carry a jeep containing a 106-mm (4.17-in) recoilless rifle and its four-man crew. The hinged rear fuselage can be replaced by a fairing making room for two one-ton cargo pallets, or for 12 stretchers plus two attendants. Other missions include tanker, ASW aircraft, navigational trainer and target tug. Features include electrically operated double-slotted flaps, scoop-type spoilers to augment lateral control, and reversing propellers.

IAI Arava of the Israel Defence Force/Air Force. Israeli Aravas are used for a variety of light transport and casevac duties, and some are converted for electronic warfare.

Specification
Type: STOL utility transport
Powerplant: two 559-kW (750-shp) Pratt & Whitney Canada PT6A-34 turboprop engines
Performance: maximum speed 326 km/h (203 mph) at 3050 m (10,000 ft); economical cruising speed 311 km/h (193 mph); initial rate of climb 393 m (1,298 ft)/minute service ceiling 7620 m (25,000 ft); range with maximum fuel 1056 km (656 miles)
Weights: empty 3999 kg (8,816 lb); maximum take-off 6804 kg (15,000 lb)
Dimensions: span 20.96 m (68 ft 9.2 in); length 13.03 m (42 ft 9 in); height 5.21 m (17 ft 1.1 in); wing area 43.68 m^2 (470.18 sq ft)
Armament: three Browning 12.7-mm (0.5-in) machine-guns, one in blister each side of forward fuselage and one pintle-mounted in tailcone in gunship configuration; 12 sonobuoys and four Mk 44 torpedoes in ASW configuration

IAI 201 Arava

Operators (military): Bolivia, Colombia, Ecuador, Guatemala, Honduras, Israel, Liberia, Mexico, Nicaragua, Salvador, US Venezuela, plus others unnanounced

Israel Aircraft Industries Kfir

Kfir (lion cub) has been developed by IAI to succeed the Dassault Mirage III in the interceptor and ground-attack roles. The basic Mirage III/5 airframe was adopted as a starting point, with a General Electric J79 turbojet – the engine used in the Phantom – replacing the SNECMA Atar employed in the Mirage series. Mating the US powerplant with the Dassault-designed airframe proved difficult, however, and IAI produced the Nesher (eagle) as an interim fighter to fill the gap until the new type was available. The Nesher was a locally built copy of the Mirage III/5, with some modifications but retaining the Atar 9C engine.

The Kfir was revealed in April 1975, when two examples were put on display. The aircraft closely resembles the Mirage 5 but has a number of differences apart from the use of a US engine and Israeli equipment. The rear fuselage is fatter and shorter than in the French-designed aircraft, with the variable exhaust nozzle protruding from the afterbody. The afterburner is cooled by air drawn from a scoop in the root of the fin. The forward section of Kfir's fuselage is larger than that of the Mirage, and the undersurface is flatter, while the nose itself has been lengthened. The leading edges of the 60° delta wing are also modified, and the strengthened landing gear uses long-stroke oleos.

The adoption of the J79-17 turbojet in place of an Atar has improved specific fuel consumption by about 20 per cent, and the high mass flow demanded by the US engine has necessitated an increase in intake inlet and duct area. The J79 also runs hotter than the French engine, which was one of the major difficulties encountered as a result of the substitution.

In July 1976 the Israeli Air Force revealed the existence of the Kfir-C2, the major external difference between this model and the original Kfir-C1 being the addition of canard surfaces. The addition of canards, slightly ahead of and above the wing, has a number of effects: it increases the lift available at a given angle of attack, allows the aircraft to operate over a greater range of angles of attack, and reduces stability because the centre of lift and centre of gravity are moved closer together. The canards have been fitted mainly to improve man-oeuvrability in combat, but they also allow the aircraft to operate from shorter runways. The Kfir-C2 has saw-teeth in the wing leading edges and small strakes along the nose, both these features complementing the canard surfaces in improving manoeuvrability. The Kfir-C1 and -C2 are otherwise identical, or nearly so, and the original aircraft are being converted to bring them up to the definitive -C2 standard. Recently the -C7 has been introduced, with uprated engine and avionics.

The Kfir carries two internally mounted 30-mm DEFA cannon with their muzzles protruding below the

The Kfir-C7 is the latest variant of this capable fighter. The Kfir now has an indigenous ECM kit as good as anything fitted elsewhere.

Israel Aircraft Industries Kfir (continued)

engine air intakes, as in the Mirage 5. The front of the barrel is fitted with specially developed gas-deflecting baffles which, according to the Israeli Air Force, allows the guns to be fired over the complete performance envelope without the risk of engine compressor stall. The aircraft can also carry a variety of external stores totalling more than 3856 kg (8,500 lb), including Rafael Shafrir Python air-to-air missiles, rocket pods, conventional or anti-runway bombs, and guided air-to-surface weapons such as Maverick, Hobos and Shrike.

The majority of Kfir's avionics is supplied by Israeli companies, although some items are built under licence from overseas firms. In many cases the original equipment has been improved on by the licensee. The Thomson-CSF Cyrano radar fitted in Mirage IIIs proved to be inadequate for the Israel Air Force's needs, so Elta has developed its EL/M-2001B ranging radar for the Kfir. This set can track low-flying aircraft even against a background of clutter and is claimed by the manufacturer to have no equivalent in the West. The EL/M-2001B feeds range data for air or ground targets into the central navigation and weapon-aiming system, which is based on an Elbit System-80 digital computer. Further information is supplied by the licence-built Singer Kearfott KT-70 inertial platform, allowing the nav-attack system to drive a head-up display supplied by Israel Electro-Optical Industries. The company based its HUD development on equipment built by Marconi Avionics for the Israeli Air Force's A-4 Skyhawks, although the unit fitted in the Kfir is of indigenous design. The HUD has two modes for delivery of air-to-ground weapons, another for air combat and a fourth for navigation. The ECM fit is one of the most comprehensive fitted to a fighter.

The Kfir is now in widespread service with the Israeli Air Force, and also has been exported to Ecuador, Colombia's Mirage 5s are being upgraded to Kfir standard by IAI. The type has seen much action over Lebanon as a fighter-bomber. The two-seat trainers in service can also be used in an offensive electronic warfare role.

Specification
Type: single-seat fighter and ground-attack aircraft
Powerplant: one General Electric J79-GE-17 (modified) turbojet rated at 8119 kg (17,900 lb) thrust with afterburning
Performance: maximum design (clean) Mach 2.2 or 2335 km/h (1,450 mph) at high altitude; tactical radius (as interceptor with two drop tanks) up to 535 km (330 miles); 11000 m (36,000 ft) 1.5 minutes
Weights: maximum combat for ground attack 14600 kg (32,120 lb); typical combat as interceptor with two Shafrir missiles 9305 kg (20,470 lb)
Dimensions: span 8.23 m (27 ft); length 15.34 m (50 ft 4 in); height 5.22 m (17 ft 1 in); 36.43 m² (329 sq ft)
Armament: two 30-mm DEFA cannon and two Shafrir AAMs or air-to-ground weapons including bombs, rockets or missiles (such as Maverick, Hobos and Shrike)
Operators: Israel, Colombia, Ecuador

Kfirs are used for fighter-bomber duties by Israel, and saw much action over Lebanon. One particular role was bombing Syrian SAM sites.

Israel Aircraft Industries IAI Lavi

The first advanced combat aircraft to be designed in Israel, the Lavi (young lion) is intended to replace the McDonnell Douglas A-4 and later the IAI Kfir, operating mainly in the attack and close support roles. It will also have considerable air combat capability, but unlike the Saab Gripen and the European Fighter Aircraft this will be a secondary role. Design was undertaken with assistance from many US companies, notably Grumman Aerospace which is producing the first 50 sets of wings and fins in advanced carbonfibre composites. About 20 per cent of the structure by weight will be of composite materials.

The configuration is the fashionable one with a delta main wing, swept canards and chin inlet, with a plain engine nozzle and no attempt to use engine thrust to augment wing lift. The wing will have leading and trailing edge flaps, giving variable camber, uses in conjunction with the powered canard and all under control of the quadruple-redundant digital fly-by-wire flight control system. The inlet will be simple, and high Mach numbers are not attempted. Instead the design is biased strongly in favour of low-level bombload, agility, small radar cross-section and very comprehensive EW (electronic warfare) installations for penetration of hostile airspace.

The Pratt & Whitney PW1120 engine is used in no other aircraft, but has the advantage of using the same core (and some other components) as the well-proved F100. It is, however, a much heavier engine than the equally powerful advanced versions of the RB199 and F404, which also offer lower fuel consumption. Internal fuel capacity of 3330 litres (732 Imp gallons) is rather less than that of the General Dynamics F-16 but greater than that of the Kfir. More remarkably, the external fuel is published as 5095 litres (1,121 Imp gallons), suggesting the carriage of two of the biggest F-15 size drop tanks on the inboard pair of wing pylons. The point must be emphasized that the Lavi is smaller than almost all other modern combat aircraft, apart from the Gripen, and it carries a modest load of attack weapons; with a reported increase in gross weight the load may be increased. On the other hand it carries it over a considerable radius and is expected to have outstanding penetration capability.

Avionics will be of the highest order, with an advanced Elta multimode radar with doppler beam-sharpening, terrin avoidance and a programmable signal processor. Electronic warfare systems include internal and external computer-managed noise and deception jammers and high-capacity payload dispensers. The seat is the Martin Baker IL10LD. First flight is expected in early 1986, with production delivery in 1989, with delivery planned during the 1990s of about 240 single-seaters and 60 dual trainers with reduced internal fuel.

Specification
IAI Lavi
Type: multi-role attack fighter
Powerplant: one 9380 kg (20,680 lb) thrust Pratt & Whitney PW1120 afterburning turbojet engine
Performance: (estimated) maximum speed, clean at high altitude Mach 1.85 or 1965 km/h (1,221 mph); low-level attack speed with eight Mk 117 bombs 997 km/h (619 mph) combat radius with same load lo-lo-lo 452 km (281 miles)
Weights: (estimated) empty 6940 kg (15,300 lb); loaded (clean) 9664 kg (21,305 lb); maximum take-off originally 17010 kg (37,500 lb), now being increased to 19200 kg (42,329 lb)
Dimensions: span 8.71 m (28 ft 6.9 in); length 14.39 m (47 ft 2.5 in); height 5.28 m (17 ft 3.9 in); wing area 32.50 m² (349.8 sq ft)
Armament: six fuselage and four underwing stations for bombs, missiles and other stores up to maximum of 2721 kg (6,000 lb), not including 4164 kg (9.180 lb) of external fuel, and two infra-red AAMs at wingtips. Note: the latest planned gross weight would allow for a weapon load of 4911 kg (10,827 lb)

Israel Aircraft Industries Nesher

As an intermediate stage in the development of an Israeli fighter derived from the Dassault-Breguet Mirage III, Israel Aircraft Industries first produced drawings and tooling for a direct copy of the French aircraft, with only detail changes. While the secret 'Black Curtain' project was under way to re-engine the Mirage with the J79 turbojet, plans of the original aircraft and its Atar 09C engine arrived covertly in Israel from French and Swiss sources. In view of the time needed to develop a satisfactory J79-Mirage it was decided as an immediate measure to manufacture the Atar-powered aircraft under the name Nesher (eagle). The decision to do so was taken in 1968 and the first Nesher flew in September 1969. The aircraft was in most respects similar to the Mirage IIICJ but incorporated simplified Israeli avionics more akin to those of the Mirage 5. When the Yom Kippur war broke out in 1973, Neshers were fully operational with the Heyl Ha'Avir. Total production exceeded 50. All were sold to Argentina, before and since the Falklands war. In 1983 survivors were fitted with inflight refuelling probes by IAI.

IAI Dagger of the Argentine air force. Daggers were used against British ships during the Falklands campaign, several being lost to Sea Harriers.

Specification
Not quoted specifically by IAI but essentially similar to Dassault Mirage IIIC
Operator: Argentina

Israel Aircraft Industries Westwind and Sea Scan

The Westwind executive jet stemmed from the Aero Commander 1121 Jet Commander. In 1967 Rockwell (owner of Aero Commander) merged with North American, and US anti-trust laws decreed that Rockwell could not market both the Jet Commander and Sabreliner. Accordingly, the Jet Commander was sold to Israel Aircraft Industries. The IAI aircraft received an 0.76 m (2 ft 6 in) fuselage 'stretch', CJ610-5 engines, an increase in weight to 8390 kg (18,497 lb) and the name Commodore Jet 1123, and later the Westwind 1123.

The 1124 Westwind with Garrett turbofan engines provided a basis for the 1124N Sea Scan maritime patrol aircraft. This has Litton LASR-2 or APS-504 search radar, a Global GNS-500A VLF/Omega navigation system, bubble windows, reversers, and provision for a MAD boom, low-light television and FLIR. External stores, including sea-rescue drop modules, can be carried on fuselage pylons, the aircraft can be fitted with ECM chaff dispensers and sonobuoy chutes, and the latest

options are winglets and two underwing pylons for missiles or other stores.

Specification
IAI 1124N Sea Scan
Type: maritime surveillance aircraft
Powerplant: two 1678-kg (3,700-lb) thrust Garrett TFE731-3-1G turbofan engines
Performance: maximum speed 872 km/h (542 mph) at 5900 m (19,355 ft); long-range cruising speed 675 km/h (419 mph); service ceiling 13700 m (44,950 ft) maximum search range 4630 m (2,877 miles)
Weights: empty 5760 kg (12,699 lb); maximum take-off 10660 kg (23,501 lb)
Dimensions: span over tiptanks 13.65 m (44 ft 9.4 in); length 16.80 m (55 ft 1.4 in); height 4.81 m (15 ft 9.4 in); wing area 28.65 m^2 (308.29 sq ft)
Armament: provision for two A/S torpedoes, two Gabriel III air-to-surface missiles, or other stores
Operators: West Germany (target

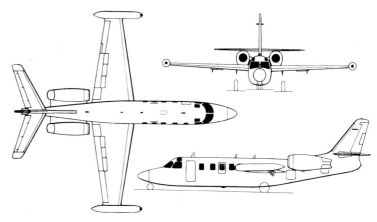

IAI Westwind 1124

towing), Honduras, Israel

The Astra is the latest version of the Westwind family. It offers unrivalled economy for VIP transport and coastal patrol.

Jodel/SAN D.140E/R

Based on the two-seat Jodel versions, the D.140 Mousquetaire was developed by Société Aéronautique Normande (SAN) as a four-seat tourer with alternative capacity for a stretcher loaded through a baggage door behind the cabin. The prototype flew on 4 July 1958. The D.140B had a revised instrument panel and foot brakes: the D.140C Mousquetaire III introduced a swept vertical tail, and the D.140E Mousquetaire IV had a larger wing, flaps and vertical tail, all-moving tailplane and improved ailerons. The D.140A had a modified airspeed indicator, and when fitted with the D.140C tail became the D.140AC. The Armée de l'Air bought 18 D.140Es for the Ecole de l'Air at Salon where they were used for recreational purposes, and 15 D.140R

Abeilles were also obtained. This version, which first flew in mid-1965, is equipped for glider and banner towing; the extensively-glazed cockpit improves rearward field of view. All models can be fitted with skis.

Specification
Jodel/SAN D.140E/R
Type: four-seat light aircraft
Powerplant: one 134-kW (180-hp) Lycoming O-360-A2A flat-four piston engine
Performance: maximum speed 255 km/h (158 mph) at sea level; cruising speed 240 km/h (149 mph) at 1800 m (5,905 ft); initial rate of climb 230 m (755 ft)/minute; service ceiling 5000 m (16,405 ft); range 1400 km (870 miles)
Weights: empty 620 kg (1,367 lb);

Jodel/SAN D.140

maximum take-off 1200 kg (2,646 ft)
Dimensions: span 10.27 m (33 ft 8.3 in); length 7.82 m (25 ft 7.9 in);

height 2.05 m (6 ft 8.7 in); wing area 18.50 m^2 (199.14 sq ft)
Armament: none
Operator: France

Kaman SH-2 Seasprite

The prototype Kaman HU-2K (later UH-2A) Seasprite shipboard utility helicopter flew on 2 July 1959, and from 1962 Kaman delivered 190 of these attractive machines each powered by a T-58 engine and with fully retractable forward-mounted main landing gears. The UH-2A and UH-2B could each carry a 1814-kg (4,000-lb) slung load or 11 passengers, and did sterling work in planeguard, SAR, fleet reconnaissance, Vertrep (vertical replenishment) and utility transport duties, operating from many surface warships as well as from shore bases. From 1967 all available Seasprites were converted to twin-T58 helicopters with full engine-out safety and generally improved performance, plus better load-carrying capability. Among many other models the most important current variant is the SH-2F (Mk 1 LAMPS, for Light Airborne Multi-Purpose System) for ship-based ASW and anti-ship missile defence with secondary SAR, observation and utility capability, in all weather conditions. With a crew comprising pilot, co-pilot and sensor operator, the SH-2F can carry full ASW gear including Canadian Marconi LN-

66HP surveillance radar, towed 'bird' for the ASQ-81(V)2 MAD, General Instruments ALR-66 passive detection receiver, ASN-123 tactical navigation system, Difar passive and Dicass active sonobuoys, and comprehensive nav/com and display systems. The 1814-kg (4,000-lb) cargo capability remains, and a 272-kg (600-lb) rescue hoist is standard. From 1973 Kaman not only delivered conversions of earlier models but also 88 new SH-2Fs, followed in 1983-4 by 18 additional machines. The Seasprite is being supplemented by the Sikorsky SH-

60B Seahawk LAMPS III, but will remain in service throughout the 1990s on the US Navy's older or smaller ships.

Specification
Kaman SH-2F Seasprite
Type: ship-based helicopter
Powerplant: two 1007-kW (1,350-shp) General Electric T58-GE-8F turboshaft engines
Performance: maximum speed 265 km/h (165 mph); cruising speed 241 km/h (150 mph); initial rate of climb 744 m (2,440 ft)/minute; service ceiling 6860 m (22,500 ft);

A Kaman Seasprite of the US Navy, carrying a pair of Mk.46 homing torpedoes.

range with maximum fuel 679 km (422 miles)
Weights: empty 3193 kg (7,040 lb); normal take-off 5806 kg (12,800 lb)
Dimensions: main rotor diameter 13.41 m (44 ft 0 in); length, rotors turning 16.03 m (52 ft 7 in); height 4.14 m (13 ft 7 in); main rotor disc area 141.26 m^2 (1,520.53 sq ft)
Armament: one or two Mk 46 torpedoes
Operator: USA (US Navy)

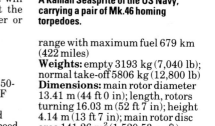

179

Kaman HH-43F Huskie

The prototype Kaman H-43B, converted from a modified HOK-1, flew on 27 September 1956. Production H-43Bs, which began to come off the line in December 1958, were powered by an 641-kW (860-hp) T53-L-1B turboshaft driving the intermeshing 'egg-beater' rotors. By mounting the engine above the cabin, the new model offered double the usable cabin space and payload of the H-43. The type was subsequently redesignated HH-43B, and export deliveries included 12 to the Burmese air force and six to the Colombian air force. Final version of the Huskie was the HH-43F with uprated T53-L-11A turboshaft and increased internal fuel-first flown in August 1964, 24 production aircraft were later supplied to Iran.

Specification
Kaman HH-43F Huskie
Type: crash rescue helicopter

Powerplant: one 858-kW (1,150-shp) Lycoming T53-L-11A turboshaft engine derated to 614 kW (825 shp)
Performance: maximum speed 193 km/h (120 mph) at sea level; cruising speed 177 km/h (110 mph); initial rate of climb 610 m (2,000 ft)/minute; service ceiling 7010 m (23,000 ft); range 811 km (504 miles)
Weights: empty 2027 kg (4,469 lb); maximum take-off with underslung load 3992 kg (8,800 lb)
Dimensions: main rotor diameter, each 14.33 m (47 ft 0 in); length of fuselage 7.67 m (25 ft 2 in); height 3.84 m (12 ft 7 in)
Armament: none
Operators: Burma, Morocco, Pakistan, Thailand, USA

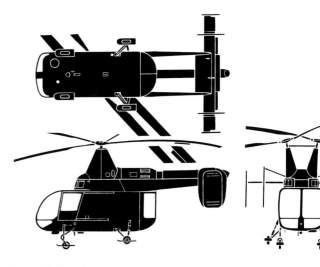

Kaman HH-43 Huskie

Kamov Ka-25 'Hormone'

Called 'Hormone' by NATO, the compact Kamov Ka-25 helicopter has appeared in various sub-types which have, since 1965, been the standard type carried aboard Soviet surface warships for defensive and offensive roles. The traditional Kamov layout with superimposed coaxial rotors reduces disc diameter, and automatic blade-folding is incorporated for stowage in small hangars. The four-legged landing gear is specially designed to operate from pitching decks, each leg having an optional quick-inflating flotation bag which gives the 'Hormone' a distinctly unusual appearance. The rear legs can be raised vertically, on their pivoted bracing struts, to lift the wheels out of the vision of the search radar always fitted under the nose. Two radars have been identified; the smaller type is carried by the 'Hormone-A' variant on ASW missions; this model also has a towed MAD 'bird', dipping sonar, electro-optical sensor (and possibly others) and an optional right side box of sonobuoys. A larger radar is fitted to the 'Hormone-B', which is believed to be able to guide the SS-N-12 'Sandbox' cruise missile fired from Soviet surface ships and, especially, submarines. Other equipment items include a cylindrical container near the rear of the cabin and a streamlined pod under the tail. In 1982 Ka-25s were seen without flotation gear but with a long ventral box housing (it is believed) a wire-guided torpedo. All Ka-25s have a large cabin normally provided with 12 folding seats additional to those for the crew of two pilots plus three systems operators. Some 460 of all variants were built by 1975, and the type continues to play an important part in Soviet

naval operations, flying from destroyers, cruisers, helicopter carriers; 18 are believed to be carried on each of the two ships *Moskva* and *Leningrad*, and the much larger *Kiev*, *Minsk* and *Novorossisk* each accommodate 16 'Hormone-A' and three 'Hormone-B'. The type has also been exported for ship- and land-based operations. The last variant is the 'Hormone-C' search-and-rescue helicopter based on the 'Hormone-A' without the latter's mission equipment. The 'Hormone' will remain in service for some years to come, despite the appearance of the Kamov Ka-27, Ka-32 and Mil Mi-14.

Specification
Type: multi-role shipboard helicopter
Powerplant: two 739-ekW (990-eshp) Glushenkov GTD-3BM turboshaft engines
Performance: maximum speed 209 km/h (130 mph); cruising speed 193 km/h (120 mph); service ceiling 3500 m (11,485 ft); range with external tanks 650 km (404 miles)
Weights: empty about 5000 kg (11012 lb); maximum take-off 7500 kg (16,535 lb)
Dimensions: main rotor diameter, each 15.74 m (51 ft 7.7 in); fuselage length 10.36 m (33 ft 11.9 in);

Now being replaced on larger vessels by the Ka-27, the Ka-25 serves as an ASW helicopter and over-the-horizon targetter, with secondary duties. The coaxial rotors enable it to fit into tight decks without having to fold rotors.

height 5.37 m (17 ft 7.4 in); main rotor disc area, total 389.16 m² (4,189.04 sq ft)
Armament: normally equipped with ventral bay or external box for two AS torpedoes, nuclear or conventional depth charges and other stores
Operators: India, Jugoslavia, Soviet Union, Syria

Kamov Ka-27 'Helix'

Early publicity for the Kamov Ka-32 was associated with civil applications, including reconnaissance from the nuclear-powered icebreakers *Arktika*, *Lenin*, *Rossiya* and *Sibir*, and all forms of transport and agricultural flying. Photographs were first taken of Aeroflot (civil) and AVMF (naval air force) examples at sea aboard the new destroyer *Udalov* in September 1981. NATO allocated the reporting name 'Helix', and the US Depart-

ment of Defense calls military versions Ka-27. Clearly an enlarged successor to the familiar Ka-25, the Ka-27 has similar three-blade coaxial rotors that each negate the torque of the other, thus removing the need for a tail rotor and giving the designer freedom to reduce overall dimensions. However, the blades of the Ka-27's rotor are of different shape and increased diameter. The fuselage has greater volume than that of the Ka-25, and it is estimated

that in a utility role the Ka-27 could carry 25 troops or substantial quantities of cargo: the civil version is described as able to lift slung loads up to 5000 kg (11,023 lb) and to carry such a load over a range of 185 km (115 miles). The ASW version, known in the West as 'Helix-A', has a large box on each side (probably for sonobuoys), a box under the tail (probably for MAD equipment), a large chin radar and extremely comprehensive avionics including

EW installations. 'Helix-B' is a targeting aircraft for anti-ship missiles, intended as a replacement for the Ka-25 'Hormone-B'. From the Soviet navy's point of view, the real advantage of the Ka-27 series is that while the overall dimensions are little altered from those of the Ka-25, permitting them to operate from existing platforms and hangars, payload and general utility have been enhanced considerably.

Specification
Kamov Ka-27 'Helix-A'
Type: ASW and multi-role naval
helicopter
Powerplant: two 1659-kW (2,225-
shp) Isotov TV3-117V turboshaft
engines
Performance (estimated):
maximum speed 260 km/h
(162 mph); range with 5-tonne
payload 185 km (115 miles)
Weights (estimated): empty
5750 kg (12,677 lb); maximum take-
off 11500 kg (25,353 lb)
Dimensions (estimated): main
rotor diameter, each 16.75 m (54 ft
11.4 in); fuselage length 11.00 m
(36 ft 1.1 in); height 5.50 m (18 ft
0.5 in); rotor disc area, total
440.70 m² (4,743.89 sq ft)
Armament: includes AS torpedoes
Operator: Soviet Union (AVMF)

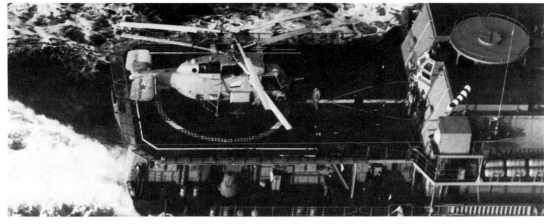

Based on the tried and tested co-axial rotor design, Kamov's Ka-27 'Helix' is entering service on important ships.

Kamov 'Hokum'

First mentioned in Western publications in mid-1984, this coaxial-rotor helicopter from the Kamov bureau, led by S.V. Mikheyev, is the first ever to be designed from the start as a helicopter fighter, its primary purpose apparently being the destruction of hostile helicopters in the neighbourhood of ground battles.

By late 1985 the 'Hokum' (NATO codename) was still virtually unknown in the West, the US Department of Defense having included a poor sketch and a few suggested figures in the 1985 edition of its *Soviet Military Power*. There are grounds for the supposition that this new helicopter, which might be the Ka-29 or -31, probably uses the same engines and rotors as the Ka-27/Ka-32 'Helix'. It has been suggested that the gross weight is in the region of 5443 kg (12,000 lb), which would fit

in with this assumption.

One of the few things that does appear to be agreed about 'Hokum' is that it has a slim fighter-type fuselage with tandem cockpits for a crew of two, and with no rear cabin. In common with other coaxial helicopters, no anti-torque tail rotor is needed, the tail being thought to resemble that of a fixed-wing aeroplane. The DoD sketch showed fixed landing gears, but it would be much more likely for the gear to retract, as in the Mil Mi-24, because high speed is an obvious requirement for a helicopter intended to catch and destroy other helicopters. There is no obvious advantage in using coaxial rotors other than overall compactness, which is an advantage in shipboard operation. This helicopter may well find a role at sea, but initially it is thought to be deployed

purely with operational manoeuvre groups of the land theatre forces, to clear the sky of NATO or other hostile helicopters and thus reduce attrition of Soviet armour and other ground forces, besides having a major effect on opposing reconnaissance assets.

One possible advantage for a fighter helicopter is that compact rotors should result in greater flight agility, especially in higher rate of roll. It is unlikely that the new Kamov uses the rigid ABC (advancing blade concept) type coaxial rotors tested in the Sikorsksy S-69 (XH-59). This helicopter, superficially very similar to the new Soviet machine, was tested by Sikorsky and the US Army at speeds up to 487 km/h (303 mph), appreciably faster than any other wingless helicopter. The only reason for doubting

use of this concept by Kamov is the Soviet preference for robust reliability rather than seemingly uncertain new technologies. With a traditional Kamov type rotor it is doubtful if speeds beyond 322 km/h (200 mph) could be attained. Despite this the Pentagon estimates 'Hokum' to have a top speed of 350 km/h (217.5 mph), with the radius of action thought to be 250 km (155 miles).

With virtually no hard numerical facts to hand there seems little point in attempting a series of guessed figures for performance or other capabilities. the engines may well be Isotov TV3-117s of some 1641 kW (2,200 shp), and armament is certain to include at least one gun as well as AAMs. It is also safe to predict that this helicopter will have night and all-weather visionics and electronic-warfare installations.

Kawasaki C-1

Produced in only modest numbers for service with the Japanese Air Self-Defence Force, the Kawasaki C-1 was one of a handful of indigenous designs which reached fruitition during the course of the 1960s and 1970s. Conceived to replace the World War II Curtiss C-46 Commando transport aircraft, the C-1 entered design under the auspices of the Nihone Aeroplane Manufacturing Company (NAMC) during 1966, although the task of assembling this type was eventually undertaken by Kawasaki at its Gifu factory. It was from Gifu that the first two prototypes made a successful maiden flight on 12 November 1970.

After initial company trials, both prototypes were duly handed over to the Japanese Defence Agency for formal evaluation, a process which was completed in March 1973 and resulted in a production go-ahead for an initial batch of 11 aircraft.

Training of personnel to operate the C-1 got under way in 1974, with deliveries of production aircraft beginning shortly before the end of the same year, initially to No. 402 Squadron at Iruma and subsequently to No. 401 Squadron at Komaki and No. 403 Squadron at Miho.

In service the type has proved to be somewhat limited with regard to payload and range, but some of the final examples feature an extra fuel tank in the wing centre-section. Despite being rather 'short-legged', the C-1 is closely tailored to JASDF

needs and can accommodate up to 60 troops in a pressurized cabin. Alternatively, when being used for paratrooping, it can take up to 45 fully equipped soldiers. Other loads include 36 stretchers and medical attendants in casualty evacuation tasks, whilst the cabin cross-section permits carriage of such items as a 2½-ton truck or a 105-mm howitzer in a freight load of 11900 kg (26,235 lb).

In addition to the basic transport model, design studies of specialized variants for inflight-refuelling, electronic warfare and weather recon-

naissance have been undertaken, but have failed to result in firm orders.

Specification
Kawasaki C-1
Type: medium-range tactical transport aircraft
Powerplant: two 6577-kg (14,500-lb) thrust Mitsubishi-built Pratt & Whitney JT8D-M-9 turbofans
Performance: cruising speed 655 km/h (407 mph) at 10670 m (35,000 ft); service ceiling 11580 m (38,000 ft); range 1300 km (808 miles) with a 7900-kg (17,416-lb)

payload
Weights: empty 24300 kg (53,571 lb); maximum take-off 45000 kg (99,206 lb)
Dimensions: span 30.60 nm (100 ft 4.75 in); length 29.00 m (95 ft 1.75 in); height 9.99 m (32 ft 9.25 in); wing area 120.5 m² (1,297.1 sq ft)
Operator: Japan

Japan's tactical transport assets rest on the C-1, which can operate from short fields. Electronic warfare versions exist.

Kawasaki (Boeing Vertol) KV107

The first Boeing Vertol 107 Model II completed by the Japanese licensee made its maiden flight in May 1962. In 1965 Kawasaki secured worldwide marketing rights and the KV107 remains in production for civil and military customers.

The principal production model is designated KV107IIA, with uprated engines. Optimised for mine countermeasures, the KV107IIA-3 operates with the Japanese Maritime Self-Defence Force (JMSDF) which purchased seven, these following two earlier KV107II-3s. The Japanese Ground Self-Defence Force (JGSDF) uses the KV107IIA-4 as its principal cargo/troop transport

helicopter, 18 following 42 KV107II-4s, one of which was configured for VIP duties. The Japanese Air Self-Defence Force (JASDF) has 15 SAR KV107II-5s, followed by 20 KV107IIA-5s. Eight basically similar helicopters were supplied to the Swedish navy in 1972-74 as the HKP-4C with Rolls-Royce Gnome H.1200 engines, Decca navigation and much Swedish equipment.

Saudi Arabia has purchased 16 KV107s, the most numerous subtype being seven KV107IIA-SM-1s for firefighting with chemical retardants or water. Four KV107IIA-SM-2s were delivered for SAR, with res-

cue hoist, medical equipment, stretcher kit and other rescue apparatus. Two VIP KV107IIA-SM-3s were delivered and three medevac KV107IIA-SM-4s were due to be handed over in 1984.

Specification
Kawasaki KV107IIA-2
Type: multirole transport helicopter
Powerplant: two 1044-kW (1,400-shp) General Electric CT58-140-1 turboshaft engines
Performance: maximum speed at sea level 254 km/h (158 mph); initial

rate of climb 625 m (2,050 ft)/minute: service ceiling 5180 m (16,995 ft); range with maximum fuel 1097 km (682 miles)
Weights: empty 5250 kg (11,574 lb); maximum take-off 9706 kg (21,398 lb)
Dimensions: main rotor diameter, each 15.24 m (50 ft 0 in); length rotors turning 25.40 m (83 ft 4 in); height 5.13 m (16 ft 10 in); rotor disc area, total 364.6 m² (3,925 sq ft)
Armament: none
Operators: Japan, Saudi Arabia, Sweden

Both JGSDF and JASDF fly the KV107, the former for cargo/assault and the latter for SAR.

Kawasaki P-2J

The Kawasaki P-2J was the outcome of a requirement for a new maritime patrol and anti-submarine aircraft for the Japanese Maritime Self-Defence Force, to provide a successor to the Lockheed P-2H Neptunes then in service. As 60 of these latter aircraft had been built by Kawasaki a derived aircraft was decided upon, instead of the more costly alternative of buying the P-3 Orion, and the prototype P-2J, a modified P-2H, flew on 21 July 1966. The JMSDF received the first of 82 production P-2Js on 7 October 1969. The P-2J's fuselage is extended by 1.27 m (4 ft 2in) to house the new avionics fit. The engines are Japanese-built General Electric T64 turboprops, but as on the P-2H auxiliary turbojets are also fitted. The normal crew is 12 comprising two pilots, seven operators in the tactical compartment and three in the ordnance room and galley aft. ASW equipment embraces AN/APS-80 search radar, MAD, ESM, sono-buoy data display system, digital data processor and an integrated data display. A searchlight is housed in the right wingtip pod.

The JMSDF shore-based anti-submarine force comprises five groups, three of which number P-2Js

in their equipment. Four of these aircraft have been converted as UP-2J target tugs, and one has been removed from the inventory and is used for research. Ironically, the replacement for the P-2J is now the P-3C Orion.

Specification
Type: long-range maritime patrol aircraft
Powerplant: two 2282-ekW (3,060-ehp) T64-IHI-10E turboprop and two

7518-kg (3,410-lb) thrust J3-IHI-7D turbojet engines licence-built by Ishikawjima-Harima
Performance: maximum cruising speed 400 km/h (249 mph); initial rate of climb 550 m (1,804 ft)/minute; service ceiling 9150 m (30,020 ft); maximum range 4450 km (2,765 miles)
Weights: empty 19280 kg (42,505 lb); maximum take-off 34000 kg (74,957 lb)
Dimensions: span, over tiptanks

The P-2J is a licence-build of the Lockheed P-2H with turboprop engines. These are being replaced by P-3C Orions.

30.87 m (101 ft 3.4 in); length 29.23 m (95 ft 10.8 in); height 8.93 m (29 ft 3.6 in); wing area 92.90 m² (1,000.0 sq ft)
Armament: anti-submarine homing torpedoes, depth charges or mines
Operator: Japan (JMSDF)

Kawasaki XT-4

To replace the Lockheed T-33A and Fuji T-1A/B in the JASDF, up to 200 examples of the XT-4 may be acquired during the late 1980s. The go-ahead came on 4 September 1981; construction of the first prototype began in April 1984 for a maiden flight in July 1985.

Bearing a resemblance to the Alpha Jet, the XT-4 will have five hardpoints for fuel or weapons. Mit-

subishi is responsible for the centre fuselage, air intakes and wing inspection doors. Fuji will contribute the rear fuselage, wings, tailplane, fin, canopy and fairings.

The Kawasaki XT-4 is one of the latest crop of indigenous jet trainers. It is equipped with hardpoints for light strike and weapons training.

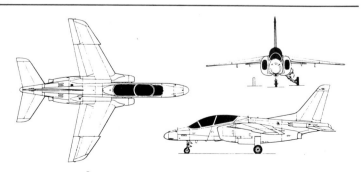

Specification
Type: jet trainer
Powerplant two 1600-kg (3,527-lb) thrust Ishikawajima-Harima XF3-30 turbofan engines
Performance: (estimated) maximum speed at optimum altitude 955 km/h (593 mph); initial rate of climb 3155 m (10,350 ft)/ minute; service ceiling 15000 m (49,210 ft); range, internal fuel 1313 km (816 miles)
Weights: (estimated) empty 3700 kg (8157 lb); maximum take-off 7500 kg (16,535 lb)
Dimensions: span 10.00 m (32 ft 9.7 in); length 12.60 m (41 ft 4.1 in); height 4.20 m (13 ft 9.4 in); wing area 21.60 m^2 (232.51 sq ft)
Armament: four underwing pylons for gun pods, practice bombs or air-to-air missiles, plus underfuselage pylon for tanks, ECM/chaff dispenser or target towing gear
Operator: none

Lockheed C-5 Galaxy

Initial US Military Air Transport Service attempts to acquire a very large strategic transport, made in 1963, centred around the CX-4 project which called for an aircraft with a maximum take-off weight of 272155 kg (600,000 lb) as well as the ability to operate from the same airfields as the C-141 StarLifter. Subsequent studies led to the CX-HLS proposal which envisaged a machine able to carry a payload of 56699 kg (125,000 lb) for 12875 km (8,000 miles) or a greater payload over a shorter distance.

Boeing, Douglas and Lockheed submissions were judged worthy of further development, and General Electric and Pratt & Whitney were awarded contracts for engines. GE's GE1/7 turbofan was selected for development as the TF39 in the summer of 1965. In October 1965 Lockheed won the aircraft competition, and this was given the designation C-5 Galaxy, construction of the first of 115 starting in August 1966. Flown on 30 June 1968, the first Galaxy was joined by seven more in a test programme not completed until 1971. Delivery to Military Airlift Command – as MATS had become – began on 17 December 1969. By this time contracts had terminated at 81, the plan to obtain 115 having succumbed to cost escalation. The 81 C-5As equipped three Military Airlift Wings, the 436th MAW at Dover AFB, Delaware, the 443rd at Altus AFB, Oklahoma, and the 60th at Travis AFB, California.

In service the wing accumulated fatigue damage much more rapidly

The C-5A can easily load helicopters and aircraft, which helps the USAF's global rapid reaction forces. Here an Army UH-1 is being loaded, no dismantling being necessary.

than anticipated. Lockheed was directed in early 1978 to manufacture two sets of wings to a virtually new design, incorporating later aluminium alloys to provide greater strength. Testing of these wings was completed in 1980 and resulted in the decision to re-wing all 77 surviving C-5As by 1987. The aircraft have also been painted in 'European One' camouflage.

In 1982 Congress approved a Lockheed proposal to produce additional aircraft (instead of C-17s). The new C-5B is similar to the C-5A but incorporates changes resulting from 12 years of service. Present planning calls for 50 C-5Bs, to be manufactured using tools retained from C-5A production, and assembly of the first of these new aircraft began in April 1984, for entry into service in December 1985. Delivery of the last C-5B is due in 1989.

Specification
Lockheed C-5B
Type: heavy logistics transport
Powerplant: four 19504-kg (43,000-lb) thrust General Electric TF39-GE-1C turbofan engines
Performance: (estimated) cruising speed at 7620 m (25,000 ft) 888-908 km/h (552-564 mph); initial rate of climb 526 m (1,725 ft)/minute: service ceiling 10895 m (35,750 ft); range with maximum payload 5526 km (3,434 miles)

The C-5A fleet is being slowly updated with strengthened wings. As each aircraft is returned for this work to be carried out, tactical camouflage is applied. The result gives the Galaxy a distinctly evil look.

Weights: (estimated) empty equipped 169644 kg (374,000 lb); maximum take-off 379657 kg (837,000 lb)
Dimensions: span 67.88 m (222 ft 8.5 in); length 75.54 m (247 ft 10 in); height 19.85 m (65 ft 1.5 in); wing area 575.98 m^2 (6,200.0 sq ft)
Armament: none
Operator: USAF (Military Airlift Command)

MILITARY AIRLIFT COMMAND

MILITARY AIRLIFT COMMAND

436ª MAW

0456

U.S. AIR FORCE

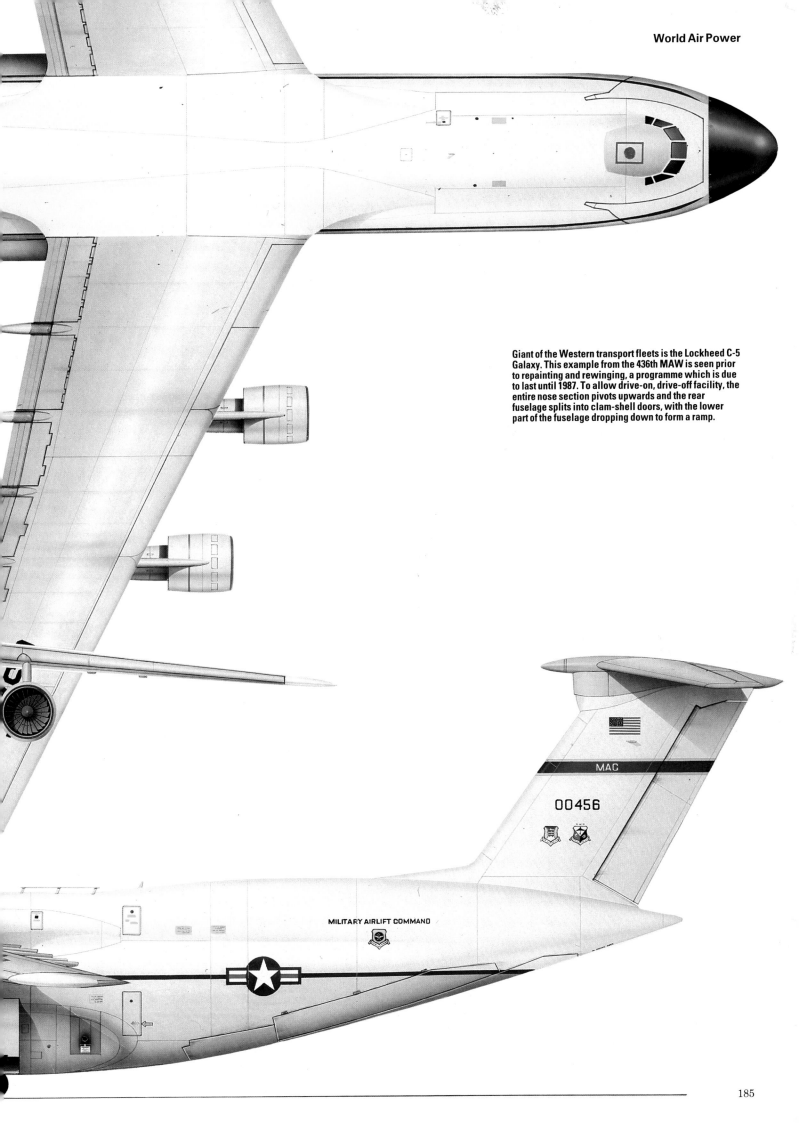

Giant of the Western transport fleets is the Lockheed C-5
Galaxy. This example from the 436th MAW is seen prior
to repainting and rewinging, a programme which is due
to last until 1987. To allow drive-on, drive-off facility, the
entire nose section pivots upwards and the rear
fuselage splits into clam-shell doors, with the lower
part of the fuselage dropping down to form a ramp.

MAC

00456

MILITARY AIRLIFT COMMAND

Lockheed C-130 Hercules

By far the most successful of all military airlifters, Lockheed's Model 382 was well named. It is one of the few types which has been in production for more than one-third of the entire period of powered flight.

Its origin came in 1951 when the USAF made the decision to acquire a fleet of turboprop transports for use by Tactical Air Command (TAC). Known as the Logistic Carrier Supporting System SS-400L, the contract for two prototypes, placed on 11 July 1951, used the designation YC-130. On 23 August 1954, the first made its maiden flight at Burbank, powered by four 2796-kW (3,750-shp) Allison T56-A1A turboprops. The first production contract was awarded in September 1952, and the first production C-130A was flown by the new Lockheed-Georgia Company at Marietta on 7 April 1955. First deliveries, in December 1956, went to the 463rd Troop Carrier Wing and the 322nd Division USAFE.

The high wing ensured that the cabin had minimal loss of capacity from the wing carry-through structure, and for the same reason the main landing gear units retracted into fairings outside the fuselage. Weather radar altered the nose profile at aircraft 30, and other modifications increased fuel capacity and strengthened the floor. Access to the hold, 12.62 m (41 ft 5 in) long and 3.12 m (10 ft 3 in) wide, is via a hydraulically-actuated ramp which takes pressurization loads.

A single AC-130A Shadow gunship was tested in Vietnam; armament included four 20-mm guns and four 7.62-mm Miniguns. Two GC-130As (later DC-130A) were converted to carry up to four drone aircraft and to launch them and control their flight. Eleven JC-130As were converted for the tracking of missiles and spacecraft, additional fuel and oil, giving endurance of 13 hours; seven were reconverted as AC-130A gunships and used in Vietnam in 1968-69. Seventeen RC-130As were equipped with TV viewfinder, cameras, mapping equipment, galley and five additional crew positions and delivered by 1959 to MAC's 1,370th Photo Mapping Wing.

The C-130B entered service with TAC on 12 June 1959. This has extra fuel, strengthened landing gear and 3020-kW (4,050-shp) T56-A-7 engines. The 230 built were supplied to the USAF, Canada, Indonesia, Pakistan and South Africa. Variants included 12 HC-130B search and rescue aircraft for the US Coast Guard, with a radio operator, two search observers and up to 44 passengers. Six C-130Bs with Fulton air-snatch satellite recovery equip-

The Royal Air Force has a large Hercules fleet which flies from Lyneham. Many have had the fuselage stretched as the C.3.

ment for the Discoverer programme flew with the USAF's 6593rd Test Sqn at Hickam AFB, Hawaii, in mid-1961. The NC-130B was converted for STOL research with two turbojet pods to provide bleed air for BLC blowing. The EC-130B ABCCC (Airborne Battlefield Command and Control Center) is a USAF version housing a giant windowless capsule with a staff of 12. RC-130B reconnaissance versions were similar to the RC-130A. The 17 WC-130B weather reconnaissance aircraft were distributed between the USAF's 53rd Sqn in Puerto Rico, the 54th on Guam and the 55th in California. The 12 C-130D wheel/ski versions of the C-130A operated in polar regions.

The C-130E introduced increased tankage, 503 being built. Derivatives included eight AC-130E gunships (with 105-mm guns) for use in Vietnam; DC-130E drone launch and control aircraft, converted by Lockheed Service Company (LASC); C-130N CAML (Cargo Aircraft Mine Layer) carrying large sea mines; a special duty EC-130E for the Coast Guard and EC-130E Coronet Solo II Elint aircraft packed with avionics and festooned with special aerials; three HC-130E SAR aircraft for the Coast Guard; and WC-130E weather aircraft for the USAF. The menacing black MC-130E Combat Talon fleet fly clandestine missions, mostly by night at treetop height, with a mass of special devices and STAR retrieval yokes to snatch up agents. The KC-130F (originally GV-1) is a Marine Corps assault transport and tanker, with outer wing hose-drum pods. The C-130F lacks the refuelling pods; both versions have been re-engined with the 3362-kW (4,508-shp) T56-A-15.

Current version is the C-130H, powered by T56-A-15s, of which well over 700 had been ordered by 1985. Variants include HC-130H long-range SAR aircraft with pick-up gear to lift persons or objects from the ground; JC-130H, four modified from HC-130H, for retrieval of space capsules on re-entry; C-130H-MP maritime patrol aircraft with a mass of sensors and very long endurance; one DC-130H modified by LASC for

drone launch and recovery; EC-130H Compass Call strategic high-power jamming platforms (several types); KC-130H inflight refuelling tankers; HC-130N SAR aircraft for retrieval of space capsules; HC-130P

helicopter inflight refuelling tankers; EC-130Q Tacamo command communication aircraft for the US Navy with 4-km (2.5-mile) trailing aerial for VLF link with submerged missile submarines; KC-

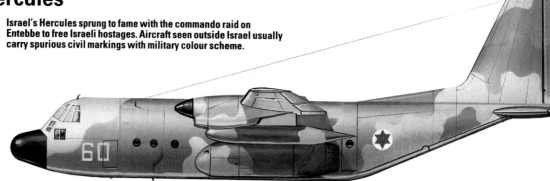

Israel's Hercules sprung to fame with the commando raid on Entebbe to free Israeli hostages. Aircraft seen outside Israel usually carry spurious civil markings with military colour scheme.

Lockheed Hercules C.Mk 3 cutaway drawing key

1 Radome
2 Weather radar scanner
3 Scanner tracking mechanism
4 Pitot head, port and starboard
5 Radome hinge
6 Radar mounting framework
7 Front pressure bulkhead
8 Downward vision windows
9 Instrument panel
10 Instrument panel shroud
11 Windscreen panels
12 Overhead switch panel
13 Co-pilot's seat
14 Cockpit eyebrow windows
15 Pilot's seat
16 Control column
17 Rudder pedals
18 Cockpit floor level
19 Nose landing gear wheel bay
20 Ground intercom socket
21 Twin nosewheels

22 Nosewheel leg door
23 Battery compartment
24 Radio and electronics racks
25 Portable oxygen bottle
26 Pilot's side console
27 Electrical system panel
28 Systems engineer's seat
29 Navigator's station
30 VHF aerial
31 Navigator's instrument panel
32 UHF aerial
33 Crew rest bunks
34 Cockpit emergency escape hatch
35 Control cable runs behind bulkhead
36 Galley unit
37 Fire extinguisher
38 Crew closet
39 Flight deck access ladder
40 Crew entry door
41 Integral airstairs

42 Cockpit section production joint double frame
43 Cargo-handling system roller conveyors
44 Main cabin bulkhead
45 Stretcher installation, maximum load 96 stretchers
46 Stretcher/troop seating mounting beam
47 Overhead equipment stowage rack
48 Cabin roof frames
49 Forward fuselage 'plug' section, 100 in (2.54 m) long
50 Aerial lead-in
51 Fuselage skin plating
52 Cabin wall trim panels
53 Forward fuselage plug section attachment double frame
54 Troop seats (stowed), maximum 92 fully equipped paratroops

55 Floor beam construction
56 Cabin window panels
57 Main cargo floor, maximum load 51,819 lb (23505 kg) on seven cargo pallets
58 Wing inspection light
59 Starboard main landing gear fairing cabin air-conditioning system
60 emergency exit window
61 Booster hydraulic reservoir
62 Air conditioning ducting
63 Foreign object damage propellor guard skin reinforcing plate
64 Main hydraulic system equipment
65 Wing root fillet
66 Handrail
67 Engine bleed air system piping
68 Fuselage/main spar attachment joint
69 Detachable leading edge section
70 Starboard inner engine nacelle
71 Engine exhaust duct
72 Allison T56-A-15 turboprop engine

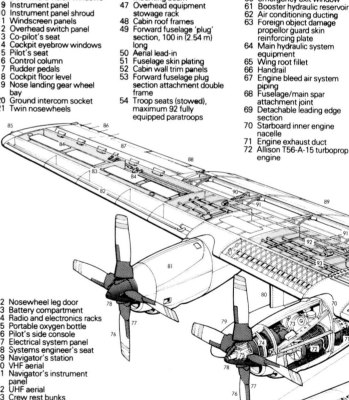

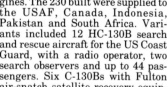

Lockheed C-130 Hercules (continued)

130R tankers for the US Marine Corps; LC-130R US Navy wheel/ski aircraft for the Antarctic; and KC-130T tankers able to refuel fighters or helicopters.

A special sub-family are the RAF's 66 C-130Ks, with UK parts and avionics. Built as Hercules C.1, long-range Falklands versions include C.1-LT2 and -LT4 with two of four cabin tanks, C.1-PLR2 with extra tanks plus FR probes and C.1(K) tankers, all with Omega and other navigation updates. The W.2 is a weather research aircraft, and the 30 C.3s have 'stretched' fuselages adding 4.57 m (15 ft), increas-

The Hercules has provided an excellent basis for many special conversions. The EC-130E has a battlefield airborne control post function, hence the large communications aerials.

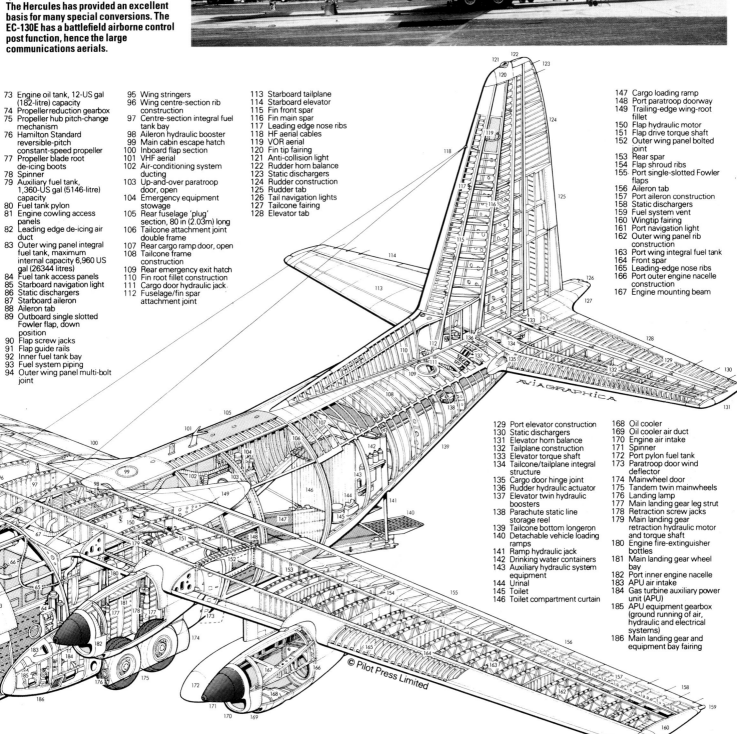

73 Engine oil tank, 12-US gal (182-litre) capacity
74 Propeller reduction gearbox
75 Propeller hub pitch-change mechanism
76 Hamilton Standard reversible-pitch constant-speed propeller
77 Propeller blade root de-icing boots
78 Spinner
79 Auxiliary fuel tank, 1,360-US gal (5146-litre) capacity
80 Fuel tank pylon
81 Engine cowling access panels
82 Leading edge de-icing air duct
83 Outer wing panel integral fuel tank, maximum internal capacity 6,960 US gal (26344 litres)
84 Fuel tank access panels
85 Starboard navigation light
86 Static dischargers
87 Starboard aileron
88 Aileron tab
89 Outboard single slotted Fowler flap, down position
90 Flap screw jacks
91 Flap guide rails
92 Inner fuel tank bay
93 Fuel system piping
94 Outer wing panel multi-bolt joint

95 Wing stringers
96 Wing centre-section rib construction
97 Centre-section integral fuel tank bay
98 Aileron hydraulic booster
99 Main cabin escape hatch
100 Inboard flap section
101 VHF aerial
102 Air-conditioning system ducting
103 Up-and-over paratroop door, open
104 Emergency equipment stowage
105 Rear fuselage 'plug' section, 80 in (2.03m) long
106 Tailcone attachment joint double frame
107 Rear cargo ramp door, open
108 Tailcone frame construction
109 Rear emergency exit hatch
110 Fin root fillet construction
111 Cargo door hydraulic jack
112 Fuselage/fin spar attachment joint

113 Starboard tailplane
114 Starboard elevator
115 Fin front spar
116 Fin main spar
117 Leading edge nose ribs
118 HF aerial cables
119 VOR aerial
120 Fin tip fairing
121 Anti-collision light
122 Rudder horn balance
123 Static dischargers
124 Rudder construction
125 Rudder tab
126 Tail navigation lights
127 Tailcone fairing
128 Elevator tab

129 Port elevator construction
130 Static dischargers
131 Elevator horn balance
132 Tailplane construction
133 Elevator torque shaft
134 Tailcone/tailplane integral structure
135 Cargo door hinge joint
136 Rudder hydraulic actuator
137 Elevator twin hydraulic boosters
138 Parachute static line storage reel
139 Tailcone bottom longeron
140 Detachable vehicle loading ramps
141 Ramp hydraulic jack
142 Drinking water containers
143 Auxiliary hydraulic system equipment
144 Urinal
145 Toilet
146 Toilet compartment curtain

147 Cargo loading ramp
148 Port paratroop doorway
149 Trailing-edge wing-root fillet
150 Flap hydraulic motor
151 Flap drive torque shaft
152 Outer wing panel bolted joint
153 Rear spar
154 Flap shroud ribs
155 Port single-slotted Fowler flaps
156 Aileron tab
157 Port aileron construction
158 Static dischargers
159 Fuel system vent
160 Wingtip fairing
161 Port navigation light
162 Outer wing panel rib construction
163 Port wing integral fuel tank
164 Front spar
165 Leading-edge nose ribs
166 Port outer engine nacelle construction
167 Engine mounting beam

168 Oil cooler
169 Oil cooler air duct
170 Engine air intake
171 Spinner
172 Port pylon fuel tank
173 Paratroop door wind deflector
174 Mainwheel door
175 Tandem twin mainwheels
176 Landing lamp
177 Main landing gear leg strut
178 Retraction screw jacks
179 Main landing gear retraction hydraulic motor and torque shaft
180 Engine fire-extinguisher bottles
181 Main landing gear wheel bay
182 Port inner engine nacelle
183 APU air intake
184 Gas turbine auxiliary power unit (APU)
185 APU equipment gearbox (ground running of air, hydraulic and electrical systems)
186 Main landing gear and equipment bay fairing

© Pilot Press Limited

ing seating from 92 to 128, paratroops from 64 to 92 and casevac stretchers from 74 to 97.

Total sales exceed 1,750 for 57 nations, including the 'stretched' L-100 series of which many have been sold to military customers.

Specification
Lockheed C-130H Hercules
Type: multirole airlift transport
Powerplant: four 3362-kW (4,508-shp) flat-rated Allison T56-A-15

turboprop engines
Performance: maximum cruising speed 602 km/h (374 mph); initial rate of climb 579 m (1,900 ft)/minute; range with maximum payload, 5 per cent reserves plus 30 minutes at sea level 3792 km (2,356 miles); range with maximum fuel, allowances as above 7876 km (4,894 miles)
Weights: empty 34686 kg (76,469 lb); maximum take-off 79379 kg (175,000 lb)

Dimensions: span 40.41 m (132 ft 7 in); length 29.79 m (97 ft 9 in); height 11.66 m (38 ft 3 in); wing area 162.12 m^2 (1,745.0 sq ft)
Armament (Gunship versions): has included 20-mm (0.79-in) and 40-mm (1.57-in) guns, a 105-mm (4.13-in) howitzer and 7.62-mm (0.3-in) Miniguns
Operators: have included Abu Dhabi, Algeria, Argentina, Australia, Belgium, Brazil, Cameroun, Canada, Chad, Chile,

Colombia, Denmark, Ecuador, Egypt, France, Gabon, Greece, Honduras, Indonesia, Iran, Iraq, Israel, Italy, Japan, Jordan, South Korea, Kuwait, Libya, Malaysia, Morocco, New Zealand, Niger, Nigeria, Norway, Oman, Pakistan, Peru, Philippines, Portugal, Saudi Arabia, Singapore, South Africa, Spain, Sudan, Sweden, Syria, Taiwan, Thailand, Tunisia, Turkey, UAE, UK, USA, Venezuela, Vietnam, North Yemen, Zaïre

Lockheed C-140 JetStar

The JetStar was produced as a private venture to meet the US Air Force UCX requirement for a utility jet for crew readiness training, navaid calibration, transport and other duties. The prototype flew on 4 September 1957 powered by two Bristol Orpheus turbojet engines. Production aircraft have four Pratt & Whitney JT12s, a high-lift leading edge, twin-wheel landing gear and reversers. External tanks at mid-span were intended to be optional, but became standard. In October 1959 the USAF selected the JetStar as the C-140. Five were used by the Special Air Missions wing, and others by the Airways and Air Communications Service for inspecting overseas navaids. Up to 10 passengers are normally accommodated. In 1973 Lockheed announced the JetStar II with four Garrett TFE731-3 turbofan engines.

Specification
Lockheed C-140 JetStar
Type: light utility jet transport
Powerplant: four 1361-kg (3,000-lb) thrust Pratt & Whitney JT12A-6A turbojet engines
Performance: maximum cruising speed 885 km/h (550 mph) at 6095 m (20,000 ft); initial rate of climb 1006 m (3,300 ft)/minute; service ceiling 10060 m (33,000 ft); range with maximum payload, 45 minutes reserve 3187 km (1,980 miles); range with maximum fuel, 45 minutes reserves 3516 km (2,185 miles)
Weights: empty 9752 kg (21,500 lb); maximum take-off 18561 kg (40,920 lb)
Dimensions: span 16.59 m (54 ft 5 in); length 18.41 m (60 ft 5 in); height 6.22 m (20 ft 5 in); wing area 50.40 m^2 (542.50 sq ft)
Armament: none

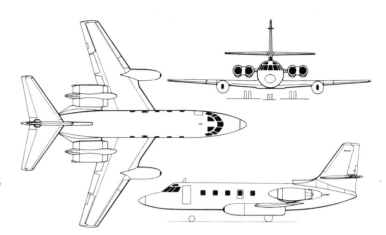

Lockheed C-140A JetStar

Operators: West Germany, Indonesia, Iran, Libya, Mexico, Saudi Arabia, US (Army and Air Force)

Lockheed C-141 StarLifter

Victory in battle often goes to the contestant who gets there 'firstest with the mostest', and the Lockheed C-141 StarLifter transport was developed to enable the United States to deploy large quantities of troops and heavy equipment very quickly indeed.

The aircraft was designed to specification SOR-182 (Specific Operational Requirement 182) issued for a turbofan-powered freighter and troop carrier for operation by the US Military Airlift Command, and was selected in a competition in which Boeing, Douglas and General Dynamics were contenders. The

transport is the flying element of the US Logistics Support System 476L, the purpose of which is to provide global-range airlift for the MAC, and strategic deployment capabilities at jet speeds for the US Strike Command, which includes the Strategic Army Corps and the Composite Air Strike Forces of Tactical Air Command.

Of conventional construction, the StarLifter is of swept-wing configuration, the wing being mounted high on top of the fuselage to minimize cabin obstruction. The four engines are mounted on pylons carrying them well below and for-

ward of the wing leading edge. A distinctive feature is the tall T-tail.

The 21.34 m (70 ft 0 in) long cabin has a maximum width of 3.12 m (10 ft 3 in) and a maximum height of 2.77 m (9 ft 1 in), and can accommodate 154 troops or 123 fully-equipped paratroops, or 80 stretchers with seats for up to 16 walking wounded or attendants. Two bunks and two seats are provided within the cabin for relief flight crew members. If required, a special pallet comprising a galley and toilet can be installed in the front of the cabin, this reducing the capacity to 120 passenger-type seats.

Two paratroop doors are provided at the aft end of the cabin. Clamshell doors and a rear ramp permit straight-in cargo loading. The ramp can be opened in flight for air-drops. Up to 149.6 m^3 (5,283 cu ft) of freight can be loaded on 10 pallets. The rollers and retaining rails for the pallets can be retracted into recesses to provide a flat floor when not in use.

The StarLifter fleet has undergone a radical alteration which entails an extra 'plug' in the fuselage to increase cargo carriage and the addition of an inflight-refuelling receptacle.

Lockheed C-141 StarLifter (continued)

Lockheed C-141B StarLifter in European One camouflage.

The StarLifter demonstrated its load-carrying potential when it established a world record for heavy cargo drops by delivering 31840 kg (70,195 lb). Several aircraft were modified to carry the Minuteman ICBM in its special transport container, a total weight of 39103 kg (86,207 lb).

The StarLifter began squadron operations with MAC in April 1965 and soon demonstrated its usefulness in war when it was used extensively to carry troops and supplies across the Pacific to Vietnam and carry wounded back to the USA. This and other operational experience indicated the need to provide the StarLifter with a flight-refuelling capability. It was frequently found, moreover, that when loaded with a bulky rather than weighty load, the aircraft had not reached its maximum weight; that is to say, it could have carried still more. Though nothing could be done to enlarge the cross-section of the StarLifter's fuselage, the latter could be lengthened.

Accordingly, in mid-1976 Lockheed was awarded a contract to develop an extended C-141 with in-flight refuelling equipment. Designated YC-141B, a converted aircraft first flew on 24 March 1977. The fuselage extension consists of a 4.06 m (13 ft 4 in) plug inserted in front of the wing and a similar 3.05 m (10 ft 0 in) plug immediately aft of the wing. At the same time refined wing-root fairings were fitted. These not only reduce drag, thus permitting high-speed and reducing fuel consumption, but also change the lift distribution, permitting the carriage of increased loads without affecting the fatige life of the wing.

The enlarged cabin can accommodate 13 standard pallets, instead of the previous 10. The US Air Force had converted all of its 277 operation StarLifters (out of 284 built) to the new configuration by 1982, in effect adding the equivalent of an extra 90 aircraft.

Specification
Type: (C-141B) long-range transport
Powerplant: four 9525-kg (21,000-lb) thrust Pratt & Whitney TF33-P-7 turbofans
Performance: maximum level speed at 7620 m (25,000 ft) over 920 km/h (570 mph); maximum cruising speed over 900 km/h (560 mph); range with maximum fuel about 8500 km (5,000 miles); range with maximum payload about 6450 km (4,000 miles)
Weights: operating weight 67970 kg (149,848 lb); maximum payload 2.25 g 40439 kg (89,152 lb); or (2.5 g) 31242 kg (68,788 lb); maximum ramp weight 156444 kg (344,900 lb)
Dimensions: span 48.74 m (159 ft 11 in); length 51.3 m (168 ft 4 in); height 11.98 m (39 ft 3½ in); wing area 299.9 m² (3,228 sq ft)
Operator: USAF

Lockheed Electra

The design of the Lockheed L-188 short/medium range turboprop airliner began in 1954. In 1955 Lockheed received an initial order for the L-188 'off the drawing board' from American Airlines, with the result that the prototype made its first flight on 6 December 1957, entering service with American and Eastern in January 1959. Of conventional layout, the L-188 Electra had large Fowler flaps, and a tricycle-type landing gear with twin wheels on each unit. The powerplant comprised four Allison 501 turboprop engines, the commercial version of the T56. The initial version accommodated 74 passengers, but later arrangements provided for a maximum of 98 passengers.

A total of 170 were built. As these aircraft were gradually replaced by jets, some were acquired by the military services of smaller nations for use in a cargo/transport role.

Argentina is the major user, with one aircraft at least configured for the Sigint role. This saw action in the Falklands.

Specification
Type: short/medium-range transport
Powerplant: four 2796-kW (3,750-shp) Allison 501-D13 or 3020-kW (4,050-shp) Allison 501-D15 turboprops
Performance: maximum speed at 3660 m (12,000 ft) 721 km/h (448 mph); cruising speed at 60705 m (22,000 ft) 652 km/h (405 mph); range with maximum payload and 3221-kg (7,100 lb) fuel reserves 4458 km (2,770 miles); range with maximum fuel plus 3785 litres (1,000 US gallons) of auxiliary

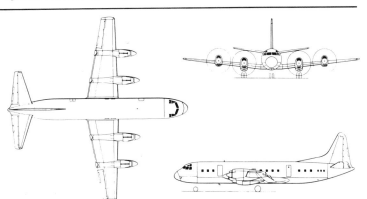
Lockheed L-188 Electra

fuel with 2 hours reserves 5568 km (3,460 miles)
Weights: empty 25991 kg (57,300 lb); empty (extra fuel) 26649 kg (58,750 lb); maximum take-off 52617 kg (116,000 lb)
Dimensions: span 30.18 m (99 ft 0 in); length 31.86 m (104 ft 6½ in); height 9.78 m (32 ft 1 in); wing area 120.77 m² (1,300 sq ft)
Armament: none
Operators: Argentina, Bolivia, Mexico, Panama

Lockheed F-104 Starfighter

Although by the early 1950s, just over a decade after the first jet aircraft had flown, NACA (the National Advisory Committee for Aeronautics, now known as NASA) had flown a series of experimental Mach 1+ aeroplanes to push back the speed and altitude frontiers, the fastest US service fighter was the Mach 0.8 North American F-86 Sabre, and the first true transonic fighter, the North American F-100 Super Sabre, had not yet flown.

This was the situation facing C. L. 'Kelly' Johnson, chief engineer of Lockheed, when in 1952 he set out to produce a fighter superior to anything being flown by the Communists over Korea. General Electric was just about to launch the J79, the engine that was to become one of the most widely used of all Western turbine powerplants.

With the promise of this engine to come, Johnson went ahead with his design. He chose Mach 2.2 as the flat-out level speed, and investigated some 300 different shapes to find one that would provide the best compromise between speed, range, manoeuvrability, and landing and take-off performance. Throughout 1953, as the bitter air war finally ended in the Korean skies, Johnson continued to interview pilots just back from combat, to find out what they wanted; meanwhile model after model went through the Lockheed wind tunnels.

Since the discovery in Germany that swept wings produced less drag than straight ones at speeds around Mach 1, virtually all designers had gone over to them. But later work by NACA showed that swept wings were actually 'draggier' at speeds

Japan is a large operator of the F-104 Starfighter. This is a two-seat TF-104J used for conversion training.

189

around Mach 2, and so Johnson chose a tiny straight wing only in 10.16 m (4 in) thick at the deepest part, and with so sharp a leading edge that it had to be covered with a protective sheet on the ground to prevent injury. The wing was heavily anhedralled to overcome the 'aileron' effect of the large rudder. To increase the lift produced by the tiny wing, high-pressure air from the engine was blown over the flaps when they were depressed for landing. The tailplane was set high on the fin in an effort to avoid pitchup, a serious aerodynamic characteristic that was known to affect jet fighters when pulling very tight turns. There was very little room for equipment, and no attempt was made to incorporate AI (Airborne Interception) radar. Most controversially, the pilot was given a downward-firing ejector seat on the grounds that a conventional upward-firing one might hit the tailplane.

On 4 March 1954 the XF-104 prototype made its first flight with simple inlets feeding a J65 engine. There were many problems to overcome, and some 50 production F-104As with advanced inlets feeding a J79 engine were assigned to the test programme in addition to the prototypes and pre-production aeroplanes. Development in fact took four years to accomplish, twice the anticipated duration, and far longer than any other US fighter up to that time. Clearance for use by the squadrons was granted in January 1958, but accidents and continued difficulties were so prolific that the F-104A was grounded three months later. In July 1958 the type was again cleared to fly.

During 1958 F-104As of the USAF Air Defense Command (responsible for the defence of the continental United States) set up international speed and altitude records. But the US Air Force was losing interest in the lightweight fighter formula, despite the efforts of Lockheed to

turn the F-104 into a workable combat aircraft, and in 1959 they were transferred to the Air National Guard, the part-time reserve organisation.

The F-104 programme by now had assumed considerable momentum, however, and the USAF was obliged to accept the next model, the greatly improved F-104C (the F-104B had been a two-seat version of the F-104A). This time, however, they went to the Tactical Air Command, where they stayed till 1965.

The F-104 story might have ended there had it not been for the decision of a group of NATO countries led by West Germany to build under licence a totally redesigned version. While the European aircraft industry was slowly regaining strength, there was certainly not enough experience to build a fighter guaranteed to match anything the Russians could put up. In the largest international programme up to then Germany, Italy, Holland, Denmark, Norway, Canada and Japan investigated a dozen or so aircraft, and in February 1959 chose what had already become the most controversial of them, the F-104, for its new multimission attack fighter, to replace a variety of earlier types such as Gloster Meteors, Lockheed Shooting Stars, North American Sabres, and Republic Thunderstreaks. Lockheed's sales tactics in the matter were to be widely criticised over the next 20 years.

So the F-104G (G for Germany, with more than 700 aircraft) was launched, to keep production lines in many countries busy for the next seven years. The Super Starfighter was the most advanced fighter anywhere at the time of its introduction; apart from being the first Mach 2+ fighter outside the USA, Britain, Soviet Union and Sweden, it had a proper fire-control radar and the world's first miniature, high-accuracy inertial navigation system for squadron service.

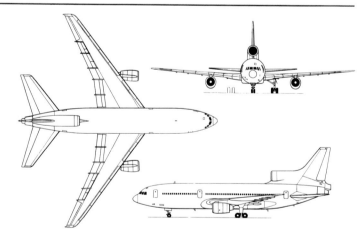

The most recent version of the F-104 family is the F-104S, an advanced interceptor for the Italian and Turkish air forces. This is basically the same as the -G model, but incorporates refinements developed over years of experience with the earlier models. But the main change was the substitution of a weapon system for air fighting rather than for ground attack. The main external differences were the addition of two wingtip-mounted Sparrow missiles (hence the 'S' in the designation), and the appearance of a pair of additional strakes under the rear fuselage. It first flew in 1968, and the 205 aeroplanes built in Italy were the last Starfighters to be built.

Of the many countries that took to F-104, only Pakistan has used the Starfighter in combat. In 1965 its Starfighters clashed with Indian fighters over the frontier between the two countries.

Lockheed's last big effort to sell the F-104 took place in 1970, proposing a version for the IFA (International Fighter Aircraft) competition, subsequently won by Northrop with its F-5E, and again a few years later in the LWF (Light-Weight Fighter) programme. F-104s soldier on with several forces around the world, but fleets are dwindling

The West Germans were the largest user of the Starfighter, but these are rapidly being replaced by Tornados. The Luftwaffe uses the Starfighter for weapons training.

rapidly, mainly being replaced by the F-16.

Specification
Type: single-seat multimission fighter
Powerplant: one General Electric J79-GE-11A of 4536-kg (10,000-lb) thrust, increasing to 7167 kg (15,800 lb) with afterburning
Performance: maximum speed 2092 km/h (1,300 mph) at 12192 m (40,000 ft); radius of action 1110 km (690 miles); service ceiling 16764 m (55,000 ft)
Weights: empty 6388-kg (14,821-lb); maximum 13054 kg (28,779 lb)
Dimensions: span 6.68 m (21 ft 11 in); length 16.69 m (54 ft 9 in); height 4.15 m (13 ft 6 in); wing area 18.22 m² (196.1 sq ft)
Armament: one 20-mm General Electric M61 six-barrel cannon, wingtip-mounted Sidewinder air-to-air missiles: various external stores to total weight of 11814 kg (4,000 lb)
Operators: Canada, Denmark, Greece, Italy, Japan, Taiwan, Turkey, West Germany

Lockheed L-1011 TriStar

During the Falklands War in 1982 the Royal Air Force identified an urgent need for additional inflight refuelling capability. Short-term answers included hurried conversion of six Vulcan bombers and four Hercules as single-point tankers. In the longer term it was recognized that the VC10 tanker fleet would not be sufficient to cope with increasing demand. It was conveniently decided to purchase and convert six Lockheed L-1011-500 TriStar airliners which had been declared surplus to requirements by British Airways.

Responsibility for the modification was entrusted to Marshall of Cambridge, four of the six aircraft being delivered to Teversham in 1983. Delivery of these as TriStar K.1s to the RAF's 216 Sqn took place in 1984-85, at which point the remaining two aircraft were handed over for conversion, after having been used with minor changes for crew training and to augment RAF capability. These will become

K(C).1s with a large cargo door. More recently two TriStar 500s have been purchased from PanAm, and these were to be converted in 1985.

All RAF TriStars have twin Flight Refuelling Mk 17T hose drum units (HDUs) in the aft fuselage, these eventually being augmented by single Mk 32 pods beneath the wings. Nine additional fuel cells are installed in the underfloor area, with a capacity of 45359 kg (100,000 lb). A refuelling probe is added above the flight deck.

Specification
Lockheed TriStar K(C).1
Type: long-range tanker/cargo aircraft
Powerplant: three 22680-kg (50,000 lb) thrust Rolls-Royce RB211-524B turbofan engines
Performance: economical cruising speed at 10670 m (35,000 ft) 890 km/h (553 mph); service ceiling 13105 m (43,000 ft); range with maximum payload and reserves

Lockheed L-1011 TriStar

9904 km (6,154 miles); range with maximum fuel and reserves 11286 km (7,013 miles)
Weights: empty 111153 kg (245,050 lb); maximum take-off 231332 kg (510,000 lb); cargo payload 42003 kg (92,600 lb)

Dimensions: span 50.09 m (164 ft 4 in); length 50.05 m (164 ft 2.5 in); height 16.87 m (55 ft 4 in); wing area 328.96 m² (3,541.0 sq ft)
Armament: none
Operator: UK (RAF)

Lockheed P-2 Neptune

Originating from design studies made in the early years of World War II, Lockheed's land-based Neptune patrol aircraft was designed, from 1947 to 1962, to represent the

foundation of the US Navy's land-based patrol squadrons. Strangely enough, the original design studies were made at a time when the US Navy had not envisaged that a land-

based patrol aircraft would be included in its inventory of operational types. The first prototype XP2V-1 flew for the first time on 17 May 1945. It was seen to be fairly

large aircraft, able to accommodate a crew of seven, and possessing a weapons bay which could carry two torpedoes or 12 depth charges, and armed with three pairs of 12.7-mm

Argentina's navy is one of the few Lockheed Neptune users left. These have a maritime patrol and electronic warfare role.

in a training role. Final Lockheed production version was the P-2H (formerly P2V-7), first flown on 26 April 1954, of which many remain in service. This was the only Neptune to have underwing auxiliary turbojets as standard on all production aircraft, plus many of the refinements introduced on P-2E and P-2F aircraft. SP-2H aircraft had Julie explosive echo-sounding and Jezebel acoustic search equipment, and LP-2Js were equipped for Arctic photo-reconnaissance. The USAF operated seven P2V-7Us in Vietnam as RB-

69A ECM test and training aircraft, on loan from the US Navy, and also acquired a small number of AP-2H aircraft for special duties. Kawasaki in Japan built 48 P-2H ASW aircraft, and has since developed a new ASW for the JMSDF under the designation P-2J, plus one UP-2J for target towing. Several airforces operate the P-Z for patrol duties. Argentina converted one for Elint duties and her force saw action in the Falklands.

Specification
Type: naval patrol bomber and ASW aircraft
Powerplant: (P-2H) two 2610-kW (3,500-hp) Wright R-3350-32W radial piston engines, plus two 1542-kg (3,400-lb) Westinghouse J34-WE-34 projects
Performance: (P-2H) maximum speed 649 km/h (403 mph) at 3050 m (10,000 ft); cruising speed 333 km/h (207 mph) at 2950 m (8,500 ft); ferry range 5930 km (3,685 miles)

Weights: (P-2H) empty 22650 kg (49,935 lb); maximum take-off 36240 kg (799,895 lb)
Dimensions: (P-2H) span 31.65 m (103 ft 10 in); length 27.84 m (91 ft 4 in); height 8.94 m (29 ft 4 in); wing area 92.90 m² (1,000 sq ft)
Armament: variations of 20-mm cannon, 12.7-mm (0.5-in) machine-guns, mines, torpedoes, depth charges, and air-to-surfce missiles
Operators: Argentine, Australia, Brazil, France, Netherlands, Portugal, UK, US Air Force, US Navy, US Naval Reserve

Lockheed P-3 Orion

Two of the six P-3F Orions delivered to the Iranian air force before the revolution are still active. These are flown from Bandar Abbas, often being used to locate targets for strike aircraft in the Gulf.

In early December 1957 the prototype of a new four-turboprop civil transport was flown by Lockheed. This had the company designation L.188, and the first deliveries of L.188A production aircraft to US airlines began in the autumn of 1958. Named Electra, about 170 were built in L.188A and L.188C versions, and these were supplied mainly to US and South American airlines.

In August 1957 the US Navy called for design proposals to meet Type Specification 146. This concerned the supply of a new advanced aircraft for maritime patrol and ASW (Anti-Submarine Wafare), and in order to save cost and, more importantly, to permit service introduction as quickly as possible, the US Navy suggested that a variant of an aircraft that was already in production would receive favourable consideration if generally suitable. Thus, Lockheed proposed a developed version of the civil Electra as its submission for the USN competition, and in April 1958 the US Navy announced that this had been selected. The initial research and development contract was awarded on 8 May 1958, and Lockheed proceeded immediately to modify the third civil Electra airframe as an aerodynamic prototype for US Navy evaluation of flight characteristics. This had a mock-up of the MAD

(Magnetic Anomaly Detection) boom as an extension of the rear fuselage and a simulated weapons-bay, and made its first flight on 19 August 1958. An operational prototype with full avionics flew for the first time on 24 November 1959, this having the designation YP3V-1, and the name of Orion was adopted for these aircraft in late 1960. The first production P3V-1 made its initial flight on 15 April 1961, and six aircraft were involved in flight testing, operational evaluation and acceptance trials before the first deliveries of production aircraft to USN Patrol Squadrons VP-8 and VP-44 began on 13 August 1962. By that time the P3V-1 had been redesignated as the P-3 Orion.

Lockheed's Model 185 retains the wings, tail unit, basic fuselage structure, powerplant and many assemblies and systems of the Electra. It differs primarily by having a fuselage which has been reduced in length by 2.24 m (7 ft 4 in) and

modified to incorporate a large weapons-bay. The change from a civil to military role involved also the provision of new avionics and other systems, including a pneumatic system for the launch of ASW stores and extra electrical power generation to cope with demand of the much-increased avionics equipment. The basic fuselage is both pressurized and air-conditioned, but the weapons-bay is excluded, and the hydraulic system includes operation of the weapons-bay doors. Mines, depth-bombs, torpedoes or nuclear weapons can be carried in the weapons-bay, and 10 underwing pylons accommodate a variety of stores.

During more than 17 years of service there has been very considerable revision of the avionics equipment, as a result of changing threats and the inevitable progressive evolution of ASW equipment over this period. In other respects there have been few changes, except that the original 3356-kW (4,500-hp) Allison

T56-A-10W turboprop was replaced by a more powerful version.

The original P-3A had what was then the advanced avionics system to equip it for an effective ASW role, for however sophisticated the weapons that such an aricraft can deploy, these are virtually useless unless the target can be identified and located. So, in addition to HF, VHF and UHF communications, the early Orions had inertial, doppler, Loran, and Tacan navigation systems, autopilot, sonobuoy signal receivers and indicators, MAD, and a modified ECM device which served as a direction finder, by detecting and locating electronic emissions from submarines.

Orions are operated normally by a crew of 10, five of these being regarded as tactical specialists who

All the US Navy's land-based anti-submarine missions are handled by the P-3 Orion. The large fleet is divided between East and West coasts.

work in a compartment within the main cabin which contains electronic, magnetic and sonic detection equipment. And because these aircraft have a patrol endurance of up to 10 hours, a large crew rest area with galley is provided in the main cabin.

The initial P-3A production aircraft are being replaced in USN squadron service progressively by new production P-3Cs. It is planned that P-3A/B aircraft released from active service will, as they become available, be used to modernize the US Navy's reserve forces, gradually replacing the Lockheed P-2 Neptunes which at present are used by the reserve.

For an aircraft which has given good service over a period of more than 17 years, it is inevitable that a number of versions and variants have evolved. The major production versions are the P-3A, -3B and -3C.

A total of 157 examples of the P-3A were built, and from the 110th aircraft these were provided with more sensitive (so-called Deltic) ASW detection equipment and improved electronic displays. Three of

The EP-3E variant is a specialized electronic intelligence gatherer. Chief task is to produce electronic 'fingerprints' of WarPac ships, analysing their radars and defensive systems.

this latter version were supplied to the Spanish Air Force. P-3Bs replaced P-3As on the production line during 1965, the new model having more powerful Allison turboprops (see data). Those which entered service with the USN were modified subsequently to allow for the carriage and deployment of AGM-12 Bullpup air-to-surface guided missiles on wing pylons. Five of this version were delivered to the RNZAF in 1966, 10 to the RAAF in 1968, and five to Norway in 1969.

Current production version is the P-3C. This has the same power plant as -3B, but has an advanced system of sensors and control equipment identified as A-NEW. Heart of the system is a digital computer which processes all ASW information, and this then becomes available for retrieval or display at any time. Under the Update and Update II pro-

grammes, P-3Cs have been given even more advanced systems. Update III provides new ASW avionics. Variants include three RP-3A special project reconnaissance aircraft, four WP-3As for weather reconnaissance, EP-3Bs for electronic reconnaissance, one RP-3D for a worldwide magnetic survey, two WP-3Ds to serve as airborne research centres, 10 EP-3E electronic reconnaissance aircraft for service with VQ-1 and VQ-2 squadrons, and six P-3Fs as long-range surveillance aircraft for the Iranian Air Force.

Specification
Type: ASW patrol/attack aircraft
Powerplant: (P-3B/C) four 3661-kW (4,910-ehp) Allison T56-A-14 turboprop engines
Performance: (P-3B/C) maximum speed at 4750 m (15,000 ft) at AUW of 47627 kg (105,000 lb) 761 km/h (473 mph); patrol speed at 457 m (1,500 ft) at above AUW 381 km/h (237 mph); mission radius 3 hours on station at 457 m (1,500 ft) at 2494 km (1,550 miles); maximum mission radius, not time on station,

Important among Western maritime patrol aircraft are the P-3 Orions operated by Norway. Their proximity to the Soviet Northern Fleet bases on the Kola peninsula enables them to keep close watch.

at maximum normal take-off weight, 3836 km (2,384 miles)
Weights: (P-3B/C) empty 27892 kg (61,491 lb); maximum normal take-off 61235 kg (135,000 lb); maximum permissible weight 64410 kg (142,000 lb)
Dimensions: span 30.38 m (99 ft 8 in); length 35.61 m (116 ft 10 in); height 10.27 m (33 ft 8½ in); wing area 120.77 m² (1,300 sq ft)
Armament: (weapons bay) one Mk-25/39/55/56 mine, or three Mk-36/52 mines, or three Mk-57 depth-bombs, or eight Mk-54 depth-bombs, or eight Mk-43/44/46 torpedoes; (underwing pylons) mines and rockets, torpedoes for ferrying, and a searchlight under the starboard wing. Maximum weapons load (P-3C) is 9070 kg (20,000 lb)
Operators: Australia, Iran, Japan, Netherlands, New Zealand, Norway, Portugal, Spain and US Navy

Lockheed CP-140 Aurora

A P-3 Orion tailored specifically to the needs of the Canadian Armed Forces, the Aurora entered service in 1980, and the 18 aircraft equip four squadrons of the Maritime Air Group. Evolution began in 1972 when Specification 15-14 called for a replacement for the CP-107 Argus. This envisaged the purchase of an existing design to handle ASW (anti-submarine warfare), ice reconnaissance, pollution control, SAR, Arctic surveillance and aerial survey. The Aurora was chosen on 21 July 1976.

It incorporates the S-3A Viking's ASW detection package, at the heart of which is a Univac AYK-14 digital computer which handles the processing of data originating from the sensors. Cabin layout has been revised, and is generally accepted as being superior to that of the P-3C. The Aurora is expected to undertake civil tasks for which a special sensor pack fits into the weapons bay.

Deliveries took place from 29 May 1980 to 10 July 1981. Three of the four squadrons (Nos 404, 405 and 415) are at Greenwood, Nova Scotia, No. 404 being responsible for training. The fourth squadron, No. 407, is at Comox, British Columbia, from where it covers the Pacific.

Specification
Type: ASW patrol aircraft
Powerplant: four 3661-ekW (4,910-eshp) Allison T56-A-14 turboprop engines
Performance: maximum speed at optimum altitude 732 km/h (455 mph); ferry range 8024 km (4,968 miles); endurance on station at 1852 km (1,151 miles) radius 8 hours 12 minutes

Weights: not known
Dimensions: span 30.38 m (99 ft 8 in); length 35.61 m (116 ft 10 in); height 10.27 m (33 ft 8.5 in); wing area 120.77 m² (1,300.0 sq ft)
Armament: mines, depth bombs, torpedoes and rockets in the weapons bay up to 2177 kg (4,800 lb), or on 10 underwing hardpoints with individual capacities varying from 277 kg (611

lb) to 1111 kg (2,450 lb)
Operator: Canada

The 18 Auroras in service with the CAF are superficially similar to the P-3 Orion, but contain the ASW gear from the S-3 Viking. Most serve on the east coast, but four fly from Comox in British Columbia.

Lockheed S-3 Viking

In 1967 the US Navy initiated a design competition for a replacement for the Grumman S-2 Tracker carrier-based ASW aircraft. When final proposals were evaluated in 1969, Naval Air Systems Command selected the submission from the Lockheed-California Company. Designated S-3A, it had been developed in partnership with Vought and Univac Federal Systems Division. Vought was responsible for the wings, tail unit, landing gear and engine pods, and Univac for the advanced digital computer. The first prototype flew on 21 January 1972, and initial deliveries were made to Squadron VS-41 on 20 February 1974. To equip 13 squadrons, each with 10 aircraft, contracts were placed for 179 production aircraft, the last delivered in 1978.

The S-3A is a shoulder-wing monoplane with wings that fold on skew hinges to overlap for carrier stowage; the vertical tail also folds. Aerodynamic features include single-slotted Fowler flaps which depress automatically at flap angles in excess of 15°, and ailerons augmented by under- and over-wing spoilers. The engines are in underwing pods. The whole aircraft is stressed for catapult launching and arrested landing. The short fuselage accommodates part of the fuel, a retractable FR probe, weapons bay and the landing gear. The pilot, co-pilot, tactical operator (Tacco) and acoustic sensor operator (Senso) have McDonnell Douglas ejection seats in the pressurized cockpit. The Viking has a comprehensive range of sonobuoys, and a MAD boom extendable from the rear fuselage. Non-acoustic sensors include the outstanding Texas Instruments APS-116 high-resolution radar, a forward-looking infra-red scanner in a retractable turret, and passive ECM wingtip pods. Accurate navigation is ensured by an advanced inertial system, augmented by doppler, Tacan and UHF/DF, and an ACLS (automatic carrier landing system) permits all-weather operation.

ES-3A electronic patrol and KS-3A tanker versions have been studied, and a single US-3A utility transport for COD (carrier on-board delivery) was flown on 2 July 1976, with a lengthened fuselage for 23 passengers or supplies. In 1982

three more US-3A conversions were put into service.

S-3A squadrons are deployed on carriers of the Atlantic and Pacific fleets, each deployment lasting approximately six months. The whole force is being updated to S-3B standard, with IBM sonics processing, expanding ESM, a new sonobuoy receiver, better radar processing, Harpoon anti-ship missiles on the wing pylons and other improvements.

Specification
Type: carrier-based ASW patrol/attack aircraft
Powerplant: 4207 kg (9,275-lb) thrust General Electric TF34-GE-400A turbofan engine
Performance: maximum speed 834 km/h (518 mph); maximum cruising speed 686 km/h (426 mph); loiter speed 296 km/h (184 mph); combat range more than 3701 km (2,300 miles)
Weights: empty 12088 kg (26,650 lb); maximum take-off 23831 kg (52,539 lb)
Dimensions: span 20.93 m (68 ft 8 in), wings folded 8.99 m (29 ft 6 in); length 16.26 m (53 ft 4 in), tail folded 15.06 m (49 ft 5 in); height 6.93 m (22 ft 9 in), tail folded 4.65 m (15 ft 3 in); wing area 55.55 m² (598.0 sq ft)
Armament: (weapons bay) four Mk 36 destructors, or four Mk 46 torpedoes, or four Mk 82 bombs, or two Mk 57 or four Mk 54 depth bombs, or four Mk 53 mines; (underwing pylons) SUU-44/A flare launchers, Mk 52, 55 or 56 mines, Mk 20-2 cluster bombs, LAU-68A, -61A, -69A or -10A/A rocket pods, Mk 20 cluster bombs, Mk 76-5 or 106-4 practice bombs, or Aero 1D auxiliary fuel tanks; (S-3B) two AGM-84B Harpoon missiles
Operator: US (US Navy)

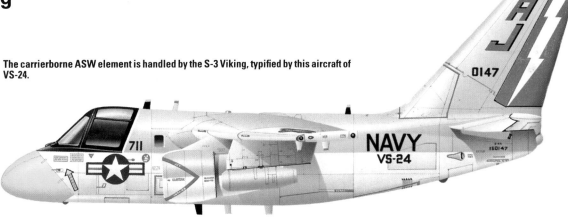

The carrierborne ASW element is handled by the S-3 Viking, typified by this aircraft of VS-24.

Seen entering the landing pattern, these VS-31 aircraft show graphically the position of the tail hook. In service, the S-3 has proved to be an easy aircraft to fly off carriers (if carrier flying can ever be called easy).

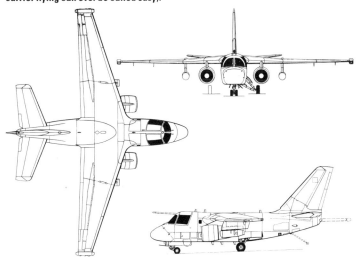

Lockheed S-3A Viking

Two specialized versions of the Viking that have been tested are the KS-3 tanker, and the US-3 COD aircraft. Here the pair are seen hooked up.

This S-3A of VS-22 displays the retractable MAD sting, which takes the detector as far away from the metal bulk of the aircraft as possible. Sonobuouys are also carried

Lockheed SR-71A

Developed from the A-12 and flown for the first time during December 1964, the Lockheed SR-71A is the world's fastest operational aircraft, approximately a dozen examples of this outstanding machine remaining active with the USAF's 9th Strategic Reconnaissance Wing at any given time. Possessing the ability to survey $260000 \ km^2$ (100,000 square miles) of the Earth's surface in just one hour, the SR-71A routinely cruises at Mach 3 at altitudes in excess of 24385 m (80,000 ft) during the course of its duties, and is able to gather a variety of data by virtue of highly classified and interchangeable photographic and electronic sensors which are installed to meet specific mission objectives.

Deliveries to Strategic Air Command began in January 1966 and it is believed that a total of 32 aircraft was built. This figure includes two examples of the two-seat SR-71B for pilot training, plus a single SR-71C two-seater built of components taken from crashed aircraft and a structural test specimen. All SR-71s are finished in special near-black paint containing microscopic iron spheres to act as an early radar absorbent skin.

Specific details of the work undertaken by the SR-71A remain shrouded in secrecy, but it is known that operations are routinely conducted from two forward operating locations by aircraft detached from the SRW's headquarters at Beale AFB, California. The first of these, at Kadena in Okinawa, normally has three aircraft attached at any time, while Mildenhall in the United Kingdom is the location of the second SR-71 detachment which usually controls the activities of two aircraft. In addition, Beale serves as the centre for crew training and may well also support operational flights using inflight refuelling to extend the range of the SR-71.

The rigours of high-speed flight at extreme altitude are such that the two crew members, comprising a pilot and a reconnaissance systems operator, both wear full pressure suits similar to those of astronauts. Indeed, selection procedures are identical to those originally used in choosing trainee astronauts, and mission briefings are akin to those of a spaceflight. At one time GTD-21 RPVs were carried pick-a-back and released for overflights of the most sensitive targets, the SR-71 acting as digital data link for the pictures and Elint.

Specification
Lockheed SR-71A
Type: two-seat all-weather strategic reconnaissance aircraft
Powerplant: two Pratt & Whitney J58-P-1 continuous turbo-ramjet engines each rated at 14752-kg (32,500-lb) afterburning thrust
Performance: maximum speed 3661 km/h (2,275 mph) or Mach 3.35 at 24385 m (80,000 ft); operational ceiling 26060 m (85,500 ft); maximum unrefuelled range 5230 km (3,250 miles) at Mach 3
Weights: empty 27216 kg (60,000 lb); maximum take-off 78018 kg (172,000 lb)
Dimensions: span 16.94 m (55 ft 7 in); length 32.74 m (107 ft 5 in); height 5.64 m (18 ft 5 in); wing area 166.76 m^2 (1,795.0 sq ft)
Armament: none
Operator: US (USAF)

An unusual feature of the SR-71 are the three-wheel main landing gear units. The tyres are painted with heat-resistant paint.

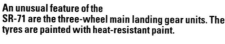

To aid braking, the SR-71 deploys a large parachute, as well as keeping the nose high for further aerodynamic braking. The SR-71 is surprisingly docile at slow speeds.

Lockheed SR-71 'Blackbird' cutaway drawing key

1 Pitot tube
2 Nose mission equipment bay
3 Detachable nose cone joint frame
4 Cockpit front pressure bulkhead
5 Rudder pedals
6 Control column
7 Instrument panel
8 Instrument panel shroud
9 Knife edged windscreen panels
10 Engine throttle levers
11 Oxygen cylinder
12 Pilot's zero-zero ejection seat
13 Upward hinged cockpit canopy cover
14 SR-71B dual control trainer variant
15 Raised instructor's rear cockpit
16 Reconnaissance systems officer's (RSO) canopy cover
17 RSO's zero-zero ejection seat
18 Side console panel
19 Cockpit environmental system equipment bay
20 Rear pressure bulkhead
21 Canopy hinge joint
22 Astro-navigation star tracker
23 Navigation and communications systems electronics
24 Nosewheel bay
25 Landing and taxiing lamps
26 Twin nosewheels
27 Torque scissor links
28 Nosewheel steering control jack
29 Nose undercarriage pivot fixing
30 Palletized reconnaissance equipment packages, interchangeable
31 Forward fuselage longeron
32 Air refuelling receptacle
33 Forward fuselage integral fuel tanks
34 Titanium skin plating
35 Fuselage chine member
36 Close-pitched fuselage frame construction
37 Forward fuselage production joint
38 Blended wing/fuselage main integral fuel tanks (JP-7 fuel)
39 Main undercarriage wheel bay
40 Three-wheel main undercarriage bogie
41 Hydraulic retraction jack
42 Starboard main undercarriage, stowed position

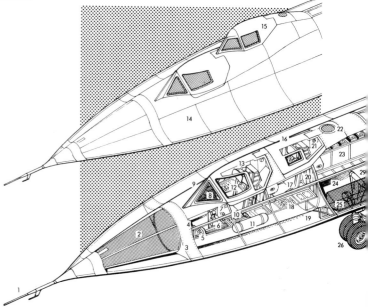

43 Corrugated titanium wing skin panelling
44 Moveable intake, conical centre-body
45 Centre-body retracted (high speed) position
46 Engine air intake
47 Automatic intake control air data probe
48 Intake suction relief doors
49 Variable suction relief doors
50 Hinged engine cowling panel

59 Aft fuselage integral fuel tanks
60 Brake parachute doors
61 Ribbon parachute stowage
62 rear fuselage longeron
63 Wing root rib
64 Inboard wing panel integral fuel tank
65 Close-pitched wing/fuselage frame construction

Apart from the red serials and stencilling, the SR-71 is covered entirely with black, radar-absorbent paint. Further radar absorption is provided by the structure, which traps radar energy within it, and the shape, which has few sharp angles to reflect weaker returns.

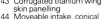

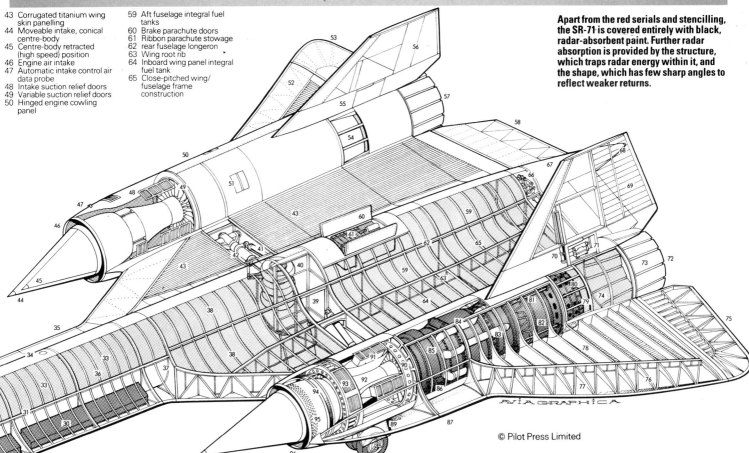

© Pilot Press Limited

51 By-pass duct blow-in doors
52 Starboard outer wing panel
53 Outboard, roll control, elevon
54 Engine bay tertiary air flaps
55 Tailfin fixed root section
56 All-moving starboard fin
57 Exhaust nozzle ejector flaps
58 Inboard, pitch control, elevon

66 Elevon mixer unit
67 Tailcone
68 Fuel jettison
69 Port all-moving tailfin
70 Fin pivot fixing
71 All-moving fin hydraulic jack
72 Port engine exhaust nozzle
73 Ejection flaps
74 Ejector mixer duct

75 Port outer elevon titanium rib construction
76 Outer leading-edge rib construction
77 Under-cambered leading edge
78 Outer wing panel construction
79 Exhaust duct tertiary doors
80 Afterburner nozzle

81 Variable-area afterburner nozzle control jacks
82 Afterburner duct
83 Compressor bleed air by-pass ducts
84 Outer wing panel/engine cowling hinge axis
85 Pratt & Whitney JT11D-20B (J58) single spool turbo-ramjet engine

86 Engine accessory equipment bay
87 Outer wing panel/nacelle chine
88 Port main undercarriage three-wheel
89 Main undercarriage leg door
90 Intake duct secondary by-pass annular louvres

91 Centre-body bleed air louvres
92 Diffuser chamber
93 By-pass duct suction relief louvres
94 Intake annular by-pass duct
95 Centre-body boundary layer bleed air holes
96 Port engine intake
97 Intake moveable centre-body

Lockheed T-33

One of the most widely used trainers of all time, the T-33A remains active in modest numbers throughout the world, being in service with no fewer than 28 air arms after 1980. A derivative of the F-80 Shooting Star, with a longer fuselage to accommodate a second cockpit, the T-33A was produced at a high rate in the 1950s, 5,691 completed by Lockheed plus 210 by Kawasaki in Japan and 656 Nene-engined CL-30 Silver Stars by Canadair. In addition, 85 were completed by Lockheed as RT-33A single-seat reconnaissance aircraft.

A major user of the 'T-bird' is still the USAF, almost 120 serving with Air Defense Tactical Air Command and the Air National Guard. These are employed as targets or electronic jamming platforms in interception exercises. Japan's JASDF still employs this type in the pilot training role. Many surviving T-33s have been modified into AG-33A attack

trainers.

Specification
Type: jet trainer
Powerplant: one Allison J33-A-35 turbojet rated at 2449-kg (5,400-lb) thrust with water injection
Performance: maximum speed at sea level 966 km/h (600 mph); speed at 7620 m (25,000 ft) 874 km/h (543 mph); initial rate of climb 1684 m (5,525 ft)/minute; service ceiling 14480 m (47,500 ft); range 2165 km (1,345 miles)
Weights: empty 3667 kg (8,084 lb); normal take-off 5427 kg (11,965 lb); maximum take-off 6551 kg (14442 lb)
Dimensions: span 11.85 m (38 ft 10.5 in); length 11.51 m (37 ft 9 in); height 3.55 m (11 ft 8 in); wing area 22.02 m² (237.0 sq ft)
Armament: (AT-33A only) two 12.7-mm (0.5-in) Browning M3 machine-guns plus provision for

bombs and rockets on underwing hardpoints
Operators: (T-33A) Canada, Colombia, Ethiopia, Greece, Guatemala, Honduras, Japan, Pakistan, Philippines, Portugal, Singapore, Taiwan, Thailand, Turkey, USA; (RT-33A) Colombia, Greece, Honduras, Jugoslavia, Pakistan, Philippines; (AT-33A)

Canada is one of the many nations which uses the venerable T-33, several being based with fighter units both at home and in Germany.

Bolivia, Burma, Colombia, Ecuador, Mexico, Nicaragua, Uruguay

---

Lockheed TR-1

A development of the notorious U-2, the TR-1 is intended to undertake tactical rather than strategic reconnaissance missions. Production at Palmdale, California, used tooling stored since completion of production of the U-2R variant in 1969.

Employing the same basic airframe and engine as the U-2R, the TR-1 incorporates new systems, the most significant being the ASARS (Advanced Synthetic Aperture Radar System), a UPD-X side-looking airborne radar possessing the ability to 'look' approximately 48 km (30 miles) into enemy territory without crossing the frontier. In this way, it is able to monitor activity while remaining (in peacetime) relatively immune to interception. The TR-1 will later employ the PLSS (Precision Location/Strike System) and the first flight with this equipment took place in 1984. Current planning calls for 10 TR-1s to be equipped with PLSS, these being used to detect and locate enemy emitters, such as ground radars, and then direct attack aircraft to them.

Initial funding in 1979 totalled $10.2 million to re-open the produc-

Lockheed TR-1A of the 95th Reconnaissance Squadron, 17th Reconnaissance Wing, based at RAF Alconbury for battlefield surveillance duties.

tion line, with a view to manufacturing 33 TR-1A single-seaters and two TR-1B two-seat trainers, as well as one ER-2 aircraft for NASA, based at Ames Research Center on Earth resources investigation. The first TR-1A flew on 1 August 1981, and was delivered to the 9th Strategic Reconnsaissance Wing at Beale AFB in the following month. The two two-seat TR-1Bs were delivered in March and May 1983, while about 20 TR-1As had been completed by late 1985. Most were initially assigned to the 9th SRW, but deployment to Europe took place in February 1983 with the arrival of the first example at RAF Alconbury, where it joined the 17th Reconnaissance

Wing which is scheduled to have a complement of 18 by late 1986.

The ASARS apparatus is housed in two wing pods, other sensors being in the modular nose, in the 'Q-bay' aft of the cockpit and in smaller areas elsewhere. TR-1s are known to have deployed non-stop from Beale to Alconbury on more than one occasion.

Specification
Type: high-altitude reconnaissance aircraft
Powerplant: one 7711-kg (17,000-lb) thrust Pratt & Whitney J75-P-13B turbojet engine
Performance: maximum cruising speed above 21335 m (70,000 ft)

more than 692 km/h (430 mph); operational ceiling 27430 m (90,000 ft); range approximately 10002 km (6,215 miles)
Weights: empty 7031 kg (15,500 lb); maximum take-off 18733 kg (41,300 lb)
Dimensions: span 31.39 m (103 ft 0 in); length 19.20 m (63 ft 0 in); height 4.88 m (16 ft 0 in); wing area approximately 92.9 m² (1,000 sq ft)
Armament: none
Operator: USAF

The TR-1A differs from the U-2R in systems and role only. Those based at Alconbury are mainly concerned with providing stand-off cross-border intelligence using SLAR imagery.

Lockheed U-2

One of the most remarkable, and politically controversial, aircraft of all time, the U-2 remains in front-line service with Strategic Air Command's 9th Strategic Reconnaissance Wing at Beale AFB, California. Today's aircraft differ considerably from those which entered service with the Central Intelligence Agency in 1957.

Brainchild of Clarence L. 'Kelly' Johnson, the U-2 was conceived in 1954 as a strategic reconnaissance and special-purpose aircraft capable of operation at extreme altitude. Authorisation to proceed with a batch of 20 was granted in late November 1954, funding and direction of 'Project Aquatone' being by the CIA. In July 1955 the first U-2 was trucked to Groom Lake, Nevada – known as Watertown Strip or 'the Ranch' – where, on 1 August 1955, Tony LeVier made the maiden flight. The U-2 showed little inclination to return to earth; after two abortive attempts LeVier succeeded, though it was found necessary to stall the aircraft to get it to stop flying. The first batch of CIA pilots were converted at Groom Lake in 1956, subsequently deploying via Lakenheath, UK, to Wiesbaden, West Germany, from where the first overflight of the Soviet Union was made on 4 July 1956. Soon afterwards, additional CIA pilots were dispatched to Incirlik, Turkey, from where further overflights were made until 1 May 1960 when Francis 'Gary' Powers was shot down near Sverdlovsk during the course of a mission which began at Peshawar, Pakistan, and which should have ended at Bodo, Norway. The subsequent furore brought an end to this phase of intelligence gathering, but the U-2 remained very important. It was employed to monitor Cuba during the 1962 missile crisis, and enjoyed a long career in South East Asia. CIA association did not cease with the Powers incident; 'Company'-furnished aircraft flown by Chinese Nationalist pilots performed numerous overflights of China until October 1974.

Numerous variants have been identified. Initial aircraft had the Pratt & Whitney J57 engine and were designated U-2A, while the U-2B was retrofitted with the more powerful J75. The U-2C had additional sensors and increased internal fuel. Five two-seat U-2Ds were built during the 1950s, while the U-2E, U-2F and U-2G all arose as modifcations. Two examples of the U-2CT trainer are still thought to be active. The only other variant positively identified is the U-2R, of 1968. This is considerably larger and heavier, and survives in a front-line capacity with SAC.

Some 113 aircraft were completed of all versions. Of these 12 were U-2Rs, but a block of serial numbers was reserved for the U-2R and at least 14 have actually been observed. However, as with many aspects of the U-2 saga, seeing may not necessarily be believing.

Specification
Lockheed U-2R
Type: high-altitude strategic reconnaissance aircraft
Powerplant: one 7711-kg (17,000-lb) thrust Pratt & Whitney J75-P-13B turbojet engine
Performance: maximum cruising

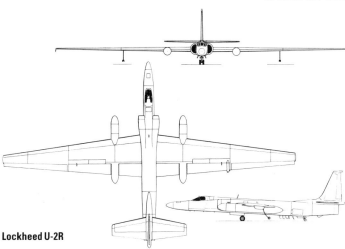

Lockheed U-2R

speed above 21335 m (70,000 ft) more than 692 km/h (430 mph); operational ceiling approximately 22860 m (75,000 ft); range about 10,002 km (6,215 miles)
Weights: empty 6850 kg (15,101 lb); maximum take-off 18597 kg (41,000 lb)
Dimensions: span 31.39 m (103 ft 0 in); length 19.20 m (63 ft 0 in); height 4.88 m (16 ft 0 in); wing area approximately 92.9 m² (1,000 sq ft)
Armament: none
Operator: USAF

The U-2R is the last variant in service, being used by the 9th SRW based at Beale AFB.

Lockheed 'Stealth' (RF-19)

Like earlier projects of the 'Skunk Works', Lockheed-California's latest product is surrounded by security which extends even to the designation. However, it is known that funding for the Lockheed contribution to 'Projectile Stealth' emanates from the Defense Advanced Research Projects Agency (DARPA), responsibility for contract management resting with the USAF Flight Dynamics Laboratory.

The 'F-19' is reported to be a single-seat fighter/reconnaissance aircraft, first flown in 1977, of which several examples had been built by 1983 for test and evaluation. This task is probably undertaken from the isolated facility at Groom Lake, Nevada. Colloquially known as 'the Ranch', it was from here that early testing of both the U-2 and A-12/F-12/SR-71 series was accomplished.

The 'F-19' is said to be similar in size to the F/A-18 Hornet, with a planform resembling the Space Shuttle, and powered by two engines in the 5443-kg (12,000-lb) class. Stealth technology minimises radar, infra-red and optical reflectivity in order to reduce vulnerability from a variety of threats, hence adoption of the acronym CSIRS (Covert Survivable Interdictor-strike Reconnaissance System) which has been associated with this aircraft.

Specification
Type: single-seat fighter/reconnaissance aircraft
Powerplant: two jet engines, each rated in the 5443-kg (12,000-lb) thrust class
Performance, Weight, Dimensions, Armament: no details available
Operator: under test by USAF

Max Holste MH.1521 Broussard

Produced initially by Avions Max Holste (merged into Aérospatiale), the Broussard STOL six-seater interested French army light aviation (ALAT) which placed a launching order. The prototype flew on 17 November 1952 and the first production aircraft on 16 June 1954. A total of 335 Broussards was built and deliveries were made to a number of countries under French military aid programmes. A few remain in service.

Specification
Type: utility transport
Powerplant: one 336-kW (450-hp) Pratt & Whitney R-985-AN Wasp radial piston engine
Performance: cruising speed 230 km/h (143 mph) at 1500 m (4,920 ft); initial rate of climb 360 m (1,181 ft)/minute; range with 600-kg (1,323-lb) payload 800 km (497 miles)
Weights: empty 1450 kg (3,197 lb); maximum take-off 2450 kg (5,401 lb)

Dimensions: span 13.75 m (45 ft 1.3 in); length 8.60 m (28 ft 2.6 in); height 2.80 m (9 ft 2.2 in); wing area 25.40 m² (273.41 sq ft)
Armament: none
Operators: Benin, Burkina Faso, Cameroun, Central African Republic, Chad, Congo, France (Armée de l'Air and ALAT), Madagascar, Mauretania, Niger, Senegambia, Togo

Max Holste MH.1521 Broussard

McDonnell Douglas A-4 Skyhawk

One of the most successful military aircraft of all time, the diminutive A-4 Skyhawk is still used widely throughout the world. Brainchild of Ed Heinemann, chief engineer with the Douglas El Segundo Division, the Skyhawk was evolved in response to a US Navy requirement for a carrier-based attack aircraft to carry a 907-kg (2,000-lb) weapon load. The Navy envisaged a turboprop aircraft weighing 13608 kg (30,000 lb), but Heinemann decided that this presented an opportunity to reverse the trend towards ever-heavier machines. The resulting aircraft tipped the scales at just half that of the specification, while for good measure it exceeded the load by a handsome margin and the speed by 100 knots. The prototype flew on 22 June 1954 and the Skyhawk quickly matured into an outstanding warplane, remaining in production for 26 years with 2,960 being built.

Early versions had the J65 engine, but the A-4E introduced the more powerful J52. A-4E production totalled 494, these being followed by the A-4F and the TA-4F and TA-4J two-seaters for Naval Air Training Command. The final new-build variant for the US armed forces was the A-4M assigned to Marine Corps light attack squadrons, and which features a braking parachute and much improved avionics.

New A-4s were supplied to Australia (A-4G and TA-4G, now stored), New Zealand (A-4K and TA-4K), Israel (A-4H, TA-4H and A-4N) and Kuwait (A-4KU and TA-4KU). Remanufactured early Skyhawks were supplied to Argentina (A-4P and A-4Q), Malaysia (A-4L) and Singapore (A-4S and TA-4S). Grumman plays a major role in refurbishing. Others were despatched to Israel during the 1973 Yom Kippur War; such aid included A-4Es, about 12 of which were passed on to Indonesia.

Twenty-three TA-4Fs were converted to OA-4M standard for forward air control (FAC) duties with the US Marine Corps. Equipment include ESM in the fin-tip and ECM aerials under the nose.

McDonnell Douglas A-4B Skyhawk of No. 5 Sqn, Argentine air force.

Specification

Type: single-seat attack bomber
Powerplant: one 5080-kg (11,200-lb) thrust Pratt & Whitney J52-P-408 turbojet engine
Performance: maximum speed with 1814-kg (4,000-lb) bombload 1040 km/h (646 mph); ferry range 3219 km (2,000 miles)
Weights: empty 4899 kg (10,800 lb); maxiumum take-off 11113 kg (24,500 lb)
Dimensions: span 8.38 m (27 ft 6 in); 12.29 m (40 ft 4 in); height 4.57 m (15 ft 0 in); wing area 24.15 m² (260.0 sq ft)
Armament: two 20-mm (0.79-in) cannon (Israeli aircraft two 30-mm/1.18-in cannon). Centreline stores station with capacity of 1588 kg (3,500 lb); two inboard stores stations with capacity of 1021 kg (2,250 lb) each; two outboard stores stations with capacity of 454 kg (1,000 lb) each. Ordnance ranging from conventional and nuclear bombs through guided missiles to air-to-air and air-to-surface rockets
Operators: Argentina, Australia (inactive), Israel, Kuwait, Malaysia, New Zealand, Singapore, US Marine Corps, US Navy (not first-line)

Right: An important role fulfilled by US Navy Skyhawks is that of potential adversary in the dissimilar air combat training (DACT) programme. The TA-4Js and A-4J in this illustration are camouflaged in the colour schemes of 'enemy' forces.

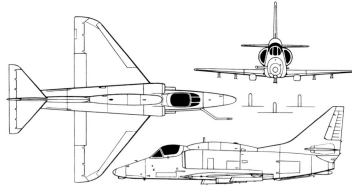

McDonnell Douglas A-4M Skyhawk

McDonnell Douglas C-9

In 1966 the USAF required a new aeromedical transport and ordered the first of 21 C-9A Nightingales in 1967. Based on the standard DC-9-30, these had JT8D-9 instead of D-7 engines and a very different interior. Three entrances were provided, two with hydraulically actuated stairways, and the third (forward) door measuring 2.06 m (6 ft 9 in) high and 3.45 m (11 ft 4 in) wide with a hydraulically-actuated ramp for loading stretchers. There is accommodation for 40 stretchers or more than 40 seated patients, and a medical team of five. An intensive-care compartment has controllable ventilation and pressure. A galley and toilets are provided fore and aft. Delivery to Scott AFB began on 10 August 1968, for MAC's 375th Aeromedical Wing.

Just before the completion of this order, in February 1973, the US Navy decided on a modified DC-9 as a fleet logistic passenger/cargo transport. On 24 April 1972 the first five of 14 C-9B Skytrain IIs were ordered. The C-9B has DC-9-30 dimensions, JT8D-9 engines, extra fuel capacity and the cargo door forward. Passengers board via front or rear ventral doors with airstairs. Normal seating capacity is 90, or 107 high-density, with galley and toilet at each end, or the C-9B can be operated in all-cargo or mixed configuration. The first flew on 7 February 1973, and all had been delivered by mid-1976. Two similar aircraft were sold to Kuwait.

Specification
Lockheed C-9B Skytrain II
Type: passenger/cargo transport
Powerplant: two 6577-kg (14,500-lb) thrust Pratt & Whitney JT8D-9 turbofan engines
Performance: maximum cruising speed 927 km (576 mph); long-range cruising speed 811 km/h (504 mph); range (LR speed, 4536-kg/10,000-lb

payload at 9145 m/30,000 ft) 4704 km (2,923 miles)
Weights: empty, cargo 27082 kg (59,706 lb), passenger 29612 kg (65,283 lb); maximum take-off 49895 kg (110,000 lb)
Dimensions: span 28.47 m (93 ft 5 in); length 36.36 m (119 ft 3.5 in); height 8.38 m (27 ft 6 in); wing area 92.97 m² (1,000.7 sq ft)
Armament: none
Operators: Colombia, Italy, Kuwait, US (USAF, USN), Venezuela

The red cross on the fin identifies this as an aeromedical evacuation C-9A Nightingale. Twenty-one such aircraft were procured for the US Air Force, each able to carry up to 40 stretcher cases.

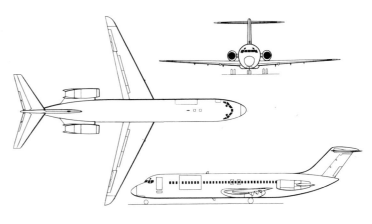

McDonnell Douglas C-9B Skytrain II

McDonnell Douglas KC-10 Extender

One of the more versatile additions to the USAF inventory in recent years, the McDonnell Douglas KC-10 Extender possesses genuine dual tanker/transport capability and, indeed, is regularly employed as such in support of major deployment by tactical fighter aircraft. Although principally concerned with aerial refuelling duties, the Extender also possesses significant airlift capacity (76825 kg/169,370 lb), and aircraft from the two squadrons which have formed to date are frequently called upon to augment MAC's fleet of strategic transport aircraft by undertaking missions to Europe and Hawaii.

Essentially a derivative of the commercial DC-10-30CF, the Extender incorporates numerous changes in order to accomplish its military duties. These include the addition of an inflight-refuelling boom with hose and drogue coupling; provision of a boom operator's station; fitment of additional fuel cells in the underfloor cargo compartments; installation of a refuelling receptacle; and provision of full military avionics equipment.

First flown in July 1980, the KC-10A production model had earlier resisted a strong challenge from Boeing, whose Model 747 had also been in contention for the ATCA (Advanced Tanker/Cargo Aircraft) programme. Following a closely fought contest, the USAF elected to

A KC-10A Extender of the 2nd Bomb Wing, operating form Barksdale AFB, Louisiana.

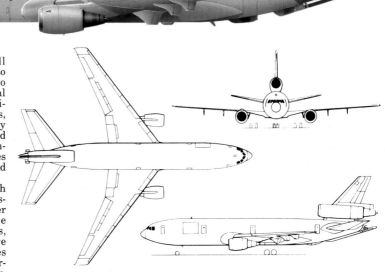

purchase the DC-10, McDonnell Douglas receiving authorization to proceed with the construction of two aircraft in November 1978. Initial proposals anticipated the acquisition of no more than 20 KC-10As, but this figure has been sharply revised in an upwards direction, and it now seems certain that the eventual total will be 60 when deliveries are completed shortly before the end of the present decade.

Since entering service with Strategic Air Command at Barksdale in spring 1981, the Extender has supported a considerable number of overseas deployments, but perhaps the most impressive demonstration of its capabilities came in October 1982 when one aircraft escorted six McDonnell Douglas F-15C Eagles on a non-stop flight from Kadena, Okinawa, to Eglin

McDonnell Douglas KC-10A Extender

AFB, Florida. During the 15-hour, 11265-km (7,000-mile) trip the KC-10A furnished fuel to the F-15s seven times and also airlifted 59 personnel and 24950 kg (55,000 lb) of support equipment.

Specification
McDonnell Douglas KC-10A Extender
Type: strategic inflight-refuelling tanker and transport aircraft
Powerplant: three 23814-kg (52,500-lb) thrust General Electric CF6-50C2 turbofans
Performance: cruising speed 908 km/h (564 mph); service ceiling 10180 m (33,400 ft); range 7032 km (4,370 miles) with maximum 76825-kg (169,370-lb) payload
Weights: empty 108891 kg (240,065 lb); maximum take-off 267620 kg (590,000 lb)
Dimensions: span 50.41 m (165 ft 4.4 in); length 55.35 m (1891 ft 7 in); height 17.70 m (58 ft 1 in); wing area 367.7 m² (3,958 sq ft)
Armament: none
Operator: USAF

The longer boom and large size of the KC-10 enables refuelling of large aircraft (such as this C-5A Galaxy) to be accomplished with great ease and safety.

McDonnell Douglas C-17

On 29 August 1981 McDonnell Douglas was chosen to proceed with development of their C-X submission, to fulfil long-range heavy-lift transportation tasks, including the movement of tanks to airstrips close to potential battle areas. But the future of the C-17 is by no means certain, following the decision to restart production of the C-5. Work is proceeding, however, with a view to introducing the type in 1992.

USAF studies identified a need for about 210 C-17s. It could airlift out-size items which at present can be accommodated only by the C-5. It will also possess short-field performance comparable with that of the C-130 Hercules. Extensive use is made of technology developed for the company's YC-15 advanced medium STOL transport. Refinements include the introduction of flaps on a swept supercritical wing incorporating winglets. It would operate from runways 914 m (3,000 ft) long and 18.3 m (60 ft) wide; inflight refuelling capability will permit global operations while a tight turn radius and independence of ground power will allow it to use airstrips close to a battle zone.

Specification (provisional)
Type: heavy-lift cargo aircraft
Powerplant: four 16783-kg (37,000-lb) thrust Pratt & Whitney PW2037 turbofan engines
Performance: maximum cruising speed at low altitude 649 km/h (403 mph); range with maximum 78109-kg (172,200-lb) payload 4450 km (2,765 miles); ferry range 9262 km (5,755 miles)
Weights: empty 117480 kg (259,000 lb); maximum take-off 258548 kg (570,000 lb)
Dimensions: span 50.29 m (165 ft 0 in); length 53.09 m (174 ft 2 in); height 16.84 m (55 ft 3 in); wing area 353.02 m² (3,800.0 sq ft)
Armament: none
Operator: none

McDonnell Douglas F-4 Phantom II

With 5,195 built in 20 years, the F-4 Phantom II is one of the most successful fighter aircraft in aviation history. Examples will probably be in operational service more than 50 years after the maiden flight.

Development got under way during 1953, and close contact with the US Navy resulted in a contract for a fighter designated F3H-G. Redesign in 1954 provided additional fuel, a second seat and more capable fire-control system. Flown as the XF4H-1 prototype on 27 May 1958, it soon became apparent that McDonnell had come up with a real winner. The 21 F-4A pre-production examples were followed by the first production batch, leading to the first operational version, the F-4B, with more powerful engines, a taller canopy and larger radar dish. Deliveries to the Navy began in June 1961 and to the Marine Corps one year later. In 1961-62 the USAF conducted a comparative study and then a formal evaluation of an F-4A configured for the strike/attack role. This confirmed the Phantom's remarkable capabilities, and a minimum-change version known initially as

The Phantom II is still a highly potent air-defence fighter and continues to fly with nations around the world. This low-visibility grey RAF Phantom FGR.2 is based in Germany where it helps defend Western airspace as part of NATO.

McDonnell Douglas F-4 Phantom II (continued)

A small force of Phantoms continues to serve with the Islamic Republic of Iran air force, committed to the war against Iraq. Missions have included attacks on shipping in the Persian Gulf.

The F-4F equips four front-line West German air force units, this example serving with JG 74 'Molders'. Aircraft are undergoing modification to allow a greater range of air-to-air missiles to be carried under the 'Peace Rhine' conversion programme.

Formidable use of the Phantom by the Israeli air force in the skies over Lebanon has included AGM-45 Shrike-equipped aircraft knocking out Syrian surface-to-air missile (SAM) sites.

Cameras for the RF-4 photo-recce models of the Phantom are housed in the nose. This is a West German RF-4E from a fleet which is undergoing improvements which include better cameras, infra-red linescan and the ability to carry weapons.

Below: An excellent underview of an RAF Phantom displays a menacing combination of armament including gun-pod, semi-recessed Sky Flash and paired Sidewinder AAMs.

the F-110A, changed to F-4C post-September 1962, was ordered into production for the USAF in April 1962. McDonnell delivered 583 F-4Cs, some being passed to Spain, followed by 825 F-4Ds more extensively changed for the land-based attack mission, some of which went to Iran, Spain and South Korea.

The next variant, the F-4E, was produced in the greatest numbers. This was fitted with an integral M61A1 gun under the nose, more powerful engines, a solid-state APQ-120 radar and an additional fuel cell in the rear fuselage. The USAF received the first of 1,003 in October 1967, later introducing a powerfully slatted wing for enhanced manoeuvrability. This version was exported to Greece, Iran, Israel, South Korea and Turkey. In addition, 140 F-4EJs went to the Japanese Air Self Defence Force, the first two built at St Louis, the next 11 supplied as kits to Mitsubishi, and the last 127 built in Japan. West German industry shared in 175 F-4Fs, with simpler avionics and no Sparrows, for the Luftwaffe. Following delivery of the last F-4B, the US Navy ordered 522 F-4Js with AWG-10 radar, more powerful Dash-10 engines, additional fuel capacity, slatted tailplane, drooping ailerons and many other changes.

The UK took delivery of 52 F-4Ks (Fleet Air Arm FG.1) and 118 F-4Ms (RAF FGR.2), partly built from British components and with the Rolls-Royce Spey turbofan of 9305 kg (20,515 lb) thrust. Both marks now serve solely with the RAF, which has also bought 15 refurbished ex-US Navy F-4Js.

Three variants were very well equipped for the reconnaissance role. First came the RF-4C, a few of the USAF's 505 being transferred to Spain. The US Marine Corps bought 46 RF-4Bs while 146 RF-4Es were exported.

Under the US Navy's CILOP (Conversion in Lieu of Procurement) programme, 228 F-4Bs became F-4Ns with new systems and generators, and close to 300 F-4Js were brought to F-4S configuration with manoeuvring slats and further updated systems. The USAF rebuilt 116 F-4Es with special avionics and weapons as F-4G 'Wild Weasel' SAM suppression aircraft, while F-4Es and some RF-4Cs are receiving the Pave Tack FLIR/laser pod.

By 1 July 1979 McDonnell Douglas had delivered 5,068 aircraft, including the 11 kits for Japan. Mitsubishi delivered the 5,195th and last Phantom II in 1981. Some 3,500 F-4s are in service, and at least 1,500 will probably be active in the year 2000. Modernisation would enhance their capabilities, and the chief programme has been evolved by Boeing in co-operation with Pratt & Whitney.

This involves re-engining with the PW1120 bypass jet, a complete avionics update and the fitment of either a conformal fuel tank or weapons carriage pallet. Flight testing is expected to begin in October 1985, both at Boeing, Wichita and in Israel. The Luftwaffe's F-4Fs are being rebuilt with pulse-doppler 'look-down' radars and other updates. Japan hopes to fit a close support weapons system with a Westinghouse APG-66 pulse-doppler radar, Kaiser/VDO head-up display (HUD), inertial navigation system and radar warning receiver. This could be installed in about 100 F-4EJs from 1987. Meanwhile, McDonnell Aircraft was in 1984 offering its own Enhanced Phantom II update programme which, though not evident externally (apart from a one-piece windscreen), leaves the aircraft with structure, fuel system and avionics good for the year 2000.

Specification
McDonnell Douglas F-4E Phantom II
Type: multi-role fighter
Powerplant: two 8119-kg (17,900-lb) afterburning thrust General Electric J79-GE-17 turbojet engines
Performance: maximum speed, clean at 10975 m (36,000 ft) 2301 km/h (1,430 mph); cruising speed with full internal fuel, four AIM-7s and two external tanks at 10060 m (33,000 ft) 921 km/h (572 mph); initial rate of climb, clean 15179 m (49,800 ft)/minute; service ceiling, clean 17905 m (58,750 ft); typical combat radius, hi-lo-hi mission with four AIM-7s and two external tanks 837 km (520 miles)
Weights: empty 14448 kg (31,853 lb); take-off, four AIM-7s plus two external drop-tanks 24410 kg (53,814 lb); maximum take-off 28030 kg (61,795 lb)
Dimensions: span 11.68 m (38 ft 4 in); length 19.20 m (63 ft 0 in); height 5.00 m (16 ft 5 in); wing area 49.24 m² (530.0 sq ft)
Armament: one M61A1 20-mm (0.79-in) cannon with 640 rounds, four AIM-7 Sparrow AAMs plus various combinations of ordnance on four wing pylons up to 5888 kg (12,980 lb)
Operators: (F-4C) Spain, USAF; (F-4D) South Korea, USAF; (F-4E) Greece, Israel, Japan, South Korea, Turkey, USAF, West Germany; (F-4F) West Germany; (F-4G) USAF; (F-4J/S) UK, USMC, USN; (F-4K/M) UK; (F-4N) USN; (RF-4B) USMC; (RF-4C) Spain, USAF; (RF-4E) Greece, Iran, Israel, Japan, Turkey, West Germany

Tactical reconnaisance for USAF Europe is provided by the RF-4C. This aircraft flies with the 1st TRS, 10th TRW from RAF Alconbury. Note the wrap-round lizard scheme.

McDonnell Douglas F-4E Phantom II cutaway drawing key

1 Starboard tailplane
2 Static discharger
3 Honeycomb trailing edge panels
4 Tailplane mass balance weight
5 Tailplane spar construction
6 Drag chute housing
7 Tailcone/drag chute hinged door
8 Fuselage fuel tanks vent pipe
9 Honeycomb rudder construction
10 Rudder balance
11 Tail warning radar fairing
12 Tail navigation light
13 Fin tip antenna fairing
14 Communications antenna
15 Fin rear spar
16 Variable intensity formation lighting strip
17 Rudder control jack
18 Tailplane pivot mounting
19 Tailplane pivot seal
20 Fixed leading edge slat
21 Tailplane hydraulic jack
22 Fin front spar
23 Stabilator feel system pressure probe
24 Anti-collision light
25 Stabilator feel system balance mechanism
26 Tailcone cooling air duct
27 Heat resistant tailcone skinning
28 Arresting hook housing
29 Arresting hook, lowered
30 Starboard fully variable exhaust nozzle
31 Rudder artificial feel system bellows
32 Fin leading edge
33 Ram air intake

34 Fuselage No 7 fuel cell, capacity 84 US gal (318 l)
35 Engine bay cooling air outlet louvres
36 Arresting hook actuator and damper
37 Fuel vent piping
38 Fuselage No 6 fuel cell, capacity 213 US gal (806 l)
39 Jet pipe shroud construction
40 Engine bay hinged access doors
41 Rear AIM-7E-2 Sparrow air-to-air missile
42 Semi-recessed missile housing
43 Jet pipe nozzle actuators
44 Afterburner jet pipe

45 Fuselage No 5 fuel cell, capacity 180 US gal (681 l)
46 Fuel tank access panels
47 Fuel system piping
48 Tailplane control cable duct
49 Fuselage No 4 fuel cell, capacity 201 US gal (761 l)
50 Starboard engine bay construction
51 TACAN aerial
52 Fuselage No 3 fuel cell, capacity 147 US gal (556 l)
53 Engine oil tank
54 General Electric J79-GE-17A turbojet engine
55 Engine accessories
56 Wing rear spar attachment
57 Mainwheel door
58 Main undercarriage wheel well
59 Lateral control servo actuator
60 Hydraulic accumulator
61 Lower surface airbrake jack
62 Flap hydraulic jack
63 Starboard flap
64 Honeycomb control surface construction
65 Starboard aileron
66 Aileron power control unit
67 Flutter damper

68 Spoiler housing
69 Wing tank fuel vent
70 Dihedral outer wing panel
71 Rear identification light
72 Wing tip formation lighting
73 Starboard navigation light
74 Radar warning antenna
75 Outer wing panel construction
76 Outboard leading edge slat
77 Slat control linkage
78 Slat hydraulic jack
79 Outer wing panel attachment
80 Starboard wing fence
81 Fuel vent system shut-off valves
82 Top of main undercarriage leg

83 Outboard pylon attachment housing
84 Inboard slat hydraulic jack
85 Starboard outer pylon
86 Mainwheel leg door
87 Mainwheel brake discs
88 Starboard mainwheel
89 Starboard external fuel tank capacity 370 US gal (1400 l)
90 Inboard leading edge slat, open

91 Slat hinge linkages
92 Main undercarriage retraction jack
93 Undercarriage uplock
94 Starboard wing fuel tank, capacity 315 US gal (1192 l)
95 Integral fuel tank construction
96 Inboard pylong fixing
97 Leading edge ranging antenna
98 Starboard inboard pylon
99 Twin missile launcher
100 AIM-9 Sidewinder air-to-air
101 Hinged leading edge access panel
102 Wing front spar
103 Hydraulic reservoir
104 Centre fuselage formation lighting
105 Fuselage main frame
106 Engine intake compressor face
107 Intake duct construction
108 Fuselage No 2 fuel cell, capacity 185 US gal (700 l)
109 Air-to-air refuelling receptacle, open

110 Port main undercarriage leg
111 Aileron power control unit
112 Port aileron
113 Aileron flutter damper
114 Port spoiler
115 Spoiler hydraulic jack
116 Wing fuel tank vent pipe
117 Port outer wing panel
118 Rearward identification light
119 Wing tip formation lighting
120 Port navigation light
121 Radar warning antenna
122 Port outboard leading edge slat
123 Slat hydraulic jack
124 Wing fence
125 Leading edge dog tooth
126 Inboard leading edge slat, open
127 Port external fuel tank capacity 370 US gal (1400 l)
128 Inboard slat hydraulic jack
129 Port wing fuel tank, capacity 315 US gal (1192 l)
130 Upper fuselage light
131 IFF antenna

132 Avionics equipment bay
133 Gyro stabiliser platform
134 Fuselage No 1 fuel cell, capacity 215 US gal (814 l)
135 Intake duct
136 Hydraulic connections
137 Starter cartridge container
138 Pneumatic system air bottle
139 Engine bleed air supply pipe
140 Forward AIM-7 missile housing
141 Ventral fuel tank, capacity 600 US gal (2271 l)
142 Bleed air louvre assembly, lower
143 Avionics equipment bay
144 Variable intake ramp jack
145 Bleed air louvre assembly, upper
146 Radar operator's Martin-Baker ejection seat
147 Safety harness

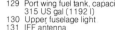

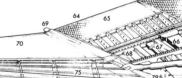

The F-4G 'Wild Weasel' variant is a dedicated tactical electronic warfare aircraft tasked with locating and destroying hostile air-defence electronic systems. Weaponry includes Shrike and Maverick air-to-surface missiles.

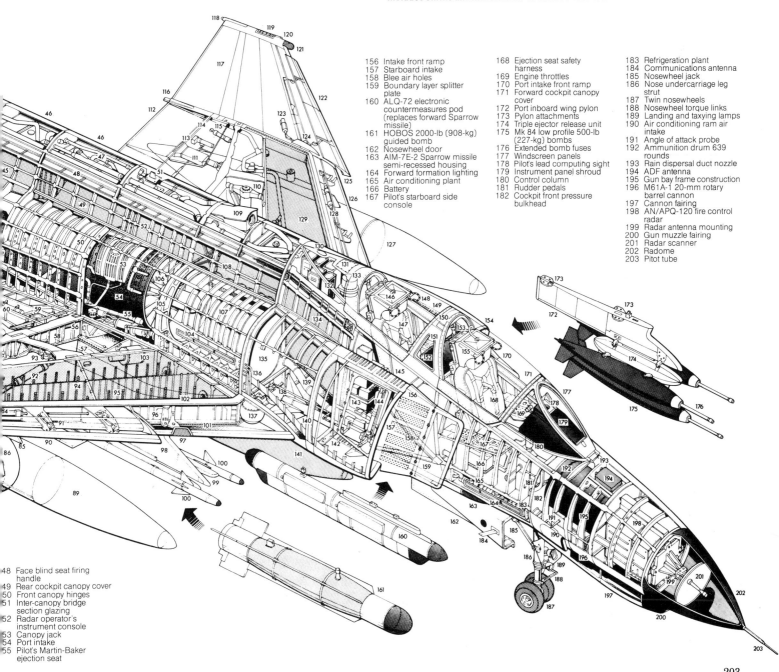

156 Intake front ramp
157 Starboard intake
158 Blee air holes
159 Boundary layer splitter plate
160 ALQ-72 electronic countermeasures pod (replaces forward Sparrow missile)
161 HOBOS 2000-lb (908-kg) guided bomb
162 Nosewheel door
163 AIM-7E-2 Sparrow missile semi-recessed housing
164 Forward formation lighting
165 Air conditioning plant
166 Battery
167 Pilot's starboard side console

168 Ejection seat safety harness
169 Engine throttles
170 Port intake front ramp
171 Forward cockpit canopy cover
172 Port inboard wing pylon
173 Pylon attachments
174 Triple ejector release unit
175 Mk 84 low profile 500-lb (227-kg) bombs
176 Extended bomb fuses
177 Windscreen panels
178 Pilot's lead computing sight
179 Instrument panel shroud
180 Control column
181 Rudder pedals
182 Cockpit front pressure bulkhead

183 Refrigeration plant
184 Communications antenna
185 Nosewheel jack
186 Nose undercarriage leg strut
187 Twin nosewheels
188 Nosewheel torque links
189 Landing and taxying lamps
190 Air conditioning ram air intake
191 Angle of attack probe
192 Ammunition drum 639 rounds
193 Rain dispersal duct nozzle
194 ADF antenna
195 Gun bay frame construction
196 M61A-1 20-mm rotary barrel cannon
197 Cannon fairing
198 AN/APQ-120 fire control radar
199 Radar antenna mounting
200 Gun muzzle fairing
201 Radar scanner
202 Radome
203 Pitot tube

48 Face blind seat firing handle
49 Rear cockpit canopy cover
50 Front canopy hinges
51 Inter-canopy bridge section glazing
52 Radar operator's instrument console
53 Canopy jack
54 Port intake
55 Pilot's Martin-Baker ejection seat

McDonnell Douglas F-15 Eagle

Now spearheading the defence of the Western world, and likely to stay there well into the next century, the McDonnell Douglas F-15 Eagle has a flight performance unsurpassed by any other fighter. It is probably the only US fighter capable of catching the Soviet Union's very fast and high-flying Mikoyan-Gurevich MiG-25, and indeed very early in its career was being openly talked about as the 'Foxbat killer'. Israel and Iran are two countries that have raged impotently as this formidable Soviet spy-plane and long-range interceptor has streaked high along their borders (and even over their territory), photographing military installations. With typical early-warning times quite long, as they have been in the mid-1970s, it has been immune from the efforts of the McDonnell Douglas F-4 Phantoms to bring them down. In at least one case Sparrow missiles from Israel F-4s launched against 'Foxbats' operating from Libya have fallen impotently into the sea.

But all this has now changed. F-15s are now operational in Europe, Israel and Saudi Arabia, and the Mach 3/24385-m (80,000-ft) cruise performance of the MiG-25 will no longer be adequate to protect it over these areas. Although armed with the same Sparrow and Sidewinder combat missiles as its predecessor, the Phantom, the F-15 behaves as a much more powerful 'first stage', giving it considerably greater speed and height at launch.

In the early 1960s, while the fabulous F-4 was still fresh to the US Navy (its sponsor and first customer), American defence experts were beginning to plan a fighter to follow it. Both the US Navy and US Air Force wanted an air-superiority aircraft, and many people saw the possibility of a common design. But as time went by the two services evolved substantially different requirements and with the example of the disastrous General Dynamics F-111A/F-111B fresh in their minds the planners let the two services have their own ways.

In September 1968, the USAF commissioned three companies to produce competitive designs. They were McDonnell Douglas, Fairchild and North American, and in December 1968 the first-named was declared the winner. Undoubtedly its F-4 background helped, but all three companies put up extraordinarily detailed schemes. The new fighter was designated the F-15 Eagle, and the first of 20 test aeroplanes made its initial flight in July 1972. The type entered service in November 1974 and the first squadron was declared operational in January 1976.

Such is the importance of the F-15 in Europe that less than two years later three squadrons had been de-

ployed to Germany to face the growing fleet of Soviet and eastern bloc warplanes ranged along the East German border. Export F-15s are now also finding their way to Israel, Saudi Arabia and Japan. So potent was the Eagle considered that the decision to permit export to Israel was long held up because of political implications in the Arab world.

In July 1979 the F-15 saw its first combat application in the skies over the Lebanon. The aircraft of the Israeli *Heyl Ha'Avir* flying in support of ground attacks by Kfirs and F-4s engaged Syrian MiG-21s. These successfully destroyed several Syrian aircraft and themselves suffered no losses. More recently the F-15s claimed 40 MiG-21 and -23 kills for no loss in hectic fighting over the Bekaa. MiG-25s have also been downed by Israeli Eagles, while Saudi machines have destroyed 2 Iranian F-4s.

A thrust/weight ratio far greater than used in any previous fighter was specified. This was necessary for two reasons: to permit a rate of climb sufficient to catch any intruding MiG-25 with its Sparrow and Sidewinder anti-aircraft missiles, and to out-turn any likely foe in combat in order to bring its gun to bear.

The missiles were already standard armament on a number of US aircraft, but a new gun was to be specially designed for the F-15 to make it more lethal. The 25-mm GAU-7 rotating-barrel gun was to have a considerably greater killing power than the 20-mm calibre M61 weapon used in the F-4E Phantoms and Lockheed F-104 Starfighters. A competition to build the gun was held between General Electric and Philco-Ford, and won by the latter company, but both firms ran into such trouble with the new caseless ammunition that the programme was abandoned and the USAF had to fall back on the well tried M61.

To exploit fully the new fighter designs being studied by the US Air Force and US Navy, the Defense Department decided the time was right for new propulsion systems, and accordingly a third competition

A trio of F-15A Eagles display the very large wing area, tapered forward fuselage and prominent powerplant housings characteristic of the aircraft.

was held, between Pratt & Whitney and General Electric, the two top US engine companies. The former's F100 engine, after some initial problems, is shaping up to be a worthy successor to the General military powerplant in the western world. As with most combat aircraft designed since the late 1960s, twin engines rather than one were chosen for better chances of survival.

The decision to carry only one crew member (its naval contemporary, the F-14, has a crew of two) meant that the F-15 had to be easy to fly and automated as far as possible.

In particular it meant that the Hughes APG-63 pulse-Doppler radar had to be easy to operate and read; pilots in combat or scanning the skies cannot afford to be looking down into the cockpit all the time.

In a word, the F-15 is designed for 'seat-pants' operation: the pilot flies the machine instinctively, his eyes constantly scanning the sky and see-

This F-15J Eagle serves with the Japan Air Self-Defence Force.

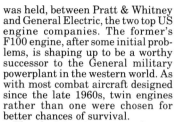

02-8801

McDonnell Douglas F-15C Eagle cutaway drawing key

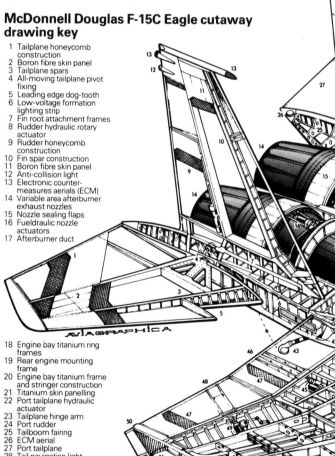

1 Tailplane honeycomb construction
2 Boron fibre skin panel
3 Tailplane spars
4 All-moving tailplane pivot fixing
5 Leading edge dog-tooth
6 Low-voltage formation lighting strip
7 Fin root attachment frames
8 Rudder hydraulic rotary actuator
9 Rudder honeycomb construction
10 Fin spar construction
11 Boron fibre skin panel
12 Anti-collision light
13 Electronic counter-measures aerials (ECM)
14 Variable area afterburner exhaust nozzles
15 Nozzle sealing flaps
16 Fueldraulic nozzle actuators
17 Afterburner duct
18 Engine bay titanium ring frames
19 Rear engine mounting frame
20 Engine bay titanium frame and stringer construction
21 Titanium skin panelling
22 Port tailplane hydraulic actuator
23 Tailplane hinge arm
24 Port rudder
25 Tailboom fairing
26 ECM aerial
27 Port tailplane
28 Tail navigation light
29 ECM aerial
30 Radar warning aerials
31 Boron fibre skin panelling
32 Fin leading edge
33 Port air system equipment bay
34 Forward engine mounting
35 Engine mounting frame
36 Bleed air system ducting
37 Engine support link
38 Engine bay fireproof bulkhead
39 Pratt & Whitney F100-PW-100 afterburning turbofan engine
40 Starboard air system equipment bay
41 Engine bleed air primary heat exchanger
42 Heat exchanger ventral exhaust duct
43 Retractable runway arrester hook
44 Wing trailing edge fuel tank
45 Flap hydraulic jack
46 Starboard plain flap
47 Flap and aileron honeycomb panel construction
48 Starboard aileron
49 Aileron hydraulic actuator
50 Fuel jettison pipe
51 Aluminium honeycomb wing tip fairing
52 Low-voltage formation lighting
53 Starboard navigation light
54 ECM aerial
55 Westinghouse ECM equipment pod
56 Outboard wing stores pylon
57 Pylon attachment spigot
58 Cambered leading edge ribs
59 Front spar
60 Machined wing skin/stringer panels
61 Outboard pylon fixing
62 HF flush aerial
63 Leading edge fuel tank
64 Inboard pylon fixing
65 Wing rib construction
66 Starboard wing integral fuel tank, total internal fuel load, 13,455-lb (6103-kg)

ing only one instrument, the head-up display. This instrument is a cathode-ray tube fed with information from the radar, showing the actual position of the target in the sky, its range, closing speed, missile safe firing distance and all the other information the pilot needs to attack the target.

In addition to picking up echoes from targets flying at the same altitude or higher (where echoes can only be produced by other aircraft), the radar can 'see' targets 'silhouetted' against the ground. This is immensely more difficult owing to the far greater strength of the ground returns compared with even the largest targets. It calls for highly sophisticated signal-processing techniques, using complex 'software' (computer programming) methods to get the most information out of the faint target returns. Improvements in software in fact permit great improvements in radar performance without physical modifications to the equipment itself.

Meanwhile, taking a lesson from the Vietnam war, the designers have added a tail-warning radar to indicate – the traditional blind spot and worry of the air-combat pilot.

Specification
Type: single-seat air-superiority fighter
Powerplant: two Pratt & Whitney F100-PW-100 turbofan engines, each of 10976-kg (23,800-lb) thrust with afterburning.
Performance: maximum speed 1482 km/h (921 mph) at low altitude, Mach 2.5 at altitude; maximum radius of action, 4631 km (2,878 miles) with three 600 US gallon tanks; maximum rate of climb 12192 m/min (40,000 ft/min); service ceiling 19203 m (63,000 ft)
Weights: empty 12700 kg (28,000 lb); maximum take-off 25401 kg (56,000 lb)
Dimensions: span 13.05 m (42 ft 10 in); length 19.43 m (63 ft 9 in); height 5.63 m (18 ft 8 in); wing area 56.5 m² (608 sq ft)

Armament: one 20-mm M61A1 rotating-barrel gun and up to eight air-to-air missiles (normally four AIM-7F Sparrow II and four AIM-9L Sidewinder) under guidance of APG-63 pulse-Doppler radar with search range of 241 km (150 miles)

Very much at the heart of air defence forces in the United States, the F-15 has significantly upgraded force effectiveness.

Operators: United States, Israel, Saudi Arabia, Japan

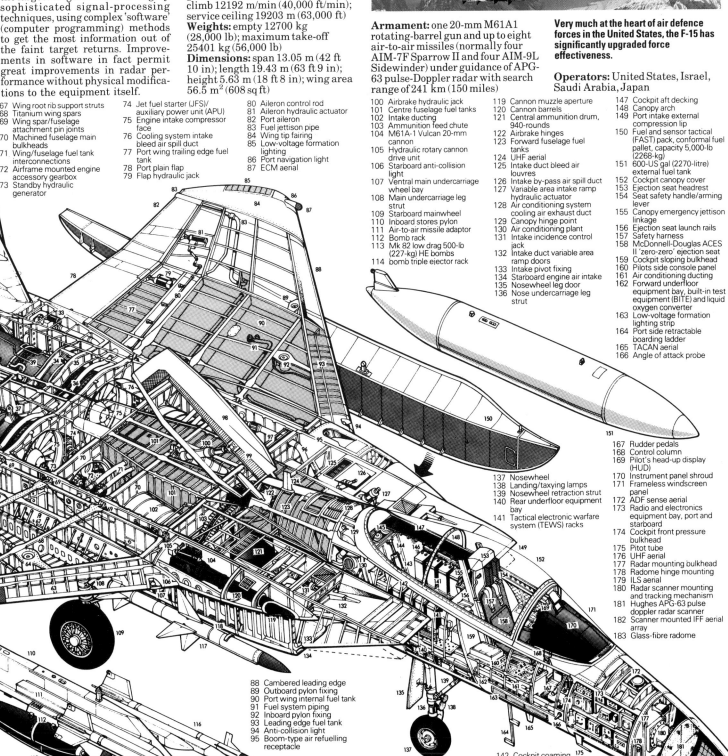

67 Wing root rib support struts
68 Titanium wing spars
69 Wing spar/fuselage attachment pin joints
70 Machined fuselage main bulkheads
71 Wing/fuselage fuel tank interconnections
72 Airframe mounted engine accessory gearbox
73 Standby hydraulic generator
74 Jet fuel starter (JFS)/ auxiliary power unit (APU)
75 Engine intake compressor face
76 Cooling system intake bleed air spill duct
77 Port wing trailing edge fuel tank
78 Port plain flap
79 Flap hydraulic jack
80 Aileron control rod
81 Aileron hydraulic actuator
82 Port aileron
83 Fuel jettison pipe
84 Wing tip fairing
85 Low-voltage formation lighting
86 Port navigation light
87 ECM aerial
88 Cambered leading edge
89 Outboard pylon fixing
90 Port wing internal fuel tank
91 Fuel system piping
92 Inboard pylon fixing
93 Leading edge fuel tank
94 Anti-collision light
95 Boom-type air refuelling receptacle
96 Bleed air duct to air conditioning plant
97 Control rod runs
98 Dorsal airbrake, open
99 Airbrake glass-fibre honeycomb construction
100 Airbrake hydraulic jack
101 Centre fuselage fuel tanks
102 Intake ducting
103 Ammunition feed chute
104 M61A-1 Vulcan 20-mm cannon
105 Hydraulic rotary cannon drive unit
106 Starboard anti-collision light
107 Ventral main undercarriage wheel bay
108 Main undercarriage leg strut
109 Starboard mainwheel
110 Inboard stores pylon
111 Air-to-air missile adaptor
112 Bomb rack
113 Mk 82 low drag 500-lb (227-kg) HE bombs
114 bomb triple ejector rack
115 Missile launch rail
116 AIM-9L Sidewinder air-to-air missile
117 AIM-7F Sparrow air-to-air missile
118 Sparrow missile launcher unit
119 Cannon muzzle aperture
120 Cannon barrels
121 Central ammunition drum, 940-rounds
122 Airbrake hinges
123 Forward fuselage fuel tanks
124 UHF aerial
125 Intake duct bleed air louvres
126 Intake by-pass air spill duct
127 Variable area intake ramp hydraulic actuator
128 Air conditioning system cooling air exhaust duct
129 Canopy hinge point
130 Air conditioning plant
131 Intake incidence control jack
132 Intake duct variable area ramp doors
133 Intake pivot fixing
134 Starboard engine air intake
135 Nosewheel leg door
136 Nose undercarriage leg strut
137 Nosewheel
138 Landing/taxying lamps
139 Nosewheel retraction strut
140 Rear underfloor equipment bay
141 Tactical electronic warfare system (TEWS) racks
142 Cockpit coaming
143 Rear pressure bulkhead
144 Canopy jack
145 Cockpit pressurization valves
146 Structural space provision for second crew member (F-15D)
147 Cockpit aft decking
148 Canopy arch
149 Port intake external compression lip
150 Fuel and sensor tactical (FAST) pack, conformal fuel pallet, capacity 5,000-lb (2268-kg)
151 600-US gal (2270-litre) external fuel tank
152 Cockpit canopy cover
153 Ejection seat headrest
154 Seat safety handle/arming lever
155 Canopy emergency jettison linkage
156 Ejection seat launch rails
157 Safety harness
158 McDonnell-Douglas ACES II 'zero-zero' ejection seat
159 Cockpit sloping bulkhead
160 Pilots side console panel
161 Air conditioning ducting
162 Forward underfloor equipment bay, built-in test equipment (BITE) and liquid oxygen converter
163 Low-voltage formation lighting strip
164 Port side retractable boarding ladder
165 TACAN aerial
166 Angle of attack probe
167 Rudder pedals
168 Control column
169 Pilot's head-up display (HUD)
170 Instrument panel shroud
171 Frameless windscreen panel
172 ADF sense aerial
173 Radio and electronics equipment bay, port and starboard
174 Cockpit front pressure bulkhead
175 Pitot tube
176 UHF aerial
177 Radar mounting bulkhead
178 Radome hinge mounting
179 ILS aerial
180 Radar scanner mounting and tracking mechanism
181 Hughes APG-63 pulse doppler radar scanner
182 Scanner mounted IFF aerial array
183 Glass-fibre radome

© Pilot Press Limited

In the hands of Israeli air force pilots the
F-15 Eagle has proved king of the skies in
the Middle East, proving its fighting
potential in the heat of battle against
Syrian forces. This example has a deadly
combination of Sidewinder and Sparrow
AAMs for close- and medium-range
attack respectively. Note the 'kill'
markings ahead of the national insignia.

695

McDonnell Douglas F/A-18 Hornet

Intended as a successor to the ubiquitous McDonnell Douglas F-4 Phantom and a replacement for the sturdy Vought A-7 Corsair, the McDonnell Douglas F/A-18 Hornet is steadily entering squadron service with the US Navy and US Marine Corps, a total of 1,366 being required for these two air arms. Plans are that this agile interceptor and strike aircraft will equip six USN and 12 USMC squadrons in the fighter role, and a further 24 and 20 units respectively in attack roles. The Hornet has performed well during its operational evaluation, turning in high serviceability, although its critics claim that the operational radius with full weapon load falls below specification. and that the project is emerging as far more costly than at first intended. Hence there are attempts in some quarters to cut back on procurement.

Coming into the world of naval operations by an unusual route, the Hornet traces its history back to the Northrop YF-17, which was passed over by the USAF in its Lightweight Fighter competition in favour of the General Dynamics F-16 Fighting Falcon. Seeing potential in the design, the US Navy adopted the YF-17, but commissioned McDonnell Douglas to develop a larger, heavier variant with increased fuel capacity, leaving Northrop as a mere associate contractor, responsible for only 40 per cent of production of the resultant F/A-18 Hornet.

Making considerable use of advanced graphite/epoxy composites in its airframe, the Hornet is optimized for high manouevrability, featuring leading-edge root extensions (LERX), leading-edge flaps and drooping aileron, plus a fly-by-wire control system. An inertial navigation system and head-up display are included in the advanced avionics suite, making the Hornet the most accurate of any carrier-based aircraft in the attack role. Provision is included for additonal sensors, such as forward-looking infra-red (FLIR), and for a broad range of current and projected weapons, giving a claimed greater 'growth' potential than the F-16. For this reason, and because of the additional security of its two engines, the Hornet has been chosen in recent competitions by the air arms of Canada, Australia and Spain. All these nations have recently operated carriers, but have no ships that can fly the Hornet.

Specification
McDonnell Douglas F/A-18A Hornet
Type: single-seat carrier-based combat/strike aircraft
Powerplant: two 7257-kg (16,000-lb) afterburning thrust General Electric F404-GE-400 turbofans
Performance: maximum speed (clean) Mach 1.8; maximum speed at intermediate power Mach 1.0; combat ceiling about 15240 m (50,000 ft); combat radius (fighter mission, unrefuelled) more than 740 km (460 miles); ferry range (unrefuelled) more than 3700 km (2,300 miles)
Weights: take-off, fighter 15234 kg (33,585 lb); take-off, fighter escort

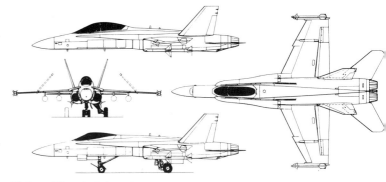

McDonnell Douglas F/A-18A Hornet (top: TF/A-18A)

Above: This Canadian Hornet displays the two-seat configuration of the TF/A-18A trainer, excellent visibility being afforded to both pilots.

Below: With greater all-round capability than any previous single-seat combat aircraft, the Hornet promises much for the US Navy and Marine Corps through to the 21st century.

15875 kg (35,000 lb); maximum
take-off 21319 kg (47,000 lb)
Dimensions: span 11.43 m (37 ft
6 in); length 17.07 m (56 ft 0 in);
height 4.66 m (15 ft 3½ in); wing
area 37.16 m² (400 sq ft)
Armament: carried on nine
external weapon stations up to a
maximum capacity of 7711 kg
(17,000 lb) 6214 kg (13,700 lb) for
high-g missions; weapons can
include Sidewinder, Sparrow or
AIM-120 air-to-air missiles, bombs,
air-to-surface missiles and rockets;
sensor pods and fuel tanks can also
be carried on these weapons
stations; inbuilt armament
comprises one M61 20-mm six-
barrel cannon in the nose
Operators: Australia, Canada,
Spain, United States

With the removal of the nose-mounted
gun and replacement by a sensor
package, the Hornet becomes an
effective recce platform. Designated RF-
18, the aircraft can quickly be converted
back to fighter/attack configuration.

McDonnell Douglas AV-8B Harrier II

For all its remarkable attributes,
the original BAe Harrier was no
more than a third-generation version of a technology demonstrator
which was to have led to a far more
advanced combat aircraft. However,
when in 1975 the British government showed no further interest in
advancing the design, McDonnell
Douglas in the USA decided to proceed with an improved Harrier, having acquired production and development rights for the aircraft as part
of the agreement under which AV-
8As were bought by the US Marine
Corps. Well pleased with their aircraft, the USMC supported work on
an upgraded aircraft, and the resultant McDonnell Douglas YAV-8B
aerodynamic prototype made its
first flight in St Louis in 1978.

With a completely new super-
critical-section wing (the main box
produced entirely in carbonfibre
composite with slotted trailing-edge
flaps), a Pegasus engine with improved front nozzles, and a raised
cockpit to improve the pilot's view,
what is now termed the AV-8B
Harrier II can carry twice the weapon load of the Harrier GR.3, and
has enormous additional advantages. Avionics are totally new, with
an advanced internal ECM system
(Zeus for the RAF) and the Hughes
Angle/Rate Bombing System. The
wing pylons have been increased in
number from four to six (US
Marines) or eight (RAF) to accommodate much greater loads of more
varied weapons. Most important is
the effortless flight of a stable and
completely safe aircraft, with no
pilot hazards or restrictions.

Harrier IIs are now in full service
with the USMC, which plans to obtain 336 for re-equipment of AV-8A
units and five more squadrons currently flying McDonnell Douglas A-
4 Skyhawks. All these will regularly
undertake shipboard training. In
addition, the RAF has placed an
order for 60, designated as the
Harrier GR.5, to augment or replace
Harrier GR.3s operating in Germany, while a further 12 will be
used by the Spanish navy from the
new carrier *Principe de Asturias*,
flying from its 'ski-jump' ramp for
take off at maximum weight. The
RAF's GR.5s are being assembled by
BAe, which also has a 25 per cent
share in manufacture of airframes

An AV-8B Harrier II of the US Marine Corps, with which it is set to fill the light-attack
role.

for the USMC and third parties.
In 1987 McDonnell expects to fly
the first TAV-8B tandem dual
trainer. This bears no direct kinship
with the BAe two-seater, and an
initial 27 are needed by the USMC to
support the Harrier II training programme. It will have limited
weapon-training capability.

Specification
**McDonnell Douglas AV-8B
Harrier II**
Type: single-seat close-support
ship- or land-based attack aircraft
Powerplant: one Rolls-Royce
Pegasus vectored-thrust turbofan

engine of 9607 kg (21,180 lb) thrust
(F402-404A) or 9752 kg (21,500 lb)
thrust (F402-406), or RAF GR.5 of
9866 kg (21,750 lb) thrust (Pegasus
Mk 105)
Performance: maximum speed,
high altitude Mach 1.1; maximum
speed at sea level 1093 km/h
(679 mph); operational radius hi-lo-
hi, short take-off run with seven
bombs plus tanks 1112 km (691
miles)
Weights: empty 5861 kg
(12,922 lb); maximum take-off, VTO
8868 kg (19,550 lb), STO 13494 kg
(29,750 lb)
Dimensions: 9.25 m (30 ft 4 in);

length 14.12 m (46 ft 4 in); height
3.55 m (11 ft 7.75 in); wing area
21.37 m² (230.0 sq ft)
Armament: two cannon pods
(USMC one 25-mm/0.98-in GAU-12/
U in one pod with ammunition in the
other, RAF two 25-mm Aden), plus
up to nine pylons for total weapon
load of 711 kg (17,000 lb)
Operators: Spain and UK (on
order), US (USMC)

The AV-8B has entered service with the
Marine Corps, one of the first units being
VMAT-203, shown here.

Messerschmitt-Bölkow-Blohm BO 105

Though it is expensive for its size, the MBB BO 105 has matured as a small helicopter of exceptional performance, agility, capability and safety. Construction of three prototypes (the first with two Allison 250-C18 turboshafts and a conventional hinged rotor, and the other two with Allison or MAN-Turbo 6022 turboshafts and hingeless main rotor units) began in 1964, the prototypes all flying in 1967 and being followed by two pre-production helicopters with uprated Allison 250-C20 turboshafts. The first of these flew on 11 January 1971 and was followed by production BO 105C helicopters. The BO 105 was the first small helicopter to offer full twin-engine safety, and all versions are available with the all-weather avionics and very comprehensive equipment. A particular feature of all versions is the rigid main rotor with a hingeless (except for the feathering hinge) hub of forged titanium carrying efficient blades of glassfibre-reinforced plastics. In the passenger role most versions seat five, though there is a lengthened six-seater and MBB in partnership with Kawasaki of Japan is also producing the 8/10-seat BK 117. Versions are being assembled in the Philippines, Indonesia and Spain, but the biggest military customer had been Federal Germany itself. The Heer (army) has 227 BO 105M (VBH) observation machines with many advanced features. A prototype is investigating further types of all-weather sights and displays. The Heer has also deployed a further 212 of the BO 105P type as the PAH-1 (anti-armour helicopter No. 1). These have six HOT anti-tank missiles, a stabilized all-weather sight above the cabin, Doppler navigation and numerous items for battlefield protection. Each army corps has an anti-tank PAH regiment with two squadrons of 28 helicopters operating in four flights of seven. A further 21 are reserved for special duties with the 6th Panzergrenadier Division. The type is also the basis of the BO 105/Ophelia (Optique Platforme

The Royal Netherlands army operates two dozen BO 105Cs, though they are flown by air force crews.

Practising 'nap-of-the-earth' flying are West German army BO 105Ps, armed with triple sets of Euromissile HOT anti-tank rounds mounted on the outriggers.

HELIcoptère Allemande) advanced experimental model. This has a mast-mounted sight (forward-looking infra-red and TV sensors and a laser rangefinder) in a spherical mounting above the rotor head, and head-up/head-down displays in the cabin. The type also has provision for helmet-mounted sights and displays, and began flight trials in 1981.

Specification
MBB BO 105P (PAH-1)
Type: anti-tank helicopter
Powerplant: two 313-kW (420-hp) Allison C20B turboshafts
Performance: maximum continuous speed 210 km/h (130 mph); mission endurance with 20-min reserve 1 hour 30 mins
Weights: empty 1322 kg (2,915 lb); maximum take-off 2400 kg

(5,291 lb)
Dimensions: main rotor diameter 9.84 m (32 ft 3⅓ in); fuselage length (plus tail rotor) 8.56 m (28 ft 1 in); height 3.0 m (9 ft 10 in); main rotor disc area 78.65 m² (846.6 sq ft)
Operators: Bahrain, Brunei, Colombia, West Germany, Indonesia, Iraq, Lesotho, Mexico, Netherlands, Nigeria, Philippines, Sierra Leone, Spain, Sudan

---

MBB/Kawasaki BK 117

During the 1970s MBB in Germany sought to develop a new helicopter from its BO 105 (the BO 107) while in Japan Kawasaki worked on the similar KH-7. In 1977 the two companies reached an agreement on joint development of a helicopter whose designation, like the helicopter itself, combines portions of both. MBB is responsible for both rotors, the important hydraulically boosted flight control system, the tailboom and tail and various minor parts. Kawasaki is responsible for the fuselage, KH-7 type transmission, fuel and electrical systems and various equipment items.

The first BK 117 was assembled at Gifu, Japan, for ground test. No. 2 flew in Germany on 13 June 1979 and No. 3 in Japan on 10 August 1979. Production helicopters began to appear from late 1981, with the civil, paramilitary (eg police) and military order books growing healthily. Though there is no duplication of manufacture (except that from 1985 PT Nurtanio of Indonesia is making the complete helicopter under licence) MBB and Kawasaki each have their own assembly line.

Like the BO 105, the advanced hingeless main rotor has glassfibre blades attached to a forged titanium

hub. Extensive sandwich panel construction, some using Kevlar composites, is used in the pod/boom fuselage. Seating can be provided for up to 10 passengers apart from the pilot, and many special role kits are offered. By late 1985 about 60 helicopters in this family had been delivered, to many kinds of customer. Numerous weapon and sensor fits have been studied, but so far the emphasis has been upon transport, including medevac and SAR configurations.

At the 1985 Paris airshow the partners (mainly in this case MBB) exhibited a new military variant, the BK 117A-3M. This was displayed with comprehensive all-weather navaids, Racal EW systems, provision for mast-mounted and helmet-mounted sights, a digital weapon control system and (as a suggested weapon fit) a Lucas chin turret with 12.7-mm (0.5-in) gun and eight HOT 2 anti-armour missiles.

Specification
MBB/Kawasaki BK 117
Type: multi-role transport helicopter
Powerplant: two 441-kW (592-shp) Avco Lycoming LTS101-650B-1 turboshaft engines

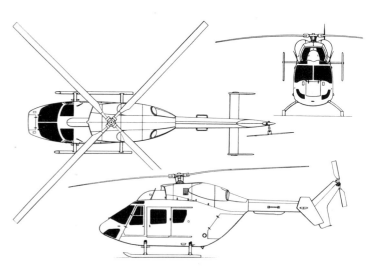

Messerschmitt-Bölkow-Blohm/Kawasaki BK 117.

Performance: maximum cruising speed 251 km/h (156 mph); economic cruising speed 230 km/h (143 mph); range at sea level with seven passengers, baggage and standard fuel 500 km (311 miles)
Weights: empty, seven seats 1658 kg (3,655 lb); maximum take-off 2850 kg (6,283 lb), with slung

load 3000 kg (6,614 lb)
Dimensions: main rotor diameter 11.0 m (36 ft 1.1 in); length (ignoring rotors) 9.98 m (32 ft 8.9 in); height 3.36 m (11 ft 0.3 in)
Armament: none, but see text for A-3M version on offer
Operator: Ciskei

Mikoyan-Gurevich MiG-15UTI 'Midget'

The Mikoyan-Gurevich MiG-15 'Fagot' survives in large numbers in its MiG-15UTI 'Midget' two-seat trainer version. Deliveries of the fighter started in 1949; China, Czechoslovakia and Poland built MiG-15s, and the latter two countries converted many single-seaters after the MiG-15 was phased out of first-line service. The MiG-15UTI then moved out of its original role as a conversion trainer and became the Eastern bloc's standard and advanced trainer. Even today it is found all over the world, mainly in third-world countries.

Specification
Type: advanced trainer
Powerplant: one 2700-kg (4,952-lb) thrust Klimov VK-1 turbojet engine
Performance: maximum speed 1076 km/h (669 mph) at 12,000 m (39370 ft); initial rate of climb 2760 m (9,055 ft)/minute; service ceiling 15500 m (50,855 ft); ferry range 2000 km (1,243 miles)
Weights: empty 3400 kg (7,496 lb); maximum take-off 5785 kg (12,754 lb)
Dimensions: span 10.08 m (33 ft 0.9 in); length 10.86 m (35 ft 7.6 in) height 3.40 m (11 ft 1.9 in); wing area 20.60 m² (221.74 sq ft)
Armament: usually one 12.7 mm (0.5-in) UBS gun; underwing hardpoints for slipper tanks or up to 500 kg (1,102 lb) of stores
Operators: Albania, Algeria, Angola, Bulgaria, Cuba, Czechoslovakia, Egypt, East Germany, Finland, Guinea, Hungary, Iraq, Mauretania, Mongolia, North Korea, Romania, Sri Lanka, Tanzania, Vietnam, North Yemen, South Yemen

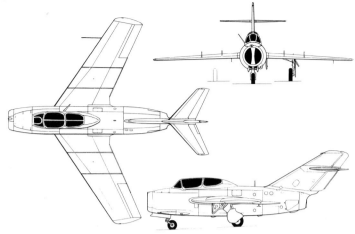

Mikoyan-Gurevich MiG-15UTI 'Midget' trainer.

Mikoyan-Gurevich MiG-17 'Fresco'

Mikoyan-Gurevich MiG-17F 'Fresco-C' of the Arab Republic of Egypt air force.

The MiG-17 was an evolutionary development of the MiG-15, primarily aimed at achieving higher Mach numbers; the MiG-15 became unstable at Mach numbers in excess of 0.92, and was a poor gun platform above Mach 0.88, so many of its adversaries could elude it in a dive. The prototype flew in January 1950, but two months later it was destroyed in a fatal accident. An extensively revised second aircraft, the SI-2, made its maiden flight in the second half of 1950, and became the prototype for the MiG-17.

The primary difference between the two aircraft was a more sharply swept wing, but there were many more changes. The tail surfaces were more sharply swept, and the rear fuselage was more gently tapered. The powerplant of the early MiG-17 was similar to that of the contemporary MiG-15bis, as was the mixed-calibre cannon armament.

Production of the MiG-17 was authorized in mid-1951. Deliveries began in late 1952, so the new aircraft was not available in time to take part in the Korean War, where the MiG-15bis distinguished itself. In fact, the improvements offered by the original MiG-17 were probably not enough to warrant interrupting production of the MiG-17 in wartime. The same applied to the near-contemporary MiG-17P 'Fresco-B', equipped with a nose-mounted radar giving some night/all-weather capability.

Major production only started when the Klimov bureau developed an augmented version of the VK-1 engine, which could not be mated to the MiG-15. The basic reheated day fighter-bomber was designated MiG-17F, and most of the subsequent production aircraft were MiG-17Fs or essentially similar aircraft; the only other variants to be produced in quantity were limited night/adverse interceptors: the MiG-17PF 'Fresco-D', with an improved version of the radar equipment fitted to the original MiG-17P, and the MiG-17PFU 'Fresco-E', which in 1956 became the first missile-armed fighter to enter Soviet service.

Full-scale production of the MiG-17 in the Soviet Union lasted only five years, before the type was superseded by the supersonic MiG-19 and MiG-21, but it has been estimated that more than 6,000 were built in that time, of which some 5,000 were MiG-17Fs. It was the most numerous type in Frontal Aviation's tactical air defence unit well into the 1960s, and remained in service until the late 1970s with training and reserve units. In addition, the MiG-17 was built under licence in Poland, under the designation LIM-5P, by the WSK-Mielec works, and at the Shenyang plant in China.

The MiG-17's claim to historical importance rests on its use by North Vietnamese Air Force (NVAF) units between 1965 and 1973. When the USAF began to plan a bombing campaign against North Vietnam in 1964, the NVAF had no fighter force; this situation began to change in late 1964, with the arrival of MiG-15s and J-5s from China. Soviet-supplied MiG-17s arrived in the following year, and along with further J-5s these were the most important NVAF fighters throughout most of the war. The MiG-15 was little used, while the MiG-21 was not available in large quantities. It came as a severe shock to the USAF and US Navy to discover that the MiG-17 was a serious match for their modern, radar-equipped, automated supersonic fighters, and this experience was instrumental in shaping the F-14, F-15 and F-16.

The number of MiG-17s in service has dwindled rapidly over the past few years, but it remains in use with Soviet combat-training units, non-Soviet Warsaw Pact reserve and fighter-bomber regiments, and many third-world air forces, including almost a dozen African air arms.

Specification
Mikoyan-Gurevich MiG-17F 'Fresco-C'
Type: single-seat fighter-bomber
Powerplant: one Klimov VK-1F turbojet of 2600 kg (5,730 lb) dry thrust and 3380 kg (7,450 lb) augmented thrust
Performance: maximum speed 1145 km/h (710 mph) at 3000 m (10,000 ft) or Mach 0.94; initial climb in 1.8 minutes to 5000 m (16,500 ft); service ceiling 16600 m (54,500 ft); ferry range 1980 km (1,230 miles) with full external fuel
Weights: empty weight 3930 kg (8664 lb); internal fuel approx. 907 kg (2,000 lb); normal take-off weight 6075 kg (13,400 lb)
Dimensions: span 9.63 m (31 ft 7 in); length overall 1.26 m (36 ft 11 in); height 3.8 m (12 ft 6 in); wing area 22.6 m² (243 sq ft)
Armament: two 23-mm NR-23 cannon with 80 rounds each, and one 37-mm N-37 with 40 rounds, in lower forward fuselage; originally four, later six underwing hardpoints for bombs or unguided rockets; points immediately outboard of wheel wells normally used for fuel tanks; later, two AA-2 'Atoll' missiles on outer pylons
Operators: Afghanistan, Algeria, Angola, Bulgaria, Congo, Cuba, Czechoslovakia, East Germany, Egypt, Equatorial Guinea, Guinea, Madagascar, Mali, Mauretania, Mongolia, Mozambique, North Korea, Poland, Romania, Somalia, Soviet Union, Syria, Vietnam North Yemen, South Yemen

This Mozambique air force MiG-17 defected to South Africa. It is seen being escorted by a Dassault-Breguet Mirage F.1AZ.

Mikoyan-Gurevich MiG-19 'Farmer' (J-6)

The Mikoyan-Gurevich MiG-19 'Farmer', the world's first production supersonic fighter, remained into production at least into 1984, and still forms the backbone of the Chinese tactical air force in its Shenyang J-6 version. In the hands of Pakistan air force pilots it has proved its worth against considerably more modern and costly opponents, with agility in combat which would do credit to a contemporary air-superiority fighter. Another good feature is the hard-hitting gun armament, with much greater projectile weight and muzzle velocity than most western 30-mm weapons.

Development started in the late 1940s, with a requirement for a new fighter designed around the newly developed Lyulka AL-5, the Soviet Union's first large axial-flow turbojet. Disappointing progress with this powerplant led to the decision to redesign the Mikoyan prototype around two small-diameter Mikulin AM-5s. The first aircraft, the I-350(M), was distinguished by a T-tail, but was destroyed in flight testing as a result of tailplane flutter. The I-350(M) was completed with a low-set tailplane, and was flown in later 1952. It was soon followed by the production MiG-19F with afterburning AM-5Fs, the first version to go supersonic in level flight, in early 1953. The initial MiG-19F and limited all-weather MiG-19PF were replaced from early 1955 by the MiG-19S, with an all-moving tailplane, refinements to flying-control and systems, and RD-9 engines (the AM-5 improved by the Tumansky bureau). The two-seat MiG-19UTI was never put into production in the Soviet Union, pilots finding little difficulty in converting from the MiG-15UTI.

The MiG-19P, with Izumrud radar in an intake bullet fairing and the inlet lip, led to the Soviet Union's first missile-armed fighter, the MiG-19PM with four K-5M 'Alkali' beam-riding air-to-air missiles replacing the guns. Production of the MiG-19 was transferred to Czechoslovakia in 1958, the Aero works producing some 850 aircraft in 1958-61. In 1961 the Chinese Shenyang works produced the first examples of an unlicensed copy of the MiG-19S, designated J-6. By 1984 several thousand J-6s had been built in China at Shenyang and Tianjin, including a few J-6Xin with pointed central radome (replacing the J-6B derived from the MiG-19PM).

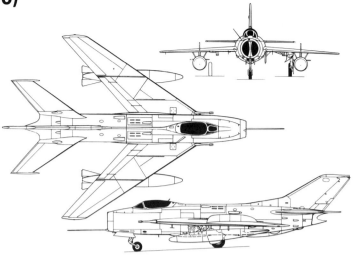

Mikoyan-Gurevich MiG-19S 'Farmer'.

Specification
Mikoyan MiG-19S (J-6)
Type: single-seat fighter-bomber
Powerplant: two 3250-kg (7,165-lb) afterburning thrust Tumansky RD-9B (J-6: Shenyang WP-6) turbojets
Performance: maximum speed, clean 1450 km/h (901 mph) at 10000 m (32,810 ft), with external tanks 1150 km/h (715 mph) at 10000 m (32,810 ft); initial rate of climb 6900 m (22,638 ft)/minute; service ceiling 17500 m (57,415 ft); ferry range, external tanks 2200 km (1,367 miles)
Weights: empty 5172 kg (11,402 lb); maximum take-off 8900 kg (19,621 lb)
Dimensions: span 9.20 m (30 ft 2.2 in); length, excluding pitot tube 14.90 m (48 ft 10.6 in); height 3.90 m (12 ft 9.5 in); wing area 25.00 m² (269.11 sq ft)
Armament: typically three 30-mm (1.18-in) NR-30 cannon each with 80 rounds, two rocket pods (often UV-16-57) on underwing pylons; Pakistani J-6s have AIM-9 Sidewinders; JZ-6, two wing guns only; J-6B radar equipped interceptor with four AAMs replacing guns; JJ-6 fuselage gun only
Operators (* = J-6 variants): Afghanistan, Albania*, Bangladesh*, Bulgaria, China, Cuba, Egypt*, Iraq*, Kampuchea*, North Korea*, Pakistan*, Tanzania*, Uganda, Vietnam*, Zimbabwe

---

Mikoyan-Gurevich MiG-21 'Fishbed'

Still in production almost 30 years after its first flight, the MiG-21 stands alongside the F-4 as one of the most important warplanes of the supersonic era. Its main attribute – simplicity and low cost – has never been lost, although the capability of the type has been steadily improved.

The design originated in a requirement for a short-range interceptor with a Mach 2 speed, issued in December 1953. (At the time, the Mach 1.3 MiG-19 was still experiencing severe control problems.) No autonomous target-detection capability was required, because the interception would be made under close ground control, and range was considered secondary to rapid climb and simplicity.

The Soviet Central Hydrodynamic Institute (TsAGI) offered two configurations for the design: a development of the MiG-19 layout,

Mikoyan-Gurevich MiG-21FL of No. 8 Sqn 'The Eight Pursoot', Indian Air Force.

with a thinner wing and more slender fuselage, and a more radical design combining a small delta wing with a conventional tailplane. TsAGI also specified that the inlet should be a nose-mounted full-cone type. Although this called for long ducts to the engine, the inlet itself was light, efficient and unaffected by airframe aerodynamic effects.

Mikoyan's designs to this requirement were so small that Western intelligence agencies over-estimated their size by 40-50 per cent until the early 1960s. In fact, the first prototypes to fly used what was virtually half the powerplant of the MiG-19 (itself not a large aircraft) – a single uprated Tumansky RD-9Ye. These were the swept-wing Ye-50 and the tailed-delta Ye-4, the latter flying in December 1955. The Tumansky R-11, around which the aircraft were designed, became available some months later, and the swept-wing Ye-2A and the delta Ye-5 flew in May and June 1956 respectively. At the end of the year, the Ye-5 was judged superior and selected for further development and production. After 30-36 months of testing and redesign, a production configuration was frozen, and the first MiG-21s were delivered in late 1958.

The first major production version

Mikoyan MiG-21MF cutaway drawing key

1 Pitot-static boom
2 Pitch vanes
3 Yaw vanes
4 Conical three-position intake centrebody
5 'Spin Scan' search-and-track radar antenna
6 Boundary layer slot
7 Engine air intake
8 'Spin Scan' radar
9 Lower boundary layer exit
10 Antennae
11 Nosewheel doors
12 Nosewheel leg and shock absorbers
13 Castoring nosewheel
14 Anti-shimmy damper
15 Avionics bay access
16 Attitude sensor
17 Nosewheel well
18 Spill door
19 Nosewheel retraction pivot
20 Bifurcated intake trunking
21 Avionics bay
22 Electronics equipment
23 Intake trunking
24 Upper boundary layer exit
25 Dynamic pressure probe for q-feel
26 Semi-elliptical armour-glass windscreen

27 Gunsight mounting
28 Fixed quarterlight
29 Radar scope
30 Control column (with tailplane trim switch and two firing buttons)
31 Rudder pedals
32 Underfloor control runs
33 KM-1 two-position zero-level ejection seat
34 Port instrument console
35 Undercarriage handle
36 Seat harness
37 Canopy release/lock
38 Starboard wall switch panel
39 Rear-view mirror fairing
40 Starboard-hinged canopy
41 Ejection seat headrest
42 Avionics bay
43 Control rods
44 Air conditioning plant
45 Suction relief door
46 Intake trunking
47 Wingroot attachment fairing
48 Wing/fuselage spar-lug attachment points (four)
49 Wing fuselage ring frames
50 Intermediary frames
51 Main fuselage fuel tank
52 RSIU radio bay
53 Auxiliary intake

54 Leading-edge integral fuel tank
55 Starboard outer weapons pylon
56 Outboard wing construction
57 Starboard navigation light
58 Leading-edge suppressed aerial
59 Wing fence
60 Aileron control jack
61 Starboard aileron
62 Flap actuator jack
63 Starboard blown flap – SPS (sduva pogranichnovo sloya)
64 Multi-spar wing structure
65 Main integral wing fuel tank
66 Undercarriage mounting/ pivot point
67 Starboard mainwheel leg
68 Auxiliaries compartment
69 Fuselage fuel tanks Nos 2 and 3
70 Mainwheel well external fairing
71 Mainwheel (retracted)
72 Trunking contours
73 Control rods in dorsal spine
74 Compressor face
75 Oil tank
76 Avionics pack
77 Engine accessories

78 Tumansky R-13 turbojet (rated at 14,550-lb/6600-kg thrust with full reheat)
79 Fuselage break/transport joint
80 Intake
81 Tail surface control linkage
82 Artificial feel unit
83 Tailplane jack
84 Hydraulic accumulator
85 Tailplane trim motor
86 Fin spar attachment plate
87 Rudder jack
88 Rudder control linkage
89 Fin structure

90 Leading-edge panel
91 Radio cable access
92 Magnetic detector
93 Fin mainspar
94 RSIU (radio-stantsiya istrebitelnaya ultrakorotkykh vol'n – very-short-wave fighter radio) antenna plate
95 VHF/UHF aerials
96 IFF antennae

97 Formation light
98 Tail warning radar
99 Rear navigation light
100 Fuel vent
101 Rudder construction
102 Rudder hinge
103 Braking parachute hinged bullet fairing

104 Braking parachute stowage
105 Tailpipe (variable convergent nozzle)

was the MiG-21F 'Fishbed-C' with greater power and other refinements. Deliveries started in 1959; the type was built under licence in Czechoslovakia and India, and a few examples supplied to China before the Sino-Soviet rift in 1960 provided the basis for the Shenyang J-7.

By the time the first MiG-21F regiments were operational, however, a considerably redesigned limited-all-weather version was under development. The initial MiG-21PF 'Fishbed-D' replaced the MiG-21F on Soviet production lines in 1962-63, was the start of a series of gradual improvements. Developments included the MiG-21FL export model, built in India; the MiG-21PFS, with blown flaps replacing the original Fowler flaps, and the later MiG-21PFM with an improved ejection system and other modifications.

The next major step was the introduction in 1965 of the MiG-21PFMA, fitted with an uprated engine, a better radar, a deeper dorsal spine, two extra wing pylons and many other improvements. Later MiG-21PFMAs, and the generally similar export-model MiG-21M, featured an internal gun. (The original MiG-21 had two internal cannon, but one was removed on the MiG-21F, and the MiG-21PF had no guns at all.) The new gun was the highly regarded twin-barrel GSh-23, a compact and very lethal weapon. An-

other member of this generation was the MiG-21R tactical reconnaissance fighter.

The next version to appear, entering service in 1969-70, was the MiG-21MF, which consolidated the evolutionary MiG-21PFMA improvements and added the new and more powerful Tumansky R-13 engine. The MiG-21RF was the equivalent reconnaissance type.

At this time, the MiG-21's range and payload, and its lack of any real air-to-ground capability, were seen as its major shortcomings. The MiG-21SMT variant was developed to remedy the situation. On the SMT, the dorsal spine went through its third design change, reaching its thickest and highest point just aft of the wing, and accommodated an extra fuel tank. By the time the MiG-21SMT reached squadron service, in 1971, priorities were chang-

Designed as a result of lessons learnt during the Korean War, the MiG-21 is a classic amongst fighters, having a rapid reaction time, impressive sortie rates and good trans/supersonic handling. Illustrated are Polish MiG-21PFMs.

ing. The MiG-23 was close to entering service, and could be adapted for ground attack, while the USAF was developing a new generation of fighters that would be much more

106 Afterburner installation	123 Ventral airbrake (retracted)	138 Port inboard weapons pylon
107 Afterburner bay cooling intake	124 Trestle point	139 UV-16-57 rocket pod
108 Tailplane linkage fairing	125 ATO assembly release solenoid (front mounting)	140 Port mainwheel
109 Nozzle actuating cylinders	126 Underwing landing light	141 Mainwheel outboard door section
110 Tailplane torque tube	127 Ventral stores pylon	142 Mainwheel leg
111 All-moving tailplane	128 Mainwheel inboard door	143 Aileron control linkage
112 Anti-flutter weight	129 Splayed link chute	144 Mainwheel leg pivot point
113 Intake	130 Twin 23-mm GSh-23 cannon installation	145 Main integral wing fuel tank
114 Afterburner mounting	131 Cannon muzzle fairing	146 Flap actuator fairing
115 Fixed tailplane root fairing	132 Debris deflector plate	147 Port aileron
116 Longitudinal lap joint	133 Auxiliary ventral drop tank	148 Aileron control jack
117 External duct (nozzle hydraulics)	134 Port forward air brake (extended)	149 Outboard wing construction
118 Ventral fin	135 Leading-edge integral fuel tank	150 Port navigation light
119 Engine guide rail	136 Undercarriage retraction strut	151 Port outboard weapons pylon
120 ATO assembly canted nozzle	137 Aileron control rods in leading edge	152 'Advanced Atoll' IR-homing AAM
121 ATO assembly thrust plate forks (rear mounting)		153 Wing fence
122 ATO assembly pack		154 Radio altimeter antenna

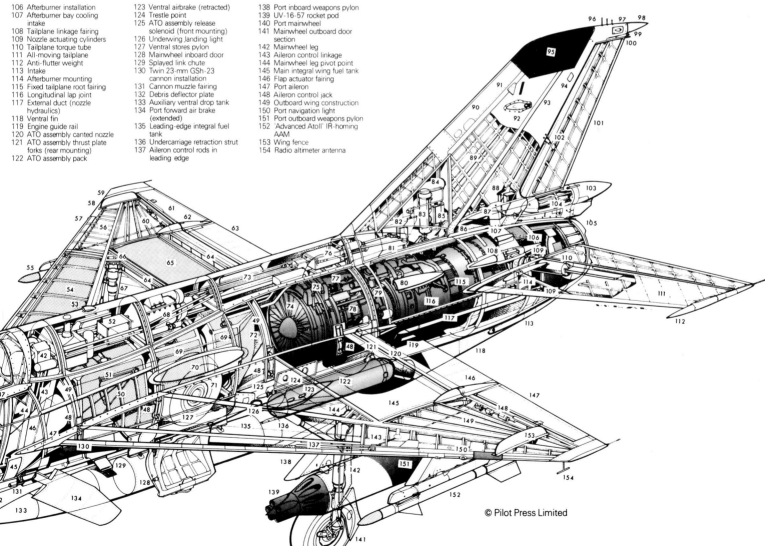

© Pilot Press Limited

Mikoyan-Gurevich MiG-21 'Fishbed' (continued)

agile than the new swing-wing type. Production of the MiG-21SMT was therfore terminated in favour of a new version optimized for the air-to-air mission.

It is probable that many of the improvements observed on this model, the MiG-bis, had been originally developed for the MiG-21SMT. They included a thoroughly revised structure with increased internal fuel capacity and longer life, and many systems changes. The most important, however, was the Tumansky R-25 engine, offering a relatively vast thrust increase combined with better handling at altitude. According to some sources, this engine was introduced with the later MiG-21Mbis; however, it is fitted to the MiG-21bis aircraft which are built by HAL in India, and the MiG-21Mbis probably features other changes.

The new MiG-21s have thrust/weight ratios in the class of the F-15, making them potentially dangerous adversaries despite their dated aerodynamic design. While the MiG-21's sustained turn rate is not up to the latest standards, instantaneous turn rates are reasonably good, thanks to the moderately loaded delta wing, and climb rate

Mikoyan-Gurevich MiG-21PFM 'Fishbed-F' serving with the Iraqi air force.

and acceleration are excellent. Increased thrust also makes it practical to carry four AAMs for combat missions: the type has been seen with radar-guided AA-2-2 'Advanced Atolls' on the outer pylons, and short-range AA-8 'Aphid' dogfight missiles inboard. Warload/radius and endurance are as poor as ever, if not worse. However, the type's primary role is tactical air defence: the fighters are scrambled on warning of an attack and vectored by ground control towards the incoming strike force, so the mission is short in time and distance. It is likely that the latest aircraft will stay in this role for some time, while the much more complex MiG-29 is assigned to more demanding missions.

Specification
Mikoyan-Gurevich MiG-21Mbis
Type: single-seat tactical air defence fighter
Powerplant: one Tumansky R-25 turbojet of approximately 5900 kg (13,000 lb) dry thrust and 9000 kg (19,850 lb) augmented thrust
Performance: maximum speed clean 2230 km/h (1,385 mph) or Mach 2.1 at 11,000 m (36,000 ft); maximum speed at sea level just over Mach 1; initial climb rate 284 m/sec (58,000 ft/min); combat radius, less than 320 km (200 miles)
Weights: empty 6200 kg (13,500 lb); normal take-off 8750 kg (19,300 lb); maximum take-off 10000 kg (22,000 lb)
Dimensions: span 7.16 m (23 ft 6 in); length overall 15.75 m (51 ft

9 in); height 4.5 m (14 ft 9 in); wing area 22.9 m^2 (247 sq ft)
Armament: one internal twin-barrel, 23-mm GSh-23 cannon, plus four wing hardpoints for up to 1500 kg (3,300 lb) of ordnance, including AA-2 'Atoll' and AA-8 'Aphid' AAMs, AS-7 'Kerry' ASMs and unguided rockets or bombs; centreline pylon for drop tank; outboard wing pylons also 'wet'
Operators: Afghanistan, Algeria, Angola, Bangladesh, Bulgaria, Cuba, Czechoslovakia, East Germany, Egypt, Ethiopia, Finland, Hungary, India, Iraq, Laos, Libya, Madagascar, Mongolia, Nigeria, North Korea, Poland, Romania, Somalia, South Yemen, Soviet Union, Sudan, Syria, US Air Force, Vietnam, Zambia

Mikoyan-Gurevich MiG-23 'Flogger'

The design of the MiG-23 started in the early 1960s, when Frontal Aviation – the Soviet tactical air defence/strike force – was taking delivery of its first MiG-21s and Su-7s. Competitive with Western contemporaries in speed and manoeuvrability, these types were badly lacking in warload, range and equipment. FA's commanders therefore put a high priority on firepower, range and equipment in defining the next tactical fighter. The requirement was issued before the air battles over Vietnam, and did not call for greatly improved agility in air combat over the MiG-21.

The Mikoyan bureau produced two prototype aircraft to the FA's specification. One of these resembled an enlarged MiG-21, fitted with two lift-jet engines in the centre-section. The other, designated E-231, was a straightforward application of the then-fashionable variable-gometry (VG) formula. It is probable that both of them made their first flights in late 1966 or early 1967; by the middle of that year, the lift-jet-equipped design had been abandoned in favour of the E-231. The E-231 was followed by a small pilot batch of closely similar 'Flogger-A' fighters.

Like contemporary Western VG aircraft, the E-231 used outboard wing pivots and powerful all-moving stabilizers to solve the trim problems inherent in variable sweep. Unlike earlier MiG designs, it had side inlets, making room for a reasonable internal fuel capacity and avionics. The outer wings were hydraulically driven from 16 deg to 72 deg sweep – an intermediate 45 deg position could be selected, but full-range, automatic control was not provided. In the interests of simplicity, there were no swivelling wing pylons, and the wing high-lift devices were confined to full-span, three-section plain flaps and drooped outboard leading edges. The type featured a unique main gear design, swinging upwards into the fuselage sides, and a large folding ventral fin.

Flight-tests of the E-231 soon revealed serious stability and control problems, as is evident from drastic modifications made to the basic design. The tailerons and dorsal fin were moved some 100 cm (40 in) rearwards, and the wings were redesigned: the new outer wings featured large leading-edge extensions, terminating in bat-like 'claws' at the junction with the fixed glove. The modifications were incorporated in the first major production version, the MiG-23MF 'Flogger-B', which entered service in 1973.

The MiG-23MF was equipped with new weapons and avionics, and represented a vast improvement over the MiG-21 in that respect. Glove pylons carried a pair of large air-to-air missiles, codenamed AA-7 'Apex' by NATO, while belly pylons were fitted for two very small, highly manoeuvrable AA-8 'Aphid' missiles. (Up to 1976, however, MiG-23MFs were often seen with launch rails for the older AA-2 'Atoll', which may have been carried as an interim weapon.) In the lower fuselage, between the intakes, was a GSh-23 cannon. The main attack radar was the Soviet Union's first pulse-Doppler fire-control radar, named High Lark by NATO. It may be partly based on Westinghouse AWG-10 technology, obtained from F-4J Phantoms shot down over Vietnam. A laser rangefinder, probably for use in air-to-air gunnery, was mounted under the nose. More antennae, housed in the tip of the fin, the ends of the 'claws' and the forward fuselage, betrayed the presence of comprehensive electronic countermeasures and surveillance equipment.

A number of improvements have been introduced on the MiG-23 during production. The original R-27 engine has been replaced by the more powerful R-29B, and provision for drop-tanks has been made beneath the outer wings. (These pylons do not swivel, so the pylons must be jettisoned for high-speed flight.) NATO applies the designation 'Flogger-G' to MiG-23MFs produced

Mikoyan MiG-23MF 'Flogger-G' cutaway drawing key

1 Pitot tube
2 Radome
3 'High Lark' radar scanner dish
4 Radar dish tracking mechanism
5 ILS antenna
6 Avionics cooling air scoop
7 Radar and avionics equipment bay
8 Ventral doppler antenna
9 Yaw vane
10 Air data probe
11 SRO-2 'Odd-Rods' IFF antenna
12 Armoured windscreen panel
13 Head-up-display
14 Instrument panel shroud
15 Radar 'head-down' display
16 Instrument panel
17 Rudder pedals
18 Angle of attack transmitter

19 Laser rangefinder housing
20 Nosewheel steering unit
21 Torque scissor links
22 Pivoted axle beam
23 Twin aft-retracting nosewheels
24 Nosewheel spray/debris guard
25 Shock absorber strut
26 Nosewheel doors
27 Hydraulic retraction jack
28 Control column

29 Ejection seat firing handles
30 Wing sweep control lever
31 Engine throttle control lever
32 Pilot's ejection seat
33 Electrically heated rear view mirror
34 Ejection seat headrest
35 Upward hingeing cockpit canopy cover
36 Canopy jack
37 Starboard air intake
38 Canopy hinge point
39 Boundary layer splitter plate
40 Boundary layer bleed air holes

41 Port engine air intake
42 Intake internal flow fences
43 Retractable landing/taxiing lamp, port and starboard
44 Temperature probe

45 Variable area intake ramp doors
46 Boundary layer bleed air ejector

47 Avionics equipment bay
48 ADF sense aerial
49 Boundary layer air duct
50 Forward fuselage fuel tank

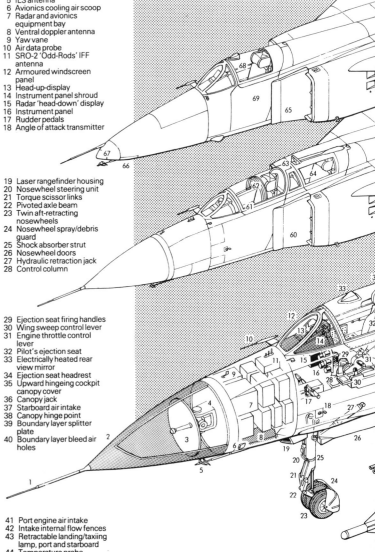

Mikoyan-Gurevich MiG-23 'Flogger' (continued)

since 1978, which are distinguished by a shorter dorsal fin and a modified nose landing gear housed under bulged doors.

Recent improvements are reported to include improved head-up displays; new MiG-23s, like modern Western fighters, dispense with a full-time radar scope in the cockpit. Since 1982, 'Flogger-G' has also been seen in a six-missile configuration, with a twin-missile adapter, probably for AA-8s, replacing the usual AA-7 rail on the glove pylons. The High Lark/AA-7 combination has also been improved; the MiG-23MF is believed to have 'some ability to engage low-flying targets'

according to the Department of Defense, and the type is now used alongside the Su-15 by Soviet PVO air defence units as well as by Frontal Aviation.

Along with the MiG-27 'Flogger-D' (see separate entry), the MiG-23

was produced at unprecedented rates. The Soviet Union's output of 'Floggers' reached a peak rate of nearly 500 aircraft a year in the mid-1970s. By the end of 1982, it was estimated that some 3,500 of the type had been built. The result was

Mikoyan-Gurevich MiG-23MF 'Flogger-B' of the East German air force.

51 Ventral cannon ammunition magazines
52 Ground power connections
53 Intake suction relief doors
54 Weapons system electronic control units
55 SO-69 Sirena 3 radar warning antennae
56 Fuselage flank fuel tanks
57 Wing glove fairing
58 Starboard Sirena 3 radar warning antennae
59 176-gal (800-litre) jettisonable fuel tank
60 Nose section of MiG-23U 'Flogger-C' two-seat tandem trainer variant
61 Student pilot's cockpit
62 Folding blind-flying hood
63 Rear seat periscope, extended
64 Instructor's cockpit

65 MiG-23BN 'Flogger-F' dedicated ground attack variant
66 Ventral radar ranging antenna
67 Laser ranger nose fairing
68 Raised cockpit section
69 Armoured fuselage side panels
70 Wing leading edge flap, lowered
71 Starboard navigation light
72 Wing fully forward (16-deg sweep) position
73 Port wing integral fuel tank, total internal fuel capacity 1,265 gal (5750 litres)
74 Full span plain flap, lowered
75 Starboard wing intermediate (45-deg sweep) position

76 Starboard wing full (72-deg sweep) position
77 Two-segment spoilers/lift dumpers
78 Non-swivelling, jettisonable wing pylon (wings restricted to forward swept position)
79 Wing glove sealing plate
80 Wing pivot bearing
81 Wing pivot box carry-through unit, welded construction
82 VHF aerial
83 Wing sweep control screw jacks
84 Fin root fillet
85 Rear fuselage fuel tank
86 Tumansky R-29B afterburning turbojet engine
87 Afterburner duct cooling air scoop

88 Cut-back fin root fillet (some 'Flogger-G' aircraft)
89 Tailplane control and hydraulic equipment bay
90 Starboard all-moving tailplane
91 Tailfin
92 Short wave ground control communications aerial
93 UHF aerial
94 ILS aerial
95 Sirena 3 tail warning radar
96 ECM aerials
97 Tail navigation light
98 Static discharger
99 Rudder
100 Rudder hydraulic actuators, port and starboard
101 Brake parachute housing
102 Split conic fairing parachute door

103 Variable area afterburner nozzle
104 Fixed tailplane tab
105 Static discharger
106 Port all-moving tailplane
107 Afterburner nozzle control jacks (six)
108 Tailplane pivot bearing
109 Tailplane hydraulic jack
110 Airbrakes (four), upper and lower surfaces
111 Airbrake hydraulic jack
112 Afterburner duct
113 Ventral fin, folded (undercarriage down) position
114 Ventral fin control jack
115 Lower UHF aerial
116 Ventral fin down position

117 Engine accessory equipment bay
118 Wing root seal
119 Port spoilers/lift dumpers
120 Flap guide rails
121 Port plain flap
122 Fixed spoiler strips
123 Static discharger
124 Port navigation light
125 Leading edge flap, lowered
126 Port wing integral fuel tank
127 Wing pylon mounting rib
128 Extended chord sawtooth leading edge
129 Port mainwheel
130 Mainwheel door/debris guard
131 Shock absorber strut
132 Hinged axle beam
133 Articulated mainwheel leg strut

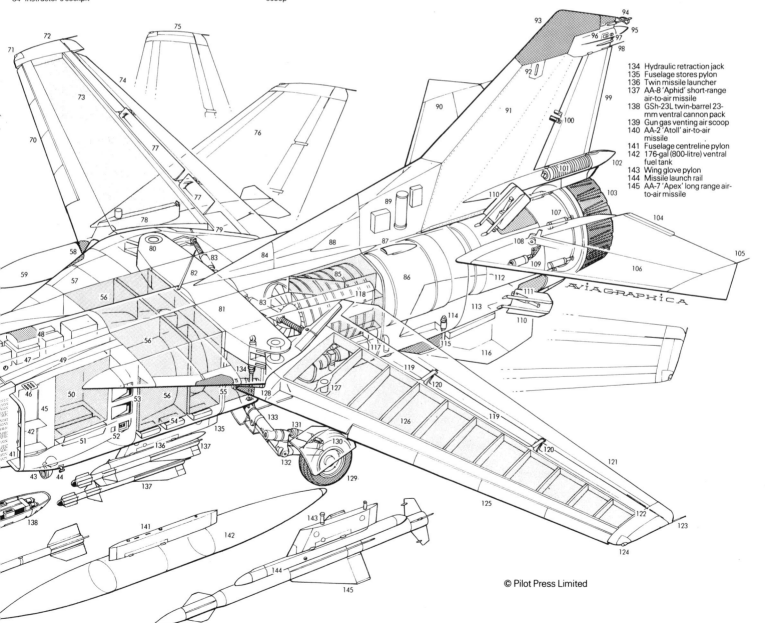

134 Hydraulic retraction jack
135 Fuselage stores pylon
136 Twin missile launcher
137 AA-8 'Aphid' short-range air-to-air missile
138 GSh-23L twin-barrel 23-mm ventral cannon pack
139 Gun gas venting air scoop
140 AA-2 'Atoll' air-to-air missile
141 Fuselage centreline pylon
142 176-gal (800-litre) ventral fuel tank
143 Wing glove pylon
144 Missile launch rail
145 AA-7 'Apex' long range air-to-air missile

Keith Fretwell

A Libyan MiG-23 'Flogger-E', armed with AA-2 'Atoll' missiles. In its first major action in the Middle East, in 1982, the MiG-23 proved no match for the F-15 and F-16, though this was largely because of great disparity in pilot tactics and skill. Israeli pilots consider the 'Flogger' inferior to the MiG-21 as a dogfighter, while the export version shown here has similar radar and weapons.

that Soviet and allied forces were able to re-equip very fully and quickly with the MiG-23 family, and it would be fair to say that the Mikoyan fighter was the main element in a rapid improvement in Warsaw Pact airpower in the late 1970s.

Other members of the MiG-23 family stem from the MiG-23MF and MiG-27. The first of these to appear was the MiG-23U 'Flogger-C', a two-seat trainer version of the MiG-23MF with a slightly raised rear cockpit and a smaller radar. The MiG-23U is used as the trainer for the MiG-27 and other attack variants as well as for the MiG-23 fighters.

Middle Eastern allies of the Soviet Union were the first export customers for the MiG-23; Egypt, Libya and Syria took delivery of their first batches in the course of 1975. All of them, however, received deliberately downgraded export models of the type. The first of these was the 'Flogger-E' fighter, which differed from the MiG-23MF in having a much smaller radar – probably related to the 'Jay Bird' system fitted to later MiG-21s – and in being armed with AA-2 'Atoll' missiles. More recently, however, export customers appear to have received the more potent MiG-23MF.

A hybrid version of the family, combining MiG-23 and Mig-27 features, was first seen in 1975 and was at first thought to be a MiG-27 development aircraft. It soon became apparent, however, that it was a strike aircraft for non-Soviet forces, and it is designated MiG-23BN. It has the nose of the MiG-27, and the same duct-mounted pylons and enlarged mainwheels. However, it has the inlets, exhaust and gun of the MiG-23. The initial version was codenamed 'Flogger-F' by NATO; the version supplied to Eastern European air forces, with additional antennae, is identified as 'Flogger-H'. This version is also being built under licence in India, alongside the 'Flogger-C' trainer.

MiG-23s have seen action in the war between Iraq and Iran, and between Syria and Israel. Syria's

'Flogger-Es' suffered heavy losses in action against the Israeli Kfirs and F-16s in June 1982, but this was ascribed to tactics rather more than technology. Generally, the MiG-23 is regarded as a well-equipped and capable aircraft, very much less agile than the F-16 but somewhat faster and, currently, carrying a heavier, longer-range armament; this imbalance exists largely because the USAF has elected to wait for the new AIM-120 AMRAAM instead of integrating the earlier Sparrow missile with the F-16. So far, though, the MiG-23 and MiG-27 have been available in far greater numbers than the F-16.

An unexpected operator of the MiG-23 is the US Air Force, which is known to maintain several of the type at a base in Nevada. Obtained from Egypt after its rift with the Soviet Union, they are used for the development of air combat tactics. Egypt has also supplied one or two MiG-23s to China.

Production of the MiG-23/27 is believed to have slowed in the last few years, making way for the introduction of more advanced types. The

MiG-29 may be destined to replace it; the DoD believes that it may be a true dual-role fighter like the F-18 or F-16, designed to perform fighter and attack missions without modifications. But the MiG-23 series unquestionably has a few years of production, and a great many years of service, still to come before the new types take over.

Specification
Mikoyan-Gurevich MiG-23MG 'Flogger-G'
Type: single-seat tactical fighter and interceptor
Powerplant: one Tumansky R-29B turbofan of 8000 kg (17,500 lb) dry thrust and 11500 kg (25,350 lb) augmented thrust
Performance: maximum speed 2500 km/h (1,550 mph) or Mach 2.35; maximum speed at sea level 1350 km/h (840 mph) or Mach 1.1; combat radius, with four AAMs and external fuel, 930 km (575 miles)
Weights: empty approximately 11300 kg (25,000 lb); normal take-off 17250 kg (38,000 lb); maximum take-off 18500 kg (41,000 lb)
Dimensions: spread span 14.25 m

The pylons visible on this MiG-23MF can carry a variety of air-to-air missiles and rocket packs. As the standard model of the 'Flogger' in Warsaw Pact service, the MiG-23MF forces are considerable.

(46 ft 9 in); swept span 8.3 m (27 ft 2 in); length including probe 18.25 m (59 ft 10 in); height 4.35 m (14 ft 4 in); wing area 37.2 m² (400 sq ft)
Armament: one twin-barrel 23-mm GSh-23 cannon in lower fuselage; two medium-range AA-7 'Apex' AAMs, or four short-range AA-8 'Aphid' AAMs, on glove pylons, plus two AA-8s under intake ducts
Operators: Algeria, Bulgaria, Cuba, Czechoslovakia, East Germany, Egypt, Ethiopia, Hungary, India, Iraq, Libya, Romania, Soviet Union, Syria, US Air Force

A variety of camouflage schemes are seen on these Indian Air Force MiG-23BN 'Flogger-H' aircraft, this model having two small avionics pods forward of the nosewheel doors.

Mikoyan-Gurevich MiG-25 'Foxbat'

One of the best known warplanes of modern times, the MiG-25 is an example of brilliant improvization to meet a menace that never fully materialized. Its remarkable use of available technology to meet an advanced requirement at moderate cost is an object lesson to all designers and planners.

The requirement for the MiG-25 emerged in the late 1950s. American designers were apparently making vast strides in the design of high-speed aircraft. The first transonic fighters had entered service in 1955, but work on a number of aircraft capable of cruising at Mach 3 at very high altitude was already well under way. These would render conventional Mach 2 interceptors useless and, quite possibly, outpace the technology of surface-to-air missiles.

The Soviet air defence organization, PVO-Strany, therefore developed a requirement for an interceptor to operate at high speeds and altitudes, carrying a powerful radar to detect and track the target (to which it would be directed by ground control) and missile armament adequate to ensure its destruction. The catch was that the aircraft would have to be cheap enough to be acquired in hundreds, and in service before the US B-70, which was then scheduled to enter service in 1964.

The final cancellation of B-70 production did not affect the programme greatly: Soviet intellignce may have been aware of the development of the Lockheed A-12 reconnaissance aircraft, and of its close cousin, the RS-71 reconnaissance-bomber. Before the new Ye-266 flew (some time in 1964), A-12s were probing Soviet airspace.

The Mikoyan team was freed from the usual central control of configuration, and sensibly copied one of the best supersonic aircraft of the day: the North American A-5 Vigilante. Even the twin fins of the Ye-266 had appeared on the Vigilante mock-up. The main differences were that the Ye-266 had a proportionally smaller wing and larger engines. The design was well thought out, and provided natural stability up to Mach 3 combined with normal low-speed behaviour.

Internally, however, the design was unique. Lacking experience with titanium, or with the characteristics of stainless steel, the designers elected to build the aircraft from arc-welded nickel-steel alloy. A major problem with any Mach 3 aircraft – sealing fuel tanks despite thermal expansion – was solved by installing a complex but quite efficient system of thin-walled

welded fuel tanks.

The engine design reflected absolute dedication to the design mission. Very large, very simple and of very low pressure ratio, and fitted with a massive afterburner, the Tumansky R-31 is optimized for high speed flight. At high Mach numbers, most of the compression is done in the inlet, while most of the combustion takes place in the afterburner. The 'core' engine is sized to provide acceleration at lower speeds. Fuel consumption, however, is phenomenally high at all times.

The avionic systems were dominated by the radar, known to NATO as Fox Fire. This was not designed for long detection range: the automated ground control system was to take the fighter within tens of miles of the target. Rather, it was designed to 'burn through' heavy jamming, and was designed using vacuum-tube technology to generate the necessary power.

Development was slow, and not entirely trouble-free; problems of high-speed stability and engine reliability were encountered, and several prototypes were lost. However, performance targets were being met by 1967 – the type appeared in public in that year, and later set a series of world records – and the aircraft entered service, as the MiG-25, in the course of 1970. Initially, it was armed with the AA-5 'Ash' developed for the Tu-28P, but this was replaced by the definitive weapon, the vast AA-6 'Acrid', in the mid-1970s.

The MiG-25 was not a complete success as an interceptor. Its sustained ceiling is much less than that of the only high-speed, high-altitude threat in existence, the SR-71, and its missiles cannot quite 'snap-up' far enough to hit the USAF aircraft. Against low-altitude penetrators, its powerful radar and missiles were ineffective. Most of the MiG-25 'Foxbat-A' interceptors built for the PVO were converted in the late 1970s and

An unusual view of a Libyan MiG-25 interceptor illustrates the massive afterburner nozzles for the Tumansky R-31 turbojet engines, these providing the 'Foxbat' with its high-speed performance.

early 1980s to the 'Foxbat-E' configuration, with AA-7 'Apex' missiles and the latest version of the High Lark radar, to augment the Soviet Union's force of look-down, shoot-down interception systems. Other 'Foxbat-A' interceptors have been supplied to allied countries, but have had no military success.

Perhaps because of the lack of high-altitude targets, some MiG-25s were completed as tactical reconnaissance aircraft. Two versions exist – the camera-equipped MiG-25R 'Foxbat-B' and the 'Foxbat-D' for electronic reconnaissance. Soviet MiG-25Rs based near Cairo were used to overfly Israel in 1971 and 1975, at first enjoying immunity from attack, but the Israelis acquired F-15s and improved Sparrow missiles, and also refined their tactics; in the 1982 air battles over the Lebanon, two Syrian MiG-25s fell victim to snap-up attacks by F-15s. The fourth production version of the type is the MiG-25U 'Foxbat-C' conversion trainer, with a student's cockpit in the space normally occupied by the Fox Fire radar. The MiG-25U is believed to be the Ye-133, which has captured a number of women's world speed records; if this is the case, it indicates that the trainer is somewhat slower than the

operational aircraft.

Development of an improved version of the MiG-25 was started in the early 1970s, to meet the threat of the B-1 bomber. Originally known as the MiG-25M, this aircraft has been redesignated MiG-31, and is described in detail below.

Specification
Type: single-seat interceptor
Powerplant: two Tumansky R-31 turbojets of 9300 kg (20,500 lb) dry thrust and 12300 kg (27,120 lb) augmented thrust
Performance: maximum speed 3010 km/h (1,870 mph) or Mach 2.82; service ceiling 24000 m (78,000 ft); initial climb rate 208 m/sec (41,000 ft/min); interception radius 400 km (250 miles)
Weights: empty approximately 20000 kg (44,000 lb); maximum take-off 37500 kg (82,500 lb)
Dimensions: span 14.0 m (46 ft); length without probe 22.3 m (73 ft 2 in); height 5.60 m (18 ft 5 in); wing area 56 m^2 (605 sq ft)
Armament: four AA-6 'Acrid' AAMs or ('Foxbat-E') AA-7 'Apex' AAMs on wing pylons
Operators: Algeria, India, Iraq, Soviet Union, Syria

Mikoyan-Gurevich MiG-25U 'Foxbat-C' two-seat trainer of the Indian Air Force.

Mikoyan-Gurevich MiG-27 'Flogger'

A specialized ground-attack version of the MiG-23 fighter, with which it shares the NATO reporting name 'Flogger', the Mikoyan-Gurevich MiG-27 differs from its companion in major respects. The most significant of these are the fixed-geometry air inlets and afterburner nozzles which indicate a simplicity of design permitted by the low-level mission requirement of only transonic speed and high agility compared with Mach 2.35 attainable at high altitudes by the MiG-23. First of the MiG-27 variants, the 'Flogger-D' features a 'duck nose' sharply tapered downwards from the windscreen and containing a laser ranger and marked-target seeker, while a six-

barrel 23-mm gun is installed in the belly. The pilot's view is improved for attack missions by a raised seat and canopy, there are heavy slabs of cockpit armour and low-pressure tyres are compatible with rough-field operation. Issued in quantity to the Soviet tactical air force, Frontal Aviation, the 'Flogger-D' has an export counterpart in the MiG-23BN 'Flogger-F' which retains the original 12500 kg (27,558 lb) R-29 afterburning turbojet (plus variable inlets and nozzle) and twin-barrel cannon. The similar MiG-23BN

The MiG-27 is distinguishable from the MiG-23BN by having small fixed splitter plates and a rotary cannon.

'Flogger-H', another 'high-speed' variant, is distinguished by two avionics pods astride the lower fuselage, forward of the nosewheel doors. In 1981 the MiG-27 'Flogger-J' was first noted, differing from the 'Flogger-D' in further nose revision, with a lip on the top and a blister fairing beneath. Podded guns on two wing pylons have barrels which can be depressed for attacking ground targets without recourse to a steep dive. Recent models, including 'Flogger-J', have wing-root strakes and kinked taileron edges. India has negotiated a production licence for 'Flogger-J' and was expected to begin manufacture in 1985.

Specification
MiG-27 'Flogger-D'
Type: single-seat ground attack and tactical nuclear strike aircraft
Powerplant: one 11500-kg (25,353-lb) thrust Tumansky R-29B afterburning turbojet engine
Performance: maximum speed Mach 1.7 at altitude; maximum speed at sea level Mach 1.1; service ceiling 16000 m (52,495 ft); combat radius, lo-lo-lo mission with four 500-kg (1,102-lb) bombs and two AA-2s 390 km (242 miles)
Weights: loaded, clean 15500 kg (34,172 lb); maximum take-off 18000 kg (39,683 kg)
Dimensions: span, fully extended, 16° sweep 14.25 m (46 ft 9 in), fully swept, 72° 8.17 m (26 ft 9.7 in);

length 16.00 m (52 ft 5.9 in); height 4.35 m (14 ft 3.3 in); wing area, unswept 27.26 m^2 (293.43 sq ft)
Armament: one fixed six-barrel 23-mm (0.91-in) gun; five weapon pylons and two rear fuselage racks for stores including AS-7 'Kerry' ASMs and self-defence AA-2 'Atoll'

AAMs up to a maximum weight of 3000 kg (6,614 lb)
Operators: Afghanistan, Algeria, Bulgaria, Cuba, Czechoslovakia, Egypt, Ethiopia, East Germany, Hungary, India, Iraq, Libya, Poland, Soviet Union, Vietnam

Distinguishing features of the MiG-27 'Flogger-D' include the window covering the nose-mounted laser rangefinder and marked target seeker (LRMTS), and a bullet-shaped antenna on each glove pylon leading-edge.

Mikoyan-Gurevich MiG-29 'Fulcrum'

Prototypes of the MiG-29 air-combat fighter were first seen by US satellites at Ramenskoye test centre in 1979, being dubbed first 'Ram-L' and later receiving the code name 'Fulcrum'. Since then the plan view has been known in the West with some accuracy, but only in 1984 was it possible to add some authenticity to the side view. One thing never in doubt was that the MiG-29 is a very formidable aircraft. Its selection by the Indian Air Force, which had previously bought the Dassault Mirage 2000 and is not unaware of the supply of General Dynamics F-16s to Pakistan, speaks for itself. As in the case of the MiG-21 almost 25 years ago, the MiG fighter was judged the best choice after prolonged evaluation.

Every inch a thoroughbred, the MiG-29 has many family trade marks and yet is in all respects a totally fresh design. In plan it has something in common with the McDonnell Douglas F/A-18 Hornet,

and is very similar in size, weight and engine power, though the Soviet engines are more powerful. The broad tapered wings have powerful variable camber, are mounted high and carry four of the six main missile pylons. From the outset the MiG-29 has been equipped with a new-technology pulse-doppler radar conferring complete look-down shoot-down capability. This fighter can operate over large areas, completely independent of the close ground control previously imposed on Soviet fighters.

It is to be expected that much of the airframe will be of composite materials, and that digital fly-by-wire flight controls will be used. The tailerons, used for both pitch and roll, are pivoted to beams projecting aft of the wing outboard of the engines. Between the engines, which are parallel as in the Grumman F-14, is room for much of the large fuel capacity. A particularly striking feature is the great depth of

the variable inlets, extending far below the bottom line of the fuselage and giving outstanding engine behaviour at extreme angles of attack. It is widely believed that this aircraft can outmanoeuvre the F-16 at similar conditions of external load (which was one of the design requirements).

Deliveries began in early 1984, by which time over 100 aircraft were flying. Both the PVO air-defence forces and the Frontal tactical air armies of theatre forces use the MiG-29, with secondary capability in surface attack. A tandem two-seat version has been forecast, for pilot conversion and EW (electronic warfare), but had not been positively identified by late 1985. So far as is known all Indian deliveries, from November 1985, are single-seaters.

Specification
Type: air-combat and multi-role fighter
Powerplant: two Tumansky R-33D

augmented turbofan engines each rated at 5100-kg (11,245-lb) thrust dry and 8300-kg (18,300-lb) with full augmentation
Performance: (estimated) maximum speed, clean at high altitude Mach 2.2 or 2337 km/h (1,452 mph); speed at sea level with six AAMs 1300 km/h (808 mph); combat radius 800 km (500 miles)
Weights: (estimated) 7825 kg (17,250 lb); maximum take-off 16500 kg (36,275 lb)
Dimensions: (estimated) span 10.25 m (33 ft 7.5 in); length 15.50 m (50 ft 10.2 in); height 5.25 m (17 ft 2.7 in)
Armament: two large guns in upper part of wing roots, one each side, with muzzles far aft of cockpit, and six AA-10 medium-range radar-guided AAMs, four under wings and two on fuselage inlet chines
Operators: India, Soviet Union

Mikoyan-Gurevich MiG-31 'Foxhound'

Dubbed 'Foxhound' by NATO, this development of the MiG-25 first flew prior to 1978 but was apparently not identified in US satellite imagery until 1983. By then the MiG-31 was in front-line regiments, and a plant at Gorkii (possibly GAZ-21) had delivered about 150 by late 1985.

At first glance the MiG-31 appears to be a tandem two-seat MiG-25, but in fact it is an almost completely new design. Whereas the MiG-25 sacrificed manoeuvrability and much else for sheer speed, the new fighter is a far more versatile and useful aircraft, though it is believed it has no provision for attack on surface targets. The airframe is stressed to much higher factors than the MiG-25, though the MiG-31 is not intended for close dogfighting. Instead it is a stand-off interceptor with full

look-down shoot-down capability, using an exceptionally powerful and capable pulse-doppler radar totally unrelated to the old-technology radar of most MiG-25s. The radar and electronic-warfare systems are managed by the backseater, who like the pilot has a zero/zero seat under an upward-hinged canopy.

The engines are almost certainly the same as those of the MiG-25M 'Foxbat-E', but with longer afterburners which may give reduced drag or higher thrust at supersonic speeds. This does not affect overall aircraft length, and what has not yet been explained is why the Western estimate of overall length should be less than that of the MiG-25, despite the fact (reported by Washington, and fairly obvious) that the fuselage ahead of the wing is longer. The

wings have leading edge root extensions but no tip pods (which in the MiG-25 contain Sirena 3A receivers and anti-flutter masses). Like the Sukhoi Su-27 this interceptor has great combat persistence, with eight large AAMs, though how these differ from the AA-10s of the Su-27 is not yet known. There is no evidence of an internal gun, which in such an aircraft seems very unlikely, and how the MiG-31 carries its eight large missiles has also not yet been made public.

Specification
Type: long-range stand-off interceptor
Powerplant: believed two 14000 kg (30,864-lb) thrust Tumansky R-31F afterburning turbojets
Performance: (estimated)

maximum speed, clean Mach 2.4 or 2251 km/h (1,585 mph); service ceiling 24400 m (80,000 ft); combat radius with eight AAMs 1500 km (932 miles)
Weights: (estimated) empty 21825 kg (48,115 lb); maximum take-off 41150 kg (90,720 lb)
Dimensions: (estimated) span 14.00 m (45 ft 11.2 in); length 23.50 m (77 ft 1.2 in), without nose pitot 22.50 m (73 ft 9.8 in); height 6.10 m (20 ft 0 in); wing area 58.0^2 (624 sq ft)
Armament: eight AA-9 long-range radar-guided AAMs
Operator: Soviet Union

Mil Mi-4 'Hound' (Z-5)

The prototype Mil Mi-4 was completed in April 1952. It adopted the layout of the Sikorsky S-55, with the powerplant in the nose and quadricycle landing gear, but added a pair of clamshell loading doors capable of admitting a small military vehicle or most light infantry weapons such as anti-tank guns. It was a far more capable and powerful transport than any Western contemporary; production Mi-4s soon introduced all-metal (instead of wooden-skinned) rotor blades and several thousand were built. An amphibious version was tested in 1959, and the Mi-4V for high altitude operation has a two-stage supercharger. The Mi-4 was also put into production at the Harbin plant in China as the Zhi-5, popularly named Syuan Fen (whirlwind). About 350 are air force transports and 50 are coastal ASW and SAR machines. Some are being re-engined with PT6T twin turboprops.

The Mi-4 became one of the first armed helicopters, with a machine-gun in the nose of the navigator's gondola and rocket pods on outriggers. With the naval AV-MF the Mi-4 carried radar, a MAD 'bird' and sonar as an anti-submarine warfare aircraft. Many have been converted as communications and command platforms for land warfare, while others have sprouted lateral Yagi aerial arrays and other antennae in the EW jamming role: Mi-4s in this latter role are identified by NATO as the 'Hound-C'.

Specification

Type: transport helicopter
Powerplant: one 1268-kW (1,700-hp) Shvetsov ASh-82V 14-cylinder radial piston engine
Performance: maximum speed 210 km/h (130 mph); at 1500 m (4,920 ft); cruising speed 160 km/h (99 mph); service ceiling 6000 m (19,685 ft); normal range 400 km (249 miles)
Weights: empty 5356 kg (11,808 lb); maximum take-off 7800 kg (17,196 lb)

Dimensions: main rotor diameter 21.00 m (68 ft 10.8 in); fuselage length 16.80 m (55 ft 1.4 in); height 5.18 m (17 ft 0 in); main rotor disc area 346.36 m^2 (3,728.31 sq ft)
Armament: one 7.62-mm (0.3-in) machine-gun in ventral gondola, and rocket or gun pods; ASW versions carry two depth charges or torpedoes
Operators: Afghanistan, Albania, Algeria, Bulgaria, China, Cuba, Czechoslovakia, East Germany, Egypt, Ethiopia, Hungary, India, Iraq, North Korea, Mali, Mongolia, Mozambique, Poland, Romania, Somalia, South Yemen, Syria, Soviet Union, Vietnam

Features similar to those of the S-55 are visible in this view of an Afghan Republican air force Mi-4, approximately a dozen of which are in service in the continuing war in Afghanistan.

Mil Mi-4 'Hound-B' anti-submnarine warfare helicopter.

Mil Mi-6 'Hook' and Mi-10 'Harke'

Announced in 1957, the Mil Mi-6 'Hook' was the first Soviet helicopter with turbine powerplant and, until the Mi-26, was by far the largest helicopter in the world. Arising from a joint civil/military need for a heavy lift helicopter, the Mi-6 has the ability to carry light armoured vehicles in its cabin with access through clamshell doors at the rear. It can be fitted with large removable wings, to unload the rotor when cruising and allow STOL take-offs with a greater payload than can be lifted vertically. The normal load is limited to around 12000 kg (26,455 lb) but, like most modern helicopters, it can also carry external cargo although the limit imposed is usually 9000 kg (19,842 lb). Approximately 70 troops can be carried, or 41 stretcher casualties. Two flying crane derivatives, the Mi-10 and Mi-10K, are known by the NATO reporting name 'Harke'. The Mi-10 has a massive quadricycle landing gear which enables it to straddle and lift bulky loads in the order of 15000 kg (33,069 lb); the Mi-10K has a shorter landing gear, and an aft-facing crew station beneath the nose from which a member of the crew monitors load positioning. Only small numbers of Mi-10s were built, whereas production of the Mi-6 exceeded 800, about half of which remain in Soviet military service.

Specification
Mil Mi-6 'Hook'
Type: heavy transport helicopter
Powerplant: two 4101-kW (5,500-shp) Soloviev D-25V turboshaft engine
Performance: maximum speed 300 km/h (186 mph); cruising speed 250 km/h (155 mph); service ceiling 4400 m (14,435 ft); range with 12000-kg (26,455-lb) payload 200 km (124 miles); range with 4000-kg (8,818-lb) payload 1000 km (621 miles)
Weights: empty 27240 kg (60,055 lb); maximum, vertical take-off 42500 kg (93,696 lb)
Dimensions: main rotor diameter 35.00 m (114 ft 10 in); fuselage length 33.18 m (108 ft 10.3 in) span, detachable wings 15.30 m (50 ft 2.4 in); height 9.17 m (30 ft 1 in); main rotor disc area 962.12 m^2 (10,356.46 sq ft)
Armament: optional manually-aimed 12.7-mm (0.5-in) DShK gun in the nose
Operators (military): Algeria, Angola, Egypt, Ethiopia, Iraq, Peru, Somalia, Syria, Vietnam

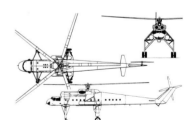

Heavy-lift duties among the Egyptian air force helicopter force are assigned to a small force of Mi-6 'Hooks'.

Mil Mi-10 'Harke'

Mil Mi-8 and Mi-17 'Hip'

First flown in 1961, the Mi-8 is used widely by all Warsaw Pact forces, chiefly in Mi-8T multirole variants used by tactical units. It is broadly comparable with the Sikorsky S-61 family, although all versions are much heavier and their usefulness is increased by rear doors for bulky loads such as artillery and missile launchers. Typical loads include 28 equipped troops or 12 stretchers.

The Mi-8's armament is intended for fire-suppression during assault landings under fire. It is large for aggressive use in combat, and less

The highly successful Mil Mi-8 'Hip' transport and assault helicopter family includes the basic transport 'Hip-C' seen here with 128×57-mm rockets in externally-mounted packs.

manoeuvrable than a specialized attack helicopter. However, a broadside of unguided missiles may prove effective in keeping the defenders' heads down while the helicopters land and unload their cargoes, by far the most dangerous point of a mission. There are many specialized versions, including communications relay platforms, electronic-intelligence variants and ('Hip-J and -K') tactical jammers. Mi-8 minesweepers were ferried to Egypt in 1974 to help clear the Suez Canal. The Mi-8 is not normally operable from ships.

Today the Mi-8 and more powerful Mi-17 family far outnumber all other large helicopters in worldwide use. With powers and weights considerably greater than those of the S-61 Sea King family, they are doing a tremendous job all over the world, and most of those exported have been paid for. Total production from Kazan and Ulan-Ude is about 11,000, and was continuing in 1985.

Specification
Mil Mi-17 'Hip-H'
Type: transport helicopter
Powerplant: two 1417-kW (1,900-shp) Isotov TV3-117MT turboshaft engines

Performance: maximum speed 240 km/h (155 mph); maximum cruising speed 240 km/h (149 mph); service ceiling 3600 m (11,810 ft); range, standard fuel and 5 per cent reserves 465 km (289 miles)
Weights: empty 7100 kg (15,653 lb); maximum take-off 13000 kg (28,660 lb)
Dimensions: main rotor diameter 21.29 m (69 ft 10.2 in); length rotors turning 2535 m (83 ft 2 in); height 4.75 m (15 ft 7 in); main rotor disc

area 356.0 m² (3,832.08 sq ft)
Armament: normally up to four 16× 57-mm (2.24-in) rocket pods on fuselage pylons; assault versions up to four triple UV-16-57s, for 192 rockets, plus four AT-2 'Swatter' anti-tank rockets, plus one 12.7-mm (0.5-in) gun aimed from cockpit
Operators (Mi-8/-17): Algeria, Angola, Bangladesh, Bulgaria, China, Cuba, Czechoslovakia, East Germany, Egypt, Ethiopia, Finland, Guinea-Bissau, Hungary, Iraq,

Front-line resupply and casevac missions are the main duties for the Pakistani Mil Mi-8Ts, a general-purpose model which can carry freight or up to 24 passengers.

Jugoslavia, Laos, Libya, Mali, Mongolia, Mozambique, Nicaragua, North Yemen, Peru, Poland, Romania, Somalia, South Yemen, Soviet Union, Sudan, Vietnam, Zambia

Mil Mi-14 'Haze'

Also known to the design bureau as the V-14, the Mil Mi-14 'Haze' is a large twin-turbine helicopter derived from the mass-produced Mi-8. It has the more powerful TV3 type engines of the Mi-17 and Mi-24. The fuselage is different from any other Mil helicopter, and has much in common with the Sikorsky S-61R, the boat-hulled transport version of the Sea King, in that it is fitted with twin-wheel bicycle landing gear with the main units retracting into rear sponsons which also incorporate water keels. The Mi-14 was designed in the 1960s as the replacement for the piston-engine Mi-4 'Hound' as the Soviet AV-MF (naval air force) shore-based ASW and multi-mission helicopter. It it is too large for convenient operation from ships, where the Ka-25 'Hormone' and Ka-27 'Helix' are used. The numbers deployed are small, the estimate in mid-1985 being only 100 for the AV-MF. The cockpit houses the pilot and co-pilot, who have comprehensive navaids including doppler, search radar with 360° surveillance, a radar altimeter and full de-icing equipment, though there are no inlet screens on the engines, which in the large main cabin, which in

Clearly derived form the Mi-8, the Mi-14 'Haze' anti-submarine warfare helicopter has a sealed hull, and main landing-gear units which retract into the sponsons.

transport versions can be equipped with 32 seats, a mission crew of at least four sit round a tactical display served by the radar, a towed MAD (magnetic anomaly detection) 'bird' and dipping sonar. The hull is watertight and has limited amphibious capability, but the Mi-14 is not intended for sustained operations from the open sea. Some examples have additional radio aerials and a few have a rescue hoist above the large sliding door on the left side. Fuel is housed in large tanks along the sides under the main floor, and possibly also in the rear sponsons, leaving the central compartment free for use as a weapon bay with

belly doors. Details of what can be carried are not yet known, and no Mi-14 has been seen with externally mounted weapons. In addition to the basic ASW version, it is probably that a transport version exists for utility and SAR roles. Because of its size, this model would probably not be suitable for the vertrep (vertical replenishment of ships) role.

Specification
Type: shore-based ASW (possibly also anti-ship) helicopter
Powerplant: two 1641-kW (2,200-shp) Isotov TV3-117MT turboprop engines
Performance: (estimated)

maximum speed 260 km/h (162 mph); maximum cruising speed 240 km/h (149 mph); range with full mission load 500 km (311 miles)
Weights: (estimated) empty 8000 kg (17,637 lb); maximum take-off 12000 kg (26,455 lb)
Dimensions: (estimated) main rotor diameter 21.29 m (69 ft 10.2 in); length rotors turning 25.30 m (83 ft 0 in); height 5.65 m (18 ft 6.4 in); main rotor disc area 356.00 m² (3,832.08 sq ft)
Armament: certainly includes AS homing torpedoes and/or depth charges; may also include anti-ship missiles
Operators: Bulgaria, Cuba, East Germany, Libya, Poland, Soviet Union

Mil Mi-24 'Hind'

The Mil Mi-24 'Hind' is a large intimidating armed helicopter, and has been a cause for controversy and a source of puzzlement in the West since it was first observed in 1973. At first it was thought to be a straightforward armed version of the Mi-8, but it soon became clear that the new helicopter was rather smaller than its predecessor, although apparently using the same engines. The early 'Hind-A' appeared to be a conventional squad-carrying helicopter with the addition of rocket pods and missile rails. This sort of combination had been experimentally used by the US Army in Vietnam, but had led to the development of specialized armed heli-

Mil Mi-24 'Hind-A' anti-tank helicopter of the Algerian air force, one of 40-plus such machines in service.

copters with automatic turreted armament and small silhouette, designed to escort the troop carrier, while other helicopters armed with guided weapons took on the enemy

armour. The 'Hind-A', however, appeared to combine all three elements into one unwieldy package: a troop carrier with guns for self-defence, but equipped with rockets for defence suppression and anti-tank missiles for attacking enemy armour.

The conundrum of how the Mi-24 was to be used became even more

Mil Mi-24 'Hind' (continued)

perplexing with the arrival in 1975 of the 'Hind-D', adding to the earlier versions' armament a highly complex nose gun installation. The 'Hind-D' also features a more heavily protected cockpit, considerably less spacious than that of its predecessor. One fear was that the gun armament is intended for use against NATO's own anti-tank helicopters in Western Europe. By early 1985 it had been estimated that more than 1,000 'Hinds' were in service, both 'Hind-D' and 'Hind-E' (with AT-6 missile capability) being in volume production.

The Mi-24 seems to combine the powerplant and transmission of the Mi-8 with a smaller rotor and airframe, retaining the fan-cooled transmission characteristic of large Mil turbine helicopters. The cabin is considerably smaller than that of the 28-seater Mi-8, but should be able to accommodate a 12-man infantry section without difficulty, off-loading them via a large side door forward of the anhedralled stub wings. The latter carry missiles on downward tip extensions, presumably to allow easy reloading from ground level while carrying the wing spar above the cabin. The rough-field landing gear is retractable.

The 'Hind-D' forward fuselage features two tandem blown canopies on separate cockpits, reducing the chance of both crewmen being disabled with one hit. The windscreens are made of flat armour glass. In the extreme nose is a turret mounting a four-barrel gun, possily of 12.7-mm calibre, although most sources report that it is larger. Aft of the turret are two installations: a blister vry similar to that under the nose of the 'Hind-A' and a larger installation which appears to contain a sensor slaved in elevation to the gun. This may be an assisted gunsight (either infra-red or TV). A large low-airspeed probe juts from the forward (gunner's) windscreen.

Now entering service on the 'Hind-E' is a new heavy anti-tank missile designated AT-6 'Spiral' by NATO, possibly weighing as much as 90 kg (200 lb) per round and with a 10-km (6-mile) range. It is likely to be laser-guided, with semi-active seeking, rather than wire-guided like the AT-2 'Swatter' previously carried by 'Hind-D'. It is also believed to be tube-launched, and it is possible that more than four could be

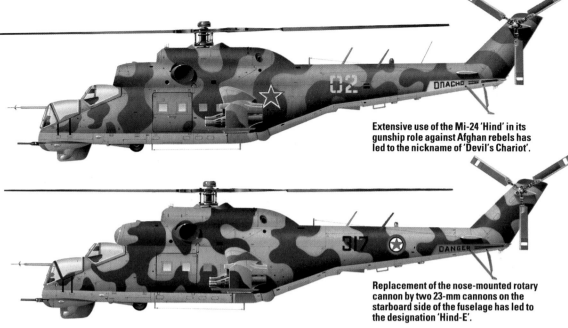

Extensive use of the Mi-24 'Hind' in its gunship role against Afghan rebels has led to the nickname of 'Devil's Chariot'.

Replacement of the nose-mounted rotary cannon by two 23-mm cannons on the starboard side of the fuselage has led to the designation 'Hind-E'.

carried on one helicopter. A new 'Hind' version with twin large calibre cannon on the starboard side of the nose is also entering service.

Performance figures for the Mi-24 are difficult to assess, but records established by Soviet women pilots in a helicopter known as the 'A-10' may give a clue. Given that the Mi-24 has as much power as the larger Mi-8, the performance of the 'A-10' is roughly what might be expected. However, its power/weight ratio is considerably less than that of the latest US armed helicopters, and with its relatively old-technology rotor system (similar to that of the Mi-8) the Mi-24 is not likely to be agile. Its large size compared with the Western ideal of a combat helicopter will also make it vulnerable to hostile fire. A surprising feature of the design, which will adversely affect its survivability, is its complete lack of infra-red signature suppression; the exhausts are open from all aspects.

The Mi-24 has been described as a 'helicopter battle-cruiser' and this may not be too bad a summing-up of what the machine does. Its main advantage is its ability to fight in several different ways: by dropping an anti-tank platoon, complete with missiles, while defending itself against ground fire with the nose gun (or in the case of the 'Hind-A', with side guns); by acting as its own escort on troop-carrying flights; or by acting as a tank-killer pure and simple, with a vast capacity for even the heaviest reload rounds. An inevitable corollary of this 'combination of all arms' in a single aircraft,

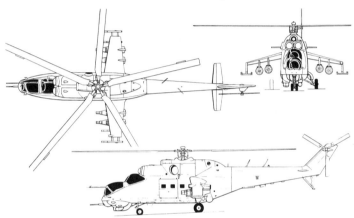

Mil Mi-24 'Hind-D' gunship helicopter

however, is that the vehicle's size and weight rule out evasive flying, and render it difficult to escape alert and well-equipped defences. The type has seen much action in Afghanistan.

Specification
Type: ('Hind-A to -C') assault helicopter and ('Hind-D') gunship
Powerplant: two 1119-kW (1,500-shp) Isotov TV-2 turboshafts
Performance: maximum speed 320 km/h (200 mph); cruising speed 260 km/h (160 mph); service ceiling 550 m (18,000 ft)
Weights: empty 6500 kg (14,000 lb); loaded 10000 kg (22,000 lb)
Dimensions: main rotor diameter 17 m (56 ft); length of fuselage 17 m (56 ft); height 4.25 m (14 ft); main rotor area 227 ^{2} (2,463 sq ft)
Armament: ('Hind-D') four-barrel cannon of 15.5- or 20-mm calibre in nose turret; (all versions) up to four pods each containing 32 57-mm rockets, plus up to four anti-tank missiles on stub wings; the 'Hind-A' has nose- and side-mounted guns
Operators: Afghanistan, Algeria, Bulgaria, Cuba, Czechoslovakia, East Germany, Ethiopia, India, Iraq, Laos, Libya, Mozambique, Nicaragua, Peru, Poland, South Yemen, USSR

The Mi-24 houses the weapons system officer in the front cockpit, while the pilot sits above and behind him. The probe ahead of the canopy frame is an air data sampling antenna.

The 'Hind's' usual weapon load includes 57-mm rocket pods. In Afghanistan, some rockets have carried chemical warheads.

Mil Mi-26 'Halo'

After M.I. Mil's death in 1970 the new bureau leader, M.N. Tishchyenko, scrapped the clumsy V-12 monster with side-by-side rotors and designed a totally new and far better machine with two of the new D-136 engines. Designated Mil Mi-26 (NATO codename 'Halo'), it could loosely be regarded as a modernized Mi-6 with engines of twice the power, and with a much bigger fuselage.

The remarkable main rotor has eight blades, each with a steel-tube spar and glassfibre/Nomex honeycomb aerofoil profile. Like the tail rotor it has electric de-icing, and though gross weight is almost 50 per cent greater than that of the Mi-6 the main rotor diameter is actually less. The well-streamlined fuselage is equipped for a flight crew of four plus a loadmaster in charge of the main hold, which is 15 m (49 ft 2.6 in) long including the rear ramp, 3.25 m (10 ft 8 in) wide and 3 to 3.17 m (9 ft 10 in to 10 ft 4.8 in) high, with comprehensive computerized loading using two electric winches. BMPs and all other tactical airborne vehicles can be carried, and there are 40 foldidng seats along the sides. Another 60 seats can be installed down the centre of the hold if necessary.

Fuel is housed beneath the floor, and under the rear of the flight deck is an APU which maintains hydraulic, electric and air-conditioning systems on the ground. The neat fixed landing gear has twin wheels; the main gears can 'kneel' individually to adjust rear door loading height and level the sill on uneven ground, and the main oleos transmit a signal of gross weight to a readout on the engineer's panel. There is provision for slung loads, and the very comprehensive avionics include weather radar, doppler and automatic hover flight control. Closed-circuit TV is provided to assist accurate positioning of slung loads.

In 1982 Mi-26s set various world records with payloads up to 25 tonnes (55,115 lb), which was lifted to 4100 m (13,451 ft). Use of a forged titanium main rotor hub and 'in-house' design of the main gearbox saved many tonnes and enabled the Mil bureau to meet the tough requirement that empty weight should not be more than half the maximum. Flight development in 1979-82 was remarkably trouble-free, and by early 1983 a military V-TA regiment was already being equipped. Since then some dozens of these outstanding helicopters have entered both V-TA and Aeroflot ser-

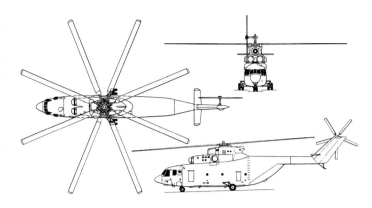

vice. There is nothing remotely in the same class in the West.

The Mi-26 'Halo' offers a significant increase to Soviet transport forces.

Specification
Type: heavy transport helicopter
Powerplant: two 8500-kW (11,400-shp) Lotarev D-136 turboshaft engines
Performance: maximum speed 295 km/h (183 mph); cruising speed 255 km/h (158 mph); hovering ceiling out of ground effect 1800 m (5,900 ft); range with maximum fuel and reserves 800 km (497 miles)

Weights: empty 28200 kg (62,170 lb); maximum take-off 56000 kg (123,459 lb)
Dimensions: main rotor diameter 32.00 m (104 ft 11.8 in); length ignoring rotors 33.727 m (110 ft 7.8 in); height to top of main hub 8.145 m (26 ft 8.7 in)
Armament: usually none
Operator: Soviet Union

Mil Mi-28

In 1983 the existence of a new Soviet armed attack helicopter was the subject of rumours in the West, and in 1984 the NATO codename 'Havoc' was allocated to this machine, which is clearly from the Mil bureau and has been identified by the US Department of Defense as the Mi-28. The US DoD produced a crude outline drawing which looked like a copy of the AH-64 Apache, but in 1985 issued a much more detailed illustration which showed several strange features.

As might be expected the whole machine, apart from the forward fuselage, appears similar to the Mi-24, but such commanality might later prove to be totally incorrect. Indeed the 'Havoc' may prove to be a much smaller and more agile helicopter than the large and massive Mi-24 series, and a British Army

expert, Maj Stewart, has said publicly that instead of using the TV3 engine the most likely powerplant is 'a new engine in the 895 to 1119 kW (1,200 to 1,500 shp) range'.

Since most information at present probably comes from satellite imagery one must accept as correct the Washington artist's positioning of the tail rotor on the right as on the first Mi-24s (but not all later versions). The main rotor has been drawn the same as on the Mi-24, but the engines are shown on the sides of the upper fuselage, indicating a totally different arrangement of free-turbine drive to the gearbox. No oil cooler duct is intended, and the two jetpipes curve up (no doubt to minimize detectable heat emission). Instead of retractable landing gear a fixed tailwheel arrangement is shown, as on Western attack heli-

copters.

Pilot and gunner are shown in stepped cockpits with flat-plate canopies, well back from the nose in a fuselage of extraordinary depth and bulk, presenting a huge target to defending sensors and spoiling forward view. On top of this huge nose is a black pimple which is clearly a radar or FLIR. In the chin position there appears to be a glazed gondola. In due course the 'Havoc', which may well prove to be something other than the Mi-28, will become better known. Much of the current US illustration may be found as wide of the mark as in similar drawings of new Soviet fixed-wing aircraft.

Specification
Type: attack helicopter
Powerplant: possibly two 1641-kW

(2,200-shp) Isotov TV3 type turboshaft engines (but see text)
Performance: maxium speed is hardly likely to be much ledss than 322 km/h (200 mph); no reliable information
Weights: no reliable information, and clearly hard to guess if engine power could be anything from 1641 to 3356 kW (2,200 to 4,500 shp)
Dimensions: no reliable information, though probably significantly smaller than the Mi-24
Armament: the Pentagon artist has shown a large single-barrel gun in a ventral turret, weapon wings of the type carried by the Mi-24 with twin AS-6 Spandrel missile tubes at the tips and inboard pylons carrying what looks like a quad HOT missile pack

Mitsubishi F-1

This close-support fighter was an adaptation of the T-2 trainer, the second and third productioon T-2s being converted to prototype F-1s. The first flew on 3 June 1975, with the second following four days later. In these aircraft the rear cockpit accommodated a fire-control system and test equipment in place of the instructor, and in summer 1975 the prototypes were delivered to the Air Proving Wing at Gifu for service testing. The first production aircraft flew 16 June 1977 and subsequently 73 were delivered by 1984.

Behind the cockpit is an avionics compartment housing a bombing computer and Ferranti inertial navigation system. The fin carries radar warning aerials and the avionics fit includes dual UHF, IFF/SIF, head-up display, radio altimeter, attitude and heading reference system, air-data computer, Tacan and a radar warning and homing system. Mitsubishi Electric supplied the multi-mode radar and the fire-control system and bombing

Mitsubishi F-1 assigned to the 3rd Air Wing, JASDF at Misawa AB.

computer.

The first F-1 unit was the 3rd *Hiko-tai*, which converted from the F-86 at Misawa AB in spring 1978. The JASDF's original plans called for an FS-X replacement to be ordered in the early 1980s, but lack of money has prevented this. The F-1s will, instead, be put through a "life-extension programme".

Specification
Type: close-support fighter
Powerplant: two 3239 kg (7,140-lb)

thrust Ishikawajima-Harima TF40-IHI-801A afterburning turbofans (licence-built Rolls-Royce Turboméca Adour)
Performance: maximum speed Mach 1.6 or 1700 km/h (1,056 mph) at 11000 m (36,090 ft); initial rate of climb 6000 m (19,685 ft); service ceiling 15200 m (49,870 ft); combat radius 1814-kg (4,000 lb) warload and external fuel 350 km (217 miles)
Weights: empty 6358 kg (14,017 lb); maximum take-off 13700 kg (30,203 lb)

Dimensions: span 7.88 m (25 ft 10.2 in); length 17.85 m (58 ft 6.8 in); height 4.39 m (14 ft 4.8 in); wing area 21.17 m² (227.88 sq ft)
Armament: one 20-mm (0.79-in) JM-61 (Vulcan) cannon; 2720 kg (5,997 lb) of external stores, including bombs, rockets, drop-tanks or Mitsubishi ASM-1 air-to-surface missiles; wingtip attachments for up to four Sidewinder or Mitsubishi AAM-1 air-to-air missiles
Operator: Japan (JASDF)

Mitsubishi MU-2/LR-1

Mitsubishi's MU-2 first flew on 14 September 1963 and remains in production. Some 50 examples serve with the Japanese armed forces. Most are based on the short-fuselage MU-2B, the JASDF receiving 27 MU-2S aircraft for the Air Rescue Wing. This variant is believed still to be in production, as is the JGSDF LR-1, used for reconnaissance and utility transport, with provision for two nose-mounted 12.7-mm (0.5-in) machine-guns and SLAR (Side-Looking Airborne Radar) or camera packages. At least 19 have been delivered and this version remains in limited production. The MU-2J is a long-fuselage model for navaid calibration; four serve with the JASDF Flight Check Group at Iruma.

Zaïre obtained at least three MU-2Bs from the USA, these being employed on light transport duties.

Specification
Mitsubishi MU-2S

Type: light SAR aircraft
Powerplant: two 533-kW (715-shp) Garrett TPE331-25A turboprop engines
Performance: maximum cruising speed at 3050 m (10,000 ft) 474 km/h (295 mph); initial rate of climb

671 m (2,200 ft)/minute; range with reserves 2100 km (1,305 miles)
Weights: empty 2560 kg (5,644 lb); maximum take-off 4560 kg (10,053 lb)
Dimensions: span 11.93 m (39 ft 1.7 in); length 10.70 m (35 ft 1.3 in);

height 3.94 m (12 ft 11.1 in); wing area 16.55 m^2 (178.15 sq ft)
Armament: none
Operators: Japan, Zaïre

Search-and-rescue duties are assigned to the Mu-2S in Japanese service.

Mitsubishi T-2

This supersonic trainer bears more than a passing resemblance to the Jaguar, and is powered by the same engines. Mitsubishi was selected as prime contractor in September 1967, and the first XT-2 flew on 20 July 1971. The flight test programme was completed in March 1974, and the T-2 entered JASDF service in 1975. All 86 aircraft had been delivered by March 1981; of these 28 are T-2s, the other 58 being T-2As fitted with a JM-61 20-mm cannon for gunnery training, and two were converted to FST-2 *kai* standard as prototypes of the F-1. The T-2A can carry three 833-litre (183-Imp gallon) droptanks, and has wingtip attachments for air-to-air missiles. Other stores can be carried on the three standard pylons if required for use in the armed training role. All have a simple ranging radar and Thomson-CSF head-up display. One T-2 has been rebuilt as a CCV (control-configured vehicle) research aircraft.

Specification
Type: two-seat advanced and weapons trainer
Powerplant: two 3239-kg (7,140-lb) afterburning thrust

Ishikawajima-Harima Heavy Industries TF40-IHI-801A turbofans (licence-built Rolls-Royce Turboméca Adours)
Performance: maximum speed Mach 1.6 or 1700 km/h (1,056 mph) at 11000 m (36,090 ft); initial rate of climb 6000 m (19,685 ft)/minute; service ceiling 1520 m (50,035 ft); maximum ferry range, external tanks 2590 m (1,609 miles)

Weights: empty 6307 kg (13,905 lb); loaded, clean 9675 kg (21,330 lb); maximum take-off 12900 lb (28,440 lb)
Dimensions: span 7.88 m (25 ft 10.2 in); length 17.85 m (58 ft 6.8 in); height 4.39 m (14 ft 4.8 in); wing area 21.17 m^2 (227.88 sq ft)
Armament: (T-2A) JM-61 (Vulcan) 20-mm (0.79-in) cannon; optional wingtip attachments for AAMs, and

Designed to meet JASDF combat training requirements, the Mitsubishi T-2 borrows much from Western trainer designs, and is currently serving as the T-2 advanced trainer and T-2A combat trainer.

three pylons beneath wings and fuselage
Operator: Japan (JASDF)

Morane-Saulnier MS.760 Paris

In January 1953 Morane-Saulnier flew the prototype MS.755 Fleuret jet trainer with side-by-side seating, which formed the basis for the MS.760 Paris, flown on 29 July 1954. A four-seat liaison aircraft, the Paris was the forerunner of the executive jet. By spring 1961 137 had been ordered, including 48 sets of components for assembly at the Argentine government factory in Cordoba; Brazil received 12 for training, 10 for liaison and eight for aerial photography. The Paris II, with 480-kg (1,058-lb) thrust Marboré VI turbojets, superseded the Paris I in production and by 1964 a total of 165 of the two series had been built, in addition to those assembled in Argentina.

France still has about 20 with the Armée de l'Air at Aix, Bordeaux, Metz and Villacoublay. The Aéronavale has nine, at Landivisiau and Lann-Bihoué on communications, radar and continuation training.

About 10 are still used by the Argentine air force.

Specification
Morane-Saulnier MS.760 Paris I
Type: four-seat liaison aircraft
Powerplant: two 400-kg (882-lb) thrust Turboméca Marboré II turbojets
Performance: maximum speed 650 km/h (404 mph) at sea level; cruising speed 570 km/h (354 mph) at 5000 m (16,405 ft); initial rate of climb 690 m (2,264 ft)/minute: service ceiling 10000 m (32,810 ft); range 1500 km (932 miles) at 7000 m (22,965 ft)
Weights: empty 1945 kg (4,288 lb); maximum take-off 3470 kg (7,650 lb)
Dimensions: span, over tiptanks 10.15 m (33 ft 3.6 in); length 10.05 m (32 ft 11.7 in); height 2.60 m (8 ft 6.4 in); wing area 18.00 m^2 (193.76 sq ft)
Armament: weapons trainer, two

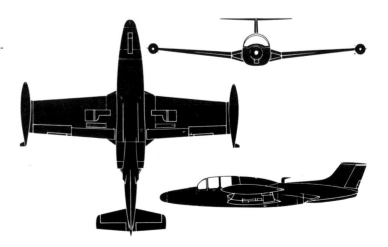

Morane-Saulnier MS.760 Paris

7.5-mm (0.295-in) machine-guns; underwing racks for two 50-kg (110-lb) bombs or four 89-mm (3.5-in) rockets

Operators: Argentina, France

Mudry CAP 10

In 1955 Auguste Mudry took the first steps towards the design of a fully-aerobatic monoplane. The side-by-side Piel Super Emeraude was taken as a basis, leading to the CAP 10 built by CAARP (*Co-opérative des Ateliers Aéronautiques de la Région Parisienne*). The first flew in August 1968 and in late 1970 production was initiated, CAARP building the fuselages and Avions Mudry at Bernay constructing the remainder and completing assembly. Construction is mainly wood, with the rear fuselage fabric-covered and plastics used for the cowling and some other parts. Delivery of 30 was made to the *Ecole de Formation Initiale du PN*, the pilot selection school at Clermont-Ferrand Aulnat; the CAP 10B is also used by the Armée de l'Air's *Equipe de Voltige Aérienne* (EVA) at Salôn-de-Provence, this model having a ventral fin and larger rudder. The Armée de l'Air has 56 in all, and the Aéronavale six. By 1985 total deliveries exceeded 200.

Specification
Type: basic and aerobatic trainer
Powerplant: one 134-kW (180-hp) Lycoming AEIO-360-B2F flat-four piston engine
Performance: maximum speed 270 km/h (168 mph); cruising speed 250 km/h (155 mph); initial rate of climb 360 m (1,181 ft)/minute; service ceiling 5000 m (16,405 ft); maximum range 1200 km (746 miles)
Weights: empty 540 kg (1,190 lb); maximum take-off, aerobatic 760-kg (1,676 lb), utility 830 kg (1,830 lb)
Dimensions: span 8.06 m (26 ft 5.3 in); length 7.16 m (23 ft 5.9 in); height 2.55 m (8 ft 4.4 in); wing area 10.85 m² (116.79 sq ft)
Armament: none
Operators: France and eight other countries including Mexico and Morocco

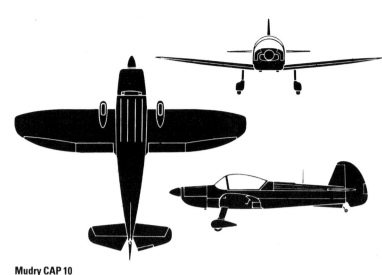

Mudry CAP 10

Mudry CAP 20

Intended solely for aerobatics, the CAP 20 is a single-seater, has a more powerful engine than the CAP 10, is stressed for g limits of +8 and −6, is slightly smaller, and the wings have no dihedral and no flaps. The prototype flew on 29 July 1969. The *Équipe de Voltige Aérienne* received six CAP 20As with lighter landing gear.

Current CAP 20LS-200 production aircraft have a more powerful engine, and CAP types are now manufactured entirely by Avions Mudry at Bernay.

Specification
Mudry CAP 20LS-200
Type: single-seat aerobatic aircraft
Powerplant: one 149-kW (200-hp) Lycoming AIO-363-B1B flat-four piston engine
Performance: maximum cruising speed 265 km/h (165 mph); initial rate of climb 840 ft (2,756 ft)/minute; endurance 2 hours
Weights: empty 480 kg (1,058 lb); maximum take-off, aerobatic 650-kg (1,433-lb)
Dimensions: 7.57 m (24 ft 10 in); length 6.46 m (21 ft 2.3 in); height 1.52 m (4 ft 11.8 in); wing area 10.47 m² (112.70 sq ft)
Armament: none
Operator: France

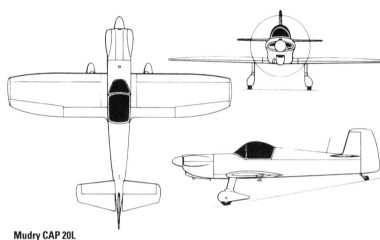

Mudry CAP 20L

Myasischev M-4 'Bison'

In 1951 V.M. Myasischev was directly ordered by Stalin to build a jet bomber to fly long-range strategic missions. the task was then beyond the state of the art, but the Myasischev M-4 (often called the Mya-4, and called *Molot*, meaning hammer, by the Soviets, and given the codename 'Bison' by NATO) proved a most successful aircraft which is still in limited service. Powered by four of the same very large turbojets as the twin-engine Tu-16 'Badger', the M-4 is a very large and impressive machine which on the ground rests on front and rear bogie landing gears, with small stabilizer gears at the wingtips. Between the main gears is the cavernous bomb bay, while all the crew ride in the pressurized nose, apart from the sixth man who mans the tail turret. The first verion, a free-fall bomber called 'Bison-A', reached regiments of the DA (Long-Range Aviation) in 1956. Altogether about 200 were built. A few were re-engined with much more powerful D-15 turbojets, and many were given different noses as 'Bison-B' and 'Bison-C' for maritime reconnaissance purposes. The bomber has a glazed nose for the bomb-aimer, and despite the extremely powerful, and very heavy, defensive armament soon showed great capability. In 1985 it was estimated that 40 remained in active service in the original role. Supporting them are another 30 rebuilt as inflight-refuelling tankers, with many modifica-

tions including a large hose drum unit in the rear of the weapon bay. 'Bison-B' was a long-range maritime reconnaissanve version, possibly rebuilt from the original bomber, with a 'solid' nose housing a mapping and ship-targeting radar with the refuelling probe above. Numerous other reconnaissance systems were installed. 'Bison-C' was an improved maritime reconnaissance version with an even larger surveillance radar (NATO name 'Puff Ball') in a more pointed nose swollen at the sides and with the refuelling probe at the tipe. These flew surveillance and electronic missions for the AV-MF (Naval Air Force).

Specification
'Bison-A'
Type: six-seat strategic bomber
Powerplant: four 9500-kg (20,943-lb) thrust Mikulin RD-3M turbojets
Performance: maximum speed at high altitude 1000 km (621 mph); service ceiling at normal loaded weight 17000 m (55,775 ft) or at maximum take-off weight 13000 m (42,650 ft); range 10700 km (6,650 miles)
Weights: empty 70000 kg (154,321 lb); normal loaded 160000 kg (352,734 lb); maximum take-off 210000 kg (462,963 lb)
Dimensions: span 50.48 m (165 ft 7½ in); length (no probe) 47.20 m (154 ft 10 in); height 14.24 m (46 ft 0 in); wing area 309.0 m² (3,326.2 sq ft)
Armament: (as built) 10 23-mm cannon in five power turrets; internal bomb bay for 15000-kg (33,068-lb) bombload
Operator: Soviet Union

A Myasischev M-4 'Bison-C' Maritime-reconnaissance aircraft, an improved version of the 'Bison-B'.

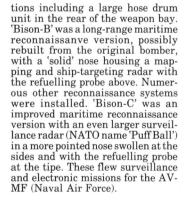

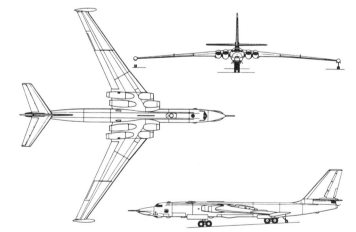

Myasischev M-4 'Bison-C'

NAMC YS-11

Developed as a twin-turboprop airliner by Nihon Aeroplane Manufacturing Co, and first flown on 30 August 1962, the YS-11 was acquired by two elements of the Japanese armed forces. The JASDF received 13; four YS-11P passenger transports were followed by a single YS-11PC passenger/cargo version and seven YS-11C pure freighters, all initially operated by the three squadrons of the Air Transport Wing from Komaki, Iruma and Miho. One YF-11FC was also purchased for navaid checking with the Flight Check Group at Iruma. With the advent of the C-1A jet transport some YS-11s have been reconfigured, including two YS-11Es with sophisticated electronics and assigned to the ECM training role. At least one has been updated to YS-11NT standard.

The JMSDF purchased 10; six YS-11Ts for ASW training had radar beneath the forward fuselage, ESM and other ASW sensors. The other four are YS-11M transports, two configured for passengers, one for mixed traffic and one for cargo. The six YS-11Ts serve with Air Training Command's 205th Kokutai at Shimofusa, the YS-11Ms being operated by the 61st Kokutai from Atsugi.

Six ex-Olympic Airways YS-11As are active in Greece, while Gabon is believed to retain one of two aircraft acquired several years ago.

Specification
Type: short range transport
Powerplant: two 2282-ekW (3,060-ehp) Rolls-Royce Dart 542-10 turboprop engines
Performance: maximum cruising speed at 4570 m (15,000 ft) 472 km/h (293 mph); initial rate of climb 372 m (1,220 ft)/minute; range with maximum payload 1110 km (690 miles); range with maximum fuel 1390 km (864 miles)
Weights: empty 14590 kg

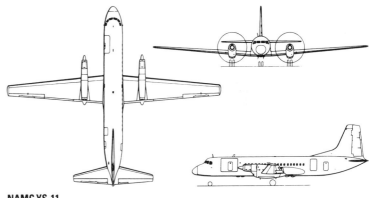

NAMC YS-11

(32,165 lb); maximum take-off 24500 (54,013 lb)
Dimensions: span 32.00 m (104 ft 11.8 in); length 26.30 m (86 ft

3.4 in); height 8.99 m (29 ft 5.9 in); wing area 94.80 m² (1,020.45 sq ft)
Armament: none
Operators: Gabon, Greece, Japan

Nanchang Q-5

Known to NATO as the 'Fantan-A' and to the People's Liberation Army Air Force as the Quiangjiji 5 (Attack Aircraft Type 5), the Q-5 was revealed in limited detail only during 1980 although conceived about a decade before as China's first 'almost indigenous' jet attack aircraft. Much of it, particularly the wing, is based on the Mikoyan-Gurevich MiG-19 already built in the People's Republic as the Shenyang J-6. Longer than its ancestor, the Q-5 differs considerably in profile by reason of a 'solid' nose and cheek air inlets made necessary by transfer of some avionics from the centre fuselage to make way for an internal weapons bay (now used for additional fuel), although the MiG's four wing strongpoints and root-mounted cannon (of newer type) are retained for close-support work. Extra fuselage pylons are also added. Powered by the same Soviet-designed, Chinese-built engines as the J-6, this considerably modified variant has a taller fin and a narrower centre fuselage, the latter an aerodynamic improvement conforming to area-rule. A nose camera is for gunnery recording only. The revised nose shape has led to speculation that a radar-equipped version is (or was) planned, but no evidence has emerged to support this suggestion, even though such equipment would improve capabilities as a tactical nuclear bomber in the apparent absence of an inertial navigation

A Nanchang Q-5 (Fantan-A) attack fighter of the Air Force of the People's Liberation Army.

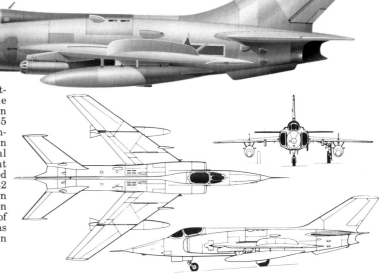

system. Interceptor Q-5s are reported to be in operation with the People's Navy, any differences in equipment being unknown. The Q-5 remains in production at the Nanchang State Aircraft Factory in Jiangxi Province, where several hundred have been built, recent (post 1981) examples having cleaned up rear fuselages. A first batch of 42 A-5s (export designation) was flown to Pakistan in 1982-3; this nation requires an eventual total of 150 of this type to equip eight squadrons and an Operational Conversion Unit.

Specification
Nanchang Q-5
Type: single-seat ground attack aircraft with interceptor capability
Powerplant: two 3250-kg (7,165-lb) thrust Wopen-6 (Tumansky R-9BF-811) afterburning turbojet engines
Performance: maximum speed 1190 km/h (739 mph) or Mach 1.12 at 11000 m (36,090 ft) and 1210 km (752 mph) at sea level; service ceiling 16000 m (52,495 ft); combat radius, maximum load, lo-lo-lo

Nanchang Q-5 (Fantan-A)

mission 400 km (249 miles), or hi-lo-hi mission 600 km (373 miles)
Weights: empty 6494 kg (14,317 lb); maximum take-off 12000 kg (26,455 lb)
Dimensions: span 9.70 m (31 ft 9.9 in); length 16.73 m (54 ft 10.7 in); height 4.51 m (14 ft 9.6 in); wing area 27.95 m² (300.86 sq ft)
Armament: two 23-mm (0.91-in)

Type 23-2 cannon (with 100 rpg) in wing roots; four wing and four fuselage pylons carrying normally 1000 kg (2,205 lb) or a maximum 2000 kg (4,409 lb) of ordnance, including AAMs, or a tactical nuclear weapon
Operators: China, Pakistan

Neiva C-42 Regente

Broadly resembling the Cessna 150/180 series, the Regente was built for the FAB (*Força Aérea Brasiliera*) in two basic versions, the C-42 and L-42. A civil prototype flew on 7 September 1961, powered by a 108-kW (145-hp) Continental O-300 engine. The C-42 was generally similar, and 80 were built as utility transports by the Sociedade Construtora Aeronáutica Neiva Ltda in 1965-68. To replace the L-6 Paulistinha (another Neiva design) and Cessna O-1 Bird Dog in the air observation post and liaison roles, Neiva had in January 1967 flown a prototype with a two/three-seat cabin with all-round glazing, and a 157-kW (210-hp) Continental IO-360-D flat-six engine. As the L-42,

40 were built in 1969-71. Most in FAB service are operated by two *Esquadroes de Ligaçao e Observaçao* (liaison and observation squadrons) which, despite their title, perform an additional COIN (counter-insurgency) role.

Specification
Neiva C-42 Regente
Powerplant: one 134-kW (180-hp) Lycoming O-360-A1D flat-four piston engine
Performance: maximum speed 220 km/h (137 mph) at sea level; initial rate of climb 210 m (689 ft)/minute; service ceiling 3600 m (11,810 ft); range with maximum fuel 928 km (577 miles)
Weights: empty 640 kg (1,411 lb);

Neiva C-42 Regente

maximum take-off 1040 kg (2,293 lb)
Dimensions: span 9.13 m (29 ft 11.4 in); length 7.21 m (23 ft 7.9 in); height 2.93 m (9 ft 7.4 in); wing area

13.45 m² (144.78 sq ft)
Armament: L-42 only has provision for light bombs, rockets or other stores on four underwing pylons
Operator: Brazil

Nord 262

A twin-engined light transport, additionally used for the training of navigators, the Aérospatiale 262 was produced by Nord-Aviation at Bourges as a pressurized turboprop successor to the Nord 260 Super Broussard (four of which remain in French military experimental use). A prototype flew on 24 December 1962, and the type entered service with Air Inter in July 1964; the last of 110 flew in November 1976. Main variants were the Nord 262A powered by Turboméca Bastan VICs, and the Nord 262C (civil) and Nord 262D Frégate (military) with 843-kW (1,130-shp) Bastan VIIA engines giving a 37 kmh (23 mph) increased in cruising speed and generally improved 'hot and high' performance. Six 262As went to the Armée de l'Air for executive transport, but were transferred to the Aéronavale in 1971 when 18 262s were supplied. The last six 262Ds went to the Armée de l'Air, initially serving the transport aircraft conversion unit. In addition to its

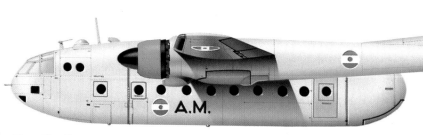

Nord 262 of the French Aéronavale, which uses them on medium-range liaison tasks.

second-hand acquisitions, the Aéronavale bought 15 262As and later added four more ex-civil aircraft, plus a single 262C. Used for communications and twin-conversion, the naval aircraft were replaced in the latter role by EMBRAER Xingus in 1983, and a dozen were converted to Nord 262E navigation trainers with appropriate equipment consoles. Another six serve trials establishments. Export military sales were restricted to three 262Cs delivered to Gabon in 1971-73, two to

Upper Volta (now Burkina-Faso) in 1974 and one to the Congo Republic in 1976 (now apparently withdrawn). Angola bought four surplus 262As from Air Algeria in 1980.

Specification
Aérospatiale (Nord) 262A
Type: 26-29-seat light transport
Powerplant: two 794-kW (1,065-shp) Turboméca Bastan VIC turboprop engines
Performance: maximum speed 385 km/h (239 mph); initial rate of

climb 378 m (1,240 ft)/minute; service ceiling 5850 m (19,195 ft); range (maximum payload) 915 km (569 miles)
Weights: basic operating 7029 kg (15,496 lb); maximum take-off 10600 kg (23,369 lb)
Dimensions: span 21.90 m (71 ft 10.2 in); length 19.28 m (63 ft 3 in); height 6.21 m (20 ft 4.5 in); wing area 55.0 m^2 (592.03 sq ft)
Armament: none
Operators: Angola, Burkina-Faso, France, Gabon

Nord Noratlas

Adopting the twin-boom layout of the Fairchild C-119, the prototype Noratlas flew on 27 November 1950. Over 200 were produced for the Armée de l'Air with deliveries beginning in 1953, and this service remains the major user. In 1956 the Federal German Republic signed a licence agreement with France to enable the Noratlas to be produced by Flugzeugbau Nord. The Luftwaffe took delivery of 186, 50 of them supplied from France, and a few have been sold to other operators.

The cargo hold measure 9.90 m (32 ft 5.8 in) in length, with a maximum width of 2.40 m (7 ft 10.5 in) and maximum height of 2.75 m (9 ft 0.3 in). Rear loading cargo doors open to the full cross-section of the hold (but have to be removed for airdropping) and there is a door to the rear on the left. Loads can include vehicles, up to 51.03 m (1,801.05 cu ft) of cargo, 36 paratroops, 45 troops, or up to 22 stretchers in the casevac role.

Specification
Type: utility transport
Powerplant: two 1521-kW (2,040-hp) SNECMA-built Bristol Hercules 739 14-cylinder radial piston engines
Performance: cruising speed 335 km/h (208 mph) at 3000 m (9,845 ft); initial rate of climb 375 m (1,230 ft)/minute; service ceiling 7500 m (24,605 ft); range 3000 km (1,864 miles)
Weights: empty 13075 kg (28,825 lb); maximum take-off 23000 kg (50,706 lb)
Dimensions: span 32.50 m (106 ft 7.5 in); length 21.96 m (72 ft 0.6 in); height 6.00 m (19 ft 8.2 in); wing

One of four Nord Noratlas transports in service with the Niger air force.

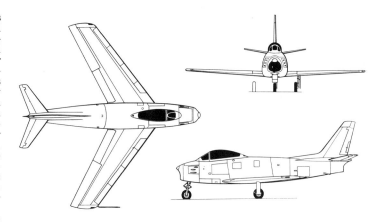

area 101.20 m^2 (1,089.34 sq ft)
Armament: none
Operators: Chad, Djibouti, France, Greece, Niger

Still serving with the French air force in modest numbers, the Noratlas performs a variety of tasks including basic cargo transport and paradropping.

North American F-86 Sabre

Designed to meet US Air Force requirements for a day-fighter which could also be used for escort duties, the North Americna F-86 Sabre as originally conceived had straight wings. However, an analysis of German research experiments in World War II indicated that a swept wing would significantly improve performance.

The first production models, designated F-86A, went into service in 1949. These aircraft had a top speed of 1093 km/h (679 mph) at sea level, a performance greatly superior to that of contemporary fighters such as the Lockheed P-80C and Gloster Meteor, the top speeds of which were below 965 km/h (600 mph).

This version saw service in the Korean War in the 1950s where its performance, together with skilful flying by experienced pilots, proved superior to that of the opposing vaunted MiG-15.

During its career, several versions of the F-86 were developed by North American, including the heavy radar-equipped F-86D interceptor,

the F-86K (the F-86D with guns instead of rockets) and the powerful F-86H fighter-bomber. Mitsubishi-built reconnaissance fighters still serve in Japan. In Canada, Canadair produced 1,815 for the Royal Canadian Air Force and other customers under the designation CL-13, many of these being powered by Orenda engines. The most advanced verison was the one developed by the Commonwealth Aircraft Corporation for the Royal Australian Air Force. These Australian Sabres were powered by Rolls-Royce Avon engines of 3400-kg (7,500-lb) static thrust and were armed with two 30-mm Aden cannon and Sidewinder air-to-air missiles.

In addition to it widespread use among NATO and British Commonwealth countries (those in Pakistan seeing action when that country found itself at war with India), Sabres were supplied to many other countries throughout the world. In fact, the demand became so great that in 1954 the F-86F was put back into production.

North American F-86E Sabre

Biggest of the overseas users was Japan, where Mitsubishi assembled and built 300 Sabres. These continued to serve in the Japanese Air

Self-Defence Force until the late 1970s. When withdrawn from front-line service Sabres continued to be used for training; the final stages of pilot tuition including no less than 290 hours on an F-86F, of which 220 hours were combat instruction.

A handful fly on in 1985, notably with Taiwan.

Specification
Type: (F-86D) single-seat fighter

Powerplant: one 2330-kg (7,650-lb) General Electric J47-GE-17 after burning turbojet
Performance: maximum design speed 1114 km/h (692 mph) at sea level; maximum speed at 12190 m (40,000 ft) 985 km/h (612 mph); combat radius with full internal fuel

446 km (277 miles); maximum rate of climb 3705 m (12,150 ft) per minute at sea level
Weights: empty 4123 kg (13,518 lb); normal take-off 5546 kg (18,183 lb); maximum take-off (area interceptor) 6092 kg (19,975 lb)
Dimensions: span 11.32 m (37 ft

1½ in); length 12.28 m (40 ft 3 in); height 4.57 m (15 ft); wing area 26.75 m² (287.9 sq ft)
Armament: twenty-four 70-mm (2.75-in) Mighty Mouse rockets
Operators: Argentina, Bolivia, Philippines, South Korea, Taiwan, Tunisia, Uruguay, Yugoslavia

North American F-100 Super Sabre

Conceived as a private-venture successor to the F-86 Sabre, the Sabre 45 – a reference to the degree of wing sweep – showed sufficient promise for a USAF contract for two prototypes, followed on 11 February 1952 by an order for a batch of 23 F-100As. Flown on 25 May 1953, the F-100 day air superiority fighter entered service with Tactical Air Command in autumn 1954. After major accidents the wings and fin were extended, but it was not until yaw and pitch dampers were introduced on the F-100C version that shortcomings were finally overcome.

Production of 2,249 ended in 1959, and the type consolidated its position as TAC's principal fighter-bomber. In addition to the F-100C and advanced F-100D, the F-100F two-seat combat proficiency trainer was also built, retaining full combat capability and providing the basis for the first 'Wild Weasel' SAM (surface-to-air missile) suppression aircraft.

The F-100 served with distinction

Now virtually gone from operational service, the F-100 still has some life left with the Turkish air force.

in South East Asia, generally employed on close air support missions. France also used the type originally in Algeria and Djibouti. Large numbers (mostly stored) are in Turkey, and a few in Taiwan. In the USA many are being converted by Sperry into QF-100 pilotless drones for realistic close air-combat training.

Specification
North American F-100D Super Sabre

Type: fighter-bomber
Powerplant: one 7711-kg (17,000-lb) afterburning thrust Pratt & Whitney J57-P-21A turbojet engine
Performance: maximum speed at 10975 m (36,000 ft) 1390 km/h (864 mph); initial rate of climb 1250 m (4,100 ft)/minute; operational ceiling 14540 m (47,700 ft); combat radius 853 km (530 miles)
Weights: empty 9525 kg (21,000 lb); maximum take-off 15800 kg (34,832 lb)

Dimensions: span 11.81 m (38 ft 9 in); length, including probe 14.33 m (47 ft 0 in); height 4.95 m (16 ft 2.7 in); wing area 35.79 m² (385.20 sq ft)
Armament: four Pontiac M-39E 20-mm (0.79-in) cannon, each with 200 rounds, plus up to 3193 kg (7,040 lb) of tactical air/ground ordnance on six hardpoints
Operators: (F-100C/D/F) Taiwan, Turkey, (QF-100D/F) USA

North American T-6 Texan

The North American T-6 is a two-seat advanced trainer, officially named Texan but better remembered by Commonwealth pilots as the Harvard.

The aircraft was introduced in 1938 and was similar to and eventually replaced the USAAF's North American BC-1A basic combat trainer. The pilot and instructor sat in tandem enclosed cockpits, the low sill of which gave an excellent view. Complete dual flight and engine controls were fitted in each cockpit. The powerful and reliable Pratt & Whitney Wasp air-cooled radial engine gave a sprightly performance, and the forgiving but responsive flying controls combined to make the T-6 an ideal training aircraft. Its distinctive rasping noise, not heard by the occupants, was caused by the high tip-speed of the direct-drive propeller.

Early aircraft had a conventional aluminium-alloy monocoque fuselage, but in 1941, because of possible shortages of strategic materials, the structure was extensively redesigned to eliminate the use of aluminium-alloy and high-alloy steels. The wings, centre-section, fin, and the flying control surfaces were made of spot-welded low-alloy steel, while the side panels of the forward fuselage and the entire rear fuselage were made of plywood. However, when the anticipated shortages did not materialize, the original method of construction was reintroduced.

Harvards were first delivered to the Royal Air Force in 1938, and remained a standard trainer at flying training schools for over 16 years. After their withdrawal in 1955 they continued in service with university air squadrons of Home Command as communications aircraft. Many Harvards also saw service in an operational armed role

The classic Harvard fulfils training duties with the South African Air Force.

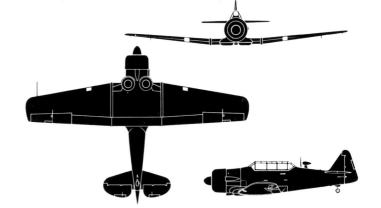

North American T-6 Texan

against the Mau Mau in Kenya and terrorists in Malaya.

The primary role of the T-6 as an armed trainer makes it suitable also for light ground attack duties. Because of this, and its strong construction, reliability and ease of flying, the T-6 continued to be used by numerous countries throughout the world.

Specification
Type: two-seat advanced trainer
Powerplant: one 409-kW (550-hp) Pratt & Whitney Wasp R-1340-AN-1 air-cooled radial
Performance: maximum speed 341 km/h (212 mph); cruising speed 272 km/h (170 mph); service ceiling 6560 m (21,500 ft); range 1400 km (870 miles)
Weights: empty 1888 kg (4,158 lb); gross 2550 kg (5,617 lb)
Dimensions: span 12.9 m (42 ft 0 in); length 9.0 m (29 ft 6 in); height 3.5 m (11 ft 8½ in); wing area 23.6 m² (253.7 sq ft)
Armament: two forward-firing 7.62-mm (0.30-in) machine-guns and one 7.62-mm (0.30-in) machine-gun on flexible mounting in rear

cockpit; in addition, underwing rockets and light bombs can be carried for weapon training and close-support duties
Operators: Angola, Argentina, Bolivia, Brazil, Canada, Chile, Dominican Republic, France, Greece,

Guatemala, Haiti, Honduras, India, Indonesia, Italy, Japan, Laos, Mexico, Morocco, New Zealand, Nicaragua, Pakistan, Paraguay, Portugal, Salvador, South Africa, Spain, Taiwan, Thailand, Tunisia, United Kingdom, Uruguay, Venezuela, Zaïre, Zambia

North American T-28 Trojan

The first T-28 flew on 26 September 1949, and delivery of 1,194 to USAF Air Training Command began in late 1950. In 1952 the US Navy ordered the T-28B with the more powerful 1063-kW (1,425-hp) Wright R-1820. Following 489 T-28Bs, North American built 299 T-28Cs with an arrester hook. Hundreds of surplus T-28s were modified by many companies to serve as ground attack aircraft, while others were made in France (as the Fennec) for this role.

Specification
Type: trainer and light attack aircraft
Powerplant: one 1063-kW (1,425-

hp) Wright R-1820 Cyclone 9-cylinder radial piston engine
Performance: maximum speed 552 km/h (343 mph); range 1706 km (1,060 miles) at 3050 m (10,000 ft)
Weights: empty 2914 kg (6,424 lb); maximum take-off 3849 kg (8,486 lb)
Dimensions: span 12.22 m (40 ft 1 in); length 10.06 m (33 ft 0 in); height 3.86 m (12 ft 8 in); wing area 24.90 m² (268.0 sq ft)
Armament: six underwing hardpoints suitable for bombs, machine-gun packs and rockets
Operators: Argentina, Dominican Republic, Ethiopia, Honduras, Laos, Philippines, Nicaragua, Thailand, Uruguay, Zaïre

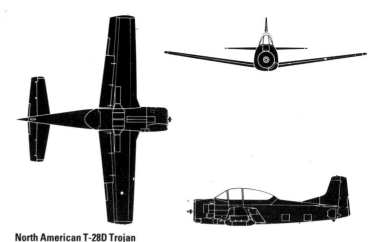

North American T-28D Trojan

North American Rockwell OV-10 Bronco

In August 1964 North American's NA-300 design was announced as the winner of a competition for a light armed reconnaissance aircraft for the US Marine Corps and Army. Production aircraft began to enter service in 1967, the USAF also deciding to acquire 157, used operationally in Vietnam at the beginning of 1968. The Marine Corps received 114, with deliveries to VMO-5 beginning on 23 February 1968. To enhance their capabilities 15 OV-10As were equipped under the 'Pave Nail' programme with a stabilized night periscope, laser rangefinder/target illuminator, Loran receiver and co-ordinate converter; these aircraft proved valuable for locating targets at night, which could then be illuminated for attack by laser-seeking missiles.

Other versions include six OV-10Bs supplied to the Luftwaffe for target towing, plus 12 OV-10B(Z) boosted by a 1338-kg (2,950-lb) thrust General Electric J85-GE-4 turbojet. The Royal Thai air force received 40 OV-10Cs, and Indonesia and Venezuela each received 16 similar OV-10Fs and OV-10Es respectively.

Two OV-10As were modified as YOV-10D NOGS (Night Observation/Gunship System) aircraft to provide the US Marine Corps with a night capability. Special equipment included a 20-mm M97 gun turret beneath the rear fuselage, a forward-looking infra-red sensor and laser target illuminator in a turret beneath an extended nose, and five pylons for a wide range of stores. Subsequently, 17 USMC aircraft were converted into OV-10D NOS (night observation surveillance) aircraft with developed equipment and avionics as installed in the NOGS prototypes, plus IR-suppressed 776-kW (1,040-hp) T76-G-420/421 engines.

Specification
North American Rockwell OV-10A Bronco
Type: multi-purpose/counter-insurgency aircraft
Powerplant: two 533-ekW (715-

North American Rockwell OV-10E Bronco of the Venezuelan air force.

Another export customer for the Bronco is Indonesia, which operates 16 OV-10Fs.

ehp) Garrett T76-G-416/417 turboprop engines
Performance: maximum speed, at sea level without weapons 452 km/ (281 mph); initial rate of climb at 5443 kg (12,000 lb) AUW 2073 m (6,800 ft)/minute; combat radius, maximum weapon load 367 km (228 miles); ferry range, with auxiliary fuel 2225 km (1,382 miles)
Weights: empty 3127 kg (6,893 lb); normal take-off 4494 kg (9,908 lb); maximum take-off 6552 kg (14,444 lb)
Dimensions: span 12.19 m (40 ft 0 in); length 12.67 m (41 ft 7 in); height 4.62 m (15 ft 2 in); wing area 27.03 m² (291.0 sq ft)
Armament: four weapon attachment points beneath the sponsons and one beneath the centre fuselage with a combined capacity of 1633 kg (3,600 lb); two 7.62-mm (0.3-in) M60C machine-guns in each sponson; USMC aircraft can carry one AIM-9D Sidewinder beneath each wing
Operators: West Germany, Indonesia, South Korea, Morocco, Thailand, USA (USAF, USMC), Venezuela

With its excellent forward air control (FAC) and counter-insurgency (COIN) abilities, the Bronco has found favour with several nations including Thailand.

Northrop F-5 Freedom Fighter

In 1954 the US government initiated a study to determine requirements for a simple fighter to be supplied to friendly nations via the Military Assistance Program (MAP). Northrop worked on the requirement as the N-156; in 1956 the USAF showed interest in a supersonic trainer derived from this

Northrop RF-5A of No. 13 Sqn, No. 1 Wing, Royal Thai air force.

(T-38) and developed in parallel with the private venture N-156C, which made its first flight on 30 July 1959. In April 1962 the US Secretary of Defense approved USAF selection of the N-156C as the required 'FX' fighter. This was designated F-5, the F-5A prototype flying in May 1963. The two-seat F-5B fighter/trainer entered service on 30 April 1964, four months before the F-5A, with the 4441st Combat Crew Training Squadron at Williams AFB. The first F-5As were delivered to Williams in August 1964.

The airframe used new structural and aerodynamic features. These included leading edge and trailing edge flaps, area-ruled fuselage with two airbrakes on the undersurface, and in some versions manoeuvring flaps. Rocket-powered ejection seats were provided. The primary interception weapons comprised two nose-mounted 20-mm M39 guns, and two AIM-9 Sidewinder missiles on wingtip launchers. One underfuselage and four underwing pylons permitted the carriage of nearly three tons of weapons, these including a wide variety of bombs and rockets.

To evaluate the combat potential of the F-5A, 12 (later 18) aircraft were deployed to South East Asia in October 1965, under the codename 'Project Skoshi Tiger'. The aircraft were diverted from the MAP and provided with an inflight refuelling probe, armour, jettisonable pylons, additional avionics and camouflage paint.

Versions of the F-5 include the basic F-5A and two-seat F-5B, the F-5G being the designation of Royal Norwegian air force F-5As with extending nose leg, drag chute and manoeuvring flaps. CF-5A/D are the versions of the F-5A/B respectively that were built for the Canadian Armed Forces by Canadian and Netherlands companies (the Royal Netherlands air force aircraft being designated NF-5A/B), all of those having very advanced features. The CASA-built Spanish versions are known in Spain's Ejercito del Aire, as the C-9 and CE-9 respectively. The RF-5A reconnaissance version carries four KS-92A cameras, mounted in the nose, and Norwegian and Spanish equivalents have the respective designations RF-5G and CR-9.

Specification
Northrop F-5A
Type: tactical fighter
Powerplant: two 1851-kg (4,080-lb) afterburning thrust General Electric J85-GE-13 turbojet engines
Performance: maximum speed Mach 1.4 or 1487 km/h (924 mph) at 10975 m (36,000ft); initial rate of climb 8758 m (28,700 ft)/minute; service ceiling 15390 m (50,500 ft); combat radius, maximum payload with allowance for 5 minutes combat at sea level 314 km (195 miles)
Weights: empty 3667 kg (8,085 lb); maximum take-off 9379 kg (20,677 lb)
Dimensions: span 7.70 m (25 ft 3 in); length 14.38 m (47 ft 2 in); height 4.01 m (13 ft 2 in); wing area 15.79 m² (170.0 sq ft)
Armament: two 20-mm (0.79-in) M39 guns and two AIM-9B Sidewinders; plus AGM-12B Bullpup ASMs; LAU-3/A and LAU-10/A rockets; Mk 81, Mk 82, Mk 83, Mk 84, M-117 and BLU-1/B bombs, to overall total of 2812 kg (6,200 lb)
Operators: Canada, Greece, Iran, Libya, Morocco, Netherlands, Norway, Philippines, South Korea, Spain, Taiwan, Thailand, Turkey, Venezuela

The popularity of the F-5 has not been restricted to smaller nations: several of the major air arms have made good use of its versatility, including Canada. This CF-116 has a camera-nose for photo-reconnaissance duties.

Northrop F-5E Tiger II

Conceived as a private venture, the F-5E is an improved F-5A Freedom Fighter incorporating a more powerful J85 engine, an integrated fire control system, extra fuel and a modified wing offering enhanced manoeuvrability. It was first flown in March 1969, entered the IFA (International Fighter Aircraft) competition to find a successor to the F-5A, and emerged as the winner with a contract covering development and production of 325 aircraft being placed on 6 December 1970. The first production F-5E flew on 11 August 1972 and soon became operational with the 425th Tactical Fighter Training Squadron at Williams AFB, Arizona, tasked with the training of pilots from recipient air arms.

The Tiger II benefited from combat experience in South East Asia which revealed that manoeuvrability was usually of greater value than speed. Consequently, the F-5E's wing incorporated elecrically-operated full-span leading edge manoeuvring flaps, as well as LEXs (leading edge extensions) to improve airflow at high angles of attack. Wing area was increased and an extendible nose leg introduced to increase the angle of attack on take-off and thus reduce field length. A drag 'chute and hook are standard. Later aircraft have a 'shark-nose' radome and larger LEXs.

Deliveries to overseas air arms began in 1973, with over 1,350 completed by late 1984. The US Navy and Air Force also use the F-5E in the 'aggressor' role in the teaching of air combat manoeuvring. The Tiger II has been the subject of licence agreements with KAL of South Korea and with AIDC of Taiwan,

Northrop F-5E Tiger II of No. 1 Sqn, Brazilian air force, at Santa Cruz AFB.

Every good fighter design should have a trainer derivative, and the F-5 is no exception. This F-5F serves with No. 11 Sqn, Royal Jordanian air force.

An F-5E of Grupo 7, Fuerza Aérea de Chile, based at Cerro Moreno. The Chilean air force has become highly efficient despite international arms embargoes, and the F-5E Tiger is its most potent asset. The F-5E has freed Chile's Hunters for ground attack duties.

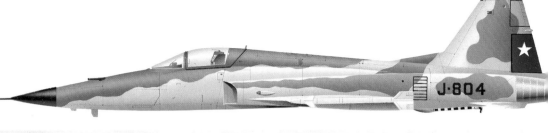

The RF-5E Tigereye is an armed recce version of the well proven F-5E, and 12 have been ordered by Malaysia, as seen here, and Saudi Arabia. Flexibility is the keynote of Tigereye, which can use interchangeable pallets containing different sensors.

and it has been assembled in Switzerland by the Federal Aircraft Factory.

The F-5F has dual combat/training capability. To accommodate the second cockpit the fuselage was lengthened by 1.08 m (3 ft 6.5 in), one gun being deleted although the fire-control system is retained. Flown on 25 September 1974, about 200 F-5Fs are used by most F-5E customers. The RF-5E TigerEye flew in January 1979, demonstrat-

ing day and night reconnaissance capability. Two were ordered by Malaysia and 10 by Saudi Arabia. The sensors are on pallets; three combinations of cameras and infrared linescan are available, and TigerEye may later have Elint (Electronic Intelligence) receivers. Weapon capability is retained, though with only one gun.

Specification
Type: tactical fighter
Powerplant: two 2268-kg (5,000-lb) afterburning thrust General Electric J85-GE-21 turbojet engines
Performance: maximum speed at 10975 m (36,000 ft) 1743 km/h (1,083 mph); initial rate of climb 10516 m (34,500 ft)/minute; service ceiling 15790 m (51,800 ft); combat radius (maximum fuel, two AIM-9E

Sidewinders, five minutes full-power combat at 4570 m/15,000 ft) 1056 km (656 miles); combat radius (maximum fuel, 2359-kg/5,200-lb weapon load, two AIM-9E Sidewinders, five minutes full-power combat at sea level) 222 km (138 miles)
Weights: empty 4410 kg (9,723 lb); maximum take-off 11214 kg (24,722 lb)
Dimensions: span 8.13 m (26 ft 8 in); length 14.45 m (47 ft 4.75 in); height 4.06 m (13 ft 4 in); wing area 17.28 m^2 (186.0 sq ft)
Armament: two M39A2 20-mm (0.79-in) cannon each with 280 rounds, two AIM-9 Sidewinder AAMs on wingtip launchers, plus up to 3175 kg (7,000 lb) of ordnance including bombs, rockets and air-to-surface missiles on five hardpoints

Northrop F-5E Tiger II

Operators: Bahrain, Brazil, Chile, Ethiopia, Indonesia, Iran, Jordan, Kenya, South Korea, Malaysia, Mexico, Morocco, Philippines, Saudi Arabia, Singapore, Sudan, Switzerland, Taiwan, Thailand, Tunisia, North Yemen, US

Northrop F-20 Tigershark

Although available for export to friendly nations under the terms of the Foreign Military Sales (FMS) programme, and despite the fact that it possesses some commonality with the F-5E, Northrop's F-20A Tigershark has still to secure its launch order. By early 1984 Northrop had invested some $600 million in the F-20, a sum which it seems increasingly unlikely to recoup.

Following extensive in-house studies, which showed that adoption of a single F404 engine in place of two J85s would result in an aircraft with far superior performance, the decision to go ahead with a prototype was taken in January 1980. This flew as the F-5G on 30 August 1982. Redesignated in recognition of their significant differences, three F-20As demonstrated excellent reliability, Nos 2 and 3 having digital avionics and a revised canopy. Sadly, No. 1 crashed in Korea in October 1984.

Specification
Type: single-seat multi-role fighter
Powerplant: one 7711-kg (17,000-lb) afterburning thrust General Electric F404-GE-100 turbofan engine
Performance: maximum speed at altitude approximately Mach 2; initial rate of climb 16093 m (52,800 ft)/minute; combat ceiling 16675 m (54,700 ft); radius on combat air patrol mission (three external tanks, two AIM-9 missiles, 1 hour 37 minutes on patrol and 20 minutes reserves at sea level) 555 km (345 miles)
Weights: empty 5089 kg (11,220 lb); maximum take-off 12474 kg (27,500 lb)
Dimensions: span 8.13 m (26 ft 8 in); length, excluding probe 14.17 m (46 ft 6 in); height 4.22 m (13 ft 10.25 in); wing area 17.28 m^2 (186.0 sq ft)
Armament: two M39 20-mm (0.79-

in) cannon each with 450 rounds; stores stations (centreline, four underwing and two wingtip) can accommodate in excess of 3765 kg (8,300 lb), including Harpoon, Maverick and Sidewinder missiles
Operator: none

The Northrop Tigershark programme has been dogged by misfortune, with two fatal crashes, and resistance from potential customers who would prefer the more glamourous F-16, despite the many advantages offered by Tigershark.

Northrop T-38A Talon

As mentioned in the description of the Northrop F-5A/B, after two years of private development Northrop's N-156 resulted in the supersonic T-38 trainer. The first of six prototypes made its initial flight on 10 April 1959. The first production T-38A Talon was delivered for service with the USAF's 3510th Flying Training Wing at Randolph AFB, on 17 March 1961. Many pupils must have wondered what they were taking on. But despite its high performance, its stalling speed was reason-

Northrop T-38 Talon of the Força Aerea Portugesa.

able at 235 km/h (146 mph), control surfaces were hydraulically powered, directional and longitudinal

stability augmentors were installed, and the Talon was designed to land safely with one aileron inoperative.

For over 20 years the T-38 has consistently maintained the best safety record of any supersonic aircraft in

USAF service.

When production ended in 1972, 1,187 T-38As had been delivered to the USAF. The US Navy acquired five from the Air Force. NASA obtained 24 from Northrop to serve as flight readiness trainers for astronauts. In addition, 46 were supplied through the USAF for use by Luftwaffe student pilots in the United States: these retain USAF insignia. The AT-38 is an attack trainer modification with special avionics. Portugal acquired some ex-USAF aircraft.

Specification
Type: two-seat supersonic basic trainer
Powerplant: two 1746-kg (3,850-lb) afterburning thrust General Electric J85-GE-5A turbojet engines

Performance: maximum speed Mach 1.3 or 1381 km/h (858 mph) at 10975 m (36,000 ft); initial rate of climb 9144 m (30,000 ft)/minute; service ceiling 16335 m (53,600 ft); range, maximum fuel and 20 minutes reserves at 3050 m (10,000 ft) 1835 m (1,140 miles)
Weights: empty 3250 kg (7,164 lb); maximum take-off 5361 kg (11,820 lb)
Dimensions: span 7.70 m (25 ft 3 in); length 14.14 m (46 ft 4.5 in); height 3.92 m (12 ft 10.5 in); wing area 15.79 m^2 (170.0 sq ft)
Armament: none
Operators: Colombia, West Germany, Portugal, Turkey, USA (USAF, USN, NASA)

For nearly 25 years, the T-38 has provided supersonic training for the USAF. It has an unrivalled safety record.

Panavia Tornado ADV

A Panavia Tornado F.Mk 2 of No. 229 OCU, the RAF's first Tornado ADV unit.

Having 80 per cent commonality with the interdictor/strike Tornado, the ADV – Air Defence Variant – is optimized for long-range missions against high- and low-level threats. Susceptible to 'back door' attacks from bombers based on Soviet Arctic airfields as well as from over the North Sea, the UK has to intercept intruders well before they achieve landfall. This requires the Tornado ADV to work in close collaboration with the NATO Airborne Early Warning Force, especially the RAF's Nimrod AEW.3s. Its Marconi AI.24 Foxhunter pulse-doppler radar can track several targets simultaneously up to a distance of well over 185 km (115 miles). Under the fuselage are four BAe Sky Flash radar-guided AAMs with a range of over 40 km (25 miles) even against low-flying targets in a high-ECM environment. These are backed by two AIM-9L Sidewinders on the inboard sides of the single pylon under each wing. To accommodate the tandem AAMs the ADV is 1.36 m (4 ft 5.5 in) longer than the IDS variant, and this makes room for an additional 909 litres (200 Imp gallons) of fuel and extra avionics. The fixed inboard parts of the wings have 68° sweep instead of 60°, to give additional chord, and they house the forward radar warning receivers repositioned from the fin. Armament is completed by a single gun in the fuselage. Possessing good short-field performance, the ADV can operate beyond 645 km (400 miles) from its base in all weathers, and engage a rapid succession of targets attempting to defend themselves by low-level approaches and/or jamming. A prototype has patrolled for 2 hours 20 minutes at a distance of 602 km (374 miles) from home during an unrefuelled sortie lasting 4 hours 13 minutes, and a retractable inflight refuelling probe in the left side of the nose gives even greater capability.

Although a stand-off interceptor, the aircraft has good manoeuvrability and can get the better of a Phantom in close combat. It is also a star performer in a tail-chase, having demonstrated an ability to fly low-level at 1481 km/h (920 mph), this being 185 km/h (115 mph) beyond the structural or aerodynamic limits at that height of prospective targets (including Western ones). Its supersonic acceleration is better than the IDS, and is improved further from the 19th aircraft by Mk 104 engines with extended afterburners. The first of three prototypes flew on 27 October 1979, and production aircraft followed from March 1984, having been ordered in batches of 18, 52 and 92 (including eight, 18 and 13 dual trainers) for a total of 165. These are known to the RAF as Tornado F.2, or F.2T for the trainer version. Crew conversion began at RAF Coningsby in early 1985, followed by replacement of Lightnings and some Phantoms as squadrons are re-equipped. Two F.2 units will be based at Leuchars, Scotland, and three at Leeming, Yorkshire, leaving four Phantom squadrons at Wattisham and in RAF Germany as the balance of Britain's interceptor force. Oman and Saudi Arabia have ordered 8 and 24 aircraft respectively.

Specification
Panavia Tornado ADV (F.2)
Type: long-range interceptor
Powerplant: two 8600-kg (18,960-lb) thrust Turbo-Union RB199-34R Mk 104 augmented turbofan engines
Performance: maximum speed Mach 2.27 at alitude, or 1481+ km/h (920+ mph) at low level; take-off run, fully armed, 762 m (2,500 ft); on-task CAP 2 hours at 555-740 km (345-460 miles) from base
Weights: empty about 14288 kg (31,500 lb); maximum take-off, clean 21546 kg (47,500 lb)
Dimensions: span 13.90 m (45 ft 7.2 in) spread, or 8.60 m (28 ft 2.6 in) swept; length 18.06 m (59 ft 3 in); height 5.70 m (18 ft 8.4 in)
Armament: one 27-mm (1.06-in) IKWA-Mauser cannon; four BAe Sky Flash (later AIM-120A) and two Bodenseewerk AIM-9L Sidewinder AAMs
Operators: UK (RAF)

Early problems with Tornado's Foxhunter radar have now been largely overcome, and the aircraft shows great promise.

Panavia Tornado IDS

The most sophisticated combat aircraft of its size in the world, the Tornado now serves with four air arms in three countries, considerably boosting NATO's operational efficiency. It can claim to be not only the fastest and most 'unstoppable' attack aircraft in the world but also the most accurate, RAF examples having swept the board at the 1984 USAF Strategic Air Command 'Giant Voice' competition – an unprecedented achievement.

Capable of carrying the wide variety of weaponry in service with Alliance members, it represents a fine example of European co-operation and may prove to be the basis of further collaborative projects. Its manufacturer, Panavia Aircraft GmbH, is owned jointly by British Aerospace (42.5 per cent), Germany's MBB (42.5 per cent) and Aeritalia of Italy (15 per cent), these three countries being the founder purchasers.

The Tornado was designed for close air support and battlefield interdiction, counter-air interdiction and strike, air superiority, air-defence interception, naval strike and reconnaissance, most of these roles being performed (though not all simultaneously) by the Tornado IDS (Interdictor/Strike). British air defence requirements are met by the Tornado ADV, described separately. Key attributes of the aircraft are its ability to fly under automatic control at a mere 61 m (200 ft) above ground level, faster than any other aircraft, in all weathers and deliver ordnance with pinpoint accuracy. With variable-geometry wings and powered by an amazingly compact and fuel-efficient engine – also the subject of international collaboration – the Tornado has a very low gust response in the low-level regime, and can be flown lower and further than any other Western aircraft before crew discomfort impairs efficiency. Nine prototypes and six pre-series aircraft flew in 1974-79, portions coming from unduplicated sources for assembly at Warton (UK), Manching (Germany) and Turin (Italy). The first production order, for 40, was placed in July 1976, and increased with tri-national batches of 110, 164, 144, 119 and 63 by January 1984 to the planned total of 640, to which will be added four pre-series aircraft refurbished to operational standard. This total includes 125 dual-pilot trainers, with full combat capability, and many of these are based at RAF Cottesmore with the Trinational Tornado Training Establishment. TTTE converts crews from the RAF, Luftwaffe, Marineflieger and Aeronautica Militare Italiana, and received its first aircraft in July 1979.

RAF designation is Tornado GR.1, or in dual-pilot form, Tornado GR.1T. Orders for 220 include 49 trainers (one of them a converted pre-series aircaft), all for delivery by December 1986. Weapons training is undertaken by each country, RAF crews qualifying at the TWCU (Tornado Weapons Conversion Unit) at RAF Honington. In wartime the instructors and aircraft of the TWCU would take to the front line as No. 45 Sqn. Also at Honington, No. 9 Sqn formed in June 1982, followed at Marham by Nos 617 and 27 – all ex-Vulcan units. Thereafter, deliveries were to RAF Germany for Nos 15, 16 and 20 Sqns at Laarbruch, and Nos 31, 14 and 17 at Brüggen, in 1983-85 for replacement of Buccaneers and Jaguars. At a later stage, No. 9 is to move to

Brüggen, and Laarbruch-based No. 2 Sqn will convert to a reconnaissance version, with two infra-red linescan units in place of its guns. A further reconnaissance unit will form in Britain, giving 11 operational squadrons in all. RAF Tornados carry nuclear weapons, free-fall retarded or laser-guided bombs and two Sidewinder AAMs for self-defence. The highly-effective Hunting JP.233 airfield attack system of two species of strewn 'bomblets', dispensed from giant underfuselage pods, will be followed in 1987 by the BAe Alarm anti-radiation missile. Outer pylons carry a Marconi Sky Shadow jamming pod and a Philips BOZ-107 chaff/flare dispenser. An inflight-refuelling probe retracts into an optional pack attached on the right of the cockpit.

German Tornados comprise 228 for the Luftwaffe and 96 Marineflieger aircraft for a total of seven wings, of which the navy's MFG 1 at Schleswig/Jagel was the first to equip, in 1982. Assigned to naval strike over the Baltic, MFG Tornados are armed with two or four MBB Kormoran anti-ship missiles beneath the fuselage and will later get the Texas Instruments Harm anti-radiation weapon. Half of MFG 2 at Eggebeck carry centreline reconnaissance pods. Luftwaffe units are JBG 38 (the weapons unit) at Jever, JBG 31 at Norvenich, JBG 32 at Memmingen – all completed by 1985 – JBG 33 at Buchel and JBG 34 at Memmingen. In the area-denial role their main weapon is the ungainly MW-1 belly container with its stocks

of laterally-ejected sub-munitions, but Hughes Mavericks would be used for pinpoint attacks. Italian Tornados also have this weapon in prospect, together with the reconnaissance pod and Kormoran missiles. Deliveries of 100 aircraft, including 12 trainers and a refurbished strike variant, began in 1982 and the first recipient (for weapons training) was the 154° Gruppo of 6° Stormo at Ghedi. In 1984 156° Gruppo of 36° Stormo at Gioia del Colle was re-equipped for tasks including naval atack, followed by 155° Gruppo/51° Stormo at Istrana/Treviso. Italy will assign 36 Tornadoes to reserve. Although Greece failed to select the Tornado in its 1984 combat aircraft competition, Saudi Ara-

An Italian air force Panavia Tornado IDS allocated to the Tri-National Tornado Training Establishment based at Cottesmore, England.

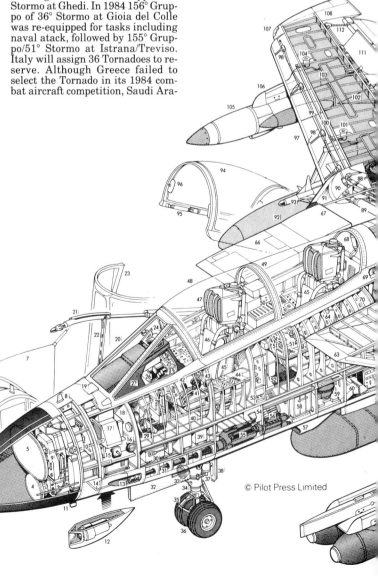

All three original Tornado users provide aircraft and instructors for the TTTE, training new Tornado crews from all nations. Here a German Tornado touches down at Cottesmore.

© Pilot Press Limited

anavia Tornado GR.Mk 1 cutaway drawing key

Air data probe
Radome
Lightning conductor strip
Terrain following radar antenna
Ground mapping radar antenna
Radar equipment bay hinged position
Radome hinged position
IFF aerial
Radar antenna tracking mechanism
Radar equipment bay
UHF/TACAN aerial
Laser Ranger and Marked Target Seeker (Ferranti), starboard side
Cannon muzzle
Ventral Doppler aerial
Angle of attack transmitter
Canopy emergency release
Avionics equipment bay
Front pressure bulkhead
Windscreen rain dispersal air ducts
Windscreen (Lucas-Rotax)
Retractable, telescopic, inflight refuelling probe
Probe retraction link
Windscreen open position, instrument access
Head-up display, HUD (Smiths)
Instrument panel
Radar 'head-down' display
Instrument panel shroud
Control column
Rudder pedals
Battery
Cannon barrel
Nosewheel doors
Landing/taxiing lamp
Nose undercarriage leg strut (Dowty-Rotol)
Torque scissor links
Twin forward-retracting nosewheels (Dunlop)
Nosewheel steering unit
Nosewheel leg door
Electrical equipment bay

40 Ejection seat rocket pack
41 Engine throttle levers
42 Wing sweep control lever
43 Radar hand controller
44 Side console panel
45 Pilot's Martin-Baker Mk 10 ejection seat
46 Safety harness
47 Ejection seat headrest
48 Cockpit canopy cover (Kopperschmidt)
49 Canopy centre arch
50 Navigator's radar displays
51 Navigator's instrument panel and weapons control panels
52 Foot rests
53 Canopy external latch
54 Pitot head
55 Mauser 27-mm cannon
56 Ammunition feed chute
57 Cold air unit ram air intake
58 Ammunition tank
59 Liquid oxygen converter
60 Cabin cold air unit
61 Stores management system computer
62 Port engine air intake
63 Intake lip
64 Cockpit framing
65 Navigator's Martin-Baker Mk 10 ejection seat
66 Starboard engine air intake
67 Intake spill duct
68 Canopy jack
69 Canopy hinge point
70 Rear pressure bulkhead
71 Intake ramp actuator linkage
72 Navigation light
73 Two-dimensional variable area intake ramp doors
74 Intake suction relief doors
75 Wing glove Krüger flap
76 Intake bypass air spill ducts
77 Intake ramp hydraulic actuator
78 Forward fuselage fuel tank

79 Wing sweep control screw jack (Microtecnica)
80 Flap and slat control drive shafts
81 Wing sweep, flap and slat central control unit and motor (Microtecnica)
82 Wing pivot box integral fuel tank
83 Air system ducting
84 Anti-collision light
85 UHF aerials
86 Wing pivot box carry-through, electron beam welded titanium structure
87 Starboard wing pivot bearing
88 Flap and slat telescopic drive shafts
89 Starboard wing sweep control screw jack
90 Leading-edge sealing fairing
91 Wing root glove fairing
92 External fuel tank, capacity 330 Imp gal (1500 litres)
93 AIM-9L Sidewinder air-to-air self-defence missile
94 Canopy open position
95 Canopy jettison unit
96 Pilot's rear view mirrors
97 Starboard three-segment leading-edge slat, open
98 Slat screw jacks
99 Slat drive torque shaft
100 Wing pylon swivelling control rod
101 Inboard pylon pivot bearing
102 Starboard wing integral fuel tank
103 Wing fuel system access panels
104 Outboard pylon pivot bearing

105 Marconi 'Sky-Shadow' ECM pod
106 Outboard wing swivelling pylon
107 Starboard navigation and strobe lights
108 Wing tip fairing
109 Double-slotted Fowler-type flaps, down position
110 Flap guide rails
111 Starboard spoilers, open
112 Flap screw jacks
113 External fuel tank tail fins
114 Wing swept position trailing edge housing
115 Dorsal spine fairing
116 Aft fuselage fuel tank
117 Fin root antenna fairing
118 HF aerial
119 Heat exchanger ram air intake
120 Starboard wing fully swept back position
121 Airbrake, open
122 Starboard all-moving tailplane (taileron)
123 Airbrake hydraulic jack
124 Primary heat exchanger
125 Heat exchanger exhaust duct
126 Engine bleed air ducting

127 Fin attachment joint
128 Port airbrake rib construction
129 Fin heat shield
130 Vortex generators
131 Fin integral fuel tank
132 Fuel system vent piping
133 Tailfin structure
134 ILS aerial
135 Fin leading edge
136 Forward passive ECM housing
137 Fuel jettison and vent valve
138 Fin tip antenna fairing
139 VHF aerial
140 Tail navigation light
141 Aft passive ECM housing
142 Obstruction light
143 Fuel jettison
144 Rudder
145 Rudder honeycomb construction
146 Rudder hydraulic actuator (Fairey Hydraulics)
147 Dorsal spine tail fairing
148 Thrust reverser bucket doors, open

149 Variable area afterburner nozzle
150 Nozzle control jacks (four)
151 Thrust reverser door actuator
152 Honeycomb trailing edge construction
153 Port all-moving tailplane (taileron)

154 Tailplane rib construction
155 Leading-edge nose ribs
156 Tailplane pivot bearing
157 Tailplane bearing sealing plates
158 Afterburner duct
159 Airbrake hydraulic jack
160 Turbo-Union R.B.199-34R Mk 101 afterburning turbofan engine

161 Tailplane hydraulic actuator
162 Hydraulic system filters
163 Hydraulic reservoir (Dowty)
164 Airbrake hinge point
165 Intake frame/production joint
166 Engine bay ventral access panels
167 Engine oil tank
168 Rear fuselage fuel tank
169 Wing root pneumatic seal
170 Engine driven accessory gearboxes, port and starboard (KHD), airframe mounted
171 Integrated drive generator (two)
172 Hydraulic pump (two)
173 Gearbox interconnecting shaft
174 Starboard side Auxiliary Power Unit, APU (KHD)
175 Telescopic fuel pipes
176 Port wing pivot bearing
177 Flexible wing sealing plates
178 Wing skin panelling
179 Rear spar
180 Port spoiler housings
181 Spoiler hydraulic actuators
182 Flap screw jacks
183 Flap rib construction
184 Port Fowler-type double-slotted flaps, down position
185 Port wing fully swept back position
186 Wing tip construction
187 Fuel vent
188 Port navigation and strobe lights
189 Leading-edge slat rib construction
190 Marconi 'Sky-Shadow' ECM pod
191 Outboard swivelling pylon
192 Pylon pivot bearing
193 Front spar
194 Port wing integral fuel tank
195 Machined wing skin/stringer panel
196 Wing rib construction
197 Swivelling pylon control rod
198 Port leading-edge slat segments, open
199 Slat guide rails
200 External fuel tank
201 Inboard swivelling pylon
202 Inboard pylon pivot bearing
203 Missile launch rail
204 AIM-9L Sidewinder air-to-air self-defence missile
205 Port mainwheel (Dunlop), forward retracting
206 Main undercarriage leg strut (Dowty-Rotol)
207 Undercarriage leg pivot bearing
208 Hydraulic retraction jack
209 Leg swivelling control link
210 Telescopic flap and slat drive torque shafts
211 Leading-edge sealing fairing

212 Krüger flap hydraulic jack
213 Main undercarriage leg breaker strut
214 Mainwheel door
215 Landing lamp
216 Hunting JP 233 Airfield Attack Weapon (two, side-by-side)

217 Submunitions compartments (30 SG357 runway penetration bombs and 215 HB876 area denial weapons in each JP 233)
218 Port shoulder pylon
219 Fuselage shoulder pylon (two)
220 ML twin stores carriers
221 Hunting BL 755 cluster bombs (eight)
222 Mk 83 high speed retarded bomb
223 Mk 13/15 454-kg (1,000-lb) HE bomb

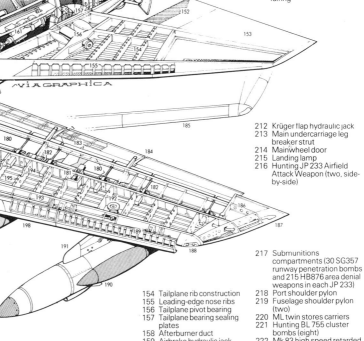

bia emerged at the same time as a strong prospect for an export order. In 1985, it announced an order for 48 aircraft, the first 20 of which will be delivered from the RAF production line. Further contracts are also in prospect from Britain and Germany for 40 each. The German aircraft would be for defence-suppression, with Harms and jamming equipment – possibly the ALQ-99E suite, as fitted to the USAF's EF-111A Raven.

Specification
Panavia Tornado IDS
Type: all-weather multi-role combat aircraft
Powerplant: two 7675-kg (16,920-lb) thrust Turbo-Union RB199-34R Mk 103 augmented turbofan engines
Performance: maximum speed Mach 2.2 clean at high alitude, or 1112 km/h (691 mph) with full external stores; time to 9145 m (30,000 ft) under 2 minutes; radius of action 1389 km (863 miles) with heavy weapon load
Weights: empty 14091 kg (31,065 lb); maximum take-off approximately 27215 kg (60,000 lb)
Dimensions: span 13.90 m (45 ft 7.2 in) spread, or 8.60 m (28 ft 2.6 in) swept; length 16.70 m (54 ft 9.5 in); height 5.70 m (18 ft 8.4 in)

Armament: two 27-mm (1.06-in) Mauser cannon, and over 8165 kg (18,000 lb) of stores on four wing and three tandem fuselage hard points
Operators: Italy, UK, West Germany (Marineflieger and Luftwaffe)

No. IX Sqn, based at RAF Honington, was the first RAF Tornado squadron, converting to the aircraft from Avro Vulcans. Tornado is NATO's finest strike and attack aircraft, able to deliver its warload with pinpoint accuracy by day or night even in the most appalling weather conditions, skimming over the terrain at below tree-top height. RAF Tornados have taken the prizes in both the USAF bombing contests they have entered.

Keith Fretwell

Pazmany PL-1B Chiensou

Originally a two-seat amateur-built aircraft originating in California, the Pazmany PL-1 was selected in 1968 by the Chinese Nationalist air force in Taiwan for production on the island as a first step in creating a local Aircraft industry. Construction of the first military prototype, the PL-1A, began in June 1968 under the supervision of Colonel C.Y. Lee of the Nationalist air force. It was built at the Aeronautical Research Laboratory, Taichung. First flight was on 26 October 1968 and the aircraft was demonstrated to Generalissimo Chiang Kai-shek four days later. Two more prototypes were built the following year, and the decision was made to build a series under the designation PL-1B Chienshou for use as primary trainers for Nationalist air cadets.

The PL-1B is a cantilever low-wing monoplane, closely resembling the PL-1A, but with wider cockpit, larger rudder and more powerful Lycoming O-320-E2A engine in place of the original 93-kW (125-hp) O-290-D. Fuel is carried in wingtip tanks. The wing is an all-metal single-spar structure in one piece with leading-edge torsion box. Piano-hinged ailerons and flaps are also of metal. The undercarriage is of fixed tricycle configuration. The two cockpit seats are side-by-side with dual control under a rearward-sliding one-piece 'bubble' canopy. The angular cantilever tail has a swept-back fin and rudder and a one-piece all-moving horizontal surface with anti-servo tab/trim tab.

Apart from 50 PL-1Bs built for the Nationalist air arm, the very similar PL-2 was ordered by South Korea, South Vietnam and Thailand.

Improving relations between the West and the Peoples Republic of China have led the Nationalist Chinese to rely more heavily on indigenous products, like the PL-1B Chienshou, a useful primary trainer.

Specification
Type: two-seat primary trainer
Powerplant: one 112-kW (150-hp) Lycoming O-320-E3A four-cylinder horizontally-opposed air-cooled piston engine
Performance: maximum speed at sea level at maximum take-off weight 241 km/h (150 mph); economical cruising speed 185 km/h (115 mph) at sea level; stalling speed with flaps down 87 km/h (54 mph); rate of climb at sea level 488 m (1,600 ft) per minute; range with maximum fuel 652 km (405 miles); take-off run 171 m (560 ft); landing run 167 m (550 ft)
Weights: empty equipped 431 kg (945 lb); maximum take-off 653 kg (1,440 lb)

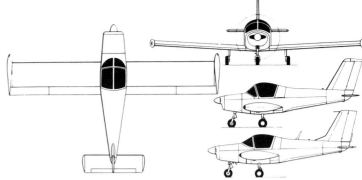

Pazmany PL-1B Chiensou

Dimensions: span 8.53 m (28 ft 0 in); length 5.98 m (19 ft 8⅛ in); height 2.24 m (7 ft 4 in); wing area 10.78 m² (116 sq ft)
Armament: none
Operators: South Korea, Taiwan

Piaggio PD-808

Designed by the El Segundo Division of Douglas Aircraft, the PD-808 was built in Italy by Piaggio. The Italian Ministry of Defence agreed to pay one-third of the development costs, buy two prototypes and provide test facilities, and in 1965 placed an order for 25, of which 22 were delivered. The prototype flew on 29 August 1964. Four versions were delivered to the Italian air force: they comprise the six-seat PD-808 VIP, four of these VIP transports serving with the air staff flying unit at Rome; eight PD-808 TA nine-seaters built for navigation training or transport; four PD-808 RMs equipped for radio and navaid calibration, and six PD-808 ECM electronic countermeasures aircraft carrying two pilots and three equipment operators. The PD-808 ECMs equip the 71 *Gruppo Guerra Elettronica* at Practica de Mare, Rome.

Specification
Type: twin-jet utility transport
Powerplant: two 1361-kg (3,000 lb) thrust Rolls-Royce Viper 525 turbojet engines
Performance: cruising speed 722 km/h (449 mph) at 12500 m (41,010 ft); initial rate of climb 1650 m (5,413 ft)/minute; service ceiling 13715 m (44,995 ft); range, with 45 minutes reserves 2128 km (1,322 miles)
Weights: empty 4830 kg

(10,648 lb); maximum take-off 8165 kg (18,000 lb)
Dimensions: span 11.43 m (37 ft 6 in); length 12.85 m (42 ft 2 in); height 4.80 m (15 ft 9 in); wing area 20.90 m² (225.0 sq ft)
Armament: none
Operator: Italy

A Piaggio PD.808 ECM aircraft of the 71 Gruppo Guerra Eletronica, Aeronautica Militare Italiano, based at Practica de Mare, Roma.

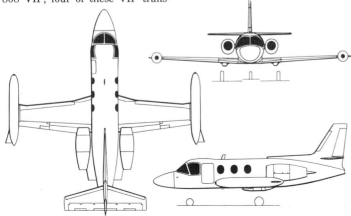

Piaggio PD.808TF

Pilatus P-3

The Pilatus P-3-01 prototype aerobatic trainer made its first flight on 3 September 1953. The Swiss *Flugwaffe* considered it underpowered and the company accordingly built the P-3-02 for service evaluation, powered by a 179-kW (240-hp) Lycoming GO-435-C2AS six-cylinder air cooled inverted-vee engine. Satisfactory tests led to an order for a service test series of 12 machines, serialled A-802 to A-813. Most of the machines were still in first-class condition 20 years after going into service. They were followed by a series of 60 P-3-05 aircraft, serials A-814 to A-873, which were powered by the same engine as the evaluation series. Although the Swiss authorities wanted more power, no suitable powerplant was available. Deliveries began in 1958.

The P-3 is a low-wing cantilever monoplane with NACA Series 649-wing section. The wing is a single-spar aluminium-alloy structure with split flaps and ailerons. The fuselage is an all-metal semi-monocoque. The angular tailplane is a cantilever all-metal structure. The nosewheel of the tricycle landing gear retracts backwards into the fuselage, and the mainwheels retract inwards.

Pupil and instructor are seated in tandem under a large glazed canopy with sliding sections. Dual controls are provided. Equipment can include oxygen apparatus, a 7.62-mm (0.3-in) machine-gun in a pod below the left wing, racks for two light practice bombs below the right wing, a single training rocket launcher under each wing, or a camera gun. An R/T transceiver is standard.

The P-3 has been used for *ab initio* and intermediate training aerobatics (with inverted-flight capability for short periods), night flying, instrument flying, and weapon training.

In order to improve flight characteristics the Swiss air arm has modified all P-3s by the addition of a long vertical fin on the rear underside of the fuselage. As a result a propensity for aircraft to go into a 'flat spin' during training was overcome.

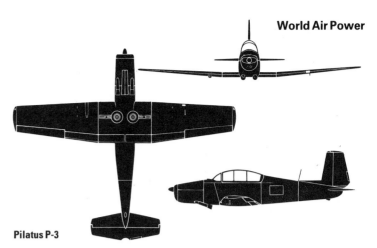

Pilatus P-3

Specification
Type: two-seat intermediate trainer
Powerplant: one 194-kW (260-hp) Lycoming GO-435-C2A six-cylinder horizontally opposed air-cooled engine
Performance: maximum speed 310 km/h (193 mph) between sea level and 2000 m (6,562 ft); economic cruising speed 255 km/h (158 mph); range 750 km (466 miels); maximum rate of climb at sea level 420 m (1,378 ft) per minute; service ceiling 5500 m (18,135 ft)
Weights: empty 1190 kg (2,623 lb); maximum take-off 1530 kg (3,373 lb)
Dimensions: span 10.4 m (34 ft 1.4 in); length 8.75 m (28 ft 8.5 in); height 3.05 m (10 ft 0 in); wing area 16.5 m^2 (1776.6 sq ft)
Armament: provision for one 7.62-mm (0.3-in) machine-gun, two practice bombs, or two rocket launchers
Operators: Brazil, Switzerland

Pilatus PC-6

The PC-6 Porter light transport has excellent STOL qualities. Almost 500 have been built, of which about 150 are military. It will carry 10 passengers in addition to the pilot, eight parachutists or two stretchers and three attendants. Roles include supply-dropping, SAR (with optional radar in a wing pod), aerial survey and photography, crop spraying, water bombing, glider towing and rainmaking. In addition to doors for both pilot and front passenger (or co-pilot, if optional dual controls are fitted), the Porter has a large, rearward-sliding door on the right side and double doors on the left. First flown on 4 May 1959, the original PC-6 Porter was powered by a 261-kW (350-hp) Lycoming GSO-480 piston engine. Early customers included the Swiss Air Force, which bought 12, and Burma, where about three remain in service. After producing the PC-6/340 and PC-6/340-H2 with increased weight, Pilatus offered the PC-6-A1/H2 Turbo Porter with a 522-kW (700-shp) Turboméca Astazou XII. When fitted with a 410-kW (550-shp) Pratt & Whitney Canada PT6A-20 the aircraft became a PC-6B1/H2, while a 429-kW (575-shp) Garrett TPE331-25D produced the PC-6C1/H2. The B1 and C1 were also produced under licence in the US by Fairchild Hiller, whose Armed Porter can carry machine-gun pods on the inner wing pylons and light bombs on the outer hardpoints. Uprated engines have resulted in further variants, of which the PC-6/B2-H2 is current. Switzerland obtained six Turbo Porters and converted its earlier aircraft to the same standard in 1980-81, all now operated by Leichte Fliegerstaffel for SAR/mountain rescue. Austria has 11 with the Flachenfliegerstaffel of Fliegerregiment I at Tulln-Langenlebarn, while the US Army garrison in Berlin flies two as UV-20A Chiricahuas. Peru bought at least 17, of which about nine remain, and the Australian Army 19, 14 survivors being based at Oakey with 173 (General Support) Sqn and the School of Army Aviation. Deliveries have recently been made to Colombia (for the military airline SATENA), Iraq and Malaysia.

Specification
Pilatus PC-6/B2-H2 Turbo Porter
Type: STOL utility transport
Powerplant: one 410-kW (550-shp) flat-rated Pratt & Whitney Canada PT6A-27 turboprop engine
Performance: maximum cruising speed 259 km/h (161 mph); initial rate of climb 387 m (1,270 ft)/minute; service ceiling 8535 m (28,000 ft); range on internal fuel 1050 km (652 miles)
Weights: empty 1218 kg (2,685 lb); maximum take-off 2200 kg (4,850 lb)
Dimensions: span 15.13 m (49 ft 7.7 in); length 10.90 m (35 ft 9.1 in); height 3.20 m (10 ft 6 in); wing area 28.80 m^2 (310.0 sq ft)
Armament: none
Operators: Angola, Argentina, Australia, Austria, Bolivia, Burma,

Pilatus PC-6/B2 Turbo Porter

Chad, Colombia, Ecuador, Iran, Iraq, Jugoslavia, Malaysia, Peru, Switzerland, Sudan, Thailand, USA (Army)

A Pilatus PC-6 Turbo-Porter of the Fuerza Aerea Colombiana, which use its Porters for transport duties with the air-force run national airline, Satena.

Pilatus PC-7 and PC-9

The PC-7 Turbo Trainer resulted from a company decision to introduce the Pratt & Whitney Canada PT6A-20 turboprop to the Pilatus P-3 basic trainer. The prototype flew on 12 April 1966, but generated little interest until the 'fuel crisis' sent jet trainer costs rocketing. Pilatus produced a second prototype powered by a 410-kW (550-shp) flat-rated PT6A-25 before the first production PC-7 flew in August 1978. Capable of performing aerobatics as well as basic and transition training, it found a ready market, aided by certification to FAR Pt 23 and US MIL Spec. It can undertake IFR and tactical training, and with six underwing hardpoints is a useful COIN aircraft. Some 350 have been produced, the first being 17 for Burma, followed by 24 for Bolivia's Colegio Militar de Aviación. The largest fleet is operated by Mexico, which bought 55, all except the first eight for COIN duties with six squadrons (Escuadrones 201 and 203-207 at Cozumel, La Paz, El Ciprés, Mérida, Puebla and Cuidad Ixtapec respectively). Guatemala has 12, used against guerrilla forces, and the Chilean navy's VT-4 at Quilpué-El Belloto has 10 for pilot training. Both Iraq (52) and Iran (undisclosed, but believed at least 36) train pilots on PC-7s for their continuing war, while other users include Abu Dhabi (14), Angola (12) and Malaysia (44). The Austrian Pilotenschule at Zelt-

The Royal Malaysian Air Force operates 44 Pilatus PC-7s in the training role, replacing a mixed fleet of Bulldogs and Cessna 172s.

weg received six in 1983-84, and the Swiss Air Force obtained 40 for Fliegerschule I-Teil at Magadino. Civilian operators have bought a further four.

Improvement plans in a 1982 scheme for the PC-7 received a spur when the RAF announced the AST.412 requirement, as a result of which the Pilatus PC-9 was offered with backing from BAe. Flown on 7 May 1984, the PC-9 has numerous differences for better performances and handling. Most obvious is the raised rear cockpit with stepped

Martin-Baker CH11 ejection seats and updated avionics including CRT and HUD displays. An airbrake is fitted, mainwheels are enclosed when retracted, and the aerofoil section is changed for operation at the higher speeds possible with the uprated engine. Initial deliveries are planned in December 1985, an initial batch of six being built in anticipation of orders. The PC-7 and PC-9 are complementary and will continue in parallel manufacture.

Specification
Pilatus PC-9
Type: turboprop basic trainer
Powerplant: one 708-kW (950-shp) flat-rated Pratt & Whitney Canada PT6A-62 turboprop engine
Performance: maximum speed 593 km/h (368 mph); initial rate of climb 1219 m (4,000ft)/minute; service ceiling 12190 m (40,000 ft); range 1112 km (691 miles)
Weights: empty 1610 kg (3,549 lb); aerobatic take-off 2200 kg (4,850 lb); maximum take-off, with stores 3200 kg (7,055 lb)

Dimensions: span 10.12 m (33 ft 2.4 in); length 10.05 m (32 ft 11.7 in); height 3.26 m (10 ft 8.3 in); wing area 16.39 m^2 (176.43 sq ft)
Armament: optional light weapons
Operators: (PC-7) Abu Dhabi, Angola, Austria, Bolivia, Burma, Chile, Guatemala, Iran, Iraq, Malaysia, Mexico, Switzerland

Piper L-4/L-18/L-21/U-7

In 1941 the US Army began a procurement programme that was to give it a large number of Piper Cubs for the artillery spotting and liaison role, these being operated in several variants under overall L-4 designations. After the end of World War II standard Piper Cub 95s were acquired for supply by the US to friendly nations, 105 L-18Bs going to the Turkish army, and 108 of 838 similar L-18Cs to other nations. In 1951 the US Army procured 150 Piper PA-18 Super Cubs with 93 kW (125 hp) engines, these being designated L-21A; a further purchase of 568 with 101 KW (135 hp) engines became designated L-21B. In 1962 L-21A/Bs remaining in service were redesignated U-7A and U-7B respectively, but all have since been withdrawn (some being sold or given to other users).

Specification
Piper L-18B
Type: two-seat liaison aircraft
Powerplant: one 67-kW (90-hp) Continental C90-8F flat-four piston

engine
Performance: maximum speed 177 km/h (110 mph); cruising speed 161 km/h (100 mph); range 402 km (250 miles)
Weights: empty 363 kg (800 lb); maximum take-off 680 kg (1,500 lb)
Dimensions: span 10.74 m (35 ft

3 in); length 6.82 m (22 ft 4.5 in); height 2.03 m (6 ft 8 in); wing area 16.63 m^2 (179.0 sq ft)
Armament: none
Operators (L-18/U-7): Argentina, Iran, Israel, Nicaragua, Norway, Sweden, Thailand, Uganda, Uruguay

Large numbers of Piper Super Cubs remain in service with a variety of military and civilian operators; this colourful example is an Israeli air force aircraft. Most surviving Cubs are used as hacks, helicopters having taken over their original role.

Piper PA-23 Apache/Aztec

The PA-23 Apache was the first of the post-war four-seat light twins to be launched on the US market. Progressive development resulted in an additional seat, swept vertical tail and 186-kW (250-hp) in place of 112 to 119-kW (150 to 160-hp) engines, the outcome being the Model PA-23-250 Aztec sold in 1960-81. The largest military order was placed by the US Navy which procured 20 UO-1 (later U-11A) for communications within the Continental USA. The Armée de l'Air bought the Aztec for support of the south Pacific nuclear test range.

Specification
Piper PA-23 Aztec
Type: five-seat transport
Powerplant: two 186-kW (250-hp) Lycoming IO-540-C4B5 (optional

TIO-540-C1A turbocharged) flat-six piston engines
Performance: maximum cruising speed 338 km/h (210 mph); initial rate of climb 454 m (1,490 ft)/minute; service ceiling 5365 m (17,600 ft); range 1947 km (1,210 miles)
Weights: empty 1330 kg (2,933 lb); maximum take-off 2359 kg (5,200 lb)
Dimensions: span 9.21 m (30 ft 2.5 in); length 11.34 m (37 ft 2.5 in); height 3.15 m (10 ft 4 in); wing area 19.28 m^2 (207.56 sq ft)
Armament: none
Operators: Argentina, Cameroon, Costa Rica, France, Malagasy, Mexico, Nigeria, Peru, Senegal, Spain, Tanzania, Uganda, Venezuela, Zambia

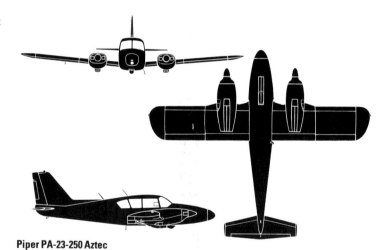

Piper PA-23-250 Aztec

Piper PA-24 Comanche

With the PA-24 Comanche of 1956 Piper introduced a relatively high-performance four-seat all-metal aircraft with retractable landing gear and a 134-kW (180-hp) Lycoming O-360-A1A engine. In 1958 Piper introduced the Comanche 250 with a 186-kW (250-hp) Lycoming O-540-A1A5 engine and fuel capacity increased from 189 to 227 litres (50 to 60 US gallons). This was superseded in August 1958 by the Comanche 260, with a 194-kW (260-hp) Lycoming IO-540 engine and revised landing gear. The most powerful of the

Comanches, the Model 400, received Type Approval on 28 January 1964. Some structural changes were made to allow operation at increased gross weight and higher speeds, resulting from the installation of a 298-kW (400-hp) Lycoming IO-720-A1A engine. Fuel capacity was increased to 379 litres (100 US gallons) and a new tail unit with all-moving horizontal surfaces was introduced. Small numbers of most versions were bought by several air forces.

Specification
Piper PA-24-260B Comanches B
Type: four-seat light monoplane
Powerplant: one 194-kW (260-hp) Lycoming IO-540-D/E flat-six piston engine
Performance: maximum cruising speed 293 km/h (182 mph) at 2135 m (7,000 ft); service ceiling 6095 m (20,000 ft); range at maximum cruising speed with standard fuel 1162 km (722 miles)
Weights: empty 784 kg (1,728 lb); maximum take-off 1406 kg (3,100 lb)

Dimensions: span 10.97 m (35 ft 11.75 in); length 7.62 m (25 ft 0 in); height 2.29 m (7 ft 6 in); wing area 16.54 m^2 (174.0 sq ft)
Armament: none

Piper PA-28

The PA-28 Cherokee first flew on 14 January 1960 and is an all-metal semi-monocoque design with a low constant-chord wing and fixed landing gear. Initial versions had a 112 or 119-kW (150 or 160-hp) engine, but power rose to 175 kW (235 hp) in 1964. Military examples have been used for training and communications. The Cherokee-140 is aimed at the low-cost trainer market, with a 104-kW (140-hp) engine. Principal users have been the Tanzanian air force and Royal Jordanian air academy. In 1973 the PA-28-151 Warrior introduced a wing of higher aspect ratio with tapered outer panels. This was also used on the PA-28-236 Dakota (described in the specification) which has been built from Piper kits by the maintenance division of the Chilean air force for use as a trainer and light transport.

Specification
Piper PA-28-236 Dakota
Type: four-seat basic trainer/light communications aircraft
Powerplant: one 175-kW (235-hp) Lycoming O-540-J3A5D flat-six piston engine
Performance: maximum cruising speed 269 km/h (167 mph); initial rate of climb 338 m (1,110 ft)/minute; service ceiling 5335 m (17,500 ft); range 1508 km (937 miles)
Weights: empty 730 kg (1,610 lb); maximum take-off 1361 kg (3,000 lb)
Dimensions: span 7.53 m (24 ft

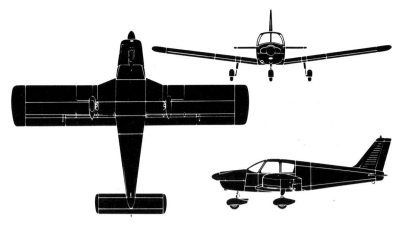

Piper PA-28-180 Cherokee

8.5 in); length 10.78 m (35 ft 4.5 in); height 2.20 m (7 ft 2.5 in); wing area 15.79 m² (170.0 sq ft)

Armament: none
Operators: Chile, Costa Rica, Finland, Jordan, Libya, Tanzania

---

Piper PA-30 Twin Comanche

This light twin was produced as a smaller six-seat companion to the Aztec. The prototype flew on 7 November 1962, and 1,996 were built. In 1970 Piper introduced the PA-39 Twin Comanche C/R, with counter-rotation engines and optional fifth and sixth seats, 144 being built. The main military user is the Spanish navy but one flew in RAF markings on RAE research programmes.

Specification
Piper PA-30 Twin Comanche
Type: four-seat light communications aircraft
Powerplant: one 119-kW (160-hp) Lycoming IO-320-B flat-four piston engine
Performance; maximum cruising speed 319 km/h (198 mph); initial rate of climb 445 m (1,460 ft)/minute; service ceiling 5670 m (18,600 ft); range 1931 km (1,200 miles)
Weights: empty 1015 kg (2,238 lb); maximum take-off 1633 kg (3,600 lb)
Dimensions: span 10.97 m (36 ft 0 in); length 7.67 m (25 ft 2 in); height 2.50 m (8 ft 2.5 in); wing area 16.54 m² (178.0 sq ft)
Armament: none
Operators: Chile, Finland, France, Nigeria, Panama, Spain

Piper PA-31 Navajo

The PA-31 Navajo is a low-wing monoplane of all-metal construction. The prototype first flew from Piper's Lock Haven headquarters on 30 September 1964, and went through numerous detailed changes before receiving its FAA Type Certificate in February 1966. The Navajo was offered with two power options—the standard version with two Lycoming IO-540-M1A5 engines and the Turbo Navajo with two turbocharged Lycoming TIO-540-A1As. In practice, the majority of customers bought the Turbo Navajo and the normally-aspirated version was eventually dropped from production. The Turbo Navajo was joined in the product line by the Navajo C/R in 1975. Improved performance was offered in this aircraft through the use of 242-kW (325-hp) Lycoming TIO-540-F2BD engines, and these incorporated the counter-rotating concept which allows both propellers to turn inwards thus eliminating many of the torque effects on piston twins. Some 1,430 examples of the Navajo, Turbo Navajo and Navajo C/R had been sold by December 1978.

The Navajo had been designed as the basis for a family of aircraft and the next move was the introduction of pressurization. The PA-31P Pressurized Navajo appeared in 1970, and 248 were sold by the time it was withdrawn in 1978. In its turn, the PA-31P permitted Piper to develop the PA-31T Cheyenne, powered by two Pratt & Whitney PT6A turboprops, which is now available in three separate versions, each offering a different choice of power and seating capacity. The PA-31P and PA31T have both been commercially successful but military use has been limited. One Pressurized Navajo was used for a time by the Escuadron 912 of Spanish air force, but no Cheyennes had been delivered to military users by the end of 1978.

The other alternative open to Piper was stretching the Navajo's fuselage, and this came about in September 1972 when the PA-31-350 Chieftain appeared. This proved to be the most successful model, substantially outselling the Turbo Navajo. Engine power increased to 261-kW (350-hp) and two extra passengers could be carried, with baggage accommodated in large nose and rear cabin compartments. Navajos became popular as light transport aircraft with several air arms, although the type was not selected by any of the US military forces. It was generally procured 'off-the-shelf' with either a utility or executive interior, and the largest operator has been the French Aéronavale which received 12 aircraft in 1973 and 1974. These replaced the Dassault MD312 Flamants which had given good service since the late-1940s. Navajos were allocated to a number of *escadrilles de servitude* for use on communications within metropolitan France. The aircraft allocated to *Escadrille* 3S have been applied to instrument training duties, and the VIP *Section de Liaison de Dugny* has three Navajos to support the Paris naval headquarters. Other users of the type include the Syrian air force, which bought two Navajos equipped with aerial survey cameras for map-

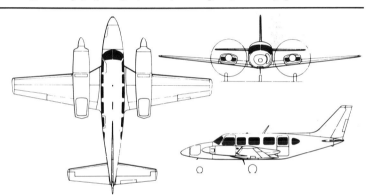

Piper PA-31 Navajo Chieftain

ping work, and the Kenyan and Nigerian air forces which both use Chieftains.

Specification
Type: eight-seat light communications aircraft
Powerplant: two Lycoming turbocharged 231-kW (310-hp) TIO-540-A1A six-cylinder horizontally-opposed piston engines
Performance: maximum speed 418 km/h (260 mph); normal cruising speed at 75 per cent power at 7163 m (23,500 ft) 398 km/h (247 mph); long-range cruising speed at 45 per cent power at 7315 m (24,000 ft) 291 km/h (181 mph);

maximum range at long-range cruising speed with 45 minutes reserves 2480 km (1,550 miles); rate of climb 425 m (1,395 ft) per minute; service ceiling 8016 m (26,300 ft)
Weights: empty 1741 kg (3,842 lb); maximum take-off 2945 kg (6,500 lb)
Dimensions: span 12.40 m (40 ft 8 in); length 9.94 m (32 ft 7½ in); height 3.96 m (13 ft 0 in); wing area 21.3 m² (229 sq ft)
Operators: Argentina (Navy), Chile (Navy), France (Navy), Kenya, Nigeria, Spain, Syria United Kingdom

One of two Piper Navajos delivered to the Kenyan air force for miscellaneous transport and communications duties.

Piper PA-34 and PZL M-20

The PA-34 Seneca light twin was introduced as a general utility model in 1971 to replace the Twin Comanche. With 149-kW (200-hp) engines, it featured a large entry door on the left with a supplementary cargo door, and the long nose had a baggage compartment. The PA-34-200T Seneca II of 1975 brought turbocharged engines, additional rear windows and an increase in useful load. This was also built under licence by EMBRAER in Brazil as the EMB-810 and by PZL-Mielec in Poland as the PZL M-20 Mewa with 164-kW (220-hp) PZL-Franklin 6A-350C-1L/R engines. The current PA-34-220T Seneca III has a single-piece windshield and increased power. The main military user is Brazil which has over 30 U-7s for communications and instrument training with the 8th Grupo de Aviação.

Specification
Piper PA-34-220T Seneca III
Powerplant: one Continental TSIO-360-KB and one LTSIO-360-KB counter-rotating flat-six piston engine, each of 164-kW (220-hp)
Performance: maximum cruising speed 360 km/h (224 mph); initial rate of climb 427 m (1,400 ft)/minute; certificated ceiling 7620 m (25,000 ft); range, with supplementary tanks 1844 km (1,146 miles)
Weights: empty 1294 kg (2,852 lb); maximum take off 2155 kg (4,750 lb)
Dimensions: span 11.85 m (38 ft 10.5 in); length 8.72 m (28 ft 7.3 in); height 3.02 m (9 ft 10.9 in); wing area 19.39 m² (208.70 sq ft)
Armament: none
Operators: Argentina, Brazil, Costa Rica, Jordan, Libya, Nicaragua, Pakistan

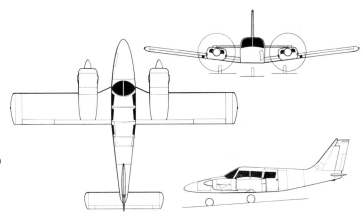

Piper PA-34 Seneca

PZL-Mielec (Antonov) An-28

The Antonov design bureau in the Soviet Union derived this STOL twin-turboprop transport from the An-14 in the early 1960s. Development was protracted, and the final choice of engine and greatly improved airframe were not finalized until 1975. Series production was assigned to WSK-PZL Mielec in Poland, the engines being produced by WSK-PZL Rzészow.

The An-28 (NATO codename 'Cash') is a refined and attractive machine with outstanding STOL performance. Construction is all-metal, except for the carbon composite skin of the long-span double-slotted flaps and fabric-covered ailerons. Flight controls are manual, but flaps, aileron spoilers (which automatically open asymmetrically following engine failure), brakes and nosewheel steering are hydraulically operated. Remarkably, the small engines provide sufficient bleed air to de-ice the entire wing and twin-finned tail, the latter having an upside-down tailplane with inverted fixed slat. Automatic slats over the outer wings make the An-28 stall-proof, even with the stick held fully back.

The two-seat cockpit has bulged side widows and electric windshield de-icing. The unpressurized main cabin is 5.26 m (17 ft 3 in) long and about 1.70 m (5 ft 7 in) wide and high. The main doors for passenger and cargo loading are the downward-opening rear doors, which can admit small vehicles. Normal passenger accommodation is for 17, arranged mainly 1+2, with all seats able to fold against the wall for carrying cargo. Maximum payload is 2000 kg (4,409 lb), and the interior can be equipped for medevac, parachute training, firefighting, navigation training, photogrammetric survey and many other duties.

Mielec's initial batch numbered just 15 aircraft, the first being completed in early 1984. It is expected that very large numbers will be ordered by Aeroflot, and smaller numbers are expected to be built for Warsaw Pact air forces and for export. Apart from its exceptional safety the An-28 offers unrestricted full-payload operation from the highest and hottest airstrips. It is possible that in due course ski and float versions may make their appearance.

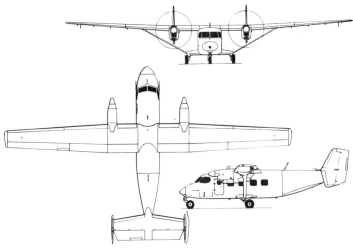

PZL-Mielec An-28 'Cash'

Specification
PZL-Mielec An-28
Type: STOL utility transport
Powerplant: two 716-kW (960-shp) PZL-10S (Glushenkov TVD-10B) free-turbine turboprop engines driving reverse-pitch propellers
Performance: economic cruising speed 337 mph/h (209 mph); take-off run 260 m (853 ft) at maximum weight; landing run 170 m (558 ft); service ceiling 6000 m (19,685 ft); range with maximum payload 560 km (348 miles)
Weights: empty 3750 kg (8,267 lb); maximum 6500 kg (14,330 lb)
Dimensions: span 22.06 m (72 ft 4.5 in); length 13.10 m (42 ft 11.7 in); height 4.90 m (16 ft 0.9 in); wing area 39.72 m² (427.56 sq ft)
Armament: none

PZL-Mielec M-26 Iskierka

The Mielec factory designed the M-26 Iskierka (little spark) to compete against PZL-Okecie's Orlik. As its name implies it is intended as the basic pilot trainer for the PWL (Polish air force), from which pupils will progress to the turbojet-powered PZL Mielec TS-11 Iskra (spark). Compared with its rival it has less power, lighter weight and a bigger wing, and thus is likely to be a much more docile and less-demanding machine. On the other hand it would be unable to take pupils so far down the syllabus as the Orlik and the jump at transition to the jet would be appreciably greater.

Like the Orlik the Iskierka is an all-metal low-wing tandem aircraft with retractable tricycle landing gear. Few details and no photograph had been released by late 1985, though first flight was due in 1984. It is possible that Mielec had second thoughts and delayed the prototype in order to re-engine it with a much more powerful turboprop in an attempt to win against the Orlik. One of the few known decisions is that the airframe uses major components similar to those of the M-20

Mewa, which is the Piper Seneca built under licence. The unpressurized cockpit has fixed seats and a single very large moulded canopy hinged open to the right. It is not known if the three-blade constant-speed propeller will be a Hartzell or a PZL-Okecie product. The Mewa can have optional rubber-pneumatic de-icing boots on the wing and 'slab' tailplane, and this will probably carry over to the Iskierka, together with the manually operated slotted flaps and hydraulically retracting landing gear.

Specification
PZL Mielec M-26
Type: military basic pilot trainer
Powerplant: originally intended to be one 153-kW (205-hp PZL-F (Franklin) 6A-350 flat-six air-cooled piston engine (but see text)
Performance: no information released
Weights: no information released
Dimensions: span 9.60 m (31 ft 6 in); length 8.30 m (27 ft 2.8 in); height 2.90 m (9 ft 6 in)
Armament: none at present planned

PZL-Mielec TS-11 Iskra/Iskra 2

The first Polish-designed jet aircraft is the PZL-Mielec TS-11 Iskra (spark) trainer. Surprisingly, in view of the Warsaw Pact's rigid policy of standardization, it was adopted by the Polish air force instead of the Pact's otherwise universal Aero L-29.

Design work on a jet advanced trainer to replace the TS-8 Bies started under the direction of Tadeusz Soltyk. The first TS-11 took to the air in December 1959. In the autumn of 1961 a later TS-11 participated in a 'fly-off' competition with the L-29 and the Soviet Union's Yakovlev Yak-30. The L-29 emerged as the winner, and was duly adopted as standard equipment by the Warsaw Pact.

However, Poland elected to continue development of the TS-11, partially because of performance improvements anticipated from the PZL-Rzeszów SO-1 turbojet. This powerplant was not available for early Iskras, which flew with the 780-kg (1,720-lb) HO-10.

Of conventional, all-metal structure, the TS-11 has an unswept, mid-mounted wing, with manual ailerons, double-slotted flaps and slatted airbrakes, inboard of a fence. The pod-and-boom fuselage is built in three sections. An avionics bay and a 23-mm cannon occupy the nose section. Tandem pressurized cockpits have an upward-hingeing canopy and lightweight ejection seats. A tapered boom, beneath which the engine exhausts, carries the manually-controlled tail. The landing gear is of the tricycle type, and designed for rough-field operation.

The engine is mounted under the wing. Later production Iskras have the SO-3 turbojet, which offers a longer period between overhauls. The fuel is housed in a fuselage tank of 500 litres (110 Imperial gallons), a 70-litre (15.5-Imperial gallon) collector tank (between the cockpit section and the engine bay) and two wing tanks each of 315-litre (69-Imperial gallon) capacity.

The Iskra carries its cannon low on the right side, and also has provision for various combinations of underwing stores. Up to 100 kg (220 lb) of bombs, rocket pods or 7.62-mm (0.3-in) machine-gun pods can be carried on each of two pylons (increased to four on later models). A single-seat attack prototype has been tested. This has an uprated engine and a 2000-litre (44-Imperial

gallon) increase in fuel capacity. A reconnaissance prototype mounts three cameras in the aft cockpit.

Pre-production Iskras, powered by the HO-10, reached the Polish air force in 1964. However, it was not until 1967 that the Iskra entered service in appreciable numbers. The type serves at the Central Flying School in the basic training role and as a weapons trainer.

In 1974 the Indian air force placed an order for 50 Iskras to replace the de Havilland Vampire. These were delivered between October 1975 and March 1976. They serve with the Fighter Training Wing at Hakimpet, whose role is to provide advanced pupils with 180 hours of jet instruction. It is anticipated that the HAL HJT-16 Kiran will eventually oust the Iskra at Hakimpet.

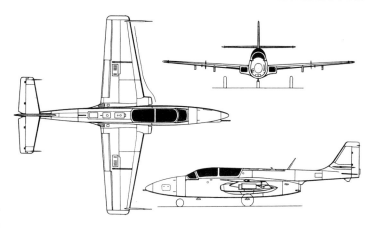

The Iskra has seen good service as the basic trainer for the Polish air force. Others were supplied to India, where they are due for replacement by the HAL Kiran.

Specification
Type: jet trainer
Powerplant: one 1000-kg 92,205-lb) PZL-Rzeszów SO-3 turbojet
Performance: maximum speed at 5000 m (16,400 ft) 720 km/h (447 mph); initial rate of climb 840 m (2,756 ft) per minute; ceiling

11140 m (36,550 ft); range with maximum internal fuel 1160 km (720 miles)
Weights: empty equipped 2495 kg (5,500 lb); loaded 3185 kg (7,020 lb); maximum take-off 3810 kg (8,400 lb)
Dimensions: span 10.06 m (33 ft

0 in); length 11.15 m (36 ft 5 in); height 3.5 m (11 ft 5½ in); wing area 17.5 m² (188.37 sq ft)
Armament: one 23-mm NS-23 cannon; 400 kg (880 lb) of underwing stores, or rocket and gun pods
Operators: India, Poland

PZL Warszawa-Okecie PZL-130 Orlik

Poland's WSK-PZL Warszawa-Okecie has been responsible for many important trainer and agricultural aircraft. In ths late 1970s Poland decided to pit the design team against WSK-PZL Mielec to see which could produce the better basic pilot trainer, with piston engine. Mielec chose a PZL-F engine, but Okecie selected a radial derived from the Soviet AI-14. Designated PZL-130 Orlik (eaglet), the Okecie aircraft was first flown on 12 October 1984.

An attractive all-metal machine, the Orlik seats pupil and instructor in tandem reclining seats with electric adjustment of height and longitudinal position. Cockpit instrumentation is planned to resemble that of high-performance combat aircraft, though all instruments are of the traditional electromechanical type. There is no provision for armament, the requirement being purely for pilot instruction, including aerobatic manoeuvres.

Controls are conventional, all movable surfaces having large trim tabs. Slotted flaps are fitted, and though pneumatic operation was studied it was finally decided to instal a hydraulic system to operate the flaps and landing gear, as in the PZL-Mielec TS-11 Iskra to which most pupils would proceed. The tricycle landing gear has wide track, with the main wheels carried outboard of the legs which retract inwards. Low-pressure tyres are fitted to permit operation from unpaved surfaces. Equipment includes communications radio, night lighting and provision for blind-flight instrument training.

At least two prototypes were built, for competitive evaluation in 1985

against the M-26 Iskierka. As this book went to press no winner had been announced. PZL Warszawa-Okecie has studied a turboprop development, but this would probably depend upon the PZL Rzeszów plant going ahead with a turboprop version of the 298-kW (400-hp) GTD-350 helicopter engine. The Czech M601 turboprop, in the 522-kW (700-hp) class, is much too powerful.

Specification
PZL Warszawa-Okecie PZL-130 Orlik
Type: basic pilot trainer
Powerplant: one 216-kW 290-hp WSK-PZL Kalisz AI-14RD nine-cylinder air-cooled radial piston engine
Performance: maximum cruising speed 330 km/h (205 mph); service ceiling 7000 m (22,970 ft); take-off run 300 m (984 km/h); range 1460 km (907 miles)

The Orlik is unusual among modern aircraft in having a radial engine. Training is the only role envisaged.

Weights: empty 950 kg (2,094 lb); loaded, normal 1300 kg (2,866 lb); maximum take-off 1500 kg (3,307 lb)
Dimensions: span 8.00 m (26 ft 3 in); length 8.45 m (27 ft 8.7 in); height 4.00 m (13 ft 1.5 in); wing area 12.20 m² (131.32 sq ft)
Armament: none

PZL-104 Wilga/Gelatik

Designed as a successor to the Yak-12 and its Polish version, the PZL-101 Gawron, the Wilga 1 all-purpose light aircraft made its maiden flight on 24 April 1962, its powerplant being a 134-kW (180-hp) WN-6B radial engine. Problems with this aircraft led to a thorough redesign of

the fuselage and tail and the resulting Wilga 2 powered by a 145-kW (195-hp) WN-6RB engine was flown on 1 August 1963. A third prototype, with a 168-kW (225-hp) Continental O-470, flew on 30 December of that year. A well-glazed cabin accommodates the pilot plus three pas-

sengers or an equivalent weight of cargo, and there are large doors to simplify loading. About 360 Wilgas were built in Poland, and the main military version in Polish service is the Wilga 32. Licence production in Indonesia of a version named the Gelatik 32, powered by the Conti-

nental O-470, totalled 56. Both types are used largely for liaison, parachute training and glider towing. Later versions include the Wilga 40 with a detachable freight container and the Wilga 43 with a slab tailplane.

Specification
PZL Wilga 32
Type: general-purpose STOL aircraft
Powerplant: one 194-kW (260-hp) Ivchenko AI-14R 9-cylinder radial piston engine
Performance: maximum speed 205 km/h (127 mph); economic cruising speed 135 km/h (84 mph); initial rate of climb 265 m (869 ft)/minute; service ceiling 3,680 m (12,075 ft); range 630 km (391 miles)

Weights: empty 737 kg (1,625 lb); maximum take-off 830 kg (1,830 lb)
Dimensions: span 11.14 m (36 ft 6.6 in); length 8.10 m (26 ft 6.9 in); height 2.50 m (8 ft 2.4 in); wing area 15.50 m^2 (166.85 sq ft)
Armament: none
Operators Wilga): include Poland

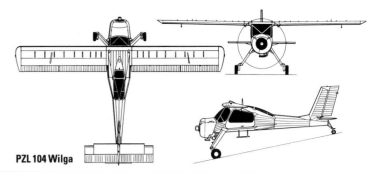

PZL 104 Wilga

PZL Swidnik Mi-2

Known to NATO as 'Hoplite', this twin-turbine helicopter was developed by the M.I.Mil bureau in the Soviet Union in the late 1950s and flown in September 1961. Distantly derived from the piston-engined Mil Mi-1, the Mi-2 offered much greater cabin volume and payload, higher performance and in almost all conditions engine-out safety. Following successful certification production was assigned to the Polish industry, and WSK-PZL Swidnik completed the first production machine in 1965.

Construction is basically all light alloy, though most of the later main rotors have a duralumin extruded spar with trailing edge pockets of bonded glassfibre. Main and tail rotors can have electric de-icing, as can the pilot's windscreens, and engine inlets can be de-iced by hot engine bleed air. The simple landing gears, with twin nosewheels, are fixed; a customer option is the addition of skis. Fuel is carried in a single 600-litre (132-Imp gallon) cell under the floor, and there is provision for carrying two 230-litre (52.6-Imp gallon) tanks on the sides of the fuselage. These tanks have to be in lieu of armament, which is carried by some versions.

Normally the cockpit sits the pilot alone on the left side, while the rear cabin, 2.27 m 7 ft 5.4 in) long, 1.20 m (39 ft 11.2 in) wide and 1.40 m (4 ft 7.1 in) high, usually has a central bench unit seating three passengers facing forward and three more facing aft. If necessary two extra seats can be added on the right side. Alternatively the cabin can be equipped to carry 800 kg (1,764 lb) of cargo, or four stretchers and an attendant, or two stretchers and two

PZL Swidnik Mi-2 of the Polish air force

seated casualties. A dual-control side-by-side cockpit is another option, and further equipment can include a central hook rated at 800 kg for a slung load, a 120 kg (264 lb) electric rescue hoist, weather radar in the nose, and for some tactical versions a comprehensive RWR (radar warning receiver) system and even ECM dispensers and jammers.

By late 1985, when production was almost complete, Swidnik had delivered about 4,500 Mi-2s in various sub-types of military and civil customers in many countries, including the Soviet Union and several Warsaw Pact air forces. Progressive updating has introduced increasing proportions of composite materials, an uprated hoist of 260 kg (573 lb) capacity and numerous special role kits. The Allison-engined Kania, Kitty Hawk and Taurus developments are civil only.

Specification
PZL Swidnik Mi-2
Type: light multi-role helicopter.
Powerplant: two 331-kW (444-shp)

PZL Swidnik Mi-2 'Hoplite'

PZL Rzeszów (Isotov) GTD-350P turboshaft engines
Performance: maximum cruising speed, clean 200 km/h (124 mph); hovering ceiling out of ground effect 1000 m (3,280 ft); range at low level with maximum payload and reserves 170 km (106 miles)
Weights: empty, passenger version 2402 kg (5,296 lb); maximum take-off 3700 kg (8,157 lb)

Dimensions: main rotor diameter 14.50 m (47 ft 6.9 in); length ignoring rotors 11.40 m (37 ft 4.8 in); height 3.75 m (12 ft 3.6 in); main rotor disc area 165.13 m^2 (1,777.5 sq ft)
Armament: some versions have lateral pylons for rocket launchers, gun pods or four AT-3 Sagger anti-armour missiles
Operators: Egypt, Poland

PZL Swidnik W-3 Sokól

The first of five prototypes of this helicopter flew at Swidnik on 16 November 1979. Though based on the Mi-2 and Kania family the W-3 Sokól (falcon) is appreciably larger and much more powerful. It is the first large turbine-engined helicopter to be designed in Poland.

The fuselage remains a light-alloy stressed-skin structure, but the main and tail rotors (now with four and three blades, respectively) have completely new blades of glassfibre construction, electric de-icing and hydraulically boosted flight controls. The main rotor blades have tapered tips, and a Salomon type hanging vibration absorber to give particularly smooth travel and low vibration in the cabin. The engines have electronic control, automatic torque sharing and further special features for reducing vibration. Fuel is housed in a group of underfloor cells with a combined capacity of

1700 litres (374 Imp gallons), giving much greater range than the Mi-2 series. The landing gears are particularly neat, with single levered-suspension legs on the main units and with the option of attaching skis or floats.

Unlike the Mi-2 the standard cockpit has two seats, and dual controls are an option. The main cabin is 3.20 m (10 ft 6 in) long and 1.60 m (5 ft 3 in) wide, giving far greater floor area than in the Mi-2. In the standard passenger configuration there are four rows each of three seats (12 in all), all quickly removable for loading up to 2000 kg (4,409 lb) of cargo. The limit for external slung load is 2100 kg (4,630 lb). Among the role equipment options is an electric rescue hoist of 150 kg (331 lb) capacity. Avionic equipment includes a two-axis autopilot and autostabilization system, plus normal equipment for

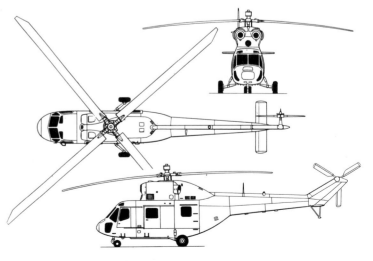

PZL Swidnik W-3 Sokól

night and bad-weather flying.

In 1985 production of this attractive helicopter was just beginning. No armed version had then been announced, but customers may be expected to include many air forces, probably including the forces of the Soviet Union, because there is no clause in the Glushenkov licence restricting production of the engine

(which is basically the same as that of the Ka-25 'Hormone') to civil application. No NATO name has been announced.

Specification
PZL Swidnik W-3 Sokól
Type: multi-role transport helicopter
Powerplant: two 640-kW (858-shp)

PZL (Glushenkov) 10W turboshaft engines, each with an emergency engine-out rating of 735-kW (986-shp)
Performance: maximum cruising speed 235 km/h (146 mph); hovering ceiling out of ground effect 1850 m (6,070 ft); range with maximum payload and reserves 680 km (422 miles)

Weights: empty, multi-role 3630 kg (8,003 lb); maximum 6400 kg (14,110 lb)
Dimensions: main rotor diameter 15.70 m (51 ft 6.1 in); length ignoring rotors 14.10 m (46 ft 3.1 in); height 4.12 m (13 ft 6.2 in); main rotor disc area 193.59 m^2 (2,083.89 sq ft)
Armament: none yet disclosed

Soviet Union 'Ram-M'

Least-known of all modern Soviet aircraft, a prototype of this high-altitude fixed-wing design was reported to have been seen on US satellite imagery in 1982. It was observed on Ramenskoye test airfield, and the resulting 'Ram-M' designation is the only one known in

the West. Nothing further has been reported, and as no NATO code name has been revealed it is probably fair to deduce that the type has not yet entered operational service.

Early reports stated that it was 'in the class of the U-2', which probably implied that it had a single jet

engine, and that it had a curious tail with a tailplane (horizontal stabilizer) joining the tops of twin fins (which were presumably carried on twin booms). There have been rumours that it is rather larger than the Lockheed U-2, with a span in excess of 30.50 m (100 ft). The

design bureau with the greatest experience of very high altitude reconnaissance aircraft is that of Yakovlev, but it would be unwise to assume that 'Ram-M' is a Yak design and premature to guess at a specification.

Republic F-84F Thunderstreak

Today only about a dozen F-84Fs are still active with the Greek Elliniki Aeroporia, these being the last airworthy examples of 2,711 made by Republic and General Motors 30 years ago. Equipping the 348th Mira at Larissa, they operate alongside about 18 RF-84F Thunderflash reconnaissance aircraft, this squadron serving as a 'finishing school' for newly qualified fighter pilots.

Greece is the last operator of the F-84F thunderstreak, a dozen flying from Larissa on weapons training duties.

Specification
Republic F-84F Thunderstreak
Type: fighter-bomber
Powerplant: one 3275-kg (7,220-lb) thrust Wright J65-W-3 turbojet engine
Performance: maximum speed at sea level 1118 km/h (695 mph); initial rate of climb 1920 m (6,300 ft)/minute; service ceiling 13505 m (44,300 ft); combat radius, clean 724 km (450 miles); ferry

range 3444 km (2,140 miles)
Weights: load, clean 8772 kg (19,340 lb); maximum take-off 10977 kg (24,200 lb)
Dimensions: span 10.24 m (33 ft

7.25 in); length 13.23 m (43 ft 4.75 in); height 4.39 m (14 ft 4.75 in); wing area 30.19 m^2 (325.0 sq ft)
Armament: six 12.7-mm (0.5-in)

Colt-Browning M3 machine-guns, plus up to 2722 kg (6,000 lb) of external ordnance, including bombs and rockets
Operator: Greece

Robin HR.100/250

Avions Pierre Robin builds an extensive range of low-wing light aircraft. The HR.100 was the company's first all-metal type in production, the prototype flying on 3 April 1959. The HR.100/250 was put into production in 1975 as a trainer for the Armée de l'Air. About 18 were delivered to the Centre d'Essais en Vol during 1976-77, and two are operated by the Aéronavale from Hyères in the communications role.

Specification
Type: four-seat lightplane
Powerplant: one 186-kW (250-hp) Lycoming IO-540-C4B5 flat-six piston engine

Performance: cruising speed 285 km/h (177 mph); initial rate of climb 324 m (1,063 ft)/minute
Weight: maximum take-off 1400 kg (3,086 lb)
Dimensions: span 9.08 m (29 ft 9.5 in); length 7.59 m (24 ft 10.8 in); height 2.71 m (8 ft 10.7 in); wing area 15.10 m^2 (162.54 sq ft)
Armament: none
Operator: France

France is the only operator of the Robin HR.100. It uses a small number of HR.100/250s, with retractable landing gear, for training.

Rockwell B-1B

Entering service with USAF Strategic Air Command's 96th Bomb Wing at Dyess AFB, Texas, in 1985, Rockwell's B-1B is a derivative of the B-1A which was cancelled by the Carter administration in June 1977. Limited testing of the four B-1A prototypes was allowed to continue, and in October 1981 President Reagan rescued Rockwell's bomber from near-limbo and ordered it into production as the B-1B, present planning calling for the delivery of 100.

The B-1B takes advantage of the many developments since the B-1A

first flew in December 1974. Most important is the adoption of some 'stealth' technology, changes to the engine intakes and other areas and RAM (radar-absorbent material) coatings significantly reducing radar cross-section and thus rendering it less likely to be detected. The radar signature is just one per cent that of the Boeing B-52.

The variable-geometry wing of the B-1 allows the type to operate at high speeds and low altitude, vastly reducing buffet with wings fully swept.

The first B-1B contracts were signed on 20 January 1982. One, for $1,317 million, covered the development phase including flight testing and two modified B-1As. The other, for $886 million, covered the first B-1B and long-lead items for ensuing B-1Bs. The modified No. 2 B-1A resumed flight trials on 23 March 1983, earmarked to explore handling (notably stability and control), perform weapons-system tests and evaluate certain new features. The No. 4 B-1A resumed flight status in summer 1984 – when, through no fault of the aircraft, the No. 2 crashed from low level – incorporating the remainder of the planned B-1B improvements, especially including defensive and offensive avionics systems.

Originally expected to fly in March 1985, the first B-1B was rolled out ahead of schedule on 4 September 1984 and was in the air on 18 October 1984. Deliveries to the USAF began in June 1985, and IOC (Initial Operational Capability) with 15 aircraft of the 96th BW was planned for August 1986 but may be

achieved earlier. Thereafter, deliveries at four per month are to continue to June 1988, though an order for 100 even more stealthy B-1Cs has long been discussed.

The B-1Bs can carry weapons ranging from free-fall nuclear bombs through AGM-69 SRAMs (Short-Range Attack Missiles) and AGM-86B (Air-Launched Cruise Missile) to conventional 'iron' bombs. Three internal bays are provided, a 9.53 m (31 ft 3 in) double bay ahead of the wing box and a single 4.57 m (15 ft 0 in) bay aft. In the conventional role 84 227-kg (500-lb) Mk 82 bombs or 24 907-kg (2,000-lb) Mk 84 bombs can be carried internally, and eight stores stations beneath the fuselage are compatible with either nuclear or conventional weapons.

Although changes have been incorporated in the airframe, including a massive increase in fuel capacity (and weight) it is the avionics which represent the most significant advance. Combining the best features of the offensive avionics of the B-52 and B-1A, as well as equipment

similar to that of the newest fighters, the B-1B has advanced forward-looking and terrain-following radars, a highly accurate Singer Kearfott inertial navigation system, an AFSATCOM (Air Force Satellite Communications) link and a doppler-radar altimeter. Defensive avionics include the Eaton AIL AN/ALQ-161 system which should enable the B-1B to penetrate existing and forecast enemy defences until well into the 1990s. This uses phased-array antennae (aerials) and is capable of identifying and dealing with many radar threats simultaneously and almost instantaneously.

Specification
Rockwell B-1B
Type: strategic heavy bomber
Powerplant: four General Electric F101-GE-102 turbofan engines, each 'in the 13608 kg (30,000 lb) thrust class'
Performance: maximum high-altitude speed approximately Mach 1.25; low-level penetration speed – at intended 61 m (200 ft) –

The B-1B incorporates much 'stealth' technology, and has a much reduced radar cross-section over the B-1A.

exceeds 966 km/h (600 mph); unrefuelled range approximately 11998 km (7,455 miles)
Weights: empty about 78018 kg (172,000 lb); maximum take-off 216364 kg (477,000 lb)
Dimensions: span, spread 41.67 m (136 ft 8.5 in), swept 23.84 m (78 ft 2.5 in); length 44.81 m (147 ft 0 in); height 10.36 m (34 ft 0 in); wing area approximately 181.16 m² (1,950.0 sq ft)
Armament: (nuclear) internal capacity for up to eight AGM-86B ALCMs or 24 AGM-69 SRAMs on rotary launchers, 12 B28 or B43 free-fall nuclear bombs or 24 B61 or B83 bombs, plus (external) 14 ALCMs or eight B28 or 14 B43/61/83; (conventional) up to 84 Mk 82 227-kg (500-lb) or 24 Mk 84 907-kg (2,000-lb) bombs, plus (external) 14 Mk 84 or 44 Mk 82 bombs
Operator: USAF (Strategic Air Command)

Rockwell/Aero Commander Shrike Commander

Model 520 and 560 Aero Commanders of the 1950s were sold to the US Army and Air Force, but have since been retired. In June 1958 Aero Commander was taken over by Rockwell and development included the 680FL Grand Commander with a stretched 11-seat interior. The smaller Commanders included the 680 with 254-kW (340-hp) Lycoming GSO-480A engines and the 500B with fuel-injected 186-kW (250-hp) Lycoming IO-540 engines. The Model 500S Shrike introduced a squared-off tail and revised nose; production ceased in 1979. Most

military deliveries were single aircraft used as VIP and general light transports, though Argentine aircraft have been used in the SAR role.

Specification
Rockwell Model 500S Shrike Commander
Type: 5/7-seat light twin
Powerplant: two 216-kW (290-hp) Lycoming IO-540-E1B5 flat-six piston engines
Performance: maximum cruising speed 348 km/h (216 mph); initial rate of climb 408 m (1,340 ft)/minute; service ceiling 5915 m

(19,400 ft); range, standard fuel no reserves 1526 km (948 miles)
Weights: empty 1972 kg (4,348 lb); maximum take-off 3062 kg (6,750 lb)
Dimensions: span 14.95 m (49 ft 0.5 in); length 11.15 m (36 ft 7 in); height 4.42 (14 ft 6 in); wing area 23.96 m² (255.0 sq ft)
Armament: none
Operators: Argentina (500U), Benin, Burkina-Faso, Colombia, Greece (680FL), Honduras, Indonesia (500A, 560A, 680FL), Iran (681B), Ivory Coast (500B), South Korea (520, 560F), Mexico

Rockwell/Aero Commander

(500S), Nicaragua (500S), Pakistan (680E), Venezuela (680)

Rockwell Turbo Commander

Rockwell's Model 680FLP pressurised Grand Commander was the basis for a range of turboprop business aircraft. The 680T was flown on 31 December 1964, powered by two Garrett TPE331-43s. The production 680T, 680V and 680W all have dual controls and eight seats as standard, the door being situated just behind the flight deck on the left. The main wheels retract rearward and rotate 90° to lie flat. The 690 of 1971 has an extended centre section and higher powered TPE331s; 15 were used by the Iranian army, navy and gendarmerie as

staff transports. Gulfstream Aerospace produced the models 690B, 695 and 695A, and now offer their Special Mission Aircraft (SMA) with a large upward-hinging right side door for use in the medevac role, and the maritime-patrol SMA Commander with an auxiliary fuel tank, nose-mounted APS-504 Litton search radar and underwing Vinten Vicon 70 surveillance pod.

Specification
Rockwell Turbo Commander 690B (IFR version)
Type: light transport

Powerplant: two 522-kW (700-shp) Garrett TPE331-5-251K turboprop engines
Performance: maximum cruising speed 526 km/h (327 mph); initial rate of climb 860 m (2,820 ft)/minute; service ceiling 9995 m (32,800 ft); maximum range 2718 km (1,689 miles)
Weights: empty 3054 kg (6,733 lb); maximum take-off 4683 kg (10,325 lb)
Dimensions: span 14.22 m (46 ft 8 in); length 13.52 m (44 ft 4.25 in) height 4.56 m (14 ft 11.5 in); wing area 24.71 m² (266.0 sq ft)

Rockwell Turbo Commander

Armament: none
Operators: Argentina, Iran, Pakistan

Rockwell International T-2 Buckeye

In late 1956 North American received a contract from the US Navy for 26 jet trainers. The first T-2A Buckeye flew on 31 January 1958, and initial deliveries went to NAS Pensacola, Florida. Production of the T-2A totalled 217, and were used mainly by squadrons VT-7 and VT-9 at NAS Meridian, Missouri, which catered for students selected for continuation training after completing 35 hours on the T-34. Modification of two T-2As began soon after production of that series ended in 1960. The major change was replacement of the Westinghouse J34 engine by two 1361-kg (3,000-lb) thrust Pratt & Whitney J60-P-6 turbojets side by side in the lower fuselage. A total of 97 of the resulting T-2Bs were built, the 34th and subsequent aircraft having additional fuel capacity. Most extensively built was the T-2C; the first flew on 10 December 1968, and 231 T-2Cs had been delivered by the end of 1975. From 1990 they will be replaced by the BAe/McDonnell Douglas T-45 Hawk. The T-2Ds of the Venezuelan air force differ in their avionics and by deletion of the carrier landing capability. T-2Es of the Hellenic air force also have avionics variations, but their major difference lies in six wing stores stations with a combined capacity of

Now due for replacement, the Rockwell T-2 Buckeye has served the US Navy well in the basic jet trainer role since 1960. This example is from VT-23.

1588 kg (3,500 lb), plus protection from small arms fire.

Specification
Rockwell International T-2C
Type: jet trainer
Powerplant: two 1338-kg (2,950-lb) thrust General Electric J85-GE-4 turbojet engines
Performance: maximum speed 853 km/h (530 mph) at 7620 m (25,000 ft); initial rate of climb 1798 m (5,900 ft)/minute; service ceiling 13870 m (45,500 ft); maximum range 1722 km (1,070

miles)
Weights: empty 3681 kg (8,115 lb); maximum take-off 5983 kg (13,191 lb)
Dimensions: span over wingtip tanks 11.62 m (38 ft 1.5 in); length 11.67 m (38 ft 3.5 in); height 4.51 m (14 ft 9.5 in); wing area 23.69 m² (255.0 sq ft)
Armament: can include gun packs, practice bombs, rockets and target towing-gear to a maximum of 290 kg (640 lb)
Operators: Greece, USA (USN), Venezuela

Rockwell T-2C Buckeye

Rockwell (North American) T-39 Sabreliner

In August 1956 the USAF specified a requirement for a general utility/trainer aircraft, then identified as UTX, signifying utility/trainer experimental, and in sending out its requests for proposals stipulated that interested manufacturers would be required to design, build and fly a prototype as a private venture. At that time North American Aviation had more or less completed the design study of a small pressurized jet transport aircraft, and was thus in a position to offer this design to the US Air Force with but few changes to make it capable of meeting their specification. So, on 27 August 1956, the company announced that it was to build a prototype of a small turbine-powered transport under the name Sabreliner.

The original design had placed the engines in the wing roots, but as detail design proceeded during early 1957 the configuration was changed to a rear-engine layout, with two turbojet engines attached to the sides of the rear fuselage. Construction was virtually complete in May 1958, but the first flight was delayed until four months later, as the result of the non-availability of suitable engines. It was with two 1134-kg (2,500-lb) General Electric J85 turbojet engines that the prototype flew on 16 September 1958, and the type completed its USAF flight test evaluation at Edwards AFB, California in December. Early in 1959 North American received an initial production order for seven aircraft.

The first of these, which by then had the USAF designation T-39A, made its initial flight on 30 June 1960. This had two Pratt & Whitney J60-P-3 turbojets of increased power, and some internal changes, and initial deliveries for the Air Training Command, on 4 June 1961, went to Randolph AFB. Subsequent contracts brought total order for the T-39A to 143, and these were delivered for service with the Air Training Command, Strategic Air Command, Systems Command, and

Rockwell CT-39 Sabreliner of the United States Air Force.

to the Headquarters of the USAF for command duties. From June 1967 the USAF also took delivery of a number of T-39As, which had been modified with strengthened landing gear and provided with seven, instead of four, passenger seats.

In the period February-June 1961, six aircraft designated T-39B were delivered to the Tactical Air Command for training duties at Nellis AFB, Nevada. These were equipped with a doppler navigation system and the NASARR all-weather search and range radar which was installed in the Republic F-105, and were used to train crews who were to fly the Thunderchief.

The designation T3J-1, subsequently T-39D, was allocated to 42 Sabreliners ordered from North American in 1962 by the US Navy. Required for the training of maritime radar operators, these had Magnavox radar systems installed, and delivery to the Naval Air Training Command HQ, at NAS Pensacola, Florida, began in August 1963. The US Navy acquired also seven Series 40 commercial Sabreliners, under the designation CT-39E, for high-priority transport of passengers, ferry pilots and cargo, and since 1973 has procured 12 of the longer-fuselage Sabreliner 60s under the designation CT-39G. These are used by both the US Marine Corps and US Navy for fleet tactical support duties. Under the

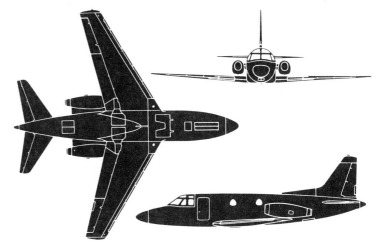

Rockwell T-39 Sabreliner

designation T-39F, a number of USAF T-39As were modified to make them suitable for the training of Wild Weasel ECM operators for service with the USAF's F-105Gs and McDonnell Douglas F-4Gs.

Specification
Type: twin-engine utility transport/trainer
Powerplant: (T-39A) two 1361-kg (3,000-lb) thrust Pratt & Whitney J60-P-3 turbojets
Performance: (T-39A) maximum speed 958 km/h (595 mph) at

10970 m (36,000 ft); cruising speed 727 km/h (452 mph) at 12190 m (40,000 ft); design range 2776 km (1,725 miles)
Weights: (T-39A) empty 4218 kg (9,300 lb); maximum take-off 8056 kg (17,760 lb)
Dimensions: span 13.54 m (44 ft 5¼ in); length (T-39A/B/D) 13.34 m (43 ft 9 in); length (CT-39E/G) 14.30 m (46 ft 11 in); height 4.88 m (16 ft 0 in); wing area 31.78 m² (342.05 sq ft)
Armament: none
Operators: Jordan, Mexico, Sweden, United States

RTAF-4 Chandra

In 1975 the Royal Thai Air Force set up an Office of Aeronautics and Aircraft Design at Bangsue. The RTAF had previously designed three unsuccessful types, only one of which was built, but the OAAD determined to create aircraft that could serve in the RTAF. Wisely the first of its designs was a derivative of a long-established basic trainer, the de Havilland Canada DHC-1 Chipmunk.

Designated RTAF-4, and named Chandra (Moon), the main difference lies in replacement of the Gipsy Major 10 engine by a more powerful American flat-four. Though this had little effect on side area, it was also decided to replace the distinctively curved 'de Havilland' vertical tail by a new fin and rudder of rather greater area, having a simpler straight-tapered shape with a small added dorsal fin. There are many other mostly minor changes, one being the provision of a new framed sliding canopy shaped like the blown Canadian pattern and built like the severe British canopy.

The RTAF-4 prototype had a number of minor differences from the standard chosen for the production Chandra, 13 of which were converted. They have proved satisfactory, and continue in use as basic pilot trainers at Don Muang airbase. The specification is generally as for the Chipmunk T.Mk 10, but with higher flight performance resulting from use of the 134-kW (180-hp) Avco Lycoming IO-360H four-cylinder air-cooled opposed engine and variable-pitch propeller.

RTAF-5

By far the largest aircraft-design task attempted in south-east Asia, this unusual machine was to some degree influenced by the Rockwell OV-10 (which equips RTAF No. 12 Sqn), though it is lighter and has about one-quarter of the engine power. The basic objectives were to create an all-Thai aircraft that would be genuinely useful and to try to combine in one aircraft the ability to fly FAC (forward air control) and close-support missions, while also being able to serve as an economical advanced trainer.

Of conventional light-alloy construction, the RTAF-5 has the tail unit carried on twin booms which project forward of the wing to accommodate the main landing gear. The mid/high wing has a fixed leading edge, slotted flaps and glassfibre tip tanks supplementing the main integral tanks in the centre section, where the pusher turboprop engine is mounted. Flaps and landing gear are electrically operated, and flight controls are manual. The crew (pupil and instructor, or pilot and backseater) sit in stepped seats under a large canopy with upward-opening double-hinged panels. Comprehensive avionics are fitted, and should an FAC version be built it would have EW/ECM provisions as well as the armament fit. In any case a gunsight is expected to be installed in the training version.

First flight was on 5 October 1984. By mid-1985 various small modifications had been made, for example to the engine inlet, and the landing gear has been cleared to retract. The RTAF hopes eventually to put this aircraft into service in numbers.

Specification
RTAF-5
Type: advance trainer, FAC and close-support aircraft
Powerplant: one 313-kW (420-shp) Allison 250-B17C turboprop engine
Performance: maximum speed at sea level 389 km/h (242 mph); maximum cruising speed 333 km/h (207 mph); take-off to 15 m (50 ft) 540 m (1,772 ft); maximum range 869 km (540 miles)
Weights: empty 1645 kg (3,627 lb); maximum take-off 2086 kg (4,599 lb)
Dimensions: span over tanks 9.86 m (32 ft 4.2 in); length 8.76 m (28 ft 8.9 in); height 3.05 m (10 ft 0 in); wing area 15.66 m² (168.57 sq ft)
Armament: provision for four underwing pylons, inners rated at 68 kg (150 lb) each and outers at 45 kg (100 lb) each; these would carry light attack stores, flares and possibly sensors

Saab 105

Developed as a private venture, the Saab 105 was intended as a trainer and light ground-attack aircraft that would be able also to handle other roles. These included liaison and executive transport (for which the side by side ejection seats can be replaced by four fixed seats), reconnaissance, survey and ambulance. The first of two prototypes flew on 29 June 1963, and the following year the Saab 105 was ordered into production for the Swedish air force, the first of 150 flying on 27 August 1965. These Sk 60s are powered by two Turboméca Aubisque turbofans and entered service in the spring of 1966 with F5, the flying training school at Ljungbyhed. Most were later given armament hardpoints, gunsights and associated equipment. The Sk 60B has ground-attack as its primary role, and the Sk 60C is equipped for photographic reconnaissance, having a Fairchild KB-18 camera in the nose, while retaining a ground-attack capability.

Some 75 Sk 60As serve in the training role with F5, pupils progressing from the Sk 61 Bulldog and flying some 160 hours on the jet before qualifying. In an emergency these trainers would comprise the equipment of five light ground-attack squadrons. A squadron of F21 based at Lulea, Sweden's most northerly air base, operates a mixture of Sk 60B attack and Sk 60C reconnaissance aircraft. F20, the air force college at Uppsala, also operates these variants, and second-line users include a staff liaison flight (with the four-seat version) and F13M who undertake target towing and weapon testing.

A development powered by J85 turbojets, the Saab 105XT first flew on 29 April 1967. Fuel capacity was increased to 2050 litres (451 Imp gallons), two 500-litre (110-Imp gallon) underwing drop tanks also being available. In addition to enhanced performance the 105XT had improved avionics, and a strengthened wing allowed the underwing load to be increased to 2000 kg (4,409 lb). This version can perform interception and target-towing roles. Infra-red guided missiles such as Sidewinder are carried for day interceptor duties. Forty aircraft designated Saab 105Ö were built for the Austrian air force and in late 1984 these were still the only jet aircraft in Austrian service. They equip four Staffeln, comprizing the Überwachungsgeschwader (surveillance wing) with two Staffeln operating in the air defence role, a task for which the 105Ö is ill-suited and a replacement has long been required. The remaining two Staffeln operate in the light-attack role as part of the Jagdbombergeschwader, one of the Staffeln being a weapon-training unit and the Düsenflugstaffel (jet conversion squadron) also flies the type. It is intended to operate the Saab 105Ö until the end of the 1980s.

Specification
Saab 105Ö
Type: light attack aircraft
Powerplant: two 1293-kg (2,850-lb) thrust General Electric J85-GE-17B turbojet engines
Performance: maximum speed 970 km/h (603 mph) at sea level; climb to 10000 m (32,810 ft) 4 minutes 30 seconds; service ceiling 13000 m (42,650 ft); range 2400 km (1,491 miles) at 13100 m (42,980 ft)
Weights: empty 2550 kg (5,622 lb); maximum take-off 6500 kg (14,330 lb)
Dimensions: span 9.50 m (31 ft 2 in); length 10.70 m (35 ft 1.3 in); height 2.70 m (8 ft 10.3 in); wing area 16.30 m² (175.46 sq ft)
Armament: provision for up to 2000 kg (4,409 lb) of underwing stores
Operators: Austria, Sweden

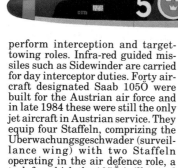
Saab 105 (Sk60) of the Swedish air force.

Austria has been a regular customer of Saab products, and operates 32 Saab 105s for air defence, ground attack and training.

Saab 35 Draken

Although designed initially as a bomber interceptor, the Draken (dragon) was developed for a wide variety of roles including ground attack and reconnaissance. The first prototype made its maiden flight in October 1955, and the J35A entered service with the Swedish air force in March 1960, armed with four Rb24 (Sidewinder) AAMs and two Aden 30-mm cannon. The last version for the Swedish air force was the J35F with a Hughes weapon system comprizing a pulse-doppler radar, automatic fire-control system and Falcon air-to-air missiles, built under licence in Sweden by LM Ericsson and other companies. The missiles were constructed by Saab, the radar-guided AIM-26A, known as the Rb27 in Swedish service, and the infra-red AIM-4D (Rb28). A J35F normally carries a combination of the two types. Many J35As were converted into Sk35C two-seat trainers. Nearly 550 Drakens have seen service with the Swedish air force, and the production run was extended by export orders received just as the line was about to be closed. Denmark ordered fighter-bomber, trainer and reconnaissance versions of the Saab 35X export model and designated them F-35, TF-35 and RF-35 respectively. The 35X is similar to the J35F but has a larger internal fuel capacity and can carry up to 4500 kg (9,921 lb) of external armament. The RF-35s have the FFV 'Red Baron' night reconnaissance pod. Drakens have also been assembled under licence by Valmet in Finland for the nation's Ilmavoimat, bringing the total production run to 606 aircraft. All three countries will continue to operate Drakens in front-line use throuogut the 1980s.

Specification
Saab Draken 35X
Type: single-seat fighter-bomber
Powerplant: one 8000-kg (17,637-kg) thrust Volvo Flygmotor RM6C (licence-built Rolls-Royce Avon 300) afterburning turbojet
Performance: maximum speed Mach 2 (2112 km/h/1,312 mph) at 12200 m (40,025 ft); initial rate of climb 10500 m (34,450 ft)/minute; radius of action (hi-lo-hi) with two 454-kg (1,000-lb) bombs and two drop tanks 1000 km (621 miles); ferry range with maximum internal and external fuel 3250 km (2,019 miles)
Weights: take-off, clean 11400 kg (25,133 lb); maximum take-off 16000 kg (35,274 lb)
Dimensions: span 9.40 m (30 ft 10.1 in); length 15.35 m (50 ft 4.3 in); height 3.90 m (12 ft 9.5 in); wing area 49.20 m (529.60 sq ft)
Armament: two 30-mm (1.18-in) Aden cannon and up to 4500 kg (9,921 lb) of external stores including Rb24 Sidewinder AAMs,

pods containing 19 75-mm (2.95-in) rockets, or 12 135-mm (5.31-in) Bofors air-to-ground rockets, nine 454-kg (1,000-lb) bombs or 14 227-kg (500-lb) bombs, and other weapons
Operators: Denmark, Finland, Sweden

Saab J35F Draken

Denmark has 16 F-35 Drakens for fighter/ground attack duties with No. 724 Sqn at Karup. The other Karup-based Draken unit, No. 429, has the RF-35 reeconnaissance variant. All Danish Drakens are undergoing modernization to bring their avionics fit up to contemporary standard.

Saab JAS 39 Gripen

Following long study Saab-Scania AB and the Royal Swedish Air Board decided in 1979 that the next-generation combat aircraft to follow the System 37 Viggen should be a remarkably small air-combat fighter with attack and reconnaissance capability. Its original designation of JAS resulted from the words fighter/attack/reconnaissance in Swedish. In 1980 the Swedish government approved go-ahead on project definition and initial engineering design, since when the progress has been rapid. In May 1982 the Swedish government approved an overall programme for 115 single-seat Gripens (griffin) and 25 tandem two-seaters with dual control and reduced fuel. In June 1982 contracts were signed for the five prototypes and first 30 production aircraft.

Aerodynamically and structurally the Gripen follows the fashion in being a canard delta with a plain engine nozzle and no attempt to use installed engine thrust for direct lift.

Smaller and lighter than almost all other modern combat aircraft, the Gripen will have a ratio of thrust to weight that (unlike its rivals) will exceed unity on all except the most heavily loaded missions. The engine is a part-Swedish development of the most powerful version of the General Electric F404, fed from simple but efficient lateral inlets. Partners throughout the West are collaborating in the programme.

British Aerospace Warton has designed the carbonfibre composite wings, and has built the first shipset. Lear Siegler supply the triplex redundant fly-by-wire flight control system (flown on a Viggen in 1982), with Moog power units for the elevons and Lucas 'geared hinges' driving the leading-edge flaps. AP Precision Hydraulics supply the fuselage-mounted landing gears, Martin-Baker the S10LS seat, British Aerospace Dynamics the environmental control systems and Microturbo the gas-turbine APU. Honeywell supply the laser-inertial

navigation system, Ericsson and Ferranti are collaborating on the multi-mode pulse-doppler radar and the FLIR (forward-looking infra red) sensor pod carried under the right inlet duct, and Hughes the wide-angle HUD (head-up display) with diffraction optics. Most of the instrument panel area is occupied by three SRA colour electronic displays, the cockpit naturally being influenced by that of the Hornet. PEAB will be heavily involved in the extremely comprehensive EW (electronic warfare) installations, which will be mainly internal but will also include podded or scabbed external jammers and dispensers.

First flight is due in 1987, with entry into service being scheduled for 1992. A second Viggen is to be used to help develop avionics and weapon-aiming systems. Sweden carefully studied alternatives to this programme, and the possibility of collaborating with another country (made difficult by the country's strong neutral stance). There is

every indication this will be the first Swedish aircraft to achieve significant foreign sales.

Specification
Saab JAS 39 Gripen
Type: multi-role combat aircraft
Powerplant: one 8165-kg (18,000-lb) thrust Volvo Flygmotor/GE RM12 augmented turbofan engine
Performance: maximum speed, supersonic down to sea level, close to Mach 2 at height; take-off and landing strip length 'well below 1000 m' (3,280 ft)
Weights: 'normal maximum loaded' 8000 kg (17,637 lb)
Dimensions: span (ignoring AAMs) 8.00 m (26 ft 3 in); length 14.00 m (45 ft 11.2 in)
Armament: one 27-mm Mauser BK27 gun; four underwing pylons for very wide range of stores or tanks including RBS15F attack missiles; wingtip rails for AAMs including RB71 Sky Flash and RB24 Sidewinder

Saab MFI-17 Supporter, PAC Mushshak

The MFI-17 stemmed from an all-metal two-seat light aircraft designed in 1958 by Bjorn Andreasson. Intended for use in the close-support role, and as a general trainer, its six underwing points can carry air-to-ground rockets, wire-guided missiles, various gun pods or supply containers. The two seats are side by side. The Supporter has good low-speed handling and manoeuvrability in combat missions, and is suitable for third world air arms. The Zambian air force bought 20 as basic trainers and Denmark uses it for training and artillery observation. In Pakistan it has been built in considerable numbers since 1976 as the Mushshak (proficient) at the air force's Kamra factory: production was scheduled to continue after 1985.

Specification
Saab MFI-Supporter
Type: two/three-seat light trainer/close support aircraft

Powerplant: one 149-kW (200-hp) Lycoming IO-360-A1B6 flat-four piston engine
Performance: maximum speed 235 km/h (146 mph); cruising speed 208 km/h (129 mph); initial rate of climb 325 m (1,066 ft)/minute; service ceiling 5190 m (17,030 ft); range with 10 per cent reserves 1120 km (696 miles)
Weights: empty 625 kg (1,378 lb); maximum take-off 1000 kg (2,205 lb)
Dimensions: span 8.85 m (29 ft 0.4 in); length 7.00 m (22 ft 11.6 in); height 2.60 m (8 ft 6.4 in); wing area 11.90 m^2 (128.09 sq ft)
Armament: various light attack stores on underwing hardpoints
Operators: Denmark, Norway, Pakistan, Sweden

The Saab MFI-17 Supporter is a useful and extremely agile spotting/liaison aircraft. Denmark is one of two countries (along with Pakistan) that has bought the MFI-17 from Sweden.

Saab 37 Viggen

The Viggen forms part of the Swedish air force's System 37, a complete weapon system including support facilities which is integrated into the nation's STRIL 60 air defence network. The Viggen was developed to replace a variety of earlier types in attack, interception, reconnaissance and training missions, and the first of seven prototypes flew on 8 February 1967. The AJ37 attack variant began to replace A32A Lansens in mid-1971. The combination of a large delta wing and flapped foreplanes, tandem-wheel landing gear stressed for no-flare landings, and a reverser, allows the Viggen to operate from short runways and lengths of roadway, thus vastly increasing survivability in wartime. The RM8A en-

gine is a development of the Pratt & Whitney JT8D-22 turbofan built under licence by Volvo Flygmotor, with a Swedish-designed afterburner, and a thrust-reverser which cuts in automatically as the nosewheel strikes the ground. The use of automatic speed-control equipment and an advanced HUD (head-up display) allow the aircraft to land on narrow tracks only 500 m (1,640 ft) long.

The AJ37's navigation and fire-control system is based on a Saab CK-37 miniaturized digital computer, which relieves the pilot of much of the workload. The CK-37 takes its input from the LM Ericsson search and attack radar, and from the air data computer, doppler radar, radio altimeter and other

sensors, the resulting information being shown on a Marconi Avionics HUD. The AJ37 carries no built-in armament, but can be equipped with a variety of air-to-air and air-to-surface weapons.

Saab also built two specialized reconnaissance versions, the SF37 and SH37. The first to enter service was the SH37, in mid-1975. This is a sea-surveillance version, using a nose-mounted surveillance radar similar to that in the AJ37 and a pod-mounted forward-looking long-range radar as its main sensors. Other equipment can include the FFV 'Red Baron' infra-red reconnaissance pod or others containing active or passive ECM (electronic countermeasures) equipment. An auxiliary fuel tank may be mounted

beneath the fuselage, and lightweight air-to-air missiles, such as Sidewinder, can be carried for self-defence. The SF37, which entered service in 1977, has no radar but carries a total of nine cameras looking forward, vertically downward and sideways, together with a FLIR (forward-looking infra-red) camera in the nose, and illumination equipment for operations at night. All cameras are controlled by the central digital computer, and when all are working they produce 75 photographs per second, which contains as much data as about 50 black-and-white TV cameras operating simultaneously. The SF37 is operated in conjunction with a System 37 intelligence platoon, which includes a mobile evaluation centre with briefing, processing, evaluation and interpretation facilities. The SH37 and SF37 are similar to the basic AJ37, as is the tandem-seat Sk 37 trainer apart from a taller fin. Production of the four versions totalled 180.

The latest version, the JA37 interceptor, is a redesigned aircraft. Its RM8B turbofan produces more thrust, giving improved climb and manoeuvrability. There are numerous airframe structural and control surface changes. An inertial navigation system, the Singer Kearfott KT-70L, is one of several new sensors which feeds the same company's central digital computer, built under licence by Saab as Computer 107. Another innovation is the LM Ericsson PS-46 pulse-doppler radar, which allows two JA37s on fighter patrol along the coast to survey as much airspace as a whole squadron of earlier aircraft. The new cockpit has three displays which can show electronic maps giving the location of air bases and anti-aircraft batteries, together with a tactical plot either received by radio link from the ground or derived directly from the JA37's radar. A Smiths Industries HUD allows the pilot to operate his radar and weapons while continuing to look forward outside the aircraft.

These four AJ37s from F6 and F15 show the 'bullet' incorporated into the wing. These contain ECM gear, and form a natural 'dogtooth' to enhance performance.

The SK37 is the two-seat conversion trainer. Due to the extra side area imposed by the second cockpit, the fin has had to be extended. Full attack capability is retained, although fuel tankage is slightly reduced.

The SH37 is a dedicated maritime reconnaissance and attack aircraft, carrying a reconnaissance pod on the centreline. These are employed around Swedish coasts, monitoring and following all manner of shipping.

The JA37 carries an underbelly pack containing an Oerlikon KCA cannon of high power. The gun fires 0.36-kg (0.79-lb) projectiles at a rate of 1,350 rounds per minute and at a velocity of 1050 m (3,445 ft)/second. These shells have as much penetrating power after 1500 m (4,921 ft) of flight as a conventional 30-mm round has as it leaves the muzzle of an Aden or DEFA cannon, and the small trajectory drop eases the problems of sighting in tight manoeuvres. Deliveries of JA 37s to the Swedish air force began in late 1978, and by 1985 more than half the 149 of this model were in service. Wings F4, F13, F17 and F21 were then mainly converted to the JA 37. Whereas earlier Viggens had a complex camouflage colour scheme, the JA is all-grey or (from 1984) all-white.

Latest Viggen variant is the JA37 dedicated interceptor. This features the extended fin of the SK37, and carries BAe Sky Flash and locally-built Sidewinders.

Specification
Saab JA37 Viggen
Type: interceptor
Powerplant: one 12770-kg (28,153-lb) thrust Volvo Flygmotor RM8B augmented turbofan
Performance: maximum speed Mach 2.0 (2112 km/h; 1,312 mph) at 12200 m (40,025 ft), or Mach 1.1 (1335 km/h; 830 mph) at low level; time to 10000 m (32,810 ft) from brake release less than 1 minute 40 seconds; estimated ceiling 15200 m (49,870 ft); tactical radius with external armament, hi-lo-hi 1000 km (621 miles), lo-lo-lo 500 km (311 miles)
Weights: take-off with normal armament 17000 kg (37,479 lb)
Dimensions: span 10.60 m (34 ft 9.3 in); length 16.40 m (53 ft 9.7 in); height 5.90 m (19 ft 4.3 in); wing area 46.00 m² (495.16 sq ft)
Armament: one 30-mm (1.18-in) Oerlikon KCA (305K) cannon and four BAe Sky Flash (RB71) and/or Rb24 Sidewinder AAMs
Operator: Sweden

The excellent turn rate and slow landing speed of the Viggen are attributable to the large canard and wing area, producing a low wing loading. This has enabled the Viggen to operate regularly from stretches of road, a ploy which would be very useful should Sweden come under attack.

The Viggen possesses one of the most
intricate camouflage schemes ever seen
on a warplane, with four colours applied
to a complicated splinter pattern. This
AJ37 attack version carries a pair of
RB75 precision attack missiles on the
fuselage pylons, and a pair of Saab
RB.04E anti-ship missiles on the wing
pylons. The Viggen has no internal gun,
but can carry a gun pod on the centreline
in place of the more usual fuel tank.

Scottish Aviation Bulldog

Originally designed as a military trainer version of the civil Pup, the Beagle Bulldog was produced by Scottish Aviation (later BAe) from 1971 until 1982. An all-metal stressed-skin machine with fixed tricycle landing gear, the Bulldog has side-by-side dual controls, with rear space for an observer or 100 kg (220 lb) of other load. The canopy slides to the rear, and other features include cockpit heating, electrically driven slotted flaps, hydraulic wheel brakes (or optional skis), comprehensive avionics for communication and navigation, an optional glider tow hook and provision for armament as detailed in the specification.

All production Bulldogs have the same basic engine (though the AEIO-360 was offered for customers requiring 20 seconds inverted flight at full power), driving a Hartzell constant-speed propeller. Four removable metal tanks in the wings hold 145.5 litres (32 Imp gal). No oxygen, de-icing system or ejection seats were provided, and the Bulldog gained its popularity from its combination of low price and low operating costs, robust simplicity and flawless qualities in the primary training role.

The chief order for the initial version was for 78 placed by Sweden, which designates the Bulldog as the Sk 61, 20 of these being used as liaison machines by combat units. Malaysia bought 15 Model 102s and Kenya five 103s. In 1973 production switched to the Series 120, with increased aerobatic capability at maximum weight and a fully aerobatic clearance at increased weight. Orders for this comprised 130 Model 121s for the RAF, designated Bulldog T.1, and 12 Model 122s for Ghana, 37 Model 123s for Nigeria, a

Model 124 demonstrator, 22 Model 125s for Jordan, six Model 126s for Lebanon, nine Model 127s for Kenya, two Model 128s for Hong Kong, one Model 129 for a civil customer in Venezuela, and six Model 130s for Botswana.

In 1974 a retractable-gear Series 200 was flown, but this was not put into production.

Specification
Bulldog Model 120 series
Type: piston-engined primary trianer and light COIN aircraft
Powerplant: one 149-kW (200-hp) Avco Lycoming IO-360-A1B6 six-cylinder air-cooled opposed piston engine
Performance: maximum speed at sea level 241 km/h (150 mph); economical cruising speed 194 km/h

Chief customer for the Bulldog has been Sweden, which operates 58 for the air force as basic trainers, and a further 20 for the army for observation. The local designation is Sk.61.

(121 mph); initial rate of climb 315 m (1,034 ft) per minute; service ceiling 4875 m (16,000 ft); range with maximum fuel 1000 km (621 miles)
Weights: empty 649 kg (1,430 lb); maximum take-off 1066 kg (2,350 lb)
Dimensions: span 10.06 m (33 ft 0 in); length 7.09 m (23 ft 3 in); height 2.28 m (7 ft 6 in); wing area 12.02 m² (129.4 sq ft)
Armament: provision for 290 kg (640 lb) hung on four underwing hardpoints, including bombs up to 50 kg (110 lb), machine-gun pods, wire-guided missiles, grenade-

130 Bulldogs were supplied to the RAF, for use by a variety of units. Important among these are the University Air Squadrons.

launchers, supply containers, survival equipment or leaflet dispensers
Operators: Botswana, Ghana, Hong Kong, Kenya, Lebanon, Malaysia, Nigeria, Sweden, UK

SEPECAT Jaguar

Developed jointly by BAC in Britain and Breguet in France, the SEPECAT Jaguar was intended originally as a light attack and training aircraft with supersonic performance. Development resulted in a machine so capable that it seemed pointless to use it only as a trainer, and the tandem dual-control versions are fully combat-ready (though not fitted with the full spectrum of avionics and weapons as is the single-seater). The small engine was developed by Turboméca and

Rolls-Royce, two being installed in a manner reminiscent of the layout developed for the McDonnell Douglas F-4 Phantom. The wing was mounted high to give good access underneath for large tanks and weapons, and is fitted with leading edge slats, full-span double-slotted flaps and powered spoilers for roll control. The rudder and slab tailplanes are also powered. The twin-wheel rough-field main gears retract forward into the fuselage, ahead of the perforated airbrakes. Much of

Jaguar International cutaway drawing key

1 Nose profile (Maritime Strike Variant)
2 Thomson-CSF Agave dual-role (air-air, air-ground) radar
3 Ferranti Type 105 Laser Ranger
4 Pitot tube
5 'Wedge-profile' optical sighting windows
6 Ferranti Laser Ranging and Marked Target Seeker
7 Total pressure probe (both sides)
8 Electronics cooling air duct
9 Air-data computer
10 Radio altimeter
11 Power amplifier
12 Avionics access doors
13 Waveform generator
14 Cooling air intake
15 Marconi Avionics nav/attack system equipment
16 Landing/taxiing lamps
17 Nosewheel leg door
18 Towing lug
19 Nosewheel forks
20 Nosewheel
21 Steering jacks
22 Nose undercarriage leg strut
23 Artificial feel control units
24 Rudder pedals

33 Pilot's side console panel
34 Martin-Baker Mk 9 'zero-zero' ejection seat
35 Seat and parachute, combined safety harness
36 Honeycomb cockpit side panel
37 Plexiglass cockpit canopy cover (upward opening)
38 Ejection seat headrest
39 Canopy struts
40 Cockpit pressurization valve
41 Rear pressure bulkhead
42 Gun muzzle blast trough
43 Battery and electrical equipment bay
44 Port engine intake
45 Gun gas vents
46 Spring-loaded secondary air intake doors
47 Boundary layer bleed duct
48 Forward fuselage fuel tank (total system capacity 924 Imp gal/1200 litres)
49 Air conditioning unit
50 Secondary heat exchanger
51 Starboard engine air intake
52 VHF homing aerials
53 Heat exchanger intake/exhaust duct
54 Cables and hydraulic pipe ducting

62 Main undercarriage hydraulic lock strut
63 Leading-edge-slat drive motors and gearboxes
64 Fuel system piping
65 Wing panel centreline joint
66 Anti-collision light
67 IFF aerial
68 Wing/fuselage forward attachment joint
69 Starboard wing integral fuel tank
70 Fuel piping provision for pylon mounted tank
71 Overwing missile pylon
72 Missile launch rail
73 Matra 550 Magic air-to-air missile
74 Starboard leading-edge slat
75 Slat guide rails
76 Starboard navigation light
77 Tacan aerial
78 Flap guide rails and underwing fairings
79 Outboard double-slotted flap
80 Starboard spoilers
81 Inboard double-slotted flap
82 Flap honeycomb construction
83 Flap drive shaft and screwjacks
84 Spoiler control links

25 Instrument panel shroud
26 Retractable in-flight refuelling probe
27 Windscreen panels
28 Smiths Electronics head-up display
29 Instrument panel
30 Smiths FS6 head-down navigational display
31 Control column
32 Engine throttles

55 Intake/fuselage attachment joint
56 Duct frames
57 Integrally-stiffened machined fuselage frames
58 Ammunition tank
59 30-mm Aden cannon
60 Ground power supply socket
61 Mainwheel stowed position

85 Wing/fuselage aft attachment joint
86 Heat exchanger air scoop
87 Control runs
88 Air conditioning supply ducting
89 Fuselage fuel tank access panels

SEPECAT Jaguar (continued)

the structure is covered in honeycomb-stabilized skin, about 40 per cent of the exterior being access panels.

Single-seat versions are equipped normally with an inflight refuelling probe. Three 1200-litre (264-Imp gallon) drop tanks can be carried on the centreline and inboard wing pylons. Most of the air-conditioning system is grouped along the top of the finely profiled fuselage, the primary heat-exchanger being situated in a large bulge ahead of the fin, a position dictated by area-rule considerations to give minimum drag. Two oblique ventral fins are added below the main engine access doors to increase side area of the rear fuselage. A retractable arrester hook is standard.

A common requirement for the Jaguar was drafted by the British (RAF) and French (Armée de l'Air) in 1965, calling for single-seat attack and dual-control trainer versions to enter service with the

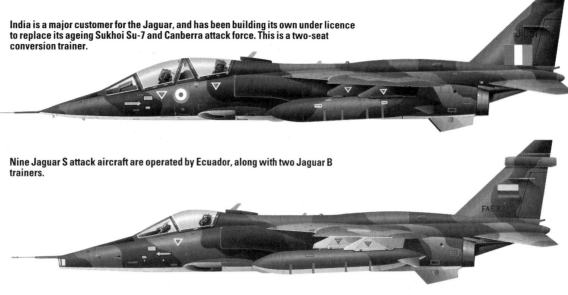

India is a major customer for the Jaguar, and has been building its own under licence to replace its ageing Sukhoi Su-7 and Canberra attack force. This is a two-seat conversion trainer.

Nine Jaguar S attack aircraft are operated by Ecuador, along with two Jaguar B trainers.

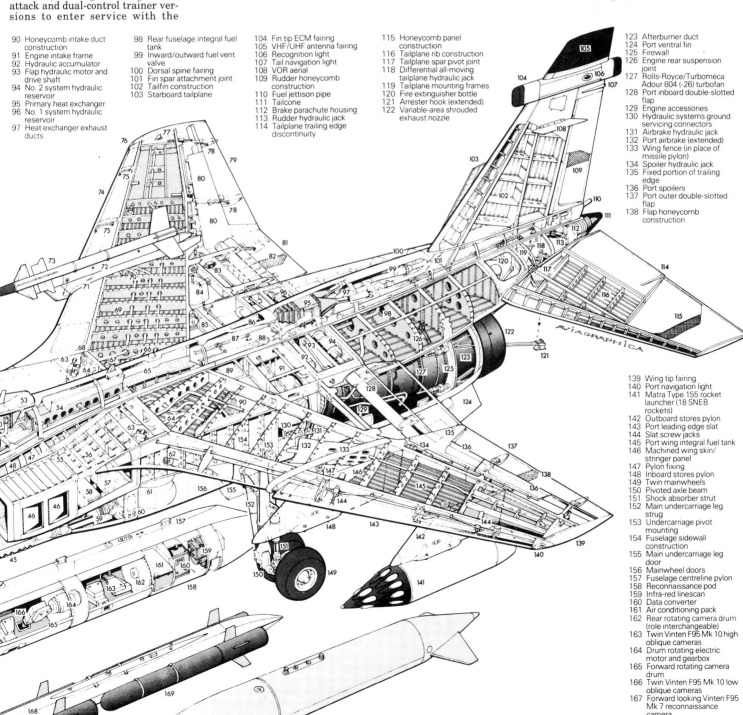

90 Honeycomb intake duct construction
91 Engine intake frame
92 Hydraulic accumulator
93 Flap hydraulic motor and drive shaft
94 No. 2 system hydraulic reservoir
95 Primary heat exchanger
96 No. 1 system hydraulic reservoir
97 Heat exchanger exhaust ducts
98 Rear fuselage integral fuel tank
99 Inward/outward fuel vent valve
100 Dorsal spine fairing
101 Fin spar attachment joint
102 Tailfin construction
103 Starboard tailplane
104 Fin tip ECM fairing
105 VHF/UHF antenna fairing
106 Recognition light
107 Tail navigation light
108 VOR aerial
109 Rudder honeycomb construction
110 Fuel jettison pipe
111 Tailcone
112 Brake parachute housing
113 Rudder hydraulic jack
114 Tailplane trailing edge discontinuity
115 Honeycomb panel construction
116 Tailplane rib construction
117 Tailplane spar pivot joint
118 Differential all-moving tailplane hydraulic jack
119 Tailplane mounting frames
120 Fire extinguisher bottle
121 Arrester hook (extended)
122 Variable-area shrouded exhaust nozzle
123 Afterburner duct
124 Port ventral fin
125 Firewall
126 Engine rear suspension joint
127 Rolls-Royce/Turboméca Adour 804 (-26) turbofan
128 Port inboard double-slotted flap
129 Engine accessories
130 Hydraulic systems ground servicing connectors
131 Airbrake hydraulic jack
132 Port airbrake (extended)
133 Wing fence (in place of missile pylon)
134 Spoiler hydraulic jack
135 Fixed portion of trailing edge
136 Port spoilers
137 Port outer double-slotted flap
138 Flap honeycomb construction
139 Wing tip fairing
140 Port navigation light
141 Matra Type 155 rocket launcher (18 SNEB rockets)
142 Outboard stores pylon
143 Port leading edge slat
144 Slat screw jacks
145 Port wing integral fuel tank
146 Machined wing skin/stringer panel
147 Pylon fixing
148 Inboard stores pylon
149 Twin mainwheels
150 Pivoted axle beam
151 Shock absorber strut
152 Main undercarriage leg strug
153 Undercarriage pivot mounting
154 Fuselage sidewall construction
155 Main undercarriage leg door
156 Mainwheel doors
157 Fuselage centreline pylon
158 Reconnaissance pod
159 Infra-red linescan
160 Data converter
161 Air conditioning pack
162 Rear rotating camera drum (role interchangeable)
163 Twin Vinten F95 Mk 10 high oblique cameras
164 Drum rotating electric motor and gearbox
165 Forward rotating camera drum
166 Twin Vinten F95 Mk 10 low oblique cameras
167 Forward looking Vinten F95 Mk 7 reconnaissance camera
168 Matra Durandal, 430-lb (195-kg) penetration bomb
169 Pylon attachment shackles
170 264 Imp gal (1200 litres) auxiliary fuel tank

Armée de l'Air in 1972 and the RAF in 1973. The first prototype flew on 8 September 1968, and all variants were developed within the allotted time and budget. The first was the Jaguar E (Ecole or school) trainer which entered the inventory of the Centre d'Expériences Aériennes Militaires at Mont de Marsan in May 1972. This has pupil and instructor seated on Martin-Baker Mk 4 seats which cannot be used safely at speeds below 167 km/h (104 mph). In most other respects, including armament, it is identical to the next version to enter service, the Jaguar A. The Jaguar A (Appui or attack) is the Armée de l'Air single-seat version, with a Mk 9 zero/zero seat, pointed nose without sensors, and simple nav/attack system based on a doppler radar and twin-gyro platform navigation system. Two DEFA 553 30-mm guns are installed in the centre fuselage, and a total external load of up to 4536 kg (10,000 lb) can be carried on centre-line and four wing pylons, including the AN.52 tactical nuclear bomb of the Armée de l'Air, Belouga cluster bombs, AS.37 anti-radar Martel, Durandal runway-piercing missiles, retarded bombs, SNEB rocket packs and Magic air-to-air missiles. The Martin Marietta/Thomson CSF Atlis TV target acquisition and laser-designation pod was added in 1978, and a simple reconnaissance pod and Super Cyclope IRLS (infra-red linescan) can also be fitted.

Britain's Jaguars have inertial nav/attack systems, a HUD (head-up display), projected map display, radar height, laser ranger (in a 'chisel' nose) and an ARI.18223 radar warning receiver installation near the top of the fin. Called Jaguar S, this version is designated Jaguar GR.1 by the RAF and 165 were delivered by 1978. The guns are 30-mm Adens, and all navigation and weapon delivery is integrated in a Navwass system controlled by a digital computer. The usual cluster bomb is the BL.755, and a specially-designed BAe multi-sensor reconnaissance pod can be carried flush under the centreline, with five optical cameras and IRLS all linked to Navwass. From 1978 all these aircraft were refitted with Adour 104 engines of greater thrust. The RAF trainer is the Jaguar B, designated

T.2 by the RAF, of which 37 were delivered. This has one gun and no nose laser, but retains Mk 9 seats.

For export various options have been added, including more powerful engines, overwing pylons for Magic, AIM-9L or other dogfight missiles, Agave nose radar with Ferranti 105S laser in a small fairing below the nose, low-light TV and new weapons options including Harpoon and Kormoran for use against ships. The first two export sales were for aircraft of standard type, though with uprated engines. The third, a large contract with India involving 116 aircraft, took advantage of most of the new options. After 45 Jaguars had been supplied from the joint production by BAe and Dassault-Breguet, assembly of 40 and finally complete licence-manufacture of a further 31 took place in India. As almost 'three-quarters of the total requirement' was made in India this also included the engines, and also involved a buy-

back by Britain of parts made in India.

All Jaguar export sales were gained by BAe in the teeth of opposition from the French 'partner'. Despite this a further order, for 18 Jaguar Internationals from Nigeria, was gained in 1983, when a further 12 were delivered to Oman.

Specification
SEPECAT Jaguar International (late Indian, Oman and Nigerian versions)
Type: single-seat attack aircraft and two-seat trainer
Powerplant: two 4205-kg (9,270-lb) afterburning thrust Rolls-Royce/Turboméca Adour Mk 811 turbofan engines
Performance: maximum speed Mach 1.6 (1700 km/h; 1,056 mph) at 11000 m (36,090 ft), or Mach 1.1 (1350 km/h; 839 mph) at sea level; typical attack radius with weapons and no external fuel, hi-lo-hi 852 km (529 miles); ferry range 3525 km

This Omani Jaguar demonstrates the type of flying at which the type excels. The RAF has consistently shown other NATO partners the value of this aircraft, where its low-level agility and speed have made it difficult to catch, while it can deliver its weapons with accuracy.

(2,190 miles)
Weights: empty 7000 kg (15,432 lb); maximum take-off 15700 kg (34,613 lb)
Dimensions: span 8.69 m (28 ft 6.1 in); length single-seat 16.83 m (55 ft 2.6 in), two-seat 17.53 m (57 ft 6.2 in); height 4.89 m (16 ft 0.5 in); wing area 24.18 m² (260.28 sq ft)
Armament: single-seat, two 30-mm (1.18-in) Aden or DEFA 553 cannon or two-seat, one Aden gun on port side; centreline and four underwing pylons for mixed stores and two overwing pylons for two air-to-air missiles to a maximum load of 4763 kg (10,500 lb)
Operators: Ecuador, France, India, Nigeria, Oman, UK

Shenyang J-5

The Soviet Union's 1950s policy of licensing aircraft manufacture to allied countries extended to China. Chinese-assembled MiG-17Fs began to appear in 1956, and by 1959 the Shenyang aircraft factory was building the MiG-17F and VK-1F without outside assistance.

The Sino-Soviet rift of 1960 did not interrupt production of the type; rather, China's isolation from any other source of military technology, and the political hostility towards the 'intellectual' community, which caused domestic technology to stagnate, tended to prolong production of the basically obsolete design. MiG-

17 manufacture is believed to have continued into the 1970s.

The standard MiG-17F is designated J-5; an equivalent to the MiG-17PF, probably reverse-engineered together with its radar equipment from examples supplied by the Soviet Union, is the J-5 Jia (or J-5A). Additionally, the Shenyang organization has developed a trainer version of the J-5, with an instructor's seat behind the pilot, no afterburner and a single cannon. Some of these aircraft have a ranging radar, similar to that of the J-5A, but no search radar. Some J-5s have been exported in an effort to gain foreign currency.

Specification
Shenyang J-5
Type: single-seat fighter-bomber
Powerplant: one Chinese-built Klimov VK-1F turbojet of 2600 kg (5,730 lb) dry thrust and 3380 kg (7,450 lb) augmented thrust
Performance: maximum speed 1145 km/h (710 mph) at 3000 m (10,000 ft) or Mach 0.94; initial climb 1.8 min to 5000 m (16,500 ft); service ceiling 16600 m (54,500 ft); ferry range 1980 km (1,230 miles) with full external fuel
Weights: empty 3930 kg (8664 lb); internal fuel approximately 907 kg (2,000 lb); normal take-off 6075 kg

(13,400 lb)
Dimensions: span 9.63 m (31 ft 7 in); length overall 11.26 m (36 ft 11 in); height 3.8 m (12 ft 6 in); wing area 22.6 m² (243 sq ft)
Armament: two 23-mm NR-23 cannon with 80 rounds each, and one 37-mm N-37 with 40 rounds, in lower forward fuselage. Four underwing hardpoints for bombs or unguided rockets; points immediately outboard of wheel wells normally used for fuel tanks.
Operators: Albania, China, Sudan, Tanzania, Vietnam

Shenyang J-6

Phased out of production in the Soviet Union during the late 1950s, the Mikoyan-Gurevich MiG-19 continues to be built in China under a licence agreement of January 1958. The J-6 (Jianjiji 6 or Fighter Aircraft Type 6) is normally credited to the Shenyang production facility, although a second assembly line is located at Tianjin. First of the Chinese production versions was the J-6 equivalent of the MiG-19S/SF

day fighter, this giving way to the J-6A/MiG-19PF limited all-weather interceptor and the later J-6B/MiG-19PM, the latter augmenting gun and rocket armament by AA-1 'Alkali' AAMs. An improved MiG-

19SF, known as the J-6C and identified by a brake parachute housing at the base of the fin, is currently in production, as is the locally-designed J-6Xin ('New J-6') which features a sharply-pointed radome in the engine air intake for a Chinese-developed airborne interception radar. The JZ-6 (Jianjiji Zhenchaji 6) is a MiG-19R reconnaissance version equivalent with the forward fuselage cannon replaced by a camera array. Despite the lack of Soviet production of the MiG-19UTI, Chinese requirements for a dual-control trainer were met by a local redesign to produce the JJ-6 (Jianjiji Jiaolianji 6) with its 0.84-m (2 ft 9.1-in) fuselage extension, semi-automatic seats slightly staggered in height and flat-topped side-hinged canopies. Several J-6s have been built for the Chinese army and navy air forces since 1961, while export variants (known as the F-6 and trainer FT-6) serve in countries listed below. Despite its age and short range, the J-6 is well liked by its pilots as a very tough and man-oeuvrable fighter, and stable weapons platform. Pakistan's aircraft have been improved by the addition of a third (underfuselage) fuel tank, US-designed AIM-9B/J Sidewinder

Shenyang JJ-6 trainer of the Egyptian air force.

AAMs and Martin-Baker PKD Mk 10 automatic zero-zero ejection seats.

Specification
Shenyang/Tianjin J-6C
Type: single-seat close-support fighter and day interceptor
Powerplant: two 3250-kg (7,165-lb) afterburning thrust Wopen-6 (Tumansky R-9BF-811) turbojet engines
Performance: maximum speed, clean Mach 1.45 (1540 km/h; 957 mph) at 11000 m (36,090 ft); maximum speed at low level Mach 1.09 (1340 km/h; 833 mph); initial rate of climb 9150 m (30,020 ft)/minute; service ceiling 17900 m (58,725 ft); combat radius with two 760-litre (167-Imp gallon) external tanks 685 km (426 miles); maximum range with two similar external tanks 1390 km (864 miles)
Weights: empty 5760 kg (12,699 lb); normal take-off, clean 7545 kg (16,634 lb); maximum take-off with external stores about 10000 kg (22,046 lb)
Dimensions: span 9.20 m (30 ft 2.2 in); length, excluding probe 12.60 m (41 ft 4.1 in); height 3.88 m (12 ft 8.8 in); wing area 25.00 m² (269.11 sq ft)
Armament: three internal 30-mm (1.18-in) NR-30 cannon; wing pylons for two 250-kg (551-lb) bombs or four rocket packs, plus fuel tanks
Operators: Albania, Bangladesh, Cambodia, China, Egypt, Iran, Iraq, Pakistan, Somalia, Tanzania, Vietnam, Zambia

Pakistan is a large operator of the J-6, with around 120 in service. Most of these have been fitted with launch rails for Sidewinders to give a vastly improved air-to-air capability.

Shenyang J-8

When the Chinese visited the Mikoyan-Gurevich design bureau in the late 1950s they were permitted to purchase information not only on the MiG-19 family and MiG-21F but also on the Ye-152A (NATO codename 'Flipper'), which had never gone into production. Rather surprisingly, over a period of many years, the Chinese industry at both the Shenyang and Xian factories worked on this very fast twin-engined delta design and eventually produced a Chinese version, first flown in the late 1960s.

Nothing was disclosed regarding the Jianjiji (Fighter aircraft) Type 8 (NATO codename 'Finback') until late 1984. Rather surprisingly the programme was then still being continued in a low key, despite the obsolete nature of the basic design. Compared with the J-7 (MiG-21F), the J-8 offers marginally higher speed and a little more radius of action, but still lacks modern radar and other avionics. So far as is known all J-8s so far built (a matter of tens rather than hundreds) have been use for general research and development. Some have the front-hinged frameless canopy of the J-7 while others have a separate windscreen and side-hinged canopy. All have a small inlet centrebody (Soviet Ye-152A and related prototypes having a giant cone housing search radar) and an airframe almost totally redesigned in detail. Whereas the emphasis in the Soviet original was all-weather interception, the J-8 is a day fighter/bomber with six or eight pylons (aircraft vary).

Though as this book goes to press the J-8 is still 'ticking over', the PLA air force does not expect ever to deploy it in numbers. Like the smaller J-7, examples of the J-8 are flying with a GEC Avionics HUD, two makes of Western IFF, at least two types of digital computer and a wide range of indigenous and imported weapons. None has been seen with a multi-mode radar. Not least of the puzzles surrounding this racy-looking fighter is that the Chinese spent ten years collaborating with Rolls-Royce, culminating in the successful type-test of an augmented Spey turbofan in 1980 and completion of tooling of a Spey factory at Xian in 1982. Yet this engine is too large and powerful to fit the J-8 and inadequate for a single-engined version. The US Defense Intelligence Agency gave lack of 'adequate jet engines' as the reason for the J-8 never having entered production, but the true reason is obsolescence of the basic concept.

Specification
Shenyang J-8
Type: fighter/bomber
Powerplant: two 6200 kg (13,668 lb) Chengdu WP-7 afterburning turbojets, based on Tumanskii R-11F2S-300
Performance: maximum speed, clean at high altitude Mach 2.35, or 2496 km/h (1551 mph); range with maximum internal and external fuel 2600 km (1,615 miles)
Weights: no data, but Ye-152A maximum was 14200 kg (31,305 lb)
Dimensions: not disclosed, but span probably same as Ye-152A (8.97 m/29 ft 5.1 in); length about 20.0 m (65 ft 7.4 in); wing area 28.20 m² (303.55 sq ft)
Armament: one to three guns of 23-mm or 30-mm calibre; six or eight pylons for up to four tanks and unknown quantity of ordnance including bombs, rockets, a nuclear bomb and AIM-9P3 Sidewinder or Matra Magic AAMs
Operator: China

Shin Meiwa PS-/US-1

One of the last military flying-boat/amphibians in the Western world, development of the PS-1/US-1 can be traced back to 1960. An unusual feature is the use of boundary-layer control blowing for improved control down to very low airspeed, using as power source a 932-kW (1,250-shp) T58 turbojet. A contract of January 1966 covered development of an ASW (anti-submarine warfare) flying-boat for the JMSDF (Japan Maritime Self-Defence Force). Known as the SS-2, this flew on 5 October 1967, entering service in 1973 with the JMSDF as the PS-1. Production totalled 23, survivors operating with the 23rd Kokutai, and with the 51st Kokutai, the JMSDF's principal operational test agency, both at Iwakuni.

Development of the SS-2A SAR (search and rescue) amphibian began in June 1970 and the first was flown from water on 16 October 1974 and from land on 3 December 1974. Delivery of the first US-1, as it is known by the JMSDF, took place in March 1975 but only nine had been handed over by the end of 1984. The last three are US-1As with more powerful engines, and it is intended to re-engine the original six. All serve with the 71st Kokutai, also at Iwakuni.

Capable of operating in wave heights of up to 3.0 m (9.8 ft), the US-1 can taxi into or out of the water under its own power, and can also be refuelled at sea, by an aircraft or ship. It has a crew of nine and space for 20 seated survivors or 12 stretchers. The crew of the standard PS-1 is 10: two pilots, flight engineer, navigator, two sonar operators, MAD (magnetic anomaly detector) radar and radio operators, and tactical co-ordinator.

Specification
Shin Meiwa US-1A
Type: SAR amphibian
Powerplant: four 2602-ekW (3,490-ehp) Ishikawajima-built General Electric T64-IHI-10J turboprop engines
Performance: maximum speed 522 km/h (324 mph) at 3050 m (10,500 ft); initial rate of climb 715 m (2,346 ft)/minute; service ceiling 7195 m (23,600 ft); maximum range 3815 km (2,371 miles)
Weights: empty 25500 kg (56,218 lb); maximum take-off,

water 43000 kg (94,799 lb), land 45000 kg (99,208 lb)
Dimensions: span 33.15 m (108 ft 9.1 in); length 33.46 nm (109 ft 9.3 in); height 9.95 m (32 ft 7.7 in); wing area 135.82 m² (1,462.0 sq ft)
Armament (PS-1 only): internal bay for four 150-kg (330-lb) anti-submarine bombs, smoke bombs, sonobuoys and explosive charges for use with acoustic echo-ranging equipment; pods beneath each wing for homing torpedoes; three 127-mm (5-in) rockets may be carried beneath each wingtip
Operator: Japan

One of the last flying-boats in operation, the Shin Meiwa US-1 and PS-1 are used for SAR and ASW respectively. All fly from the base at Iwakuni.

Shorts Skyvan

Shorts flew the first military Skyvan Srs 3M in 1970. With a basic layout lending itself to military operation, the Skyvan is capable of very short take-off and landing (238 m; 780 ft and 212 m; 695 ft), even under 'hot and high' conditions, and despite its workhorse shape cruises at a useful speed. The ramp door in the rear fuselage can be opened in flight to drop supplies that are as tall as 1.37 m (4 ft 6 in). Excluding the flight deck, 22.09 m³ (780.0 cu ft) of cabin volume is available on 11.15 m² (120.0 sq ft) of floor area with a usable length of 5.66 m (18 ft 7 in), entered through the rear door which is 1.98 m (6 ft 6 in) high and 1.96 m (6 ft 5 in) wide. The 3M can accommodate 22 equipped troops or 16 paratroops plus dispatcher, or 12 stretcher cases and two attendants, or up to 2359 kg (5,200 lb) of freight. Equipment includes a blister window on the port side for the dispatcher and rollers in the floor for positioning cargo. Parachuting fittings include anchor cables for static lines, inward-facing seats with safety nets, a signal light and a guard rail over the tail.

Oman is the largest user of the military Skyvan, No. 2 Sqn of the Sultan's Air Force operating 15 from Seeb with Racal ASR.360 surveillance radar. The Ghana Air Force flies six from Takoradi on tactical support, communications, coastal patrol and casevac duties. No. 121 Sqn of the Republic of Singapore Air Force uses six for search-and-rescue and anti-smuggling duties around the island. The Comando de Aviación Naval Argentina assigned five 3Ms to the Prefectura Naval Argentina (at least one was lost in 1982), this being a small force which

also operates Hughes 500M helicopters for coastguard duties. No. 2 Sqn of the Indonesian air force flies three on short-range transport work, and they are equipped to civil standard so that they can also operate for the Ministry of the Interior. Three Skyvan 3Ms are flown by the Royal Thai Border Police. A light transport squadron of the Austrian air force, III Geschwader of Fliegerregiment I, uses two 3Ms for supply and aerial survey from the base at Tulln. The Mauritian air force and army operate two aircraft jointly, from the base at Nouakchott. Three 3Ms are flown by the air wing of the Royal Nepalese Army on light transport duties, and one of them is used also by the Royal Flight. The air force of the Yemen Arab Republic

flies two in the transport role, and the Ecuadorean army uses one. Other operators include the Botswana Defence Force (2), Guyana Defence Force (2), Lesotho Police (2), Malawi Police (1) and Panama National Guard (1).

Specification
Shorts SC.7 Skyvan Series 3M
Type: light STOL utility transport
Powerplant: two 533-kW (715-shp) Garrett TPE331-201 turboprop engines
Performance: maximum cruising speed 325 km/h (202 mph) at 3050 m (10,000 ft); initial rate of climb 466 m (1,530 ft)/minute; service ceiling 6705 m (22,000 ft); range at economic cruising speed with reserves 1078 m (670 miles)

Oman is a major Skyvan operator, with 15 flying from Seeb with No. 2 Sqn. These are used for surveillance duties as well as the more usual transport role.

Weights: empty 3357 kg (7,400 lb); maximum take-off 6577 kg (14,500 lb)
Dimensions: span 19.79 m (64 ft 11 in); length, with radome 12.60 m (41 ft 4 in); height 4.60 m (15 ft 1 in); wing area 35.12 m² (378.0 sq ft)
Armament: none
Operators: Argentina, Austria, Botswana, Ecuador, Ghana, Guyana, Indonesia, Lesotho, Malawi, Mauritania, Mexico, Nepal, Oman, Panama, Senegambia, Singapore, Thailand, South Yemen

Shorts C-23A Sherpa

First flown on 23 December 1982, the Sherpa is a freighter version of the Shorts 330-200 incorporating a full-width rear ramp door; in addition to the basic freighter role it can be reconfigured as a passenger transport.

The winning submission in the US Air Force's EDSA (European Distribution System Aircraft) competition, an initial batch of 18 was ordered as the C-23A, for delivery from October 1984 to March 1986. Equipping Military Airlift Command's 10th Military Airlift Squadron at Zweibrücken AB, West Germany, these are employed primarily for the transport of aircraft spares,

especially fighter engines, between USAFE bases. The USAF is expected to take up an option for a further 48 C-23As for service elsewhere, delivered at two per month from March 1986. The C-23A can accommodate containers up to LD3 size, four LD3s being carried in the cargo configuration; alternative loads include two half-ton vehicles,

Shorts won a major order with the selection of their 330 Sherpa for the USAF. Designated C-23A in service, these fly spare parts and personnel around the USAF's European bases. More may be bought for service in other areas.

or two LD3s and nine passengers. Seat rails provide lashing points for bulk cargo. The hydraulically-operated ramp can be lowered to a chosen height, eliminating the need for specialized ground equipment, controlled from inside or outside the aircraft. There is another cargo door at the front on the left.

Specification
Type: utility transport
Powerplant: two 893-kW (1,198-shp) Pratt & Whitney Canada PT6A-45R turboprop engines
Performance: maximum cruising speed at 3050 m (10,000 ft) 351 km/h (218 mph); initial rate of climb 360 m (1,180 ft)/minute; range with 3175-kg (7,000-lb) payload and reserves 362 km (225 miles); range with 2268-kg (5,000-lb) payload and reserves 1239 km (770 miles)
Weights: empty 6680 kg (14,727 lb); maximum take-off 10387 kg (22,900 lb)
Dimensions: span 22.76 m (74 ft 8 in); length 17.69 m (58 ft 0.5 in); height 4.95 m (16 ft 3 in); wing area 42.08 m² (453.0 sq ft)
Armament: none
Operator: US Air Force

SIAI-Marchetti S.208M

Evolved from a series of popular touring aircraft, 44 examples of the S.208M were delivered in 1971-72 to the Italian Aeronautica Militare, which uses them for liaison and instrument/navigation training. The S.208M is identical with the civil S.208 except that the right cabin door can be jettisoned in emergency, the baggage compartment is accessible from the rear of the cabin, and fuel is housed in two integral wing tanks plus two wingtip auxiliary tanks.

Specification
Type: four/five-seat liaison and training aircraft
Powerplant: one 194-kW (260-hp) Lycoming O-540-E4A5 flat-six piston engine
Performance: maximum cruising speed 300 km/h (186 mph); service ceiling 5400 m (17,715 ft); range 1200 km (746 miles)
Weights: empty 780 kg (1,720 lb); maximum take-off 1350 kg (2,976 lb)
Dimensions: span 10.86 m (35 ft 7.76 in); length 8.00 m (26 ft 3 in); height 2.89 m (9 ft 5.8 in); wing area 16.09 m² (173.2 sq ft)
Armament: none
Operator: Italy

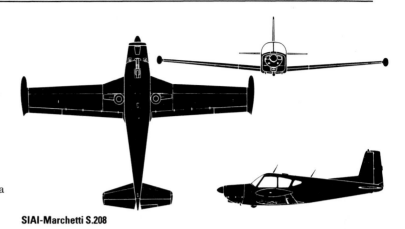

SIAI-Marchetti S.208

SIAI-Marchetti S.211

Having gained experienced of worldwide training procedures through its SF.260, SIAI-Marchetti elected to enter the light jet trainer market in the mid-1970s. The designers opted for low gross weight, low fuel consumption, an advanced wing and incorporation of new technology to reduce cost and maintenance requirements, yet give students a steep learning curve to transition direct to combat aircraft. Stepped tandem cockpits have Martin-Baker Mk 8 seats operable between 111 and 741 km/h (69 and 460 mph) at all altitudes and capable of modification to Mk 10 (zero-zero) standard. The wing is of supercritical section, with four hardpoints to enable the S.211 to be used in the light attack role. Optional avionics include doppler radar, a radar ranger and HUD (head-up display).

The prototype flew on 10 April 1981, followed by two more by early 1983. Orders and options for 90 were then claimed. The first, placed in July 1982 by Singapore, is for 10 plus 20 options, six delivered from Italy and four in kit form for assembly by Singapore Aircraft Industries which will produce the batch of 20.

Specification
Type: basic trainer and light attack aircraft
Powerplant: one 1134-kg (2500-lb) thrust Pratt & Whitney Canada JT15D-4C turbofan engine
Performance: maximum cruising speed 667 km/h (414 mph); initial rate of climb 1280 m (4,200 ft)/minute; service ceiling 12190 m (40,000 ft); endurance with reserves 3 hours 50 minutes
Weights: empty 1615 kg (3,560 lb); maximum take-off, trainer 2500 kg (5,512 lb), light attack 3100 kg (6,834 lb)
Dimensions: span 8.43 m (27 ft 7.9 in); length 9.31 m (30 ft 6.5 in); height 3.80 m (12 ft 5.6 in); wing area 12.60 m² (135.63 sq ft)
Armament: light bombs, rockets and grenade launchers, or machine-

The SIAI-Marchetti S.211 is a useful light attack aircraft and trainer. So far only Haiti, Singapore and Somalia have ordered the type.

gun or reconnaissance pods up to a total of 600 kg (1,323 lb)
Operators: Haiti, Singapore, Somalia

SIAI-Marchetti SF.260

Designed by Stelio Frati as a fast, compact and flamboyant private aircraft, the SF.260 prototype first flew on 15 July 1964. The aircraft has since found favour with a large number of air forces, used for such diverse roles as training, liaison, aerobatic practice, light attack, forward air control, fishery protection, search and rescue and maritime patrol. The 260M, a three-seat trainer developed from the civil A, flew on 10 October 1970 and introduced structural modifications including strengthened wings. May 1972 saw the first flight of the SF.260W Warrior equipped with underwing pylons, this is a light strike platform and has become popular with air forces that cannot afford to risk expensive metal in minor skirmishes. The list of mission profiles varies from long-duration sorties close to base (5 hours or so at about 80-km/50-mile range) to 5 minutes over a target 565 km (351 miles) distant. The SF.260SW Sea Warrior maritime patrol variant has enlarged tiptanks housing (left) lightweight Bendix radar and (right) photo-reconnaissance equipment in addition to fuel.
Of 36 SF.260MBs delivered 1969-

71, 33 are still in service with the Belgian air force on elementary training duties from Groetsenhoven. Burma's 10 MBs are operated in the dual strike/trainer role, and Dubai has an SF.260WD. Ireland replaced its eight Chipmunks with 11 SF.260WE Warriors and the home market, Italy, received 33 for basic training. Libya is by far the largest customer with an order for 240 260MLs. The Philippine air force's 17th attack wing has 16 WPs and the Singapore air force flies 28 MSs for basic training. Korat Air Base is home to the Royal Thai Air Force's 18 MTs, used for training, as are the 12 WTs of the Tunisian Republic's air force and the 20 MCs used by Zaire. Zambia's nine MZs are primarily for training but, like the 17 Warriors of Zimbabwe, are occasionally flown on COIN missions.

Specification
SIAI-Marchetti SF.260W Warrior
Type: trainer/tactical support aircraft

Thirty-one SF.260MBs are used by Belgium for basic training before pupils move on to Alpha Jets. Other countries use the SF.260 in the COIN role.

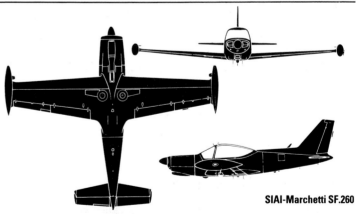

SIAI-Marchetti SF.260

Powerplant: one 194-kW (260-hp) Lycoming O-540-E4A5 flat-six piston engine
Performance: maximum speed 305 km/h (190 mph) at sea level; maximum cruising speed 280 km/h (174 mph) at 1500 m (4,920 ft); initial rate of climb 380 m (1,247 ft)/minute; service ceiling 4480 m (14,700 ft); range, self-ferry with two seats occupied and fuel reserves 1715 km (1,066 miles)
Weights: empty 830 kg (1,830 lb); maximum take-off 1300 kg (2,866 lb)
Dimensions: span, over tiptanks 8.35 m (27 ft 4.7 in); length 7.10 m (23 ft 3.5 in); height 2.41 m (7 ft 10.9 in); wing area 10.10 m² (108.72 sq ft)
Armament: two or four underwing hardpoints with a combined maximum capacity of 300 kg (661 lb); typical optional loads include one or two SIAI gun pods each with one 500-round 7.62-mm (0.3-in) FN machine-gun, two

SIAI-Marchetti SF.260WT of the Tunisian air force. These have a major training role, but could be called upon to give light support if so desired. Unguided rockets are the favoured weapon.

Simpres AL-8-70 launchers each with eight 70-mm (2.75-in) FFAR rockets, two Matra F-2 launchers each with six 68-mm (2.68-in) SNEB 253 rockets, two SAMP EU 32 125-kg (276-lb) GP bombs or EU 13 120-kg (265-lb) fragmentation bombs,

two Alkan 500B cartridge throwers, one Alkan 500B cartridge thrower and one photo-reconnaissance pod with two 70-mm automatic cameras, or two 83-litre (18.25-Imp gallon) auxiliary tanks
Operators (various military

versions): Belgium, Bolivia, Brunei, Burma, Burundi, Comoros, Dubai, Ecuador, Ghana, Ireland, Italy, Libya, Morocco, Nicaragua, Philippines, Singapore, Somalia, Thailand, Tunisia, Zaïre, Zambia, Zimbabwe

SIAI SF.600TP

The SF.600TP Canguro (kangaroo) was designed by Ing Stelio Frati as the piston-engine F.400 12-seater with a minimum of complex systems and fixed landing gear; it now incorporates turboprop engines and has optional retractable gear. The square-section fuselage provides a 5.05-m (16 ft 6.8-in) length of space, with a large sliding door on the left; a swing-tail is an option. The production of an initial batch of 20 has been authorised, but no military orders have been recorded to date. Versions include four-stretcher medevac with attendants, water and oxygen; an ECM model with chaff dispensers on underwing pylons; a navigation trainer with three student consoles and one instructor position; and paratroop transport for 12 equipped soldiers with weapons

racks in the centre of the cabin.

Specification
Type: multirole utility transport
Powerplant: two 313-kW (420-shp) Allison 250-B17C turboprop engines
Performance: maximum cruising speed 287 km/h (178 mph); initial rate of climb 462 m (1,516 ft)/minute; service ceiling 7315 m (24,000 ft); maximum range 1580 km (982 miles)
Weights: empty 1800 kg (3,968 lb); maximum take-off 3300 kg (7,275 lb)
Dimensions: span 15.00 m (49 ft 2.6 in); length 12.15 m (39 ft 10.3 in); height 4.60 m (15 ft 1.1 in); wing area 24.00 m² (258.34 sq ft)
Armament: none
Operators: none

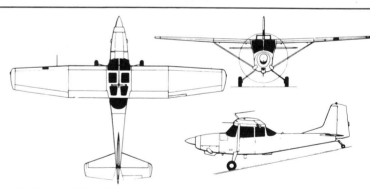

Several versions of the SIAI-Marchetti SF.600TP have been developed to attract military orders but so far none have been forthcoming. Among these versions are ECM, medevac, navigation training and paradropping aircraft.

SIAI-Marchetti SM.1019

To replace earlier observation aircraft, SIAI installed an Allison 250 turboprop in a modified Cessna O-1 Bird Dog. Standard accommodation is for two in tandem, with dual controls provided for use in the training role. The Italian army received 100 SM.1019EI aircraft for training, artillery spotting, forward air control, battlefield reconnaissance and helicopter escort.

Specification
SIAI-Marchetti SM.1019EI
Type: training and observation aircraft
Powerplant: one 198-kW (400-shp)

Allison 250B-17B turboprop engine
Performance (utility): economic cruising speed 280 km/h (174 mph); initial rate of climb 550 m (1,804 ft)/minute; service ceiling 7620 m (25,000 ft); maximum range 925 km (575 miles) at sea level
Weights: empty 690 kg (1,521 lb); maximum take-off 1300 kg (2,866 lb)
Dimensions: span 10.97 m (36 ft 0 in); length 8.52 m (27 ft 11.4 in); height 2.86 m (9 ft 4.6 in); wing area 16.16 m² (173.85 sq ft)
Armament: light tactical stores on three hardpoints can include rocket launchers, anti-personnel bombs,

SIAI-Marchetti SM.1019E

gun pods, wire-guided missiles (up to AS.12 size) or night

reconnaissance packs
Operator: Italy

Sikorsky S-55

One of the first truly useful helicopters, the S-55 served with all elements of the US forces in troop transport, airlift, SAR and communications roles for many years from 1951. At least 20 are still active, some – notably those of Chile – have been updated and fitted with turbine engines as the S-55T. The Westland Whirlwind is described separately.

Specification
Sikorsky UH-19B
Type: utility helicopter
Powerplant: one 522-kW (700-hp)

Sikorsky's venerable S-55 staggers on with a few air forces. Those of Japan, now out of service, were used for SAR duties, wearing this fetching colour scheme.

Wright R-1300-3 piston engine
Performance: maximum speed at sea level 180 km/h (112 mph); cruising speed 146 km/h (91 mph); range 579 km (360 miles)
Weights: empty 2381 kg (5,250 lb); maximum take-off 3583 kg (7,900 lb)
Dimensions: main rotor diameter 16.15 m (53 ft 0 in); fuselage length 12.88 m (42 ft 3 in); height 4.06 m

(13 ft 4 in); main rotor disc area 204.94 m² (2,206 sq ft)
Armament: none

Operators: Chile (S-55T), Dominica, Honduras, South Korea, Turkey

Sikorsky S-58/H-34

First flown on 8 March 1954, the S-58 remains operational with a number of air arms. It met a 1951 US Navy requirement for a machine to perform the ASW (anti-submarine warfare) mission, a task too great for the HO4S (S-55). Introduced to service as the HSS-1 in 1955, it also served the US Army as a transport with up to 18 seats, the Marine Corps as an assault helicopter and the US Coast Guard in the SAR role. Many were completed for friendly nations under the Military Assistance Program, while 166 were licence-built in France by Sud-Aviation for ALAT and Aéronavale use, initially in the armed role in Algeria.

Many survivors have been converted to PT6T turbine power, most notably those in military service with Indonesia and Thailand which have nine and 18 S-58Ts respective-

ly. The Westland Wessex is described separately.

Specification
Sikorsky S-58/H-34
Type: general-purpose helicopter
Powerplant: one 1137-kW (1,525-hp) Wright R-1820-84 piston engine
Performance: maximum speed at sea level 196 km/h (122 mph); maximum cruising speed 156 km/h (97 mph); range with maximum fuel and reserves 398 km (247 miles)
Weights: empty 3515 kg (7,750 lb); maximum take-off 5897 kg (13,000 lb), UH-34 series 6350 kg (14,000 lb)
Dimensions: main rotor diameter 17.07 m (56 ft 0 in); length overall 20.06 m (65 ft 10 in); height 4.36 m (14 ft 3.5 in); main rotor disc area 228.53 m^2 (2,460.0 sq ft)
Armament: none
Operators: (S-58T) Argentina,

Costa Rica, Haiti, Indonesia, Thailand; (S-58) Belgium, Haiti, Honduras, Laos, Nicaragua, Taiwan, Uruguay

One Sikorsky CH-34 is operated by the Sandinista regime's air force, having been supplied originally by the United States.

Sikorsky S-61/CH-3/SH-3/HH-3 Series

Conceived as the HSS-2 (later SH-3A) amphibious ASW (anti-submarine warfare) helicopter, the S-61 proved to be most versatile, variants being purchased by the US Navy and Air Force and many other armed services, and produced under licence in Italy, Japan and the UK, over 1,170 having been completed with production continuing in Italy and the UK. Flown on 11 March 1959, the SH-3A entered service with the US Navy in September 1961. The similar Canadian CH-124 was assembled from Sikorsky parts by United Aircraft of Canada. Mitsubishi of Japan produced the HSS-2, HSS-2A and HSS-2B for the Maritime Self-Defence Force, 118 being completed.

The SH-3D has more powerful T58-GE-10 engines and greater fuel capacity. The US Navy took 72, similar machines going to the navies of Argentina (five S-61D-4s), Brazil (six S-61D-3s) and Spain (18, later updated to SH-3H configuration). The US Navy bought two other

variants for the Washington-based Executive Flight Detachment of Marine Corps squadron HMX-1. Fitted out with VIP interiors, eight VH-3As were later replaced by 11 VH-3Ds. Efforts to update the US Navy's ASW SH-3s centred around mod-

Above: Developed for service in Vietnam, the HH-3 features a large inflight-refuelling boom for use with HC-130 Hercules. The role is combat SAR, retrieving downed airmen from behind enemy lines under cover of attack aircraft.

Below: United Aircraft of Canada assembled Sikorsky-built CH-124 Sea Kings for the CAF, which still operates 35 for ASW work. These are equivalent to the US Navy's SH-3A, but, like the American aircraft, have undergone several equipment updates.

ifications leading to the SH-3H, with lightweight sonar, active and passive sonobuoys and MAD (magnetic anomaly detection) gear; about 160 older US Navy Sea Kings had been brought to this standard by 1980. Other US Navy variants produced as conversions from SH-3As include the armed and armoured HH-3A for combat SAR (12), the mine countermeasures RH-3A (nine) and the SH-3G for utility tasks (105).

In 1962 the USAF obtained three HSS-2s on loan for resupply missions to 'Texas Tower' radar sites in the Atlantic. Three more were acquired as CH-3Bs and their value prompted a decision in November 1962 to obtain an S-61 optimized for transport. Incorporating a hydraulically-operated rear ramp the CH-3C (S-61R) also had tricycle landing gear and an APU (auxiliary power unit). It flew in June 1963, and the USAF eventually received 50, most of which were brought up to the CH-3E standard of later production (69) with uprated T58-GE-5 engines. Eight armed and armoured combat SAR HH-3Es for the Aerospace Rescue and Recovery Service proved so successful in South East Asia that CH-3Es were modified to this standard. Forty essentially similar machines, minus combat gear, were acquired by the US Coast Guard from 1968 as HH-3F Pelicans.

Sikorsky's Italian licensee, Agusta, has not only produced large numbers of SH-3D Sea Kings (many fitted with the Italian Marte antiship system with Sistel radar and Sea Killer Mk 2 missiles) but is also now the only source of the HH-3F (S-61R) and commercial Sea Kings. Westland's licensed versions are described separately.

Specification
Sikorsky SH-3D
Type: amphibious all-weather ASW helicopter
Powerplant: two 1044-kW (1,400-shp) General Electric T58-GE-10 turboshaft engines
Performance: maximum speed 267 km/h (166 mph); speed for maximum range 219 km/h (136

A Sikorsky S-61A Nuri of the Malaysian air force.

mph); service ceiling 4480 m (14,700 ft); range with 10 per cent reserves 1006 km (625 miles)
Weights: empty 5382 kg (11,865 lb); maximum take-off 8449 kg (18,626 lb)
Dimensions: main rotor diameter 18.90 m (62 ft 0 in); length rotors turning 22.15 m (72 ft 8 in); height 5.13 m (16 ft 10 in); main rotor disc area 280.47 m² (3,019.0 sq ft)
Armament: provision for 381 kg (840 lb) of weapons, including homing torpedoes
Operators: (Sikorsky) Argentina, Australia, Belgium, Canada, Denmark, Egypt, West Germany, India, Indonesia, Japan, Malaysia, Norway, Pakistan, Spain, US Air Force, Coast Guard, Marine Corps and Navy; (Agusta) Argentina, Brazil, Iraq, Iran, Italy, Libya, Morocco, Peru, Saudi Arabia, Syria; (Mitsubishi) Japan

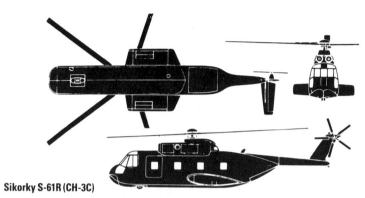

Sikorky S-61R (CH-3C)

The US Navy has been systematically updating its Sea Kings. Most are now of SH-3H standard, with new detection gear, including MAD. These are deployed on aircraft-carriers to provide ASW coverage in conjunction with the wider-ranging S-3 Vikings.

Sikorsky CH-54 Tarhe

Now relegated to second-line service with the US Army, the Sikorsky CH-54 Tarhe was evolved in response to a late 1950s requirement for a heavy-lift helicopter. Although the use of cargo hooks to lift heavy and bulky loads was a feature of the design, the Tarhe was also intended to straddle interchangeable pods. First flown in May 1962, the type proved sufficiently promising to warrant a production order for what was, by US Army standards, a relatively small quantity, together with a universal pod to be employed for various special roles. In Vietnam the

CH-54 was instrumental in the recovery of 380 aircraft and helicopters, as well as the rapid movement of bulldozers, graders and even armoured fighting vehicles. The CH-54B had more powerful engines, an uprated gearbox, high-lift rotor blades, twin mainwheels and an improved flight system. Several are still airworthy.

Specification
Sikorsky CH-54A Tarhe
Type: heavy-lift crane helicopter
Powerplant: two 3579-kW (4,800-shp) Pratt & Whitney T73-P-1

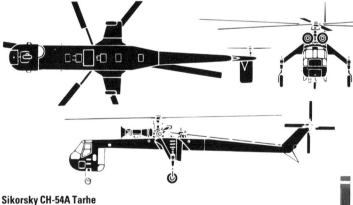

Sikorsky CH-54A Tarhe

Sikorsky CH-54 Tarhe of the United States Army. The Army currently has 72 of these monsters for heavy-lift work, most serving with the National Guard.

turboshaft engines
Performance: maximum cruising speed 169 km/h (105 mph); service ceiling 2745 m (9,000 ft); range with maximum fuel and reserves 370 km (230 miles)
Weights: empty 8724 kg (19,234 lb); maximum take-off 21319 kg (47,000 lb)
Dimensions: main rotor diameter 21.95 m (72 ft 0 in); length, rotors turning 26.97 m (88 ft 6 in); height 7.75 m (25 ft 5 in); main rotor disc area 378.24 m² (4,071.5 sq ft)
Armament: none
Operator: USA

This US Army CH-54 demonstrates the ability of the Tarhe to lift outsize items such as trucks. As well as having a cargo hook for slinging items, the Tarhe can straddle objects such as a portable building, to transport them without swinging them.

Sikorsky S-65A/H-53 Sea Stallion

In 1960 the US Marine Corps decided that it needed a more advanced assault transport helicopter. Sikorsky evolved the S-65 using the dynamic parts of the S-64A Skycrane and introduced a watertight hull to give emergency water-landing capability. Rear doors allow loading of bulky items, and hydraulic winches and a roller track facilitate the positioning of pallets, vehicles or a 105-mm howitzer and its carriage. The first prototype flew on 14 October 1964, and the first production CH-53A Sea Stallions were delivered to Marine units in September 1966, soon going to Vietnam. They proved able to operate in all weathers, had external cargo system for in-flight pickup and release without assistance from the ground, and hydraulically-folded main rotor blades and tail.

In September 1966 the USAF ordered eight for the Aerospace Rescue and Recovery Service, to be equipped similarly to the HH-3E and with T64 engines uprated from 2125 kW (2,850 shp) to 2297 kW (3,080 shp); the first of these HH-53Bs flew on 15 March 1967. The improved HH-53C has 2927 kW (3,925 shp) T64-GE-7 engines, auxiliary fuel tanks, an inflight refuelling probe, rescue hoist, and a 9072-kg (20,000-lb) external cargo hook; the first entered service on 30 August 1968. Eight HH-53Cs were modified under the Pave Low 3 programme to provide night search and rescue capability, these aircraft being equipped with FLIR (forward-looking infra-red), inertial and doppler navigation and terrain following radar.

Final version for the Marine Corps was the CH-53D with more powerful engines, one-man cargo handling, and seats for 55 equipped troops; this first entered service in March 1969. When the last was delivered, on 31 January 1972, production of the twin-engined CH-53 for the Marines totalled 265. All but the first 34 were equipped for minesweeping, but in October 1972 pro-

Above: RH-53Ds carried the US Navy's minesweeping effort for several years. These are now being replaced by the MH-53E which has far greater capability.

Right: These two RH-53s are seen prior to being painted in sand camouflage for the abortive Iranian hostage raids.

duction was initiated of 30 RH-53D MCM (mine countermeasures) helicopters, for sweeping acoustic, magnetic and mechanical mines. First delivery, to the US Navy's HM-12 Squadron, was made in September 1973. These have been followed by the MH-53E.

The designation CH-53G applied to 112 aircraft for the German army with 2927-kW (3,925-shp) T64-GE-7 engines. Two were delivered by Sikorsky on 31 March 1969; of the remainder 20 were assembled from American components by VFW in Germany, and 90 were manufactured by the German company. Two aircraft designated S-65Ö were supplied by Sikorsky in 1970 to the Austrian air force for operation as rescue aircraft in Alpine regions.

Specification
Sikorsky CH-53D
Type: heavy assault transport helicopter
Powerplant: two 2927-kW (3,925-shp) General Electric T64-GE-413 turboshaft engines
Performance: maximum speed 315 km/h (196 mph) at sea level; cruising speed 278 km/h (173 mph); range with 10 per cent reserves 414 km (257 miles)
Weights: empty 10653 kg

(23,485 lb); mission take-off 16511 kg (36,400 lb); maximum take-off 19051 kg (42,000 lb)
Dimensions: main rotor diameter 22.02 m (72 ft 3 in); length rotors turning 26.90 m (88 ft 3 in); height 7.59 m (24 ft 11 in); main rotor disc area 380.87 m² (4,099.84 sq ft)
Armament: none
Operators: West Germany, Iran, Israel, US (USAF, USMC, USN)

Germany is a large operator of the CH-53, with over 100 on charge with the Heeresflieger (army). These are designated CH-53G, and serve with MHFTR-15, -25 and -35.

Sikorsky S-70/UH-60 Black Hawk/SH-60B

Now reaching the US Army in large quantities, the UH-60A Black Hawk has been designed around crash-worthiness, with cabin, rotor-blades and fuel system reputedly capable of taking hits up to 23-mm calibre. This aircraft is used for medevac duties.

At the beginning of the 1970s the US Army began procurement of the Utility Tactical Transport Aircraft System (UTTAS), primarily a combat assault helicopter. The fly-off evaluation of competing prototypes occupied seven months of 1976, and on 23 December of that year Sikorsky's design was selected for production as the UH-60A Black Hawk. A total of 1,107 had been ordered by 1985, for delivery in 1978-88. The four-blade main rotor is of advanced design with aft-swept tips; it is designed to survive hits from 12.7-mm or 23-mm armour-piercing shells, and the hub has elastomeric bearings needing no lubrication. The fuel system is regarded as being crashworthy, as is the fuselage which is designed to survive hits from armour-piercing rounds of rifle calibre. Accommodation is provided for a crew of three and 11 troops, the pilot and co-pilot having armoured seats. Eight seats can be replaced by four stretchers or internal cargo. An external hook of 3629 kg (8,000 lb) capacity is provided for airlift of artillery and supplies. Six Black Hawks can be carried in a Lockheed C-5A Galaxy.

The EH-60A is a special ECM version carrying 816 kg (1,800 lb) of Quick Fix II electronics to detect and jam hostile transmissions. The US Army hopes to buy 77, as well as 78 of the EH-60B Sotas (stand-off target acquisition system) with a special radar. The USAF has 11 ordinary UH-60As and is buying 89 HH-60A Night Hawks packed with gear (and a refuelling probe) for

special missions and rescue, plus 66 more simple HH-60Es.

The US Navy's continuing search for advanced ASW helicopters resulted in competition for the LAMPS (light airborne multi-purpose system) Mk III configuration, won by the S-70L with Navy designation SH-60B Seahawk. These helicopters have an advanced ASW capability and serve also for ASST (anti-ship surveillance and targeting) roles, with capability for search and rescue, medevac, and vertical replenishment to ships at sea. They serve on board cruisers, destroyers and frigates for sea periods of up to three months. The SH-60B has automatic main rotor and tail pylon folding, shorter wheelbase, and naval avionics, including radar, MAD, sonar and ESM. The large APS-124 search radar is under the cockpit, and 25 tubes for launching sonobuoys are built into the left side, with four complete sets of reloads carried internally. The first of 204 Seahawks entered service in 1984.

Specification
Sikorsky UH-60A
Type: combat assault squad transport
Powerplant: two 1163-kW (1,560-hp) General Electric T700-GE-700 turboshaft engines
Performance: maximum speed 296 km/h (184 mph) at sea level;

The UH-60 received its baptism of fire during the US intervention in Grenada. This aircraft is being used for casevac.

cruising speed 269 km/h (167 mph) at 1220 m (4,000 ft); range with 30 minutes reserves 600 km (373 miles)
Weights: empty 4819 kg (10,624 lb); mission take-off 7375 kg (16,260 lb); maximum take-off 9185 kg (20,250 lb)
Dimensions: main rotor diameter 16.36 m (53 ft 8 in); length rotors turning 19.76 m (64 ft 10 in); height 5.13 m (16 ft 10 in); main rotor disc area 210.14 m^2 (2,262.0 sq ft)
Armament: provision for one or two 7.62-mm (0.3-in) M60 machine-guns

firing from opened side doors, special External Stores Support System for 16 Hellfires, or other loads
Operators: UH-60A, USA (US Army and US Air Force); SH-60B, Australia (on order), USA (US Navy)

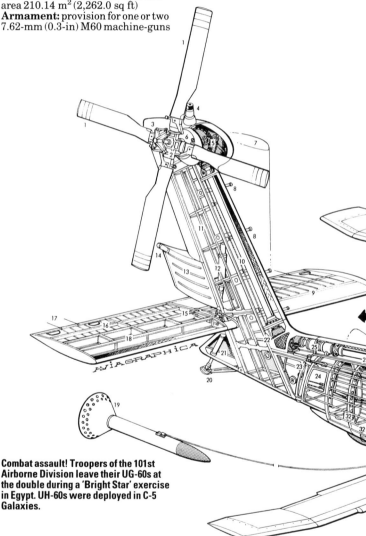

Combat assault! Troopers of the 101st Airborne Division leave their UG-60s at the double during a 'Bright Star' exercise in Egypt. UH-60s were deployed in C-5 Galaxies.

The HH-60A has been developed for the combat SAR role, complete with refuelling probe. Other missions to be undertaken by this aircraft (called Night Hawk) will be covert Special Forces missions.

Sikorsky SH-60B Seahawk cutaway drawing key

1 Graphite epoxy composite tail rotor blades
2 Lightweight cross beam rotor hub
3 Blade pitch change spider
4 Anti-collision light
5 Tail rotor final drive bevel gearbox
6 Rotor hub canted 20 degrees
7 Horizontal tailplane folded position
8 Pull-out maintenance steps
9 Port tailplane
10 Tail rotor drive shaft
11 Fin pylon construction
12 Tailplane hydraulic jack
13 Cambered trailing edge section
14 Tail navigation light
15 Tailplane hinge joint (manual folding)
16 Handgrips
17 Static dischargers
18 Starboard tailplane construction
19 Towed magnetic anomaly detector (MAD)
20 Tail bumper
21 Shock absorber strut
22 Bevel drive gearbox
23 Tail pylon latch joint
24 Tail pylon hinge frame (manual folding)
25 Transmission shaft disconnect
26 Tail rotor transmission shaft
27 Shaft bearings
28 Tail pylon folded position
29 Dorsal spine fairing
30 UHF aerial
31 Tailcone frame and stringer construction
32 Magnetic compass remote transmitters
33 MAD detector housing and reeling unit
34 Tail rotor control cables
35 HF aerial cable
36 MAD unit fixed pylon
37 Ventral data link antenna housing
38 Lower UHF/TACAN aerial
39 Fuel jettison
40 Anti-collision light
41 Tie-down shackle
42 Tailcone joint frame
43 Air system heat exchanger exhaust
44 Engine exhaust shroud
45 Emergency locator aerial
46 Engine fire suppression bottles
47 IFF aerial
48 Port side auxiliary power unit (APU)
49 Oil cooler exhaust grille
50 Starboard side air conditioning plant
51 Engine exhaust pipe
52 HF radio equipment bay
53 Sliding cabin door rail
54 Aft AN/ALQ-142 ESM aerial fairing, port and starboard
55 Tailwheel leg strut
56 Fireproof fuel tanks, port and starboard, total capacity 361 US gal (1368 litres)
57 Starboard stores pylon
58 Castoring twin tail wheels
59 Torpedo parachute housing
60 Mk 46 lightweight torpedo
61 Cabin rear bulkhead
62 Passenger seat
63 Honeycomb cabin floor panelling
64 Sliding cabin door
65 Recovery Assist, Secure and Traverse (RAST) aircraft haul-down fitting
66 Ventral cargo hook, 6,000-lb (2722-kg) capacity
67 Floor beam construction
68 Spring-loaded door segment in way of stores pylon
69 Pull-out emergency exit window panel
70 Pneumatic sonobuoy launch rack (125 sonobuoys)
71 Rescue hoist/winch
72 General Electric T700-GE-401 turboshaft engine
73 Engine accessory equipment gearbox
74 Intake particle separator air duct
75 Engine bay firewall
76 Oil cooler fan
77 Rotor brake unit
78 Engine intake ducts
79 Maintenance step
80 Engine drive shafts
81 Bevel drive gearboxes
82 Central main reduction gearbox
83 Rotor control swash plate
84 Rotor mast
85 Blade pitch control rods
86 Bi-filar vibration absorber
87 Rotor head fairing
88 Main rotor head (elastomeric, non-lubricated, bearings)
89 Blade pitch control horn
90 Lead-lag damper
91 Individual blade folding joints, electrically actuated
92 Blade spar crack detectors
93 Blade root attachment joints
94 Main rotor composite blades
95 Port engine intake
96 Control equipment sliding access cover
97 Engine driven accessory gearboxes
98 Hydraulic pump
99 Flight control servo units
100 Flight control hydro-mechanical mixer unit
101 Cabin roof panelling
102 Radar operator's seat
103 AN/APS-124 radar console
104 Tie-down shackle
105 Gearbox and engine mounting main frames
106 Maintenance steps
107 Main undercarriage leg mounting
108 Shock absorber leg strut
109 Starboard mainwheel
110 Pivoted axle beam
111 Starboard navigation light
112 Cockpit step/main axle fairing
113 Forward cabin access panel
114 Collective and cyclic pitch control rods
115 Sliding fairing guide rails
116 Cooling air grille
117 Main rotor blade glass-fibre skins
118 Honeycomb trailing edge panel
119 Titanium tube blade spar
120 Rotor blade drooped leading edge
121 Leading edge anti-erosion sheathing
122 Fixed trailing edge tab
123 Cockpit eyebrow window
124 Rear view mirrors
125 Overhead engine throttle and fuel cock control levers
126 Circuit breaker panel
127 Pilot's seat
128 Safety harness
129 Crash resistant seat mounting
130 Pull-out emergency exit window panel
131 Flight deck floor level
132 Cockpit door
133 Boarding step
134 AN/APS-124 search radar antenna
135 Ventral radome
136 Retractable landing/hovering lamp
137 Downward vision window
138 Yaw control rudder pedals
139 Cyclic pitch control column
140 Instrument panel
141 Centre instrument console
142 Stand-by compass
143 ATO/co-pilot's seat
144 Outside air temperature gauge
145 Instrument panel shroud
146 Air data probes
147 Windscreen panels
148 Windscreen wipers

149 Hinged nose compartment access panel
150 Pitot tubes
151 Avionics equipment bay
152 Forward data link antenna
153 Forward AN/ALQ 142 ESM aerial housings

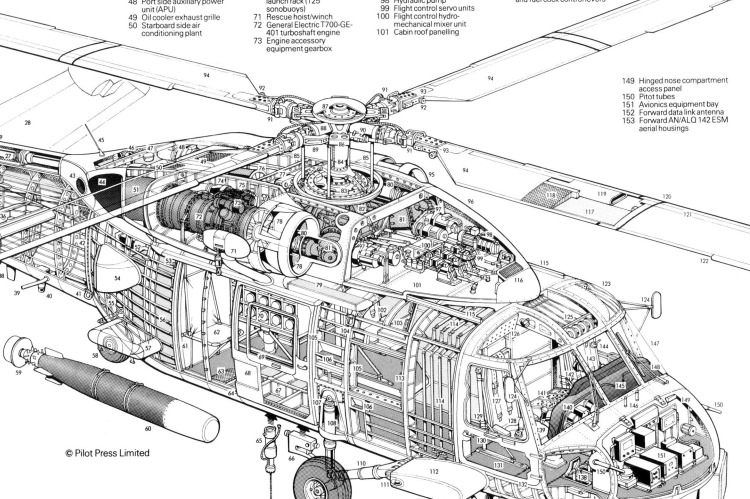

© Pilot Press Limited

Based on the same S-70 design as the UH-60, the SH-60B Seahawk is a dedicated anti-submarine warfare platform, coming complete with naval avionics (including ESM and radar), sonobuoy dispensers in the side of the fuselage MAD 'bird' towed from a pylon on the starboard side. A hoist is provided for SAR and 'plane-guard' duties while weapons pylons for torpedoes are placed on the fuselage sides. This example is from HSL-41, the Fleet Requirements Unit, based at North Island.

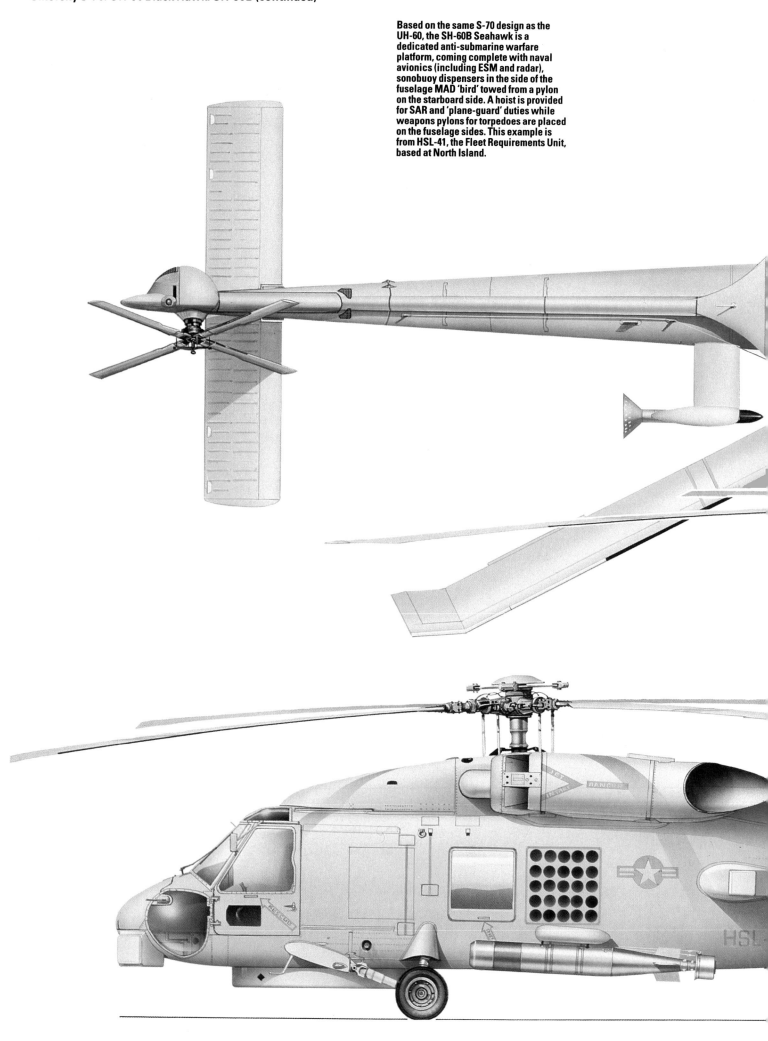

Sikorsky S-76

Aimed at the commercial market, the S-76 flew on 13 March 1977; several hundred have now been delivered , including a small number for military service. The main version is the S-76 Mk II, with a new cabin ventilation system and additional maintenance access panels. A more basic version is the S-76 Utility, 17 going to the Philippine air force, 12 of them being H-76s for counter-insurgency, troop support and medevac tasks. The H-76 led to the AUH-76, able to carry machine-guns, machine-gun or rocket pods, TOW, Hellfire, Sea Skua and Stinger missiles as well as torpedoes. A naval variant has also been proposed, to carry Ferranti Sea Spray 3 or MEL

Super Searcher radar and being able to undertake anti-ship or anti-submarine tasks.

Specification
S-76 Utility
Type: utility helicopter
Powerplant: two 485-kW (650-shp) Allison 250-C30S turboshaft engines
Performance: maximum cruising speed 286 km/h (178 mph); initial rate of climb 411 m (1,350 ft)/minute; range with 12 passengers and fuel reserves 748 km (465 miles)
Weights: empty 2540 lb (5,600 kg); maximum take-off 4672 kg (10,300 lb)
Dimensions: main rotor diameter

Sikorsky S.76

13.41 m (44 ft 0 in); fuselage length 13.22 m (43 ft 4.5 in); height, tail rotor turning 4.52 m (14 ft 9.75 in); main rotor disc area 141.21 m^2

(1,520.0 sq ft)
Armament: AUH-76 only
Operators: Brunei, Jordan, Philippines, Trinidad & Tobago

SOCATA Rallye/Guerrier

The SOCATA subsidiary of Aérospatiale took over the Rallye from bankrupt Morane-Saulnier in 1966. First flown on 10 June 1959, the Rallye has automatic slats which make it capable of very slow flight in safety. All Rallyes have a large canopy which encloses a two-, three- or four-seat cabin, most having dual controls. Many air forces purchased 'off-the-shelf'. In 1974 the French Aéronavale purchased 10 Rallye 100S spinnable two-seat trainers and six 3/4-seat 100STs used as air experience aircraft for cadets at the École Navale at Lanvéoc-Poulmic. The Armée de l'Air has five 134-kW (180-hp) Rallye MS.893E communications aircraft.

The Model 235G Guerrier has a strengthened airframe and four underwing hardpoints. It can also carry a stretcher or an underwing TV pod with direct pilot control for battlefield real-time reconnaissance. SOCATA has delivered few

Guerriers, although it is in use with its Air International Formation training group at Arcachon.

Specification
SOCATA 235G Guerrier
Type: two/four-seat light army support aircraft
Powerplant: one 175-kW (235-hp) Avco Lycoming O-540-B4B5 flat-six piston engine
Performance: maximum speed 275 km/h (171 mph); cruising speed 245 km/h (152 mph); initial rate of climb 300 m (984 ft)/minute; service ceiling 4500m (14,765 ft); maximum range with four loaded hardpoints and reserves 530 km (329 miles)
Weights: empty 764 kg (1,684 lb); maximum take-off 1298 kg (2,862 lb)
Dimensions: span 9.74 m (31 ft 11.5 in); length 7.25 m (23 ft 9.4 in); height 2.80 m (9 ft 2.2 in); wing area 12.28 m^2 (132.19 sq ft)
Armament: four underwing points

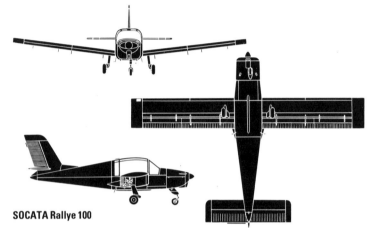

SOCATA Rallye 100

for twin 7.62-mm (0.3-in) gun pods, Matra F22 rocket launchers each with six 68-mm (2.68-in) rockets, 50-kg (110-lb) bombs or battlefield illumination, surveillance or SAR

equipment
Operators: France (100S, 100ST), Rwanda, Salvador

Soko Galeb

Produced at Mostar by Soko Vazduhoplovna Industrija, the G2-A Galeb (seagull) is a basic trainer, the design of which began in 1957 to a Jugoslav JRV (air force) specification. The first prototype flew in May 1961, and production took place in 1963-83. Britain supplies the engine, Folland Type 1-B automatic seats and some avionics. Access to the engine is obtained by removal of four bolts, allowing the rear fuselage to be detached. Designed for load factors of + 6g and − 4g, the Galeb may be used for armament training or for tactical reconnaissance with a 175-mm (6.89-in) focal length camera in the rear cockpit, backed by night photography flares on the wing pylons. Wingtip tanks are added when the aircraft is assigned to navigation training. Docile handling qualities include a stalling speed of 158 km/h (98 mph) with full flaps and the ability to operate from short grass runways. Galebs are

operated by the main JRV flying school at Mostar and by other training establishments such as Skopje.

Progressive improvements include air-conditioning, and the G2-A-E export version introduced full IFR instrumentation and a revised navigation and communications fit. This was chosen by the Libyan Arab air force for its new Air Academy at Zawia, 50 being delivered in 1975 (believed joined by more in 1983). Libyan G2-A-Es fit between the primary stage on SF.260s and advanced instruction with the L-39 Albatros. The first export customer was Zambia, which received six G2-As in 1971 plus Jugoslav instructors; their replacement at Mbala is the Super Galeb.

Specification
Soko G2-A Galeb
Type: armed basic jet trainer
Powerplant: one 1134-kg (2,500-lb) thrust Rolls-Royce Viper 11 Mk 22-6

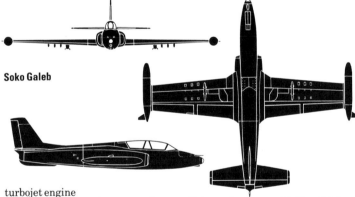

Soko Galeb

turbojet engine
Performance: maximum speed 812 km/h (505 mph); initial rate of climb 418 m (1,371 ft)/minute; service ceiling 12000 m (39,370 ft); range with tip tanks 1240 km (771 miles)
Weights: empty 2620 kg (5,776 lb); maximum take-off, basic trainer 3488 kg (7,690 lb), attack aircraft 4178 kg (9,211 lb)

Dimensions: span 10.47 m (34 ft 4.2 in); length 10.34 m (33 ft 11.1 in); height 3.28 m (10 ft 9.1 in); wing area 19.43 m^2 (209.15 sq ft)
Armament: two internal 12.7-mm (0.5-in) machine-guns; light bombs and rockets on two underwing hardpoints
Operators: Jugoslavia, Libya and Zambia

Soko Jastreb

The Jastreb (hawk) represents a minimum-change conversion of the G2-A Galeb to a single-seat ground attack aircraft, a metal fairing covering what would be the rear cockpit. The front cockpit has a canopy hinged to the right and a Folland 1-B seat and oxygen, but pressurization and air-conditioning were offered only for export. The engine is

uprated and has self-contained starting. Take-off performance can be boosted by two 454-kg (1,000-lb) thrust JATO rockets, and braking is assisted by a parachute. Improved navigation and communications equipment is fitted and there are forward-facing cameras in each tip tank to complement the optional centre-fuselage installation. Local

airframe strengthening has been incorporated, particularly for heavier weapon loads.

In JRV service the standard model is designated J-1, and some 100 remain from production launched in 1968. Jastreb bases include Pula and Zadar, the former housing a weapons training unit using some of its aircraft in the target-towing and

air combat manoeuvring roles. The export J-1-E has the same improved avionics as the G2-A-E Galeb, although the six operated by Zambia are to J-1 standard. The TJ-1 Jastreb restores the rear cockpit, but loses the fuselage camera. Used for operational conversion and proficiency training, the TJ-1 first flew in 1974, and about 20 are in service

Soko Jastreb (continued)

with attack squadrons operating the J-1. Replacement with the Soko/IAR-93 is beginning.

Specification
Type: single-seat attack aircraft
Powerplant: one 1361-kg (3,000-lb) thrust Rolls-Royce Viper Mk 331 turbojet engine
Performance: maximum speed 820 km/h (510 mph); initial rate of climb 1260 m (4,134 ft)/minute; service ceiling 12000 m (39,370 ft); range with tip tanks 1520 km (944 miles)
Weights: empty 2820 kg (6,217 lb); maximum take-off 5100 kg (11,244 lb)

Soko Jastreb of the Jugoslavian air force.

Dimensions: span over tip tanks 11.68 m (38 ft 3.8 in); length 10.88 m (35 ft 8.3 in); height 3.64 m (11 ft 11.3 in); wing area 19.43 m² (209.15 sq ft)

Armament: three Colt-Browning 12.7-mm (0.5-in) guns; inboard underwing attachments for bombs of up to 250-kg (551-lb), launchers for 12 rockets of 57-mm (2.24-in) calibre

or 45-kg (100-lb) photo flares; six other pylons each for one 127-mm (5-in) rocket
Operators: Jugoslavia, Libya, Zambia

Soko Kraguj

The Kraguj, a simple COIN aircraft, was designed at the Aeronautical Research Establishment in Belgrade, and manufacture was undertaken by the state-owned Soko factory at Mostar. A prototype flew in 1966, and about 30 were built subsequently for ground-attack units of the Jugoslav air force. Able to operate from grass fields or unprepared strips, with a ground run of 120 m (394 ft) or less, the Kraguj entered service in 1968, mainly as an elementary weapons trainer.

Specification
Type: single-seat light close-support aircraft
Powerplant: one 254-kW (340-hp)

Lycoming GSO-480-B1A6 flat-six piston engine
Performance: maximum speed 275 km/h (171 mph) at sea level; initial rate of climb 480 m (1,575 ft)/minute; range with maximum fuel 800 km (497 miles)
Weights: empty 1130 kg (2,491 lb); maximum take-off 1625 kg (3,583 lb)
Dimensions: span 10.64 m (34 ft 10.9 in); length 7.93 m (26 ft 0.2 in); height 3.00 m (9 ft 10.1 in); wing area 17.00 m² (182.99 sq ft)
Armament: one 7.7-mm (0.303-in) machine-gun in each wing; two inboard underwing hardpoints each able to carry a 100-kg (220-lb) bomb, 150-litre (33-Imp gallon) napalm

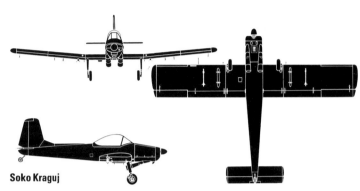

Soko Kraguj

tank or 12-round rocket pack; and four outboard underwing hardpoints each able to carry one 57-mm (2.24-

in) or 127-mm (5-in) rocket
Operator: Jugoslavia

Soko G-4 Super Galeb

Despite its name, the Super Galeb is a totally new aircraft, bearing a close superficial similarity to the BAe Hawk. It has a respectable light attack capability and so entered service with the JRV flying school at Mostar in 1982 as a replacement for both the Galeb and Lockheed T-33, and is expected to supplant the TJ-1 for weapons training. Future JRV pilots will fly only the UTVA-75 primary trainer and Super Galeb before converting to the MiG-21U. The prototype flew on 17 July 1978; one of the few subsequent changes was incorporation of a slab tailplane with pronounced anhedral. The Super Galeb was shown to the West at the 1983 Paris Salon surrounded by a variety of Jugoslav-developed weapons. Export customers have not yet been identified.

Specification
Type: basic trainer and light attack aircraft

The Super Galeb is a totally new aircraft, and will largely replace its earlier namesake. Light strike can be added to the trainer role if required.

Powerplant: one 1814-kg (4,000-lb) Rolls-Royce Viper Mk 632 turbojet engine
Performance: maximum speed at 6000 m (19,685 ft) 910 km/h (565 mph); initial rate of climb 1800 m (5,906 ft)/minute; absolute ceiling 15000 m (49,215 ft); combat radius lo-lo-lo 300 km (186 mph)
Weights: empty 3250 kg (7,165 lb); maximum take-off, training 4760 lb (10,494 lb), attack 6330 kg (13,955 lb)
Dimensions: span 9.88 m (32 ft 5 in); length 11.86 m (38 ft 11 in); height 4.28 m (14 ft 0.5 in); wing area 19.50 m² (209.9 sq ft)
Armament: up to 1350 kg (2,976 lb) of ordnance which can comprise S-8-16 cluster bombs, KPT-150 bomblet dispensers, L-57-16MD 57-mm

(2.24-in) and L-128-04 128-mm (5.04-in) rocket pods, VRZ-57 training rockets, and a ventral pod containing either a Jugoslav-built

GSh-23L twin-barrel 23-mm (0.91-in) cannon pod or a reconnaissance pack
Operator: Jugoslavia

Soko/CNIAR Orao/IAR-93

A unique example of co-operation in military aircraft design by countries on opposite sides of the Iron Curtain, the Orao was made possible by the unusual latitude allowed to Romania by its Soviet ally in the Warsaw Pact, and the communist persuasion of non-aligned Jugoslavia. Built in the latter country as the Soko Orao (eagle) and in Romania as the CNIAR (Centrul National al Industreiei Aeronautice Române) IAR-93, the aircraft was designed by a joint team and developed under a programme known as JuRom, the name indicating the partner countries. Prototypes assembled by Soko and CNIAR made

their first flights within minutes of each other on 31 October 1974, and these were followed by a pair of two-seat variants, both flown on 29 January 1977. During 1978 deliveries began of a pre-production batch of 15 to each country, after which the initial series model, known as the Orao 1 and IAR-93A, entered service. These early versions each have a pair of non-afterburning Viper Mk 632 engines of 1814 kg (4,000 lb) thrust, but after a short run of single- and two-seat aircraft to this standard (20 of them for Romania) manufacture began of the definitive Orao 2 and IAR-93B, which feature a licence-built after-

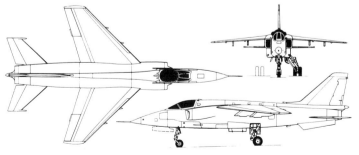

Soko/CNIAR (JuRom) Orao/IAR-93

burner and structural changes such as integral wing fuel tanks and a honeycomb rudder and tailplane. Both versions have a tandem dual trainer model, with similar opera-

tional capability to the single-seat version, for advanced training and weapons instruction. All variants have Martin-Baker Mk 10 automatic seats, but while single-seaters

SOKO/CNIAR Orao/IAR-93 (continued)

have a powered clamshell canopy, the trainers have canopies manually hinged to the right side. Romania has ordered 165 IAR-93Bs; Jugoslav plans are believed to be similar but had not been announced by 1985. The Orao/IAR-93 is limited to the close support role by its lack of radar or inertial navigation, but low-level interception is a secondary duty.

Specification
CNIAR IAR-93B

Type: single-seat close-support fighter with two-seat derivative
Powerplant: two 2268-kg (5,000-lb) afterburning thrust Rolls-Royce Viper Mk 633-47 turbojet engines
Performance: maximum speed 1160 km/h (721 mph) at sea level; initial rate of climb 3960 m (12,992 ft)/minute; service ceiling

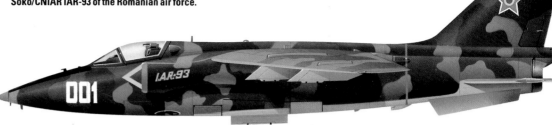

Soko/CNIAR IAR-93 of the Romanian air force.

12500 m (41,010 ft); mission radius, hi-hi-hi, with four 250-kg (551-lb) bombs and auxiliary fuel 530 km (329 miles)
Weights: empty 5900 kg (13,007 lb); maximum take-off 10097 kg (22,260 lb)

Dimensions: span 9.62 m (31 ft 6.7 in); length, single-seat 14.90 m (48 ft 10.6 in); height 4.45 m (14 ft 7.2 in); wing area 26.00 m^2 (279.87 sq ft)
Armament: two 23-mm (0.91-in) twin-barrel cannon with 200 rpg;

one centreline and four wing pylons carrying five 250-kg (551-lb) bombs, or equivalent loads including rocket pods, reconnaissance pods, munition dispensers or tanks
Operators: Jugoslavia (Orao), Romania (IAR-93)

Sukhoi Su-7 'Fitter'

Though criticized for its poor payload/range capabilities, the obsolescent Sukhoi Su-7 has the saving graces of excellent handling qualities, good low-level gust response, manoeuvrability, simplicity and

unbreakable toughness. Remaining in service with 15 air arms, although almost replaced within Soviet frontline units, it has seen action on several occasions during wars in the Middle East and Indian sub-continent. First flown in 1955, the aircraft was intended as an interceptor to shoot down the F-100 and similar fighters. It entered service four years later in its Su-7B form, under the NATO reporting name 'Fitter-A', in a totally different role, rapidly establishing itself as the standard fighter-bomber of the Soviet air force and some Warsaw Pact allies. Three progressively improved models followed, but had insufficient changes to merit a revised Western designation. In the Su-7BM, underwing stores pylons were doubled to four, an uprated engine installed (its take-off power being boosted, if re-

Sukhoi Su-7 'Fitter-A' of the Egyptian air force.

quired, by two JATO bottles), and a revised internal cannon installed with incrased muzzle velocity. The aircraft also introduced a radar warning receiver in the tail and two duct fairings running along the spine. Rough-field operation was provided for in the Su-7BKL, its main wheels carrying soft-field steel skis or skids, and whose large low-

Indian pilots have proved the worth of the elderly Sukhoi, taking their aircraft to war against Pakistan. These are still in service but are sadly being rapidly replaced by more modern types such as the SEPECAT Jaguar.

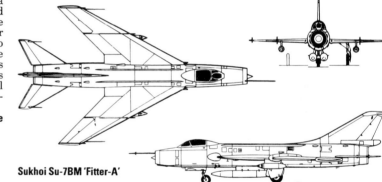

Sukhoi Su-7BM 'Fitter-A'

Sukhoi Su-7 (continued)

pressure nosewheel tyre is betrayed by a blistered door to its bay. Further changes of detail were incorporated in the later Su-7BMK, but little could be done to moderate the demands of the thirsty AL-7F engine, which on full afterburner at sea level would consume the entire 2941 litres (647 Imp gallons) of internal fuel in a little over eight minutes. Even so, fuel capacity is reduced in the operational trainer versions (Su-7UM and Su-7UMK, known to NATO as 'Moujik') to make way for a second seat, despite a slight lengthening of the fuselage.

Specification
Sukhoi Su-7BMK 'Fitter-A'
Type: single-seat ground-attack fighter
Powerplant: one 10000-kg (22,046-lb) afterburning thrust Lyulka

Around 70 'Fitter-As' serve in the ground attack role with the Czech air force. These will be replaced by Su-25 'Frogfoots' over the next few years.

AL-7F-1 turbojet engine
Performance: maximum speed 850 km/h (528 mph) at sea level without afterburning, or 1350 km/h (839 mph) with afterburning; initial rate of climb 9120 m (29,921 ft)/minute; service ceiling 15150 m (49,705 ft)
Weights: empty 8620 kg (19,004 lb); normal loaded 12000 kg (26,455 lb); maximum take-off

13500 kg (29,762 lb)
Dimensions: span 8.93 m (29 ft 3.6 in); length, including probe 17.37 m (56 ft 11.9 in); height 4.57 m (14 ft 11.9 in); wing area 27.60 m² (297.09 sq ft)
Armament: two 30-mm (1.18-in) NR-30 cannon (with 70 rpg) in wing roots; six weapons pylons; two underfuselage and two beneath the inner wings each carry up to 500 kg

(1,102 lb) of stores, plus two under the outer wings each carrying up to 250 kg (551 lb); weapon load reduced to 1000 kg (2,205 lb) when two 600-litre (132-Imp gallon) drop tanks are carried on fuselage pylons
Operators: Algeria, Czechoslovakia, Egypt, Hungary, India, Iraq, Poland, Romania, Soviet Union, Vietnam

Sukhoi Su-15 'Flagon'

The Sukhoi Su-15 demonstrates the Soviet Union's practice of constantly improving a basic design over a period of years to produce a highly effective definitive production aircraft. For ten years it has been the main interceptor in service with the IA-PVO air defence force, with over 850 estimated to be operational. However, PVO re-equipment needs are now being met by the MiG-23S, MiG-29, MiG-31 and Su-27, and the Su-15 is now being replaced. By 1985 numbers had dropped to below 700.

The Su-15 traces its origin back to a requirement issued in the early 1960s for a supersonic interceptor with higher flight performance and better radar then the Su-11, which was then under development. The Su-15 prototypes, tested in 1964-65 by V. Ilyushin (son of the design bureau leader), appear to have combined twin Tumansky R-11 afterburning engines, Skip Spin radar and AA-3 'Anab' missiles of the Yakovlev Yak-28P with the virtually unchanged delta wing and tail of the Su-9/-11; there was a new fuselage and a single large braking parachute. The two engines were fed by variable-geometry side inlets with auxiliary inlet doors. The Su-15 thus represented a very low-risk development, making use of a large proportion of components from existing aircraft.

A pre-series batch of Su-15s made their appearance at the 1967 Domodedovo air display, but this basic type was not built in large numbers. Development of the type was already under way, as indicated by the presence at Domodedovo of an experimental STOL version with three lift-jets installed in the centre section of the fuselage. The sweep of the outer wing panels was reduced, to give enhanced lift and lateral control, especially at low airspeeds. Although the Soviet Union apparently abandoned research into jet-lift STOL aircraft, the wing planform of this prototype foreshadowed

Sukhoi Su-15 'Flagon-F' armed with AA-3 'Anab' missiles.

the compound sweep of later Su-15 variants.

The first major production version of the Su-15, called 'Flagon-D' by NATO, entered service in the late 1960s. It introduced more powerful engines and compound sweep on the outer wing panels, joined to the inner wing by a very short unswept section. This 'soft dog-tooth' is reminiscent of that incorporated in the wing of the Ilyushin Il-62 airliner. It is likely that the modification to the Su-15 wing was intended to improve low-speed handling and to reduce landing speeds from the high level of the earlier sub-types.

'Flagon-E', with more powerful R-13 engines, was also said to have 'improved electronics', but these did not impart full capability against low-flying targets, nor compatibility with the AA-7 'Apex' missile. The final production version, identified as 'Flagon-F' by NATO, is distinguished by a radome of curved contours and offering lower drag, which replaces the conical radome of earlier aircraft. Two-seat Su-15U conversion trainers are identified as 'Flagon-C', and these have the cockpits at the same level, the instructor having a large periscope. Like all late single-seat versions, this has twin nosewheels.

The Su-15 has been based only in the Soviet Union, deployed mainly in the medium- to high-altitude intercept role, which it shares with the MiG-25 and -31. Like previous single-seat Soviet interceptors, it operates in a closely ground-controlled environment. It was thought that the pilot need never see the target except on head-down radar and flight-director displays, but this was certainly not the case with the

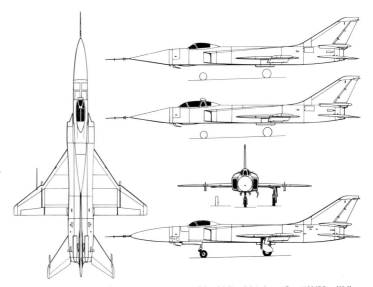

Sukhoi Su-15 'Flagon-F' (upper view: 'Flagon-D', middle side view: Su-15U 'Moujik').

Su-15 that shot down the KAL 747 in 1983. In its primary role, the Su-15's attributes include great speed and climb, large internal fuel capacity, and extremely mature engines and AAM systems. If a Mach 2 fighter can be called a workhorse, the Su-15 is one, and it is likely to see many more years' service with the IA-PVO.

Specification
Sukhoi Su-15 'Flagon-E'
Type: all-weather interceptor
Powerplant: two 7500-kg (16,535-lb) afterburning thrust Tumansky R-13-F2S-300 turbojet engines
Performance: maximum speed, with 4 AAMs at medium to high

altitudes 2445 km/h (1,519 mph), or 1100 km/h (684 mph) at sea level; service ceiling 20000 m (65,615 ft); combat radius 725 km (450 miles)
Weights: empty 12500 kg (27,558 lb); normal loaded 18000 kg (39,683 lb); maximum take-off 20000 kg (44,092 lb)
Dimensions: span 10.50 m (34 ft 5.4 in) length 21.50 m (70 ft 6.5 in); height 5.00 m (16 ft 4.9 in); wing area 35.70 m² (384.28 sq ft)
Armament: two or four air-to-air missiles, normally AA-3-2 'Advanced Anab' plus two AA-8 'Aphid'; two 23-mm (1.18-in) GSh-23L gun pods
Operator: Soviet Union

Sukhoi Su-17/20/22 'Fitter'

When a version of the Sukhoi Su-7 with variable-geometry outer wing panels (the Su-17G) was revealed in 1967, it was labelled 'Fitter-B' by NATO and dismissed as a research version of an unimpressive ground-attack fighter. Not until the mid-1970s did it dawn on the West that the modification, together with a more powerful but fuel-efficient engine and new avionics, had resulted in a vastly improved aircraft with

doubled weapon load, 30 per cent greater range and substantially better short-field take-off characteristics. So successful has been the aircraft that numerous versions are in service with Frontal Aviation, the

Soviet naval air arm, Warsaw Pact air forces and client countries abroad.

Sukhoi Su-17 variants based in Eastern Europe and the USSR have been progressively improved from the basic 'Fitter-C' initial production model, firstly to the Su-20M 'Fitter-D' with undernose terrain-avoidance radar and a marked-target seeker in the inlet centre-body. An Su-20U conversion trainer, the 'Fitter-E', parallels the 'Fitter-C' except for a slightly drooped forward fuselage and lack of a left wing root gun, while the Su-22U 'Fitter-G'

operational trainer has a taller, straight-topped fin, small ventral fin, and a marked-target seeker. Newest of the single-seat variants is the Su-22BM 'Fitter-H' which has revised fin, augmented sensors and other avionics, and a deep dorsal fairing behind the canopy, presumably for extra fuel tanks. Export versions of the 'Fitter-C' have a reduced avionics fit and are designated Su-20, but when the Su-22 'Fitter-F' appeared as a 'Fitter-D' counterpart, its bulged rear fuselage revealed a change of engine to the 11500-kg (25,353-kg) thrust Tumansky R-29B afterburning turbojet for even better performance. A 'Fitter-H' counterpart, the Su-22BKL 'Fitter-J' is similarly powered and identified by a small ventral fin and more angular dorsal fin. Su-22s are also employed as interceptors with AA-2 'Atoll' AAMs. A Tumansky-powered two-seater has also been noted in Soviet service.

Specification
Sukhoi Su-17 'Fitter-C'
Type: single-seat ground attack fighter

Powerplant: one 11200-kg (24,692-lb) afterburning thrust Lyulka AL-21F-3 turbojet engine
Performance: maximum speed Mach 2.17 or 2300 km/h (1,429 mph) at altitude, or Mach 1.05 (1285 km/h; 798 mph) at sea level; initial rate of climb 13800 m (45,276 ft)/minute; service ceiling 18000 m (59,055 ft); combat radius with 2000 kg (4,409 lb) of stores, hi-lo-hi mission 630 km (391 miles), lo-lo-lo mission 360 km (224 miles)
Weights: empty 10000 kg (22,046 lb); loaded, clean 14000 kg (30,865 lb); maximum take-off 17700 kg (39,022 lb)
Dimensions: span, extended (28° sweep) 14.00 m (45 ft 11.2 in), fully swept (62°) 10.60 m (34 ft 9.3 in); length 18.75 m (61 ft 6.2 in); height 4.75 m (15 ft 7 in); wing area 40.10 m^2 (431.65 sq ft)
Armament: two 30-mm (1.18-in) NR-30 cannon in wing roots (with 70 rpg); four underwing and four fuselage weapon pylons for up to 4000 kg (8,818 lb) of ordnance, including tactical nuclear weapons and AS-7 'Kerry' ASMs; AA-2 and AA-2-2 'Atoll' self-defence AAMs
Operators: Afghanistan, Algeria, Angola, Czechoslovakia, Egypt, West Germany, Iraq, Libya, Peru, Poland, Soviet Union, Syria, Vietnam, North and South Yemen

Latest version of the big Sukhoi is the 'Fitter-J' with bulged rear fuselage (with Tumansky engine) and larger dorsal spine (to reduce drag and hold more fuel). The fin has a squared-off end, giving this variant a distinctive look. Other 'Fitter-Js' have been exported, notably to Libya.

Sukhoi Su-24 'Fencer'

It is evident that VVS procurement has been strongly influenced by programmes of the USAF, and to an unprecedented degree many modern Soviet tactical aircraft are counterparts of US designs. In the case of the Sukhoi Su-24 (NATO codename 'Fencer') the inspiration was the General Dynamics F-111, even to the extent of seating the crew side-by-side. Perhaps by chance, the Sukhoi design avoided all the very severe development problems encountered by the US type. Though its main gears retract into the fuselage they do not inhibit the carriage of extremely heavy weapon loads on fuselage pylons, and as the Soviet aircraft is both lighter and much more powerful than most F-111s it has better all-round performance.

It is no exaggeration to claim that this aircraft has been more feared than any other Soviet attack aircraft in history. Just as the NATO reporting name 'Foxbat' caused ripples of concern in the Pentagon in the mid/late 1960s, so did the name 'Fencer' cause furrowed brows in the 1970s. In fact, though triggered by the F-111, the Su-24 has a much closer counterpart in the Panavia Tornado. All-round capability of the two aircraft is uncannily similar, though more advanced technology in the Western aircraft, especially in the engine, results in it being significantly smaller. More to the point, the earlier timing of the Soviet type resulted in its entering service almost a decade earlier than the Tornado, so that by late 1985 the number in service was at least 700.

Basic geometry is that refined at TsAGI in 1960-62 and also used by the MiG team on the Ye-231 (MiG-23) family. The Sukhoi team was instructed to duplicate the MiG on roughly twice the scale of gross weight, using two engines instead of one. The result has always been a

The Sukhoi Su-24 'Fencer' carries pylons under the swinging section of the wings for weapons carriage.

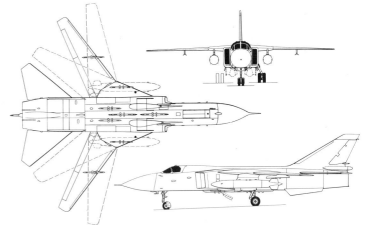

Sukhoi Su-24 'Fencer' of Frontal Aviation.

special aircraft, used only from major paved airfields and never offered to Warsaw Pact nations or foreign clients. EW (electronic warfare) installations are probably the most comprehensive in the world in their era, though the wealth of aerials that cover the airframe have not all been identified.

The wing is structurally mounted high on the broad fuselage, with the swing-wing pivots carried outboard on small fixed gloves. Each outer wing is driven electro-hydraulically, early Su-24s having settings of 16°, 45° and 68° selected manually. Later models may have a Mach/sweep programmer. Each wing panel has full-span hydraulically operated slats and double-slotted flaps, the inner sections of flap being locked except at minimum sweep. Differential spoilers augment the tailerons for roll control, act symmetrically as airbrakes and open automatically as lift dumpers after landing. The vertical tail is large, and augmented by large ventral fins on the chines of the extremely wide flat-bottomed rear fuselage.

The large afterburning turbojet or augmented turbofan engines had not been positively identified by late 1985. There are grounds for supposing that there are Su-24s with two types of engine. It is generally believed in the West that most have the Lyulka AL-21F, but absence of visible smoke often suggests an alternative engine. There have certainly been variations in inlet shape and nozzle geometry, the majority of current aircraft, called 'Fencer-B',

having almost vertical nozzle lips and a rear fuselage smoothly curved around the jetpipes (early Su-24s, called Fencer-A', having had a rectangular box-like rear section).

Not least of this aircraft's impressive attributes is its internal fuel capacity, never estimated at less than 13,000 litres (2,860 Imp gallons), augmented by two drop tanks usually carried on the fixed glove pylons and estimated at a unit size of 2,000 litres (440 Imp gallons) or sometime 2,500 litres (550 Imp gallons). Under the outer wings are two pivoting pylons usable at all wing angles, while the fuselage has two tandem and two side-by-side pylons. Usually all eight pylons are fitted with identical crutched ejector racks. Sensors are predictably diverse. All Su-24s have a multimode main radar, terrain-avoidance or terrain-following radar, forward-looking infra red and a laser ranger. In one of the first US assessments a decade ago the Su-24 was said to be able 'to deliver ordnance in all weather within 55 m (180 ft) of target'. The latest, so-called 'Fencer-C', version has a forest of probes on or above the tip of the nose, more instrumentation and/or EW aerials carried below the nose and twin blister fairings extending from the inlet lips to the roots of the wing gloves. It has been surmised these are the same target-illuminating CW radars as carried on the lower flanks of the nose of various MiG-

23BNs and MiG-27s. Still unresolved is the puzzle of the two large fairings under the fuselage, large portions of which actually form part of the twin airbrakes. One if not both is a gun, but their visibly different appearance has caused controversy.

Specification
Sukhoi Su-24
Type: all-weather interdictor/attack aircraft
Powerplant: believed to be two 11200 kg (24,691 lb) Lyulka AL-21F-3 afterburning turbojets (but see text)
Performance: (estimated) maximum speed, clean at high altitude Mach 2.18 or 2315 km/h (1,439 mph); service ceiling 17500 m (57,400 ft); combat radius hi-lo-hi with 2000 kg (4,409 lb) bomb load 1800 km (1,118 miles)
Weights: (estimated) empty 18100 kg (39,900 lb); maximum take-off 39500 kg (87,080 lb)
Dimensions: (estimated) span, spread 17.25 m (56 ft 7 in), swept 10.00 m (32 ft 9.7 in); length, excluding probe 21.29 m (69 ft 10 in); height 5.50 m (18 ft 0 in); wing area, spread 46.45 m² (500.0 sq ft)
Armament: at least one cannon in fuselage, and possibly two guns of different types; up to 10000 kg (22,050 lb) of fuel and ordnance carried on eight fuselage and wing pylons
Operator: Soviet Union

Sukhoi Su-24 'Fencer'

Sukhoi Su-25 'Frogfoot'

Expansion of Soviet tactical air strength during the 1970s kept the West on the lookout for signs of new combat aircraft, satellite reconnaissance of the Ramenskoye test centre revealing a number of designs in the early stages of flight development. Amongst these was an aircraft resembling the Northrop A-9, and as the A-9 was unsuccessful in the competition which brought the Fairchild A-10 Thunderbolt II into USAF ser-

vice, the Soviet machine's close-support role was readily apparent. Known originally to NATO as the Ram-J (tenth new type seen at Ramenskoye), it was identified subsequently as the Sukhoi Su-25 and

Sukhoi Su-25 'Frogfoot' of the Soviet air force in Afghanistan.

issued with the fighter-classification reporting name 'Frogfoot'. By 1982 a trials squadron was operating in Afghanistan against tribesmen trying to resist the Soviet occupation. This opportunity has been taken to

'Frogfoots' are being supplied to Warsaw Pact countries, the first confirmed being Czechoslovakia (illustrated) and Hungary. It is very likely that Poland and East Germany have also been supplied with this close-support fighter.

develop operational techniques, including co-ordinated low-level attacks by Mil Mi-24 helicopter gunships and Su-25s in support of ground troops. Smaller than the Thunderbolt, and with engines estimated at much lower thrust, the Su-25 is thought to carry less weight of ordnance than its American counterpart, despite having ten weapons pylons. Flight performance may be better, and avionics probably include laser ranging and comprehensive weapon delivery systems, as well as very complete EW (electronic warfare) systems, partly in large wingtip pods; a multimode radar is probably not fitted. Deliveries to operational units of Frontal Aviation in the western USSR and eastern Europe took place at a high rate from 1983, these aircraft being pro-

duced by the manufacturing plant at Tbilisi.

Specification
Type: single-seat close-support aircraft
Powerplant: two non-afterburning turbojets each of up to 4100-kg (9,039-lb) thrust
Performance: maximum speed about 880 km/h (547 mph); combat radius about 550 km (342 miles)
Weights: maximum take-off 16300 kg (35,935 lb)

Dimensions: span 15.50 m (50 ft 10.2 in); length 14.50 m (47 ft 6.9 in)
Armament: one multi-barrel cannon beneath the centre fuselage; 10 hardpoints for some 4000 kg (8,818 lb) of ordnance
Operator: Soviet Union

Sukhoi Su-27 'Flanker'

In 1980 a new and impressively large twin-engined fighter was seen on the Ramenskoye test airfield, receiving the temporary designation 'Ram-J'. In 1982 its designation was announced by the US Department of Defense as the Sukhoi Su-27, the NATO reporting name being 'Flanker'.

Since then not a great deal more has been learned about this aircraft, which follows standard Soviet practice in using the same aerodynamic geometry as another type but on a different scale. In this case the partner is the Mikoyan-Gurevich MiG-29 and the Sukhoi aircraft is the larger. Probably incorporating advanced alloys and fibre-reinforced composites in its construction (though not to the same degree as Western fighters) the Su-27 has two widely spaced augmented jet engines, with tankage between them. The fixed wing is mounted well aft, with tapered and slightly swept outer portions joined to an enormous inner section with an acutely swept leading edge, extending forward to the windscreen of the single-seat cockpit. Twin vertical tails and slab tailplanes are carried on cantilever beams extending along the outer sides of the engines. One of the few major recognition features distinguishing this fighter from the MiG-

29 is that the engine nozzles and fuselage extend well aft of the tail, the opposite of the MiG-29.

It is not yet known whether the Su-27 has the same extremely deep slab-sided inlets as the MiG-29 but it is probable. Some artist's impressions show an enormous wing with curved streamwise tips of Küchemann type, while others show a smaller wing with square tips carrying anti-flutter bodies or AAM rails. The following specification is based on US Department of Defense estimates, and no indication had been given by late 1985 on how the armament is arranged. One DoD estimate gives it the same armament (six 'AA-10' AAMs) as the MiG-29, in which case one wonders why this much bigger aircraft was built. Both have full look-down shoot-down capability, similar speed and adequate radius of action. Again, while two large guns appear in most artwork showing the MiG-29, no guns have been suggested for the Su-27, though on the basis of available evidence the Sukhoi ought to be the most agile close-combat aircraft in the world, beating even the MiG-29.

Suffice to report that in late 1984 it was said to be 'nearing deployment' and will probably be used by frontal theatre units and also by the

The first photographs of the Su-27 show it to have rounded wingtips, although production aircraft may have missile rails.

IA-PVO home defence forces. It is also thought in Washington to be the most likely base of a carrier version for embarkation in the giant new Soviet nuclear carrier.

Specification
Sukhoi Su-27
Type: long-range air-combat fighter
Powerplant: two augmented turbofans in 13150 kg (29,000 lb) class, probably Tumansky R-29Bs
Performance: (estimated) maximum speed Mach 2.35 or

2495 km/h (1,550 mph); speed at sea level Mach 1.1 or 1347 km/h (847 mph); combat radius 1500 km (932 miles)
Weight: (estimated) maximum take-off 28000 kg (61,730 lb)
Dimensions: (estimated) spoan 14.00 m (45 ft 11 in); length 21.00 m (69 ft 0 in); height 5.50 m (18 ft 0 in)
Armament: probably includes two advanced guns; six or eight AAMs, said to be of AA-10 type. No mention has been made of surface-attack weapons

Swearingen (Fairchild) Merlin IIIA

Swearingen Aircraft rebuilt a Beech Queen Air with a new streamlined, pressurized fuselage and two PT6A turboprop engines, designating it Merlin IIA. This led to the heavier and more powerfu IIIA with many airframe changes and improved performance. First military customer was the Belgian air force, whose Pembroke C.51s needed costly resparring; six Merlins were delivered

in mid-1976 as replacements with the 21 Smaldeel, a mixed transport squadron based at Brussels-Melsbroek. The Argentines took delivery of four aircraft from 1978 for use in the communications role. Fairchild Industries has since replaced the Merlin IIIA by the Fairchild 300 (SA227) but no military orders have been announced.

Specification
Swearingen SA226T Merlin IIIA
Type: 11-seat light transport
Powerplant: two 626-kW (840-shp) Garrett TPE331-3U-303G turboprop engines
Performance: maximum cruising speed 523 km/h (325 mph) at 4875 m (16,000 ft); initial rate of climb 771 m (2,530 ft)/minute; service ceiling 8,810 m (28,900 ft);

range, 45 minutes reserves 4603 km (2,860 miles)
Weights: empty 3357 kg (7,400 lb); maximum take-off 5670 kg (12,500 lb)
Dimensions: span 14.10 m (46 ft 3 in); length 12.85 m (42 ft 2 in); height 5.13 m (16 ft 10 in); wing area 25.78 m² (277.50 sq ft)
Armament: none
Operators: Argentina, Belgium

Swearingen Merlin/Metro

Swearingen (now Fairchild-Swearingen) used its Merlin business aircraft to develop a 19-passenger pressurized commuter airliner designated SA-226TC Metro, flown

on 26 August 1969. It had a circular-section fuselage with airstair door ahead of the wings and cargo hatch in the rear. An executive equivalent was designated SA-226AT Merlin

IV. Both have had detail changes, including new 4-blade Dowty propellers, increased power and strengthened landing gear, current models (SA-227PC Metro IIIC and SA-

227AT Merlin IVC) being known as the Fairchild 400. Both are in military service for passenger (up to 23 passengers and two crew), medevac (six slung stretchers with up to 10 seats for passengers or attendants) or maritime surveillance (under-fuselage Litton APS-504 radar, MAD tailboom, bubble windows, Bendix multiband scanner and Zeiss RMKA 8.5/23 vertical camera). An alternative Electronic Surveillance model has a SLAMMR antenna (aerial) on the roof, radar and communications pods under the forward and rear fuselage and low-band communications intercept receivers on the outer wings. A projected special missions aircraft, in ASW and maritime patrol versions, was announced in September 1982 under the name Air Sentry.

Specification
Fairchild 400
Type: transport and special-missions aircraft
Powerplant: two 820-kW (1,110-shp) Garrett TPE331-14UA/UB-801G turboprop engines
Performance: maximum operating speed 555 km/h (345 mph); initial rate of climb 717 m (2,353 ft)/minute; service ceiling 10085 m (33,090 ft); range 3125 km (1,942 miles)
Weights: empty 4393 kg (9,686 lb); maximum take-off 7484 kg (16,500 lb)
Dimensions: span 17.37 m (57 ft 0 in); length 18.09 m (59 ft 4.25 in); height 5.08 m (16 ft 8 in); wing area 28.71 m^2 (309.0 sq ft)
Armament: none
Operators: (Merlin/Metro) Argentina, Chile, Mexico, South Africa, Sweden, Thailand

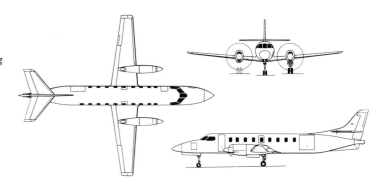

Swearingen (Fairchild) Merlin IV

Transall C.160

One of the handful of types which have achieved the rare distinction of being reinstated in production, the Transall C.160 is again in process of delivery to the French air force, albeit in somewhat improved form.

Originally conceived as a replacement for the Nord Noratlas, which equipped transport units of France's Armée de l'Air and West Germany's Luftwaffe, the C.160 was one of the first successful joint European aerospace ventures, being produced by a consortium of companies which was collectively known as the Transport Allianz group. Indeed, the name and designation chosen for the resulting machine reflected the origins of the project, for the initial quantity to be acquired was set at 160 (50 C.160F aircraft for France and 110 C.160D aircraft for West Germany) whilst the name was merely a contraction of Transport Allianz. Members of the original production group included Nord-Aviation, Hamburger Flugzeugbau (HFB) and Vereinigte Flugtechnische Werke (VFW), these joining forces at the beginning of 1959.

Three prototypes were built in all, one by each of the three major partners in this venture, and the first of these made a successful maiden flight on 25 February 1963. They were followed by six pre-production specimens from May 1965 whilst production-configured C.160s began to emerge in the spring of 1967, deliveries getting under way soon afterwards, and by the time manufacture ceased in 1972 a total of 169 had been built. In addition to the 160 supplied to the two principal partners, nine more of a variant known as the C.160Z were sold to South Africa, the only other air arm to operate the original type being that of Turkey, which took delivery of 20 C.160T aircraft (former Luftwaffe examples) in the early 1970s.

Subsequently, at the end of the 1970s, it was decided to reopen the production line in France, that country's air force ordering 25 more examples which differ from their predecessors by virtue of additional fuel capacity and improved avionics gear. Range limitations have been partly resolved by the extra centre-section fuel tank, but the newest C.160s also feature inflight-refuelling capability in the form of a probe above the cockpit. Maximum pay-

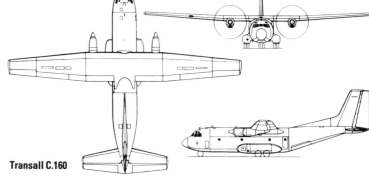

Transall C.160

load is 1600 kg (35,275 lb), while 93 troops or 88 paratroops can be accommodated.

Specification
Transall C.160
Type: medium tactical transport aircraft
Powerplant: two 4,549-eKW (6,100-ehp) Rolls-Royce RTy.20 Tyne Mk 22 turboprops.
Performance: maximum speed 592 km/h (368 mph) at 5000 m (16,405 ft); service ceiling 8230 m (27,000 ft); range 4800 km (2,983 miles) with 8000-kg (17,637-lb)

payload
Weights: empty 28760 kg (63,405 lb); maximum take-off 51000 kg (112,435 lb)
Dimensions: span 40.00 m (131 ft 3 in); length 32.40 m (106 ft 3.5 in); height 12.37 m (40 ft 7 in); wing area 160.0 m^2 (1,722.3 sq ft)
Operators: France, West Germany

Transall production has been restarted to provide the French air force with 25 extra transports. These have inflight-refuelling capability.

Tupolev Tu-16 'Badger'

The development in the early 1950s of the Mikulin bureau's massive AM-3 (RD-3) turbojet marked the end of the Soviet Union's dependence on Western engine technology. Rated at 8200-kg (18,078-lb) thrust in its initial version, the AM-3 made possible the design of new bombers with fewer engines than their Western counterparts, being twice as powerful as most contemporary Western engines. Known as the Tupolev Tu-88 or Avicraft N, the prototype of the Tu-16 'Badger' series was flown in 1952; it was thus a contemporary of the British V-bombers rather than the American Boeing B-47. Like many Soviet aircraft, in its design philosophy it was a mixture of the conservative and the radically new. The fuselage was virtually that of the piston-engined Tu-85, and the structure, systems and defensive armament were all based on the Tu-4 (B-29). The wing set the pattern for future Tupolev aircraft, with a high degree of sweep and a bogie landing gear retracting rearwards into trailing edge pods. The layout offered a sturdy and simple gear, left the centre-section free for weapon stowage and offered wide track; in addition, the pods helped to delay the transonic drag-rise. An apparent anomaly was the fixed forward-firing gun, retained to the present day.

The Tu-88 was ordered into production for the DA (Long-Range Aviation) and entered service in 1955. Later versions were equipped with the uprated AM-3M, and most of the type were eventually fitted with inflight refuelling equipment. The system fitted to Tu-16s so far seen is unusual, involving a tip-to-tip connection; other Soviet aircraft have nose probes. Some Tu-16s were completed as specialized tankers with a tip-hose or belly hose-reel.

Soviet production of the Tu-16 probably gave way to the Tu-20 and Tu-22 in the late 1950s, after about 2,000 had been delivered. In the 1970s the type was put back into production at Xian in China, as a replacement for the Tu-4. The Chinese designation of the Tu-16 is H-6. In 1955 the Tu-16 also formed the basis for the Tu-104, the first

Soviet jet airliner.

A new lease of life for the Tu-16 came with the rising power of the Soviet navy in the early 1960s. Tu-16s were steadily transferred from the DA to the AVMF (Soviet Naval Aviation), and became that service's first missile-carriers. The first missile-armed variant, the 'Badger-B' of 1961, carried two turbojet-powered AS-1 'Kennel' missiles under the wings; these aircraft were converted to 'Badger-Gs', with the more advanced AS-5 'Kelt' rocket missile: a number of these were delivered to Egypt and saw missile-firing action against Israel in 1973. The 'Badger-C', also seen in 1961, carries a large supersonic AS-2 'Kipper' missile, guided with aid from a powerful radar installation replacing the glazed nose. Most of the 200 survivors carry AS-6 'Kingfish' on wing pylons. A similar radar is featured by the 'Badger-D' maritime reconnaissance aircraft, together with an array of radomes indicating an electronic surveillance capability. The 'Badger-E' is a reconnaissance conversion of 'Badger-A'. 'Badger-F' is similar but has specialized electronic intelligence (Elint) pods under each wing and a plethora of aerials. 'Badger-G Modified' carries AS-6 'Kingfish' Mach-3 missiles on wing pylons and has unknown de-

vices on the nose and a large belly radome. 'Badger-H' is a stand-off ECM jammer with powerful emitters and the ability to dispense an estimated 9 tonnes (19,842 lb) of chaff. 'Badger-J' is a more advanced jammer, with noise jammers in A- to I-bands in a canoe fairing under the fuselage. 'Badger-K' is an electronic reconnaissance version with large receiver blisters ahead of and behind the weapons bay.

About 350 Tu-16s are still in service with the AVMF, including tankers as well as strike aircraft. The reconnaissance and Elint aircraft are likely to be replaced in first-line service by the better-optimized turboprop Ilyushin Il-38 and Il-18, while the missile-armed variants are being replaced by Tu-22Ms. Already these outstanding big twin-jets have served for 30 years, far surpassing the record of any Western aircraft in this category, early B-52s being withdrawn in the 1960s and 1970s.

Specification
Type: strategic bomber, ECM, jammer platform, Elint aircraft, flight-refuelling tanker, missile launcher and reconnaissance aircraft
Powerplant: two 9500-kg (20,944-

lb) thrust Mikulin AM-3M (RD-3M) turbojet engines
Performance: maximum speed Mach 0.91 or 992 mph (616 mph); cruising speed Mach 0.8 or 850 km/h (528 mph); service ceiling 14000 m (45,930 ft); maximum range, no weapons 6400 km (3,977 miles)
Weights: empty 36000 to 37200 kg (79,366 to 82,012 lb); maximum take-off 7200 kg (158,733 lb)
Dimensions: span 32.93 m (108 ft 0.5 in); length 34.80 m (114 ft 2.1 in); height 10.80 m (35 ft 5.2 in); wing area 164.65 m² (1,772.34 sq ft)
Armament: ('Badger-C') one AS-2 'Kipper' or two AS-6 'Kingfish', ('Badger-G') two As-5 'Kelt', (G Modified) two AS-6 'Kingfish'; bomber versions have provisions for 9000 kg (19,842 lb) of internal stores: all versions have seven 23-mm (1.18-in) NR-23 cannon: one fixed in forward fuselage, two in tail and two each in forward dorsal and rear ventral barbettes
Operators: China, Egypt, Iraq, Indonesia (in storage), Libya, Soviet Union

The Soviet naval aviation (AV-MF) has around 320 Tu-16s in service for strike, tanking and specialized roles. This is a 'Badger-C' anti-shipping strike aircraft, complete with pylons for missiles.

Xian H-6 (licence-built Tu-16) of the AFPLA (Chinese air force).

Tupolev Tu-16 'Badger-B' of the Indonesian air force, carrying AS-1 'Kennel' missiles. These are now in storage.

Tupolev Tu-22 'Blinder'

Development of this large supersonic bomber, under the Tupolev bureau designation Tu-105, was initiated in 1955-56 following rejection of the Tu-98. By that time it was clear that the effectiveness of Western air-defence systems was improving rapidly, and that supersonic missile-armed all-weather interceptors and medium/high-altitude SAMS would be in large-scale service by the end of the decade. The Tu-16, with its modest cruising altitude and defensive gun turrets was already obsolescent for penetration of hostile territory. The design objectives of the Tu-105 were to produce an aircraft with performance generally comparable to the Tu-16, but with greatly increased penetration altitude and speed, combined with high speed at low level.

The Tu-105 was thus designed to incorporate supersonic dash at high altitude without excessively penalizing subsonic efficiency. The engines were mounted high on the rear fuselage in slim cowlings to avoid the weight and drag penalties of the long inlet ducts of the Tu-98 (but these returned on the Tu-22M). The wing was of the acutely swept type also used on the Tu-102 (Tu-28P) interceptor, the compound-sweep layout giving minimum subsonic and low-speed penalty. Unlike the Tu-98, the fuselage, wing, landing gear pods and engine nacelles were positioned and designed in accordance with the rule. As in the Tu-98, elimination of defensive armament except the tail cannon saved weight and volume, and the crew remained at three, seated in tandem downward-ejecting seats. For the first time in a Soviet aircraft, bombing/navigation radar displaced the glazed nose.

The source of the engines has long been a mystery, but it appears they are Kolesov VD-7s, similar to the engines fitted to the Myasishchev M-50. The nacelles were fitted with plain inlets; production Tu-22s have translating inlets which slide forward at low speeds to open an auxiliary annular aperture.

The Tu-105 probably flew in 1959, in time for 10 (including one with a missile) to be demonstrated at Tushino in 1961. The type's debut was a complete surprise to Western intelligence, but progress in defence systems had come to the point where a supersonic dash and higher cruise altitude were not, in practice, much of an advantage over the subsonic Tu-16. In addition, the decision to rely solely on large numbers of enormous ICBMs for strategic attack had already been taken in the Soviet Union, leading to the tem-

Tupolev Tu-22 'Blinder-A' of the Libyan air force.

Tupolev Tu-22 'Blinder-C' maritime reconnaissance aircraft of the AV-MF.

porary cessation of bomber development, and bringing transfer of many aircraft to Soviet naval Aviation (AVMF).

There was, however, a continuing role for the Tu-22 in the precision strike and missile-carrying role, and approximately 170 of this type were delivered to Long-Range Aviation (DA) units from about 1964. These were of the variant known to NATO as 'Blinder-B', with a bomb bay modified to accept the AS-4 'Kitchen' air-to-surface missile, as well as free-fall weapons. The designation 'Blinder-A' was applied to the nine non-missile aircraft seen at Tushino in 1961. The AS-4 is the Soviet equivalent of the defunct British Hawker Siddeley Blue Steel, rocket-powered and with a 200-mile (320-km) range.

The main AVMF variant of the Tu-22 in current service is the maritime reconnaissance/Elint 'Blinder-C'. The 'Blinder-B' and AS-4 do not appear to be in AVMF service. In the reconnaissance role the 'Blinder-C' offers payload and flight-refuelled range similar to those of the Tu-16, but far higher speeds and reduced mission times. Only some 50 are in AVMF service. Both the DA and AVMF operate a few 'Blinder-D' (probably Tu-22IU) conversion trainers, with a separate second cockpit above and behind the standard (pupil's) cockpit.

Despite its great size and capability the Tu-22 has not found an important niche in the Soviet forces. Its main role has been maritime reconnaissance and all forms of strike in the European region, but its importance has diminished with the introduction of the Sukhoi Su-24 into Frontal Aviation and the Tu-22M 'Backfire' into the DA and AVMF. The original swing-wing Tu-

Tupolev Tu-22 'Blinder-A'

22M 'Backfire-A' is believed to have been a rebuild of existing Tu-22s.

The small batches of 'Blinder-Bs' supplied to Libya and Iraq (without AS-4 missiles or nuclear weapons) were probably surplus to DA requirements. The Iraqi aircraft have been used in action against Kurdish insurgents, and one Libyan machine at least against Tanzania, in support of Uganda. The Tu-22 has thus become one of the few large jet bombers to drop bombs in anger.

Specification
Tupolev Tu-22 'Blinder-C'
Type: bomber, reconnaissance and maritime strike aircraft
Powerplant: probably two 1500-kg (33,069-lb) afterburning thrust Kolesov VD-7 turbojet engines
Performance: maximum speed Mach 1.5 or 1600 km/h (994 mph) at 11000 m (36,090 ft); cruising speed Mach 0.85 or 900 km/h (559 mph); service ceiling with afterburning 18300 m (60,040 ft); maximum range, subsonic 6500 km (4,039 miles); unrefuelled tactical radius, high with 4000 km (249 miles) supersonic dash 2800 km (1,740 miles)
Weights: empty 40000 kg (88,185 lb); internal fuel 36000 kg (79,366 lb); maximum take-off 85000 kg (187,393 lb)
Dimensions: span 27.70 m (90 ft 10.6 in); length 40.53 m (132 ft 11.7 in); height 10.67 m (35 ft 0 in); wing area 145.00 m² (1,560.82 sq ft)
Armament: one 23-mm (1.18-in) cannon in radar-directed tail barbette, and about 10000 kg (22,046 lb) of internal stores or, 'Blinder-B', one AS-4 'Kitchen' cruise missile
Operators: Iraq, Libya, Soviet Union

Tupolev Tu-22M/Tu-26

From the outset the Tu-22, despite its great size and fuel capacity, was limited in operational value by its long field length and moderate combat radius. By the mid-1960s the TsAGI refinement of a variable-geometry swing-wing scheme for retroactive application to existing aircraft showed how to overcome both problems. Designation Tu-22M was applied by VVS to modified Tu-22s fitted with new pivoting outer wings set to angles from 20° to 55°. The first two prototypes started conversion in 1966 and by 1973 there appear to have been at least 12 such rebuilds, assigned the NATO name of 'Backfire', later changed to 'Back-

fire-A'. They are believed to have equipped a VVS combat unit, and these aircraft may still be in service with the Dalnaya Aviatsya.

To reduce drag and achieve many other advantages the Tupolev bureau offered a much more extensive redesign, which was accepted as the definitive aircraft for new construction in the gigantic Kazan

Tupolev Tu-26 'Backfire-B' of the AV-MF.

manufacturing complex. Called Tu-22M by the Soviet Union during the SALT 2 talks, this appears to have been misleading; the actual VVS number is believed to be Tu-26. NATO name is 'Backfire-B'.

This aircraft retains much of the Tu-22M wing, with a very large fixed portion (far more than a mere glove), but with improved high-lift

systems, completely new inward-retracting main landing gear with the multi-wheel bogies housed in the fuselage. Another major change was to relocate the engines, of totally

different and larger type, in the rear fuselage, where they are fed by giant ducts from variable inlets on each side of the forward fuselage. The main fuselage barrel sections are virtually unchanged from the original Tu-22, but the crew compartment is completely redesigned for pilot, co-pilot, systems manager and 'radio operator'/gunner seated 2 by 2 in upward ejection seats. There is no glazed front compartment, though the main radar and flight refuelling probe are again almost the same as on the Tu-22.

The outer wings pivot from 20° to 65°, and are fitted with full-span powered slats, powered outboard ailerons (depressed with the flaps), double slotted flaps and spoilers which serve as airbrakes and lift dumpers (but are not believed to augment roll control). The tail unit is similar in principle and drive systems to the Tu-22, though the tailplanes are of different shape. Extreme care has been taken to maximize internal fuel capacity, using very advanced techniques, to such a degree that total mass has been increased from the 36 tonnes of the Tu-22 to 64 tonnes. After one inflight refuelling an early 'Backfire-B' was observed to remain on flight test for a further 10 hours, and missions lasting 17 or even 18 hours in all (usually with one refuelling) are not uncommon. For some reason US assessments at first disregarded the possibility that this bomber might be any threat to the USA, and the rundown of what was then Aerospace Defense Command continued. During SALT 2 talks the Russians eagerly confirmed this belief, and even removed the flight-refuelling probes to try to give the impression that 'Backfire-B' is, as they call it, a 'theatre aircraft'

rather than an intercontinental one. As it takes about 15 minutes to replace each probe the exercise seemed hardly worthwhile.

Today, 'Backfire-B' is recognized as having plenty of intra-theatre, or intercontinental, capability. It certainly poses a threat to almost the whole continental United States, whose air defence units have received McDonnell Douglas F-15s. The main duties of this bomber are, however, concerned with the European theatre, with the Atlantic, with the Pacific and with China and Japan. In many kinds of role it serves with enormous value: level bombing, low attack, strategic reconnaissance, Elint, anti-ship attack, missile guidance and every kind of maritime support role. Of course, the EW (electronic warfare) and ECM installations may be expected to be outstandingly comprehensive, though fundamentally older in technology than the offensive/defensive avionics of the Rockwell B-1B. Good photographs reveal at least 33 aerials (antennas) all over the airframe, most of them still not positively identified.

By late 1985 about 320 'Backfires' were in service. A few are of the so-called 'Backfire-C' type with distinctive wedge inlets, vertical in the front view and similar to those of the Mikoyan-Gurevich MiG-25/20/31. This may result from an attempt to improve cruise efficiency rather than gain a trivial amount in dash Mach number. Most of the aircraft routinely intercepted appear to belong to the naval AV-MF, which has possibly 120. The other 200 belong to the DA and to three main Air Armies charged with overland operations.

Specification
Tupolev Tu-22M/Tu-26
Type: multi-role long-range attack and reconnaissance aircraft
Powerplant: two augmented turbofans, believed to be 20,000 kg (44,090 lb) Kuznetsov NK-144 variants
Performance: (estimated) maximum speed, clean at high altitude Mach 1.92 or 2040 km/h (1,267 mph); speed at low level Mach 0.9; maximum unrefuelled combat radius 5470 km (3,400 miles)

Weights: (estimated) empty 46000 kg (101,400 lb); fuel 64000 kg (141,090 lb); ordnance 12000 kg (26,455 lb); maximum take-off 122000 kg (269,000 lb)
Dimensions: (estimated) span, spread 34.45 m (113 ft 0 in), swept 26.21 m (86 ft 0 in); length 42.50 m (139 ft 5.2 in); height 10.06 m (33 ft 0 in); wing area 170.0 m (1,829.9 sq ft)
Armament: twin 23 mm guns in radar-directed tail turret; attachments for one, two or three cruise missiles recessed into belly and on pylons inboard of wing pivots, currently AS-4 'Kitchen' or AS-6 'Kingfish' and with 'AS-X-15' predicted imminently; alternatively up to 12000 kg (26,455 lb) of free-fall stores of varied types including bombs and mines, all carried externally
Operator: Soviet Union

Caught over the Baltic by a Swedish air force fighter, this AV-MF 'Backfire-B' has extra weapons racks mounted on the fuselage under the leading edge. Rear defence is handled by a pair of remotely-controlled 23-mm cannon.

Tupolev Tu-95 and Tu-142

For over 30 years the giant swept-wing turboprop aircraft known to NATO as 'Bear' has roamed the Northern hemisphere, disproving

Tu-142 'Bear-F' of the AV-MF, used for anti-submarine patrol.

most of the accepted tenets of aircraft design. Though driven by propellers it has the dash speed of a jet, and though seemingly cumbersome and likely to be mechanically unreliable it is loved by its crews and, unbelievably, remains in production and continues to spawn new variants.

The original Tu-95 flew in 1954 and entered service with the Dalnaya Aviatsya (long-range aviation) as a strategic bomber. Defended by three turrets each with twin 23 mm guns, it carried two cumbersome nuclear bombs internally, or 10000 kg (22,046 lb) of conventional bombs. With a usual flight crew of seven in pressurized compartments at nose and tail (in most versions also aft of the wing), it was configured for missions lasting 25 hours. The giant turboprops drive eight-blade counter-rotation propellers, which in cruising flight have the blades set at an extraordinarily coarse pitch for good fuel economy. Subsequent versions have augmented tankage and a flight refuelling probe for endurance measured in days.

The original bomber, 'Bear-A', accounted for almost all the original run of an estimated 300, completed by 1961. Originally allotted the service designation Tu-20 this and other DA models were later referred to by the design bureau number of Tu-95. A few of this model remain in service.

'Bear-B', first seen in 1961, was originally configured to carry the giant ASM known as 'Kangaroo', with 800 kilotonne nuclear warhead or very large conventional warhead, and with the bomber's glazed nose replaced by a huge surveillance radar (NATO 'Crown Drum'). An estimated 110 remain in use, but rebuilt as 'Bear-G'. 'Bear-C' could also carry 'Kangaroo' but had augmented avionics, including a large blister on both sides of the rear fuselage.

'Bear-D' has for 20 years been by far the most common variant encountered by Western forces. Used by the naval AV-MF for multi-sensor reconnaissance over almost all sea areas north of the equator, it is distinguished by its two large ventral blisters, that under the glazed nose housing an I-band 'Mushroom'

navigation/mapping radar and that under the former weapon bays the surveillance radar dubbed 'Big Bulge', one of the largest ventral radars ever installed. It has blisters on both sides of the rear fuselage, pods on the tips of the tailplane, a longer inflight refuelling probe, larger gun fire-control radar and augmented EW (electronic warfare) installations. Some have the tail turret replaced by the long EW-packed tail fairing caried by the Tu-126 'Moss'.

'Bear-E' is again operated mainly by the AV-MF and is a straightforward rebuild of the original bomber as a maritime reconnaissance aircraft. It has all the usual blisters and bulges, apart from 'Mushroom' and 'Big Bulge', but has the weapon bays occupied by extra fuel tankage and a very large detachable pallet on which are mounted up to seven optical cameras (one with about 3.05 m/ 10 ft focal length) and infra red linescan equipment. A small chin radar (as used on many Tu-16 'Badgers') assists navigation, and in most examples there are two tandem blisters under the forward weapon bay, as carried by the AV-MF's 'Badger-D'.

All the foregoing stem from the original production batch of bombers. Over 100 remain in use by the VVS (air force) in various forms, though no tanker or AWACS versions have been positively identified. Those used by the naval AV-MF appear to be redesignated as Tu-142s, though this may refer only to the three new-build versions described below.

'Bear-F', first noticed by the West in 1973 but planned in the mid-1960s, was a new series built at Taganrog at the rate of about 12 per year; it is a substantially redesigned aircraft for long-range ASW (anti-submarine warfare). The wing is restressed for increased weights and greater fuel capacity, while the fuselage is lengthened (mainly ahead of the wing). The pressurized compartment aft of the wing is extended and rearranged as an ASW sensor and sonobuoy area, with a galley and crew rest cabin, the upper and lower turrets, side blisters and other equipment being removed. The rear weapon bay houses AS weapons, and apparently further sonobuoys, while the front bay houses a surveillance radar of I/J band type (NATO name not disclosed). There are no tailplane tip fairings, but a MAD (magnetic anomaly detector) sensor projects aft

from the top of the fin. Some 'Bear-Fs' have, or had, lengthened main-gear fairings, and all have bulged nose-gear bays. By late 1985 production was probably complete at about 70.

'Bear-G' is a rebuild of the 'Kangaroo'-carrying '-B' and '-C' versions. The original missiles still exist but have been placed in store. In their place the carrier aircraft (Tu-95s) have been reconfigured to carry the supersonic AS-4 'Kitchen'. There have been associated changes and updates in avionics, and it would appear likely to add two more 'Kitchen' missiles on inboard wing pylons as fitted to 'Bear-H.'

'Bear-H', the only current production version, is a new Tu-142 variant specially produced to carry the new cruise missile dubbed 'AS-X-15', with a configuration said by the US Department of Defense to resemble the AGM-109 Tomahawk, and with a range of 3000 km (1,874 miles). Two of these missiles, with wings folded, are carried on each of the deep pylons under the inboard wings. So far as is known all 'X-15' missiles have one or more nuclear warheads. The carrier aircraft as a large surveillance radar in the nose; DoD artists have drawn it as resembling 'Crown Drum', with a flight refuelling probe above. They have also shown acute anhedral on the wings and tailplane, but their artwork is notoriously poor.

Specification
Tupolev Tu-95 'Bear-F'
Type: multi-role (see text) strategic aircraft

Tupolev Tu-142 'Bear-D' cutaway drawing key

1 Fixed inflight-refuelling probe
2 Observers/bomb aimer's compartment nose glazing
3 Nose radome
4 Avionics equipment bay, port and starboard
5 'Odd-Rods' IFF aerials
6 Cockpit enclosure, pilot and co-pilot
7 Nose undercarriage pivot mounting
8 Retractable landing/taxiing lamps, ports and starboard
9 Nosewheel steering jacks
10 Aft retracting twin nosewheels
11 Nosewheel doors
12 Pitot tubes
13 Cockpit roof escape hatch
14 Forward pressurized crew compartment
15 Observation hatch
16 Wing root attachment joint
17 Wing centre section carry-through
18 Inboard wing panel
19 Starboard engine nacelles
20 AV-60N eight-bladed contra-rotating propellers
21 Propeller spinners
22 Wing fences
23 Outboard wing panel
24 Wing tip lighting
25 Starboard aileron
26 Aileron tab
27 Outboard Fowler-type flap, lowered
28 Flap guide rails
29 Nacelle tail fairing
30 Aerodynamically extended tail fairing (some 'Bear-F' aircraft)
31 Mainwheel doors
32 Inboard Fowler-type flap, lowered
33 Satellite communications antennae
34 ADF sense aerial
35 Circular section unpressurized fuselage
36 Retractable dorsal cannon barbette, 2 x 23-mm cannon, remotely controlled
37 Fuselage profile 'Bear-E' & 'F'
38 Raised cockpit section
39 Extended crew compartment fuselage plug
40 Search radar
41 Two-section weapons bay ('Bear-F')
42 Weapons bay camera and reconnaissance pallet ('Bear-E')
43 Fin root fillet
44 Starboard tailplane

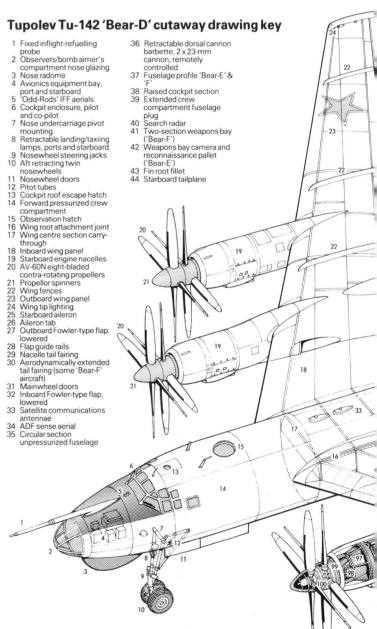

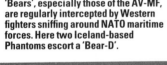

Powerplant: four 11033-kW (14,795 ehp) Kuznetsov BNK-12MV turboprop engines
Performance: maximum speed with full weapon load at high altitude 925 km/h (575 mph); range with full weapon load 12550 km (7,800 miles); unrefuelled combat radius 8285 km (5,150 miles)
Weights: empty typically 90000 kg (198,400 lb); maximum 188000 kg

(414,460 lb)
Dimensions: span 51.10 m (167 ft 8 in); length, with probe 49.50 m (162 ft 5in); height 12.12 m (39 ft 9 in); wing area 310.50 m (3,342.30 sq ft)
Armament: two 23 mm guns in tail turret; unknown quantity of AS torpedoes, mines and other stores (original bombload 10000 kg (22,046 lb)

Although not the most numerous in service, the 'Bear-D' is the most regularly seen by Western forces. It is employed for missile targetting, Elint and general maritime reconnaissance.

45 Tailfin
46 Short-wave ground control communications antennae
47 HF aerial cable
48 Fin tip aerial fairing
49 Magnetic Anomaly Detector (MAD) boom ('Bear-F')
50 Rudder
51 Rudder tab
52 Sensor equipment tail fairing (some 'Bear-D' aircraft)
53 I-band tail warning radar
54 Tail gunner's compartment

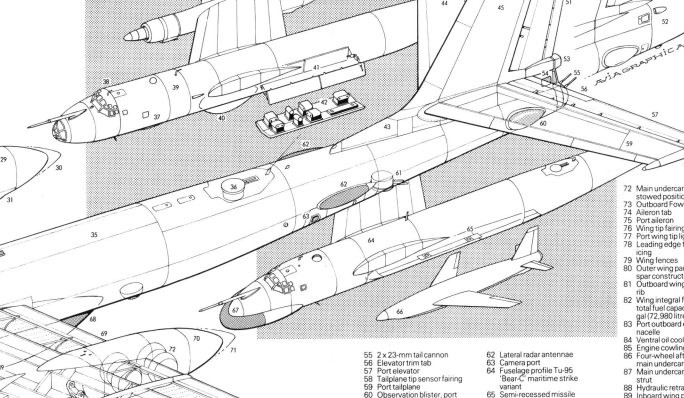

55 2 x 23-mm tail cannon
56 Elevator trim tab
57 Port elevator
58 Tailplane tip sensor fairing
59 Port tailplane
60 Observation blister, port and starboard
61 Ventral 2 x 23-mm cannon barbette, remotely controlled
62 Lateral radar antennae
63 Camera port
64 Fuselage profile Tu-95 'Bear-C' maritime strike variant
65 Semi-recessed missile housing
66 'Kangaroo' air-to-surface missile
67 'Crown-Drum' nose radome
68 Ventral X-band surveillance radar
69 Port inboard Fowler-type flap
70 Nacelle tail fairing
71 Extended tail fairing profile

72 Main undercarriage stowed position
73 Outboard Fowler-type flap
74 Aileron tab
75 Port aileron
76 Wing tip fairing
77 Port wing tip light
78 Leading edge thermal de-icing
79 Wing fences
80 Outer wing panel three-spar construction
81 Outboard wing panel joint rib
82 Wing integral fuel tanks, total fuel capacity 16,540 gal (72,980 litres)
83 Port outboard engine nacelle
84 Ventral oil cooler
85 Engine cowling panels
86 Four-wheel aft retracting main undercarriage bogie
87 Main undercarriage leg strut
88 Hydraulic retraction jack
89 Inboard wing panel four-spar construction
90 Engine fire extinguisher bottle
91 Jetpipe, twin outlets
92 Bifurcated jet pipe
93 Engine bearer struts
94 Main engine mounting ring frame/firewall
95 Kuznetsov NK-12MV turboshaft engine
96 Engine accessory equipment
97 Engine air intake
98 Propeller reduction gearbox
99 Engine cowling annular air intake
100 Propeller hub pitch change mechanism
101 Port contra-rotating propellers

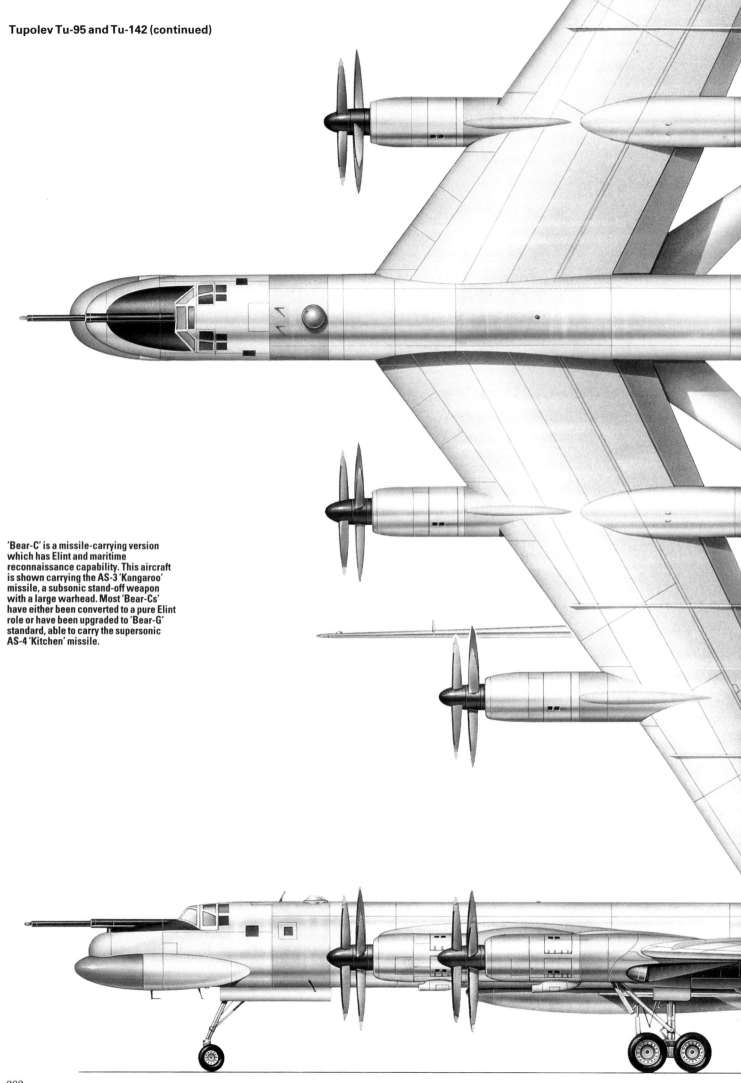

'Bear-C' is a missile-carrying version which has Elint and maritime reconnaissance capability. This aircraft is shown carrying the AS-3 'Kangaroo' missile, a subsonic stand-off weapon with a large warhead. Most 'Bear-Cs' have either been converted to a pure Elint role or have been upgraded to 'Bear-G' standard, able to carry the supersonic AS-4 'Kitchen' missile.

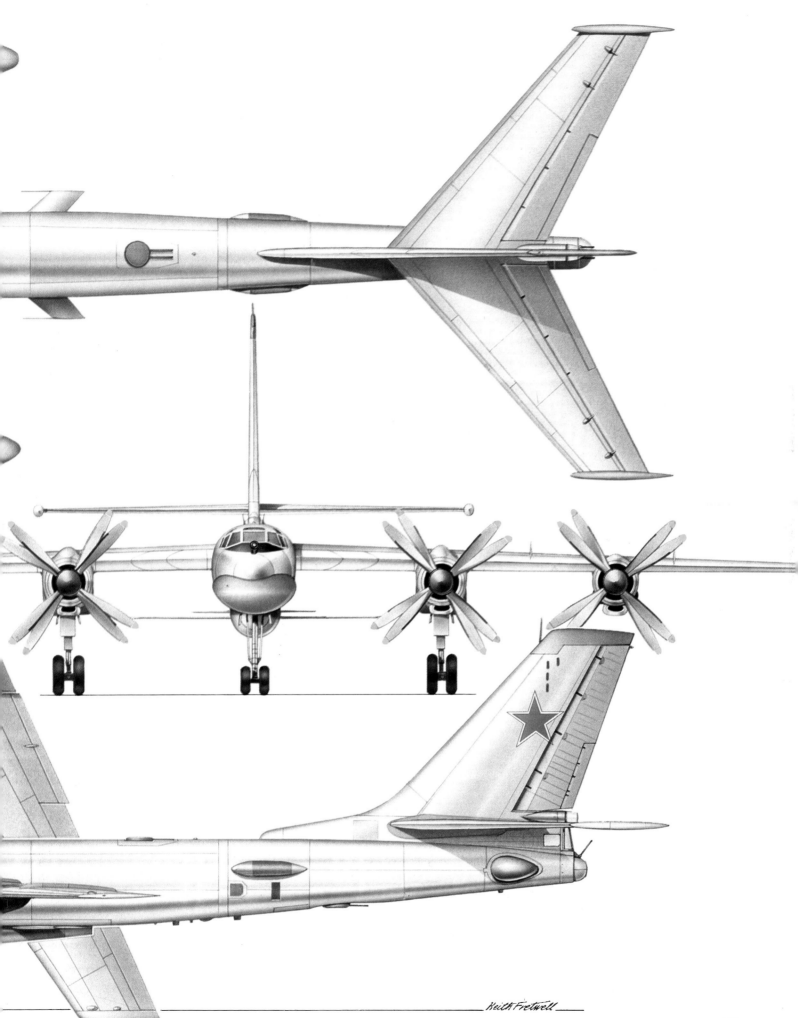

Keith Fretwell

Tupolev Tu-124 'Cookpot'

After enjoying limited success as an airliner, the Tupolev Tu-124 'Cookpot', a scaled-down version of the Tu-104, has remained in service as a VIP and government transport, for which its small size and relatively good field performance render it suitable. It was designed in the late 1950s to meet an Aeroflot requirement for an Ilyushin Il-14 replacement and to operate from small and unprepared fields. The prototype, flown in June 1960, was the first airliner to be designed for turbofan engines. Fuselage diameter is smaller than on the Tu-104, the main landing gears are shorter and the wing-loading is lower. The wing has double-slotted trailing edge flaps and overwing spoilers for better short-field performance, aided by a large airbrake beneath the centre-section. Initial Tu-124s have 44 seats, but the Tu-124V seats 56; VIP versions are the 36-seat Tu-124K and 22-seat Tu-124K2.

Specification
Type: passenger transport
Powerplant: two 5400-kg (11,905-lb) thrust Soloviev D-20P turbofan engines
Performance: cruising speed Mach 0.82 or 870 km/h (541 mph); service ceiling 11700 m (38,385 ft); range with maximum payload of 6,000 kg (13,228 lb) 1259 km (777 miles); range with 30-35 passenger 2100 km (1,305 miles)
Weights: empty 22900 kg (50,486 lb); maximum take-off 37500 kg (82,673 lb)
Dimensions: span 25.55 m (83 ft 9.9 in); length 30.58 m (100 ft 3.9 in); height 8.08 m (26 ft 6.1 in); wing area 120.00 m² (1,219.71 sq ft)

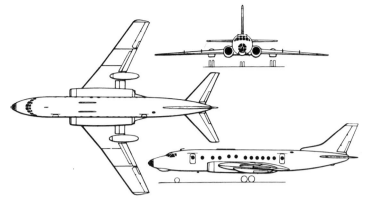

Tupolev Tu-124 'Cookpot'

Armament: none
Operators: (governments of)

Czechoslovakia, East Germany, India, Iraq, Soviet Union

Tupolev Tu-126 'Moss'

Like the Ilyushin Il-38, this aircraft marks the Soviet Union's first entry into a new field, that of airborne early warning and control (AEW/C). In the same way, too, it is not competitive with closely contemporary Western systems, which could draw on decades of operational experience and development.

Clearly a relation of the Tu-95 family, the Tu-126 is based on the bomber's commercial cousin, the Tu-114. Developed in parallel with the bomber, the Tu-114 was the largest, longest-range airliner of its day. It was heavier and more powerful than the original Tu-95, and featured an increase in wing area, in the shape of increased flap chord. Its career was not particularly successful. It flew in 1955, but development was slow, and it was not until April 1961, six years after its first flight, that the Tu-114 entered service with Aeroflot. It had been in service for less than two years when its planned replacement, the Il-26, made its first flight. Only 30 Tu-114s were built, all for Aeroflot, and they were replaced by Il-62s on the main international routes by 1969.

In the early 1960s, the USAF and British bomber arms abandoned their old high-altitude tactics in favour of low-level, 'under the radar' penetration. Development of tha Soviet AEW/C system was a natural response to this new threat, and the Tu-114 was a logical carrier aircraft: in production, common to other service types, and with a large payload and very long endurance. The Tu-126 prototype probably flew in the second half of 1967, and appeared in a documentary film made in the following year. It became operational in 1970-71.

The Tu-126 radar, known to NATO as 'Flap Jack', occupies an 11 m (36 ft) diameter rotodome, carried above the rear fuselage on a single long-chord pylon; a ventral fin and a refuelling probe are fitted, and the retractable tailwheel of the Tu-114 is replaced by fixed wheels on the ventral fin. There are numerous other antennas, presumably connected with electronic support measures (ESM) and communications systems. The airframes show no sign of major surgery, and were probably specially built.

The Tu-126 has never been highly regarded in the West. Its radar is, apparently, not advanced enough to pick out low-flying aircraft against ground clutter, and can do so only to a limited extent over water. The 'Flap Jack' radar does not use pulse-Doppler techniques – in any case, the vast eight-blade steel propellers would feed so many false signals into such a radar that its processor would be overwhelmed – but probably uses some earlier form of moving-target indication. Only ten Tu-126s are believed to remain in service, and they are probably on the point of being replaced by the far superior Ilyushin 'Mainstay-A'.

Specification
Tupolev Tu-126 'Moss'
Type: Airborne early warning and control aircraft

Powerplant: four Kuznetsov NK-12MV turboprops developing 11200 kW (15,000 shp)
Performance: maximum speed 870 km/h (540 mph) at 9000 m (30,000 ft); economic cruising speed 608 km/h (465 mph) at 10700 m (35,000 ft); normal mission endurance 20 hours; maximum range 9650 km (6,000 miles); service ceiling 12500 m (41,000 ft)
Weights: empty 95,250 kg (210,000 lb); maximum take-off weight 188000 kg (415,000 lb)
Dimensions: span 51.1 m (167 ft 8 in); length excluding probe 55.2 m (188 ft 1 in); height 16.05 in (52 ft 8 in); wing area 311 m² (3,348 sq ft)
Operator: Soviet Union

Tu-126 'Moss' of the Soviet air force, showing the large rotodome mounted above the rear fuselage.

Tupolev Tu-128 'Fiddler'

When first seen in 1961, this large supersonic twin-jet was thought by Western observers to be a Yakovlev design. In fact it was the Tupolev Tu-28 long-range surveillance fighter, from which was derived the Tu-28P 'Fiddler' interceptor. The Tupolev bureau numbers for these two types were Tu-102 and Tu-128. In many respects the largest fighter in the world, and certainly the biggest and most powerful ever put into service, the Tu-128 has a long fuselage with enormous fuel capacity to handle PVO (air defence force) missions covering vast areas of the Soviet frontier. The original Tu-28 was intended to operate almost without ground help, but the Tu-128 is assisted by ground radars and defence systems which guide it towards hostile aircraft. Then the extremely large 'Big Nose' I/J-band radar takes over until either a radar- or an IR-homing AA-3 missile can be fired. A pair of each type of AAM is carried, and no other interceptor has been seen armed with these large weapons. The weight of this fighter is spread by bogie landing gears which in flight retract backwards into fairings typical of Tupolev aircraft of the era. Capability against low-flying aircraft may have been improved since 1980, because not even the Su-27 (Ram-K) or an interceptor version of Su-24 could offer equal area defence, but the 100-odd still in use were being withdrawn in 1983.

Specification
Tu-128 'Fiddler'
Type: long-range interceptor
Armament: four AA-3 'Ash' AAMs, two radar-guided and two IR-homing
Powerplant: two afterburning turbojets, almost certainly Lyulka AL-21F-3 each rated at 11000-kg (24,250-lb) thrust
Performance: maximum speed at high altitude 1900 km/h

Tupolev Tu-128 'Fiddler'

Tu-128 'Fiddler', armed with four AA-5 'Ash' missiles.

(1,200 mph) or Mach 1.8; service ceiling 20000 m (65,615 ft); radius at high altitudes with four AAMs 1250 km (777 miles)
Weights: (estimated) empty 24500 kg (54,012 lb); maximum take-off 40000 kg (88,183 lb)
Dimensions: span 18.10 m (60 ft 0 in); length 27.20 m (89 ft 3 in); height 7.0 m (23 ft 0 in); wing area 80.0 m² (860 sq ft)
Armament: four AA-5 'Ash' long-range air-to-air missiles, carried in IR and SARH pairs.
Operators: Soviet Union

The Tu-128 is the largest fighter in service, using its vast range to overcome any performance deficiencies. This aircraft has two mighty 'Ash' missiles: four is a more normal load, in two pairs of IR- and SARH—guided versions. These aircraft defend the vast areas in the north of the Soviet Union.

Tupolev Tu-134

Soviet commercial aircraft have never matched standards set in the West, or the standards of design set by Soviet military types, and the Tupolev Tu-134 airliner is no exception to the rule. Superficially similar to the DC-9 and One-Eleven, it is heavier, carries fewer passengers in a less comfortable cabin, and has a much higher landing speed.

The Tu-134 is a direct descendant of the Soviet Union's first jet airliner, the Tu-104. Derived from the Tu-16 bomber, the Tu-104 was followed by the scaled-down Tu-124. The latter, though, was built in limited numbers and suffered from noise and vibration in the cabin. The design of a revised version, with a similar fuselage, modified wing and a rear engine installation, began in 1962, and the Tu-124A flew some time in 1963, emerging as a significantly larger and heavier aircraft than the original. It was redesignated Tu-134 before its official unveiling in the following year.

Entering service in 1967, the Tu-134 was steadily improved with the addition of thrust reversers, an APU

and other equipment. It was superseded in production by the Tu-134A, with a slightly longer fuselage and other refinements; early in the Tu-134A production run, a solid nose replaced the bomber-type navigator's glazed nose. The final version, the Tu-134B, was introduced just before production ended in the late 1970s. The aircraft remains in use as a VIP transport for several air forces.

Specification
Tupolev Tu-134

Type: airliner and VIP transport
Powerplant: two 6800 kg (15,000 lb) thrust Soloviev D-30 turbofans
Performance: cruising speed 849 km/h (530 mph) at 11000 m (36,000 ft); range with maximum payload 2400 km (1,500 miles); service ceiling 11900 m (39,000 ft); take-off field length 2180 m (7,150 ft); landing field length 2050 m (6,730 ft)
Weights: empty 27500 kg (60,500 lb); maximum take-off 45,000 kg (99,000 lb)

Hungary is one of the nations which uses the Tu-134 as a VIP transport. Soviet aircraft are used for staff transport.

Dimensions: span 29 m (95 ft 2 in); length 34.9 m (114 ft 8 in); height 9 m (29 ft 7 in); wing area 127 m² (1,370 sq ft)
Operators: Angola, Bulgaria, Czechoslovakia, East Germany, Hungary, Mozambique, Poland, Soviet Union, Vietnam

Tupolev Tu-? 'Blackjack'

This massive supersonic bomber, by a useful margin the heaviest warplane ever designed, is currently undergoing flight tests and may be in service by 1987. It was first observed in late 1981, and was assigned the temporary reporting name RAM-P. It is essentially the Soviet Union's equivalent to the Rockwell B-1B.

'Blackjack' is designed to replace the Tu-95 and remaining Mya-4s in the penetration/strike role, and to supplement the advanced 'Bear-H' as a cruise-missile carrier. In overall layout, it is similar to the B-1B, indicating that their mission profiles are also similar. Placing the engines in pods close to the centre of gravity allows the entire fuselage

volume to be used for fuel and weapons and makes it possible to use fuel transfer to compensate for changes in wing sweep. The mid-set tailplane, also common to both aircraft, is necessary to keep the tailplane away from the heat and vibration of the engine exhaust.

Apart from size – 'Blackjack' is 60 per cent heavier than the B-1, in a similar configuration – the main difference between the two aircraft is the wing planform. The Soviet aircraft's wing pivots are further outboard than the highly loaded pivots of the B-1, the entire wing structure being similar to that of the 'Backfire'.

The Department of Defense estimates that 'Blackjack' has a similar

combat radius to the B-1, but it probably carries a larger weapon load due to its greater size. Its armament will include new cruise missiles, now under development, as well as nuclear gravity weapons.

Specification
Tupolev Tu-? 'Blackjack'
Type: strategic bomber and cruise missile carrier
Powerplant: four 20000 kg (44,000 lb) thrust augmented turbofans
Performance: maximum speed 2220 km/h (1,380 mph) at 12000 m (40,000 ft) or Mach 2.1; maximum speed at sea level 1040 km/h (645 mph) or Mach 0.85; maximum unrefuelled combat radius, 7300 km

(4,500 miles)
Weights: empty 118000 kg (260,000 lb); maximum take-off 285000 kg (630,000 lb)
Dimensions: swept span 36.7 m (121 ft); spread span 54 m (177 ft); length 53 m (175 ft); height 13.5 m (44 ft); wing area 370 m² (4,460 sq ft)
Armament: probable defensive radar-directed cannon/chaff dispenser in tail; up to 20 AS-X-15 air-launched cruise missiles with range of 3000 km (1,850 miles) or single large long-range supersonic ALCM with range well over 3200 km (2000 miles)
Operator: Soviet Union (operational 1987-88)

UTVA-66/60

Announced in 1968, the UTVA-66 is descended from the 1959 UTVA-56 four-seat utility aircraft, the production development of which was the UTVA-60. This all-metal aircraft, powered by a 210-kW (270-hp) Lycoming GO-480-B1A6 engine,

was produced in five versions: the U-60-AT1 basic utility model for air taxi, light freight, liaison or sporting flying; the -AT2, a dual-control version for training; the -AG, an agricultural aircrAft; the -AM ambulance variant with accommodation for two stretchers loaded via an upward-hinged rear cabin canopy;

and the U-60H, a seaplane with a strengthened fuselage, flown in October 1961 using Edo floats, although production models used Yugoslav-built floats. The seaplane was available in the same versions as the landplane, and examples of both types served in the Yugoslav air force.

Development continued with the UTVA-66 with the more powerful 239-kW (320-hp) Lycoming engine. The new model was also available in utility and glider-towing configuration (UTVA-66), ambulance (-66AM) and seaplane (-66H), and utilized a number of UTVA-60 components. The wings were the same size but leading-edge slats were added, tail surfaces were enlarged and a servo tab plus a controllable trim tab were fitted to the elevator. Landing-gear shock absorbing was improved, and a larger fuel tank with 250-litre (55-gallon) capacity was fitted. A substantial number of UTVA-66s are in service with the Yugoslav air force.

Specification
Type: single-engine utility aircraft
Powerplant: one 239-kW (320-hp) Lycoming GSO-480-B1J6 piston engine
Performance: maximum speed at sea level 230 km/h (143 mph); maximum cruising speed 230 km (143 mph); initial rate of climb at sea level 270 m (885 ft) per minute; service ceiling 6700 m (22,000 ft); take-off distance to 15 m (50 ft) 352 m (1,155 ft); landing run from 15 m (50 ft) 181 m (594 ft); range 760 km (466 miles)
Weights: empty 1250 kg (2,756 lb); loaded 1814 kg (4,000 lb)
Dimensions: span 11.40 m (37 ft 5 in); length 8.38 m (27 ft 6 in); height 3.20 m (10 ft 6 in); wing area 18.08 m² (194.50 sq ft)
Operators: Jugoslavia

UTVA-75

A replacement for the Aero 3 as the Jugoslav air force basic trainer, the UTVA-75 was designed and built by a partnership consisting of UTVA, Prva Petoletka and two Belgrade Institutes. Design began in 1974 and two prototypes were flown in 1976. More than 100 had been delivered to civilian clubs by mid-1984, and possibly as many more to the air force. In addition to its use as a basic trainer, with side-by-side seating, the UTVA-75 is also capable of glider towing. Construction is all-metal, and racks beneath the wings can carry light weapons loads, including rockets, for armament training.

Specification
Type: trainer and utility transport
Powerplant: one 134-kW (180-hp) Lycoming IO-360-B1F flat-four piston engine
Performance: maximum speed 215 km/h (134 mph); cruising speed 165 km/h (103 mph); initial rate of climb 270 m (886 ft)/minute; service ceiling 4500 m (14,765 ft); range 800 km (497 miles)
Weights: empty 685 kg (1,510 lb); maximum take-off 960 kg (2,116 lb)
Dimensions: span 9.73 m (31 ft 11.1 in); length 7.11 m (23 ft 3.9 in); height 3.15 m (10 ft 4 in); wing area 14.63 m² (157.48 sq ft)
Armament: two wing pylons, each of 100-kg (220-lb) capacity, for gun pods, bombs or two-barrel rocket launchers
Operator: Jugoslavia

Valmet L-70 Miltrainer

The prototype L-70 (originally Leko-70) flew on 1 July 1975, some two-and-a-quarter years after receipt of a contract to design a new trainer for the Finnish Ilmavoimat. Valmat delivered 30 in 1981-82, the service name being Vinka (a cold wind).

A side-by-side all-metal aircraft, it is designed and built for 8,000 hours fatigue life and is fully aerobatic. Two more seats can be installed at the rear, or a stretcher and one medical attendant can be carried in addition to the pilot, or 280 kg (617 lb) of cargo. The L-70 can be equipped for glider or target towing, crop-spraying or dusting, photography, or the transport of 300 kg (661 lb) of stores on four pylons.

Specification
Type: trainer
Powerplant: one 149-kW (200-hp) Lycoming AEIO-360-A1B6 flat-four piston engine
Performance: maximum speed 235 km/h (146 mph) at sea level; cruising speed 222 km/h (138 mph) at 1525 m (5,005 ft); initial rate of climb 342 m (1,122 ft)/minute; service ceiling 5000 m (16,405 ft); range 950 km (590 miles)
Weights: empty 767 kg (1,691 lb); maximum take-off, aerobatic

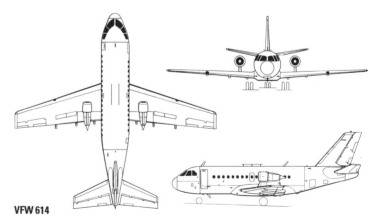

1040 kg (2,293 lb), normal 1250 kg (2,756 lb)
Dimensions: span 9.63 m (31 ft 7.1 in); length 7.50 m (24 ft 7.3 in); height 3.31 m (10 ft 10.3 in); wing area 14.00 m² (150.70 sq ft)
Armament: four wing attachments for 50-kg (110-lb) bombs, twin 7.62-mm (0.3-in) or similar gun pods, pods for 18×37-mm (1.46-in) or

The Valmet L-70 is the basic trainer used by the Finnish air force. Thirty L-70s serve, their stable handling and full aerobatic capability making them excellent for all levels of basic training.

6×68-mm (2.68-in) rockets, or camera, rescue or other pods
Operator: Finland

Valmet L-70 Vinka

VFW 614

The VFW 614 was designed by Vereinigte Flugtechnische Werke in West Germany, developed with government assistance and produced by VFW-Fokker following the formation of a joint venture company. Participants included MBB of Germany, Fokker-VFW in Holland and Belgium's Fairey and SABCA. The overwing engines were developed by Rolls-Royce and SNECMA, though the latter withdrew. The fuselage is pressurized and seats up to 44 passengers, plus a flight crew of two. The first prototype flew on 14 July 1971 but the programme was abandoned in 1978. Three were ordered for the Luftwaffe and delivered in April, June and August 1977 to the Flugbereitschaftstaffel (special air mission squadron) at Cologne-Bonn airport, where they joined a mixed fleet (707, JetStar, Hansa Jet, Skyservant and UH-1D) on official communications. They are to be replaced in about 1988 by larger aircraft.

Specification
Type: short-haul transport
Powerplant: two 3302-kg (7,280-lb) thrust Rolls-Royce M45H Mk 501 turbofan engines
Performance: maximum speed 735 km/h (457 mph); initial rate of climb 945 m (3,100 ft)/minute; service ceiling 7620 m (25,000 ft); range with 40 passengers and reserves 1205 km (749 miles)
Weights: empty 12180 kg (26,852 lb); maximum take-off 19950 kg (43,982 lb)
Dimensions: span 21.50 m (70 ft 6.5 in); length 20.60 m (67 ft 7 in); height 7.84 m (25 ft 8.7 in); wing area 64.00 m² (688.91 sq ft)

VFW 614

Armament: none
Operator: West Germany (Luftwaffe)

Vickers Viscount

The world's first turboprop transport was designed by Vickers-Arm-strong, beginning in the month that World War 2 ended, the prototype flying on 16 July 1948. Successive versions had increased power, weight, span and length. The last of 459 Viscounts was delivered in

1964, and at least five continue to fly in military markings. In Britain, single examples of the Types 837 and 838 serve with the Radar Research Squadron at Royal Aircraft Establishment, Bedford, as testbeds for avionics with prominent bulges and pods. In South Africa No. 21 Sqn at Zwartkop has a Type 781D with VIP interior for official communications (backed by BAe 125s). The Lao People's Liberation Army Air Force has a 768D believed still to be airworthy. Turkey acquired three Type 794s, two of which fly with the 8th VIP transport unit at Yesilköy, providing a service for the air warfare school.

Specification

Vickers Viscount 810
Type: 52/70-seat airliner
Powerplant: four 1342-kW (1,800-shp) Rolls-Royce Dart Mk 525 turboprop engines
Performance: maximum cruising speed 575 km/h (357 mph); service ceiling 7620 m (25,000 ft); range with maximum payload 2776 km (1,725 miles)
Weights: empty 18854 kg (41,565 lb); maximum take-off 32885 kg (72,500 lb)
Dimensions: span 28.56 (93 ft 8.5 in); length 26.11 m (85 ft 8 in); height 8.15 m (26 ft 9 in); wing area 89.46 m² (963.0 sq ft)
Armament: none
Operators: Laos, South Africa, Turkey, UK

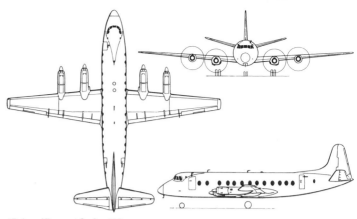

Vickers Viscount Series 800

Vought A-7 Corsair II

In May 1963 the US Navy announced the VAL competition to replace the A-4 Skyhawk in light attack squadrons. LTV Aerospace (Vought) was announced winner on 11 February 1964, with an aircraft derived from the F-8 Crusader, but shorter and with a fixed-incidence wing with less sweep. Structure was strengthened to carry heavy weapon loads, and a non-afterburning turbofan was installed, initially the Pratt & Whitney TF30-P-6.

Flown on 27 September 1965, the A-7A joined fleet squadrons only a year later, in October 1966, and made its combat debut at the beginning of December 1967 over North Vietnam. After delivering 199 A-7As production switched to 196 A-7Bs with a slightly more powerful TF30. In October 1966 the USAF ordered the A-7D with totally different avionics for the tactical attack mission, a boom receptacle instead of a folding FR probe and power provided by a variant of the Rolls-Royce Spey. Vought built 459 and many are still active with the Air National Guard. The Navy introduced even better avionics and the M61 gun, but retained the TF30 in 67 A-7Cs be-

fore the main run of 596 A-7Es with the TF41-A-2 engine, these being the most powerful Corsairs of all. Entering service in July 1969, the A-7E was the principal US Navy light attack aircraft, now fast being replaced by the F/A-18 Hornet. It

Below: Portugal is a major operator of the Vought A-7, with 44 A-7Ps serving with Esquadra 302 and Esquadra 304 at Monte Real. Although the Corsair II possesses air-to-air capability, the Portuguese machines are used for strike duties.

Above: Still serving in large numbers aboard US Navy carriers, the Vought A-7E has been the major light strike element since 1969. It is now being replaced by the McDonnell Douglas F/A-18 Hornet. These VA-46 'Clansmen' aircraft carry empty triple ejector racks.

Vought A-7 Corsair II (continued)

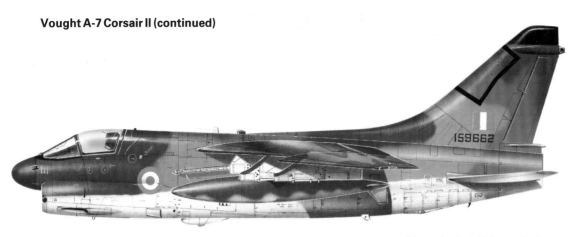

Left: Greece operates 48 A-7H and 5 TA-7H from Souda and Larissa on attack duties alongside F-104 Starfighters.

Right: Corsair IIs are used mainly by the Air National Guard within the USAF, having been replaced by Fairchild A-10s in the front-line units. Nevertheless, the A-7 is still a most useful close support and attack aircraft, with good load-carrying ability and accuracy. Here a USAF A-7D lets fly with unguided rockets.

Inflight-refuelling is commonly practised beteen US Navy aircraft using the 'buddy-buddy' system. The tanker aircraft carries a special pod with hose and drogue which is extended when required. The A-7 is sometimes used in this role, as demonstrated by these aircraft from VA-105 (tanker) and VA-37.

still equips (late 1984) some 20 squadrons as well as a handful of Reserve units. Updating, notably by the provision of a FLIR (Forward-Looking Infra-Red) weapon-aiming pod, has ensured that they remain operationally effective.

Other Corsair II variants include the Navy's TA-7C two-seat combat proficiency trainer (60 conversions of A-7Bs and A-7Cs), the A-7H land-based version for Greece (60 plus five two-seat TA-7Hs), the A-7K two-seat trainer for the Air National Guard (31 built and the ultimate new-build version) and the A-7P for Portugal (44 refurbished A-7As, plus six TA-7P two-seaters).

Specification
Vought A-7E Corsair II
Type: single-seat attack aircraft
Powerplant: one 6804-kg (15,000-lb) thrust Allison/Rolls-Royce TF41-A-2 turbofan engine
Performance: maximum speed at sea level 1112 km/h (691 mph); ferry range, maximum fuel 4604 km (2,861 miles)
Weights: empty 8669 kg (19,111 lb); maximum take-off 19051 kg (42,000 lb)
Dimensions: span 11.81 m (38 ft 9 in); length 14.06 m (46 ft 1.5 in); height 4.90 m (16 ft 0.75 in); wing area 34.84 m^2 (375.0 sq ft)
Armament: one Vulcan M61A-1 cannon with 1,000 rounds in the fuselage, and in excess of 6804 kg (15,000 lb) of stores can be carried on two fuselage stations and six underwing pylons, including bombs, rockets, gun pods, air-to-air and air-to-surface missiles
Operators: Greece, Portugal, US Air Force, US Navy

Vought A-7K Corsair II cutaway drawing key

1 Radome
2 Radar scanner dish
3 Radar tracking mechanism
4 Pitot tubes
5 Rain dispersal air ducts
6 AN/APQ-126(V) forward looking radar transmitter/ receiver
7 Cooling air louvres
8 Engine air intake
9 ILS aerial
10 Forward radar warning antenna
11 'Pave Penny' detector unit
12 Intake duct framing
13 Boron carbide (HCF) cockpit armour panelling
14 Armoured front pressure bulkhead
15 Rudder pedals
16 Control column
17 Instrument panel shroud
18 AN/AVQ-7(V) head-up-display (HUD)
19 Windscreen panels
20 Cockpit canopy cover, hinged to starboard
21 Ejection seat canopy breakers
22 Face blind firing handle
23 Seat safety lever
24 Starboard side console panel
25 Pilot's Douglas Escapac 1-C2 ejection seat
26 Port side console panel
27 Engine throttle control
28 Boarding steps
29 Cannon muzzle blast trough
30 Retractable boarding ladder
31 Taxying lamp
32 Nosing undercarriage shock absorber leg strut
33 Levered suspension axle beam
34 Twin nosewheels
35 Nosewheel doors
36 Cannon barrels
37 Rear seat boarding steps
38 Canopy emergency release
39 Angle of attack sensor
40 Rear seat control linkages
41 Rear instrument panel shroud
42 2nd pilot/instructor's ejection seat
43 Cockpit coaming
44 Rear throttle lever
45 Ammunition feed and link return chutes
46 M61A-1 Vulcan, 20-mm rotary cannon
47 Gun gas spill duct
48 Liquid oxygen container
49 Emergency hydraulic accumulator
50 Electronics system built-in test panel
51 Ventral doppler aerial
52 Port radio and electronics equipment bay
53 Cooling air extractor fan
54 Forward fuselage fuel cells, total internal fuel capacity 1425 US Gal
55 Fuselage stores pylon, 500-lb capacity
56 Wing front spar/fuselage attachment joint
57 Aileron control rod
58 Cockpit rear pressure bulkhead
59 Ammunition drum, 500-rounds
60 Air refuelling, lights
61 TACAN antenna
62 Wing centre section carry-through structure
63 Transformer rectifier
64 Wing skin panel centreline joint strap
65 Universal air refuelling receptacle
66 Starboard wing integral fuel tank
67 Fuel system piping
68 Pylon attachment hardpoints
69 Inboard leading edge flap, down position
70 Flap hydraulic actuators
71 Centre wing pylon, 3500-lb capacity
72 AIM-9 Sidewinder air-to-air missile
73 Missile launch rail
74 Fuselage missile pylon
75 Snakeye, 500-lb retarded bomb
76 Multiple ejector rack
77 Mk.82 500-lb H.E. bombs
78 Outboard wing pylon, 3500-lb capacity
79 Leading edge dog-tooth
80 Wing fold hydraulic jack
81 Outer wing panel hinge joint
82 Outboard leading edge flap
83 Hydraulic flap actuators
84 Starboard navigation light
85 Wing tip fairing
86 Formation light
87 Outer wing panel folded position

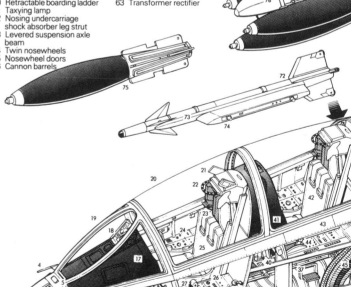

For conversion and weapons training, the US Navy procured 60 TA-7C. Most of these were conversions of earlier models with an extra seat in a lengthened cockpit. This VA-174 aircraft drops a pair of practice 'slick' bombs.

88 Starboard aileron
89 Aileron hydraulic jack
90 Fuel jettison pipe
91 Starboard single-slotted trailing edge flap, down position
92 Flap hydraulic jacks
93 Starboard spoiler, open position
94 Spoiler hydraulic actuator
95 Upper formation light
96 Anti-collision light
97 Control rod linkages

98 Read spar/fuselage attachment joint
99 Gravity fuel filler cap
100 Rear fuselage fuel cell
101 Dorsal spine fairing
102 Engine compressor intake
103 Intake centre fairing
104 Fuselage upper longeron
105 Rear fuselage frames

106 Hydraulic reservoir
107 Vertical tail control rods
108 Fin rod fillet
109 Vertical tail feel trim unit
110 Vertical tail autopilot controller
111 Rudder feel control unit
112 Tailfin construction
113 Flush VHF aerial
114 Starboard all-moving tailplane
115 Fin leading edge ribs
116 Di-electric fin tip aerial fairing
117 UHF/IFF aerial
118 VOR aerial
119 Tail navigation light
120 Tail radar warning antenna (electronic countermeasures, ECM)
121 Rudder construction
122 Rudder hydraulic actuator
123 Fin attachment post
124 Detachable tail cone
125 Jet pipe
126 Engine exhaust nozzle
127 Port all-moving tailplane construction
128 Tailplane spar box
129 Leading edge ribs
130 Tailplane pivot fixing
131 Tailplane control lever arm
132 Tailplane hydraulic actuator
133 Backup tailplane control interconnecting yoke
134 Rear engine mounting
135 Rolls Royce Allison TF41-A-2 non-afterburning turbofan
136 Fuselage lower longeron
137 Ventral chaff dispenser

138 Engine bay access panels
139 Boron carbide (HFC) engine bay armour
140 Arrester hook
141 Hook hydraulic actuator
142 Engine accessory gearbox
143 Main engine mounting trunion
144 Hydraulic accumulators
145 Position of strike camera, mounted on starboard side
146 Fuel vent mast
147 Port spoiler
148 Flap hinge arm
149 Flap hydraulic actuator
150 Flap rib construction
151 Port single-slotted trailing edge flap

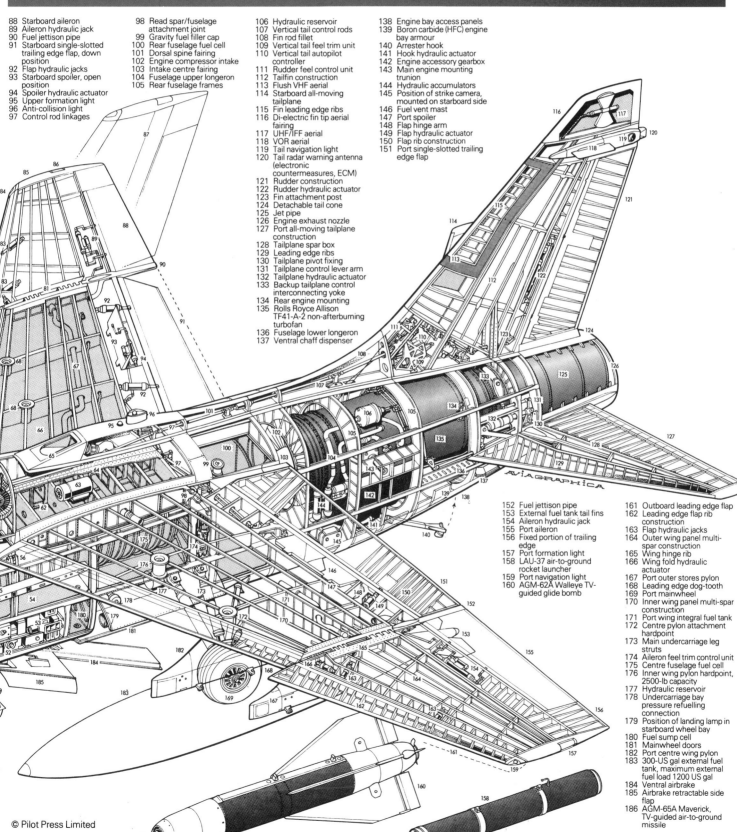

152 Fuel jettison pipe
153 External fuel tank tail fins
154 Aileron hydraulic jack
155 Port aileron
156 Fixed portion of trailing edge
157 Port formation light
158 LAU-37 air-to-ground rocket launcher
159 Port navigation light
160 AGM-62A Walleye TV-guided glide bomb

161 Outboard leading edge flap
162 Leading edge flap rib construction
163 Flap hydraulic jacks
164 Outer wing panel multi-spar construction
165 Wing hinge rib
166 Wing fold hydraulic actuator
167 Port outer stores pylon
168 Leading edge dog-tooth
169 Port mainwheel
170 Inner wing panel multi-spar construction
171 Port wing integral fuel tank
172 Centre pylon attachment hardpoint
173 Main undercarriage leg struts
174 Aileron feel trim control unit
175 Centre fuselage fuel cell
176 Inner wing pylon hardpoint, 2500-lb capacity
177 Hydraulic reservoir
178 Undercarriage bay pressure refuelling connection
179 Position of landing lamp in starboard wheel bay
180 Fuel sump cell
181 Mainwheel doors
182 Port centre wing pylon
183 300-US gal external fuel tank, maximum external fuel load 1200 US gal
184 Ventral airbrake
185 Airbrake retractable side flap
186 AGM-65A Maverick, TV-guided air-to-ground missile

Vought F-8 Crusader

Conceived in response to a US Navy requirement of 1951 for a supersonic air-superiority fighter, the Crusader first flew on 25 March 1955, exceeding Mach 1 on its maiden flight. Orders for production F8U-1s (later F-8As) followed swiftly and this initial version attained operational status with VF-32 in March 1957.

Progressive refinement of an outstanding design, paying particular attention to the radar and weapons control system, resulted in several variants, and by the time production ceased in January 1965 1,281 had been completed, as well as the RF-8A reconnaissance derivative. A remanufacturing project in the late 1960s resulted in earlier models acquiring greater weapons capability as the F-8H (89 rebuilt F-8Ds), F-8J (136 F-8Es), F-8K (87 F-8Cs) and F-8L (61 F-8Bs); 73 RF-8As were given a strengthened airframe, ventral fins and new equipment, becoming RF-8Gs which still served in 1984 with one Reserve squadron at Washington.

France purchased 42 F-8E(FN)s with French AAMs and special high-lift systems to fit them for service aboard the small Aéronavale carriers. The 20 or so survivors are soon to be retired as they run out of hours. Much later, the Republic of the Philippines took advantage of a cut-price offer when it purchased 25 surplus F-8Hs in the late 1970s for service with the 7th tactical fighter squadron at Basa AB. All were from long-term storage and remanufactured by Vought before shipment.

Flotille 12F, home-based at Landivisiau, is the last naval user of the excellent F-8 Crusader. These are designated F-8E(FN).

Vought F-8H of the 7th Tactical Fighter Squadron, Philippines Air Force, based at Basa.

Specification
Vought F-8E Crusader

Type: single-seat all-weather fighter
Powerplant: one 8165-kg (18,000-lb) afterburning thrust Pratt & Whitney J57-P-20A or -420 turbojet engine
Performance: maximum speed at 10975 m (36,000 ft) 1802 km/h (1,120 mph); initial rate of climb 6400 m (21,000 ft)/minute; service ceiling 17680 m (58,000 ft); range 1770 km (1,100 miles)
Weights: empty 9038 kg (19,925 lb); normal loaded 12701 kg (28,000 lb); maximum take-off 15422 kg (34,000 lb)
Dimensions: span 10.72 m (35 ft 2 in); length 16.61 m (54 ft 6 in); height 4.80 m (15 ft 9 in); wing area 35.52 m² (350.0 sq ft)
Armament: four 20-mm (0.79-in) Colt Mk 12 cannon each with 144 rounds, plus up to four AIM-9 Sidewinder air-to-air missiles, eight 227-kg (500-lb) bombs or Zuni rockets, or two Bullpup air-to-surface missiles; F-8E(FN) is

Vought F-8E Crusader

compatible with Matra R.530 AAM
Operators: France, Philippines.

Westland Lynx (Army)

Launched as part of the Anglo-French helicopter agreement of February 1967, the Westland Lynx is an extremely modern and versatile machine. Its design is wholly of Westland origin, but the production of the type is shared in the ratio of 70/30 between the UK and France, in the form of the nationalized Aérospatiale concern. One of the primary French responsibilities is the forged titanium hub, a one-piece structure for the four-blade semi-rigid main rotor which is one of the most important features of the design. All versions of the Lynx have advanced digital flight controls plus all-weather avionics, and no previous helicopter can equal the type for agility and all-weather one-man operation. The origins of the design lie with the WG.13 proposal, which was schemed in general-purpose naval and civilian applications. But so versatile did the design appear that the concept was expanded to land-based tactical operations, in which the type's agility and performance would prove a very considerable asset. The first prototype of the Lynx flew on 21 March 1971, and the six prototypes were used exhaustively for all aspects of the certification programme, for trials and for record-breaking. The second production model was the Lynx AH.Mk 1 battlefield helicopter for the British army. This first flew on 11 February 1977, and the type was cleared for service introduction at the end of 1977. Since that time the

The Lynx in its element: armed with TOW missiles and sporting a high-mounted sight, the Lynx can destroy tanks from behind the cover of trees.

Westland Lynx (Army) (continued)

Lynx has built up an enviable reputation as a versatile battlefield helicopter, being able to carry up to 12 troops in addition to a crew of two, or 907 kg (2,000 lb) of internal freight, or a slung load of 1361 kg (3,000 lb), or a wide assortment of weapons including eight TOW anti-tank missiles aimed with a stabilized sight mounted in the flightdeck roof. The chief distinguishing feature of the land-based Lynx is its skid landing gear, the naval Lynx having wheeled tricycle landing gear.

Westland is producing variants of the basic design, including the current WG.30 and the future Lynx 3. The WG.30 is designed for civil and military applications, and is in essence a 'big-fuselage' Lynx with uprated powerplant and larger-diameter main rotor, and can carry up to 22 troops. The Lynx 3 is a more advanced armed development of the Lynx AH.Mk 1; among its features are the ability to carry HOT, TOW or Hellfire anti-tank missiles (with

The British Army has adopted the Lynx as its main battlefield helicopter, using it for both combat assault and anti-armour.

reloads in the cabin, a technique pioneered with the Lynx AH.Mk 1), and to carry the Stinger air-to-air missile for self-defence or the destruction of enemy helicopters over the battlefield. Sensors proposed for the Lynx 3 are either a china- or mast-mounted package of target-acquisition and night-vision items. With a maximum take-off weight of 5443 kg (12,000 lb), the Lynx 3 will also be able to lift 14 troops over a range of 105 km (65 miles). The pro-

totype should fly by 1986, with deliveries beginning shortly after this.

Specification
Westland Lynx AH.Mk 1
Powerplant: two 671-kW (900-shp) Rolls-Royce Gem 41 turboshafts, each flat-rated to 559 kW (750 shp)
Performance: maximum speed 259 km/h (171 mph); range 540 km (336 miles) with a full load of troops
Weights: empty, equipped for anti-tank strike 3072 kg (6,772 lb);

maximum take-off 4536 kg (10,000 lb)
Dimensions: main rotor diameter 12,802 m (42 ft 0 in); length overall, rotors turning 15.163 m (49 ft 9 in); height 3.66 m (12 ft 0 in); main rotor disc area 128.69 m^2 (1,385.35 sq ft)
Armament: weapons can include a 20-mm cannon, rocket pods, or various types of air-to-surface missile including HOT, TOW, and AS.11
Operators: Qatar, UK

Westland Lynx (Navy)

On technical grounds the Westland Lynx is the best medium shipboard helicopter in the world, and from the original WG.13 concept (which formed part of the Anglo-French helicopter agreement of 1967) has come a series of uprated versions which not only bring in additional missions but also greatly enhanced capabilities. Though all are similar in terms of overall dimensions, the later versions have increased power for operation at greater weights, and the recent development of upgraded models of the Westland 30 (the large-fuselage variant) has opened the way to machines which essentially equal the mission capability of the Sikorsky SH-60B, but in a smaller and more compact helicopter. The original Lynx HAS.2 for the Royal Navy was actually the first production variant to fly, in February 1976. Powered by two 559-kW (750-shp) Gem 2 engines, this has a gross weight of 4309 kg (9,500 lb), yet carries a crew of two (three in the ASW or SAR roles) plus all equipment for a wide range of shipboard missions including ASW, SAR, ASV (anti-surface vessel) search and strike, reconnaissance, troop transport (typically 10 troops), fire support, communication and fleet liaison and vertrep (vertical replenishment) duties.

Equipment of all these models includes a search radar, which in the 60 Lynx HAS.2s of the RN is the Ferranti Seaspray; the equivalent machines of the French Aéronavale have the OMERA-Segid ORB 31W. In the ASW search role other sensors can include Bendix or Alcatel dipping sonars or a TI MAD (magnetic anomaly detector). The basic Lynx has one of the world's most advanced flight control systems which, in conjunction with comprehensive nav-aids, makes possible precision flying in even the worst weather, as was amply proved during over 3,000 hours of combat operations off the Falklands in 1982. During this campaign the new Sea Skua anti-ship missile was also brought into action for the first time. Though other missiles can be carried, the Sea Skua is the most effective in the world for this mission, and up to four can be fired and guided automatically by radar homing even in conditions of zero visibility.

In 1979 the Royal Netherlands Navy began receiving the upgraded Lynx Mk 27, first of a Mk 2 family with Gem 41-1 engines and weights ranging from 4763-4990 kg (10,500-11,000 lb). The most recent batches for the RN and Aéronavale are of this new standard which offers greater capability. Lynx 3 development to 5443 kg (12,000 lb) has been completed, and Lynx 4 will take the weight to 6577 kg (14,500 lb) using a five-blade rotor.

Specification
Westland Lynx HAS.2
Type: multi-role shipboard helicopter

Powerplant: two 559/671 kW (750/900-shp) Rolls-Royce Gem 2 turboshaft engines
Performance: maximum cruising speed 232 km/h (144 mph); time on ASW hover at 93 km (58 miles) radius 2 hours; ferry range 1046 km (650 miles)
Weights: empty 2740 kg (6,040 kb); maximum take-off, early production 4309 kg (9,599 lb), later production 4763 kg (10,500 lb)
Dimensions: main rotor diameter 12.80 m (42 ft 0 in); length rotors turning 15.16 m (49 ft 9 in); height rotors turning 3.60 m (11 ft 9.75 in); main rotor disc area 128.71 m^2 (1,385.45 sq ft)

Armament: two Mk 44, Mk 46 or Sting Ray homing AS torpedoes, or four BAe Sea Skua anti-ship missiles, or two Mk 11 depth charges, in each case with full mission equipment
Operators: Argentina, Brazil, Chile, Denmark, France, West Germany, the Netherlands, Nigeria, Norway, Qatar (Police), UK (RN)

Twenty-two Lynxes serve with the Royal Netherlands navy on a variety of duties, including SAR, VIP transport and anti-submarine warfare.

The Lynx HAS.2 is one of the most complete naval helicopters, being able to carry out virtually all tasks required. This aircraft carries four Sea Skua anti-ship missiles.

France's Lynxes differ from other naval Lynxes by having French equipment, notably the Omera-Segid ORB-31-W radar and Alcatel dunking sonar. French weapons are also carried, although this aircraft carries a pair of US-built Mk 44 torpedoes. Aéronavale Lynxes are flown from ships for anti-submarine protection, vertical replenishment and cross-decking, SAR and general observation. Six are on strength.

269 MARINE

DANGER→

269

Westland Scout

This five/six-seat helicopter flies liaison, casualty evacuation, air/sea rescue, ground attack, reconnaissance, training and light freight missions. Equipped as an ambulance it carries two stretchers internally and two on external panniers. Normal seating is for five, with three on a rear bench seat and two in front. The prototype flew on 20 July 1958, followed by a more powerful variant on 9 August 1959. The Scout entered service with the British Army in 1963, and over 50 examples are still operated by the AAC (Army Air Corps). Forming part of the 400-strong AAC helicopter fleet, the Scouts have seen extensive use in Northern Ireland on anti-terrorist duties. In the Falklands campaign many were fitted urgently with a wide range of combat equipment, including IR-exhaust suppressors, armour, stabilized sights and night vision equipment; they carried various weapons.

The AAC has put the Scout's versatility to wide use, the aircraft having filled almost every role, except heavy lift: for anti-tank duty the Scout carries a roof-mounted sight and SS.11 missiles. Distinguishing features of the Scout against the maritime Wasp are skid landing gear and low-set horizontal stabilizers. A few Scouts were sold for export, including two to the Royal Australian Navy for operation from survey ships. Three went to the Jordanian air force, one for the personal use of King Hussein. The

The Scout has been used by the Army Air Corps for many roles. This aircraft is being used to transport troops, while the starboard skids have a stretcher case mounted on them for casevac. For anti-armour duties, the Scout uses a sight mounted above the cockpit.

government of Uganda bought two, and a further two went to the Bahrain State Police.

Specification
Westland Scout AH.1
Type: five/six-seat general purpose helicopter
Powerplant: one 511-kW (685-shp) Rolls-Royce Nimbus 102 free-turbine turboshaft engine
Performance: maximum cruising speed 196 km/h (122 mph); initial rate of climb 509 m (1,670 ft)/minute; range with four passengers and reserves 507 km (315 miles)
Weights: empty 1466 kg (3,232 lb); maximum take-off 2404 kg (5,300 lb)
Dimensions: main rotor diameter 9.83 m (32 ft 3 in); length, rotors turning 12.29 m (40 ft 4 in); height, rotors turning 3.56 m (11 ft 8 in); main rotor disc area 75.89 m^2 (816.87 sq ft)
Armament: manually-controlled guns of up to 20-mm (0.79-in); fixed 7.72-mm (0.3-in) GPMG or other gun installations; rocket pods or guided anti-tank missiles such as SS.11
Operators: Australia, Bahrain, Jordan, Uganda, UK

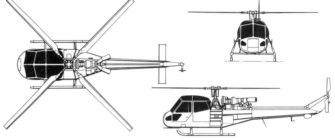

Westland Scout AH.1

Westland Sea King

In 1959, the year in which the Sikorsky S-61 helicopter first flew, Westland of the UK concluded a licence agreement and the company developed the Westland Sea King HAS.1. This became the new ASW (anti-submarine warfare) helicopter of the Royal Navy, with 56 delivered in 1969-72. Compared with the US Navy HSS-2 (SH-3), the Sea King

HAS.1 has equipment for completely autonomous operation with no help from the parent warship, including AW 391 dorsal radar, Plessey Type 195 dunking sonar and a fully fitted tactical compartment for managing a whole ASW operation. These machines have been modified to Sea King HAS.2 standard with more powerful Rolls-Royce Gnome

engines and improved equipment, and 21 Sea King HAS.2 helicopters were also newly built. The Sea King HAR.3 is the RAF search-and-rescue model with very complete and great versatiity (SAR models carry up to 22 rescuees including stretcher casualties). The Sea King HC.4 is the version of the Commando for the Royal Navy (used for

Royal Marine assault transport) with the shipboard features, such as folding blades and tail, but with simple fixed landing gear and suitable for 27 troops or 2722 kg (6,000 lb) of cargo. The Sea King HAS.5 is the current Royal Navy ASW model, with dramatically uprated avionics, all Sea King HAS.2s being converted to this standard; 17 were built new and after the Falklands (when Sea Kings flew almost non-stop in terrible weather) nine more were ordered. Key items in the Sea King HAS.5s avionics suite are the MEL Sea Searcher surveillance radar with a radome of considerably different shape and size, provision for the launch of passive sonobuoys, LAPADS (Lightweight Acoustic Processing and Display System) by GEC Avionics for the quicker and more precise handling of acoustic data, and a better display system. To permit the installation of the extra equipment, the cabin has been extended aft by 1.83 m (6 ft). The Royal Navy has received eight Sea King AEW helicopters in response to a need appreciated in the Falklands campaign of 1982. This model has Thorn EMI Searchwater radar with its antenna in an inflated and swivelling radome projecting from the right side of the fuselage. Two were modified HAS.2s and the other six converted HAS.3s. They have comprehensive Elint and communications systems. Westland has exported ASW and SAR Sea Kings to

An urgent need for shipborne AEW capability has been met by the Sea King AEW.3, which carries Thorn EMI Searchwater radar in an inflatable radome which can swivel upwards for ground operations. So far, eight aircraft have been converted from older machines to this standard.

Westland Sea King (continued)

eight countries, including the up-graded Sea King Mk 50 with Bendix sonar for Australia. The Commando transport, based on the Sea King, is described separately.

Specification
Type: ASW and multi-role helicopter
Powerplant: two 1238 kg (1,660-shp) Rolls-Royce Gnome H.1400-1 turboshaft engines
Performance: cruising speed at maximum weight 208 km/h (129 mph); initial rate of climb 616 m (2,020 ft)/minute; range with standard fuel 1230 km (764 miles)
Weights: empty, ASW 6202 kg (13,672 lb); maximum take-off 9525 kg (21,000 lb) or, AEW version 9707 kg (21,400 lb)
Dimensions: main rotor diameter 18.90 m (62 ft 0 in); length rotors turning 22.15 m (72 ft 8 in); height rotors turning 5.13 m (16 ft 10 in);

Westland Sea King HAR.3 of No. 202 Sqn, RAF.

main rotor disc area 280.47 m² (3,019.08 sq ft)
Armament: extremely comprehensive ASW sensors and systems plus up to four Mk 46 torpedoes or Mk 11 depth charges
Operators: Australia, Belgium, Egypt, West Germany, India, Norway, Pakistan, UK (RN)

Below: Equipped with nose search radar, Norway's Sea Kings sport distinctive Day-Glo patches for visibility.

Right: Australia operates the Westland Sea King for anti-submarine and SAR duties.

Westland Commando

Optimized for tactical military duties, the Westland Commando evolved from the Westland Sea King anti-submarine warfare helicopter, itself a development of the original Sikorsky SH-3 Sea King which is widely used by the US Navy.

Lacking specialized ASW equipment, the Commando was originally conceived mainly for the export market and has achieved some success in this field, examples being purchased by Egypt and Qatar. A basically similar machine, known as the Sea King HC.Mk 4, has also been acquired by the Royal Navy and this presently serves with two Commando squadrons, being employed purely in the assault role.

Featuring a fixed tail-wheel landing gear and lacking the sponsons of the Sea King, the Commando can carry up to 28 fully-armed troops as well as a crew of two. In addition, it possesses the ability to carry underslung loads of up to 3629 kg (8,000 lb), whilst provision for fitment of armament also exists, the

Among Egypt's Commandoes is this aircraft configured for VIP duties. Others serve as combat assault aircraft.

parent company having evaluated a variety of weapons including guns, rocket pods and missiles. A particularly versatile helicopter, the Commando is equally adept at assault, casualty evacuation, logistical support, combat search and rescue and strike tasks.

Unlike the Commando, which is principally intended for use from land bases, the Sea King HC.Mk 4 is frequently deployed aboard aircraft-carriers and thus possesses folding main rotor blades and tail unit to facilitate stowage aboard ship. Airlift capability is slightly less than that of the Commando, at 27 troops or 3402 kg (7,500 lb) of cargo slung externally, but the type more than proved its value in the Falklands during the 1982 campaign, playing a major part in supporting the rapid advance of British troops on Port Stanley. A total of 15 Sea King HC.Mk 4s was obtained, these routinely being used to airlift Royal Marines and possessing the ability to operate in all kinds of environment, ranging from the heat and humidity of the tropics to the bitter cold of the Arctic circle.

Specification
Westland Sea King HC.Mk 4
Type: tactical transport/assault helicopter
Powerplant: two Rolls-Royce Gnome H.1400-1 turboshafts each rated at 1238 kW (1,660 shp)
Performance: cruising speed at sea level 208 km/h (129 mph); range with maximum payload and reserves 444 km (276 miles); ferry range with maximum fuel 1460 km (907 miles)
Weights: empty 5,544 kg (12,222 lb); maximum take-off

9525 kg (21,000 lb)
Dimensions: main rotor diameter 18.90 m (62 ft 0 in); length rotors turning 22.15 m (72 ft 8 in); height 5.13 m (16 ft 10 in); main rotor disc

area 280.47 m^2 (3,019 sq ft)
Armament: a wide variety of cannon, machine-guns, rockets and missiles can be carried
Operators: Egypt, Iraq, Qatar

Operating from ships of the Royal Navy, the Sea King HC.4 is used for marine assault purposes. As well as troop transport to the combat zone, the aircraft has excellent heavy lift qualities.

Westland Wasp

Though its development can be traced back to the Saro P.531, flown in 1958, the Westland Wasp emerged in October 1962 as a highly specialized machine for flying useful missions from small ships, such as frigates and destroyers with limited deck pad area. The missions are ASW (anti-submarine warfare) and general utility, but the Wasp is not sufficiently powerful to carry a full kit of ASW sensors as well as weapons, and thus in this role relies on

Developed from the same origins as the Scout, the Wasp HAS.1 performs as many roles, but from the deck of a ship. Light anti-ship attacks can be carried out using the AS.12 missile, as demonstrated here. Other roles are liaison, VIP transport, observation, limited ASW, SAR and casevac.

the sensors of its parent vessel and other friendly naval forces, which tell it where to drop its homing torpedoes. In the anti-surface vessel role the Wasp is autonomous, and though it has no radar it can steer the AS.12 wire-guided missiles under visual conditions over ranges up to 8 km (5 miles). Other duties include SAR (search and rescue), liaison, VIP ferrying, casualty evacuation with two internally carried stretchers, ice reconnaissance and photography/survey. The cockpit is equipped for bad-weather operation with auto-stabilization, radar altimeter, beacon receivers, UHF radio and UHF homer, and in RN service limited EW provisions. The stalky quadricycle landing gear has wheels that castor so that, while the machine can be rotated on deck,

it cannot roll in any direction even in a rought sea. Sprag (locking) brakes are fitted to arrest all movement. Provision was made for various hauldown systems such as Beartrap to facilitate alighting on small pads in severe weather. Deliveries to the Royal Navy began in 1963, and a few were active in Operation 'Corporate' in the South Atlantic right at the end of their active lives when most had been replaced in RN service by the Lynx. Wasp HAS.1s flew 912 combat sorties during which they made 3,627 deck landings. Most were used in reconnaissance and utility missions, though several operated in the casevac role. Three engaged the Argentine submarine *Santa Fe* and holed its conning tower with AS.12 missiles (which passed clean through before exploding).

Specification
Type: light multi-role ship-based helicopter
Powerplant: 529-kW (710-shp) Rolls-Royce Nimbus 503 turboshaft engine
Performance: maximum speed with weapons 193 km/h (120 mph); cruising speed 177 km/h (110 mph); range 435 km (270 miles)
Weights: empty 1566 kg (3,452 lb); maximum take-off 2495 kg (5,500 lb)
Dimensions: main rotor diameter 9.83 m (32 ft 3 in); length overall 12.29 m (40 ft 4 in); height 3.56 m (11 ft 8 in); main rotor disc area 75.89 m^2 (816.87 sq ft)
Armament: two Mk 44 torpedoes or two AS.12 anti-ship missiles
Operators: Brazil, Indonesia, New Zealand, South Africa, UK (RN)

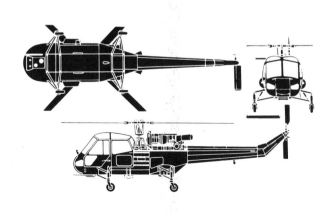

Westland Wasp HAS.1

Westland Wessex

Despite the fact that it is based upon the early 1950s vintage Sikorsky S-58, the Westland Wessex is still used extensively by the British armed forces, examples serving with the Royal Air Force and Fleet Air Arm on a variety of tasks; these encompass air-sea rescue, commando assault and logistical support. Acquired initially to fulfil the Royal Navy's urgent need for an anti-submarine warfare helicopter, the Wessex was developed by Westland with a single Gazelle turbine engine, and entered service with the FAA as the Wessex HAS.1 during 1960. The next major variant was the Wessex HC.2 for the RAF, this having a totally different nose housing a Coupled Gnome twin powerplant and being able to carry a payload of 16 troops, or a slung load of 1814 kg (4,000 lb), or seven stretchers when operating in the casevac role. Improved ASW equipment brought about the Wessex HAS.3, called 'The Camel' by virtue of the humped dorsal radome, many of the original HAS.1s being retrofitted. Two Wessex HC.4s for The Queen's Flight followed, the final British version being the Wessex HU.5, used by the Royal Marines as an assault transport; the Mks 4 and 5 were basically similar to the Mk 2 except for furnishings.

All UK versions were used extensively during the Falklands campaign, taking part in operations on East Falkland and also being involved in lesser actions at South Georgia. A single HAS.3 single-handed crippled the submarine *Santa Fe*, while HU.5s shuttled round the clock during the invasion of the main islands. The Royal Aus-

Operating from the smaller ships of the Royal Navy, Wessex HAS.3s provide ASW coverage as well as more general duties. This aircraft is 'Humphrey', the Wessex which crippled Argentina's *Santa Fe*, seen off South Georgia.

This Wessex HU.5 is typical of those used by the Royal Marines for assault purposes. Now being replaced by Sea King HC.4s and Lynxes in this role, this mark performed sterling work during the Falklands campaign.

tralian Navy uses the Gazelle-engined Mk 31B, other military operators using versions based on the twin-turbine HC.2.

Specification
Westland Wessex HU.5
Type: tactical transport and assault helicopter
Powerplant: one Rolls-Royce Coupled Gnome 101/111, with two 1007-kW (1,350-shp) engines limited to a combined output of 1156 kW (1,550 shp)
Performance: maximum speed 212 km/h (132 mph); cruising speed 195 km/h (121 mph); service ceiling 4300 m (14,100 ft); range with standard fuel 628 km (390 miles)
Weights: empty 3927 kg (8,657 lb); maximum take-off 6123 kg (13,500 lb)
Dimensions: main rotor diameter 17.07 m (56 ft 0 in); length of fuselage 14.74 m (48 ft 4.5 in); height 4.93 m (16 ft 2 in); main rotor disc area 228.81 m² (2,463.0 sq ft)
Armament: provision for machine-guns and anti-tank missiles
Operators: Australia, Brunei, Ghana, Iraq, UK (RAF, RN, RM)

Above: No. 28 Sqn is based at Kai Tak airport in Hong Kong, and is tasked with utility duties and general policing around the territory. Among the services provided by the squadron's Wessex HC.2s is firefighting, with underslung water-hoppers.

Westland Whirlwind

Westland Aircraft Ltd at Yeovil, England, acquired licence-rights to build the Sikorsky S-51 in 1947, and this marked the beginning of an association with Sikorsky aircraft which continues in 1986.

The potential of the larger Sikorsky S-55 was such that Westland lost little time in obtaining licence-rights to build military versions for the British armed forces, and for export within specified geographical areas.

The prototype of Westland's Whirlwind HAR.1 (HAR = helicopter, air rescue) for the Royal Navy first flew on 15 August 1953, and initial deliveries of 10 production examples for the Fleet Air Arm began to enter service with No. 705 squadron soon after.

In 1960 Westland designed the uprated Whirlwind Series 3 with turbine power. First of the Series 3

Whirlwinds entered service with the RAF's No. 225 Squadron, at RAF Odiham, on 4 November 1961. These were powered by the Bristol Siddeley (now Rolls-Royce) Gnome, a licence-built version of the General Electric T58 turboshaft, and with this powerplant the Whirlwind gained a new lease of life. Subsequently, piston-engined Whirlwinds were returned to the factory

for conversion to turbine power, Royal Navy HAS.7s becoming redesignated HAS.9s.

A handful are still in Brazilian navy service in 1985.

Specification
Type: (HAR.10) light utility helicopter
Powerplant: one 783-kW (1,050-shp) Rolls-Royce Gnome H.1000 turboshaft
Performance: maximum speed 171 km/h (106 mph); cruising speed 167 km/h (104 mph); range 483 km (300 miles)
Weights: empty equipped 2246 kg (4,952lb); maximum take-off 3629 kg (8,000 lb)
Dimensions: main rotor diameter 16.15 m (53 ft 0 in); tail rotor diameter 2.72 m (8 ft 11 in); length (rotors turning) 19.00 m (62 ft 4 in); height 4.76 m (15 ft 7½ in); main-rotor disc area 188.2 m² (2,026 sq ft)
Armament: can include a homing torpedo or Nord AS.11 anti-tank missiles
Operators: Brazil, Nigeria, Qatar, UK

Westland Whirlwind Series 3 of the Brazilian navy, one of the last users of the type.

Westland 30

Developed from the Lynx, the Westland 30 is a two-crew transport helicopter with a larger fuselage, extra fuel, a new automatic flight-control system and revised electronics. The deeper, broader cabin (13.0 m³/460 cu ft volume) will carry 14 fully-equipped troops, 19 passengers in airline seats, a smaller number of VIPs in luxury surroundings or can have various configurations for cargo. The prototype flown on 10 April 1979 was representative of the Series 100 helicopter, powered by 846-kW (1,135-shp) Rolls-Royce Gem 41-1 engines. After deliveries for short-range airline operations, the type was replaced by the Series 100-60 with 940-kW (1,260-hp) Gem 60-3s. In September 1983 improved performance, especially in 'hot and high' conditions, became available from the Series 200 with 1277-kW (1,712-shp) General Electric CT7-2B engines. In prospect is a Series 300, still CT7-powered but with higher maximum weight and speed, a five-blade composite rotor, developed dynamic components and reduced noise and vibration. Westland is the first helicopter manufacturer to offer an EFIS (Electronic Flight Instrumentation System), with four CRTs replacing traditional instruments as a standard fitment. Attempts to sell the Westland 30 to

The Westland 30 combines the flight systems of the Lynx with a larger, more capacious fuselage which is capable of carrying far more in the way of troops or cargo. Despite its good load-carrying qualities, none have been sold to military customers yet.

military operators have yet to be successful. Attention is currently concentrated on the RAF's AST.404 specification for between 75 and 125 Wessex and Puma replacements to enter service from 1989. The Westland 30-404 submission is based on the Series 300, with a maximum weight of 7267 kg (16,020 lb) and the option to change to Rolls-Royce/Turboméca RTM 322 engines. Westland is also offering for export the TT30 Tactical Transport, a minimally-equipped version for IFR operation, although capable of being outfitted to higher standards at customer option.

Specification
Westland 30 Series 100-60
Type: multirole transport
Powerplant: two 940-kW (1,260-shp) Rolls-Royce Gem 60-3 turboshaft engines
Performance: maximum cruising

speed 219 km/h (136 mph); hovering ceiling out of ground effect 884 m (2,900 ft); range with maximum payload 472 km (293 miles)
Weights: empty 3167 (6,982 lb); maximum take-off 5806 kg (12,800 lb)
Dimensions: main rotor diameter

13.31 m (43 ft 8 in); length rotors turning 15.90 m (52 ft 2 in); height rotors turning 4.72 m (15 ft 6 in); main rotor disc area 139.14 m² (1,497.7 sq ft)
Armament: none
Operators: civil airlines only by 1986

Yakovlev Yak-11

Although phased out by the Soviet Union and the Warsaw Pact forces, the Yakovlev Yak-11 'Moose' continues in service with several users. Designed as an advanced combat trainer, the Yak-11 which entered service in 1947 was almost a 1944-vintage Yak-3 fighter with a low-powered engine. Production totalled 3,850 aircraft, a further 707 being made in Czechoslovakia as the C-11. These tough and simple machines have pneumatic landing-gear and flap actuation and are fully aerobatic. A few continue in use as transitional aircraft between the basic trainer and the MiG-15UTI.

Specification
Type: advanced trainer

Powerplant: one 425-kW (570-hp) Shvetsov ASh-21 7-cylinder radial piston engine
Performance: maximum speed 460 km/h (286 mph) at 2250 m (7,380 ft); service ceiling 7100 m (23,295 ft); range 1290 km (802 miles)
Weights: empty 1900 kg (4,189 lb); maximum take-off 2400 kg (5,291 lb)
Dimensions: span 9.40 m (30 ft 10.1 in); length 8.50 m (27 ft 10.6 in); height 3.28 m (10 ft 9.1 in); wing area 15.40 m² (165.77 sq ft)
Armament: provision to carry one 12.7-mm (0.5-in) UBS machine-gun and two 50-kg (110-lb) practice bombs
Operators: Afghanistan, Albania,

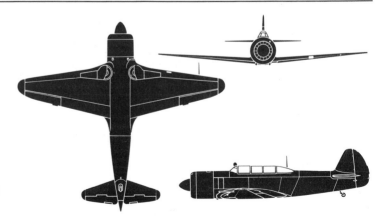

Angola, Bangladesh, Cambodia, China, Czechoslovakia, Mongolia, Somalia, Vietnam, North Yemen

Yakovlev Yak-11 'Moose'

Yakovlev Yak-18

The Yakovlev Yak-18 'Max' basic trainer has shown extraordinary longevity; itself a development of the UT-2 of 1934, it remained under active development into the 1970s and its replacement, the Yak-52, bears a close resemblance to it. The prototype flew in 1946, with the tail-wheel landing gear of its predecessors and the same 'helmet-type' cowling over the five-cylinder M-11FR radial engine. The tricycle landing gear of later versions was introduced on the Yak-18U of 1954, but the main production version was the Yak-18A, introduced in 1957, which added a more aerodynamically efficient cowling over the much more powerful AI-14 engine. The aircraft is of metal construction with fabric covering. From it have stemmed many special competition aerobatic versions. The prototype Yak-18P of 1949 had the single cockpit in the aft position, while the initial production aircraft had a forward cockpit. The pilot was moved aft of the wing again in the Yak-18PM of 1965, which was strengthened to accept +/−6g, and it was this aircraft which started the run of Soviet success in international aerobatics. The Yak-18PS of 1970 reverted to tail-wheel landing gear, and led to the Yak-50, -52, -53 and -55. The latest four-seat Yak-18T, first seen in 1967, has a wider centre section and a cabin-type fuselage. Over 6,700 Yak-18 trainers were built; the Chinese CJ-6 is a derivative.

Specification
Yakovlev Yak-18A
Type: two-seat basic trainer
Powerplant: one 194-kW (260-hp) Ivchenko AI-14R 9-cylinder radial piston engine
Performance: maximum speed 263 km/h (163 mph) at sea level; service ceiling 5060 m (16,600 ft); range 710 km (441 miles)
Weights: empty 1025 kg (2,260 lb); maximum take-off 1315 kg (2,899 lb)

Until recently, the Yak-18 has been used in a variety of marks as the main aerobatic aircraft in the Soviet Union; they are flown by the air force for aerobatic training. This is a Yak-18PM.

Dimensions: span 10.60 m (34 ft 9.3 in); length 8.35 m (27 ft 4.7 in); height 3.35 m (11 ft 0 in); wing area 17.00 m² (182.99 sq ft)
Armament: none
Operators: Afghanistan, Albania, Angola, Bangladesh, Bulgaria, East

Yakovlev Yak-18A 'Max'

Germany, Guinea Republic, North Korea, Mali Republic, Mauritania, Mongolia, Romania, Vietnam

---

Yakovlev Yak-28

Yakovlev Yak-28P 'Firebar' of the IA-PVO (air defence organization), Soviet air force, armed with AA-3 'Anab' missiles.

This family of supersonic combat aircraft has no direct equivalent in the West. Fast and rugged, with generous and usable internal volume, Yak-28s have seen service as bombers, reconnaissance and electronic countermeasures aircraft and fighters, and remain in service for most of those missions.

The Yak-28 started life as a modernized and faster version of the Yak-25/27 with more powerful engines, improved aerodynamics and a better internal arrangement. In the course of development and flight testing, between 1960 and 1963, it emerged as a new aircraft.

Two basic versions of the type were developed, a glazed-nose tactical version and a solid-nosed interceptor. The first of the former group were the Yak-28I 'Brewer-B' and 'Brewer-C' light bombers, which formed the main long-range element of Frontal Aviation until they were replaced by the Su-24 in the late 1970s. They were followed by the Yak-28R 'Brewer-D' reconnaissance aircraft, which was first observed in 1969. The final version in this series, probably produced by modifying Yak-28Is, is the 'Brewer-E' electronic countermeasures (ECM) escort aircraft, with a powerful ECM suite installed in the weapon bay. The reconnaissance and ECM versions were still in service in late 1984, and will probably be replaced by Su-24 variants in the late 1980s.

The Yak-29P 'Firebar' fighter may have entered service around 1964, but the definitive configuration, with a more pointed radome and other changes, was not seen until 1967. Generally similar in size and weight to the Su-15, and equipped with the same 'Skip Spin' radar, the Yak-28P was not as fast in level flight or climb, but probably offered a greater operational radius and endurance, and the advantage of a two-man crew. The Yak-28P remains in service, but is expected to be retired by 1988. Also in service is the Yak-28U 'Maestro', a non-operational two-seat trainer with a completely separate student's cockpit in the nose.

Specification
Yakovlev Yak-28P 'Firebar'
Type: two-seat interceptor
Powerplant: two Tumansky R-11 turbojets of 4600 kg (10,140 lb) dry thrust and 6200 kg (13,670 lb) augmented thrust
Performance: maximum speed 2000 km/h (1,240 mph) or Mach 1.87; service ceiling 16500 m (55,000 ft); initial climb rate 142 m/sec (28,000 ft/min); interception radius 900 km (360 miles)
Weights: empty approximately

13600 kg (30,000 lb); maximum take-off 22000 kg (50,000 lb)
Dimensions: span 12.95 m (42 ft 6 in); length 23.17 m (76 ft); height 3.95 m (13 ft); wing area 37.6 m² (405 sq ft)
Armament: two AA-3 'Anab' AAMs and two AA-2 'Atoll' AAMs on wing pylons
Operator: Soviet Union

Showing its unusual bicycle-type landing gear, this Yak-28 is the 'Brewer-D' reconnaissance variant. Sensors are carried in interchangeable pods in the space left where the weapons bay is in other marks. A few are still in service.

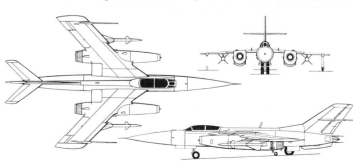

Yakovlev Yak-28P 'Firebar'

Yakovlev Yak-38 'Forger'

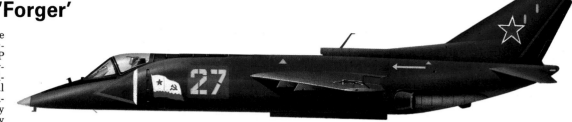

Operated from the three multi-role warships *Kiev, Minsk* and *Novorossiysk*, the Yakovlev Yak-36MP 'Forger' is the Soviet Union's equivalent of the BAe Sea Harrier, although only in the most general terms. From the 1917 revolution until recent times, the Soviet navy remained a small force with purely local defence responsibilities, but since 1956 Admiral Gorshkov has made efforts to develop a 'blue-water navy' which rivals the United States for mastery of the seas. As evidenced by the Yak-38 'Forger-A' and its two-seat trainer version, the Yak-38UV 'Forger-B', the USSR still has a long way to go, though its first giant carrier was in 1985 fast nearing completion.

First seen in July 1976, when the class-name ship *Kiev* sailed into the Mediterranean, the Yak-38 appears to be intended more for the development of techniques than for combat use, although its stated missions are operational. The four underwing pylons carry all weapons externally, giving anti-ship and ground-attack potential, and visual air combat (especially against NATO ocean-patrol/ASW aircraft and helicopters) is carried out with guns and AAMs. Unlike the Sea Harrier, the Yak-38 has no radar except a ranging unit, but most remarkable of all is its powerplant. Two lift engines mounted aft of the cockpit are enclosed by doors when not in use, while the two rear nozzles of the main engine turn downwards to give lift. There is a small toe-in on all four jet pipes to equalize fore-and aft

Yakovlev Yak-38 'Forger-A' of the Soviet navy.

thrust components. This arrangement, used previously in the VAK-191B and other types, enables the main engine to be smaller and more fuel-efficient than in a single-engine machine, but it carries certain penalties. The greatest is that the Yak-38 is incapable of a Harrier-type rolling (or ski-jump) take-off to maximize weapon loads. When seen in operation aircraft take-off and land vertically, the latter with a degree of precision which indicates ship-based (probably laser) guidance. About 12 'Forgers' are based on each ship. It is doubtful that they will be deployed on any other vessels, though they might be compatible with *Moskva* and *Leningrad*. Development and refinement has been progressive over the past eight years, and it is noteworthy that in 1984 the official Western estimate of the weapon load was doubled.

These 'Forgers' show the open intakes for the forward lift engines. These become deadweight when the aircraft is in level flight but allow the main engine to be smaller than in the Sea Harrier.

Specification
Yakovlev Yak-38 'Forger-A'
Type: single-seat VTOL shipboard strike fighter
Powerplant: one 8151-kg (17,985-lb) thrust vectoring turbojet (said to be a Lyulka AL-21) in the rear fuselage, and two 3572-kg (7,875-lb) thrust Koliesov lift turbojets in the forward fuselage
Performance: maximum speed Mach 0.95 or 1010 km/h (628 mph) at high altitude, or Mach 0.8

(978 km/h; 608 mph) at sea level; initial rate of climb 4500 m (14,764 ft)/minute; service ceiling 12000 m (39,370 ft); combat radius (lo-lo-lo mission with maximum stores) 240 km (149 miles)
Weights: empty 7485 kg (16,502 lb); maximum take-off 11700 kg (25,794 lb)
Dimensions: span 7.32 m (24 ft 0.2 in); length 15.50 m (50 ft 10.2 in); height 4.37 m (14 ft 4 in);

Around 12 'Forgers' are based on each carrier, providing both air defence and attack for the battle fleet.

wing area 18.50 m² (199.14 sq ft)
Armament: four underwing pylons for up to 3600 kg (7,937 lb) of stores, including AA-8 'Aphid' air-to-air missiles, air-to-surface or anti-ship missiles, GSH-23L gun pods, rocket launchers and bombs
Operators: Soviet Union (AV-MF)

Yakovlev Yak-40

The Yakovlev Yak-40 'Codling' was designed in 1964 to meet an Aeroflot requirement for a modern aircraft to replace its old piston-engined transports. The Yakovlev bureau chose the unusual combination of jet engines and an unswept wing, the need being for good STOL performance rather than speed. Manual controls are employed, and the aircraft has

been designed for utmost simplicity. About 1,000 were built, small numbers being used by various military services.

Specification
Type: 32-seat STOL transport or VIP aircraft
Powerplant: three 1500-kg (3,307-lb) thrust Ivchenko AI-25 turbofan

engines
Performance: maximum speed 560 km/h (348 mph) at 7300 m (23,950 ft); long-range cruising speed 470 km/h (292 mph); range, with 1700-kg/3,748-lb payload or 19 passengers 2100 km (1,305 miles), with 30 passengers 1450 km (901 miles)

Yakovlev Yak-40 'Codling'

Weights: empty 10263 kg (22,626 lb); maximum take-off 16000 kg (35,274 lb)

Dimensions: span 25.00 m (82 ft 0.3 in); length 20.30 m (66 ft 7.2 in); height 6.50 m (21 ft 3.9 in); wing

area 70.00 m² (753.50 sq ft)
Armament: none
Operators (governments):

Bulgaria, East Germany, Madagascar, Poland, Soviet Union, Yugoslavia, Zambia

Yakovlev Yak-52

Prospective aircrew for the Soviet air forces are screened and evaluated through the nationwide paramilitary training organization, DOSAAF. After many years during which it operated Yak-11 and Yak-18 trainers originally developed for the regular air forces in the 1950s, DOSAAF is now taking delivery of a purpose-designed low-cost basic trainer, the Yak-52.

The Yak-52 is a member of a family of light aircraft developed since 1975. The first of these was the Yak-50, a single-seat competitive aerobatic aircraft based on modified Yak-18 components. The Yak-52, announced in late 1978, differs in having two seats in tandem and a semi-retractable tricycle landing gear – presumably, in order to avoid having to instruct new recruits in irrelevant 'tail-dragger' techniques. The wheels are fully exposed even when the gear is up, so that a wheels-up landing does not damage the aircraft.

All production of the Yak-52 is undertaken by IAv Bacau in Romania, which started deliveries of the type in 1980. Other members of the family include the single-seat Yak-53, otherwise similar to the Yak-52, which is also built in quantity in Romania, and the mid-wing Yak-55 aerobatic type which was entered in world competitions in 1984.

Specification
Yakovlev Yak-52
Type: two-seat basic trainer
Powerplant: one Vedeneev M-14P nine-cylinder radial piston engine of 269 kW (360 hp)
Performance: maximum speed 285 km/h (177 mph); service ceiling 6000 m (19,900 ft); initial climb rate 10 m/sec (1,970 ft/min); range 550 km (340 miles)
Weights: empty 1000 kg (2,205 lb); maximum take-off 1290 kg (2844 lb)
Dimensions: span 9.5 m (31 ft 2 in); length 7.67 m (25 ft 2 in); height 2.95 m (9 ft 8 in); wing area 15 m² (162 sq ft)
Operator: Soviet Union

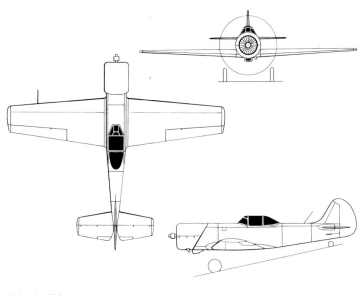

Yakovlev Yak-50

Zlin 42

Following its success with the Z.26 series, in which the two occupants were seated in tandem, Zlin designed and built a side-by-side trainer, the Z.42, which entered production in 1971. The 42M has a revised fin and constant-speed propeller, production of this variant beginning in 1974. A development is the Zlin 142, with 157-kW (210-hp) M337AK engine, about 150 being delivered in 1981-84.

Specification
Type: two-seat trainer
Powerplant: one 134-kW (180-hp) Avia M137 AZ inverted 6-inline piston engine
Performance: maximum speed

225 km/h (140 mph) at sea level; cruising speed 215 km/h (134 mph) at 600 m (1,970 ft); initial rate of climb 312 m (1,024 ft)/minute; service ceiling 4250 m (13,945 ft); range 530 km (329 miles)
Weights: empty 645 k (1,422 lb); maximum take-off, normal 970 kg (2,138 lb) aerobatic 920 kg (2,028 lb)
Dimensions: span 9.11 m (29 ft 10.7 in); length 7.07 m (23 ft 2.3 in); height 2.69 m (8 ft 9.9 in); wing area 13.15 m² (141.55 sq ft)
Armament: none
Operators: Bulgaria, Czechoslovakia, Hungary

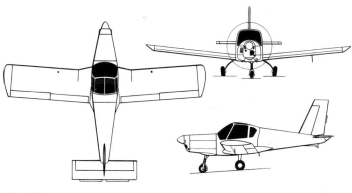

Zlin Z.142

Zlin 326

The Zlin Z.26 Trener was designed as a tandem primary trainer to meet specifications for Czech civil and military flying schools, the prototype flying in 1947 with a 78-kW (105-hp) Walter Minor 4 engine. Following competitive evaluation, the wooden Z.26 entered production and was designated C-5 by the Czech air force, which received first deliveries. The all-metal Z.126 replaced the earlier model in production from 1953, and this in turn was supplanted by the Z.226 with the 119-kW (160-hp) Walter Minor 6-III. The Z.226B was a glider-tug which flew in 1955, followed the next year by the Z.226T Trener-6. a fully-aerobatic single-seat version, the Z.226A Akrobat, also flew in 1956. This name was retained for the similar variant, the Z.326, a derivative of the Z.226T with retractable landing gear and other improvements; this entered production in 1959 and by 1965 1,540 examples of the family had been built. The Z.526F appeared in 1966 as the Trener-Master; for the single-seat version the name Akrobat as again used; with a one-piece canopy the latter became the

Z.526AS Akrobat Special, and for advanced aerobatics the Z.526AFS Akrobat. The Z.526L of 1972 was fitted with a 149-kW (200-hp) Lycoming AIO-360-B1B engine, and final development was the Z.726 Universal, similar to the Z.526F but with shorter-span wings and a 134-kW (180-hp) Avia M137 engine.

Direct military use of the Zlin 26 series seems to have been confined to four countries. The Cuban air force received 60 Z.226/326 aircraft; East Germany uses the same types alongside Soviet trainers; the Força Popular Aérea de Libertaçao de Moçambique had seven Z.326; and the Czech air force used the Z.526. Indirect military use includes Hungary where, although aircraft are civil-registered, they belong to the state and are used to train both civil and military pilots.

Specification
Type: trainer and aerobatic aircraft
Powerplant: one 119-kW (160-hp) Walter Minor 6-III inverted 6-inline piston engine
Performance: maximum speed 243 km/h (151 mph) at sea level;

cruising speed 212 km/h (132 mph); initial rate of climb 264 m (866 ft)/minute; service ceiling 4750 m (15,585 ft); range 580 km (360 miles) or, with tiptanks, 980 km (609 miles)
Weights: empty 650 kg (1,433 lb); maximum take-off, normal 975 kg (2,150 lb), aerobatic 910 kg

(2,006 lb)
Dimensions: span 10.60 m (34 ft 9.3 in); length 7.80 m (25 ft 7.1 in); height 2.06 m (6 ft 9.1 in); wing area 15.45 m² (166.31 sq ft)
Armament: none
Operators: Cuba, Czechoslovakia, East Germany, Mozambique

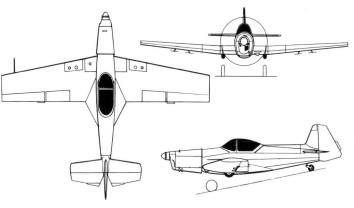

Zlin Z.526AFS Akrobat

This F-15 Eagle carries a typical air-to-air missile load of four AIM-9L Sidewinders and a similar number of AIM-7F Sparrows. These two missiles are the most widely used in the Western world.

The World's Air-launched Missiles

Aérospatiale AS.12 and AS.15TT

The SS.12 was originally developed by Nord as a multi-purpose weapon. The air-launched version of the SS.12, the AS.12, was first produced in 1960 for carriage by the French navy's Dassault Etendard and Aérospatiale Super Frelon aircraft. However, the missile soon became one of the main air-to-surface weapons carried by maritime patrol aircraft and ASW helicopters such as the Lockheed Neptune, Breguet Atlantic, Aérospatiale Alouette, Westland Wessex and Westland Wasp. The solid-propellant AS.12 used a basic command wire-guidance system to cruciform fins with optical tracking via a set of flares in the rear of the missile, the major disadvantage being that the aircraft has to remain within relatively close range of the target whilst the missile is under guidance, and is thus highly vulnerable to enemy defensive fire. During the 1982 Falklands war the Royal Navy fired a number of AS.12s from Wasp helicopters at the Argentinian submarine *Santa Fé* off South Georgia; these did some damage to the fin and pressure hull and helped

Saudi Arabia has helped pay for the development of the AS.15TT missile as part of a massive arms deal that they negotiated with France.

prevent the boat from submerging. The AS.12 is being replaced in Royal Navy service by the Sea Skua. The SS.12 has also been used as an anti-ship weapon on fast attack craft. Countries operating the AS.12 include Argentina, Chile, France, Iran, Iraq, Kuwait, Turkey, the UK, and the United Arab Emirates.

In the French Navy the AS.12 is being replaced by the helicopter-launched solid-propellant command-guided AS.15TT (*tous temps*, or all-weather) missile. This relies on a radio link to cruciform rear fins for azimuth guidance, both the missile and target being tracked by the launch helicopter's Thomson-

CSF Agrion 15 radar. The cruise height is between 3 and 5 in (10 and 16 ft) above the water, and is maintained by a radar altimeter. At around 300 m (33 yards) from the target the missile is commanded to descend to wave-top height in order to ensure a hit. The first complete test firing took place in October 1982, and Saudi Arabia has helped to pay for most of the development work. Both ship-launched and coastal-defence versions are also under development, and the only current customers are France and Saudi Arabia.

Specification

AS.12
Dimensions: length 1.87 m (6 ft 1.6 in); diameter 21.0 cm (8.25 in); span 65.0 cm (2 ft 1.6 in)
Weights: total round 76 kg (168 lb); warhead 28.4 kg (62.6 lb)
Performance: speed low subsonic; range 8 km (5 miles)

Specification
AS.15TT
Dimensions: length 2.16 m (7 ft 1 in); diameter 18.5 cm (7.3 in); span 56.4 cm (1 ft 10.2 in)
Weights: total round 96 kg (212 lb); warhead 29.7 kg (65.5 lb)
Performance: speed high subsonic; range 16 km (10 miles)

Aérospatiale AS.30 and AS.30 Laser

Essentially a scaled up AS.20, the Aérospatiale AS.30 started life in 1958 as the Nord 5401. In 1960 the basic AS.30 was fielded aboard the Dassault Mirage III fighter-bomber to meet a French air force requirement for an air-to-surface missile (ASM) that could be launched without the carrier coming within 3 km (1.86 miles) of the target. With a range of over 10 km (6.2 miles) and a terminal CEP of less than 10 m (32.8 ft), the original version required the operator to keep the missile aligned to the target with a joystick and tracking flares on the rear of the missile's body; a radio link was used to transmit corrective guidance commands to the onboard autopilot for course corrections. In 1964 an improved variant with a TCA semi-automatic guidance system and new flip-out fins entered production and service only for the French air force. This AS.30 TCA employs a SAT tracker unit for the

The AS.30 has been operational for nearly 25 years, and in that time has seen a continual enhancement of its capability. With its Mach 1.5 speed, the AS.30 has a range of some 11.25 km (7 miles).

continuous monitoring of an IR flare on the missile's rear, the pilot having only to keep the target centred in his attack sight. Over 3,870 AS.30s were built for a number of nations, operators being France, India, Israel (now out of service), Peru, South Africa, Switzerland, UK (now out of service) and West Germany. Of these South Africa has used the missile under operational conditions to attack and damage the abandoned and drifting oil tanker *Wafra* off her coastline, the launch aircraft in this case being the BAe Buccaneer S.Mk 50.

To enhance the weapon's capabilities into the 1990s, Thomson-CSF and Aérospatiale began work in

1974 on a laser-guided version. The target designation pod is the Automatic Tracking Laser Illuminator System (or Atlis), whilst the laser seeker head is called Ariel. By 1980 the first homing trials with pre-production rounds were under way, and in late 1983 the first deliveries of production AS.30 Laser rounds were made to the French air force for use on its SEPECAT Jaguar fighter-bombers, and several other as yet unidentified countries have since adopted the type.

Specification
AS.30 series
Type: air-to-surface missile

Dimensions: length (AS.30) 3.839 or 3.885 m ((12 ft 7.1 in or 12 ft 9 in), or (AS.30 Laser) 3.65 m (11 ft 11.7 in); span 1.0 m (3 ft 3.4 in); diameter 0.34 in (13.4 in)
Launch weight: 520 kg (1,146.4 lb)
Propulsion: solid-propellant rocket motor
Performance: speed Mach 1.5; range 11.25 km (7 miles); CEP (AS.30 and AS.30 TCA) less than 10 n (32.8 ft) or (AS.30 Laser) about 2.0 m (6.6 ft)
Guidance: AS.30 manual, AS.30 TCA semi-automatic to line-of-sight, and AS.30 Laser laser-homing
Warhead: 240-kg (529-lb) HE with impact or delay-action fuzing

Aérospatiale ASMP

The Aérospatiale ASMP (Air-Sol Moyenne Portée, or medium-range air-to-surface missile), is due to become France's main air-delivered nuclear weapon. It is powered by a

liquid-fuel ramjet system and will be used mainly against tactical targets such as road and railway bridges, transport depots, and command, control and communications facili-

ties. It will also have a semi-strategic role against hardened targets, and for this a total of 16 Dassault-Breguet Mirage IVA bombers of the Force de Frappe are being converted

to the Mirage IVP configuration to carry one round under the fuselage in place of the current AN22 60/70-kiloton yield free-fall nuclear bomb. The first of two squadrons to operate the Mirage IVP will commission in 1987.

For the tactical role with the ASMP the French air force is procuring 85 Dassault-Breguet Mirage 2000N two-seat low-altitude strike fighters, which from 1968 onwards will initially supplement and then replace the SEPECAT Jaguars at present assigned to this mission with single AN52 15-kiloton yield free-fall bombs. The French navy is also converting approximately 50 of its carrier based Dassault-Breguet

The large ASMP (medium-range air-to-surface missile) weighs 1000 kg (2,205 lb) at launch and can travel at speeds up to Mach 4.

Super Etendard fighters as launch platforms.

Missile guidance is of the preprogrammed inertial type with several flight profiles available. In general terms these profiles are believed to be similar to those available to the American AGM-69 SRAM. A total of 100 operational rounds is to be pro-

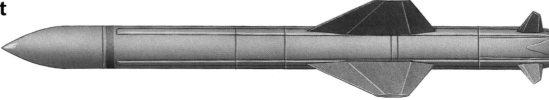

cured for the two services, and these will replace the majority of the free-fall nuclear bombs now held in stock.

Specification
ASMP
Type: tactical and/or semi-strategic air-to-surface missile

Dimensions: length 5.38 m (17 ft 7.8 in); span not known; diameter 0.96 m (3 ft 1.8 in)
Launch weight: 1000 kg (2,205 lb)
Propulsion: liquid-fuel ramjet
Performance: speed Mach 4; range 30 km (186 miles); CEP not known
Guidance: inertial
Warhead: 150-kiloton yield nuclear

Aérospatiale Exocet

Developed in the late 1960s to meet a French navy reqirement, the Aérospatiale Exocet completed its first manufacturfer's trials in mid-1972. In October of that year evaluation tests were undertaken by the French navy in conjunction with the Royal Navy and the West German Kriegsmarine. The results led to a missile improvement programme in 1973, the first production round being fired in the following year. Out of some 30 rounds fired the hit rate was 91 per cent. By early 1983 some 2,000 rounds of the various Exocet versions had been ordered by 27 customers.

The basic MM.38 round is a two-stage solid-propellant missile fitted with four cruciform wings and four tail control surfaces. The missile is stored in a rectangular box-like container-launcher. Before firing the range and bearing of the target is determined by the launch platform's own fire-control system and stored in the missile's guidance system. The missile is then launched towards the target, a low-altitude inertially-controlled profile being flown throughout the cruise phase. At around 10 km (6.1 miles) from the target's estimated location the ADAC active radar seeker is switched on and the target acquired by the missile; the seeker then locks on and the missile descends to one of

The Exocet family is the most widely used of all Western anti-ship missiles.

three pre-set sea-skimming altitudes, chosen according to the sea state and the target damage required. MM.38 operators include Argentina, Brazil, Brunei, Chile, Ecuador, France, Indonesia, Malaysia, Morocco, Nigeria, Oman, Peru, Philippines, Thailand, the UK and West Germany.

Improvements to the sustainer motor and container-launcher have resulted in the evolution of the larger MM.40 variant. This has increased range, but also allows a larger number of rounds to be carried for the same space and weight occupied by the MM.38 system. Among operators of the MM.40 are Argentina, Colombia, Ecuador, France, Morocco, Qatar, Tunisia and the United Arab Emirates. The first variant to see action, however, was the air-launched AM.39. This version evolved from a modified MM.38 (called the AM.38) that incorporated a one-second ignition delay on the motor to allow the weapon to fall clear of the launch aircraft. The first

AM.38 tests were conducted in April 1973. In 1977 the first test rounds of the shorter and lighter AM.39 were fired, with production starting in that year for the French navy and for export to Abu Dhabi, Argentina, Bahrain, Brazil, Iraq, Kuwait, Pakistan and Peru. The AM.39 was then succesfully used by Iraq in the current Gulf War from Aérospatiale Super Frelon helicopters, sinking at least three Iranian warships and damaging or destroying a number of merchant vessels and oil rig platforms. The Argentinians subsequently used the AM.39 in the Falkland Island war, together with the MM.38. The missiles fired from Dassault-Breguet Super Etendard fighters and crude shore batteries resulted in the loss of the destroyer HMS *Sheffield* and the merchant vessel *Atlantic Conveyor*, plus severe damage to the destroyer HMS *Glamorgan*. Following the success of the Super Etendard/AM.39 combination, Iraq has leased five such aircraft from the French Aéronavale

to attack Iranian oil industry targets located in the Gulf.

In the mid-1970s the French navy requested the development of an encapsulated submarine-launched version, the SM.39. In this version the missile is encased in a 5.8-m (19-ft)long capsule that weighs 1350 kg (2,976 lb) when loaded and is capable of being launched from standard torpedo tubes. Like the Sub-Harpoon the SM.39 is released from its capsule upon reaching the surface of the water, and then rapidly assumes the Exocet's sea-skimming flight profile.

Specification
AM.39
Dimensions: length 4.69 m (15 ft 4.65 in); diameter 35.0 cm (1 ft 1.75 in); span 1.004 m (3 ft 3.5 in)
Weights: total round 652 kg (1,437 lb); warhead 165 kg (364 lb)
Performance: maximum speed Mach 0.93; range 50-70 km (31-43.5 miles) depending on launch altitude

Aérospatiale SS.11

Originally developed by Nord-Aviation, the Aérospatiale SS.11 started life in 1953 as the Type 5210 and entered service with the French army in 1956. Apart from its normal ground- or vehicle-launched role, it can also be launched from a helicopter or ship. It is a manually-guided line-of-sight weapon, the operator acquiring the target by means of a telescopic sight. As soon as the missile enters his field of view after launch the operator commands it to his line of sight via a joystick control and wires, and then flies it to the target using tracking flares mounted on the rear of the missile for visual reference. From 1962 a modified S.11B1 variant was produced with transistorized firing equipment. This weapon can be fitted with a variety of warheads including the Type 140AC anti-tank, the Type 140AP02 semi-armour-piercing delay-action anti-personnel, and the Type 140AP59 anti-personnel fragmentation. Production ceased at the beginning of the 1980s after some

First developed in the 1950s, the SS.11 family has seen action in over a dozen conflicts, including the 1982 Falklands war.

179,000 rounds of the SS.11 family had been built for more than 20 countries. A modified SS.11 derivative with a much improved semi-automatic guidance system, the Harpon, was produced in some numbers for the French, West German and Saudi Arabian armies from 1967 onwards. The missile family has seen action in numerous conflicts over the years and was used most recently from British army Westland Scout helicopters against

Argentine ground positions during the recapture of the Falklands in 1982. It is also seeing regular use in the Gulf War with both the Iraqis and the Iranians. Current operators of the SS.11 are Argentina, France, India, Iran, Iraq, Italy, Tunisia, Turkey, UK, Venezuela and a number of undisclosed customers. As far as is known, the only current possessor of the Harpon type is India.

Specification

SS.11B1
Type: anti-tank missile
Dimensions: length 1.20 m (3 ft 11.25 in); diameter 16.40 cm (6.46 in); span 50.00 cm (1 ft 7.7 in)
Launch weight: 29.9 kg (65.9 lb)
Propulsion: two-stage solid-propellant rocket
Performance: range 500-3000 m (545-3,280 yards)
Warhead: see text
Armour penetration: Type 140AC 600 mm (23.62 in) and Type 140AP02 10 mm (0.4 in)

Boeing AGM-69A Short-Range Attack Missile (SRAM)

In 1964 the US Air Force began development of the Boeing AGM-

69A SRAM, for use primarily against major defensive installa-

tions deep within enemy territory whilst the launch platform remains

outside the enemy's engagement zone. The missile was also required

Boeing AGM-69A SRAM (continued)

to attack main-mission targets as well if they were suitable or if they had exceptionally heavy anti-aircraft defences. The first production round was delivered to the Strategic Air Command in 1972, the last of 1,500 being delivered three years later. Some 1,150 SRAMs currently remain in in the operational inventory. The major carrier of the type is the Boeing Stratofortress, the B-52G and B-52H models each being able to carry 20 missiles. The more usual load is six or eight SRAMs as well as four free-fall thermo-nuclear gravity bombs. The General Dynamics FB-111A can carry up to six SRAMs, but those aircraft which do carry them as part of their normal weapons load only have just two.

Four basic flight profiles can be utilized in an attack: semi-ballistic from the point of launch to the target; altimeter-controlled terrain-following; ballistic pull-up from behind screening terrain using inertially-guided flight for the terminal phase; and a combination of inertial and terrain-following. Each profile can further be enhanced by the programming into the missile's onboard guidance system of deviations in direction of up to 180°. The range depends entirely upon what launch altitude and flight profile is chosen. Once over the target the warhead can either detonate on contact to give a ground burst or at a present altitude to give an air burst nuclear explosion, the type of explo-

sion being chosen according to the target type and the damage level required. The missile's computer can be retargeted at any time up to launch. The SRAM is expected to stay in service for a number of years to come.

Specification
AGM-69A
Type: short-range air-to-surface strategic missile
Dimensions: length 4.267 m (14 ft 0 in) for internal carriage or 4.826 m (15 ft 10 in) for external carriage; span 0.762 m (2 ft 6 in); diameter 0.445 m (17.5 in)

Equipping Strategic Air Command's bomber force, the AGM-69A SRAM (Short Range Attack Missile) is mainly designed for interdiction duties deep behind enemy lines.

Launch weight: 1016 kg (2,240 lb)
Propulsion: two solid-propellant rocket motors
Performance: speed Mach 3.5; range between 56.3 and 80.5 km (35 and 50 miles) at low altitude, or between 160.9 and 221.3 km (100 and 137.5 miles) at high altitude; CEP 457 m (500 yards)
Warhead: W69 170-kiloton yield nuclear

Boeing AGM-86 Air-Launched Cruise Missile (ALCM)

The Boeing ALCM is the result of a US Air Force requirement to provide an air-launched strategic weapon for deployment on the Boeing B-52 bomber and successor designs.

The original AGM-86A was to be interchangeable with the AGM-69A SRAM on the latter's internal B-52 eight-round rotary launcher. However, because the weapon was considered to be short on range, and because of a Department of Defense's 1979 decision to hold a fly-off between the AGM-86A and the General Dynamics AGM-109 Tomahawk cruise missile, a considerably stretched version, the AGM-86B, was produced. This was some 30 per cent longer and effectively doubled the missile's range for a given warhead.

With a range in excess of 3000 km (1,850 miles), the AGM-68B gives the USAF a weapon which will enable its bombers to make strategic attacks without having to face the task of trying to penetrate modern defences.

In 1980, following a considerable delay in announcing the results, the US Air Force revealed that the AGM-86B was the chosen weapon. The first two rounds were delivered to the Strategic Air Command in 1981. The B-52G is being modified to carry 12 AGM-86B missiles on two underwing pylons whilst retaining its internal load of SRAMs and free-fall nuclear bombs. The more modern B-52H will have the same external pylon load, but the bomb bay of each aircraft is to be rebuilt to

accommodate an additional eight ALCMs on a new rotary launcher. The follow-on Rockwell B-1B will be able to contain the same internal rotary launcher and up to 14 more ALCMs will be carried on external racks. A total of 4,348 ALCMs is planned before production switches to the Advanced Cruise Missile design.

Specification
AGM-86B
Type: air-launched cruise missile

Dimensions: length 6.325 m (20 ft 9 in); span 3.658 m (12 ft 0 in); diameter 0.693 m (27.3 in)
Launch weight: 1281.4 kg (2,825 lb)
Propulsion: one turbofan
Performance: speed 805 km/h (500 mph); range 3138 km (1,950 miles); CEP between 10 and 30 m (32.8 and 98.4 ft)
Guidance: inertial with terrain contour updating
Warhead: W80-1 200-kiloton yield nuclear

British Aerospace Sea Eagle

The P3T was designed by British Aerospace to meet Air Staff Requirement 1226 issued in the early 1970s for an all-weather, night-capable, over-the-horizon range, fire-and-forget missile to replace the TV-guided AJ168 Martel missile. By 1977 the project had progressed to the definition phase with the launch trials taking place in November 1980. Full-scale development firings started in the following April, and in early 1982 a production contract was awarded for initial service deliveries in 1985. The Sea Eagle, as the P3T was named, is to arm two squadrons of maritime strike BAe Buccaneers (four missiles per aircraft), the BAe Nimrod maritime patrol force (two or four missiles per aircraft) and the Royal Navy's BAe Sea Harriers (two missiles per aircraft). In mid-1983 the Indian navy ordered the missile to equip a number of Westland Sea King Mk ß2B helicopters that it is procuring. These helicopters will carry two missiles each.

The airframe of the Sea Eagle is basically that of the Martel but fitted with an underbelly inlet for

The British Aerospace Sea Eagle has been bought by India before it has actually entered service with the Royal Navy. India will operate it from a batch of new Westland Sea King helicopters.

the Microturbo TR1-60 turbojet propulsion unit. Guidance is initially by an onboard autopilot wih the target's last known position and speed stored in its microprocessor memory; the pre-set cruise-phase flight profile and altitude are maintained by the autopilot and a Plessey radar altimeter, working through cruciform rear fins, whilst the terminal target-acquisition phase is undertaken by a highly sophisticated Marconi Space and Defence Systems active radar seeker. This guidance package allows both salvo attacks on a single target with the missiles

attacking from different directions, and the overflying of one target so that a second and more desirable target can be attacked.

A ship-launched version of Sea Eagle is also being developed to meet a requirement from the Royal Navy for an MM.38 Exocet replacement on its surface ships. Known as the Sea Eagle SL (formerly P5T), the missile is carried in a lightweight container-launcher and will be fitted with solid-propellant booster motors to bring it up to flight speed. This arrangement also allows its launch from the Westland Sea King or

Westland/Agusta EH101 helicopters. It is probable that other longer-range variants are also to be developed.

Specification
BAe Sea Eagle
Dimensions: length 4.14 m (13 ft 7 in); diameter 40.0 cm (1 ft 3.75 in); span 1.20 m (3 ft 11.75 in)
Weights: total around 550-600 kg (1,213-1,323 lb); warhead 150-200 kg (331-441 lb)
Performance: maximum speed Mach 0.9+; range 50-100 km (31-62 miles)

British Aerospace Sea Skua

Designed as the new-generation helicopter-launched anti-ship missile for use against small and agile missile-armed surface craft at ranges in excess of their missiles, the British Aerospace Sea Skua (formerly CL-834) is currently replacing the obsolete Aérospatiale AS.12 wire-guided missile in Royal Navy service. Up to four Sea Skuas are

carried by the RN's Westland Lynx, the target being illuminated by the helicopter's own frequency-agile I-band Ferranti Sea Spray surveillance and target-tracking radar to allow the semi-active homing seeker on the missile to pick up the reflected energy and so guide the weapon. The Sea Skua is treated as a round of ammunition and needs only

minimal maintenance checks. The missile uses BAJ Vickers solid-propellant boost and sustainer motors when launched for powered flights. Its cruise height is preset to one of four sea-skimming altitudes according to the sea state encountered. Once the Sea Skua has been launched, its altitude is maintained by a British Aerospace-built TRT

radio altimeter and an autopilot, working through the cruciform canard fins, until a position is reached near the target. Here a command instruction from the launch platform or from the missile's onboard guidance system orders the missile to climb to let the Marconi Space and Defence Systems semi-active radar homing head lock on.

British Aerospace Sea Skua (continued)

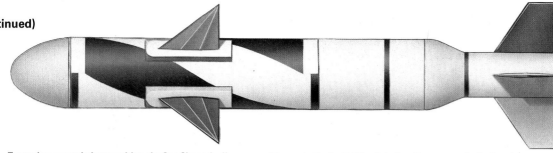

The first trials using the complete guidance system were conducted in December 1979. However, the missile was committed to combat during the 1982 Falklands war before it was officially declared operational. The missile was fired on four occasions, and scored eight hits out of eight rounds launched, sinking two Argentinian vessels and damaging two others. The semi-armour-piercing warhead proved quite effective in penetrating the target's hulls and superstructure.

A surface-launched version (for use from small craft, hovercraft and coastal-defence batteries) is also under consideration. British Aero-

Treated as a round of ammunition, the Sea Skua missile was used in combat in the Falklands before it was actually declared operational. Surprisingly, no export sales have yet been announced for this excellent missile.

space has also matched the missile to helicopters other than the Lynx, and has offered it as part of the armaments package of the BAe Coastguarder.

Specification
BAe Sea Skua
Dimensions: length 2.85 m (9 ft 4 in); diameter 22.2 cm (8.75 in); span 62.0 cm (2 ft 0.4 in)

Weights: total round 147 kg (325 lb); warhead 35 kg (77 lb)
Performance: maximum speed Mach 0.9+; range 20 km (12.4 miles)

Euromissile HOT

The long-range HOT is designed to be fired from vehicles, helicopters and fixed static positions against tanks and APCs.

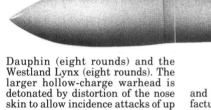

The Euromissile HOT is the heavyweight spin-stabilised tube-launched wire-guided counterpart to the MILAN for use from dug-out positions, vehicles and helicopters. Planned as the direct replacement for the SS.11, the HOT has automatic command to line of sight guidance with an IR tracking system. All the operator has to do is to keep his optical tracking sight on the target ·to ensure a hit. This guidance system allows a very rapid gathering of the missile to the line of sight after launch, thus enabling a very good short-range engagement envelope.

The helicopter types that have been fitted include the MBB PAH1 (six rounds), the Aérospatiale SA 341 and SA 342L Gazelle (four or six rounds), the Aérospatiale SA 361H

Dauphin (eight rounds) and the Westland Lynx (eight rounds). The larger hollow-charge warhead is detonated by distortion of the nose skin to allow incidence attacks of up to 65°. The warhead is said to be capable of penetrating the armour of all known battle tanks in frontal attacks. The Syrians used HOT against the Israelis in the 1982 'Peace for Galilee' war from Gazelle helicopters on up to 100 occasions, and credit the system with destroying a sizeable number of Israeli tanks

and APCs. Euromissile, the manufacturer, states that as at early 1984 14 countries have ordered 52,907 missiles.

Specification
HOT
Type: anti-tank missile
Dimensions: length 1.275 m (4 ft 2.2 in); diameter 16.5 cm (6.5 in); span 31.2 cm (12.28 in)

Launch weight: missile 23.5 kg (51.8 lb) and missile in launch tube 32 kg (70.55 lb)
Propulsion: solid-propelled booster/sustainer rocket
Performance: range 75-4250 m (82-4,650 yards)
Warhead: 6-kg (3.2-lb) hollow-charge HE
Armour penetration: 800 mm (31.5 in) or more

Hughes AGM-65 Maverick

Only 2.49 m (8 ft 2 in) long and weighing from 210 kg (463 lb), Maverick is the smallest fire-and-forget missile in the USAF's tactical inventory. It has been manufactured in a variety of guidance systems, including IIR and laser.

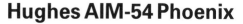

Smallest of the fully-guided launch-and-leave ASMs for the US services, the Hughes AGM-65 Maverick was originally a US Air Force programme but has now been adopted for both the US Navy and Marine Corps. The basic centroid TV homing version, the AGM-65A, entered US Air Force service in January 1972, and at least 30 were fired in combat during that year in Vietnam; another 69 were fired by Israel pilots against Arab targets during the 1973 Yom Kippur War. Although scoring 87 hits, the missile was found to be severely limited in use because of the low magnification of its TV camera, which forced pilots to close to well within the maximum launch range in poor weather just to see the target clearly enough to achieve a lock-on. To overcome this problem the AGM-65B scene-magnification version was next produced. This has the TV image magnified to twice its previous size and made clearer, thus enabling the pilot to identify the target, lock-on the missile and fire it much more quickly and at a greater slant range than that of the AGM-65A.

The follow-on version was the AGM-65C for the US Marine Corps, which was laser-guided for use in the close-support role against targets designated by ground-based or airborne laser designators. The latter can be any of the Pave Knife, Pave Penny, Pave Spike or Pave Tack systems, or even a compatible non-US designator.

This weapon was superseded in 1982 by the AGM-65E, which has a 136-kg (300-lb) penetrating blast fragmentation warhead with a three-position selectable fuze delay.

In May 1977 Hughes began development for the US Air Force of the AGM-65D with an imaging IR seeker that enables the Maverick to lock on at a range at least twice the distance otherwise possible in European areas in mist and rain or at

night. It will be the standard missile used with the LANTIRN night and adverse-weather detection system now being fitted to USAF General Dynamics F-16 Fighting Falcon fighter and Fairchild A-10 Thunderbolt II attack aircraft.

The US Navy will adopt the AGM-65F, which is essentially the same as the AGM-65D but with the warhead and fuze of the AGM-65E and modified guidance software to give maximum effect against surface ships. Weapons of the AGM-65 series are fielded by Egypt, Greece, Iran, Israel, Morocco, Saudi Arabia, Singapore, South Korea, Sweden, Switzerland, Taiwan, Turkey, US Air Force, US Navy, US Marine Corps and West Germany. Others are believed to be negotiating for the type.

Specification
AGM-65 Maverick
Type: air-to-surface missile
Dimensions: length 2.489 m (8 ft 2 in); span 0.719 m (2 ft 4.3 in); diameter 0.305 m (12 in)
Launch weight: 210 kg (463 lb) except AGM-65E/F 287.4 kg (633.6 lb)
Propulsion: solid-propellant rocket
Performance: speed subsonic; range between 0.9 and 24.2 km (0.56 and 15 miles); CEP about 1.5 m (5 ft) except AGM-65C/E which is less than this figure
Guidance: (AGM-65A/B) TV imaging, (AGM-65C/E) laser-homing and (AGM-65D/F) imaging IR
Warhead: 56.7-kg (125-lb) HE shaped charge except AGM-65E/F 136.1-kg (300-lb) HE penetrating blast fragmentation

Hughes AIM-54 Phoenix

Development of the long-range Hughes AIM-54 Phoenix, under the designation AAM-N-11, was started by Hughes Aircraft in 1960 as a follow-on from the 213-km (132-mile) range GAR-9 (later AIM-47A) Falcon derivative that had been intended to arm the defunct US Air Force's Mach 3.2 North American F-108 interceptor. The new missile was to be matched to the US Navy's

General Dynamics F-111B interceptor with the pulse-Doppler AWG-9 radar system. Such a combination flew in March 1969 and successfully engaged two targets simultaneously during a test trial. With the cancellation of the F-111B programme the Grumman F-14 Tomcat was substituted as the carrier for the radar system/missile combination. First deliveries of the AWF-9 took

place in 1970, with production of the AIM-54A beginning in 1973. In 1977 a modified missile, the AIM-54B, was placed in limited production with final deliveries being made in 1980. At the same time as production of the AIM-54B began, the development of a completely updated version, the AIM-54C, was started to meet new operational requirements and deny the Soviets

any advantage gained by the compromising of the AIM-54A in Iran. The AIM-54C has all-digital electronics, a new strapdown inertial reference system, a solid-state radar, and considerably enhanced ECCM systems. Following successful firing trials the missile entered production in 1981, first deliveries being made to the US Navy in October of that year.

The Phoenix is designated to provide a single-aircraft air-defence cover that exceeds 12,000 square miles (31000 km²) from altitudes just above sea level to 2481- m (81,400 ft) in the AIM-54A version or to 30490 m (100,000 ft) in the AIM-54C version. With six missiles carried, the Tomcat can simultaneously launch and engage up to six individual aircraft or missile targets at maximum missile range and various altitudes. Coupled with the Grumman E-2C Hawkeye AEW aircraft, the Tomcat/Phoenix combination provides the major American carrier battle groups with effective defence against the Soviet navy's aircraft and missile attack units.

Flight control is exercised by cruciform rear fins, and during its cruise phase the Phoenix adopts a sample-data SARH mode of guidance, the AWH-9 providing target illumination for each missile. At a range of some 20 km (12.5 miles) from the target, the missile switches to its own nose-mounted planar pulse-Doppler radar for the terminal phase of the flight. The missile's own radar can also be used for direct guidance in short-range engagements rather than using the aircraft's own radar and, in the latest missile version, it is said that the SARH mode of guidance can now be used throughout the engagement to defeat countermeasures. Further missile improvements are also under investigation for the 1990s.

The Hughes Phoenix is the most capable air-to-air missile available to the West. It is employed solely on the Grumman F-14 Tomcat, whose AWG-9 radar is specially integrated with Phoenix operations.

Specification
AIM-54
Dimensions: length 4.01 m (13 ft 1.8 in); span 0.925 m (3 ft 0.4 in); body diameter 0.38 m (1 ft 3 in)
Launch weight: 447 kg (985 lb)
Performance: speed AIM-54A Mach 4.3 and AIM-54C Mach 5; range AIM-54A 3.9-136 km (2.4-85 miles) and AIM-4C 3.5-148 km (2.2-92 miles)
Warhead: 60-kg (132-lb) HE fragmentation with delay action, impact and active-radar proximity fuze

Hughes AIM-120A Advanced Medium Range Air-to-Air Missile (AMRAAM)

The Hughes AIM-120A AMRAAM is the American part of a NATO development and programme for the new generation of air-to-air missiles. Designed as an all-weather weapon to attack targets from any approach angle and at both visual and beyond-visual engagement ranges, the AIM-120A is similar to the Sparrow and the British Sky Flash in using a small explosive charge to generate sufficient force to eject it downwards and clear of the aircraft during launch. The carrier itself is fitted with a search radar with track-while-scan facilities that utilizes a radio data link to keep the missile's inertial reference system updated with the target's position, thereby allowing it to reach a point where the missile's own active terminal radar seeker can take over the final interception phase. If the target attempts to jam the mid-course guidance link, the missile can either switch to a home-on-jam mode or change repeatedly between this mode and the active radar seeker as a counter ECM tactic. Multiple launches against either a single target or up to eight individual targets will be possible from a launch platform. Flight control is achieved by moveable tail-mounted fins, whilst the mid-fuselage fins are fixed.

The US armed forces require a total of some 24,000 rounds (split in the approximate ratio of two to one for the Air Force over the Navy and Marine Corps), with the first to enter service during Fiscal Year 1986 on the US Air Force's General Dynamics F-16C/D Fighting Falcon multi-role fighter. A European consortium, led by BAe Dynamics, will manufacture the weapon for the NATO air forces in Europe. In the United Kingdom it is planned initially to arm the Royal Navy's updated BAe FSR.Mk 2 Sea Harrier, with the Royal Air Force's air defence Panavia F.Mk 2 Tornado fighter to follow. The West German Luftwaffe will also arm its McDonnell Douglas F-4F Phantom fleet with the missile. Although the AIM-120A is due to enter series production in America this year, it is believed that the programme is running into problems with the US Congress over costs.

Specification
Length: 3.65 m (143.7 in)
Diameter: 0.178 m (7 in)
Wing span: 0.526 m (20.7 in)
Launch weight: 148 kg (325.6 lb)
Performance: range 75 km (46.875 miles); speed Mach 4+
Warhead: 22 kg (48.4 lb) HE-fragmentation

Hughes BGM-71 TOW

The Hughes BGM-71 TOW (Tube-launched Optically-tracked Wire-guided) heavy anti-tank missile for helicopter- or vehicle-launch application entered the design phase in 1962, the first guided firings taking place in 1968. Two years later TOW entered service, and by the summer of 1972 had seen its first combat firings when it was used against North Vietnamese tank units. During the 1973 Yom Kippur War the Israelis had it delivered as part of the arms lift by the USA, and by 1984 the missile had been used in many conflicts all over the world. On the strength of its operational success the TOW has become the West's most numerous ATGW with over 350,000 units so far produced for more than 25 countries. As in most contemporary systems, all the operator has to do is keep the cross hairs of his optical sight on the target, an IR sensor tracking the signal from the missile to permit the calculation of correction commands which are automatically sent via the guidance wire link. In order to improve the lethality of the warhead of infantry units' TOW missiles, a two-stage upgrade programme was adopted: the first phase was marked by the procurement of a warhead of 127-mm (5-in) diameter fitted with a telescopic nose probe fuze that pops out when the missile is in flight to

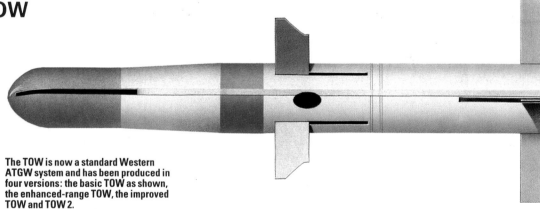

The TOW is now a standard Western ATGW system and has been produced in four versions: the basic TOW as shown, the enhanced-range TOW, the improved TOW and TOW 2.

give an optimum stand-off penetration capability, the missile fitted with this warhead being known as the Improved TOW; the second stage saw the introduction of the TOW 2 missile, which has a 152-mm (6-in) diameter warhead fitted with telescopic nose probe, improved digital guidance and a new propulsion system. All these improvements are also being retrofitted to the helicopter and armoured vehicle TOW launcher systems. The TOW version sometimes described as the Extended-Range TOW is the basic TOW variant that entered production from 1976 onwards with increased range capabilities. The TOW missile is very widely used, typical operators being Bahrain, Canada, Denmark, Egypt, Finland, Greece, Iran, Israel, Italy, Japan, Jordan, Kenya, Kuwait, Lebanon, Luxembourg, Morocco, the Netherlands, North Yemen, Norway, Oman, Pakistan, Portugal, Saudi Arabia, South Korea, Taiwan, Thailand, Turkey, United Arab Emirates, UK, USA and West Germany.

Specification
TOW
Type: anti-tank missile
Dimensions: length 1.174 m (3 ft 10.2 in) for basic model, 1.555 m (5 ft 1.2 in) for Improved TOW with probe, and 1.714 m (5 ft 7.5 in) for TOW 2 with probe; diameter 15.2 cm (6 in); span 34.3 cm (13.5 in)
Launch weight: 22.5 kg (49.6 lb) fo basic model, 25.7 kg (56.65 lb) for Improved TOW, and 28.1 kg (61.95 lb) for TOW 2
Propulsion: two-stage solid-propellant rocket
Performance: range 65-3000 m (70-3,280 yards) for pre-1976 models, and 65-3750 m (70-4,100 yards) for post-1976 models
Warhead: 3.9-kg (8.6-lb) shaped-charge HE for basic and Improved TOW models, and 5.9-kg (13-lb) shaped-charge HE for TOW 2 model
Armour penetration: 600 mm (23.62 in) for basic model, 700 mm (27.56 in) for Improved TOW, and 800 mm (31.5 in) or more for TOW 2

International Advanced Short Range Air-to-air Missile (ASRAAM)

In contrast to the fast pace of the American AIM-120A AMRAAM project, the European part of the new air-to-air missile generation, the Advanced Short Range Air-to-air Missile (ASRAAM), has only just completed its feasibility study stage. Being developed by the BBG company (formed recently by BAe Dynamics and the West German firm Bodenseewerke Geratetechnik), the new missile is now in the definition phase, with service entry due sometime in the 1990s to replace the heavier Sidewinder as the standard NATO dogfight missile. It is projected to have a wingless configuration and to be around 2.5 m (98.4 in) long and 0.15 m (5.9 in) in diameter. It will also have greatly improved flight envelope and manoeuvrability characteristics in comparison with the latter Sidewinder models, and will introduce a new guidance package into the short-range missile field. This will involve the fitting of an inertial reference unit for the initial stages of the flight and an advanced infra-red seeker for the terminal phase. This mixture considerably reduces the missile's vulnerability to IR jamming or flare-dispensing decoy systems. The ability to lock on to a target after launch will also be aided by the introduction into service of the helmet-mounted target acquisition sighting systems that are currently under study.

Each ASRAAM produced will be attached to a re-usable Missile Support Unit (MSU) that is essentially a shallow rail running along its top; the upper surface of the rail is fitted with attachment points that are compatible with all Sidewinder launcher installations. A fairing at the aft end of the rail houses the interface units required to mate the missile to the aircraft systems. The MSU and missile are both powered by the aircraft's own electrical system through these connections. As part of the design work BBG are studying other potential applications for the weapon; these include use as a future point defence surface-to-air missile and a self-defence anti-radiation defence suppression missile. Due to the projected high kill probability, which is a consequence of the ASRAAM's inherent direct-hit ability, a modern minimum-sized HE fragmentation warhead is believed to be under development together with a relatively simple fuzing system. However, because of the long timescale involved it is possible that America might well decide to withdraw from the overall NATO programme and go it alone to produce her own short range air-to-air missile, based on new technology and improved versions of the well-proven Sidewinder components.

Matra R530

Development of the Matra R530 began in 1957, and the missile is still in service today after having been produced in both IR and semi-active radar homing (SARH) variants. The R530 is normally carried on Dassault-Breguet Mirage III and F.1 interceptors, although the French navy uses the type on its Vought F-8E(FN) Crusaders. Approximately 4,400 were produced, and the system has been sold to 14 countries. It has seen combat with the Argentine, Iraqi, Israeli and Pakistani air forces, but its success rate is not believed to have been high. In the IR type the homing contains the SAT Type AD3501 all-aspect (including head-on) seeker, whilst the alternate SARH version uses an EMD Type AD26 seeker guided by the Cyrano series radars of the Mirage family. A slightly different SARH seeker is used on the French navy

The Matra R530 is used with the Mirage family, although the later members use the Super R530.

missiles, which have to be compatible with the Crusader's radar. Normally one of each missile type is carried to improve the hit probability, although it is possible for the head to be exchanged at squadron level to meet the operational requirements. Propulsion is by either a Hotchkiss Brandt/SNPE Antoinette or a higher-performance SNPE Madelain dual-thrust solid-propellant rocketmotor, and cruciform delta wings and tail fins are used for flight control. The R530 is currently used by Argentina, Australia, Brazil, Colombia, Egypt, France (air force and navy), Iraq, Jordan, Pakistan, South Africa, Spain, Venezuela and one other. Israel and Lebanon no longer operate the R530.

Specification
R530 (SARH)
Dimensions: length 3.284 m (10 ft 9.3 in); span 1.103 m (3 ft 7.4 in); body diameter 0.263 m (10.35 m)
Launch weight: 192 kg (423.3 lb)
Performance: speed Mach 2.7; range 18 km (11.2 miles)
Warhead: 27-kg (59.5-lb) continuous-rod or pre-fragmented HE with impact, delay and proximity fuzes

Specification
R530 (IR)
Dimensions: length 3.198 m (10 ft 5.9 in); span 1.103 m (3 ft 7.4 in); body diameter 0.263 m (10.35 in)
Launch weight: 193.5 kg (426.6 lb)
Performance: speed Mach 2.7; range 18 km (11.2 miles)
Warhead: 27-kg (59.5-lb) continuous-rod or pre-fragmented HE with impact, delay and proximity fuzes

Matra Super R530

Essentially a development of the Matra R530, the Matra Super R530 started life in 1971 to meet the higher speed- and altitude-performance requirements of the new generation of French interceptors The Super R530F entered service in 1980 and uses an EMD Super AD26 semi-active radar homing head with Matra autopilot and proportional navigation system matched to the Cyrano IV radar of the Dassault-Breguet Mirage F.1. It can be used to engage targets either above or below the launch platform by vertical separations of up to 7000 m (22,965 ft) or 9000 m (29,530 ft) in the latest version. The latter is an improved model, the Super R530D, with a modified seeker head matched to the RDI/RDM radars carried by

the Dassault-Breguet Mirage 2000 family. Service entry date for the Super R530D is expected to be 1986. To power the missiles Thomson-Brandt developed the Angèle rocket motor with a dual-composition solid propellant that has a much higher specific impulse capability than previous motors produced by the company. Flight control is exercised by four cruciform tail fins and four very low aspect body wings fitted midway along the missile's length. The missile body is constructed from steel and steel honeycomb and this, coupled with the use of a ceramic radome housing, allows the Super R530 to withstand speeds of up to Mach 4.6 and manoeuvres of up to 6 g at an altitude of 25000 m (82,020 ft); at altitudes up to 17000 m (55,775 ft) it can pull up to 20 g during an interception. At least 10 countries have now placed orders for the Super R530F, these being France, Kuwait, Libya, Morocco, Iraq and five unspecified countries.

Specification
R530F

The Super R530 differs greatly from the R530 externally, but owes a great deal to its predecessor inside.

Dimensions: length 3.54 m (11 ft 7.4 in); span 0.9 m (2 ft 11.4 in); body diameter 0.263 m (10.35 in)
Launch weight: 250 kg (551 lb)
Performance: speed Mach 4.6: range R530F 35 km (21.75 miles), and R530D 50-60 km (31-37.3 miles)
Warhead: HE fragmentation with radar proximity fuze and weighing more than 30 kg (66 lb).

Matra Super R550 Magic

Developed by Matra as a direct competitor to the US AIM-9 Sidewinder, the Matra R550 Magic began development as a company project in 1968, receiving official sanction via a French air ministry contract in the following year. The first fully guided round was launched on 11 January 1972 from a Gloster Meteor during a test launch against a CT-20 drone target. Further trials were then undertaken against manoeuvring targets until the missile was accepted into service in 1975. Since then both the French navy and air force have accepted the missile into their inventories, and more than 7,000 missiles have been ordered for them and by 18 other countries. The Magic has been cleared for use on the IAI Dagger, the Dassault-

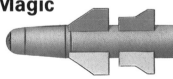

The French Sidewinder equivalent is found in the inventories of most Mirage operators.

Breguet Mirage III, 5 and F.1, the Vought F-8 Crusader, the SEPE-CAT Jaguar, the BAe Sea Harrier, the Mikoyan-Gurevich MiG-21, the Dassault-Breguet Super Etendard, the Macchi M.B.326K, and the Dassault-Breguet/Dornier Alpha Jet.

In mid-1983 it was revealed that a **Magic Mk 2** was under development, with production to start in 1984. The Magic Mk 2 employs a new solid-propellant rocket motor, a new and improved IR homing head that includes head-on engagement capability with seeker slaving, and a new electro-magnetic proximity fuze.

The basic Magic Mk 1 can be launched from almost any target aspect (except head on) within a 140° engagement envelope at all altitudes up to 1800 m (59055 ft). Above this height certain launch limitations apply. The missile can engage targets in a dogfight situation at around 300-m (985-ft) range and can be fired from an aircraft flying at over 1300 km/h (808 mph) and in 6-g manoeuvre. The IR guidance used is a SAT Type AD3601 homing head with a liquid nitrogen-cooled lead sulphide seeker element. Propulsion is by an SNPE Roméo single-stage composite double-base rocket motor, and cruciform canard fins are used for flight control. The Magic Mk 1 was used by Argentina in the Falklands war on her Mirage

IIIEA interceptors without any success, but Iraq has scored a number of kills with it on her Mirage F.1E and MiG-21 fighters in the Gulf War against Iran. The Magic is currently used by Abu Dhabi, Argentina,

Chile, Ecuador, Egypt, France (air force and navy), Greece, India (air force and navy), Iraq, Jordan, Kuwait, Libya, Morocco, Oman, Pakistan, Peru, Saudi Arabia, Spain and South Africa.

Specification
Magic Mark 1
Dimensions: length 2.75 m (9 ft 0.27 in); span 0.66 m (2 ft 2 in); body diameter 0.157 m (6.18 in)
Launch weight: 78,8 kg (198 lb)
Performance: speed Mach 3.0;

range 0.32-10 km (0.2-6.25 miles)
Warhead: 12.5-kg (27.56-lb) conventional rod/fragmentation type, of which 6 kg (13.2 lb) is HE with delay action impact and passive IR proximity fuzes

MBB Kormoran

Designed to meet a 1964 West German naval air arm (Marineflieger) requirement, the MBB Kormoran was originally based on a French Nord design, the AS.34, using the inertial guidance system from the defunct AS.33 project. However, following the creation of the German MBB aerospace consortium, the missile became a major project with help from the French concern Aérospatiale. The weapon was given a new and more sophisticated guidance package, and the first flight trials were undertaken on 19 March 1970, the first production rounds being delivered in December 1977. By the middle of the following year MFG 2, equipped with Lockheed F-104G Starfighters, was fully operational with the missiles at Eggbeck. Normally two missiles are carried under the wings of the F-104G, whilst the newly-introduced Panavia Tornados of the Marineflieger can carry four, although a maximum of eight is possible. The Kormoran is

also operated by the Italian air force.

After release from the launch aircraft, two double-propellant SNPE Prades boost motors burn for about one second, then the main SNPE Eole IV sustainer motor cuts in to provide thrust for a further 100 seconds of powered flight. For the initial cruise phase of the mission a Stena/Bodenseewerk inertial guidance platform coupled with a modified TRT radio altimeter is used to hold the missile on course (using the cruciform rear fins) at a height of about 30 m (100 ft). Near the estimated target location the missile is commanded to descend to its wavetop attack altitude by the auto-pilot, and the Thomson-CSF two-axis radar, acting in either a pre-set

active or passive mode, searches for, acquires and then locks on to the enemy vessel. The missile then strikes the target just above the waterline, the warhead (with 56 kg/123.5 lb) of explosive, delay-action fuze and 16 radially-mounted charges) explodes deep within the hull to maximize the damage caused.

The development of a Kormoran Mk 2 version for the Marineflieger has begun, and this is expected to have an improved radar seeker with enhanced ECM resistance, a longer range and a heavier and more destructive warhead.

Designed in the late 1960s, the first production rounds of the Kormoran were not delivered until 1977. A Mk 2 version is already under development for the West German navy to arm its Tornado strike aircraft during the 1990s.

Specification
Kormoran
Dimensions: length 4.40 m (14 ft 5.25 in); diameter 34.4 cm (1 ft 1.5 in); span 1.00 m (3 ft 3,4 in)
Weights: total round 600 kg (1,323 lb); warhead 165 kg (364 lb)
Performance: maximum speed Mach 0.95; range 37 km (23 miles)

McDonnell Douglas AGM-84A/RGM-84A Harpoon

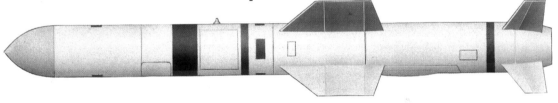

The Harpoon in its many forms will become the mainstay of the US Navy's anti-ship missile inventory until the end of the century.

In 1967, following the sinking of the Israeli destroyer *Eilat* by an SS-N-2 'Styx', the US Navy began to show serious interest in developing its own anti-ship missile. The result was a formal proposal that led to the McDonnell Douglas Harpoon, but an interim alternative was also sought. This became the Fireflash SSM, based on the BQM-34A Firebee target and reconnaissance drone. By late 1971, however, the Fireflash was dropped as it was rapidly becoming a serious competitor to Harpoon rather than just an interim weapon. In the meantime Harpoon had begun life in the Fiscal Year 1968 programme as a 92.5-km (57.5-mile) range anti-ship missile (designated AGM-84A) suitable for air launch. By 1970 Harpoon's capabilities had been extended to ship-launch applications (as the RGM-84A), and in January 1971 McDonnell Douglas was selected as the prime contractor. The final variant, the encapsulated torpedo tube-launched Sub-Harpoon version for submerged launch from submarines, was started in 1972 and subsequently replaced the Submarine Tactical Missiles (STAM) programme that was cancelled in the following year.

The Harpoon development programme always emphasized simplicity and low technical risk. The missile, boosted by a solid-propellant rocket and sustained by a Teledyne CAE J402 turbojet, is usually fired in a pre-set Range and Bearing Launch (RBL) mode, turning on its Texas Instruments two-axis active-radar terminal only at the last

moment to acquire the target without giving it time to instigate evasive measures. The frequency-agile radar can be set for large, medium or small acquisition windows that determine the range from the target at which the radar is activated. The smaller the window the more precise the initial target data must be, and the less the chance that the missile will be defeated by defensive ECM techniques in its terminal flight phase. Initial guidance in the flight is undertaken either by Lear Siegler or a Northrop three-axis strapdown altitude reference system with a Honeywell radar altimeter, working through cruciform rear fins. The alternative launch technique is the Bearing Only Launch (BOL) mode, in which the missile is fired on the target bearing and the radar is activated early in the flight, scanning 45° to each side of the missile's bearing to search for a target. If no target is acquired after a suitable time on the initial bearing, the missile switches to a pre-set search pattern. In either launch mode, once the target is detected and the seeker achieves a lock-on, the missile climbs rapidly in a pop-up man-

oeuvre and dives onto it. The newer Block IB and Block IC missiles now being built have a range increase of 15 per cent and a sea-skimming terminal attack profile. The latter was first adopted for the Royal Navy's Sub-Harpoon missiles. A Block II missile with a greater than 190-km (118-mile) range, variable flight profiles and improved ECM resistance is expected in the late 1980s for service in the 1990s.

It is reckoned that one Harpoon will destroy an 'Osa', 'Komar', 'Matka' or 'Nanuchka' class missile boat; two will disable a frigate; four will knock out a missile cruiser; and five will destroy a 'Kirov' class nuclear-powered battlecruiser or a 'Kiev' class carrier. The warhead is a 227-kg (500-lb) penetration-blast type fitted with time-delay contact and proximity fuses. On surface warships the Harpoon is either carried in its own cyclindrical container-launcher or carried in the missile magazine of a Tartar/Standard SM-1 launcher. On some frigates two of the boxes in the ASROC ASW missile-launcher have been converted to fire the Harpoon. Aircraft normally carry two missiles

under their wings. A new vertical-launch system for ships, capable of firing the Harpoon among other missiles, is currently under development for newer generations of USN vessels.

The Harpoon series is in widespread service with the USA and its allies, operators of the shipborne version including Australia, Denmark, Greece, Iran, Israel, Japan, the Netherlands, Saudi Arabia and Turkey. The USA deploys all three versions, and other operators of the air-launched model are Japan and the UK, while Sub-Harpoon is additionally found in the inventories of Australia and the UK.

Specification
McDonnell Douglas Harpoon
Dimensions: length 4.58 m (15 ft 0.75 in) for RGM-84A and Sub-Harpoon, and 3.84 m (12 ft 7 in) for AGM-84A; diameter 34.3 cm (1 ft 1.5 in); span 91.4 cm (3 ft 0 in)
Weights: total round 667 kg (1,470 lb) for RGM-84A and Sub-Harpoon, and 522 kg (1,150 lb) for AGM-84A; warhead 227 kg (500 lb)
Performance: maximum speed Mach 0.85; range 120 km (75 miles)

Rafael Shafrir and Python

Development of the Raphael Shafrir Mk 1 infra-red homing missile began in 1961. By 1965 the Shafrir had surpassed its main rival, the AIM-9

Sidewinder, in most respects. This was due mainly to the greater body diameter, which allowed significant design improvements to be made to

the internal installations that increased missile lethality. The Shafrir Mk 1 became the initial pre-production version, whilst the defi-

nitive operational model, the Shafrir Mk 2, entered squadron service in 1969. Since then the Shafrir has seen extensive use in combat against the Arab air forces and has been used on the Israeli air force Dassault-Breguet Mirage IICJs, IAI Neshers and IAI Kfirs. To date it has destroyed over 200 aircraft and has a claimed single-shot kill probability of around 65-70 per cent. The Shafrir is a 'see-and-shoot' missile; when a target is detected within firing range, audio and visual signals alert the pilot, who then fires the round. Once this happens the IR guidance system takes over and the missile is then independent. During the 1982 Falklands war the Argentines used Shafrirs on their IAI Daggers in the initial air battles of 1 May, but hit nothing as the missiles were launched at long range and outside their engagement envelope. Other Shafrir operators are Chile, Colombia, South Africa, Taiwan, Turkey and three others.

Rafael Python 3, as found on most Israeli fighters, especially the IAI Kfir.

In the same year the Israelis used against the Syrian air force what was thought to be a Shafrir Mk 3 version in their invasion of Lebanon, but this subsequently turned out to be the development version of the Python air-to-air missile, which had first been revealed at the 1981 Paris air show. The Python has a more sophisticated and sensitive wider-look IR guidance seeker than the Shafrir, and is capable of being used from all engagement aspects (including head-on) at much longer ranges. Operational deployment of the definitive Python production version was expected in late 1983 or early 1984. Both missiles use moving canard surfaces for flight control purposes.

Specification
Shafrir Mk 2
Dimensions: length 2.47 m (8 ft 1.24 in); span 0.52 cm (1 ft 8.47in); body diameter 0.16 m (6.3 in)
Launch weight: 95 kg (205 lb)
Performance: speed Mach 2.5;

range 0.5-5 km (0.31-3.1 miles)
Warhead: 11-kg (24.25-lb) pre-fragmented HE with contact, delay and proximity fuses

Specification
Python
Dimensions: length about 3.0 m (9 ft 10.1 in); span not available
Launch weight: 120 kg (264.6 lb)
Performance: speed Mach 3; range 0.5-15 km (0.31-9.3 miles)
Warhead: improved version of type fitted on Shafrir

Raytheon AIM-7 Sparrow series

The Sparrow I radar beam-riding air-to-air missile began life in 1946 as Sperry Gyroscope's Project Hot Shot for the US Navy. By 1951 full engineering development had begun and in 1956 the weapon had entered service. In 1955 Douglas started limited development of a Sparrow II with active radar-homing, but this was cancelled in the following year. In the meantime Raytheon started development of the Sparrow III variant which used essentially the same missile body as the Sparrow II but with a semi-active radar homing system. The Sparrow III entered service with the US Navy in 1958 and has been progressively modified ever since. In 1960 the US Air Force also adopted the SARH Sparrow to arm its new McDonnell F-110 (later F-4) Phantom interceptor. The main production model, the AIM-7E (over 25,000 built), was followed into service by a redesigned version, the AIM-7F, in 1977 for use on the McDonnell Douglas F-15 Eagle and F-18 Hornet fighters. In comparison with the AIM-7E the new missile has a considerably enhanced performance envelope, a more powerful rocket motor and a smaller all solid-state guidance system with conical-scan seeker head. To improve its own AIM-7E missiles' performance the UK in 1969 embarked on the development of an I-band monopulse seeker head with markedly improved ECCM features in compari-

The Sparrow family has been the West's premier SARH air-to-air missile since the early 1960s.

son with the original American seeker. The new missile was put together by British Aerospace as the Sky Flash, and entered service with the Royal Air Force in 1978 initially for the McDonnell Douglas Phantom FG.Mk 1 and FGR.Mk 2 and then the Panavia Tornado F.Mk 2 when the latter comes into service. In 1982 the US manufacturer of the Sparrow (Raytheon and GD Pomona) switched to building the latest version, the AIM-7M, which is an AIM-7F with a new seeker head comparable in operation and performance with the type on the Sky Flash.

The Sparrow saw considerable combat use in Vietnam with the US Air Force and US Navy, and also with Iran against Iraq during the current Gulf War and with Israel against the Arabs in the post-1967 conflicts. The results in each were not exactly inspiring, and the Israelis still maintain that they prefer to use short-range missiles and cannon rather than SARH missiles. The Sparrow, in all its versions, is used by Canada, Greece, Iran, Israel, Italy, Japan, Saudi Arabia,

South Korea, Spain, Turkey, UK, USA (air force, navy and marine corps) and West Germany. The British Sky Flash derivative is operated by Sweden (Rb71) and UK.

Specification
AIM-7E
Dimensions: length 3.66 m (12 ft 0 in); span 1.02 m (3 ft 4in); body diameter 0.203 m (8 in)
Launch weight: 205 kg (451.9 lb)
Performance: speed Mach 4.0; range 30-37 km (18.6-23 miles) depending on the launch platform's speed
Warhead: 29.5-kg (65-lb) HE fragmentation with contact, delay action and proximity fuses

Specification
AIM-7F
Dimensions: length 3.66 m (12 ft 0 in); span 1.02 m (3 ft 4in); body diameter 0.203 m (8 in)
Launch weight: 228 kg (502.6 lb)
Performance: speed Mach 4.0; range 44-70 km (27.3-43.5 miles) depending on the launch platform's radar capability and speed

Warhead: 39-kg (86-lb) HE fragmentation with contact, delay action and proximity fuzes

Specification
AIM-7M
Dimensions: length 3.68 m (12 ft 0.88 in); span 1.02 m (3 ft 4in); body diameter 0.203 m (8 in)
Launch weight: 228 kg (502.6 lb)
Performance: speed Mach 4.0; range 44-70 km (27.3-43.5 miles) depending on the launch platform's radar capability and speed
Warhead: 39-kg (86-lb) HE fragmentation with contact, delay action and proximity fuses

Specification
Sky Flash
Dimensions: length 3.68 m (12 ft 0.88 in); span 1.02 m (3 ft 4in); body diameter 0.203 m (8 in)
Launch weight: 193 kg (425.5 lb)
Performance: speed Mach 4.0; range 30-38 km (18.3-23 miles) depending on the launch platform's speed
Warhead: ??.5-kg (??-lb) HE fragmentation with contact, delay action and proximity fuzes

Raytheon AIM-9 Sidewinder series

Originally developed by the US Naval Weapons Center, China Lake, in the late 1940s, the AIM-9 Sidewinder is still in production in 1984 as a viable short-range air-to-air missile. The first IR-guided round was launched on 11 September 1953, production rounds entering US service in May 1956. The original versions were restricted in use to close-range rear attacks at medium to high altitudes in good visibility with a claimed 70 per cent single-shot kill probability. The early Sidewinders proved this claim in action only two years later, when in October 1958 the Chinese Nationalist air force fired a number of rounds from North American F-86 Sabres against Communist Chinese Mikoyan-Gurevich MiG-15s and MiG-17s during the brief conflict over the islands of Quemoy and Matsu. By 1962 the first major pro-

duction variant, the AIM-9B Sidewinder, was in widespread service with NATO air forces, and further improvements were planned. The US Navy for a number of years used the solitary SARH variant, the AIM-9C, on its Vought F-8 Crusaders coupled to the fighter's APQ-94 series radar. A follow-on IR guided version, the AIM-9D, proved so successful in US Navy service that it formed the starting point for most of the subsequent modifications to give both air-and land-launched variants. New solid-state electronics coupled with expanded seeker head acquisition capabilities allowed large numbers of the earlier versions to be remanufactured as missiles with vastly improved performances, whilst entirely new versions were also developed. For instance, the AIM-9L that entered production in 1977 adopted an entirely new all-

aspects IR seeker together with an improved warhead, fuze and rocket motor. Each new version incorporates technology that has been learnt from the lessons gained in combat use with Iran, Israel, Pakistan, the USA, Taiwan and the United Kingdom. It was the AIM-9L on Royal Navy British Aerospace Sea Harriers during the Falklands war of 1982 that allowed these aircraft to destroy 19 confirmed Argentine aircraft. All versions produced use cruciform canard fins for flight control and have solid-propellant rocket motors.

Currently about to enter service is the AIM-9M, a companion to the AIM-9L with the same dimensions and performance, but weighing in at 86 kg (189.6 lb) thanks to the use of an improved seeker with better IR counter-countermeasures and lock-on capability against strong IR

backgrounds. Other modern Sidewinders are rebuilt older models, the AIM-9N being the AIM-9B/E upgraded to AIM-9J standard, and the AIM-9P comprising AIM-9B/E/J and new-build missiles to the approximate standard of the AIM-9J/L series with smokeless propellant.

Operators of the Sidewinder series are Argentina, Australia, Belgium, Brazil, Canada, Chile, Denmark, Egypt, France, Greece, Indonesia, Iran, Israel, Italy, Japan, Jordan, Kenya, Kuwait, Malaysia, Mexico, Morocco, Netherlands, North Yemen, Norway, Pakistan, Philippines, Portugal, Saudi Arabia, Singapore, South Korea, Spain, Sudan, Sweden, Switzerland, Taiwan, Thailand, Tunisia, Turkey, UK (air force and navy), USA (air force, navy and marines), Vietnam and West Germany.

Specification
AIM-9B
Dimensions: length 2.83 m (9 ft 3.4 in); span 0.559 m (1 ft 10in); body diameter 0.127 m (5 in)
Launch weight: 70.4 kg (155.2 lb)
Performance: speed Mach 2.5; range 3.2 km (2 miles)
Warhead: 4.54-kg (10-lb) HE fragmentation with passive IR fuze

Specification
AIM-9J
Dimensions: length 3.07 m (10 ft 0.87 in); span 0.559 m (1 ft 10in); body diameter 0.127 m (5 in)
Launch weight: 78 kg (172 lb)
Performance: speed Mach 2.5; range 14.5 km (9 miles)
Warhead: 10.2-kg (22.5-lb) HE

fragmentation with passive IR or active radar fuze

Specification
AIM-9L
Dimensions: length 2.85 m (9 ft 4.2 in); span 0.63 m (2 ft 0.8 in); body diameter 0.127 m (5 in)
Launch weight: 83.5 kg (188 lb)
Performance: speed Mach 2.5; range 17.7 km (11 miles)
Warhead: 10.2-kg (22.5-lb) HE fragmentation with passive IR or active radar fuze

The AIM-9L is the latest major version of this important missile, and was used to great effect by the Royal Navy in the Falklands. These missiles are being loaded on to a Sea Harrier.

Rockwell AGM-114A Hellfire

Although officially described as America's next-generation anti-armour weapon, the AGM-114A Hellfire can actually be used against all types of land hard targets. The missile began life in the late 1960s as a US Army design concept that progressed to early engineering development flights in 1972. The US Army then chose Rockwell to undertake the full engineering development phase from 1976 onwards, with actual operational testing of the missile in 1980-1. The missile is now in full production, with Martin Marietta as the second-source supplier. Initial operational capability is due in Fiscal Year 1985 as part of the armament of the US Army's new

Hughes AH-64A Apache attack helicopter; up to 16 Hellfires will be carried on four wing hardpoints. The US Marine Corps will also to retrofit their current fleet of Bell AH-1T SeaCobra attack helicopters with the missile, and will buy new build Cobras already configured for the system. Other airborne launch platforms have been investigated, including the Fairchild A-10A Thunderbolt II close-support aircraft of the United States Air Force.

The seeker fitted to Hellfire is of the semi-active laser-homing type, and as such is not limited to the direct line-of-sight attack of other airborne launched ATGW; it can be fired without seeker lock-on to fly

over an obstacle, and will then commence a search for any reflected signals from a ground-based laser designator illuminating the target. The seeker will then automatically lock on and guide the missile via the onboard microprocessor logic system and canard flight controls to impact. For direct fire the launch platform has its own laser designator system. Using both launch modes, the Hellfire has successfully been fired against mobile manoeuvring targets simulating armoured vehicles of various types.

Other guidance modes are planned for the missile; these are believed to include an air defence suppression seeker for use against

Soviet battlefield SAMs and self-propelled AA guns equipped with radars, and an imaging infra-red seeker for use against camouflaged targets. The new guidance modes will also allow the Hellfire to retain the ability to engage moving targets.

Specification
AIM-9B
Dimensions: length 1.626 m (64 in)
Diameter: 0.177 m (7 in)
Wing span: 0.33 m (13 in)
Launch weight: 43 kg (94.6 lb)
Performance: range 0.3-5.5 km (0.1875-3.4375 miles); speed Mach 1.17; CEP 1.479 m (5 ft)
Warhead: 9 kg (19.8 lb) HE hollow-charge

Saab-Bofors Rb05A

The Saab-Bofors Rb05A is a simple, manually-controlled radio-command weapon for carriage on a wide variety of launch platforms. It is intended mainly for use against land and sea targets, but it may also in certain circumstances be used in an air-to-air role against such targets as a hovering helicopter. The airframe consists of a pointed cylindrical body with long-chord cruciform wings and aft-mounted cruciform control surfaces. A liquid-propellant rocket motor is centrally located, and the electrical preheating of the round is undertaken by the carrier platform.

Once the missile has been launched from a height of 20-50 m (65-

Armed with an HE blast fragmentation warhead, the Mach 1+ Rb05A missile may be used in a limited air-to-air role as well as in its primary function as a tactical air-to-ground weapon.

165 ft), the aircraft climbs to around 300-400 m (985-1,310 ft) and the pilot manually guides the weapon by lining it up on the target with the visual aid of rear-mounted tracking flares. Any control signals required are passed to the Rb05A via the radio link to the missile's onboard receiver. Once the missile is in the target's vicinity, a proximity fuze detonates the HE blast-fragmentation warhead.

The Rb05A is used by the Swedish air force's Saab AJ37 Viggen attack and Saab 105 light strike and trainer aircraft. Production started in the early 1970s and ceased in 1977. A more sophisticated version, the Rb05B with electro-optical TV homing, was to have been built, but this was cancelled when the Swedes bought the Hughes AGM-65A Maverick in its place.

Specification

Rb05A
Type: air-to-surface missile
Dimensions: length 3.60 m (11 ft 9.7 in); span 0.80 m (2 ft 7.5 in); diameter 0.30 m (11.8 in)
Launch weight: 305 kg (672.4 lb)
Propulsion: liquid-propellant rocket
Performance: speed Mach 1+; range 9 km (5.6 miles); CEP less than 10 m (32.8 ft)
Guidance: manual radio command
Warhead: HE blast fragmentation

Saab-Bofors RBS 15

The contract for the RBS 15 was awarded in July 1979 to the Saab-Bofors Missile Corporation. The missile will arm the 'Spica' and later classes of missile craft in place of the Norwegian Penguin Mk II. In August 1982 Saab-Bofors announced a development contract for the RBS 15F air launched version. This is to be used to arm the Royal Swedish air force's fleet of Saab Viggen attack aircraft and the new-generation Saab JAS for anti-ship duties.

The RBS 15 is housed (with wings folded) in a container-launcher, and consists of three sections: the forward part contains the PEAB fully digital Kuband frequency-agile pulsed active radar seeker and its associated micro-processor and electronics, the middle portion contains the FFV blast-framentation war-

The RBS 15 ship-launched anti-ship missile. An air-launched version is under development.

head with both delay-action and proximity fuzes, and the aft section houses the Microturbo TRI-60 turbojet sustainer. The RBS 15 is also fitted with strap-on booster motors located on the launch-phase stabilizer fins; these boosters burn for about three seconds and are then jettisoned. The RBS 15F has no booster motors. The flight altitude is

controlled throughout by a radio altimeter, and course by an autopilot working through the cruciform rear wings.

Specification
Saab-Bofors RBS 15 and RBS 14F
Dimensions: length 4.35 m (14 ft 3.25 in); diameter 50.0 cm (1 ft

7.7 in); span 1.40 m (4 ft 7.1 in)
Weights: total round 780 kg (1,720 lb) for RBS 15 and 598 kg (1,318 lb) for RBS 15F; warhead not known
Performance: maximum speed high subsonic; range 75 km (46.6 miles)

Soviet AA-2 'Atoll' series

'Atoll' is the NATO reporting name assigned to a Soviet air-to-air missile family that in its first-generation form closely resembled the early Sidewinders, primarily because a sample of the latter had been obtained by the Communist Chinese air force in 1958, when one of its Mikoyan-Gurevich MiG-15s flew home with a dud Sidewinder lodged in its fuselage after an air battle with Taiwanese fighters. According to American intelligence sources, there are four versions, the AA-2a, AA-2b, AA-2c and AA-2d. It is highly likely that the AA-2a and AA-2b are IR and SARH versions of the first-generation model, whilst the AA-2c and AA-2d are the corresponding versions of the slightly more reliable second-generation AA-2-'Advanced Atoll' variant. The early IR model has seen widespread combat service and its performance is reckoned to be relatively poor. The

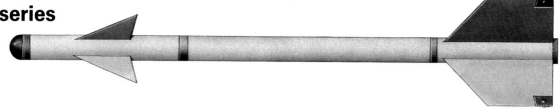

AA-2 'Atoll', the Soviet Sidewinder copy. A radar-homing version is also in service.

'Atoll' is standard armament for the early (two missiles) and late (four missiles) model MiG-21 'Fishbed' interceptor, and has been widely exported. Other aircraft seen with the missile include the MiG-17 'Fresco' (2), MiG-19 'Farmer' (2), MiG-23 'Flogger' (4), Sukhoi Su-20/22 'Fitter' (2), Sukhoi Su-11 'Fishpot' (2) and the Yakovlev Yak-28P 'Firebar' (2). In the Soviet air force and several Warsaw Pact countries the AA-2 is being replaced by the AA-8 'Aphid'. Current operators of the 'Atoll' series are Afghanistan,

Albania, Algeria, Angola, Bangladesh, Bulgaria, China (own version), Cuba, Czechoslovakia, East Germany, Egypt, Ethiopia, Finland, Hungary, India (licence built), Iraq, Jugoslavia, Laos, Libya, Madagascar, Mongolia, Mozambique, Nigeria, North Korea, North Yemen, Peru, Poland, Romania, Somalia, South Yemen, Sudan, Syria, Tanzania, USSR, Vietnam, and Zambia.

**Specification
AA-2 series**

Dimensions: length IR types 2.80 m (9 ft 2.24 in) and SARH types 3.10 m (10 ft 2.05 in); span 0.53 m (1 ft 8.87 in); body diameter 0.12 m (4.72 in)
Launch weight: IR types 70 kg (154.3 lb) and SARH types 75 kg (165.3 lb)
Performance: speed Mach 2.5; range IR 1st generation 5.7 km (3.54 miles) and 2nd generation 8 km (5 miles); range SARH 1st generation 8 km (5 miles) and 2nd generation 10 km (6.2 miles)
Warhead: 6 kg (13.2 lb) HE fragmentation with impact and delay action fuzes

Soviet AA-6 'Acrid'

Designed in the late 1950s and early 1960s as a pure long-range medium- to high-altitude bomber killer, the solid-propellant AA-6 'Acrid' entered service with the Soviet air force in the early 1970s as the largest air-to-air missile fielded by any nation to date. It is the nearest Soviet equivalent to the American Phoenix missile and is the primary armament for the Mikoyan-Gurevich MiG-25 'Foxbat' Mach 2.8 interceptor, which carries four missiles on underwing pylons (two of the IR version on the inner pylons and two of the SARH version on the outer pylons). The GCI-controlled 'Foxbat' uses its own 'Fox Fire' radar to illuminate the target for the SARH missiles, which have an inertial or autopilot midcourse

The giant 'Acrid' is the largest air-to-air missile in service. It is carried almost exclusively by the MiG-25 'Foxbat'.

guidance phase. As with nearly all Soviet missiles, the standard practice is to ripple-fire the 'Acrids' in pairs (IR version followed by SARH version) at a single target to improve the kill probability. The 'Acrid' has also been exported outside the Soviet Union to those countries which have the 'Foxbat-A' interceptor in service, namely Algeria, Iraq, Libya and Syria. Improvements

have now been made, and it is expected that the stand and improved versions of the 'Acrid' will serve on through the early 1990 before being phased out.

**Specification
AA-6 'Acrid'**
Dimensions: Ir version 5.80 m (19 ft 0.35 in) and SARH version 6.29 m (20 ft 7.64 in); span 2.25 m

(7 ft 4.58 in); body diameter 0.40 m (1 ft 3.75 in)
Launch weight: IR version 750 kg (1653.4 lb) and SARH version 800 kg (1763.7 lb)
Performance: speed Mach 4.5; range IR version 25 km (15.53 miles) and SARH version 70 km (43.5 miles)
Warhead: 90-kg (198.4-lb) HE blast fragmentation with contact and proximity fuzes

Soviet AA-7 'Apex'

The solid-propellant low- to medium-altitude AA-7 'Apex' was first revealed in the late 1970s as a third-generation Soviet equivalent to the medium-range AIM-7 Sparrow missile that had been compromised during the Vietnam War. Since then American intelligence reports have identified two versions, the AA-7s and AA-7b that correspond to the IR (Soviet designation R-23I) and SARH versions respectively. Instead of the more standard single SARH seeker head, the AA-76 (Soviet designation R-23R) has four fin-like blade receiver antennae just aft of the nose that work on the interferometer principle. Flight control is achieved by a set of control fins at the tail around the rocket motor exhaust pipe, four much larger fixed delta wings being located approximately half way along the missile body itself.
The normal launch platform for the 'Apex' is the Mikoyan-Gurevich

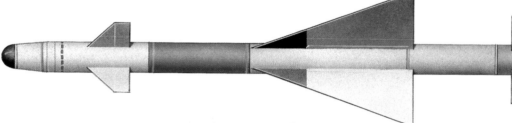

'Apex' is the Soviet Sparrow equivalent, but comes in IR as well as SARH versions. One of each type is usually carried.

MiG-23 'Flogger' family with two missiles (one of each type) on the underwing glove hardpoints. The radar associated with the Soviet air force 'Flogger-B/G' versions is the limited look-down search-and-track J-band 'High Lark' set, which illuminates the target for missile-guidance purposes when it is in a continuous-wave mode. Apart from technology from the American Sparrow, the 'Apex' probably contains components from the 1950s developed AA-5 'Ash' IR and SARH missile

family. The 'Ash' still equips the long-range Tupolev Tu-28 'Fiddler' interceptor. What is clear from analysis of Soviet air-to-air missile systems, however, is that they are very often developed for specific aircraft types rather than to the normal Western philosophy of developing a missile for a role and then fitting it to every suitable launch platform available. Current operators of the AA-7 are Czechoslovakia, East Germany, India, Libya, Syria and the USSR.

**Specification
AA-7 'Apex'**
Dimensions: length IR versions 4.20 m (13 ft 9.35 in) and SARH version 4.60 m (15 ft 1.1 in); span 1.40 m (4 ft 7.12 in); body diameter 0.223 m (8.78 in)
Launch weight: IR version 300 kg (661 lb) and SARH version 320 kg (705.5 lb)
Performance: speed Mach 3.5; range IR version 20 km (12.5 miles) and SARH version 55 km (34.2 miles)
Warhead: 40-kg (88-lb) HE fragmentation with contact and proximity fuses

Soviet AA-8 'Aphid'

The eventual replacement for the 'Atoll', the short-range AA-8 'Aphid' (Soviet designation R60) was revealed concurrently with the AA-7 'Apex' in the mid-1970s. Although only photographs of an IR-guided version have been released so far, there are persistent reports of a SARH version for use on the 'High

Lark' radar-equipped Mikoyan-Gurevich MiG-23 'Flogger' and late-generation MiG-21 'Fishbed' interceptors. The AA-8 is one of the smallest guided air-to-air missiles yet built and is known to be carried on the MiG-23 (two or four missiles on under-fuselage racks, plus two AA-7s), second- and third-generation

MiG-21 (usually two IR-guided AA-8s with 'Advanced Atoll' SARH missiles on the outer pylons), Sukhoi Su-15 'Flagon' (two IR on underfuselage pylons, plus two AA-7s) and the Soviet navy's Yakovlev Yak-36MP 'Forger-A' (usually two IR type on outer wing pylons). Apart from the USSR, AA-8 operators are Cuba,

Czechoslovakia, East Germany, India, Libya and Syria. Reports from the Middle East indicate that Syria has already used the AA-8 as a dogfight missile in the air battles against Israel during the 1982 invasion of Lebanon with negligible results to date.
Designed specifically for the dog-

fight role, the AA-8 is a highly manoeuvrable missile with four delta wings at the tail and four canard delta control surfaces, the latter immediately aft of four nose-mounted fixed aerodynamic surfaces that may both aid in the guidance

phase (by being an ECCM system) and enhance manoeuvring capabilities.

Specification
AA-8 'Aphid'
Dimensions: length IR version

2.15 m (7 ft 0.65 in) and SARH version 2.35 m (7 ft 8.52 in); span 0.40 m (1 ft 3.75 in); body diameter 0.12 m (4.72 in)
Launch weight: IR version 55 kg (121.25 lb) and SARH version 60 kg (132.3 lb)

Performance: speed Mach 3.0; range IR version 10 km (6.2 miles) and SARH version 15 km (9.23 miles)
Warhead: 6-kg (13.2 lb) HE blast fragmentation with contact and proximity fuses

Soviet New-generation air-to-air missiles

According to leaked American intelligence reports there are at least three Soviet fourth-generation air-to-air missile systems that have just or are about to enter service with the Soviet air force.

The AA-9 is an all-altitude all-aspects medium-range missile with snap-down capabilities for use against cruise missile targets. Possibly derived from AA-7 technology, the AA-9 appears to be the Soviet answer to the American AIM-7M

Sparrow and British Sky Flash systems. It is carried by the Mach 2.4 Mikoyan-Gurevich MiG-31 'Foxhound' on four underwing pylons, and relies on the pulse-Doppler radar of the 'Foxhound' for target data. In trials it demonstrated the ability to intercept 'cruise missile' target drones flying at around 90 m (295 ft) whilst the launch platform flew at vertical separations in excess of 6000 m (19,685 ft). Claimed maximum ranges are 70 km (43.5 miles)

at high altitude and around 40 km (24.85 miles) at low altitudes.

The AA-X-10 is in its final development phase, and the missile is due to enter service in late 1984 with the MiG-29 'Fulcrum' and Sukhoi Su-27 'Flanker' interceptors. The AA-X-10 is a medium-range all-altitude missile, with terminal homing of the active radar type and capable of snap-down attacks. A high-altitude beyond-visual-range engagement capability in excess of

35-40 km (21.75-24.85 miles) is claimed, together with a low-altitude performance range of 12.5-20 km (7.5-12.4 miles).

The AA-X-12(?) is a new IR-guided dogfight missile which is stated to be comparable in performance with the American AIM-9L Sidewinder. It is expected to enter service in the mid- to late 1980s to arm the new generation of Soviet interceptors, the 'Fulcrum' and 'Flanker', as their standard dogfight missile.

Soviet AT-2 'Swatter'

The AT-2 'Swatter' was the second of the Soviet first-generation ATGWs to be identified. It is a manually-guided command to line of sight vehicle- and helicopter-mounted system known to the Soviets as the PTUR-62 'Falanga'. It is unusual among ATGWs in having a UHF radio command guidance link with three possible frequencies for ECCM purposes. It is at its most effective when launched directly at the target but can, if required, be switched from one target to another as long as the new one is within the field of fire. The missile arms itself when it reaches 500 m (545 yards) from the launcher. In a later 'Swatter-B' version the maximum range was increased from the 3000 m (3,280 yards) of the original 'Swatter-A' to 3500 m (3,930 yards). A 'Swatter-C' version is now used on the Mil Mi-24 'Hind-A' and 'Hind-D' gunship heli-

AT-2 'Swatter', as used by Mil Mi-24 'Hind' helicopters. Four are carried on launch rails on the stub wing pylons.

copters (four rounds), with semi-automatic command to line of sight guidance and a further increase in maximum range to 4000 m (4,375 yards). All three versions are used on BRDM-1 and BRDM-2 scout car conversions in quadruple mounts. The 'Swatter' is now being replaced by more modern systems, and as far as it is known has never been used in combat, although 'Hind A' and 'Hind-D' helicopters have been seen carrying the type in Afghanistan.

Countries that use the 'Swatter' are Bulgaria, Cuba, Czechoslovakia, East Germany, Egypt, Hungary, Libya, Poland, Romania, South Yemen, Syria and the USSR.

Specification
AT-2 'Swatter'
Type: anti-tank missile
Dimensions: length 1.14 m (3 ft 8.88 in); diameter 13.2 cm (5.1 in); span 66.0 cm (2 ft 2 in)
Launch weight: 'Swatter-A'

26.5 kg (58.4 lb), 'Swatter-B' 29.5 kg (65 lb) and 'Swatter-C' 32.5 kg (71.65 lb)
Propulsion: solid-propellant rocket
Performance: range 'Swatter-A' 500-3000 m (545-3,280 yards), 'Swatter-B' 500-3500 m (545-3,825 yards) and 'Swatter-C' 250-4000 m (275-4,375 yards)
Warhead: hollow-charge HE
Armour penetration: 'Swatter-A' 480 mm (18.9 in) and 'Swatter-B/C' 510 mm (20.08 in)

Soviet AT-3 'Sagger'

The AT-3 'Sagger' is known to the Soviets as the PTUR-64 'Malatyuka', and until recently was their standard domestic and export ATGW. It was produced in three versions, the optically-tracked manually wire-guided Sagger-A', the 'Sagger-B' which came into service in late 1973 and has an improved propulsion motor to make it 25 per cent faster, and the 'Sagger-C' which is the 'Sagger-B' with semi-automatic guidance and entered service in the late 1970s. The 'Sagger' is used in a number of ways, that most often encountered being the three-man team carrying two rounds and the control unit. The 'Sagger' is also used on sextuple launchers on the BRDM-1 (with no reloads) and the BRDM-2 (with

eight reloads) tank destroyers. Single-rail launchers are fitted to the BMP-1 and BMD infantry combat vehicles (with four reloads each) for self-defence. The missile, believed to be the 'Sagger-B' and 'Sagger-C' versions, is also carried on the Mil Mi-2 'Hoplite', Mi-8 'Hip' and Mi-24 'Hind' helicopters of the Warsaw Pact and allies, whilst the Yugoslavs have mated it to the Aérospatiale SA 342 Gazelle and their indigenous BOV-1 armoured vehicle on sextuple launcher (with six reloads). Other vehicle mounts include a modified East German BTR-40 APC and the Czech OT-64 APC. The missile has seen extensive combat service with many nations. The Arabs used it against the Israelis in the War of Attrition, the 1973 Yom

Kippur War and the 1982 'Peace for Galilee' campaign. The North Vietnamese have used it against the South Vietnamese, Americans and Chinese in Indo-China, and the Iraqis have deployed the weapon extensively against Iranian targets in the Gulf War from the ground, vehicles and 'Hind-D' helicopters. Although not given the missile before their ideological split with the Soviets, the Chinese have since copied the type, and the Taiwanese have used it as the basis of their Kun Wu ATGW. The AT-3 'Sagger' is widely employed, known operators being Algeria, Angola, Bulgaria, China (unlicensed copy), Cuba, Czechoslovakia, East Germany, Egypt, Ethiopia, Hungary, India, Iraq, Israel, Libya, Mozambique,

North Korea, Poland, Romania, South Yemen, Syria, Taiwan (unlicensed copy), USSR, Vietnam and Yugoslavia.

Specification
AT-3 'Sagger'
Type: anti-tank missile
Dimensions: length 0.883 m (2 ft 10.76 in); diameter 11.9 cm (4.69 in); span not known
Launch weight: 11.29 kg (24.9 lb)
Propulsion: two-stage solid-propellant rocket
Performance: range 300-3000 m (330-3,280 yards)
Warhead: 3-kg (6.6-lb) hollow-charge HE
Armour penetration: 410 mm (16.14 in) or more

Soviet AT-6 'Spiral'

The tube-launched AT-6 'Spiral' is belived to be the first third-generation Soviet ATGW. At present it is only deployed on the Mil Mi-24 'Hind-E' assault helicopter (four rounds) but its presence on the new Mi-28 'Havoc' attack helicopter as its standard ATGW payload is expected soon. Much conjecture has arisen over its guidance system, laser homing with a fire-and-forget

capability being the most favoured suggestion. Recently, however, informed sources in the USA have indicated that the weapon has a much-improved radio command guidance unit with considerably enhanced ECCM capability than its predecessor, the AT-2 'Swatter'. This would explain the apparent absence of a laser designator on the 'Hind-E'. No country other than the

Soviet Union has fielded the AT-6, which indicates the importance of this missile in the Soviets' anti-armour force.

Specification
AT-6 'Spiral'
Type: anti-tank missile
Dimensions: length about 1.8 m (5 ft 10.86 in); diameter about 14.0 cm (5.5 in); span not known

Launch weight: 32 kg (70.55 lb)
Propulsion: dual-thrust solid-propellant rocket
Performance: range 100-7000 m (110-7,655 yards)
Warhead: 8-kg (17.6-lb) hollow-charge HE
Armour penetration: 800 mm (31.5 in) or more

Soviet air-launched anti-ship missiles

The first Soviet air-launched anti-ship missile achieved operational status with the Soviet Naval air force in the late 1950s. Designated AS-1 'Kennel' by NATO, one such missile was carried under each wing of the Tupolev Tu-16 'Badger-B' bomber. Range was limited to about 80 km (50 miles) and the missile was a relatively unsophisticated beam-rider. The warhead was of conventional high explosives. Exported to Egypt and Indonesia, the missile is no longer in service with any of its users. In 1960 a new turbojet-powered missile was introduced into service, carried under the fuselage of the dedicated missile-carrying Tu-16 'Badger-C' bomber. Used only by the Soviet naval air force, the AS-2 'Kipper' is fitted with a conventional warhead and used solely as an anti-ship weapon. It has autopilot guidance with mid-course correction capability, and carries an active radar seeker suitable for use against large targets such as carriers. The missile remains in service today.

In the following year the Long-Range Air Force introduced into service the massive turbojet-powered aircraft-shaped AS-3 'Kangaroo' missile for strategic use. Designated for use against large area targets, the missile can also be used for attacks on groups of ships. Lacking any terminal guidance system, the AS-3 more than makes up for this by carying an 800-kiloton thermonuclear warhead. Again, the missile remains in use on the Tupolev Tu-95 'Bear-B' missile-carrier.

Although seen in early 1961, the AS-4 'Kitchen' was not actually deployed with the Tupolev Tu-22 'Blinder-B' until 1965. The missile is powered by a single-stage liquid-fuel rocket, and was the first Soviet air-launched missile of multi-purpose nature. Used by the Soviet navy and Long-Range Air Force, the missile is available in anti-radar, anti-ship and strategic forms with a variety of homing systems to suit its role: all have inertial guidance, with active, passive or no terminal homing. Evidence of a continuing development programme was seen with the deployment of the Tupolev Tu-22M (or Tu-26) 'Backfire-B' bomber

AS-4 'Kitchen' arms the Tupolev Tu-26 'Backfire' bomber.

AS-6 'Kingfish' is a multi-role missile capable of carrying a nuclear warhead or a ton of high explosive.

with one missile in a recessed fuselage weapons bay or with one missile under each wing. Evaluation of available photographs and comparison with those under the 'Blinder' suggest that there are two different families of AS-4 missiles, one developed for 'Backfire' alone (the under-fuselage single-round type being nuclear-armed, and the missiles under the wings being conventional-warhead anti-ship or anti-radiation variants). The US navy believes that the AS-4 is most dangerous when launched at high altitude for a high-level flight profile and very steep terminal dive onto the target.

The replacement for the 'Kennel' was fielded in 1966, when the AS-5 'Kelt' was first seen under the wings of the Tu-16 'Badger-G' bomber. Powered by a single-stage liquid-fuelled rocket, the AS-5 is similar in appearance to the AS-1 and is fitted with a conventional warhead. The guidance system is considerably improved, and both active-radar and passive-radar homing versions were used in combat by the Egyptians during the 1973 war against Israel. Out of 25 fired only five hit their targets, the rest being shot down by Israeli air-defence systems. The AS-5 is used by both the Soviet Naval Air Force and the Long-Range Air Force.

In 1970 the Soviet navy and Long-Range Air Force brought into service the AS-6 'Kingfish' missile to complement the 'Kitchen'. The AS-6

is also a multi-role missile capable of the same roes as the AS-4 with the ame types of guidance and homing. However, it is powered by a single-stage solid-propellant motor and is carried in pairs under the wings of the 'Backfire-B' Tupolev Tu-16 'Badger-C mod' and Tu-16 'Badger-G mod'. Like the 'Kitchen' it is considered a special threat by the US Navy. Both types are operated only by the USSR.

Specification
AS-2 'Kipper'
Dimensions: length 10.0 m (32 ft 9.7 in); diameter 90.0 cm (2 ft 11.4 in); span 4.90 m (16 ft 0.9 in)
Weights: total round 4200 kg (9,259 lb); warhead 1000 kg (2,205 lb)
Performance: maximum speed Mach 1.2; range 185 km (115 miles) from a high-altitude launch

Specification
AS-3 'Kangaroo'
Dimensions: length 14.90 m (48 ft 10.6 in); diameter 1.85 m (6 ft 0.8 in); span 9.15 m (30 ft 0.25 in)
Weights: total round 11000 kg (24,250 lb); warhead 2300 kg (5,071 lb) 800-kiloton thermonuclear
Performance: maximum speed Mach 1.8; range 650 km (405 miles) from a high-altitude launch

Specification
AS-5 'Kelt'
Dimensions: length 8.60 m (28 ft

2.6 in); diameter 90.0 cm (2 ft 11.4 in); span 4.50 m (15 ft 1.1 in)
Weights: total round 3000 kg (6,614 lb); warhead 1000 kg (2,205 lb)
Performance: maximum speed Mach 1.2; range 230 km (143 miles) from a high-altitude launch and 180 km (112 miles) from a low-altitude launch

Specification
AS-4 'Kitchen'
Dimensions: length 11.30 m (37 ft 0.9 in); diameter 90.0 cm (2 ft 11.4 in); span 3.00 m (9 ft 10 in)
Weights: total round 5900 kg (13,007 lb); warhead 1000 kg (2,205 lb) high explosive or 350-kiloton nuclear
Performance: maximum speed Mach 3.5; range 460 km (286 miles) from a high-altitude launch and 300 km (186 miles) from a low-altitude launch

Specification
AS-6 'Kingfish'
Dimensions: length 10.00 m (32 ft 9.7 in); diameter 90.0 cm (2 ft 11.4 in); span 2.50 m (8 ft 2.4 in)
Weights: total round 5000 kg (11,023 lb); warhead 1000 kg (2,205 lb) high explosive or 350-kiloton nuclear
Performance: maximum speed Mach 3; range 560 km (348 miles) from a high-altitude launch and 250 km (155 miles) from a low-altitude launch

---

Soviet strategic air-to-surface missiles

At present the strategic elements of the air armies of the Soviet Union use strategic variants of the AS-3 'Kangaroo', AS-4 'Kitchen' and AS-6 'Kingfish' ASSMs. The Soviets also use versions of the last two missiles with passive radar-homing systems to destroy radars assessed as being of prime importance in the defence of targets likely to be attacked by Soviet strategic bombers.

Of the three listed systems only the AS-3 was developed solely for the strategic mission, the others also being available in anti-shipping variants for use by the Soviet naval air force. The 'Kangaroo' was based on an aircraft airframe with turbojet propulsion. The guidance is handled by an autopilot with mid-course command-correction facilities; no terminal homing system is fitted, and this lack of terminal accuracy dictates the use of an 800-kiloton yield thermonuclear warhead. The range is 650 km (404 miles) using a high-altitude supersonic flight profile before a terminal dive at the target location. Carried only by the Tupolev Tu-95 'Bear-B' and 'Bear-C' four-engine long-range bombers, the

The AS-5 'Kelt' bears a family resemblance to the AS-1 'Kennel', with the major change being from jet to rocket power. In place of the jet intake, the AS-5 has the nose (and probably guidance system) of the 'Styx' shipborne SSM.

AS-3 is gradually being replaced by the AS-4 (carried by the 'Bear-G' conversion) and by the AS-X-15. This latter is carried by the new-production 'Bear-H', which can launch a number of the low-altitude AS-X-15 cruise missiles, which each have a range of 3000 km (1,865 miles) and a 200-kiloton yield warhead.

Both the Mach-3.5 AS-4 and the Mach-3.0 AS-6 are single-stage solid-propellant missiles. The AS-4 is inertially guided to its target, whereas the AS-6 has an autopilot guidance system. In the normal high-altitude flight profile the AS-4 has a range of 460 km (286 miles) and the AS-6 of 560 km (348 miles) with very steep terminal dives. In both cases the missiles can be used on a low-altitude profile, which reduces their ranges to 300 km

(186 miles) and 250 km (155 miles) respectively. The nuclear warhead carried by both is 350-kiloton yield, although this may be exchanged for a 1000-kg (2,205-lb) HE warhead if required.

The strategic forces also use the AS-5 'Kelt' on occasion with their medium bomber units. Thought to be used in this context with a passive radar-homing seeker for defence-suppression tasks, the 'Kelt' carries only a conventional 1000-kg (2,205-lb) HE warhead. It is a Mach-1.2 liquid-propellant rocket-powered winged missile with high- and low-altitude fight profile ranges of 230 km (143 miles) and 180 km (112 miles) respectively. According to Israeli sources, the terminal dive angle is very shallow and the weapon can easily be engaged by air-defence systems.

Further strategic air-launched

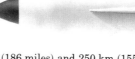

strategic missiles of higher performance are known to be in development. Of the current weapons the USSR uses all, while the conventionally-armed AS-4 is used by Iraq, and the AS-5 by Egypt and Iraq.

Specification
AS-3 'Kangaroo'
Type: air-to-surface strategic missile
Dimensions: length 14.90 m (48 ft 10.6 in); span 9.14 m (30 ft 0 in); diameter 1.85 m (6 ft 0.8 in)
Launch weight: 11000 kg (24,250 lb)
Propulsion: one turbojet
Performance: speed Mach 1.8; range 650 km (404 miles); CEP not known
Guidance: autopilot with mid-course correction
Warhead: 2300-kg (5,071-lb) nuclear with a yield of 800 kilotons

Soviet tactical air-to-surface missiles

It is notoriously difficult to obtain any reliable data on Soviet weapons of any description unless they have been captured. However, in the area of tactical air-to-ground weapons it is known that the Soviets have paralleled many of the Western equipment types, and in some specific areas such as fuel-air explosive munitions have gained a real lead in development.

The early Soviet tactical ASMs never received NATO reporting names, but from American sources it has been discovered that they were physically based on air-to-air missiles. The first real ASM to be seen by NATO was given the designation AS-7 'Kerry'. This entered service in the late 1970s with both the Soviet air force and navy, and is a single-stage radar beam-riding solid-pro-

pellant missile equivalent to the much earlier American Bullpup series. This was followed by the AS-8, which is a fire-and-forget anti-tank weapon for the Soviet attack helicopters such as the Mil Mi-24 'Hind', and perhaps the Sukhoi Su-25 'Frogfoot' battlefield support aircraft. The AS-9 was the next in the sequence, and this is reported to be a comparatively large turbojet-powered supersonic anti-radiation missile (ARM) with a range of 100 km (62 miles) and a warhead of 150 kg (331 lb) for use by bombers and strike aircraft in penetrating air defences. The more recent AS-11 and AS-12 are said to be improved versions of this missile with different homing heads and increased performance. The intervening weapon, the AS-10, is a Mach-1 solid-

propellant missile with electro-optical homing, a range of 11.1 km (6.9 miles) and a 100-kg (220-lb) warhead. The last missile for which details are available is the AS-14 (formerly known to NATO as the Advanced Tactical ASM). This is a larger version of the AS-10 with mid-course guidance and the electro-optical homing system used for the terminal phase of its maximum 40-km (25-mile) flight trajectory. Recent photographic evidence has shown it under the wings of a Mikoyan-Gurevich MiG-27 'Flogger-J' fighter-bomber.

The Soviets also have laser-guided versions of their standard FAB-500, FAB-750 and FAB-1000 GP low-drag iron bombs, together with versions of the 210-mm (8.27-in) S-21 and 325-mm (12.8-in) S-32

air-to-surface unguided rockets that have been fitted with limited visual-guidance systems. These supplement the vast array of conventional weapon types deployed by the Soviets.

Specification
AS-7 'Kerry'
Type: air-to-surface missile
Dimensions: not known
Launch weight: 1200 kg (2,646 lb)
Propulsion: solid-propellant rocket motor
Performance: speed Mach 1; range 11.1 km (6.9 miles); CEP not known
Guidance: radar beam-riding
Warhead: 100-kg (220.5-lb) HE blast fragmentation

Texas Instruments AGM-45 Shrike and General Dynamics AGM-78 Standard

Specialized anti-radar missiles (ARMs) were developed by the US Navy from 1958 onwards as a means of improving the survivability of conventional attack aircraft, either by deterring the enemy from operating his defensive radar or by directly destroying the radar's antenna. The first tactical ARM to enter production was the Texas Instruments AGM-45A Shrike in 1963. This was essentially a Sparrow AAM airframe with an enlarged blast-fragmentation warhead and a smaller rocket motor. Although used extensively by the US forces in Vietnam, by Israel against the Arabs and by the UK in the 1982 Falklands war, the Shrike has displayed a not altogether satisfactory performance as a result of design limitations associated primarily with the seeker. There are no memory circuits available, and this means that the shutdown of the radar being attacked causes the missile to go ballistic. The seeker is also rigidly mounted, so the missile must be pointed towards the target at launch, and the seeker has to be tuned before take-off to the wavelength band of the radar system under attack, otherwise it cannot pick up any emissions. A total of 13 different seekers to cover likely target systems has thus been developed. Total production for the USAF and US Navy was about 18,500 rounds, the larger number going to the former service, which uses it primarily on the McDonnell

Douglas F-4G Phantom 'Wild Weasel II' defence-suppression aircraft together with the General Dynamics AGM-78 Standard ARM.

The latter missile was contracted in 1966 because the Shrike's combat performance was found to be bad. Designated AGM-78A Standard in its initial form, it was based on the Standard shipboard SAM and initially equipped with the Shrike seeker with all its faults. Production soon shifted to the AGM-78B version with a gimballed wide-band seeker and a memory circuit that required no pretuning, thus permitting an attack even if the radar had ceased emitting signals. An AGM-78C variant was then produced for the US Air Force, the subsequent AGM-78D and AGM-78D2 models further increasing the seeker capabilities. Over 3,000 rounds had been built by the time the last delivery of a batch of AGM-78D2 missiles was made in August 1976. Ultimately both the Shrike and Standard will be replaced by the AGM-88A HARM, and current operators of the type are Israel (not certainly), South Korea, the US Air Force, US Marine Corps and US Navy. The Shrike is in slightly more widespread service, current operators including Iran, Israel, the UK, US Air Force, US Marine Corps and US Navy.

Specification
AGM-45 Shrike
Type: anti-radiation air-to-surface

missile
Dimensions: length 3.048 m (10 ft 0 in); span 0.914 m (3 ft 0 in); diameter 0.203 m (8 in)
Launch weight: 176.9 kg (390 lb)
Propulsion: solid-propellant rocket motor
Performance: speed Mach 2; range 46.5 km (28.9 miles); CEP reasonable if the target radar continues to emit
Guidance: passive radar-homing
Warhead: 65.8-kg (145-lb) HE blast fragmentation

Specification

AGM-78 Standard ARM
Type: anti-radiation air-to-surface missile
Dimensions: length 4.572 m (15 ft 0 in); span 1.092 m (3 ft 7 in); diameter 0.343 m (13.5 in)
Launch weight: 615.1 kg (1,356 lb)
Propulsion: solid-propellant rocket motor
Performance: speed Mach 2.5; range 112.65 km (70+ miles); CEP good even if the target radar ceases transmitting
Guidance: passive radar-homing
Warhead: 97.4-kg (214.7 lb) HE blast fragmentation

The much larger and more capable Standard ARM (foreground) is replacing the Shrike (background) in US service. Developed from the Standard naval SAM, the ARM has a memory circuit which enables it to attack a radar site even when it has stopped transmitting.

Texas Instruments AGM-88A High-speed Anti-Radiation Missile (HARM)

Similar in appearance to the Shrike missile it will replace, HARM (High-speed Anti-Radiation Missile) is a larger weapon with a greatly improved performance. It will also replace the Standard ARM, having similar performance with much improved electronics.

Although the Standard ARM was an improvement on the Shrike its combat performance in Vietnam was still not very inspiring as its memory circuits proved less than satisfactory. Also the Standard was five times costlier and three times heavier than the Shrike, so a requirement for a new ARM was established. The result was the Texas Instruments AGM-88A HARM, which emphasizes high speed so that any defending radar operator has only minimum warning times to 'shut down' his system; this attacker's advantage is multiplied by the fact that the launch platform does not need to execute any characteristic launch manoeuvre.

Initial development of the HARM began in late 1969 by the US Navy, but progress was halted by severe technical problems which were not

resolved until 1973. Further delays were then experienced with the Texas Instruments guidance seeker and initial production deliveries were not made until 1983. The missile has three modes of operation: the self-protection mode, in which a threat receiver on the launch platform detects a radar signal and programmes the missiles seeker before it is fired; the 'target-of-opportunity' mode, in which the sensitive seeker on the missile itself locks on to an emitting radar; and the 'prebriefed' mode, in which the missile is fired blind in the general

direction of a possible target with its seeker searching for a signal onto which the missile can home. In the last mode, failure to detect a signal initiates a programme for self-destruction. Like its two predecessors, the HARM can also be fired in a 'loft' manoeuvre to increase its range, the target being acquired on the downward portion of the trajectory. The warhead is detonated at a preset height over the target by a laser proximity fuze in order to maximize damage to the antenna and electronics.

Specification
AGM-8A
Type: anti-radiation air-to-surface missile
Dimensions: length 4.171 m (13 ft 8.2 in); span 1.118 m (3 ft 8 in); diameter 0.254 m (10 in)
Launch weight: 361.1 kg (796 lb)
Propulsion: solid-propellent rocket
Performance: speed mach 3+; range 74.4+ km (46.25+ miles); CEP very good
Guidance: passive radar-homing
Warhead: 65.8-kg (145-lb) HE blast fragmentation

Index